S0-AQK-867

TRAFFIC ENGINEERING HANDBOOK

Fourth Edition

James L. Pline

Editor

Institute of Transportation Engineers

PRENTICE HALL, *Englewood Cliffs, New Jersey 07632*

Library of Congress Cataloging-in-Publication Data

Traffic engineering handbook / Institute of Transportation Engineers ;
  James L. Pline, editor. -- 4th ed.
       p.   cm.
  Includes bibliographical references and index.
  ISBN 0-13-926791-3
  1. Traffic engineering--Handbooks, manuals, etc.   I. Pline, James
L. II. Institute of Transportation Engineers.
  HE333.T68 1992
  388.3'12--dc20   90-28690
                                                          CIP

Editorial/production supervision
  and interior design: Bayani Mendoza de Leon
Cover design: Edsal Enterprises
Manufacturing buyers: Linda Behrens/Dave Dickey

© 1992 by Prentice-Hall, Inc.
A Simon & Schuster Company
Englewood Cliffs, New Jersey 07632

First edition copyright 1941, Institute of Traffic Engineers and National Conservation Bureau.

Second edition copyright 1950, Institute of Traffic Engineers and Association of Casualty and Surety Companies.

Third edition copyright 1965, Institute of Traffic Engineers.

All rights reserved. No part of this book may be
reproduced, in any form or by any means,
without permission in writing from the publisher.

Printed in the United States of America

10 9 8 7 6 5 4 3 2 1

ISBN 0-13-926791-3

ISBN 0-13-926791-3

90000>

9 780139 267918

PRENTICE-HALL INTERNATIONAL (UK) LIMITED, *London*
PRENTICE-HALL OF AUSTRALIA PTY. LIMITED, *Sydney*
PRENTICE-HALL CANADA INC., *Toronto*
PRENTICE-HALL HISPANOAMERICANA, S.A., *Mexico*
PRENTICE-HALL OF INDIA PRIVATE LIMITED, *New Delhi*
PRENTICE-HALL OF JAPAN, INC., *Tokyo*
SIMON & SCHUSTER ASIA PTE. LTD., *Singapore*
EDITORA PRENTICE-HALL DO BRASIL, LTDA., *Rio de Janeiro*

The Institute of Transportation Engineers (ITE) is made up of more than 11,000 transportation engineers and planners in over 70 countries. These transportation professionals are responsible for the safe, efficient and environmentally compatible movement of people and goods on streets, highways, and transit systems. For more than 60 years the Institute has been providing transportation professionals with programs and resources to help them meet those responsibilities. Institute programs and resources include handbooks, technical reports, a monthly journal, professional development seminars, local, regional, and international meetings, and other forums for the exchange of opinions, ideas, techniques, and research.

For current information on Institute's programs, please contact:

INSTITUTE OF TRANSPORTATION ENGINEERS
525 School Street, S. W., Suite 410
Washington, DC 20024-2729 USA
Telephone: (202) 554-8050
Facsimile: (202) 863-5486

OTHER INSTITUTE OF TRANSPORTATION ENGINEERS
BOOKS AVAILABLE FROM PRENTICE HALL

Residential Street Design and Traffic Control
Transportation and Traffic Engineering Handbook
Transportation and Land Development
Manual of Traffic Signal Design, 2nd edition
Traffic Engineering Handbook, 4th edition
Traffic Signal Installation and Maintenance Manual
Transportation Planning Handbook

CONTENTS

PREFACE

For many years, transportation engineers and planners, as well as laypersons seeking understanding of transportation issues, have relied on the *Transportation and Traffic Engineering Handbook* for information on the state of the art of established practice in transportation engineering. When it came time to begin working on a new edition of the handbook, the Editorial Committee realized that the challenge of a single, comprehensive reference book was a formidable task. A massive amount of information would need to be included in a completely updated book, necessitating Herculean efforts in production and resulting in a publication that would be physically intimidating. As a result, the Editorial Committee initiated instead a two-volume series: The *Traffic Engineering Handbook* and the *Transportation Planning Handbook*. The volumes were planned and prepared under the close supervision of a single Editorial Committee which included the designated editor of each handbook. Although the content of each handbook was carefully reviewed for state-of-the-art content and completeness, some duplication was unavoidable and indeed necessary to ensure that each book would adequately cover those subjects deemed essential.

With the creation of this edition of the *Traffic Engineering Handbook* a link has been made to one of the most significant, if not historic, achievements of the Institute of Transportation Engineers—namely, the 1941 publication of the first edition of the *Traffic Engineering Handbook*. That book was a pioneering accomplishment, representing the first book ever to be dedicated to the subject of traffic engineering. Two subsequent editions, in 1950 and 1965, proved very successful in representing the state of the art of established practice; they served as basic references for the practicing professional. In 1976 and again in 1982, the information included in The *Traffic Engineering Handbook* was incorporated into the first and second editions of the *Transportation and Traffic Engineering Handbook* which reflected the profession's focus on all modes of surface transportation.

This book becomes the fourth edition in the original *Traffic Engineering Handbook* series and is dedicated to continuing the objective established in 1941—to collate in one volume basic traffic engineering information as a guide to best practice in the field. The purpose of the Handbook is to provide professional engineers with a basic day-to-day source of reference on the principles and proven techniques in the practice of traffic engineering. The material presented is not intended to serve as a statement of a standard of practice, but rather as guidelines for the practicing professional engineer. Although not intended to serve as a textbook for higher education in basic or advanced traffic engineering, the Handbook frequently and appropriately serves as a reference source for educators and students. Each section of the Handbook includes a list of publications that should be consulted for further reading in a specific subject area.

Continuing the precedent established by the first edition, this edition of the handbook provides a guide to the state of practice of the fundamental elements of the profession. Many of the chapters and topics cover subjects traditionally found in the handbook, but which are given a contemporary treatment to reflect the current state of the practice.

New chapters are added to each edition of the handbook to ensure comprehensive coverage of traffic engineering. This edition has a new chapter devoted to intelligent vehicle/highway systems (IVHS). IVHS activities in the United States currently include scattered research initiatives and selected demonstration projects. Other countries, such as Japan and several in Europe, already have programs that surpass the U.S. programs in scope and scale. The U.S. Congress and the Department of Transportation are currently taking action to establish federal programs to launch a coordinated, large-scale program to provide the United States with leadership in transportation for the 21st century. Future editions of the Handbook will in all likelihood see the various components of IVHS absorbed into the more traditional chapters, but at this juncture in time it was considered appropriate to present IVHS as a separate chapter.

As the Handbook is being put into print, world events are creating one of the most exciting periods in recent history. We are indeed becoming a global village, both economically and environmentally. The reunification of Germany and the prospect of an economically unified Europe lend dramatic emphasis to the opportunities facing the transportation professional. The challenges of international competitiveness, alternative energy sources, and environmental quality are at the forefront of national policy priorities. Transportation is a vital factor in all those priorities. Not since the beginning of the "interstate era" almost three decades ago have the opportunities been so exciting. The leveraging of new technology, the quest for innovative financing of transportation systems, and the opportunity for impacting those institutions or institutional arrangements that bear on the efficiency and productivity of the transportation industry are all factors affecting the costs of mobility. This *Traffic Engineering Handbook* has been prepared with a compelling urge to integrate aspects of these events and factors into the body of the text; however, the authors have been restrained by the overall objective of creating a primary reference source for the state of practice in traffic engineering. Future editions and their authors will have the rewarding responsibility of transforming the results of those exciting events into guidelines for the state of practice of the traffic engineering profession.

C. MICHAEL WALTON, *Ph.D., P.E.*
Chair
Handbooks Editorial Committee
June, 1991

ACKNOWLEDGMENTS

The planning for and completion of this book has benefitted from the many hours contributed by the persons listed below. The Institute expresses our appreciation of their dedicated service to the profession.

Work on this book began in 1985 with the appointment of the Content Advisory Committee chaired by Harold L. Michael. The committee's report to the International Board of Direction formed the basis of the deliberations of the Handbooks Editorial Committee which was charged with overseeing the drafting of the Traffic Engineering and Transportation Planning Handbooks. The Handbooks Editorial Committee, along with the respective handbook editors, determined the handbooks format, content, schedule, and selection of chapter authors. The Committee was also responsible for assuring that the content of each handbook was not in conflict with and reflected current practice. Detailed reviews of each chapter draft and subsequent revisions were made by the Panel of Chapter Reviewers whose comments assisted the authors in preparing the final manuscript. The Institute's Technical Council has provided input on content and in the selection of the Panel of Reviewers.

My role as editor was greatly assisted by the chapter authors for their expertise and extra effort in addressing the editorial comments. The ITE Headquarters staff is also gratefully acknowledged for its administrative support and coordination with the many people involved in this project, specifically Thomas W. Brahms, Juan M. Morales, Jane A. Wetz and Mark R. Norman.

James L. Pline
Editor

Handbooks Editorial Committee

Chair:
C. Michael Walton

Editors:
John D. Edwards, Jr.
James L. Pline

Associate Editors:
Wolfgang S. Homburger
William R. McGrath

Panel of Chapter Reviewers

Gerson J. Alexander
Daniel H. Baxter
R. Clarke Bennett
Howard H. Bissell
Frederick H. Blake
Daniel Brand
Thomas E. Bryer
Noel Bufe
Donald G. Capelle
John P. Cavallero, Jr.
Paul J. Claffey
Donald E. Cleveland
Mildred E. Cox

Charles W. Craig
J. Robert Doughty
Patricia H. Ehrlich
Barry W. Fairfax
Daniel B. Fambro
John J. Fruin
Charles J. Goedken
John T. Hanna
H. Milton Heywood
Thomas Hicks
Samuel Hochstein
Jack B. Humphreys

Leslie N. Jacobson
Jean M. Keneipp
C. Larry King
Melvin J. Kohn
Peter R. Korpal
Joel P. Leisch
Jeffrey A. Lindley
Martin E. Lipinski
Joseph M. McDermott
Joseph C. Oppenlander
Sheldon I. Pivnik
Alex J. Redford
Marshall F. Reed, Jr.

Robert H. Reeder
Lyle Saxton
Richard N. Schwab
Samuel J. Schwartz
Alice Snow-Robinson
Paul L. Streb
Philip J. Tarnoff
Gary K. Trietsch
John G. Viner
Vernon H. Waight
W. Scott Wainwright
Ronald C. Welke
Charles V. Zegeer

1

DRIVER AND PEDESTRIAN CHARACTERISTICS

ROBERT DEWAR, PHD., *Professor Emeritus of Psychology*

University of Calgary

Introduction

The human factors in highway transportation relate to the capabilities and limitations of the road user (driver, cyclist, pedestrian). Thus an understanding of the human element in the system is essential, as humans design, build, operate, and maintain vehicles, roads, and roadway environments (including traffic control devices). The limitations of the road user in terms of experience, impairment, physical and mental skills, motivation, and other characteristics are factors in the safe and efficient functioning of any transportation system. Characteristics of road users (motor vehicle operators, bicyclists, and pedestrians) and their interaction with the other elements of the system have all been the objects of considerable research, but they are still not well understood.

There are three major components to driving: the vehicle, the roadway/environment, and the driver. This chapter addresses the last of these, examining driver characteristics and how they interact with the other two components. On the basis of decades of research on driving and traffic safety it has become apparent that the weak link in the system is the driver. This is perhaps because humans were originally designed to walk and run, rather than to operate a motor vehicle. This limitation is reflected in the conclusion by Rumar that the driver is an "outdated human with Stone-Age characteristics and performance who is controlling a fast, heavy machine in an environment packed with unnatural, artificial signs and signals."[1]

Driver characteristics

Driver involvement*

There were approximately 162 million licensed drivers in the United States in 1987, or approximately 0.88 licensed drivers per registered motor vehicle. Table 1–1 shows that approximately 92% of males and 78% of females who are old enough to drive in their state are licensed to drive. The percentage of the population holding drivers' licenses increases through age 45 to 50 and then starts to decrease.

A recent Transportation Research Board study found that very young and very old drivers have the highest accident rates. Figure 1–1 shows the variation of traffic accident involvement rates with driver age. The high accident rates of young drivers are usually attributed to youthful aggressiveness, alcohol abuse, and lack of driving experience rather than to any deficiency in performance abilities. However, the performance capabilities that individuals need for safe driving *do* decrease with age.

The driving task

Driving is a very complex task involving a variety of skills, the most important of which are the taking in and processing of information and making quick decisions based on this information. Although a great many models and schematic representations of driving can be found, that by Wilde[2]

---

[1] K. Rumar, "Impacts on Road Design of the Human Factor and Information Systems." Proceedings of the 9th International Road Federation Meeting, Stockholm (1981), pp. 31–49.

*Information provided courtesy of Douglas Harwood, Midwest Research Institute.

[2] G.S. Wilde, "Social Interaction Patterns in Driving Behavior: An Introductory Review," *Human Factors,* 18 (1976), 477–492.

TABLE 1–1

Distribution of Licensed Drivers in the United States by Age and Sex

Age	Male Drivers			Female Drivers			Total Drivers		
	Number (thousands)	Percent of Total	Percent of Population	Number (thousands)	Percent of Total	Percent of Population	Number (thousands)	Percent of Total	Percent of Population
Under 16	48	0.1	2.7	40	0.1	2.4	89	0.1	2.6
16	900	1.1	45.7	775	1.0	41.4	1675	1.0	43.6
17	1327	1.6	68.8	1151	1.5	63.0	2478	1.5	66.0
18	1491	1.8	81.0	1301	1.7	73.0	2792	1.7	77.1
19	1599	1.9	86.1	1401	1.8	76.6	3000	1.9	81.3
(19 and under)	5366	6.4	57.1	4668	6.0	51.7	10034	6.2	54.4
20	1642	2.0	87.6	1467	1.9	79.0	3109	1.9	83.3
21	1741	2.1	92.4	1573	2.0	84.1	3314	2.0	88.2
22	1846	2.2	93.6	1685	2.2	85.6	3531	2.2	89.6
23	1988	2.4	96.5	1831	2.4	88.9	3818	2.4	92.7
24	2020	2.4	95.2	1865	2.4	88.2	3885	2.4	91.7
(20–24)	9236	11.0	93.2	8421	10.8	85.3	17658	10.9	89.3
25–29	10642	12.7	96.7	9935	12.8	90.6	20578	12.7	93.7
30–34	10300	12.3	96.7	9729	12.5	91.2	20029	12.4	93.9
35–39	9225	11.0	99.5	8793	11.3	93.0	18018	11.1	96.2
40–44	7600	9.0	99.5	7204	9.3	90.9	14804	9.1	95.1
45–49	5994	7.1	99.5	5628	7.2	89.0	11622	7.2	94.2
50–54	5161	6.1	97.7	4791	6.2	85.0	9952	6.2	91.1
55–59	5044	6.0	95.3	4656	6.0	80.0	9700	6.0	87.3
60–64	4817	5.7	95.1	4485	5.8	77.0	9302	5.7	85.4
65–69	4114	4.9	91.6	3817	4.9	70.8	7931	4.9	80.2
70 and over	6488	7.7	85.1	5702	7.3	46.3	12190	7.5	61.2
TOTAL	83989		91.6	77830		78.4	161818		84.8

SOURCE: *Highway Statistics 1987,* Report No. FHWA-PL-88-008 (Washington, DC: Federal Highway Administration, 1988), p. 32.

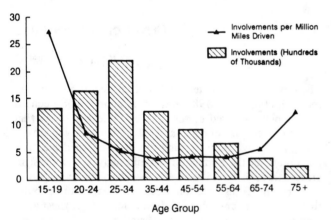

Figure 1–1. Driver involvements in accidents and involvement rates by age, 1983.
SOURCE: *Transportation in an Aging Society: Improving Mobility and Safety for Older Persons,* Special Report 218, (Washington, DC: Transportation Research Board, 1988), p. 40.

illustrates the complexity of the task well (see Figure 1–2). Among the human characteristics involved are basic skills, current physical and psychological states of the driver, and several modulating factors. Most of these characteristics will be discussed in this section.

In Wilde's model the subjective estimated danger (based on a variety of sources) is compared with the level of risk the driver is willing to accept, and decisions are made on the basis of this comparison. If the estimated danger is less than the tolerated danger level, the person will drive in a manner that may be hazardous.

The importance of correct and rapid information processing is well illustrated in the perception model proposed by

Vanstrum and Caples,[3] as seen in Figure 1–3. This model includes a "zone of committed motion projected ahead" of the vehicle and is composed of four segments: (1) distance traveled during minimum perception time; (2) distance traveled during minimum decision time; (3) distance traveled during minimum reaction time; and (4) minimum committed motion area of the vehicle after a response has been made to turn or stop. Zone 4 (minimum stopping distance) depends on vehicle speed and weight, brake efficiency, co-efficient of friction of the road surface, etc. In Figure 1–3, box *X* represents a hazard on the roadway, or a potential hazard such as a curve or railway crossing. *T*, the true point, is the last point at which action can be initiated to avoid the hazard (the point of no return). *M* is the mental point, the driver's perception of the true point, and *A* is the action point—where the driver decides he or she will take action (slow, stop, steer, accelerate). Because driving is dynamic in nature, the various points, the committed zone, and the driver's perception of these relationships change continually. The distance between points *T* and *M* is called the *perceptual error,* and *M* can be ahead of or behind point *T* (see Figure 1–4). The distance between *M* and *A* is the driver's *margin of error.* Most drivers allow some margin of error. If *A* is to the left of *T,* the situation is safe, but if the margin of error does not compensate for the perceptual error, an accident will occur.

Perceptual error can result from a number of factors—alcohol, fatigue, inexperience, inattention. According to Vanstrum and Caples, perceptual error is the proximate cause of most preventable traffic accidents. Prevention can

---

[3] R.C. Vanstrum and G.B. Caples, "Perception Model for Describing and Dealing with Driver Involvement in Highway Accidents," *Highway Research Record,* 365 (1972), 17–24.

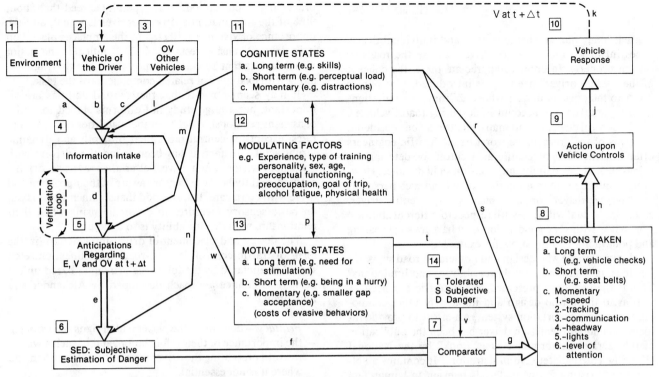

Figure 1–2. Cognitive and motivational model of driver behavior.
SOURCE: G.S. WILDE, "Social Interaction Patterns in Driving Behavior: An Introductory Review," *Human Factors*, 18, 1976. Copyright 1976 by The Human Factors Society, Inc., and reprinted by permission.

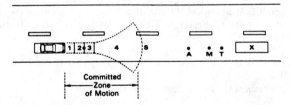

Figure 1–3. Perception model by Vanstrum and Caples.
SOURCE: R.C. VANSTRUM AND G.B. CAPLES, "Perception Model for Describing and Dealing with Driver Involvement in Highway Accidents," *Highway Research Record,* 365 (1972), 19.

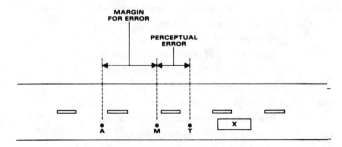

Figure 1–4. Margin of error and perceptual error.
SOURCE: R.C. VANSTRUM AND G.B. CAPLES, "Perception Model for Describing and Dealing with Driver Involvement in Highway Accidents," *Highway Research Record,* 365 (1972), 20.

be accomplished by increasing the margin of error and/or decreasing the perceptual error. The latter error can be reduced by making hazards more visible and increasing drivers' sensory and perceptual abilities. The need for good information-processing skills is evident here.

The driving task itself can also be broken down into three main elements—control, guidance, and navigation, as outlined by Alexander and Lunenfeld.[4]

Control involves the driver's interaction with the vehicle, in terms of speed and direction (accelerating, braking, steering). Relevant information comes mainly from the vehicle and its displays.

Guidance refers to maintaining a safe path and speed and keeping the vehicle in the proper lane on the road. Information comes from roadway alignment, hazards, traffic control devices, and other vehicles and pedestrians using the roadway.

Navigation means planning and executing a trip from one location to another. Navigational information comes from maps, guide signs, landmarks, etc. The control has the highest priority, in the event of an emergency, while navigation has the lowest. Performance is relatively simple and overlearned at the control level, but information handling is more complex at the other levels.

---

[4]G.J. Alexander and H. Lunenfeld, "Positive Guidance in Traffic Control" (Washington, DC: Federal Highway Administration, 1975).

The design driver

A great many human characteristics and individual differences influence the ability of drivers to use the roadway system properly. In order to appreciate these, the concept of the "design driver" needs to be introduced. This phrase refers to the range of drivers whose abilities and limitations need to be taken into account in designing roads, vehicles, traffic control devices, road maps, etc. It may be considered in terms of the "reasonable worst case." Traffic engineers often use the 85th percentile as a cutoff for determining things such as speed limits and decision sight distance. However, it must be kept in mind that there is no such person as the "average driver" or the typical 85th-percentile driver, as any individual will vary with respect to different abilities. The driver with very good vision may have average hearing and poor motor coordination, for example.

How will those who design and implement road signs, for instance, take into account driver abilities and limitations? This is the challenge presented to the traffic engineer, as well as to those who design and implement other components of the transportation system. An effort to provide information about the design driver has been the publication of the "Driver Performance Data Book" by the National Highway Traffic Safety Administration.[5] It contains a collection of existing source materials relevant to human factors in driving: response time, visual performance, auditory performance, information processing, anthropometrics, and precrash behavior. While much of the information is of indirect relevance to driving, the document is a valuable source of basic data on abilities required for the driving task.

Information processing and perception

Approximately 40% of all traffic accidents involving human error have as contributory factors difficulties in information processing or perception.[6] It is generally agreed that about 90% of the information a driver receives is visual, and the importance of vision is reflected in the emphasis placed on it in driver licensing tests. Table 1-2 indicates the major visual factors that *should* be important to the driving task, along with examples of related components of the task.

It may seem obvious that these visual abilities are all important, but a large study in California[7] failed to find a consistent relationship between these abilities and accident rates, except for central movement in depth and dynamic visual acuity. In fact, it has been found that drivers with very poor vision often have better than average driving records, mainly because they are aware of their problem and drive with extra care. It seems odd that so much stress is put on basic sensory measures in driver licensing when their relationship with driving ability is so tenuous.

The design and placement of driver information (in the form of traffic control devices; roadway delineation, etc.) should be dictated by the following principles based on the Positive Guidance approach developed by Alexander and Lunenfeld.[8]

Primacy—determine the placement of signs according to the importance of their information, and in such a way as to avoid presenting the driver with information when and where it is not essential.

Spreading—where all the information required by the driver cannot be placed on one sign or on a number of signs at one location, spread it out over space so as to reduce the information load on the driver.

[5] R.L. Henderson, ed., "Driver Performance Data Book" (Washington, DC: National Highway Traffic Safety Administration, 1987).

[6] J.R. Treat, N.S. Tumbas, S.T. McDonald, D. Shinar, R.D. Hume, R.D. Mayer, R.L. Stansifer, and N.J. Castallen, "Tri-Level Study of the Cause of Traffic Accidents," Report No. DOT-HS-034-3-535-77 (TAC) (Indiana University, 1977).

[7] R. Henderson and A. Burg, "Vision and Audition in Driving," Final Report DOT HS 801-265, National Highway Traffic Safety Administration (1974).

[8] Alexander and Lunenfeld, "Positive Guidance."

TABLE 1-2
Visual Factors in the Driving Task

Visual Factor	Definition	Related Driving Tasks
Accommodation	Change in the shape of the lens to bring images into focus	Changing focus from dashboard displays to the roadway
Static visual acuity	Ability to see small details clearly	Reading distant traffic signs
Adaptation	Change in sensitivity to different levels of light	Adjusting to changes in light upon entering a tunnel in daylight
Angular movement	Seeing objects moving across the field of view	Judging speed of cars crossing the path of travel
Movement in depth	Detecting changes in size of the image on the eye	Judging speed of an approaching vehicle
Color	Discrimination of different colors	Identification of colors of signals
Contrast sensitivity	Seeing objects that are similar in brightness to their background	Detection of dark-clothed pedestrians at night
Depth perception	Judgment of the distances of objects	Passing on two-lane roads with oncoming traffic
Dynamic visual acuity	Ability to see objects that are in motion relative to us	Reading traffic signs while moving
Eye movement	Changing the direction of gaze of the eyes	Scanning the road environment for hazards
Glare sensitivity	Ability to resist and recover from the effects of glare	Reduction in visual performance due to headlight glare
Peripheral vision	Detection of objects at the side of the visual field	Seeing a bicycle approaching from the left
Vergence	The angle between the lines of sight of the two eyes	Change from looking at the dashboard to looking at the road

Coding—where possible organize pieces of information into larger units. Color and shape coding of traffic signs accomplish this by representing specific information about the message based on the color of the sign background and the shape of the sign panel.

Redundancy—say the same thing in more than one way. The STOP sign in North America has a unique shape and message, both of which convey the message to stop. The same information may also be given with two devices (e.g., "no passing" indicated with a sign and pavement markings).

A fundamental component of this approach is the concept of *expectancy*. Drivers operate with a set of expectancies—freeway exits will be on the right side of the roadway, (on the left in Britain, Australia, etc.); advance warning will be given of hazards on the road; other drivers will obey traffic rules—and if these expectancies are violated there is the potential for an accident. Therefore, information from traffic control devices, the roadway environment, etc., must be provided when and where it is expected.

It is evident on the basis of work on accident causation that visual search, involving quick and efficient eye movements, is essential to the driving task. Rockwell and his colleagues[9] did a good deal of the pioneering work on eye movements in driving, using sign reading as one of the main tasks. They were able to determine the time spent looking at various roadway

elements, including signs, and proposed formulae to calculate the amount of time available for detecting and reading overhead highway signs. The specific distances involved depend on several factors (e.g., letter size, message complexity and familiarity, vehicle speed, traffic characteristics, driver visual acuity). Recommendations about sign message size are made based on the maximum time/distance at which the largest letter can be read, time necessary to read the message, total time the driver looks at the sign, and the time during which the driver shares sign reading with other aspects of the driving task. Rockwell suggests that data on eye movement can be helpful in evaluating highway signs and in understanding how drivers process sign information while driving.

What are the important aspects of perception and information processing? Evidence suggests that the main factors involve looking in the right place at the right time—that is, paying proper attention. Tests on selective attention have been found to correlate well with accident rates. In-depth accident investigations have shown that nearly half of the human causal factors in accidents relate to attention—for example, internal distraction, inattention, preoccupation, etc.[10] (See Figure 1–5.) This information tends to be ignored by those designing roadways, traffic control devices, driver training, and safety campaigns.

The question of whether attention can be improved through training is an interesting one. Some claim it can, but there has been little research on the issue and little effort to integrate this factor into driver training. A perceptual style referred to as *field dependence* may well be a

---

[9] V.D. Bhise and T.H. Rockwell, "Strategies in the Design and Evaluation of Road Signs through the Measurement of Driver Eye Movements." Paper presented at the Annual Meeting of the Human Factors Society, New York, 1971.

[10] Treat and others, "Tri-Level Study."

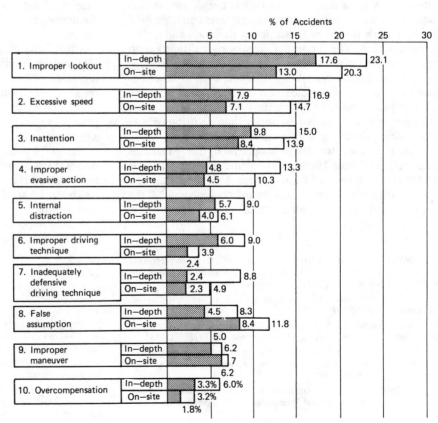

Figure 1–5. Percentage of accidents caused by specific human direct causes.
SOURCE: D. SHINAR, *"Psychology on the Road"* (New York: John Wiley, 1978), p. 119.

contributing factor here, as studies have shown that drivers who are field dependent (i.e., have difficulty sorting out relevant from irrelevant information and are easily distracted) have higher accident rates than do other drivers[11] and are poorer than average at detecting traffic signs embedded in visual scenes.[12]

As drivers gain experience they get better at scanning the roadway environment and at ignoring irrelevant information. Eye movements change from narrow scanning and looking just ahead of the vehicle to sampling more widely as driving experience increases.[13] This may be one reason why new drivers have difficulty in complex driving environments. Proper allocation of attention has become more difficult in the last decade or so because drivers are overloaded with more complex dashboards in newer vehicles and with more traffic signs to read.

A related issue is "spare mental capacity" (the ability to do a number of things at one time), which has been found to be related to accident record—those with more of this capacity have fewer accidents, as they are able to attend to and process more than one source of input at a time (i.e., they are better at time-sharing). Laboratory research[14] suggests that drivers with low spare mental capacity are more likely to have accidents.

Reaction time

A significant component in the successful processing and use of information is the speed with which this processing is done. One of the human factors often cited by traffic engineers concerned with safety is *perception-reaction time* (PRT), or how quickly a driver can respond to an emergency situation. A good deal of research has been done on this topic, but there is still not complete agreement on what PRT is appropriate for use in the design of roadways and traffic control devices. The concept of *decision sight distance* is an important part of roadway engineering, as it relates to stopping, steering, and overtaking. Adequate sight distance depends, of course, on vehicle speed and size as well as driver PRT. It plays a role in many traffic accident litigation cases where a central issue is how much time or distance was available for the driver to respond to a hazard.

The PRT used for design standards by the American Association of State Highway and Transportation Officials (AASHTO) was derived from research done some time ago and includes 1.5 sec for perception and decision and 1.0 sec for making the response, for a total of 2.5 sec.[15]

The complexities of driver PRT are often not appreciated by those who use the concept. The total time to complete an appropriate response involves the following elements: detection (is there something there?), identification (what is it?), decision (what should I do?), response execution (including movement time—e.g., getting the foot from accelerator to brake, and completing the response of pressing the brake), and time for the vehicle to respond (come to a stop or move laterally to avoid an obstacle). The last element is not a component of PRT, but is part of the total time required for the vehicle to come to a stop or move laterally to avoid a collision. All of this takes considerable time. This time is increased under conditions where the hazard is unexpected (which is most of the time in emergency situations), the driving scene is visually complex or dark, and when the driver is impaired, distracted by other components of the driving task, not paying attention, fatigued, or elderly. Hence the need to take into account not the "average" driver, but the one whose performance is much below that of most others on the road.

The complexities of driver reaction time have been described in detail by Triggs,[16] who points out that a PRT in excess of 2.5 sec should be the standard in design of roads, traffic control devices, etc. He examined PRT in a variety of driving situations ranging from braking to traffic signs and the presence of police to emergency situations (e.g., braking by a lead vehicle). The 85th-percentile reaction times to brake ranged from 1.26 sec to over 3 sec, and 7 of the 12 situations tested produced times greater than 2.5 sec, suggesting that the standard of 2.5 sec is too low.

An examination of the current standards for PRT was carried out by Hooper and McGee.[17] On the basis of available literature on the topic they determined that the current time of 2.5 sec was too short, in view of the complexities of the driving task. They suggest 3.2 sec as a more reasonable figure.

Olson[18] has also summarized some of the literature on PRT and drawn similar conclusions. The element of surprise increases PRT by about one-third to one-half, emphasizing the need to be cautious in the interpretation of much of both the laboratory and the roadway research on this topic. Surprise is a typical component of the emergencies that arise in a driving task.

Taoka[19] has also challenged the 2.5-sec PRT used by AASHTO in design considerations following a review of recent work on this topic. He points out that the research on which the 2.5-sec figure was based has definite limitations. Specifically, subjects were alert (expecting to have to make a braking response), young (usually under 30), and driving in an uncluttered environment during daylight and in good

[11] W.L. Mihal and G.V. Barrett, "Individual Differences in Perceptual Information Processing and Their Relation to Automobile Accident Involvement," *Journal of Applied Psychology*, 61 (1976), 229–233.

[12] R. Loo, "Individual Differences and the Perception of Traffic Signs," *Human Factors*, 20 (1978), 65–74.

[13] R.R. Mourant and T.H. Rockwell, "Strategies of Visual Search by Novice and Experienced Drivers," *Human Factors*, 14 (1972), 325–335.

[14] C.S. Lim and R.E. Dewar, "Cognitive Ability and Traffic Accidents." Proceedings of the 10th Congress of the International Ergonomics Association, Sydney, Australia (1988), pp. 593–595.

[15] "A Policy on Geometric Design of Highways and Streets," American Association of State Highway and Transportation Officials (Washington, DC, 1990).

[16] T. Triggs and W.G. Harris, "Reaction Time of Drivers to Road Stimuli," *Human Factors Report HFR-12* (Clayton, Australia: Monash University, 1982).

[17] K. Hooper and H. McGee, "Driver Perception-Reaction Time: Are Revisions to Current Specification Values in Order?" *Transportation Research Record*, 904 (1983), 21–30.

[18] P.L. Olson, "Driver Perception Response Time" (SAE Paper No. 890731, 1989), 67–77.

[19] G.T. Taoka, "Statistical Evaluation of Brake Reaction Time," *Compendium of Technical Papers* (52nd Annual Meeting of ITE, Chicago, August 1982), pp. 30–36.

weather. Such conditions are not representative of those found in typical driving. On the basis of his analysis of these problems and the existing research, Taoka concludes that a PRT of 3.5 sec would be more appropriate to take into account the 85th-percentile driver. Even using the 85th percentile as the design standard, there will still be 15% of drivers who will be unable to respond within this time frame.

Discussions of PRT assume the process begins when the driver first detects the object of concern. However, active search often precedes the beginning of the detection/perception phase of information processing. Alexander[20] points out the need to take this factor into account, challenging the idea that the process of responding to a hazard begins with detection. A certain amount of time is required for head and eye movement, as the driver scans the roadway environment for information. In some cases it takes on the order of seconds (not fractions of a second) for a successful search, even when the hazard, traffic sign, or whatever is readily visible. This is especially the case with older drivers and in complex environments.

On the basis of the work on PRT and its recent reevaluation and criticism, it is clear that caution must be exercised in applying this concept to the design of roadways and traffic control devices. More time than is typically allowed for emergency responses will be required by a sizable proportion of drivers. The implications for design and placement of traffic control devices, intersections, and curves are clear.

Sight distance

Directly associated with PRT is an important element in the design of roads and the location of traffic control devices—the concept of *sight distance.* There are two types—stopping sight distance (SSD) and decision sight distance (DSD). With respect to the former, AASHTO[21] states that "The minimum sight distance available on a roadway should be sufficiently long to enable a vehicle traveling at the design speed to stop before reaching a stationary object in its path." It should be at least that required for a "below-average operator or vehicle to stop in this distance." It is interesting to note that the needs of the below-average driver are mentioned, but without defining it as the 85th percentile, or some such level, as is often used in other design criteria. The SSD is the sum of the distance the vehicle travels from the time the driver sights the hazard to the instant the brakes are touched plus the distance required to stop after initial brake activation. The first is brake reaction distance and the second is braking distance. The former is determined by several factors—object distance, visual acuity, reaction time of driver, visibility of the object, and type and condition of the road. The criteria used for establishing SSD are that an object 6 inches high should be visible from a driver's eye height of 3.5 ft.

The AASHTO geometric design book indicates that, based on relevant literature, the minimum brake reaction time should be at least 1.64 sec but that, since much of the

---

[20]G.J. Alexander, "Search and Perception-Reaction Time at Intersections and Railroad Grade Crossings," *ITE Journal* 59, 11 (1989), 17–20.

[21] American Association of State Highway and Transportation Officials, *A Policy on Geometrical Design of Highways and Streets* (Washington, DC: 1990).

TABLE 1–3
Decision Sight Distances

Design Speed (MPH)	Decision Sight Distance for Avoidance Maneuver (ft)				
	A	B	C	D	E
30	220	500	450	500	625
40	345	725	600	725	825
50	500	975	750	900	1025
60	680	1300	1000	1150	1275
70	900	1525	1100	1300	1450

The following are typical avoidance maneuvers covered in the above table:
- Avoidance Maneuver A: Stop on rural road.
- Avoidance Maneuver B: Stop on urban road.
- Avoidance Maneuver C: Speed/path/direction change on rural road.
- Avoidance Maneuver D: Speed/path/direction change on suburban road.
- Avoidance Maneuver E: Speed/path/direction change on urban road.

SOURCE: *A Policy on Geometric Design of Highways and Streets* (Washington, DC: American Association of State Highway and Transportation Officials, 1990).

research was done under conditions that were better than many driving situations, the figure should be 2.5 sec to take into account the complexity of driving and account for the majority of drivers. This is still not adequate for the most complex situations, as mentioned above. Braking distance depends on a number of physical factors such as road surface condition, tire type and pressure, presence of moisture, etc.

Stopping sight distance is often inadequate in complex situations where information is difficult to perceive or where unusual or unexpected actions are needed. Drivers need time to search for, detect, and interpret information; to decide what to do; then to initiate action. "Decision sight distance is the distance required for a driver to detect an unexpected or otherwise difficult-to-perceive information source or hazard in a roadway environment that may be visually cluttered, recognize the hazard or its threat potential, select an appropriate speed and path, initiate and complete the required safety maneuver safely and efficiently."[22] The values of DSD are greater than SSD because the former gives the driver an additional margin of error (see Table 1–3). Additional distance is required in a variety of situations: visual noise, competing information, (such as traffic control devices, advertising signs, and roadway elements), poor weather conditions, etc. Advance warning from traffic signs is often necessary where adequate DSD cannot be provided by the roadway alignment.

Driving at night

Night driving presents special problems, particularly for the elderly and others with certain visual problems. Efforts at clearly defining the nighttime visual environment have not been very successful because of its complexity, making it difficult to determine the major environmental causes of night accidents.

Night myopia is a fairly common problem in which people have difficulty fixating objects on the road ahead. Where the visual field has little stimulation there is a natural tendency for the eyes to assume a fixation and accommodation distance much closer than the road (often 3 to 6 ft away). This puts objects on the road out of focus. A more obvious

---

[22] Ibid.

problem is that there is much less information available to the driver at night to judge speed of the vehicle and to detect hazards on or near the roadway. Most drivers travel on unlit roads at speeds too fast to detect hazards with the illumination available from their headlights.

A significant number of drivers encounter difficulties in night driving. About 25% of those queried in an Australian survey[23] indicated difficulties, while 6% did not drive at night for this reason. This was the case more often for women and for drivers over 70.

Dark adaptation (the adjustment of the eyes to low light levels) may take some time when one suddenly leaves a brightly lit urban area. It has been suggested that highway illumination could be reduced gradually rather than suddenly to ease this problem. Changes in color vision also occur under low illumination, making some color codes on signs less effective. Certain types of street lights will distort colors, making green signs look black and causing some cars to appear the same color as the roadway, thus reducing their conspicuity.

A major problem at night is reduction in contrast sensitivity, which makes it difficult to see even large objects when they do not stand out against the background. This problem is common among elderly people and those with medical problems such as multiple sclerosis. It is apparent that another factor is alcohol use, which in itself can impair already impoverished vision at night. Fatigue is also a factor in many nighttime accidents.

All of these factors that contribute to the difficulties of seeing the roadway environment at night make it obvious that care must be taken to design roads, traffic control devices, and vehicles so that adequate time and distance are provided for the driver to respond safely.

Fatigue

One of the human conditions that contributes to traffic accidents is driver fatigue. The extent of this problem is not known, as it is difficult to document it as a causal factor, and its effects are often combined with those of alcohol and darkness. The issue has generated a good deal of interest and controversy in recent years, as it has been identified as a significant factor in truck accidents. A study of 225 truck accident reports by the AAA Foundation for Traffic Safety indicated that fatigue was a "probable/primary cause of 41% of heavy truck accidents."[24] One of the main problems associated with truck driver fatigue is that the limits on hours of service (15 consecutive hours on the job in many areas, for example) are easy to violate and difficult to monitor.

Hulbert[25] reported that among a sample of 126 drivers nearly half admitted that they fell asleep or dozed on one or more occasions during extended periods of driving. Fatigue is a significant safety problem in long-distance and nighttime driving. Fatigue can be either task-induced (brought on by the demands of the driving task) or due to sleep deprivation. Driving under high stress conditions can also induce fatigue. Typical indices of fatigue are steering wheel reversals, changes in acceleration, lapses in attention, and subjective feelings of sore muscles and boredom. According to Nelson[26] fatigue develops in three stages: (1) realignments of the body adjustments in posture, (2) discrete symptoms such as awareness of difficulty in maintaining performance, difficulty in focusing, mental blocking, boredom, distractability, and feelings of aggressiveness or hostility, and (3) a strong desire to discontinue driving.

Unlike the effects of heavy work, the effects of driving are more subtle, making physiological measures of fatigue imprecise. Thus the subjective perceptions of drivers (measured by questionnaires) become important. Nelson[27] indicates that complaints of drivers increase linearly as a function of driving time, and that introverted drivers, as compared to extroverted ones, experience less fatigue at the beginning of a driving task but more at the end.

In a study of the combined effects of fatigue and alcohol[28] subjects operated a driver trainer until they found the task too aversive to continue. The time driven was much less for those with high blood alcohol content (BAC) levels (greater than 0.08) than for the low (less than 0.08) or no alcohol content. However, alcohol use resulted in a greater tolerance for personal stress. It appears that drinking drivers may be aware of distress that precedes feelings of fatigue, but do not understand its significance. Awareness of these negative subjective states might aid in reducing driver error.

Alcohol and drugs

The single greatest "human condition" factor that influences traffic safety is alcohol impairment. A great deal has been written on this subject and considerable effort has been put into alcohol countermeasure programs. The proportion of accidents involving alcohol depends on several factors and varies from one country to another and between locations within a country. There are difficulties in making comparisons across countries because of differences in data-recording procedures and in the definition of a "fatal" traffic accident (in some countries the victim must die at the scene of the accident, in others within 48 hours, one month, or one year). With these inconsistencies, meaningful comparisons are impossible. In addition, the legal definition of "impaired" varies. For example, it is a BAC of 0.05 in Australia, 0.08 in Canada and 0.10 in the United States. Nevertheless, it is clear that alcohol impairment is a very significant problem in traffic safety. Statistics reported in an Organization for Economic Co-operation and Development (OECD) review (1978) show that the percentage of

[23] J.M. Stewart, B.L. Cole, and J.L. Pettit, "Visual Difficulty Driving at Night," *Australian Journal of Optometry,* 66 (1983), 20–24.

[24] AAA Foundation for Traffic Safety, "A Report on the Determination and Evaluation of the Role of Fatigue in Heavy Truck Accidents" (Falls Church, VA: 1985).

[25] S. Hulbert, "Effect of Driver Fatigue." In T.W. Forbes, ed., *Human Factors in Highway Traffic Safety Research* (New York: John Wiley, 1972), 288–302.

[26] T.M. Nelson, "Personal Perceptions of Fatigue." In H.C. Foot, A.J. Chapman, and F.M. Wade, eds., *Road Safety* (New York: Praeger, 1981), 121–187.

[27] Ibid.

[28] T.M. Nelson, C.J. Ladan, and D. Carlson, "Perception of Fatigue as Related to Alcohol Ingestion," *Waking and Sleeping,* 3 (1979), 115–135.

fatally injured drivers with BAC's over 0.05 were: England and Wales 43%; Canada 41%; U. S. 58%.[29]

Among the individual driver variables that determine the influence of alcohol are age and drinking experience. Younger (16–34) and older (55 and above) drivers are more likely to be involved in alcohol-related accidents than are those in the intermediate age groups. Those with less drinking experience (drink once a month or less) are also more likely to be in an alcohol-related accident. The combination of youth, inexperience in driving, and inexperience in drinking is particularly dangerous.

There are several "psychological" variables that need to be considered. It is often assumed that the main detrimental effect of alcohol is slowing of motor responses and increase in risk taking. The evidence for the latter is weak and the former is much less important than the impairment in information-processing ability. Reduction in those abilities due to alcohol consumption is reflected in the following deficits:

1. Less efficient scanning of the driving environment
2. Narrowing of visual attention and increased duration of eye fixations, resulting in longer times required to get information from the roadway scene
3. Reduced information from peripheral vision when engaged in doing another task in central vision (which is typical of the driving situation)
4. Tracking (e.g., steering) while doing another task
5. Slower recovery from glare.

The main point is that the impairment is primarily in the taking in and processing of information needed for the driving task, rather than with sensory functions, such as visual acuity, or motor skills. At very high BAC levels these functions are also impaired.

The effects of alcohol are exacerbated in combination with other variables such as darkness, fatigue, and certain drugs (including many drugs taken for medicinal purposes). Many drivers are unaware of the additional hazards associated with drinking when on medication, although they are supposed to be warned of these hazards when medications are prescribed.

The implications of information about alcohol and drug impairment for the traffic engineer may seem obscure, but some work has shown that run-off-the-road accidents on curves may be reduced by the application of specific traffic control devices—chevron alignment markers and wider pavement edge lines.[30]

Blaschke and colleagues[31] have summarized the literature on the effects of alcohol and other drugs on driving; they conclude that a variety of driving-related behaviors are impaired differentially, depending on the BAC levels (see Figure 1–6). Note that complex information processing

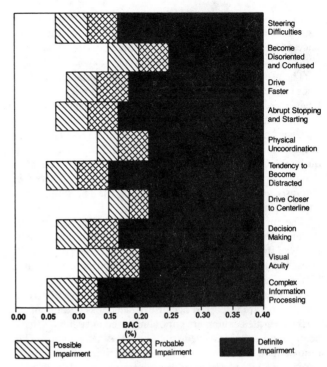

Figure 1–6. BAC at which the effects of alcohol on factors related to driving typically begin.
SOURCE: J.D. Blaschke, M.E. Dennis, and F.T. Creasy, "Physical and Psychological Effects of Alcohol and Other Drugs on Drivers," *ITE Journal,* 59 (1987), 37.

begins to deteriorate at levels as low as 0.05 and that several other factors may be impaired at levels below 0.10.

Among the perceptual abilities reduced by alcohol is sign reading, as demonstrated by a study done in South Carolina.[32] Drivers who were sober or with BAC levels of 0.08 or 0.13 drove on a test track at night and the distance at which they could read signs was measured. No differences were found between sober drivers and those at the 0.08 level, but at the 0.13 level drivers had to be about 13% closer than sober drivers to read the signs. High reflectance of the signs increased reading distance under all conditions, especially the 0.08 condition.

The influence of alcohol and sleep deprivation on vehicle control was examined by Huntley and Centybear.[33] Fine steering reversals (2 degrees), course steering reversals (12 degrees), and brake and accelerator use were recorded as subjects drove a serpentine course under the influence of 0 or 0.10 BAC, and after a normal night's sleep or 29 hours' sleep deprivation. Alcohol resulted in an increase of fine steering, accelerator use, and speed changes; however, sleep deprivation had no effect. An interesting additional finding was that the magnitude of the alcohol effect was directly related to extroversion of the driver. These researchers point out the need to take into account this personality characteristic

[29] "The Role of Alcohol and Drugs in Road Accidents," *Organization for Economic Co-operation and Development* (Paris, September 1978).

[30] I.R. Johnson, "The Effects of Roadway Delineation on Curve Negotiation by Sober and Drunk Drivers" (Australian Road Research Board Research Report No. 128, April 1983).

[31] J.D. Blaschke, M.E. Dennis, and F.T. Creasy, "Physical and Psychological Effects of Alcohol and Other Drugs on Drivers," *ITE Journal,* 57, 6 (1987), 33–38.

[32] J.A. Hicks, "An Evaluation of the Effects of Sign Brightness on the Sign-Reading Behavior of Alcohol-Impaired Drivers," *Human Factors,* 18 (1976), 45–52.

[33] M.S. Huntley and T.M. Centybear, "Alcohol, Sleep Deprivation, and Driving Speed Effects upon Control Use During Driving," *Human Factors,* 16 (1974), 19–28.

when studying the effects of alcohol on driving. The issue of alcohol use becomes complicated by the fact that the influence of both alcohol and drugs is enhanced by social context, motivation, and expectation.

The use of edgeline pavement markings, especially wider ones (8 inches as compared to 4 inches), has been shown to have a benefit on steering for impaired drivers on rural roadways at night.[34]

Drugs commonly examined as they relate to driving are amphetamines, barbiturates, cannabis, narcotics, and tranquilizers. The effects of each of these classes of drugs on driving have been reviewed in a special issue of *Accident Analysis and Prevention* (February 1976). The major conclusions will be examined briefly.

The influence of amphetamines on driving is not well documented, but laboratory data suggest that basic skills involved in the driving task are not detrimentally affected.[35] However, these drugs may induce overconfidence and thus increase risk-taking behavior. Evidence shows that amphetamines can enhance some relevant skills, especially in people who are impaired because of sleep loss or alcohol.

Barbiturates, which are commonly used as sedatives, have been found to be involved in 2% to 9% of accidents.[36] This type of drug has been found to degrade driving skills (motor, perceptual, and tracking task performance, as well as vehicle handling) in laboratory tests. Impairment is magnified when barbiturates are taken in combination with alcohol.

Derivatives of cannabis, such as marijuana and hashish, can produce strong psychological effects on the user. It appears that many people drive under the influence of marijuana, but direct evidence showing links between its use and traffic accidents is scarce. There is evidence, however, that it impairs perceptual and attentional skills and possibly tracking ability. Risk taking, as measured in driving simulators, is not impaired.[37]

Evidence related to the influence of narcotics and driving comes mainly from driving records and interviews, rather than from experiments on the topic. The relevant work suggests that narcotic users do not have poorer driving records than do drivers of similar age in the general population.[38]

With the increase in the use of tranquilizers in the last three decades, there has been an interest in their effects on driving. Research on the issue shows an increase in accident risk among users of tranquilizers. Among the functions impaired are psychomotor skills and information processing.[39] The combined effects of tranquilizers and alcohol cause greater impairment than either one by itself.

Elderly drivers

There has been a dramatic increase in research on the elderly driver in the past decade, and for obvious reasons, for the absolute numbers and proportion of older drivers have increased substantially. One result of this increased activity has been the establishment by the National Academy of Science of a Committee for the Study on Improving Mobility and Safety of Older Persons. Experts from a wide variety of disciplines were invited to provide input to the committee. Following a two-day colloquium on the subject, a major report was published with reviews of the literature on topics ranging from renewal licensing and crash protection to traffic sign conspicuity and roadway delineation.[40] This report is an excellent source of information and recommendations about older road users and their needs and limitations.

Particular concern must be addressed to the needs of older drivers. The proportion of older drivers is expected to increase dramatically over the next 40 years because of the maturing of the baby boom generation and improvements in health care. Figure 1–7 shows the expected "squaring of the population pyramid" with expected changes in the age distribution of the U.S. population over the period from 1950 to 2030. By the year 2000, 13% of the American population will be over age 65, and as much as 20% of the population could be over age 65 by the year 2030.

Statistics have indicated that accident rates increase after the age of 55 and that specific types of violations (those

[34] Anonymous, "Engineering the Way Through the Alcohol Haze," *ITE Journal* 50, 11 (1980), 12–15.

[35] P.M. Hurst, "Amphetamines and Driving Behavior," *Accident Analysis and Prevention*, 8 (1976), 9–13.

[36] S. Sharma, "Barbiturates and Driving," *Accident Analysis and Prevention*, 8 (1976), 27–31.

[37] H. Moskowitz, "Marijuana and Driving," *Accident Analysis and Prevention*, 8 (1976), 21–26.

[38] N.B. Gordon, "Influence of Narcotic Drugs on Highway Safety," *Accident Analysis and Prevention*, 8 (1976), 3–7.

[39] M. Linnoila, "Tranquilizers and Driving," *Accident Analysis and Prevention*, 8 (1976), 15–19.

[40] Transportation Research Board, "Transportation in an Aging Society," *Special Report 218* (Washington, DC: 1988).

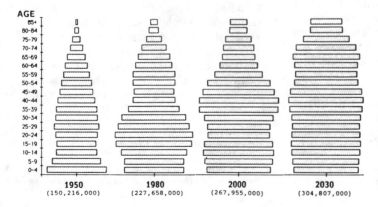

Figure 1–7. Squaring of the U.S. population pyramid, 1950–2030.

SOURCE: *Transportation in an Aging Society: Improving Mobility and Safety for Older Persons,* Special Report 218, (Washington, DC: Transportation Research Board, 1988), p. 21.

involving traffic signs, right of way, and turning left) occur. In addition, most accidents involving older drivers take place in ideal driving conditions, as the elderly drive much less at night and less in heavy traffic.

The world of the elderly driver and pedestrian is substantially different from that of the young person. The one limitation of elderly drivers, which has probably received the most attention by the traffic safety community, is their failing sensory and information-processing capacities. The abilities known to deteriorate with age as they might relate to traffic safety are:

1. decrease in visual acuity starting at age 40
2. less light gets into the eye as pupil size decreases and the lens yellows with age (about one-third of the light that enters the eye of a typical 20-year-old)
3. glare sensitivity increases and recovery takes longer
4. contrast sensitivity is poor
5. night vision is reduced
6. more time is required to change focus
7. eye movements are slower
8. medication for visual problems may interfere with vision in a variety of ways
9. the following medical problems reduce vision:
   a. glaucoma—loss of peripheral vision (drugs for it may reduce pupil size and sensitivity)
   b. cataracts—reduce light transmission and scatter light, leading to glare sensitivity
   c. macular degeneration—loss of central vision
   d. diabetes—rapid changes in refractive error of the eye; often leads to cataracts.

Several possible solutions to these problems have been suggested, including higher levels of illumination on roads, more powerful headlights, tinting the top part of windshields (rather than the entire windshield), larger and brighter traffic control devices such as pavement markings and signs, making drivers aware of poor vision and its consequences, ensuring correct prescriptions for glasses, and restricted driving licenses (e.g., daytime driving only).

The limitations associated with vision are easy targets for study, as they eventually affect all of us. Among the general population there has been little solid evidence linking sensory deficits and driving ability (as measured by accident rate). Where a relationship has been found it is usually in drivers over age 50.[41] In particular, the sensory abilities that appear to be significantly correlated with accident rate are angular movement and the speed with which targets can be detected in peripheral vision.

Other sensory abilities are not unimportant in the driving task, but they are not reliable predictors of accidents in the elderly. One reason for this is that many older drivers compensate for such limitations by driving more slowly, and less at night, in heavy traffic, and in bad weather. The abilities described above, as well as the many studied by Henderson and Burg, which did not correlate with accidents, are among the more basic information-processing skills. Higher-order abilities, such as attention and the ability to time-share tasks,

are closely related to driving, as indicated above. The abilities to attend selectively, to divide attention, and to avoid distraction are known to diminish with age.

Among the reasons why older drivers might have difficulty with information processing are less efficient scanning behavior and eye movements, diminished visual field size, difficulty in selective attention, and slower decision making. The combination of slower information processing, decision making, and reaction time obviously creates difficulties for the elderly driver, unless some compensation for these deficits is made. In addition, physical problems such as arthritis restrict neck and head movement, making shoulder checks difficult or impossible. Minor memory deficits may also play a role in driving difficulties—forgetting where to turn or what was on the traffic sign at the previous intersection, for example.

Essential sources of information from the roadway are traffic signs and other traffic control devices. Considerable work has been done on the perception of signs. It is evident from this that elderly drivers do have more difficulty than do younger ones. Two important characteristics of signs are their conspicuousness and legibility. Nighttime conspicuity was studied by Olson,[42] who pointed out that brightness is the main factor in the attention-getting qualities of a sign. How bright should signs be? In general, the brighter the better, but other factors determine the likelihood of detecting a traffic sign—visual complexity of the roadway scene, contrast with the background, sign size and placement, and attention on the part of the driver. Adequate advance warning of the presence of sign information and proper placement of signs can greatly enhance their effectiveness for older drivers. In order for a traffic sign to be noticed in a visually complex environment, its reflectivity at night must be increased by a factor of about 10 to achieve conspicuity equivalent to that found in a low-complexity environment.

The effects of age on sign legibility at night have been examined by Sivak, Olson, and Pastalan,[43] who had subjects drive toward signs and identify the orientation of the letter *E* in a normal configuration or its reverse. Legibility distances for older drivers (over age 61) were found to be 65% to 77% of those for the younger drivers (under 25). Very high contrast between letters and background were not optimal, especially for the older group. This is due to irradiation, or visual impingement of brighter areas of a sign onto the darker areas, making the sign legend less legible. This same research found that older subjects required signs to be about three times the brightness level needed by younger subjects to read the sign at the same distance.

In a review of the limitations of older drivers, Mace[44] refers to the concept of Minimum Required Visibility Distance, the distance from a traffic sign required by drivers in order to detect, understand, make a decision, and complete a vehicle maneuver before reaching the sign. This distance is

[41] Henderson and Burg, "Vision and Audition in Driving."

[42] P.L. Olson, "Problems of Nighttime Visibility and Glare for Older Drivers, *Effects of Aging on Driver Performance,* (Society of Automotive Engineers, SP-76, 1988), 53–60.

[43] M. Sivak, P.L. Olson, and L.A. Pastalan, "Effect of Driver's Age on Nighttime Legibility of Highway Signs," *Human Factors,* 23 (1981), 59–64.

[44] D. Mace, "Sign Legibility and Conspicuity." In *Transportation in an Aging Society,* Transportation Research Board, Special Report 218, (Washington, DC: 1988), 270–290.

increased for older drivers for the following reasons: slower detection time (due to distractions from irrelevant stimuli, etc.); increased time to understand unclear messages such as symbols, poorer visual acuity, and contrast sensitivity; slower decision making; visual complexity of the roadway scene; inadequate sign luminance and legend size; and low driver alertness. The specific required distance would depend on vehicle type, speed limit, and roadway conditions.

The design of many traffic control devices and roadway features assumes visual abilities and reaction times that older drivers typically are not able to meet. Perhaps the most striking of these is the standard for size of lettering on traffic signs, which for decades has been dictated by the guideline of 1 inch of letter height per 50 ft of reading distance. The visual acuity required to read print this size is approximately 20/23, while the minimum acuity for a driver's license in most jurisdictions is 20/40. Many drivers over age 75 have acuity close to this limit. A better criterion for letter size would be 1 inch per 30 ft of distance.

Older drivers are overrepresented in left-turn accidents, which reflects, in part, difficulties in judging distances and speeds of oncoming vehicles. Greater time to cross intersections while turning may be required, suggesting the need for more protected left-turn signals where there are large numbers of older drivers.

In view of the limitations experienced by many older drivers, there should be greater effort put into making them aware of their deficiencies, and into reeducation programs. Such programs should be geared to the needs of the older driver, not just a repeat of driver education given to novices. The curriculum should stress their limitations, present the material more slowly, and relate it to their driving experience. Of importance here is information that drivers may not know, such as difficulties the older driver has in estimating the passage of time (to judge speed of other vehicles), difficulty in processing a lot of information at one time, reduced night vision, and detrimental effects of medication.

The "55 Alive/Mature Driving" program in the United States has been successful in increasing driver knowledge and reducing violations and accidents. The course content includes:

1. accident experiences of the elderly
2. physical changes relating to driving performance
3. common hazards encountered
4. rules of the road
5. effects of alcohol and medications
6. adverse driving conditions
7. characteristics of other road users
8. vehicle maintenance, licensing, and insurance.

Social factors

A great many individual differences exist among drivers and pedestrians. Unfortunately, the influence of these on road-user behavior is not well understood. Social-psychological aspects have been generally ignored, as stress has been on motivation, personality, information processing, and motor skills. However, driving is very much a social behavior, as we seldom drive alone on empty roadways. Variables that are social in nature, with examples, are:

1. communication between drivers (signaling lane change, blowing horn, waving)
2. expectations about the behavior of other road users (we expect others to obey rules of the road, be courteous, and know how to drive properly)
3. imitation of other road users (drivers increase speed when others are speeding)
4. influence of the presence of others (there is more anxiety when in heavy traffic: distraction by passengers is a common problem)
5. prejudice toward other road users (belief that elderly drivers are slow and stupid and that young males are reckless drivers)
6. invasion of one's personal space (negative reactions when other drivers follow too close)
7. influence of the norms of society (social pressure to obey signs and signals; pressure from peers to drink and drive).

The elusive concept of attitudes has been the subject of much attention by those who promote traffic safety, and is generally considered to be the "cause" of many accidents, especially among young drivers. Measurement of attitudes can be difficult, as methods tend to use self-report questionnaires. Even if results of a questionnaire are valid, there is often little correlation between these "attitudes" and driving behavior. For this reason attitudes often fail as predictors of behavior. Most attitudes are learned early in life, from parents and other authority figures; therefore it is important to teach children (as young pedestrians and cyclists) the proper attitudes (through example and information) toward use of the road.

Attitudes toward driving may reflect personal values and emotions (e.g., aggression, defiance). Attitudes toward taking risks on the road may be influenced by a general lack of "regard for life," which has come about in recent decades, with the witnessing of violence and death daily in the mass media. There is also the possibility that as cars become safer (air bags, seat belts, padded dashes, collapsible steering columns, etc.) the driver takes more chances, believing that a crash will not lead to serious injury, and that property damage will be paid for by the insurance company. This idea reflects Wilde's[45] risk-compensation theory, which states that road and vehicle improvements may influence the traffic accident rate temporarily, but that the accident rate will be more or less constant over time, as road users compensate for these improvements by engaging in more risky behavior. The target level of risk for any individual (which is determined by a number of factors, such as age and motivation) will remain the same, in spite of safety measures introduced. Thus far the theory has received little support. For example, Evans[46] examined a number of short-term and long-term trends in

[45] G.S. Wilde, "Risk Homeostasis Theory and Traffic Accidents: Propositions, Deducations, and Discussion of Dissension in Recent Reactions," *Ergonomics,* 31 (1988), 441–468.

[46] L. Evans, "Risk Homeostasis Theory and Traffic Accident Data" (Report GMR-4910), *General Motors Research Laboratories,* Warren, MI (1985).

accident data and found that accidents did decline following improvements in road and vehicle design.

Drivers' perception of their own driving ability is inflated, according to a study by Svenson,[47] who asked university students from the United States and Sweden to judge their skill in driving and how safe they were as drivers. On the judgments of safety, half of the subjects believed they were among the top 20% (U.S.) or 30% (Sweden) of drivers in their respective groups. Among the U.S. drivers, 46.3% saw themselves as among the most skillful 20%, while for the Swedish drivers this figure was only 15.5%. Ninety-three percent of the U.S. sample and 69% of the Swedish sample said they were more skillful than the average driver in their group. In view of these misperceptions of driving ability, it is little wonder that many drivers take chances on the road.

Motivation

Much of the work on motivation has been directed toward young drivers and their attitudes. Basic to motivation is the reason for driving. If it is just to get from point A to point B the driving behavior may well be different than if one drives to express hostility, to show off to peers, or to be alone for a while. It has been found that teenagers often use driving to express feelings of competitiveness, initiative, independence, control over their environment, or excitement. Appeals to the thrill of driving are a major motive used to sell cars, by manufacturers who claim that they "build excitement" into their new models.

In their book on driving behavior, Näätänen and Summala[48] put considerable stress on the "extra motives" in driving, including internalized values such as aggression, competitiveness, self-esteem, and outlets for frustration. The perception of hazards may be distorted when our motive is to get somewhere in a hurry. The "hurry hypothesis" has been suggested as an explanation for many accidents (but it is difficult to confirm this, as few drivers will admit to having an accident because of this motive).

Emotions

It is difficult to make a distinction between motivation and emotion in driving, as they influence each other. Emotional disturbances influence many aspects of our lives, including driving. In-depth investigations of accidents that have involved a "psychological autopsy" (an examination of the emotional state of the driver and other psychological factors prior to the accident) often report that drivers were to some extent "impaired" by their emotions. Estimates are that between 10% and 35% of drivers in accidents are emotionally upset or tense at the time of the accident. One study showed 20% of drivers in fatal crashes were emotionally upset during the 6 hours before the accident. Such information comes from interviewing relatives, friends, and colleagues. For the majority of male drivers the problem was associated with a domestic quarrel.

McMurray[49] examined the 7-year driving records of 410 drivers who were undergoing a divorce. Accident and violation rates were found to be higher for these drivers, during the period 6 months before and 6 months after their filing for divorce, as compared with their own earlier records and with the records of the general driving population. Violation rates were greatest during the months after filing for divorce.

The psychological well-being of drivers is influenced by many factors in the transportation system, including roadway design (e.g., narrow, winding roads), traffic conditions (e.g., rush-hour congestion), weather (rain and snow), environmental conditions (air pollution), and vehicle conditions (noise and vibration). Stokols and Novaco[50] have summarized the literature on this topic and conclude that there is a positive correlation between traffic volume and increases in driver heart rate, blood pressure, and electrocardiogram irregularities. Driving simulator studies have demonstrated the increase in physical arousal associated with complexity of the driving task.

Personality

The search for personality characteristics associated with traffic accidents has been generally unsuccessful. One of the reasons for this is the lack of precision of personality tests. Accident repeaters have often been the targets of study in hope of finding something different about them. In a study examining many personality factors, it was found that there were no differences on some of these factors, while chronic traffic violators were better on 14 characteristics and control subjects were better on 11. In spite of these findings that personality variables do not correlate highly with accidents, some general characteristics do appear to be more prominent among drivers with poor driving records.

The idea that people drive as they live has been popular for some time, and the evidence suggests a relationship may exist between the two. High-risk drivers have been found to be emotionally unstable, hostile, resistant toward authority, tense, emotionally immature, and anxious. It may be difficult to think of a countermeasure for these problems, but we could at least make drivers aware that they may be at higher risk under conditions of emotional stress. Parents of young drivers could use this knowledge to judge whether their children should drive when upset or preoccupied with emotional factors.

Four theories have been postulated to explain traffic accident causation: (1) social maladjustment—poor driving is a reflection of a more general pattern of irresponsible and antisocial behavior and attitudes; (2) personal maladjustment—personal stress determines risky behavior on the road; (3) impulsive noncontrol—involves low ability to cope with risk-taking impulses and a tendency to use driving as an

[47] O. Svenson, "Are We All Less Risky and More Skillful Than Our Fellow Drivers?" *Acta Psychologica,* 47 (1981), 143–148.

[48] R. Näätänen and H. Summala, *Road-User Behavior and Traffic Accidents* (Amsterdam: North-Holland, 1976).

[49] L. McMurray, "Emotional Stress and Driving Performance: The Effect of Divorce," *Behavioral Research and Highway Safety,* 1 (1970), 100–114.

[50] D. Stokols and R.W. Novaco, "Transportation and Well-being," *Transportation and Behavior* (New York: Plenum Press, 1981), 85–130.

emotional release; and (4) information-processing deficit—poor drivers lack adequate speed and accuracy in perceptual and motor skills.

Mayer and Treat[51] gathered data relevant to these theories by administering a battery of 22 tests to 30 drivers who indicated they had three or more traffic accidents in the previous 3 years and 30 who had not been involved in any accidents over that time. The sample, however, was limited to university students, who are younger, fitter, and more intelligent than the average driver. The seven measures that discriminated the two groups were: psychopathology; number comparison (an index of clerical speed and accuracy); citizenship; school socialization; negativism; external locus of control, and attitudes toward risk taking. The last five measures fall in the social-maladjustment category, while the first is a measure of personal maladjustment.

A follow-up study by these researchers involved an in-depth investigation of accidents where interviews were conducted with 177 drivers who were at fault and 110 who were not. The former were higher on social maladjustment, which corresponds with data from the groups cited above who were given the battery of 22 tests. In addition, it was found that drivers committing errors associated with alcohol, inattention, and human conditions were more personally maladjusted than were the control group of drivers.

The concept of "accident proneness" has received a good deal of attention, as some data suggest that certain drivers have more than their share of accidents. It is thought that certain personality characteristics manifest themselves in behavior and emotional needs, which result in accidents; that is, certain personalities predispose people to dangerous behavior. There is, however, an absence of sound evidence supporting the existence of an "accident-prone" personality.[52] Certain people do have more accidents than others, but this is often due to differences in exposure to hazards or else is temporary. The notion of temporary accident proneness makes more sense than a personality trait, which is an enduring aspect of the person's identity. Various situational factors temporarily impair drivers—these include fatigue, depression, confusion, and so forth.[53]

Vehicle design

It is important that the vehicle "fits" the driver, as a pair of shoes or a coat would fit the body. However, large drivers find their vehicles crowded and difficult to enter, while others (especially small females) have difficulty reaching dashboard controls or seeing out of large cars. With the extremes in sizes of both people and vehicles, it is impossible for all cars to fit all people. A standard utilizing the 95th percentile is usually the goal in design of machines, which means that the dimensions are expected to be adequate for all but the smallest 2 1/2% and the largest 2 1/2% of users. This means there are still millions of drivers who are

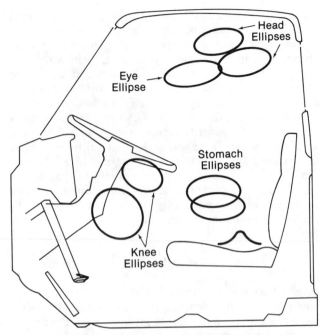

Figure 1–8. The eye ellipse and contours of the positions of other body parts in the work space of truck drivers.
SOURCE: SAE, "U.S. Truck Driver Anthropometric and Truck Work Space: Survey and Recommended Practices," SP-712, Warrendale, PA, p. 20. Reprinted with permission © 1987 Society of Automotive Engineers, Inc.

not comfortable with the dimensions of their respective vehicles.

The placement of controls and displays in vehicles is typically based on the H-point (the hip location), from which physical dimensions of the body such as leg length, reach, and eye height are determined. A major point of reference for visual information (whether from inside or outside the vehicle) is the "eyellipse," which defines the location of the driver's eye (see Figure 1–8). This height above the roadway is a factor in the design of roads and assumptions about decision sight distance. This height was recently lowered from 3.75 ft to 3.5 ft in the United States, but most roads are designed with the older criterion, which presents a safety hazard for people driving small cars. Many smaller, low-volume roads were designed several decades ago when drivers' eyes were much higher above the road than they are today, so hazards on the road could be seen at greater distances.

Parts of the vehicle often obstruct the view of the road or of displays on the dashboard. The rearview mirror is notorious for getting in the way of seeing cyclists. The location of in-vehicle displays and controls often confuses drivers, especially in unfamiliar vehicles (e.g., rented cars). A U.S. study[54] found that operating problems in rented cars occurred with climate controls (23%), shoulder harness (20%), seat belts (15%), and windshield washers (19%). In a reaction-time experiment, drivers often took considerable time finding and activating vehicle controls. Frequently the controls were not

[51] R.R. Mayer and J.R. Treat, "Psychological, Social and Cognitive Characteristics of High-Risk Drivers: A Pilot Study," *Accident Analysis and Prevention*, 9 (1977), 1–8.

[52] F.L. McGuire, "Personality Factors in Highway Accidents," *Human Factors*, 18 (1976), 433–442.

[53] Ibid.

[54] J.J. McGrath, "Driver Expectancy and Performance in Locating Automotive Controls," SP-407, SAE, Detroit (1976).

where drivers expected them to be. Such delays or the activation of the wrong controls can have disastrous consequences when driving at high speed. Greater uniformity of layout of all controls would obviously be better for drivers.

This lack of uniformity also presents problems in understanding the meaning of symbols on dials and controls. There are various versions of the same message, and some do not look like what they are intended to represent. Elderly drivers, in particular, have problems here. As more information becomes available in vehicles, the problem will get worse, unless an effort is made to standardize the presentation of symbols. There will also be the general problem of information overload for drivers, who may become distracted or confused by displays of useless information and complex electronic maps.[55] The potential for accidents attributable to distractions from using electronic maps, cellular phones, computers, intelligent vehicle information, etc., is very real, but little is known about this problem to date.

Tinted windshields are a source of debate over whether they represent a safety hazard. They reduce the light getting to the driver's eye by about 35% and red light by up to 50%.[56] This can cause problems for many drivers, especially the elderly at night.

Glare from bright objects on the outside or the inside of vehicles is also a common source of visual problems. The same applies to reflections of light-colored interiors in the windshield and veiling glare from dirty or scratched windshields.

One simple but effective way to reduce accidents is the use of daytime running lights (DRL), or having vehicle headlights on during the day. Research done in various parts of the world has consistently reported reductions in accidents (from 12% to 41%) for vehicles using DRL, as compared with control vehicles without them.[57] Motorcycles are now required in many states to use headlights at all times; they come on automatically when the ignition is turned on.

Truck drivers

The driving task, the required operator skills, and information needs are very different when it comes to the operation of large trucks. In spite of their specialized training and extensive driving experience, truckers are still susceptible to the difficulties and limitations that influence the average car driver—alcohol and drug use, fatigue, risk taking, information-processing problems, etc. Commercial vehicle operation is different in a number of ways. The purpose of the trip is to get to a specific destination, within a specific time, and by a specific (usually the shortest) route. As a result, truckers would be expected to engage in more thorough and intelligent trip planning than most automobile drivers. Route familiarity will be very high for those driving a regular route, but may be largely unfamiliar for those who deliver parcels for a large department store, move household furniture within an urban area, or are courier drivers.

With the deregulation of the trucking industry and the pressure to move goods more quickly and economically in recent years, there has been an increase in the number and size of trucks on highways. The crash rate of heavy trucks increased 15% from 1982 to 1985.[58] In a survey of 1,300 tractor-trailer operators in California, the majority said that at least 70% of those in their profession regularly break the rules. In addition, 20% reported seeing average speeds of 70 mph or more. Owner-operators were more likely to speed than were fleet operators. Perceived frequency of drug use was also high, as 36% of respondents said truckers sometimes drive under the influence of drugs.

The complexity of vehicle controls and the very different handling characteristics of a large truck make its operation more difficult for several reasons, some of which are related to highway engineering and roadway design. The increased eye height of truckers makes it possible in many situations for them to "read the road" farther in advance than can drivers of smaller vehicles. However, they require this information at a greater distance due to the need for greater stopping and maneuvering distances. The trucker may also have to take in and process more information than other drivers from the roadway and traffic control devices, as well as from inside the vehicle. The potential for information overload due to excessive information inside the vehicle is increasing for automobile drivers, as indicated earlier, but there is an even greater attention demand placed on operators of large, complex trucks, as the number of controls and displays is typically much greater in such vehicles.

The safe operation of large trucks is sensitive to sight distance at intersections (especially uncontrolled ones). According to Fambro and colleagues[59] the sight distances recommended are a function of type of traffic control, type of maneuver (crossing vs. turning), highway geometry (cross section and design speed), driver characteristics (perception and reaction time), and vehicle characteristics (length and deceleration capability). For example, the required intersection sight distances to cross a two-lane 60-mph highway are 800 ft and 1,060 ft, respectively, for single units and large combination trucks.

The truck driver characteristics that involve anthropometry have been examined by the Society of Automotive Engineers,[60] which has provided a detailed account of methods for determining truck driver physical measurements and for measuring work space in trucks. The dimensions of the vehicle itself, as well as driver-selected seat position, eye and head position and other critical dimensions, are reported in detail. Figure 1–8 illustrates the eyellipse and contours of other body parts (the top and rear of the head, left and right knees, and two stomach points) as they relate to operator work space in trucks.

[55] R.E. Dewar, "In-vehicle Information and Driver Overload," *International Journal of Vehicle Design*, 9 (1988), 557–564.

[56] M.J. Allen, *Vision and Highway Safety* (Philadelphia: Chilton, 1970).

[57] D.A. Attwood, "The Potential of Daytime Running Lights as a Vehicle Collision Countermeasure," *SAE Technical Paper Series*, No. 810190, (1981).

[58] Insurance Institute for Highway Safety," *Status Report*, 22, 11 (1987), p. 4.

[59] D.B. Fambro, J.M. Mason, and T.R. Neuman, "Accommodating Large Trucks at At-Grade Intersections." In B.L. Smith and J.M. Mason, eds., *Accommodation of Trucks on the Highway: Safety in Design* (New York: American Society of Civil Engineers, 1988), 104–122.

[60] S.A.E., "U.S. Truck Driver Anthropometric and Truck Work Space: Survey and Recommended Practices," SP-712 (Warrendale, PA: 1987).

A major concern with the safe operation of trucks is fatigue, due typically to long hours of work and stress. A study by the American Automobile Association[61] analyzed 221 truck accidents in five American states. Drivers' pre-accident activities were examined to determine whether fatigue was a primary/probable cause, contributory, or un-related. It was reported to be a primary/probable cause in 41% of the heavy truck accidents. At the time of the study the upper limit imposed by the Bureau of Motor Carrier Safety for total working time was 15 hours. Twenty of the accidents occurred between 5 and 6 A.M. and 15 between 7 and 8 A.M. According to the police reports, 15 of the accidents involved the drivers' sleeping, 11 involved "inattention," and 13 drivers drifted off the road. Only 18.4% of the drivers in these accidents had been driving for less than 9 hours.

Truck driver fatigue is significantly influenced by working irregular schedules. The natural circadian rhythm of the human body (awake during daylight hours and asleep during night hours) comes into play in lowering body temperature and alertness during night work. This, combined with the monotony of night driving, exacerbates the fatigue conditions that are often contributory factors in nighttime truck accidents.

A large nationwide interview survey of some 2,000 truck drivers examined these issues.[62] The median number of hours on duty was 9.4, while 20% of the drivers had worked 15 or more hours on at least one of the previous 6 days. About one-seventh of all driving periods were between midnight and 6 A.M. and nearly 65% of the accidents occurred in the second half of the driving period. An accident involving a dozing driver was about seven times more likely to occur on one of the early morning hours than at other times of the day.

The survey results indicate that drivers on long hauls with irregular schedules showed greater subjective fatigue, physiological stress, and performance degradation, as compared with drivers working similar numbers of hours on regular schedules. Fatigue effects became evident much earlier for drivers on irregular schedules. Similarly, bus drivers on irregular schedules suffered greater subjective fatigue and physiological stress, as compared with those on regular schedules.

In addition to the stress associated with long hours and fatigue is the physical stress resulting from vehicle vibration and poor ride quality of many trucks. Although there has been a good deal of research on whole-body vibration, the long-term effects of this on drivers are not well understood. However, it is known that vibration can cause adverse health effects, probably mediated by the central and autonomic nervous systems and the endocrine system of the body.[63]

Vibration and poor posture may combine to increase the incidence of spinal cord disorders among truck drivers.

Another environmental factor influencing physical and psychological well-being of truckers is noise. Private cars at highway speeds have a noise level in the range of 65–85 dBA, while in commercial vehicles, levels below 80 dBA are often considered "quiet," and levels as high as 90 are found in some trucks.[64] It has been shown that high noise levels can increase cardiovascular problems and that gastrointestinal disorders are a consequence of long-term exposure to high noise levels. Vigilance and memory may also deteriorate under noisy conditions. A heightened state of arousal, with accompanying development of fatigue, can result from noise because of the difficulty in shutting it out from our awareness and the need to put forth greater effort to attend to and process essential auditory information while driving.

Traffic control devices relevant to truck operators

Although most highway signs have information for most drivers, some signs are specific to truck operators (e.g., weight limits, routes for hazardous cargo, truck lanes, reduced speed limits, height limitations under bridges). Hence, they must be on the lookout for these additional sources of information.[65] The greater distance between the driver's eyes and the vehicle headlights can pose a disadvantage in reading signs at night. The retroreflective material on these signs directs light back to its source. Therefore, the greater the distance between the headlights and the driver's eyes, the less intense the sign will appear to the driver, as the light from the sign is being directed back to a location farther from the driver's eyes. The result is shorter sign legibility distance, which is a disadvantage for drivers of large vehicles, as they need such information at a greater distance than do others because of greater stopping and maneuvering distances required. Raising the height of headlights would simply increase the problem of glare for drivers of other vehicles.

A traffic control problem arises when an attempt is made to convey information about safe speed on freeway ramps; this is because the speed will have to be less for large trucks than for cars. Thus a speed sign appropriate for safe negotiation by trucks will be inappropriate for drivers of smaller vehicles. The issue of rollover of trucks with high loads at ramps is under investigation by the Federal Highway Administration in an effort to find appropriate traffic sign messages. One solution is the use of changeable message signs that present a message (preferably flashing) that the truck's speed is too high. This approach has been used at the ends of major highways and at other locations such as low bridges and sharp curves.

Guide sign information, such as route markers, destination names, and street names, must be carefully displayed for truck drivers since the consequences of getting lost (in

[61] AAA Foundation for Traffic Safety, *A Report on the Determination and Evaluation of the Role of Fatigue in Heavy Truck Accidents* (Falls Church, VA, 1985).

[62] J.C. Miller and R.R. Mackie, "Effects of Irregular Schedules and Physical Work on Commercial Driver Fatigue and Performance." In D.J. Oborne and J.A. Levis, eds. *Human Factors in Transport Research,* vol. 1 (London: Academic Press, 1980), 126–133.

[63] R.D. Pepler, C.M. Overbey, and T.J. Naughton, Jr., "Truck Ride Quality and Drivers' Health: Methodology and Development." In Ibid., vol. 2, pp. 58–66.

[64] W. Tempest and M.E. Bryan, "The Effects of Vehicle Noise on Health and Driver Performance." In Ibid., vol. 2, pp. 173–178.

[65] H. Lunenfeld, "Accommodation of Large Trucks." In B.L. Smith and J.M. Mason, eds. *Accommodation of Trucks on the Highway: Safety in Design* (New York: American Society of Civil Engineers, 1988), 89–103.

terms of safety, frustration, and delays) may be more significant for the truck operator than for others.

Pavement marking design may need to be different for truck traffic as well. Passing zone sight distances that are safe for automobiles may be too short for trucks, with their slower acceleration and greater lengths. Some jurisdictions use supplementary signs to warn truckers not to pass at specific locations marked as passing zones.

Signals can also be inappropriate for truck operation. An example here is yellow phases that are too short for trucks to stop in time (especially on downhill approaches). Such problem areas should receive special attention by those implementing traffic signals. Slow truck acceleration may disrupt traffic flow and operations, causing delay and the disruption of signal-light progression on city streets.

Construction and maintenance zones pose special problems for trucks, because closed lanes and detours may be difficult for large trucks to negotiate safely. Median crossovers seem to pose particular problems. For a good summary of these issues, see Graham.[66] One difficulty can arise at night from the use of flashing arrowboards to direct traffic. Their intensity is normally reduced for nighttime operation, but they can still create significant glare in the eyes of a truck driver, as they are often mounted at a height close to that of the driver's eyes. Another problem is the likelihood of large trucks knocking over barriers and cones (not always unintentionally). Work crews need to be alert to this problem, since other drivers could be exposed to significant danger if these traffic control devices are not in place as they should be. Truck traffic also poses a problem in construction zones as a result of the amount of dirt and moisture they deposit on signs, barriers, and beacons.

Motorcyclists

As was the case with trucks, the driving task and the associated risks are quite different for motorcycles from those for automobiles, but for more or less the opposite reasons. The small size and high acceleration capability of motorcycles make them very maneuverable, but in a collision the rider has almost no protection.

Cooper and Rothe[67] have presented a summary of the characteristics and motivations of motorcyclists, based on interviews with motorcyclists in British Columbia. Four major reasons for starting to use motorcycles were identified: a psychological drive toward risk taking and the challenge presented by the operation of these vehicles; for recreation and pleasure; as a result of incomplete social adjustment, where there is a desire to share a way of life considered to be out of the mainstream of society; and for economic reasons, as a mode of transportation. Motorcyclists are more likely than operators of other vehicles to belong to some organization whose function centers around the operation of the vehicle itself. Some see their biking activities in terms of individual freedom from the restraints of society. For them freedom and rebellion are represented by speed.

Of the total sample of 877 riders interviewed, 418 had successfully completed a motorcycle training course and 57 had failed formal training. Complete accident records were available for 98.4% of these. Results showed that nearly a quarter of the untrained riders were not properly insured, compared with fewer than 4% of trained riders. Marital and educational status were similar for the two groups, except that trained riders were more likely to have completed postsecondary education. Trained motorcyclists were more likely to be properly licensed, wear bright clothing when riding, do most of their riding on highways, and wear protective gear while riding. In addition, they were less likely to think helmet protection was overrated and they had less riding experience. There were no significant differences between formally trained and untrained riders on accident or conviction history.

The youngest riders (16–24), males, and unlicensed operators were most likely to have traffic convictions while driving other vehicles. Similarly, motorcycle convictions were most likely for the younger riders. Unmarried riders were also more likely to have such convictions.

A major reason for motorcycle accidents involving other vehicles is failure in information processing on the part of the drivers of the other vehicles. They typically fail to detect the motorcycle in time to avoid the collision. These are classical cases of the "look but fail to see" type of accident. An analysis of these difficulties and the proposed countermeasures by Wulf and colleagues[68] suggest that much of the research is based on inadequate methodology, and they conclude that the effectiveness of many of the safety measures on the roadway remains to be demonstrated. Among the reasons for information-processing failures are what Wulf and co-workers call low levels of "cognitive conspicuity," which is a function of the interests and experiences of the observer (i.e., the meaning of the stimulus). Motorcycles on public roads are relatively rare in comparison to cars and trucks, so the element of expectancy plays a role here. In addition, a number of characteristics of the motorcycle itself and the rider are determinants of conspicuity. "Bikes" are much smaller than other motor vehicles. As a result, their presence, distance, and speed are difficult to judge. Moreover, riders often wear dark clothing, which makes them somewhat inconspicuous.

Excess speed can also be a problem. It is difficult to estimate the speed and distance of a motorcycle because small objects look farther away than do large ones at the same distance. When viewing small vehicles from nearly straight on (e.g., as when starting to turn left in front of oncoming traffic) it is also difficult to judge speed and distance, as the size of the image on the eye (a major cue to distance) changes slowly until the vehicle is quite close. The rate of change of the size of the image on the eye is slower for a motorcycle than for a car.[69] In addition, there is the possibility that the A-pillar of a car or

[66] J.L. Graham, "Design Considerations for Trucks in Work Zones." In Ibid., pp. 74–88.

[67] P.J. Cooper and J.P. Rothe, "Motorcyclists: Who They Are and Why They Do What They Do," *Transportation Research Record,* 1168 (1988), 78–85.

[68] G. Wulf, P.A. Hancock, and M. Rahimi, "Motorcycle Conspicuity: An Evaluation and Synthesis of Influential Factors," *Journal of Safety Research,* 22 (1989), 153–176.

[69] P.L. Olson, "Motorcycle Conspicuity." In G.A. Peters and B.J. Peters, eds., *Automotive Engineering and Litigation,* vol. 1 (New York: Garland Law Publishing, 1984), 435–447.

truck will obscure vision of the bike coming from the right. Hence, drivers often pull out from STOP signs into the path of an oncoming motorcycle because they "did not see" it approaching.

Various measures have been proposed to increase motorcycle conspicuity. These include having bikers wear bright clothing, use of bright fairing on the vehicle and headlights for daytime, as well as retroreflective clothing and tire sidewalls for nighttime conspicuity. Although conspicuity of oncoming motorcycles at night is better than in daylight, the single headlight in darkness looks farther, and it is more difficult to judge its distance, than is the case for other vehicles. Two lights provide an indication of the vehicle's size, and hence both the distance and change in distance over time (speed) are easier for other road users to estimate.

A concern that has received attention in recent years is the increased power of motorcycle engines and the increased use of high-performance bikes. As a result of this concern many jurisdictions restrict novice riders to motorcycles with engines of a maximum size/power. In a thorough examination of the literature on this issue, conducted by the Traffic Injury Research Foundation of Canada,[70] it was concluded that there is no consistent scientific evidence of a relationship between the size of motorcycle engines and motorcycle accidents. Earlier studies showing such a relationship failed to take into account exposure (opportunity for a collision) or used inappropriate exposure indices, such as number of registered motorcycles, rather than distance traveled or conditions under which travel occurred. In addition, the characteristics of the riders (e.g., attitudes, risk-taking behavior, alcohol use) may be different among riders of different size bikes. If these factors are uncontrolled in studies of accidents, no conclusive evidence is available on the relationship between engine size and safety.

Bicyclists

The bicycle operator may have very different characteristics, and certainly has a very different task and set of dangers, than those who drive motor vehicles. Many of the former are children with limited experience in operating in traffic and little appreciation of the hazards they face on the road. Bicycles are used largely for recreation in North America, whereas they are primarily a mode of transportation in many other parts of the world, especially Asia. Thus there may be national differences in cycling behavior and the type of traffic in which cyclists travel.

In response to the rapid increase in bicycle use in North America in the 1970s, many cities built exclusive bicycle paths or bikeway systems. Many were not properly designed or evaluated. Cyclists' perceptions and use of these were studied in a national survey of bicyclists conducted on three types of bicycle facilities.[71] The types examined were: exclusive bicycle paths, restricted right-of-way on a shared facility (e.g., bike lanes on city streets) and shared right-of-way designated as such by traffic signs (sidewalk path shared by pedestrians and city streets designated as bike routes). A total of 606 cyclists were interviewed at 49 different locations in various parts of the United States.

The results showed that bike paths were considered the safest and the signed routes on streets, the least safe. The lack of protection from cars was the major problem with the latter type of facility. Twenty percent of those interviewed on signed bike routes on regular streets did not realize they were on a bicycle route. There were seen to be very few advantages of designated, signed bike routes over regular streets. Recreational riders showed a greater tolerance for gravel paths than did commuters, but riders in general were quick to point out the hazards on gravel bike paths—namely washboard surfaces and loose material on the surface.

Major factors determining cyclists' use of a sidewalk bikeway were the amount of traffic on the street, continuity of the system, and the type of cyclist. Likelihood of a favorable rating of the bike lanes increased dramatically as the width of the lanes increased from 5 to 6 ft. Allowing cars to park in the bike lanes particularly bothered cyclists. Problems included car doors opening and glass or debris in the lane, as well as encroachment into the lane by right-turning cars. Reasons for not using bike lanes were as follows: It would take the cyclist out of his or her way; there was too much traffic on the route; a lack of awareness of the bike route; too many STOP signs; and route stopped short of desired destination.

An unusual bicycle lane configuration was implemented on one-way arterial streets in Madison, Wisconsin, with the bike lane being on the right side on one street and on the left side on another street.[72] Accident data were collected over 4 years before and 4 years after the lanes were introduced (1974–1977 and 1978–1981). The accident rate increased on these streets by 18.6%, when the increase in bike traffic was taken into account. This difference was associated primarily with the bike lane on the left side of the street, and when cars were turning left. Cyclists are not accustomed to seeing cycles on their left, and cyclists are not used to traveling with other vehicles on their right. The difficulty may have been largely a novelty effect, as the problem diminished sharply after the first year of operation. Care must obviously be taken when introducing new systems that mix bicycles with motor vehicles.

The frequency and types of bicycle accidents are age-dependent. In order to understand why these differences occur it is necessary to examine cyclists' knowledge about traffic regulations, attitudes toward safety, and cognitive and motor abilities. A study done in the Netherlands looked at cyclists in five age categories: 9–11, 12–15, 16–18, 19–59, and 60–83.[73] A questionnaire survey included a test of knowledge (using photos of cyclists approaching an intersection where there might be a conflict between the cyclist and another road user), attitudes toward compliance with traffic

[70] D.R. Mayhew and H.M. Simpson, "Motorcycle Engine Size and Traffic Safety," Traffic Injury Research Foundation of Canada, Ottawa, 1989.

[71] B. Kroll and R. Sommer, "Bicyclists' Response to Urban Bikeways," *IAP Journal* (January 1976), 42–51.

[72] R.L. Smith and T. Walsh, "Safety Impacts of Bicycle Lanes," *Transportation Research Record,* 1168 (1988), 40–56.

[73] W. Maring and I. van Schagen, "Age Dependence of Attitudes and Knowledge in Cyclists," *Accident Analysis and Prevention,* 22, 2 (1990), 127–136.

rules, and opinions about car drivers, cyclists, and self. The relationship between age and scores on these parts of the questionnaire indicated an inverted-U shape, with youngest and oldest subjects being worse on knowledge, attitudes toward rules of the road, and self-reported compliance with these rules. This suggests that these aspects of safe cycling behavior may be the reasons behind the young and the old cyclists having highest accident rates. Unfortunately, these two groups may also be least aware of their cycling limitations, as they rated their own bicycle safety to be higher than did the other three age groups. In addition, all groups rated themselves as safer than other cyclists, as was the case for motor vehicle drivers, reported earlier.

The role of alcohol in motor vehicle accidents is well established. However, it is perhaps less well known that it plays a significant role in adult bicycle accidents. One study[74] of this problem measured BAC levels of 140 injured cyclists and a large control group. Alcohol was involved in 24% of the injured cyclists, but only 4% of the controls. At BACs above 0.10 the likelihood of being in an accident was 10 times that of a sober cyclist.

An additional important factor in bike accidents is the inadequacy of bicycle lighting and street illumination at night. An examination of nighttime fatalities in Australia[75] found that in 90% of these fatal accidents at night, as compared with 40% during daytime, the cyclist was hit from behind. About 80% of the nighttime accidents occurred on arterial roads and the majority were in high speed-limit zones.

Research on bicycle accidents is often hampered by a lack of adequate information. For example, data were gathered from several North Carolina hospital emergency rooms over the summers of 1985 and 1986 and compared with police-reported data.[76] Two-thirds of those treated were children under 15 and 70% were male. One-fifth of these involved collisions with motor vehicles. However, only 10% of these incidents appeared on police accident files. It is clear from this that police files miss a substantial proportion of bicycle-injury accidents.

Pedestrian characteristics

The design of roadways emphasizes the safe and efficient movement of motor vehicles. Little thought is typically given to pedestrians. At various times all users of the roadway system are pedestrians. The need to move both vehicles and pedestrians efficiently and safely at the same locations presents a significant challenge to the traffic engineer, as the competition increases between drivers and pedestrians for space on the streets. Pedestrian accidents represent a significant proportion of traffic accidents (15% to 45%, depending on the country).[77] Rates in North America are among the lowest, perhaps because there is relatively less pedestrian traffic than most other areas of the world.

Pedestrians will typically not walk more than a mile to work or half a mile to catch a bus, and 80% of their distances traveled will be less than 3,000 ft.[78] Peak times are around noon (about double the average volume at the morning and afternoon rush hours). Pedestrians often consider themselves outside the law, and enforcement typically is low. They want to get to their destinations by the shortest distance, so they jay-walk and avoid both overpasses and underpasses.

Walking speeds used for design purposes assume free flow with plenty of space for pedestrians to choose their own speeds (however, speed drops as density increases). Although it is often assumed for engineering design purposes that walking speeds are 4 ft/sec, many (especially the elderly) walk much slower than this (Figure 1–9). Six levels of service for pedestrian traffic have been identified, based on number of square ft per person.[79] (For more details on level of service, see Chapter 5 in this volume.)

Accident data have shown that the proportion of accidents associated with left-turning vehicles is nearly double that for right-turning vehicles at intersections of two one-way streets. In comparison with through maneuvers, the likelihood of a pedestrian accident during left-turning maneuvers is about four times as great. Among the contributing factors here are poor visibility from within the vehicle due to pedestrians being obscured by the A-pillar and possibly by dirt on the windshield. These problems are not present to nearly the same degree in the case of right-turning maneuvers. Habib[80] summarized statistics from the New York City area that showed that both pedestrians and drivers failed to yield the right of way with about equal frequency with right-turning maneuvers, but on left-turning maneuvers, the drivers failed to yield to the pedestrian 62% of the time, compared with a 38% failure rate for pedestrians.

The reason for the difference between left-turning and right-turning maneuvers might be explained by eye movements and fixations. A study by Shinar, McDowell, and Rockwell[81] showed that eye fixations were 3.6 degrees to the right on right curves, but almost straight ahead on left curves. Drivers generally look more at the right side than at the left side of the road.[82] Another factor is that drivers must shift attention at signalized intersections from the traffic light to the crosswalk they are about to cross as they approach the intersection. As the driver gets closer to

[74] S. Olkkonen and R. Hoknanen, "The Role of Alcohol in Nonfatal Bicycle Injuries," *Accident Analysis and Prevention*, 22, 1 (1990), 89–96.

[75] M. Hoque, "An Analysis of Fatal Bicycle Accidents in Victoria (Australia) with a Special Reference to Nighttime Accidents," *Accident Analysis and Prevention*, 22, 1 (1990), 1–11.

[76] J.C. Stutts, J.A. Williamson, and F.C. Sheldon, "Bicycle Accidents: An Examination of Hospital Emergency Room Reports and Comparison with Police Accident Data," *Transportation Research Record*, 1168 (1988), 60–71.

[77] B.M. Hiehl, S.J. Older, and D.J. Griep, "Pedestrian Safety," *OECD*, Paris, 1969.

[78] B. Pushkarev and J.M. Zupan, *Urban Space for Pedestrians* (Cambridge, MA: MIT Press 1975).

[79] J.J. Fruin, *Pedestrian Planning and Design* (New York: Metropolitan Assoc. of Urban Designers and Environmental Planners, 1971).

[80] P.A. Habib, "Pedestrian Safety: The Hazards of Left-Turning Vehicles," *ITE Journal*, 50, 4 (1980), 33–37.

[81] D. Shinar, E.D. McDowell, and T. Rockwell, "Eye Movements in Curve Negotiations," *Human Factors*, 19 (1977), 63–72.

[82] R.R. Mourant and T. Rockwell, "Strategies of Visual Search by Novice and Experienced Drivers," *Human Factors*, 14 (1972), 325–335.

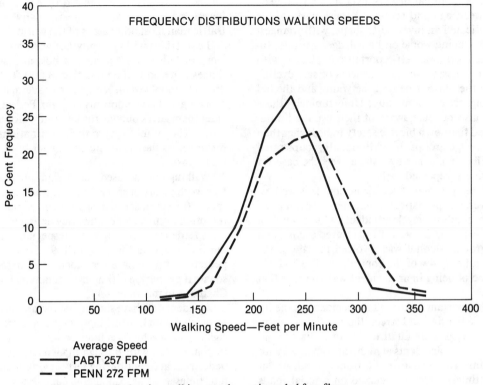

Figure 1–9. Pedestrian walking speeds—unimpeded free-flow.
SOURCE: J.J. FRUIN, *Pedestrian Planning and Design,* revised edition, Mobile, Alabama: ELEVATOR WORLD, INC. (1987), p. 40.

the location of the turn, the angle between these increases, resulting in less time to look at the crosswalk for pedestrians.

Possible solutions to this problem include modification of vehicle design (narrower A-pillars, more efficient windshield wipers), environmental changes (placement of an additional signal at the far left side of the intersection, so that both the crosswalk and the signal would be within easy view with minimal eye movement on the part of the driver), education of drivers and pedestrians to increase awareness of the issue, and recommending slower turning speeds to drivers.

The merits of using pavement markings at pedestrian crosswalks are unclear. Research done in Europe indicates that they do have safety benefits. However, a large U.S. study based on five years' worth of data at 400 intersections (each with one painted and one unpainted crosswalk) found that pedestrian accident rates were about twice as great at marked, as opposed to unmarked, crosswalks.[83]

There is clearly a need to record accidents in terms of some index of exposure to risk. One of the few studies in which this was done used an index based on the sum of the products of conflicting pedestrian-vehicle flows.[84] It showed that marked crosswalks are much safer than unmarked ones.

It has been suggested that markings could aid elderly pedestrians by keeping them walking straight across the street rather than at an angle.

Pedestrian signals

The need for traffic signals to control pedestrian movement seems apparent. Traffic-control measures to reduce pedestrian-vehicle conflicts will, of course, increase delays for both pedestrian and vehicular traffic. The engineer is thus in a conflict between providing safety and generating operational efficiency.

The basic types of pedestrian signal timing are:

1. *concurrent (standard),* where pedestrians walk concurrently with vehicle traffic
2. *early release,* where pedestrians leave the curb before traffic is allowed to turn
3. *late release,* where pedestrians must wait a portion of the green phase while traffic turns left
4. *exclusive,* where pedestrians have a protected crossing interval (scramble timing allows pedestrians to cross the intersection diagonally).

In spite of the numerous research reports on pedestrian signals, there is a good deal we do not know about their effects on safety. Some studies examined too few intersections, or too few accidents to draw meaningful conclusions. The use of several years' data is recommended. Some studies

[83] B.F. Herms, "Pedestrian Crosswalk Study: Accidents in Painted and Unpainted Crosswalks," *Highway Research Record,* 406 (1972), 1–13.

[84] R. Knoblauch, "Urban Pedestrian Accident Countermeasures Experimental Evaluation: Vol. 2, Accident Studies," U.S. Dept. of Transportation Final Report DOT HS-801-347 (February 1975).

have examined compliance, but not accidents. There is a need to demonstrate a relationship between compliance and accident rate before such compliance data can be useful. Others have examined understanding of signals or opinions about their effectiveness, which tells little about their impact on safety.

In a review of the literature on pedestrian signals, Khasnabis and colleagues[85] indicate that a number of problems have been identified with these devices. There are typically higher violation rates for scramble timing and higher compliance rates with late-release timing. Scramble timing causes greatest delays for pedestrians and vehicles, whereas delay is least for concurrent timing. Comparisons of pelican crossings (marked crosswalks with pedestrian-actuated signals) and zebra crossings (with alternate black and white stripes) generally report no difference in terms of pedestrian accidents.

Symbols have come into use on signals, as they have on signs. Robertson[86] studied preference for and understanding of symbols for pedestrian signals and recommended the "walking man" symbol for the walk phase and the hand signal for the don't-walk phase. He reported a great deal of difference in opinion about symbols and colors among the traffic engineers, adults, and children he studied. In another study, Robertson[87] examined pedestrian response to several messages—steady and flashing DON'T WALK, steady and flashing WALK, and DON'T START. The data indicate no difference between steady and flashing DON'T WALK, that DON'T START has no advantage over DON'T WALK, and that the flashing WALK is not an effective warning about turning vehicles. Robertson and Carter[88] showed that the flashing WALK and the steady WALK signals were understood by only 2.5% of a sample of 400 pedestrians. The former was used to warn of vehicles turning, but is no longer in the Manual on Uniform Traffic Control Devices. The flashing DON'T WALK has been used to indicate a clearance interval (during which pedestrians should not start across the roadway). It is clear that uncertainty of information is a major problem for pedestrians.

The effects of pedestrian signal design and timing on safety have been examined in an extensive study by Zegeer and associates[89] of 1,297 signalized intersections in 15 American cities. Of the 2,081 pedestrian accidents (over a three-year period), 1.4% were fatal; pedestrians engaged in hazardous actions in 49.2% and drivers in 41.5%.

The variables that explained the most variance in pedestrian accidents were pedestrian volume (with daily volumes of fewer than 1,200 pedestrians having many fewer accidents than those above this) and traffic volume, as would be expected. Other features that were associated with higher pedestrian accidents were the presence of buses, intersection of two-way (as opposed to one-way) streets, and wider streets (for some categories of streets with pedestrian volumes of more than 1,000 per day).

When pedestrian and traffic volumes were controlled, the data indicated that pedestrian accidents were lower where there are exclusive-timed pedestrian signals, as compared with standard-timed ones or no pedestrian signal, for intersections with moderate to high pedestrian volumes. There were no differences between intersections with standard (or concurrent) signals and those with no pedestrian signals.

The work by Zegeer and colleagues suggests the following reasons why use of pedestrian signals with concurrent timing might not be effective:

1. Compliance with pedestrian signals is generally poor, with violation rates of the DON'T WALK signal being higher than 50% in most cities.
2. The presence of a pedestrian signal may create a false sense of security, leaving pedestrians with the impression that they are fully protected.
3. The use of flashing DON'T WALK for clearance intervals is not well understood, nor is the use of the flashing WALK to indicate turning vehicles.
4. Pedestrians tend not to use pedestrian-actuated signals.
5. There is a lack of uniformity in the use of pedestrian signals across cities.

A threshold value of clearance interval exists beyond which pedestrians will ignore the signal and accept natural gaps in the traffic to cross. Thus, a clear need exists for greater uniformity in signal use and for the education of pedestrians and drivers about the meaning of pedestrian signals.

One of the conditions that leads to a significant number of pedestrian accidents is the conflict created when vehicles turn right at an intersection, especially when they are allowed to do so on a red light. Drivers are supposed to stop and yield to pedestrians in this situation, but they often fail to do so. The right-turn-on-red (RTOR) rule had been in use in the western states of the United States for some time before being introduced into most of the eastern states in the 1970s. Concern for the safety of pedestrians arose in connection with this change.

The issue was examined in a study by Preusser and co-workers,[90] who showed that a significant increase in pedestrian and bicyclist accidents occurred after the introduction of the RTOR at signalized intersections. These increases for pedestrian accidents in four jurisdictions in the United States were: 43%, 107%, 57%, and 82%. Analysis of the police reports suggested that drivers stop for a red light, look left for a gap in the traffic, and fail to see pedestrians and cyclists coming from their right. Possible countermeasures suggested by these authors were education of pedestrians to

[85] S. Khasnabis, C.V. Zegeer, and M.J. Cynecki, "Effects of Pedestrian Signals on Safety, Operations, and Behavior—Literature Review," *Transportation Research Record,* 847 (1982), 78–86.

[86] H.D. Robertson, "Pedestrian Preferences for Symbolic Signal Displays," *Transportation Engineering* (June 1972).

[87] H.D. Robertson "Pedestrian Signal Displays: An Evaluation of Word Message and Operation," *Transportation Research Record,* 629 (1977), 19–22.

[88] H.D. Robertson and E.C. Carter, "The Safety, Operational, and Cost Impacts of Pedestrian Indications at Signalized Intersections," *Transportation Research Record,* 959 (1984), 1–7.

[89] C. Zegeer, K.S. Opiela, and M.J. Cynecki, "Effects of Pedestrian Signals and Signal Timing on Pedestrian Accidents," *Transportation Research Record,* 847 (1982), 62–72.

[90] D. Preusser, W. Leaf, K. DeBartolo, R. Blomberg, and M. Levy, "The Effect of Right-Turn-on-Red on Pedestrian and Bicyclist Accidents," *Journal of Safety Research,* 13, (1984), 45–55.

be alert to the direction of movement of vehicles and to their use of turn signals and to delay crossing until the driver has seen them; better warrants for the prohibition of such turns; and an exclusive pedestrian light phase with no vehicular movement during that time.

The ITE Technical Council Committee 4A-15[91] conducted a nationwide survey of government traffic engineers (291 completed questionnaires) to examine criteria and conditions used to justify implementation of various pedestrian controls at signalized intersections. Six levels of pedestrian traffic control were defined by the committee as a starting point for asking about this issue in the survey. However, only about two-thirds of the respondents agreed with the committee's definitions of the six levels. In addition, another two-thirds of those responding indicated that they did not have policies, guidelines, or standards for providing pedestrian control. When asked about pedestrian understanding of signals, they expressed the view that only 4% of pedestrians understand the flashing WALK indication, and that 39% understand the flashing DON'T WALK signal. If they are correct, the need for public education is obvious. It was also felt that pedestrian laws are not generally enforced. The committee recommended that uniform guidelines be established for pedestrian control at signalized intersections, in view of the fact that such controls are determined primarily on the basis of engineering judgment and local preferences. As might be expected, there was seen to be a need for a national campaign to educate pedestrians about these matters.

The durations of pedestrian walk signals at many intersections are based on the assumption that the walking speed of pedestrians is 4 ft/sec (1.3 m/sec). However, a significant proportion of pedestrians walk more slowly than that. Estimates by the ITE suggest that the mean speed is 3.7 ft/sec and that 35% of pedestrians walk more slowly than the 4-sec design standard.[92] A study in Sweden[93] found that pedestrians aged 70 or older, when asked to cross an intersection very fast, fast, or at normal speed, considered fast to be less than 4 ft/sec. The 85th percentile for a comfortable speed was 2.2 ft/sec, well below the 4.0 standard often used. Walking speeds would likely be slowed even more under winter conditions with snow and heavy footwear.

Pedestrians are frequently confused about how to interpret WALK and DON'T WALK signals in the steady and flashing modes. The use of these signals is not consistent across jurisdictions. Some pedestrians, especially older ones, are sometimes confused about how to respond to the DON'T WALK signal when it comes on after they are part way across the street. Some turn around and return to the curb, thinking they are in danger if they continue across. The intent of the message is "Don't start across," but this is not at all obvious.

In an extensive study of 2,100 pedestrian accidents, Snyder[94] concluded that pedestrian search and detection failures were frequent causes of accidents. He classified pedestrian accidents into several specific types, with differential frequencies of occurrence as follows.

1. *Dart-out, first half:* A pedestrian who is not at an intersection appears suddenly from the roadside (24%). The prime example of this is the child running from behind parked cars.

2. *Dart-out, second half:* This is the same as the first type, except that the pedestrian covers half of the crossing before being struck (9%).

3. *Intersection dash:* These are similar to the dart-outs, but occur in or near a crosswalk at an intersection (9%).

4. *Multiple threat:* The pedestrian is struck by a car after other cars have stopped for the pedestrian and are blocking the view of the driver of the striking car (3%).

5. *Vehicle turn or merge with attention conflict:* Here the driver is turning or merging with traffic, and attention is directed to the traffic to look for a gap to enter when the driver hits a pedestrian who is crossing the roadway (7%).

6. *Bus-stop related:* Pedestrian crosses in front of the bus, which is blocking the view of oncoming drivers (3%). Location or design of the bus stop appears to be a major factor here.

About one in five traffic fatalities is a pedestrian,[95] costing American society about $1 billion annually. However, good information about pedestrian accidents is not readily available. Accident statistics for pedestrians typically are reported in terms of the number of accidents in specific jurisdictions or at particular types of intersections, without regard for the influence of pedestrian volume at these locations. It is rare to find data on accident *rate.* However, where rates have been calculated (by observing pedestrian volumes during selected hours of the day and using these to determine rates based on accident frequency during those hours) the data indicate that intersections with signals have lower rates than those without.

Turning vehicles, especially those turning left, present a special problem for the pedestrian. Robertson and Carter[96] examined 202 pedestrian accidents at 62 intersections and reported that 29% involved turning vehicles (17% of the vehicles entering the intersection turned); left-turning accidents accounted for 59% of the turning accidents (but only 44% of the total turns). This overrepresentation of left-turning vehicles is substantiated by Habib, who found that left-turning maneuvers were about four times as hazardous as the through movement. This problem is more acute at signalized intersections. The reason that left turns pose a greater hazard is

[91] ITE Technical Council Committee 4A-15, "Pedestrian Control at Signalized Intersections," *ITE Journal* (1986), 56, 2, 13–15.

[92] E. Hauer, "The Safety of Older People at Intersections," Special Report 218, *Transportation Research Board* (Washington, DC, 1988), 194–252.

[93] S. Dahlstedt, "Walking Speeds and Walking Habits of Elderly People," National Swedish Road and Traffic Research Institute, Stockholm, Sweden, undated.

[94] M.B. Snyder, "Traffic Engineering for Pedestrian Safety: Some New Data and Solutions," *Highway Research Record,* 406, (1972), 21–27.

[95] F. Ranck, "Walk Alert: The New National Pedestrian Program," *ITE Journal* 59, 8 (1989), 37–40.

[96] Robertson and Carter, "Safety, Operational, and Cost Impacts."

likely related to the fact that drivers, watching for a gap in oncoming traffic, will be distracted from noticing pedestrians in the crosswalk.

It often comes as a surprise to learn of the high proportion of pedestrians who are intoxicated when they are killed in traffic accidents. A report that examined 1984 data from the Fatal Accident Reporting System indicates that 40% of adult pedestrians (over 14 years of age) were intoxicated when fatally injured.[97] Of these, three out of five were at very high blood-alcohol levels.

Pedestrian accidents involving joggers seem to get more than the average amount of attention in the media, due possibly to the increase in numbers of joggers over the last several decades. Williams[98] examined 60 cases of joggers being struck by motor vehicles and concluded that only a tiny proportion of all pedestrian fatalities involve joggers. He suggests that the frequency of such accidents is greatly exaggerated. However, the magnitude of the problem is difficult to gauge, as accident reports often fail to differentiate joggers from other pedestrians. The 60 accidents examined were acquired through newspaper reports from across the United States. In these accidents 30 joggers were killed and 35 injured. No doubt the high percentage of fatalities among these is due to selective reporting—death of a jogger is bigger news than is an injury. Approximately two-thirds of the joggers were male and the average age was 24 years. Joggers were hit from behind in 31 cases. This reflected one of the most serious errors, namely jogging with, instead of against, traffic. Victims were running across the road in 12% of the cases. Just under half of the accidents occurred in darkness. The collision was due to negligence on the part of the driver in 38% of the cases.

It is evident that joggers take considerable risks if they jog on the roadway at night or in the same direction as vehicular traffic. Among the countermeasures suggested by Williams are more jogging paths, use of reflective clothing at night, running in the opposite direction to the traffic and well away from the driving lane, if jogging on the roadway. Greater awareness of each other's presence on the part of both driver and jogger is needed.

Nighttime conditions

Perhaps the most hazardous situation for pedestrians is walking on or beside the roadway at night. Pedestrians at night are, of course, much more difficult for drivers to detect, and pedestrians typically overestimate the distance at which they can be seen by approximately twice the range. In addition, clothing with low reflectivity is often worn. For example, the reflectivity of the roadway background ranges from 2% to 10%, and that of blue denim is about 3%.[99] A pedestrian in dark clothing blends in visually with the dark surroundings and relatively dark road surface.

One of the more effective ways of reducing pedestrian accidents at night is the use of clothing, patches of material, or tags that are retroreflective. Light hitting such objects returns to its source, making the object highly visible. Lamps attached to the body or carried by the pedestrian also enhance conspicuity at night. Blomberg, Hale, and Preusser[100] examined various commercially available materials and lights for their ability to increase conspicuity of pedestrians. In a closed course roadway system subjects drove an instrumented vehicle and the distances at which they could detect and identify pedestrians and cyclists were measured. The use of a leg lamp for cyclists provided the greatest detection distances, while a flashlight was detected farthest in the pedestrian condition, and about six times as far as a pedestrian wearing a white T-shirt and blue jeans.

The problem of pedestrian conspicuity was examined by Shinar,[101] who measured the actual and the estimated nighttime visibility of pedestrians. Subjects were driven toward pedestrians in dark clothing who either were wearing a retroreflective tag or were not. Pedestrians with tags were visible at about twice the distance as those without tags, under conditions of both high and low beam headlights on the subjects' cars. Subjects' estimates of their own visibility under these conditions varied in the same manner as did the actual visibility distances. Only when glare was present did pedestrians underestimate the distances at which they were visible. When subjects played the role of pedestrians they overestimated the distance at which they could be detected by twice as much in the high-beam condition and by a factor of 1.4 in the low-beam condition. Subjects were generally unaware of the effectiveness of retroreflective tags in enhancing their visibility.

Social factors

Several social factors relate to the presence and behavior of both pedestrians and drivers. Among the factors that influence walking speed are density, sex, and size of the group. Speed is reduced in higher-density conditions and when pedestrians walk in pairs, as compared to alone. When in pairs there is less agility or more potential restriction of movement.

In a field study by Malamuth, Shayne, and Pogue[102] of driver behavior at crosswalks, a female pedestrian waited to cross either alone, with a shopping cart, or with an infant stroller. Drivers stopped more frequently in the presence of the stroller than for the other conditions, reflecting the influence of social factors in stopping behavior at crosswalks.

[97] J.C. Fell and B.G. Hazzard, "The Role of Alcohol Involvement in Fatal Pedestrian Collisions," *Proceedings of the American Association for Automotive Medicine,* 2 (1985), 105–125.

[98] A.F. Williams, "When Motor Vehicles Hit Joggers: An Analysis of 60 Cases," *Public Health Reports,* 96 (1981), 448–451.

[99] P.L. Olson and M. Sivak, "Visibility Problems in Nighttime Driving." In G.A. Peters and B.J. Peters, eds., *Automotive Engineering and Litigation* (New York: Garland Law Publishing, 1984), 383–405.

[100] R.D. Blomberg, A. Hale, and D.F. Preusser, "Experimental Evaluation of Alternative Conspicuity-Enhancement Techniques for Pedestrians and Bicyclists," *Journal of Safety Research,* 17 (1986), 1–12.

[101] D. Shinar, "Actual Versus Estimated Nighttime Pedestrian Visibility," *Ergonomics,* 27 (1984), 863–871.

[102] N.M. Malamuth, E. Shayne, and B. Pogue, "Infant Cues and Stopping at the Crosswalk," *Personality and Social Psychology Bulletin,* 14 (1978), 334–336.

It seems reasonable to assume that drivers would reduce speed in the presence of pedestrians, especially children. This assumption was tested by Thompson, Fraser, and Howarth,[103] who measured vehicle speeds and distances from the curb outside five British junior high schools. The mean speed was 28.4 mph, while the posted speed limit was 30 mph. The limit was exceeded by 36% of the vehicles. The presence of children on the roadside made no difference in speed or vehicle position. However, when groups of more than 10 pedestrians were present there was a speed reduction of 1 mph. The authors suggest that drivers are inadequately prepared for the unpredictable behavior of children.

Pedestrian behavior is often a function of what other pedestrians are doing; that is, they imitate each other. An example of this is seen in a study by Russell, Wilson, and Jenkins,[104] who varied the sex, race (black and white), and number (1 or 2) of jaywalking models. Pedestrians were more likely to jaywalk in the presence of white male models and in the presence of two models.

Handicapped pedestrians

Handicapped people are much more mobile in our society than they were a few decades ago. As a result, a great many design standards for transportation facilities have been changed to meet their needs. Although the typical image of the handicapped person is someone in a wheelchair, there are a variety of other handicaps. For example, vision and hearing deficits are a safety problem as related to the detection of traffic hazards. The blind rely on the sense of touch to tell them when they have reached an intersection. Their needs are now being met by varying the texture of sidewalk surfaces and providing braille maps. Blind pedestrians also rely on hearing to get information about the speed, direction of travel, and location of vehicles.

Among the aids used for blind pedestrians are audible pedestrian signals, attached to a speaker on the light signal, which emits a sound such as beeping, chirping, or buzzing. A study of these by Wilson[105] showed that in England their use did not affect delay, but decreased by 5% the time taken for pedestrians to cross on the walk signal indication, and fewer people failed to cross completely during the walk phase. It was concluded that there were positive safety effects, including a small but significant reduction in delay to nonhandicapped people. Unfortunately, no data were obtained on the behavior of unaccompanied blind people. It should be noted that such auditory information does not always indicate which direction to cross the road, creating a potential source of confusion. Signs warning drivers of deaf or blind pedestrians are also used where there are large numbers of these people likely to cross the street.

Difficulties associated with balance, stamina, and reaction time can pose a safety problem. Among the semi-ambulatory are pedestrians with heart disease, cerebral palsy, and those who rely on the use of canes and crutches. Dexterity and walking speed are also reduced by arthritis, which is more common among elderly road users. Handicapped pedestrians are adversely affected by uneven or rough sidewalk and road surfaces, as well as by high curbs and steep ramps. Architectural guidelines have eliminated many of these problems. It is important to provide areas (and facilities such as benches or chairs) where the elderly and handicapped pedestrians who lack stamina and strength can stop and rest.

The handicapped pedestrian includes not just those with physical problems such as restricted mobility or perception, but also those temporarily disabled because they are encumbered by carrying luggage, packages, children, etc.

Handicaps are not only physical in nature but also mental. Pedestrians with low intelligence, who are illiterate, have dyslexia (a disorder in which left-right discrimination is difficult and letters are reversed), and other such problems often have difficulties understanding traffic control devices.

The following factors should be taken into account in planning pedestrian facilities for the handicapped:

1. minimum clearance for wheelchairs
2. use of nonskid, level surfaces
3. gradients (ramps, sidewalks) not more than 5% incline
4. maximum curb heights of about 10 inches
5. adequate lighting
6. good access to pedestrian-activated signal controls (some designs are difficult to operate as they have small buttons or require excessive amounts of force to activate)
7. reduction of physical barriers such as street furniture, flower boxes, mail boxes, and garbage cans
8. adequate time to cross the street
9. proper orientation or wayfinding aids such as signs and maps (difficulty in finding one's way, as either a pedestrian or a driver, can lead to confusion and frustration)
10. good snow-clearing facilities.

In spite of the many countermeasures available for pedestrian safety, this continues to be a problem. Implementation of educational campaigns, especially for children and the elderly, has met with some success and is more frequent in places such as Europe where pedestrian volumes are high compared with North America.[106]

Child pedestrians

The behavior of child pedestrians is different in a number of important ways from that of adults. The child's conception of safety is poorly formulated and his or her schema for critical behaviors such as crossing the street is not well developed. Accident statistics show that road accident rate is a function of age among young pedestrians, being greatest for those in the 3- to 8-year range (see Figure 1–10). Boys consistently have more accidents than do girls.

[103] S.J. Thompson, E.J. Fraser, and C.F. Howarth, "Driver Behavior in the Presence of Child and Adult Pedestrians," *Ergonomics,* 28 (1985), 1469–1474.

[104] J.C. Russell, D.O. Wilson, and J.F. Jenkins, "Informational Properties of Jaywalking Models as Determinants of Imitated Jaywalking: An Extension to Model Sex, Race, and Number," *Sociometry,* 39 (1976), 270–273.

[105] D.G. Wilson, "The Effects of Installing an Audible Signal for Pedestrians at a Light Controlled Junction," *TRRL Laboratory Report 917* (1980).

[106] S. Sandels, "People in Traffic," *Safety Education* (Summer 1970), 20–24.

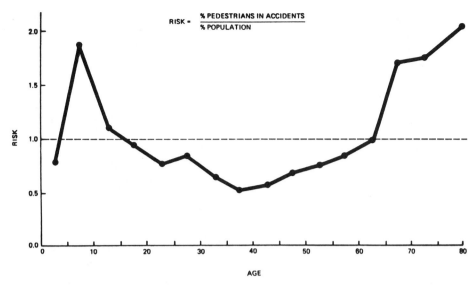

Figure 1-10. Pedestrian intersection accident risk by age, based on exposure. SOURCE: H.D. ROBERTSON AND E.C. CARTER, "The Safety, Operational, and Cost Impacts of Pedestrian Indications at Signalized Intersections," *Transportation Research Record,* 959 (1984), 4.

The sudden increase after age 3 arises as mobility and exploratory behavior increase dramatically at this age. Dashing-out accidents (usually from behind parked vehicles) are the most common type. Child pedestrian accidents are most common in the following situations:[107]

1. residential areas and where there is no traffic control
2. when the child is running
3. when the driver's vision is blocked.

The reasons for the relatively high accident rate among young pedestrians are varied. Sandels[108] has extensively examined the behaviors and cognitions of children and suggests that the necessary degree of maturity for safe behavior is reached between the ages of 9 and 12. A brief examination of the limitations and characteristics of these little road users will help to understand the problem. The following factors appear to contribute to the child pedestrian problem:[109]

1. Their small stature makes it difficult for them to evaluate the traffic situation correctly.
2. They have difficulty distributing their attention, and they are therefore easily preoccupied or distracted in hazardous traffic situations.
3. They have difficulty discriminating right from left.
4. They have difficulty in correctly perceiving the direction of sound and the speed of vehicles.
5. Many youngsters believe that the safest way to cross the street is to run.

6. Many children believe it is safe to cross against the red light.
7. Children have a poor understanding of the use of traffic control devices and crosswalks.

Another factor in avoiding danger on the road is the speed with which a pedestrian can recognize and react to the situation. Visual reaction time decreases with age in children, by a factor of about three between the ages of 4 and 17.[110] Auditory reaction time is also slower for younger children. Attention span is shorter in children than in adults. The ability to attend to more than one thing at a time increases from about age 5. This no doubt accounts in part for the large number of child accidents involving running onto the road in pursuit of objects such as balls into the path of moving vehicles.

A related skill, which also develops with age, is the ability to scan the visual environment for information. Finding relevant information in a complex environment is difficult for many adults as well as for most children. Difficulties with visual tasks for youngsters appear to be a function of limitations in speed of eye movements, attention, and memory, rather than visual acuity. A related limitation for children is that they are less able than adults to see peripheral movement (that is, to see out of the corner of the eye).

Reiss examined knowledge and perceptions of young pedestrians in a survey of 933 school students aged 5 to 14. Kindergarten children were interviewed individually, while the others were tested in groups. An amazing proportion of the younger subjects indicated that they would cross the street when the traffic signal was red.[111] Even among teenagers about 20% would commit such a violation, although they would likely do so under "safer" conditions than would the

[107] M.L. Reiss, "Knowledge and Perceptions of Young Pedestrians." Paper presented at the 56th Annual Meeting of the Transportation Research Board, January 1977.

[108] S. Sandels, *Children in Traffic,* (London: Paul Elek, 1968).

[109] Reiss, "Knowledge and Perceptions."

[110] Ibid.

[111] Ibid.

younger pedestrians. Rural youngsters, as compared to urban and suburban children, were more likely to cross the road against a red light. Younger students were also more likely to cross at unprotected corners, take a route that avoids traffic, and experience fear where there were no safety patrols or guards. Boys were more likely than girls to think it is safer to run than to walk across the street, to run across the street when there is a break in the traffic, and to indicate that when a child is struck "nothing happens to them."

Rules of the road are not readily understood by children, which contributes to many errors on the part of young pedestrians. Although they may be able to follow the rules of games, young children (ages 6 to 8) are not fully aware of the role that rules play in regulating behavior. The purpose and function of rules become appreciated at about the age of 10. In common use now is the "safety city," with miniature streets, signals, and vehicles to teach young children the basis of traffic safety.

The difficulty that children have judging distance of vehicles was examined by Zwahlen.[112] Ten children and 10 adults made judgments about whether different distances, from 180 to 210 ft, were greater or smaller than a standard distance (200 ft). The two distances were indicated by the location of cars. Although the two groups did not differ in their judgments of the distance, the variability of their judgments was greater for the children. In addition, there were no significant differences between the two groups in number of head movements toward the test stimulus nor in average judgment time. These findings suggest that children in the age group 6 to 13 have not yet reached a high degree of consistency in making distance judgments.

Size constancy—the ability to judge accurately the size of objects independent of their distance from the observer—is known to develop with age, as the child learns to interpret different environmental cues to distance. Young children have been shown to have relatively poor size constancy, suggesting that they could easily overestimate the distance of vehicles, which look very small when they are far away because they project a small image on the eye.

In addition to accurate perception of distance, it is essential for pedestrians to judge velocity of vehicles correctly. Auditory, as well as visual, information provides cues here. Salvatore,[113] in a developmental study of velocity perception, showed that auditory cues and vehicle size may be more important than visual velocity cues for children. He also showed that older children are more likely to make correct judgments of slow and medium velocities of oncoming vehicles, but that correct judgment of fast velocities is inversely related to age. Girls age 11 to 14 were found to overestimate vehicle speeds, compared with boys the same age. These gender differences were not found in 5- and 6-year-olds. In general, the younger children have great difficulty with any but the simplest situations.

Pedestrians often make assumptions about what drivers can see and can do in a conflict situation. Children often believe that adults will always be kind to them, so drivers will be able to stop instantly if a child is in danger. Their perception of driver behavior and cause and effect is not well developed.[114]

A large proportion of children's accidents are the result of unsafe or illegal actions on the part of the child. Local traffic engineers and police are often called upon to assist in the planning and implementation of traffic safety programs for schools. Special treatments for children involve use of crossing guards, safe route-to-school programs, signing and pavement markings in school zones, and speed enforcement in these areas. In general, crossing guards and increased enforcement are the best measures for child pedestrian safety.

A good deal of effort has been put into making the roadway environment near schools and the behavior of school children safer. The ITE Technical Committee 4A-1 has published a report on recommended practice for school trip safety programs.[115] The members outline six phases, which constitute a thorough and effective means of establishing a safer pedestrian environment for school children.

1. *Setting up the school trip safety process:* The organization of the program requires the establishment of a committee including representatives from law enforcement, engineering, school administration, teachers, and parents. This committee (one in each area, not one for each school) develops and reviews policy, maintains public relations, recommends action, and oversees the operation of the safety program.

2. *Route deficiency identification:* An inventory of roadway and pedestrian facilities is made; traffic studies are conducted to determine vehicular volume and speeds, gap adequacy, accident frequency, and vehicle-pedestrian conflicts. Studies are done to establish pedestrian volume, and characteristics such as age, size, mental capabilities, arrival times, reaction time, and walking speed. Other factors include sight distance, lighting, speed limits, roadway width, and traffic control devices in use. Criteria are then established for acceptable levels of these factors.

3. *School walk route map:* A designated route map, reflecting the various considerations outlined above, is developed and produced. These are issued to children and parents.

4. *Selection of control measures:* Deficiencies in the routes are identified and improvements made, as feasible. Deficiencies to be taken into account include gaps available, sight distances, ages of child pedestrians, vehicular volume and speeds, conditions after dark, land use, existing traffic controls, and special conditions such as high crime rates. Traffic control incorporates signs, signals, pavement markings, grade separation, and crossing guards. Education and enforcement programs are also important elements of the control measures.

[112] H.T. Zwahlen, "Distance Judgment Capabilities of Children and Adults in a Pedestrian Situation," *3rd International Congress on Automotive Safety Proceedings,* 2 (Supplement) (1974), 38-1–38-23.

[113] S. Salvatore, "The Ability of Elementary and Secondary School Children to Sense Oncoming Car Velocity," *Highway Research Record,* 10 (1973), 264–275.

[114] Sandels, *Children in Traffic.*

[115] ITE Technical Committee 4A-1, "School Trip Safety Program Guidelines," Institute of Transportation Engineers, Washington, DC, 1984.

5. *Implementation of program:* Successful implementation of such programs depends on public acceptance and adequate funding, as well as the cooperative effort of all parties involved. Users (children and their parents) need to be properly informed of the program and its implementation.

6. *Reevaluation:* Annual reevaluations are recommended, as conditions and personnel change and facilities deteriorate over time. These include engineering studies of traffic and pedestrian volume and collection of accident data.

Elderly pedestrians

Being a pedestrian can be a hazardous activity for the elderly for a number of reasons, including limited vision and hearing, slower reaction time, reduced walking speed, and prejudice on the part of drivers toward older pedestrians. Statistics indicate that people over age 70 are more likely to be involved in a severe pedestrian accident than are younger people (see Figure 1–10). This is due in part to greater vulnerability of older people because of physical frailty, more easily broken bones, longer recovery time, etc. The elderly tend to walk a great deal because they have a lot of free time; also, it is good exercise and an inexpensive mode of travel for short trips. The elderly are more law-abiding than the general population. They may in fact be too trustworthy of traffic signals and of drivers when it comes to crossing the street.

As the population of road users ages, the needs of the pedestrian must take age factors into account. Winter[116] reviewed the demographics of the older driver and pedestrian and the special needs associated with their limitations. There are a number of physical, cognitive, psychological, and environmental factors that influence their performance and ability to learn traffic safety programs.

There is a popular conception that drivers have less respect for older pedestrians, and this seems to be borne out by the research. Howarth and Lightburn[117] indicate that pedestrians, including the elderly and children, take the evasive action in the event of a conflict (91% of the time in distant encounters and 100% in close ones). Drivers were most likely to take no evasive action in the case of pedestrians over 60 years of age. This finding is also supported by Zuercher,[118] who found more aggressive driver behavior toward pedestrians dressed to look like old people.

Safety countermeasures

In spite of the large number of accidents, there has not been a pedestrian safety effort to match that for seat belt use or drinking and driving. The Walk Alert program is an effort of the National Safety Council (NSC), Federal Highway Administration (FHWA), National Highway Transportation Safety Administration (NHTSA), and more than 100 community and service organizations, initiated in 1986. Some earlier studies of safety campaigns directed at specific pedestrian accident types have been shown to be effective in reducing pedestrian traffic accidents.[119] However, the messages developed were never used at a national level. The Walk Alert program involves the implementation of these programs at the national level in the United States.

Even though numerous campaigns for safety of pedestrians have been developed and used over the years, most seem to be ineffective. An examination in 1986 by a National Safety Council panel of experts of 274 pieces of pedestrian safety material, audiovisuals, and programs revealed that only 16% were adequate for use in a national program. The remainder were "inadequate, incomplete, inaccurate, or misleading."[120] Some engineering or physical change in the facility is often the most appropriate countermeasure for safety; however, we need to know more about how these interact with and influence pedestrian behavior.

As a large proportion of pedestrian accidents are at midblock, the restriction of parking at curbs in areas with many pedestrians is a possible remedy, but it may not be practical as a widespread measure. Increased street lighting has also been found to reduce pedestrian accidents at night.

Habib[121] recommends the following measures to reduce pedestrian accidents due to left-turning vehicle maneuvers: improve sight angles from vehicles by modifying them to place the driver closer to the center of the vehicle or farther from the A-pillar that obstructs vision; make A-pillars smaller; design windshield wipers to remove dirt and moisture more effectively from edges of the windshield; mount traffic signals on the far left side of the sidewalk (to increase the likelihood of pedestrians being in drivers' field of vision as they turn; allow drivers to judge walking speeds more accurately and increase eye contact between drivers and pedestrians); remove obstacles such as vegetation, signs, and street furniture from within 50 ft of the intersection; educate drivers to be aware of the problem and to keep better lookout for pedestrians.

Physical barriers are an obvious solution, but agile young people often climb over them. Similarly, grade-separated crossings can be very effective, but pedestrians tend not to use them if the additional time to cross the road is more than a small fraction of the time to cross at street level.

Many countermeasures for pedestrian safety have been suggested. A search of the relevant literature on pedestrian safety by Zegeer[122] revealed several treatments used to alleviate the problem. Use of one-way streets can reduce the complexity of crossings for pedestrians, who need to look only one way. In addition, drivers can devote more attention to pedestrian traffic, as vehicle traffic is all going in the same

[116] D.J. Winter, "Needs and Problems of Older Drivers and Pedestrians: An Exploratory Study with Teaching/Learning Implications," *Educational Gerontology,* 10 (1–2) (1984), 135–146.

[117] C. Howarth and A. Lightburn, "How Drivers Respond to Pedestrians and Vice Versa." In D. Oborne and J. Levis, eds., *Human Factors in Transport Research* (London: Academic Press, 1980), 363–370.

[118] R. Zuercher, "Communication at Pedestrian Crossings," Proceedings of the International Conference on Pedestrian Safety, Haifa, 1977, 115–118.

[119] Ranck, "Walk Alert."

[120] Ibid. p. 38.

[121] Habib, "Pedestrian Safety."

[122] C.V. Zegeer, "Feasibility of Roadway Countermeasures for Pedestrian Accident Experience," *Pedestrian Impact Injury and Assessment,* P-121 (Warrendale, PA: Society of Automotive Engineers, 1983), 39–49.

direction as they are. A study of 1,297 intersections in 15 American cities indicated lower pedestrian accidents at intersections of two one-way streets than at those involving two-way streets.[123] In the 2,081 accidents examined (over a 3-year period) pedestrians were engaged in hazardous actions in 49.2% and drivers in 41.5% of them. As a large proportion of pedestrian accidents are at mid-block, the restriction of parking at curbs in areas with many pedestrians is a possible remedy, but it may not be practical as a widespread measure. Increased street lighting has also been found to reduce pedestrian accidents at night. In spite of the reduced pedestrian traffic at night, about a third of these accidents occur in darkness. One remedy for this is street lighting. Well-lit locations decrease pedestrian accidents[124] and increase gap acceptance for crossing roads. The high cost of street lighting would seem to be justified where pedestrian volumes at night are high.

An important part of the effort to improve pedestrian safety is the education of drivers. They are usually unfamiliar with the limitations and the typical behaviors of children in traffic. They seldom signal their presence, or if they do they assume the pedestrian perceives them and understands what to do. Drivers tend to believe they have the right of way, as roads are for motor vehicles. They often assume that if pedestrians glance in their direction they see the vehicle and understand the hazard. It has been found that drivers are less likely to slow down for a pedestrian crossing the street if the pedestrian looks at the driver. Drivers apparently assume it is up to the pedestrian to take evasive action in the case of a conflict.

REFERENCES FOR FURTHER READING

ALLEN, M.J., *Vision and Highway Safety.* (Philadelphia: Chilton, 1970).

EVANS, L., AND R. SCHWING (Eds.), *Human Behavior and Traffic Safety.* (New York: Plenum Press, 1985).

FORBES, T.W. (Ed.), *Human Factors in Highway Traffic Safety Research.* (New York: John Wiley, 1972).

GOEDKIN, C. (Ed.), *Highway Safety Forum.* (New York: American Society of Civil Engineers, 1985).

NÄÄTÄNEN, R., AND H. SUMMALA, *Road-user Behavior and Traffic Accidents.* (Amsterdam: North Holland, 1976).

ROTHENGATTER, J.A., AND R.A. DE BRUIN, *Road Users and Traffic Safety.* (Wolfeboro, NH: Van Gorcum, 1987).

SHINAR, D., *Psychology on the Road: The Human Factor in Traffic Safety.* (New York: John Wiley, 1978).

---

[123] Zegeer and others, "Effects of Pedestrian Signals."

[124] A.J. Harris and A.W. Christie, "Research on Two Aspects of Street Lighting," *Public Lighting,* 19 (1954).

2

TRAFFIC AND VEHICLE OPERATING CHARACTERISTICS

Douglas W. Harwood, P.E., *Principal Traffic Engineer*

Midwest Research Institute

This chapter reviews two of the basic elements of traffic operations: vehicle operating characteristics and traffic characteristics. The discussion of vehicle operating characteristics addresses vehicle types and dimensions; design vehicles; turning radii and offtracking; resistance to motion; power requirements; acceleration and deceleration performance; vehicle operating costs; and transit and buses. The discussion of traffic characteristics addresses traffic volumes, traffic composition, speed characteristics, and spacing and headway characteristics. The chapter emphasizes the highway mode of transportation with some coverage of urban transportation as well.

Overview of highways and vehicles

Highways provide the predominant mode of transportation in the United States and much of the world. Table 2–1 shows the extent of the highway system in the United States, which consists of over 3,800,000 miles of public roads and streets. Approximately 56% of these highways are paved. Over 50,000 miles (1%) of this system consist of interstate highways and other freeways and expressways with full access control. The U.S. road and street system also includes an additional 350,000 miles (9%) of highways functionally classified as arterials, 800,000 miles (21%) classified as collectors, and 2,662,000 miles (69%) classified as local roads.

Highway traffic consists of a vast fleet of personal vehicles for transporting individuals and small groups and commercial vehicles for transporting freight and larger groups of passengers. In 1987, there were an estimated 184 million motor vehicles registered in the United States, including 137,300,000 passenger cars (75%), 41,100,000 trucks (22%),

4,900,000 motorcycles (2.7%), and 600,000 buses (0.3%). Vehicle ownership per capita in the United States is the highest in the world, as shown in Table 2–2. This table also shows that the number of passenger cars per capita in 1988 varies from 0.38 in North and Central America to 0.01 in Africa. The national average is 0.56 passenger cars per capita in the United States. Figure 2–1 shows the distribution of passenger car ownership per capita by state in the United States in 1987, including both personal and commercial automobiles.

Vehicle operating characteristics

The following section summarizes the operating characteristics of vehicles that should be considered in design and operation of the highway system. Key elements of the discussion include vehicle types and dimensions, design vehicles, turning radii and offtracking, resistance to motion, power requirements, acceleration performance, deceleration performance, vehicle operating costs, and transit and buses.

Vehicle types and dimensions

A variety of motor vehicle types—including passenger cars, light trucks and vans, single-unit trucks, combination trucks, buses, recreational vehicles, and motorcycles—operate on American highways. These vehicle types each have unique size, weight, and operational characteristics, and the specific mix of vehicle types expected to be present on a highway facility must be considered in highway design and traffic engineering.

TABLE 2-1
Public Road and Street Mileage in the United States by Functional Class and Surface Type—1987

Area Type and Functional Class	Unpaved Mileage				Paved Mileage					Total Mileage
	Unimproved	Graded and Drained	Gravel or Stone	Total	Low Type	Intermediate Type	High Type Flexible	High Type Rigid	Total	
Rural roads										
Interstate highways	0	0	0	0	0	1,041	19,186	12,884	33,111	33,111
Other principal arterials[a]	0	0	0	0	770	2,838	67,075	10,036	80,719	80,719
Minor arterial	0	0	141	141	3,922	15,364	120,736	7,091	147,113	147,254
Major collector	557	4,297	51,167	56,021	72,943	93,063	204,727	8,659	379,392	435,413
Minor collector	2,297	11,589	93,794	107,680	50,816	65,059	69,526	1,718	187,119	294,799
Local road	132,581	213,812	1,146,574	1,492,967	298,210	149,105	217,212	15,048	679,575	2,172,542
TOTAL—Rural	135,435	229,698	1,291,676	1,656,809	426,661	326,470	698,462	55,436	1,507,029	3,163,838
PERCENT—Rural	4.3	7.3	40.8	52.4	13.5	10.3	22.1	1.8	47.6	100.0
Urban roads										
Interstate highways	0	0	0	0	1	48	5,391	5,777	11,217	11,217
Other freeways and expressways	0	0	0	0	3	62	3,878	3,447	7,390	7,390
Other principal arterials	0	0	46	46	586	3,161	40,499	6,191	50,437	50,483
Minor arterial	17	79	446	542	3,650	10,947	53,756	6,089	74,442	74,984
Collector	30	144	871	1,045	6,323	17,908	47,532	4,052	75,815	76,860
Local road	1,149	4,773	34,682	40,604	71,566	143,134	190,925	43,025	448,650	489,254
TOTAL—Urban	1,196	4,996	36,045	42,237	82,129	175,260	341,981	68,581	667,951	710,188
PERCENT—Urban	0.2	0.7	5.1	5.9	11.6	24.7	48.2	9.7	94.1	100.0
TOTAL—Rural and urban	136,631	234,694	1,327,721	1,699,046	508,790	501,730	1,040,443	124,017	2,174,980	3,874,026
PERCENT—Rural and urban	3.5	6.1	34.3	43.9	13.1	13.0	26.9	3.2	56.1	100.0

[a] Includes some non-Interstate freeways and expressways

SOURCE: *Highway Statistics 1987,* Report No. FHWA-PL-88-008 (Washington, DC: Federal Highway Administration, 1988), p. 109.

TABLE 2-2
Vehicle Ownership per Capita, Continents and Selected Countries

Location	Passenger Cars per Capita 1978	Passenger Cars per Capita 1988	Vehicles per Capita 1978	Vehicles per Capita 1988
Africa	0.013	0.013	0.019	0.020
Asia	0.012	0.017	0.020	0.029
Europe	0.154	0.200	0.182	0.238
North and Central America	0.370	0.385	0.476	0.526
Oceania	0.323	0.435	0.400	0.556
South America	0.056	0.063	0.071	0.077
World average	0.071	0.083	0.091	0.106
Australia	0.385	0.455	0.476	0.588
Brazil	0.063	0.077	0.077	0.083
Canada	0.417	0.476	0.526	0.625
France	0.333	0.400	0.385	0.455
Germany, West	0.357	0.476	0.370	0.526
Italy	0.303	0.417	0.323	0.526
Japan	0.185	0.250	0.294	0.435
Spain	0.179	0.270	0.213	0.323
USSR	0.025	0.056	0.050	0.083
United Kingdom	0.256	0.370	0.294	0.435
United States	0.526	0.556	0.667	0.769

SOURCE: Motor Vehicle Manufacturers Association, *World Motor Vehicle Facts,* 1989; and previous handbooks.

Passenger cars and light trucks. Passenger cars and light trucks provide the overwhelming majority of personal transportation on American highways. Passenger cars are two-axle, four-tire automobiles, generally with seating for two to six passengers. Passenger cars typically weigh between 1,500 and 4,000 lb. Figure 2–2 illustrates the typical range of several key dimensions for passenger cars.

Light trucks with four tires, including pickup trucks and vans, were originally intended for commercial use, but they are being increasingly used for personal transportation. Light trucks generally have gross vehicle weights under 10,000 lb. Recent data indicate that sales of light trucks have increased to the point where they constitute 32% of total personal transportation sales.

Single-unit trucks. Single-unit trucks, also referred to as straight trucks, are trucks in which the cargo area and the power unit are mounted on a common frame and cannot be separated. They include two-axle, six-tire trucks; three-axle trucks; and some four-axle trucks. Single-unit trucks are used for transporting commercial goods and generally have gross weights between 10,000 and 40,000 lb. Single-unit trucks are most commonly used for short-haul and local pickup and delivery operations. The configuration of a single-unit truck is illustrated in Figure 2–3.

Combination trucks. Combination trucks consist of a power unit or tractor and one or more trailers, hence the name *tractor-trailer.* The tractors and trailers that make up a combination truck are joined at hitch points where the units can rotate relative to one another and are, therefore, said to be *articulated.* Combination trucks are used extensively in local and long-haul commercial goods transportation. They generally have gross vehicle weights up to 80,000 lb and may have even higher gross weights in some jurisdictions.

Two general types of trailers are used in combination trucks. A *full trailer* has one or more axles at both its front and rear ends; a full trailer is towed by the tractor or a preceding trailer but can transfer no weight to the previous unit. A *semitrailer* has one or more axles near its rear end, but has no front axles; the front end of the semitrailer rests on the rear end of the tractor or preceding trailer and transfers part of its weight to that unit.

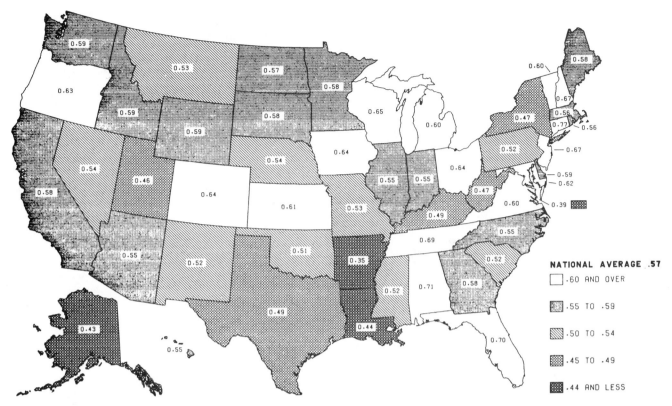

Figure 2–1. Distribution of private and commercial passenger car ownership per capita in the United States.
SOURCE: *Highway Statistics 1988,* Report No. FHWA-PL-88-008 (Washington, DC: Federal Highway Administration, 1988), p. 16.

Figure 2–3 illustrates several common types of combination trucks. Combination trucks are classified by the number and length of trailers used. Types of combination trucks include:

Single-trailer trucks

o Tractor plus one semitrailer—typically has three, four, or five axles. The five-axle version is the workhorse of the commercial trucking industry, the so-called 18-wheeler. Until 1982, most tractor-semitrailer combinations used 40- or 45-ft trailers; since the passage of the 1982 Surface Transportation Assistance Act (STAA), 48-ft trailers have become the industry standard and 53-ft trailers are becoming more common.

o Straight truck plus full trailer—a single-unit truck with a cargo area towing a full trailer with a second cargo area.

Double-trailer trucks

o Tractor plus semitrailer plus full trailer—known as a "double" or "double-bottom" truck. Three types of double combinations are in common use:

Twin-trailer truck—a tractor plus a 27- or 28-ft semitrailer plus a 27- or 28-ft full trailer.

Rocky Mountain double—a tractor plus a 45-ft semitrailer plus a 28-ft full trailer. Generally permitted only in some western states.

Turnpike double—a tractor plus a 45-ft semitrailer plus a 45-ft full trailer. Permitted only on some toll roads.

Triple-trailer trucks

o Tractor plus semitrailer plus two full trailers—permitted in a few states

The types of cargo area configurations used for semitrailers and full trailers include enclosed vans, cargo tanks, flatbed trucks, and dump trucks (bulk solid carriers).

Buses. Buses are used to transport passengers. Generally, any vehicle that can transport more than 15 passengers is classified as a bus. Most buses are single-unit vehicles with two or three axles, although some public transit systems use articulated buses up to 60 ft in length. Three types of buses with different functional uses are generally considered in traffic engineering analyses: local transit buses; school buses; and intercity buses. The characteristics of transit buses are addressed later in this chapter.

Recreational vehicles. Recreational vehicles include passenger cars with trailers, pickup trucks with camper

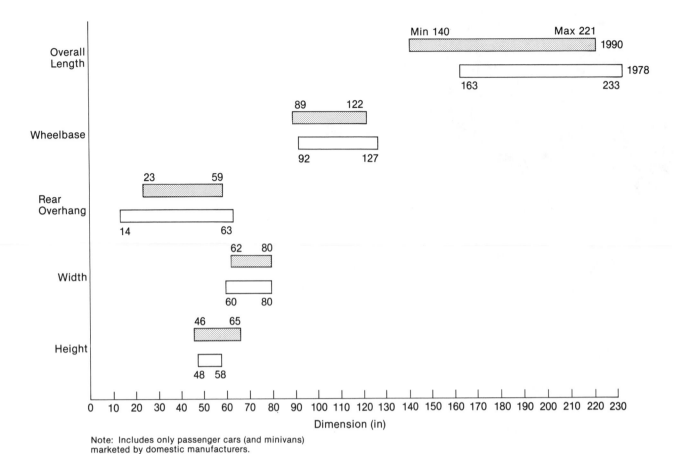

Figure 2–2. Changes in passenger car dimensions.
SOURCE: MOTOR VEHICLE MANUFACTURERS ASSOCIATION, *Parking Dimensions of 1990 Model Passengers Cars;* and previous editions of the same publication.

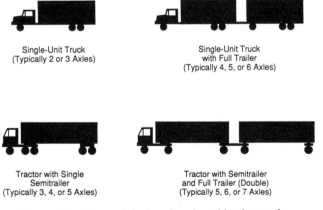

Figure 2–3. Types of single-unit and combination trucks.

bodies, specially equipped vans, and motor homes. While recreational vehicles may be physically and operationally similar to some passenger cars, light trucks, and buses, they are distinguished by their recreational purpose, which influences the times and places at which they are used. Recreational vehicles often perform poorly because they are more heavily loaded than other vehicles of similar size.

Motorcycles. Motorcycles are small, two- or three-wheel vehicles that carry one or two passengers seated on the frame of the vehicle and not enclosed by any passenger compartment. A sidecar may be added to carry additional passengers. Some motorcycles are also used to pull small trailers.

Design vehicles

Design vehicles are selected motor vehicles with dimensions and/or operating characteristic of such a critical nature that they influence or control the design and operation of the highway. Part of the highway design process is the selection of the appropriate design vehicle(s) for a given condition.

There are many assumptions concerning design vehicles imbedded in current geometric standards and criteria. These include driver eye height, vehicle headlight and taillight height, passenger car braking ability, and acceleration characteristics of passenger cars as well as trucks and other heavy vehicles. As the characteristics of the vehicle fleet have evolved over the years, geometric design criteria have been adjusted.

Current geometric design policies for highways are presented in the American Association of State Highway and Transportation Officials (AASHTO) publication *A Policy for Geometric Design of Highways and Streets.*[1] The design vehicles used in the AASHTO policy, and their key dimensions, are covered in Chapter 6, "Roadway Geometric Design." The choice of design vehicle for a specific roadway or intersection is influenced by the functional classification of the highway, by the highway system on which it is located, and by the proportions of the various types and sizes of vehicles expected to use the facility. A single-unit truck is generally the smallest vehicle used in design of public roadways. On most highways, one of the tractor-semitrailer combination design vehicles should be considered in design, particularly where turning roadways are bordered by curbs or islands. Even if tractor-semitrailer trucks use a facility only occasionally, the pavement should be sufficiently wide to accommodate them, particularly for turning movements.

The 1982 STAA legislation required highway agencies to allow the operation of larger trucks on a national truck network of highways designated by the secretary of transportation. Under the provisions of the 1982 STAA, states must permit tractor-semitrailer combination trucks with trailers up to 48 ft in length and double-trailer trucks with trailers up to 28.5 ft in length on the national truck network. In addition, the 1982 STAA included "grandfathering" provisions that require states that already allowed semitrailers longer than 48 ft in length to continue to allow them. Thirty-four states allow 53-ft semitrailers on at least some facilities, either under the STAA "grandfathering" provisions, under permit, or on specific toll roads.

Other design vehicle controls include vehicle height, which determines minimum vehicle clearances. Recommended values for minimum vertical clearance are:

- Passenger vehicles (for parking garages)   8 ft
- Buses (for bus terminals and parkways)   12.5 ft
- Trucks   14 ft

On some highway systems and at some specific locations, provision for particularly high loads is made by utilizing minimum vertical clearances of 15 to 17 ft.

Heavy vehicle operating characteristics also control the design of alignment. The weight/power ratio of a truck affects its ability to accelerate or maintain speed on upgrades.

Turning radii and offtracking

Two factors that limit the maneuverability of larger vehicles, including trucks, are turning radii and offtracking. Trucks cannot turn as sharply as can passengers cars and they also have greater offtracking. Both factors are significant in the design of intersections, and offtracking is also considered in the design of pavement width on horizontal curves. Turning radii of larger vehicles are also considered in the design of cul-de-sacs on residential streets. Current ITE guidelines for subdivision streets recommend the use of a 50-ft minimum

radius for cul-de-sacs to accommodate turns by fire and sanitation trucks.[2]

The minimum turning radii of each of the current AASHTO design vehicles are presented in Chapter 6, "Roadway Geometric Design." These data indicate that passenger cars can turn at much sharper radii than can trucks or buses.

While the difference in turning radii between passenger cars and trucks becomes significant only on very sharp turns, their differences in offtracking are significant in the design of virtually every intersection. Offtracking is the phenomenon by which the rear wheels of a vehicle do not follow the same path as the front axles. Offtracking occurs in all types of vehicles, but is relatively small in passenger cars. At low speeds, the rear axle of the truck follows a path inside the path of the front axle. Low-speed offtracking is primarily a function of the radius of the path and the spacings of the truck axles and the hitch points that join the tractor and trailers. A key feature in geometric design is the swept-path width for a vehicle making a turn. Swept-path width is the maximum difference in path between the rear inside axle and the front outside axle of the vehicle. Since low-speed offtracking is typically calculated along the vehicle centerline, the swept-path width can be calculated by adding half the tractor steering axle width plus half the rear trailer axle width (typically 7.58 ft) to the offtracking. In some cases, the front outside corner of the truck tractor or semitrailer may track outside of the front outside axle and an additional increment may need to be included within the swept-path width.

Table 2–3 shows the low-speed offtracking of several vehicle types for different radii and angles of turn.[3] Low-speed offtracking and swept-path width for particular maneuvers at intersections can be estimated from published turning templates for specific design vehicles (see Chapter 6) and from computer models.

Low-speed offtracking is slightly increased by pavement cross-slope (superelevation). For example, Table 2–3 shows that a 48-ft single-semitrailer truck would have 16.9 ft of offtracking making a 90°, 50-ft radius turn on a level pavement. This same truck could develop additional offtracking up to 0.40 ft when making the same turn on a pavement with a cross-slope of 0.06 ft/ft.[4] This small effect is not incorporated in published turning templates used for design purposes, such as those shown in Chapter 6.

Consideration of low-speed offtracking by itself is adequate for design of low-speed urban intersections. On higher-speed facilities, a second offtracking term that increases with the square of vehicle speed becomes important. This second, or high-speed, offtracking term is in the opposite sense to the low-speed term and makes the rear of the truck move back toward the outside of the curve. For example, if a 48-ft single-semitrailer truck is traversing a 1,000-ft radius horizontal curve with superelevation of 0.06 ft/ft at 10 mph, it will offtrack by 1.34 ft toward the inside of the curve. At 59 mph, the rear axle will exactly follow the

[1] *A Policy on Geometric Design of Highways and Streets* (Washington, DC: American Association of State Highway and Transportation Officials, 1990).

[2] *ITE Recommended Guidelines for Subdivision Streets: Recommended Practice* (Washington, DC: Institute of Transportation Engineers, 1984).

[3] D.W. Harwood and others, *Truck Characteristics for Use in Highway Design and Operation,* Report Nos. FHWA-RD-89-226 and -227 (Washington, DC: Federal Highway Administration, December 1989).

[4] Ibid.

TABLE 2–3

Truck Offtracking for Selected Combinations of Turn Radius and Turn Angle

	Maximum Offtracking (ft)*								
Turn Radius (ft):	50			100			300		
Turn Angle (deg):	60	90	120	60	90	120	60	90	120
Design vehicle									
Single-semitrailer truck with 37-ft trailer	9.3	11.8	13.3	6.0	6.5	6.6	2.1	2.1	2.1
Single-semitrailer truck with 45-ft trailer	12.1	15.5	—	8.0	9.0	9.4	2.9	2.9	2.9
Single-semitrailer truck with 48-ft trailer	13.0	16.9	—	8.8	10.0	10.5	3.3	3.3	3.3
Single-semitrailer truck with 48-ft trailer and long tractor	13.4	17.4	—	9.1	10.4	10.8	3.4	3.4	3.4
Double-trailer truck with cab-over-engine tractor	9.2	11.3	12.6	5.8	6.1	6.2	1.9	1.9	1.9
Double-trailer truck with cab-behind-engine tractor	9.6	11.9	13.4	6.0	6.4	6.4	2.1	2.1	2.1

*Add 7.58 ft to entries to get maximum swept-path width.

SOURCE: D.W. HARWOOD and others, *Truck Characteristics for Use in Highway Design and Operation,* Report Nos. FHWA-RD-89-226 and -227 (Washington, DC: Federal Highway Administration, December 1989).

front axle because of the high-speed offtracking effect. At higher speeds, the rear axle of the truck will track outside of the front axle.[5] High-speed offtracking is not generally significant in the design of low-speed urban intersections, but it is a potentially important issue in determining the need for pavement widening on horizontal curves.

Resistance to motion

The forces that must be overcome by motor vehicles if they are to move are rolling, air, grade, curve, and inertial resistance. Grade acts as a retarding force only when vehicles are on upgrades and inertia acts only when speed increases are involved. When vehicles are being stopped or slowed, all these resistances except downgrades and inertia help braking action. An additional resistance to motion during deceleration in gear is provided by engine compression forces.

Rolling resistance. Rolling resistance results from the frictional slip between tire surfaces and the pavement; flexing of tire rubber at the surfaces of contact; rolling over rough particles (i.e., stones or broken asphalt); climbing out of road depressions; pushing wheels through sand, mud, or snow; and internal friction at wheel, axle, and drive shaft bearings and in transmission gears. The rolling resistance force for a passenger car on a smooth pavement can be determined as follows:[6]

$$R_r = (C_{rs} + 2.15\, C_{rv}\, V^2)W \qquad (2.1)$$

where: R_r = rolling resistance force (lb)
C_{rs} = constant (typically, 0.012 for passenger cars)
C_{rv} = constant (typically 0.65×10^{-6} sec²/ft² for passenger cars)
V = vehicle speed (mph)
W = gross vehicle weight (lb)

TABLE 2–4

Rolling Resistances of Passenger Cars

	Rolling Resistance (lb/ton of vehicle weight)			
Vehicle Speed (mph)	Smooth Pavement	Badly Broken and Patched Asphalt	Dry, Well-Packed Gravel	Loose Sand
20	25	29	31	35
30	27	34	35	40
40	29	40	50	57
50	31	51	62	76
60	34	—	—	—

SOURCE: A.D. ST. JOHN and D.R. KOBETT, *Grade Effects on Traffic Flow Stability and Capacity,* National Cooperative Highway Research Program Report 185 (Washington, DC: Transportation Research Board, 1978); and P. CLAFFEY, *Running Costs of Motor Vehicles as Affected by Road Design and Traffic,* National Cooperative Highway Research Program Report 111 (Washington, DC: Highway Research Board, 1971).

Table 2–4 shows the rolling resistance for a passenger car on a smooth pavement based on Equation (2.1). The table also shows the higher rolling resistances of passenger cars for surfaces of lower quality. These estimates of rolling resistance for lower quality surfaces are based on the work of Claffey,[7] who measured the increased fuel consumption for vehicle operation on lower quality surfaces.

Rolling resistance for trucks can be estimated as:[8]

$$R_r = (C_a + 1.47\, C_b\, V)\, W \qquad (2.2)$$

where: R_r = rolling resistance force (lb)
C_a = constant (typically 0.2445 for trucks)
C_b = constant (typically 0.00044 sec/ft for trucks)
V = vehicle speed (mph)
W = gross vehicle weight (lb)

Air resistance. Air resistance is composed of the direct effect of air in the pathway of vehicles, the frictional force of air passing over the surfaces of vehicles (including the

---

[5] Ibid.

[6] A.D. St. John and D.R. Kobett, *Grade Effects on Traffic Flow Stability and Capacity,* National Cooperative Highway Research Program Report 185 (Washington, DC: Transportation Research Board, 1978).

[7] P. Claffey, *Running Costs of Motor Vehicles as Affected by Road Design and Traffic,* National Cooperative Highway Research Program Report 111 (Washington, DC: Highway Research Board, 1971).

[8] Ibid.

undersurface), and the partial vacuum behind the vehicle. The air resistance force for a motor vehicle can be estimated as:[9]

$$R_a = 0.5 \ \frac{(2.15 \ \rho \ C_D \ A \ V^2)}{g} \qquad (2.3)$$

where: R_a = air resistance force (lb)
ρ = density of air (0.002385 lb/ft^3 at sea level; less at higher elevation)
C_D = aerodynamic drag coefficient
A = frontal cross-sectional area (ft^2)
V = vehicle speed (mph)
g = acceleration of gravity (32.2 ft/sec^2)

The aerodynamic drag coefficient (C_D) is typically 0.5 for the average passenger car on the road today. The average passenger car currently being marketed has a drag coefficient of 0.4. Some production vehicles have drag coefficients as low as 0.3 and passenger car prototypes with drag coefficients as low as 0.15 have been built.[10] Historically, trucks have had aerodynamic drag coefficients ranging from 0.5 to 0.8.[11] The typical truck on the road today has a drag coefficient of about 0.5, and the drag coefficients of trucks are also expected to decrease in the future.[12]

Grade resistance. Grade resistance is the force acting on a vehicle because it is on an incline. It equals the component of the vehicle's weight acting down the grade. Equation (2.4) gives the grade resistance force:

$$R_g = \frac{W \ G}{100} \qquad (2.4)$$

where: R_g = grade resistance force (lb)
W = gross vehicle weight (lb)
G = gradient (%)

Curve resistance. Curve resistance is the force acting through the front-wheel contact with the pavement needed to deflect a vehicle along a curvilinear path. This force is a function of speed because the faster an object is moving, the greater the force required to change its direction. The curve resistance force is determined as:

$$R_c = 0.5 \ \frac{(2.15 \ V^2 W)}{g \ R} \qquad (2.5)$$

where: R_c = curve resistance force (lb)
V = vehicle speed (mph)
W = gross vehicle weight (lb)
g = acceleration of gravity (32.2 ft/sec^2)
R = radius of curvature (ft)

---

[9] Ibid.

[10] W.D. Glauz, "Future Changes to the Vehicle Fleet." In *Relationship Between Safety and Key Highway Features: A Synthesis of Prior Research,* State of the Art Report 6 (Washington, DC: Transportation Research Board, 1987).

[11] St. John and Kobett, *Grade Effects.*

[12] D.W. Harwood and W.D. Glauz, "Future Driver and Vehicle Characteristics and Their Influence on Highway Design for Safety." In T.A. Hall, ed., *Engineering 21st Century Highways,* (New York: American Society of Civil Engineers, 1989).

Inertial resistance. Inertial resistance is the force that must be overcome to change speed. It is a function of vehicle weight (regardless of type of vehicle) and the rate of acceleration or deceleration. It may be computed from the following equation:

$$R_i = \frac{W a}{g} \qquad (2.6)$$

where: R_i = inertial resistance force (lb)
W = gross vehicle weight (lb)
a = acceleration rate (ft/sec^2)
g = acceleration of gravity (32.2 ft/sec^2)

Power requirements

Power. Power is the time rate of doing work, and the maximum power that an engine can deliver is a measure of its performance capability. Power is typically expressed in units of horsepower (1 hp = 550 ft-lb/sec). The power actually used by a motor vehicle for propulsion can be determined from Equation (2.7):

$$P = 0.00267 \ R \ V \qquad (2.7)$$

where: P = power actually used (hp)
R = sum of resistances to motion ($R_r + R_a + R_g + R_c + R_i$) (lb)
V = vehicle speed (mph)

The maximum power output available for propulsion at a given engine speed equals the maximum gross brake horsepower at the flywheel for that engine speed less the power consumption of engine accessories, such as the alternator, automatic transmission, power steering, and air conditioner. For passenger cars with typical accessories, maximum horsepower available for propulsion at 60 mph is about 50% of the manufacturer's nominal engine horsepower rating. For large trucks, approximately 94% of the manufacturer's rated horsepower is available for propulsion. These estimates can be used to examine maximum acceleration rates and maximum speeds on grades for nominal engine horsepower in relation to engine speed and reliable values of resistances (particularly rolling and air resistance).

Empty weights and nominal horsepower ratings representative of major categories of motor vehicles are given in Table 2–5.

Weight/power ratio. Weight/power ratios are useful for indicating the overall performance characteristics of vehicles, particularly for making approximate performance comparisons among different vehicle types. The weight/power ratio (the gross weight of the vehicle in pounds divided by the power in horsepower) is a measure of the ability of a vehicle to accelerate and maintain speed on upgrades. Because weight is a rough indicator of resistance to motion, the higher the weight/power ratio, the lower the acceleration performance of the vehicle. A low weight/power ratio means high performance because it reflects a high ratio of power capability to motion resistance.

TABLE 2-5
Representative Motor Vehicle Weight and Power

Motor Vehicle Category	Empty Weight with Driver Aboard (lb)	Nominal Power (hp)
Passenger car	3,400	105
Large pickup truck	4,200	175
Two-axle, six-tire truck	10,000	175
Tractor-semitrailer	25,000	325

SOURCE: Estimates from various sources including W.D. GLAUZ, "Future Changes to the Vehicle Fleet: Effect on Highway Safety." In *Relationship between Safety and Key Highway Features* (Washington DC: Transportation Research Board, 1987); D.W. HARWOOD and W.D. GLAUZ, "Future Driver and Vehicle Characteristics and Their Influence on Highway Design for Safety." In *Engineering 21st Century Highways* (New York: American Society of Civil Engineers, 1989); and previous handbooks.

Figure 2-4 shows historical and projected data for the weight/power ratios of passengers cars and light trucks. The figure illustrates that downsizing of passenger cars for increased fuel economy has reduced weight/power ratios over time. Most of this reduction took place in the late 1970s and early 1980s. Very little change in weight/power ratios of passenger cars has taken place during the 1980s as average passenger car weight has remained constant at about 3,200 lb and engine horsepowers have changed little. Current trends suggest that passenger car weights and horsepowers may increase during the 1990s.[13]

Truck weight/power ratios vary widely. Vehicle weight depends on the weight of the load carried. For large trucks and truck combinations, the load can vary from zero to more than twice the vehicle's weight. Furthermore, the power available for propulsion depends on engine condition and size, transmission arrangement, and engine speed. Truck sizes and weights have increased steadily, most recently as a result of the 1982 STAA, but truck engine horsepowers have increased

Figure 2-4. Historical and projected data for distribution of passenger car weight/power ratio.
SOURCE: *Highway Capacity Manual*, Special Report 209 (Washington, DC: Transportation Research Board, 1985), p. 2–16.

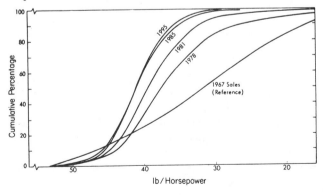

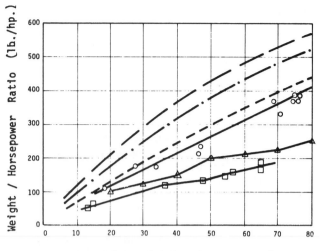

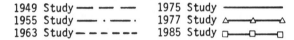

1949 Study — — —	1975 Study ————
1955 Study —·——·—	1977 Study △—△—△
1963 Study ———————	1985 Study □—□—□

Figure 2–5. Trend in weight/power ratios of trucks from 1949 to 1985.
SOURCE: *A Policy on Geometric Design of Highways and Streets,* Washington, DC: American Association of State Highway and Transportation Officials, 1990. Used by permission.

more than weights. Thus, truck weight/power ratios have steadily decreased over the past 40 years, as shown in Figure 2–5.[14] Current warrants for truck climbing lanes are based on a design truck with a weight/power ratio of 300 lb/hp, which represented a typical low-performance truck in the mid-1970s.[15] The most recent available data from a 1984 Federal Highway Administration study show that the average weight/power ratio of trucks in 1984 was approximately 175 lb/hp and only 12.5% of trucks had weight/power ratios greater than 250 lb/hp.[16]

The weight/power ratio is important in defining the hill-climbing and speed maintenance ability of the vehicle. Table 2–6 provides 1979 data on actual sustained operating speeds on upgrades in California for a variety of vehicle types. Also shown is the speed below which the slowest 12.5% of vehicles traveled. Based on the frequency of the vehicle types in the traffic streams, these data can be used to estimate the impedance to traffic flow of heavy vehicles and the proportion of potential passing maneuvers. Additional information

---

[14] *A Policy on Geometric Design,* 1990; and Harwood and others, *Truck Characteristics.*

[15] *A Policy on Geometric Design,* 1990; and G.F. Hayhoe and J. G. Grundman, *Review of Vehicle Weight/Horsepower Ratio as Related to Passing Lane Criteria,* Final Report of the National Cooperative Highway Research Program Project 20-7, Task 10, Transportation Research Board, unpublished, October 1978.

[16] T.D. Gillespie, *Methods for Predicting Truck Speed Loss on Grades,* Report No. FHWA/RD-86/059 (Washington, DC: Federal Highway Administration, November 1985); and Harwood and others, *Truck Characteristics.*

[13] Ibid.

TABLE 2-6

Vehicle Speeds on Upgrades

Vehicle Type	+1.78% Average Speed[a]	+1.78% 12.5 %ile Speed[a]	+3.00% Average Speed	+3.00% 12.5 %ile Speed	+4.00% Average Speed	+4.00% 12.5 %ile Speed	+5.00% Average Speed	+5.00% 12.5 %ile Speed	+6.00% Average Speed	+6.00% 12.5 %ile Speed	+7.00% Average Speed	+7.00% 12.5 %ile Speed
Compact sedan			60	55					57	48		
Standard sedan			62	56					60	52		
Pickup truck			60	54					58	52		
Pickup with canopy			57	52					54	46		
Pickup with camper			53	43					47	38		
Light 4-wheel drive			58	53								
U.S. van			59	54					56	51		
VW van			54	46								
Vehicle/light trailer			53	44								
Vehicle/travel trailer			51	43					40	30		
Motor home			53	46					44	35		
Intercity bus			54	46								
Two-axle truck	54	43	49	40	44	34	40	27	34	23	34	22
Three-axle truck[b]	56	49	48	39	43	31	38	26	33	19	33	23
Four-axle truck[b]	55	48	48	34	42	31	38	28	32	21	26	17
Five-axle truck[b]	55	47	46	34	42	30	35	23	35	21	30	15

[a]Sustained average and 12.5 percentile speeds observed in California for selected vehicle types on upgrades.
[b]Trucks shown as having three or more axles include both straight trucks and combination trucks.

SOURCE: P.Y. CHING AND F.D. ROONEY, "Truck Speeds on Grades in California," California Department of Transportation, June 1979.

on vehicle weight/power ratios and speed-maintenance capabilities as a factor in highway design and vehicle operation can be found in another California study on truck operations on upgrades[17] and in the 1990 AASHTO publication *A Policy on Geometric Design of Highways and Streets*.[18]

Acceleration performance

Information on vehicle acceleration capabilities is needed for evaluation of passing zone lengths on two-lane highways and determination of minimum lengths of acceleration lanes at interchanges. Normal roadway acceleration rates are a factor in designing cycle lengths of traffic signals, in computing fuel economy and travel time values, and in estimating how normal traffic movement is resumed after a breakdown in traffic flow patterns.

Maximum acceleration rates. Typical maximum acceleration rates for passenger cars and tractor-semitrailer combination trucks are shown in Table 2–7 for acceleration from a standing start to speeds from 10 mph to 50 mph. These maximum acceleration rates are expressed as a function of weight/power ratio. Maximum level-road acceleration rates for passenger cars and for tractor-semitrailer combinations over the same range of weight/power ratio are given in Table 2–8 for 10-mph increases in speed at running speeds of 20 mph to 50 mph.

Maximum acceleration rates for operation on upgrades from 2% to 10% are presented in Table 2–9. These data were

TABLE 2-7

Maximum Acceleration from Standing Start

Vehicle Type	Weight-to-Power Ratio (lb/hp)	Typical Maximum Acceleration Rate on Level Road (ft/sec^2) 0 to 10 mph	0 to 20 mph	0 to 30 mph	0 to 40 mph	0 to 50 mph
Passenger car	25	9.3	8.9	8.5	8.2	7.8
	30	7.8	7.5	7.2	6.8	6.5
	35	6.8	6.5	6.2	5.9	5.5
Tractor-semitrailer	100	2.9	2.3	2.2	2.0	1.6
	200	1.8	1.6	1.5	1.2	1.0
	300	1.3	1.3	1.2	1.1	0.6
	400	1.3	1.2	1.1	0.7	—

Passenger car acceleration rates based on performance equations on p. 6 of A.D. ST. JOHN and D.R. KOBETT, *Grade Effects on Traffic Flow Stability and Capacity*, National Cooperative Highway Research Program Report 185 (Washington, DC: Transportation Research Board, 1978). Vehicle horsepowers based on data in Table 2–5 for pickup trucks at 25 lb/hp and for passenger cars at 30 and 35 lb/hp.

Truck acceleration based on data from T.D. HUTTON, "Acceleration Performance of Highway Diesel Trucks," Paper No. 70664 (Society of Automotive Engineers, 1970) shown in Figure 2–7.

TABLE 2-8

Maximum Acceleration for 10-mph Increments

Vehicle Type	Weight-to-Power Ratio (lb/hp)	Typical Maximum Acceleration Rate on Level Road (ft/sec^2) 20 to 30 mph	30 to 40 mph	40 to 50 mph	50 to 60 mph
Passenger car	25	7.8	7.1	6.3	5.6
	30	6.5	5.8	5.2	4.5
	35	5.6	5.0	4.4	3.8
Tractor-semitrailer	100	2.1	1.5	1.0	0.6
	200	1.3	0.8	0.5	0.4
	300	1.0	0.6	0.3	—
	400	0.9	0.4	—	—

Based on same data as in Table 2–7.

---

[17]R.E. Nail and others, *Speed Trends of Five-Axle Trucks on Grades in California* (Sacramento: California Department of Transportation, January 1985).

[18]*Policy on Geometric Design*.

TABLE 2–9

Maximum Acceleration on Upgrades

Speed Change	Passenger Car (30 lb/hp)					Tractor-Semitrailer (200 lb/hp)				
	Level	2%	4%	6%	10%	Level	2%	4%	6%	10%
0 to 20 mph	7.5	6.9	6.2	5.6	4.3	1.6	1.0	0.3	a	a
20 to 30 mph	6.5	5.9	5.2	4.6	3.3	1.3	0.7	a	a	a
30 to 40 mph	5.8	5.2	4.5	3.9	2.6	0.8	0.2	a	a	a
40 to 50 mph	5.2	4.6	3.9	3.3	2.0	0.5	a	a	a	a
50 to 60 mph	4.5	3.9	3.2	2.6	1.3	0.4	a	a	a	a

[a] Truck unable to accelerate or maintain speed on grade. Derived from maximum acceleration rates on level terrain as shown in Tables 2–7 and 2–8 using Equation (2.8).

developed from the values in Tables 2–7 and 2–8 using the following approximate relationship:[19]

$$a_{GV} = a_{LV} - \frac{G\,g}{100} \qquad (2.8)$$

where: a_{GV} = maximum acceleration rate at speed V on grade (ft/sec^2)

a_{LV} = maximum acceleration rate at speed V in level terrain (ft/sec^2)

G = gradient (%)

g = acceleration of gravity (32.2 ft/sec^2)

[19] St. John and Kobett, *Grade Effects*.

The relationships between time traveled and speed achieved for passenger cars accelerating at their maximum rate from a standing stop are given in Figure 2–6 for operation on a level road and on a 6% upgrade. Data are for the range of passenger car weight/power ratios described in Tables 2–7 to 2–9. Similar data for accelerations by trucks with a range of weight/power ratios are shown in Figure 2–7.

Normal acceleration rates. Maximum acceleration rates are seldom used in normal driving. Speed vs. distance relationships for observed normal acceleration by passenger cars from speeds of 0 to 50 mph are illustrated in Figure 2–8. These speed vs. distance relationships represent acceleration rates of approximately 3.5 ft/sec^2 and less, which are less than 65% of the maximum acceleration rates found in Table 2–7. Figure 2–8 represents acceleration rates observed when drivers were not influenced to accelerate rapidly. They are typical of passenger cars starting up after a traffic signal turns green and those used for passing on four-lane highways.

Figure 2–9 illustrates the speed vs. distance, speed vs. time, and distance vs. time relationships for normal acceleration from a stop, such as at an intersection. The speed vs. distance relationship in Figure 2–9 is equivalent to that for zero initial speed in Figure 2–8. The speed vs. time and distance vs. time relationships can be derived directly from the speed vs. distance relationship.

Figure 2–6. Speed vs. time curves for maximum acceleration by passenger cars.

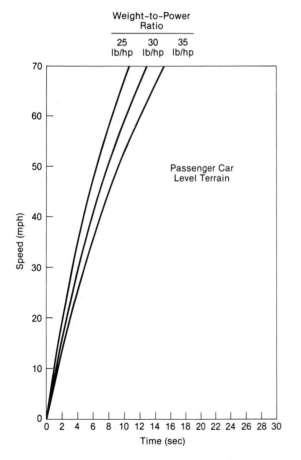

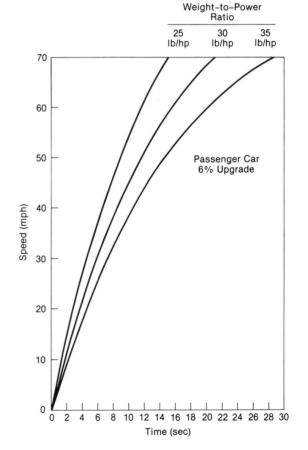

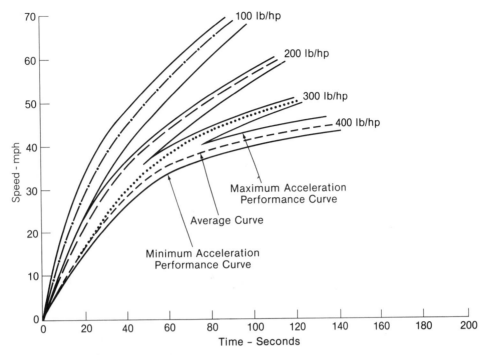

Figure 2–7. Speed vs. time curves for acceleration by trucks with various weight/power ratios.
SOURCE: T.D. HUTTON, "Acceleration Performance of Highway Diesel Trucks," Paper No. 70664 (Society of Automotive Engineers, 1970).

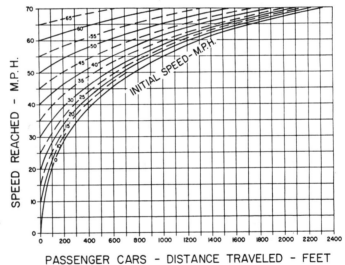

Figure 2–8. Normal acceleration distances for passenger cars on level grade.
SOURCE: *A Policy on Geometric Design of Highways and Streets,* 1990. American Association of State Highway and Transportation Officials, Washington, DC. Used by permission.

Passing zones on two-lane highways. The length of road needed to complete a passing maneuver on a two-lane highway is a function of the vehicle acceleration rate. This is because the more quickly the vehicle can accelerate while passing, the shorter the road length traversed during passing. The acceleration rates for the initial phase of a passing maneuver recommended by AASHTO are 2.05 ft/sec^2 for an average speed of 34.9 mph; 2.10 ft/sec^2 for 43.8 mph; and 2.16 ft/sec^2 for 62.0 mph.[20]

---

[20] *Policy on Geometric Design.*

Deceleration performance

Some deceleration of motor vehicles occurs automatically when the accelerator pedal is released because of the retarding effect of the resistance to motion, including engine compression forces. For controlled deceleration and for maximum rates of deceleration, however, vehicle brakes are used to restrain vehicle motion.

Deceleration without brakes. Deceleration rates without brakes are much greater at high than at low running

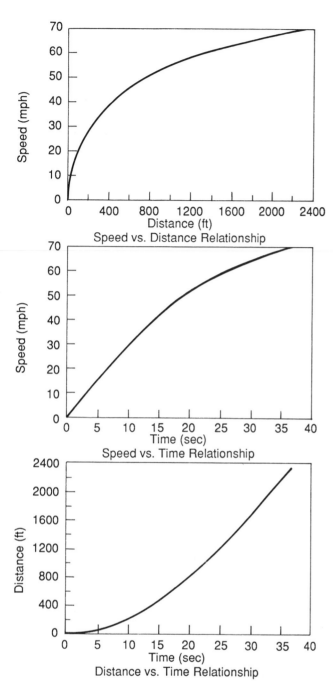

Figure 2–9. Speed, time, and distance relationships for normal acceleration from a stop by passenger cars.

speeds because the resistances to motion, particularly air resistance, are greater. This is important in planning for the control of high-speed traffic. For example, at speeds of 70 mph anything that causes a driver's foot to be removed from the accelerator pedal will result in a rapid drop in speed of about 3.2 ft/sec^2 without a brake light warning to alert following motorists.[21]

An increase in one or more of the resistances to motion will cause the vehicle to decelerate automatically unless compensated for by an immediate increase in the throttle opening. For instance, at locations where a level or descending road changes to an upgrade, or where a straight road enters a sharp curve, vehicles will decelerate appreciably unless the driver depresses the accelerator pedal enough to offset the effect of the added resistance.

Deceleration with brakes. Information on the deceleration rates of motor vehicles with braking (both maximum rates and observed rates for normal deceleration) is needed by traffic engineers. Maximum rates are used for estimating minimum stopping sight distances in emergencies. Normal deceleration rates provide a basis for estimating reasonable time and road lengths for stops at signs and signals where frequent normal stops are necessary.

Maximum deceleration rates. Retardation forces developed in brake drums or discs determine the braking deceleration rates of motor vehicles as long as slippage does not occur between the pavement and tire surfaces. When the applied braking force cannot be carried to the pavement without skidding, the wheels of a vehicle become locked and deceleration rates are determined by the effective coefficient of friction at the tire-pavement contact surface. This coefficient of friction is a function of pavement type, tire condition, and whether or not the pavement surface is wet or dry. Representative values of tire-pavement friction coefficients are given in Table 2–10, together with values recommended for highway design purposes. Because braking systems in good order can usually provide more braking force than can be carried to the pavement, maximum deceleration rates in locked-wheel braking depend primarily on this coefficient of friction between tire and pavement surfaces. The tire-pavement friction coefficient is numerically equivalent to the maximum deceleration rate in locked wheel braking expressed in units of the acceleration of gravity (g). Thus, a tire-friction coefficient of 0.5 corresponds to a locked-wheel deceleration rate of 0.5 g or 16.1 ft/sec^2.

Table 2–10 also includes the minimum distances for braking to a stop that correspond to each tire-pavement friction coefficient presented. The recommended AASHTO criteria for stopping sight distance are based on the braking distances shown for a wet-surface plus an allowance of 2.5 sec for perception-reaction time.[22]

While locked-wheel braking forms the basis for current design criteria, it is not a desirable braking mode. Vehicles, especially trucks, are prone to loss of control in locked-wheel braking because the cornering friction is minimized when the wheels are locked. Antilock brake systems can reduce the potential for loss of control and increase the amount of tire-pavement friction that can be used in controlled braking maneuvers without locking the wheels of a vehicle. Passenger cars are increasingly being equipped with microprocessor-controlled antilock brake systems to keep wheels from locking during braking maneuvers. Trucks in Europe are commonly

[21] E.E. Wilson, "Deceleration Distances for High-Speed Vehicles." In *Proceedings of the 20th Annual Meeting* (Washington, DC: Highway Research Board, 1940).

[22] *Policy on Geometric Design.*

TABLE 2–10
Passenger Car Friction Coefficients and Stopping Distances

Surface Description	Tire-Pavement Friction Coefficient			Braking Distance (ft)		
	Dry Surface[a]		Wet Surface[b]	Dry Surface[a]		Wet Surface[b]
	New Standard Tires	Badly Worn Tires	Recommended by AASHTO for Design[c]	New Standard Tires	Badly Worn Tires	Recommended by AASHTO for Design[c]
	Running Speed = 11 mph					
Dry bituminous concrete	0.74	0.61	—	5	7	—
Sand asphalt	0.75	0.66	—	5	6	—
Rock asphalt	0.78	0.73	—	5	6	—
Portland cement concrete	0.76	0.68	—	5	6	—
	Running Speed = 20 mph					
Dry bituminous concrete	0.76	0.60	0.40	18	22	33.3
Sand asphalt	0.75	0.57	0.40	18	23	33.3
Rock asphalt	0.76	0.65	0.40	18	21	33.3
Portland cement concrete	0.73	0.50	0.40	18	27	33.3
	Running Speed = 30 mph					
Dry bituminous concrete	0.79	0.57	0.35	38	53	85.7
Sand asphalt	0.79	0.48	0.35	38	63	85.7
Rock asphalt	0.74	0.59	0.35	41	51	85.7
Portland cement concrete	0.78	0.47	0.35	38	64	85.7
	Running Speed = 40 mph					
Dry bituminous concrete	0.75	0.48	0.32	71	111	166.7
Sand asphalt	0.75	0.39	0.32	71	137	166.7
Rock asphalt	0.74	0.50	0.32	72	107	166.7
Portland cement concrete	0.76	0.33	0.32	70	162	166.7
All pavements Running Speed = 50 mph			0.30	—	—	277.8
Running Speed = 60 mph			0.29	—	—	413.8
Running Speed = 70 mph			0.28	—	—	583.3

[a]T.E. Shelbourne and R.L. Sheppe, "Skid Resistance Measurement of Virginia Highways," Research Report 5-5, Highway Research Board (Washington, DC: 1948), 62–80.
[b]*A Policy on Geometric Design of Highways and Streets* (Washington, DC: American Association of State Highway and Transportation Officials, 1990).
[c]Based on AASHTO maximum stopping sight distance criteria for design speed equal to running speed.

equipped with antilock brake systems, and this European technology may soon be in use in the United States.[23]

Normal deceleration rates. Maximum deceleration rates are seldom used by drivers except in emergency situations. Deceleration rates up to 10 ft/sec^2 are reasonably comfortable for passenger car occupants. Observed normal (nonemergency) deceleration to speeds from 0 to 50 mph for passenger cars on dry pavements is represented by the speed-distance relationships shown in Figure 2–10. As will be shown in Chapter 9, "Traffic Signals," vehicle clearance intervals at traffic signals (yellow or yellow-plus-all-red time) are determined based on a comfortable deceleration rate of 10 ft/sec^2.

Vehicle operating costs

Table 2–11 presents estimates of average passenger car fuel consumption by manufacturer for automobiles sold in the United States. Table 2–12 shows average passenger car operating costs in the United States over time since 1950.

For engineers to plan and design highway facilities that are compatible with economical vehicle operation, they must know how the fuel consumption and vehicle operating costs shown in Tables 2–11 and 2–12 vary with road geometry, surface conditions, traffic flows, and speed variations. The

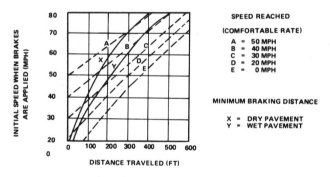

Figure 2–10. Deceleration distances for passenger cars approaching intersections.
SOURCE: *A Policy on Geometric Design of Highways and Streets,* 1990. American Association of State Highway and Transportation Officials, Washington, DC. Used by permission.

information on vehicle operating costs presented in the following paragraphs and in Tables 2–13 through 2–20 was obtained from *Report 111* of the National Cooperative Highway Research Program (NCHRP).[24] These tables are based on data predating the fuel price changes and vehicle emission and fuel consumption rate regulations of the 1970s and are the most comprehensive available. More recent spot checks of certain vehicle types and road conditions indicate that the changes in relative differences in fuel consumption rates

[23]R.W. Radlinski and S.C. Bell, *NHTSA's Heavy Vehicle Brake Research Program—Report No. 6: Performance Evaluation of a Production Antilock System Installed on a Two-Axle Straight Truck,* Report No. DOT HS 807 046; and Harwood and others, *Truck Characteristics.*

[24]Claffey, *Running Costs of Motor Vehicles.*

TABLE 2–11

Fuel Consumption Ratings by Vehicle Manufacturer (mi/gal)[a]

Vehicle Manufacturer	Model Year								
	1974	1976	1978	1980	1982	1984	1986[b]	1988[b]	1989[b]
Domestic fleets									
American Motors	16.3	18.4	18.6	22.3	24.3	36.0	33.7	—	—
Chrysler	13.9	16.3	18.4	22.3	27.6	27.8	27.8	28.4	27.9
Ford	14.2	16.6	18.4	22.9	25.0	25.8	27.0	26.4	26.5
General Motors	12.1	16.6	19.0	22.6	24.6	24.9	26.6	27.6	27.2
Import fleets									
Fuji (Subaru)	25.7	29.7	29.4	28.7	32.0	33.3	31.9	31.9	32.1
Honda	31.1	32.6	33.7	30.1	33.9	35.2	33.7	32.0	31.5
Nissan	24.0	27.0	26.8	32.2	31.2	32.5	30.3	30.4	30.4
Toyota	22.5	25.0	26.8	28.3	30.9	33.5	32.7	32.6	31.9
Volksvagen-Audi[c]	25.9	27.5	27.2	32.3	33.4	29.1	29.8	30.5	30.2

[a]Data are sales-weighted fleet averages.
[b]Data for 1986 and later years are preliminary estimates.
[c]Includes Porsche for 1984 and prior years.

SOURCE: Compiled by Motor Vehicle Manufacturers Association from:

(1974–77) U.S. Environmental Protection Agency (Murrell, Foster, Bristor), *Passenger Car and Light Truck Fuel Economy Through 1980,* Society of Automotive Engineers Paper No. 800853.

(1978–89) U.S. Department of Transportation, September 1, 1989.

TABLE 2–12

Passenger Car Operating Costs, United States[a]

	1950	1960	1970	1979	1986	1989
Variable costs (cents/mi)						
Gas and oil	2.14	2.62	2.76	4.11	4.48	5.20
Maintenance	0.68	0.79	0.68	1.10	1.37	1.90
Tires	0.46	0.49	0.51	0.65	0.67	0.80
Total	3.28	3.90	3.95	5.86	6.52	7.90
Fixed costs ($/yr)						
Fire and theft insurance[b]	15.79	30.38	44.00	74.00	86.00	109.00
Collision insurance[c]	—	—	102.00	168.00	191.00	245.00
Property damage and liability insurance[d]	59.71	109.76	154.00	241.00	232.00	309.00
License and registration	15.47	22.40	24.00	90.00	130.00	151.00
Depreciation[e]	442.05	646.00	729.08	942.00	1320.00	2094.00
Finance charge	—	—	—	296.00	637.00	626.00
Total	533.02	808.54	1053.08	1811.00	2596.00	3534.00
Total variable and fixed costs ($)						
At 10,000 mi/yr[f]	861.02	1198.54	1448.08	2397.00	2959.00	3820.00
At 20,000 mi/yr[e]	1189.02	1588.54	1843.08	3188.00	4239.00	5588.00
Cost per mile (cents)						
At 10,000 mi/yr	8.61	11.99	14.48	23.97	29.60	38.20
At 20,000 mi/yr	5.95	7.94	9.22	15.94	21.20	27.90

[a]Cars specified: 1950, "car in $2,000 price class"; 1960, Chevrolet, 8-cylinder, Bel-Air 4-door sedan; 1970, Chevrolet, 8-cylinder, Impala 4-door hardtop; 1979, Chevelle, 8-cylinder, Malibu Classic 4-door hardtop; 1986 and 1989, Chevrolet Celebrity, 6-cylinder, 4-door sedan.
[b]$100 deductible in 1979 and later years.
[c]$100 deductible through 1979; $250 deductible in later years.
[d]Property damage and liability insurance coverage: 1950, $15,000/$30,000; 1960, $25,000/$50,000; 1970 to date, $100,000/$300,000.
[e]Depreciation based on four-year/60,000-mile retention cycle; for mileage in excess of 15,000, an additional depreciation allowance of $41.00 per thousand in 1979; $68.00 per thousand in 1986; $95.00 per thousand in 1989.
[f]For 1986 and later years, depreciation based on six-year/60,000-mile retention cycle.

SOURCE: American Automobile Association, "Your Driving Costs," various issues.

TABLE 2–13

Fuel Consumption as Affected by Speed and Gradient, Passenger Cars

Uniform Speed (mph)	Gasoline Consumption on Upgrades (gal/mi)					
	Level	2%	4%	6%	8%	10%
10	0.072	0.087	0.103	0.121	0.143	0.179
20	0.050	0.070	0.086	0.104	0.128	0.160
30	0.044	0.060	0.078	0.096	0.124	0.154
40	0.046	0.062	0.078	0.096	0.124	0.156
50	0.052	0.070	0.083	0.104	0.130	0.162
60	0.058	0.076	0.093	0.112	0.138	0.170
70	0.067	0.084	0.102	0.122	0.148	0.180

The composite passenger car represented here reflects the following vehicle distribution: large cars, 20%; standard cars, 65%; compact cars, 10%; small cars, 5%.

The values in this table are for a tangent, high-type pavement and free-flowing traffic. They should be increased by about 20% for speeds of 30 mph and 50% for speeds of 50 mph when operation is on badly broken and patched asphalt pavement. They should also be increased for operation on horizontal 5-degree curves; increase by about 3% at 30 mph and 30% at 60 mph; on 10-degree curves, increase by about 20% at 30 mph and 100% at 50 mph.

SOURCE: P. CLAFFEY, *Running Costs of Motor Vehicles as Affected by Road Design and Traffic,* National Cooperative Highway Research Program Report 111 (Washington, DC: Highway Research Board, 1971).

related to operating conditions are consistent with the values in these tables. While the data in Tables 2–13 through 2–20 are out of date, the only comparable procedure in the AASHTO *Manual on User Benefit Analysis for Highway and Bus Transit Improvements*[25] is also out of date. The latter procedures are currently being updated in NCHRP Project 7-12.[26]

The mix of vehicles considered in Tables 2–13 through 2–19 is specified in the notes to Table 2–13 and should be reviewed to assess its applicability to any particular analysis. It has been found that when panel trucks, pickups, and vans are classified as passenger vehicles in accordance with recent usage trends, the values are usable and consistent with current fuel consumption rates. In addition, the assumptions that must be made for most analyses concerning speed distributions, acceleration practices, and so on, are sources of far greater uncertainty. The effect of these factors, as well as estimates of future fuel consumption rates, should be reviewed for each situation and new base values should be established for the categories in Table 2–13 where appropriate.

Fuel consumption. Vehicle fuel consumption is a major item of operating expense and is strongly influenced by road and traffic conditions. Table 2–13 presents the fuel consumption rates of the composite car for operation at various speeds on level roads and on gradients. The composite car reflects the vehicle distribution given in the first footnote of Table 2–13.

Table 2–14 gives the fuel consumed by the composite passenger car for stop-and-go and speed-change cycles in

[25] *A Manual on User Benefit Analysis of Highway and Bus Transit Improvements 1977* (Washington, DC: American Association of State Highway and Transportation Officials, 1978).

[26] "Microcomputer Evaluation of Highway User Benefits," Project 7-12, National Cooperative Highway Research Program, currently underway.

TABLE 2–14

Excess Fuel Consumption for Stop and Slowdown Cycles, Passenger Cars

Running Speed (mph)	Excess Gasoline Consumed by Amount of Speed Reduction Before Accelerating Back to Speed					
	10 mph	20 mph	30 mph	40 mph	50 mph	60 mph
10	0.0016	—	—	—	—	—
20	0.0032	0.0066	—	—	—	—
30	0.0035	0.0062	0.0097	—	—	—
40	0.0038	0.0068	0.0093	0.0128	—	—
50	0.0042	0.0074	0.0106	0.0140	0.0168	—
60	0.0046	0.0082	0.0120	0.0155	0.0190	0.0208

The composite passenger car represented here reflects the following vehicle distribution: large cars, 20%; standard cars, 65%; compact cars, 10%; small cars, 5%.

Excess fuel consumption for stop-go cycles at given running speeds.

SOURCE: P. CLAFFEY, *Running Costs of Motor Vehicles as Affected by Road Design and Traffic,* National Cooperative Highway Research Program Report 111 (Washington, DC: Highway Research Board, 1971).

excess of that for continued operation at the given running speeds. Tabular values for stop-and-go cycles (the upper values in each column of Table 2–14) do not include fuel consumption for stopped delays. Fuel consumption while stopped may be computed by using the composite passenger car idling fuel consumption rate of 0.58 gal/hr.

Table 2–15 is similar in form to Table 2–13, but it provides fuel consumption rates for a two-axle, six-tire, single-unit truck with an average gross vehicle weight of 12,000 lb.

Table 2–16 is similar in form to Table 2–14 and gives the excess fuel consumption for speed changes for the same two-axle, six-tire, single-unit truck considered in Table 2–17. Average idling fuel consumption while stopped is 0.65 gal/hr for this vehicle.

TABLE 2–15

Fuel Consumption as Affected by Speed and Gradient, Two-Axle, Six-Tire Trucks

Uniform Speed (mph)	Gasoline Consumption on Upgrades (gal/mi)					
	Level	2%	4%	6%	8%	10%
10	0.074	0.120	0.175	0.225	0.289	0.357
20	0.059	0.112	0.167	0.214	0.295	0.394
30	0.067	0.121	0.181	0.232	0.305	—
40	0.082	0.141	0.210	—	—	—
50	0.101	0.159	—	—	—	—
60	0.122	—	—	—	—	—

The composite two-axle, six-tire truck represented here reflects the following vehicle distribution: two-axle trucks at 8,000 lb gross vehicle weight (GVW): 50%; two-axle trucks at 16,000 lb GVW: 50%.

Operation is in the highest gear possible for the grade and speed (fourth, third, or second). When vehicle approach speed exceeds the maximum sustainable speed on upgrades, speed is reduced to this maximum as soon as the vehicle gets on the grade.

The values in this table are for a tangent, high-type pavement and free-flowing traffic. They should be increased by about 7% for speeds of 30 mph and 20% for speeds of 50 mph when operation is on badly broken and patched asphalt pavement. They should also be increased for operation on horizontal 5-degree curves; increase by about 3% at 30 mph and 23% at 50 mph; on 10-degree curves, increase by about 21% at 30 mph and 43% at 40 mph.

SOURCE: P. CLAFFEY, *Running Costs of Motor Vehicles as Affected by Road Design and Traffic,* National Cooperative Highway Research Program Report 111 (Washington, DC: Highway Research Board, 1971).

TABLE 2–16

Excess Fuel Consumption for Stop and Slowdown Cycles, Two-Axle, Six-Tire Trucks

| Running Speed (mph) | Excess Gasoline Consumed by Amount of Speed Reduction Before Accelerating Back to Speed (gal) | | | |
	10 mph	20 mph	30 mph	40 mph
10	0.0036	—	—	—
20	0.0073	0.0097	—	—
30	0.0080	0.0148	0.0173	—
40	0.0096	0.0167	0.0226	—

The composite two-axle, six-tire truck represented here reflects the following vehicle distribution: two-axle trucks at 8,000 lb gross vehicle weight (GVW): 50%; two-axle trucks at 16,000 lb GVW: 50%.

Excess fuel consumption for stop-go cycles at given running speeds.

SOURCE: P. CLAFFEY, *Running Costs of Motor Vehicles as Affected by Road Design and Traffic,* National Cooperative Highway Research Program Report 111 (Washington, DC: Highway Research Board, 1971).

TABLE 2–17

Tire Wear Cost Factors (cents/mi), Passenger Cars

| Uniform Speed (mph) | High-Type Concrete | | | High-Type Asphalt | | | Dry, Well-Packed Gravel |
	Tangent	5°	10°	Tangent	5°	10°	Tangent
20	0.09	0.11	0.17	0.27	0.32	0.51	1.03
30	0.19	0.43	0.82	0.36	0.82	1.56	1.05
40	0.29	2.17	4.83	0.43	3.22	7.16	1.07
50	0.32	4.44	14.34	0.45	6.25	20.10	1.10
60	0.31	7.64	—	0.46	16.34	—	—
70	0.30	—	—	0.44	—	—	—
80	0.27	—	—	0.43	—	—	—

The composite passenger car represented here reflects the following vehicle distribution: large cars, 20%; standard cars, 65%; compact cars, 10%; small cars, 5%.

Tire costs were computed by using a weighted average cost of $119 for a set of four new, medium-quality tires based on the following unit tire costs by vehicle type (as noted in the Northeastern states in 1969): large cars, $35 per tire; standard-size cars, $30 per tire; compact cars, $25/tire; small cars, $15/tire.

There are approximately 1,500 grams of usable tire tread in 80% of passenger car tires. This weight of usable tire tread was also recorded for the tires in the tire-wear test.

Tire-wear costs are based on 1969–1970 prices. See text for method of update.

SOURCE: P. CLAFFEY, *Running Costs of Motor Vehicles as Affected by Road Design and Traffic,* National Cooperative Highway Research Program Report 111 (Washington, DC: Highway Research Board, 1971).

Tire wear. Tire wear costs in Tables 2–17 to 2–19 are based on 1969–1970 prices as indicated in the notes. Prices must be adjusted to current or forecast conditions in accordance with the objectives of the analysis. Such adjustments are affected by three principal variables: (1) the change in the price of tires for each vehicle type; (2) the change in durability of tires in general use; and (3) the change in vehicle mix. For example, if study-year prices for the tires of the four vehicle types referred to in Table 2–17 are $80, $65, $55, and $45; if the vehicle mix is "large (including vans, campers, pickups, and panels) 20%, "standard" 20%, "compact" 20%, and small 40%; and if average tire life has increased from 25,000 to 40,000 miles, then an adjustment factor is computed as follows:

$$\frac{(80 \times 20 + 65 \times 20 + 55 \times 20 + 45 \times 40)}{100 \times 40,000}$$

$$\div \frac{(35 \times 20 + 30 \times 65 + 25 \times 10 + 15 \times 5)}{100 \times 25,000} = 1.08$$

Thus, all values in Table 2–17 would be increased by 8% for this example analysis. As with changes in fuel consumption, it is expected that future research will provide refinements in many operating cost factors. For credibility of analytic results, the effects of new relationships and sensitivity should be related to the policy, planning, and design objectives of the analysis.

The excess tire-wear costs for stop-and-go and 10 mph speed-change cycles are shown in Table 2–18 for the composite passenger car for a series of running speeds on high-type road surfaces.

Tire-wear cost values for the representative single-unit truck for straight road operations at 45 mph for curve operations and for operation on major urban arterials are given in Table 2–19.

Maintenance costs. Maintenance costs for vehicle components should be estimated for known or assumed study-year conditions. These costs are expected to change significantly and are influenced by inflation and changes in design related to fuel conservation. In 1970, passenger car maintenance costs were estimated to be 1.15 cents/mi compared to over 4.0 cents/mi for 1979.[27, 28] Although not

TABLE 2–18

Excess Tire Wear Costs for Speed Changes, Passenger Cars

| Running Speed (mph) | Cost of Tire Wear on Four Tires (cents per cycle) | | | |
| | Stop-Go Speed Change Cycles | | 10-mph Slowdown Cycles | |
	Concrete	Asphalt	Concrete	Asphalt
20	0.10	0.30	0.04	0.10
30	0.30	0.60	0.08	0.15
40	0.58	0.85	0.09	0.14
50	0.72	1.10	0.09	0.14
60	0.80	1.20	0.08	0.12
70	0.85	1.25	0.08	0.12

The composite passenger car represented here reflects the following vehicle distribution: large cars, 20%; standard cars, 65%; compact cars, 10%; small cars, 5%.

Tire costs were computed by using a weighted average cost of $119 for a set of four new, medium-quality tires based on the following unit tire costs by vehicle type (as noted in the Northeastern states in 1969): large cars, $35 per tire; standard-size cars, $30 per tire; compact cars, $25/tire; small cars, $15/tire.

There are approximately 1,500 grams of usable tire tread in 80% of passenger car tires. This weight of usable tire tread was also recorded for the tires in the tire-wear test.

Tire-wear costs are based on 1969–1970 prices. See text for method of update.

SOURCE: P. CLAFFEY, *Running Costs of Motor Vehicles as Affected by Road Design and Traffic,* National Cooperative Highway Research Program Report 111 (Washington, DC: Highway Research Board, 1971).

---

[27] Claffey, *Running Costs of Motor Vehicles.*

[28] J.E. Ullman, *Cost of Owning and Operating Automobiles and Vans 1979* (Washington, DC: Federal Highway Administration, May 1979).

TABLE 2–19

Tire Wear Costs per Axle (cents/mi), Two-Axle, Six-Tire Trucks

Type of Operation and Road Surface	Rear Axle- Four Tires	Front Axle- Two Tires
Uniform speed of 45 mph on high-type concrete	0.40	0.10
25–30 mph on four-lane major urban arterial with high-type concrete surface (3–4 stops per mi)	1.96	0.28
25 mph on 30° curve with high-type surface	10.80	1.30
25 mph on 60° curve with high-type surface	108.00	9.20

Tire costs were computed by using a weighted average cost of $120 for a medium-quality, 10-ply 8.25 × 20 transport tire (1970 price in the Northeastern states). Each tire has approximately 4,500 grams of usable tread before recapping is necessary. Value of tire carcass when recapped was assumed to be $20.

Tire wear costs are based on 1969–1970 prices. See text for method of update.

SOURCE: P. CLAFFEY, *Running Costs of Motor Vehicles as Affected by Road Design and Traffic,* National Cooperative Highway Research Program Report 111 (Washington, DC: Highway Research Board, 1971).

directly comparable, the 1970 cost for pickup trucks was 1.42 cents/mi and the 1979 cost for passenger vans was 5.28 cents/mi. The maintenance cost per mile for over-the-road trucks is less than for pickups and vans, because they travel much greater mileage per year and have very different design and operating characteristics.

Oil consumption. Oil consumption results from oil contamination during use and oil loss through leakage and combustion. The combined oil consumption rates for contamination, leakage, and combustion for passenger cars and two-axle, six-tire trucks are given in Table 2–20 for operation in free-flowing traffic on dust-free (high-type) roads.

Depreciation. The magnitude of motor vehicle depreciation cost, when defined as the quotient resulting from dividing the difference between original cost and scrap value by lifetime mileage, depends largely on nonhighway factors (new- and used-car market values and user travel desires).

TABLE 2–20

Engine Oil Consumption Rates (quarts/1,000 mi)[a]

Speed (mph)	Passenger Car[b]	Two-Axle, Six-Tire Truck[c]
30	0.97	2.77
35	0.97	2.77
40	1.12	2.84
45	1.28	3.00
50	1.45	3.16
55	1.64	3.33
60	1.78	3.50

[a]Tests conducted in free-flowing traffic on high-type roads. Oil consumption includes both oil for oil changes and additions between oil changes. Minimum trip length = 10 miles.
[b]An 8-cylinder Chevrolet sedan with an engine displacement of 283 cu in represented the typical passenger car.
[c]A truck with a 6-cylinder engine (engine displacement of 351 cu in) represented the typical two-axle, six-tire truck.

SOURCE: P. CLAFFEY, *Running Costs of Motor Vehicles as Affected by Road Design and Traffic,* National Cooperative Highway Research Program Report 111 (Washington, DC: Highway Research Board, 1971).

Because the work of the traffic engineer has little effect on these factors, vehicle depreciation cost factors are not included here. The used-car-market price book generally sets remaining value up to about 5 years. A write-off somewhere in the fifth to tenth years is commonly set.

Congestion costs. Highway congestion conditions affect vehicle operating costs for fuel and oil consumption, tire wear, and maintenance principally as a result of the speed changes and stopped delays associated with congestion. The frequency and severity of the speed changes and the duration of stopped delays caused by congestion vary widely from one congested location to another. When it is desired to determine vehicle operating costs caused by congestion at a given location, the recommended procedure is to observe, at that location, both the frequency of speed changes by ranges of speed change and the frequency of vehicle stops by range of stop duration and to compute the resulting fuel and oil consumption, tire wear, and maintenance costs by using the speed-change and idling operating costs data given above. Similarly, the fuel conservation potential of alternative traffic control strategies and geometric designs can be estimated.

Transit and buses

Transit vehicles are the rolling stock operating on guideways or highways used to transport passengers.[29] Transit vehicles include both conventional highway vehicles, like those discussed earlier in this chapter, and special-purpose vehicles dedicated to transit use. The following major categories of transit vehicles are commonly used:

1. *Van:* a passenger vehicle on an automobile or light truck chassis, propelled by an internal combustion engine, with a capacity for 6 to 15 persons.
2. *Minibus:* a vehicle less than 25 ft long, propelled by an internal combustion engine, with a capacity for more than 15 persons.
3. *Transit bus:* a vehicle more than 25 ft long usually propelled by a diesel engine. It is usually designed for frequent-stop service with front and center doors, low-back section, and without luggage-storage compartments or restroom facilities that are common on intercity buses.
4. *Trolley bus:* a transit bus propelled by electricity obtained from overhead wires.
5. *Articulated bus:* a diesel or electric transit bus with a permanently attached semitrailer, with full interior passenger circulation.
6. *Double-deck bus:* a bus with two levels.
7. *Dual-powered bus:* a transit bus that can be propelled by electricity or by a diesel engine.
8. *Street car or light-rail vehicle (LRV):* an electrically propelled rail vehicle operated singly or in trains on shared or semiexclusive right of way.
9. *Rail transit car:* an electrically propelled vehicle usually operated in trains on exclusive right-of-way.

[29]Material in this section on transit and buses was prepared by Herbert S. Levinson. For more details, see Chapter 10, "Urban Mass Transit Systems." In the *Transportation Planning Handbook*.

TABLE 2–21

Characteristics of Typical Transit Vehicles, United States and Canada

Type of Vehicle or Train	Length (ft)	Width (ft)	Typical Capacity[a]			Remarks
			Seats	Standees[b]	Total	
Minibus-short haul	18–25	6.5–8.0	15–25	0–15	15–40	
Transit bus	30.0	8.0	36	19	55	Example: General
	35.0	8.0	45	25	80	Motors, RTS II,
	40.0	8.5	83	32	85	1978
Articulated transit bus	55.0	8.5	66	34	100	Chicago-AM General-MAN
	59.7	8.5	73	37	110	AM General-MAN
Street car	46.7	9.0	59	40–80	99–139	P.C.C.[c]
Light rail car train	151.2	8.7	128	248–272	376–400	San Diego—6-axle car, 2-car train (DU-WAG)
	142.0	8.8	104	250–356	354–460	Boston—6-axle car, 2-car train (Boeing Vertol)
Rail rapid transit train	605.0	10.0	500	1,300–1,700	1,800–2,200	10-car train, IND New York
	600.0	10.0	576	1,224–1,664	1,800–2,240	8-car train, R-46 cars, New York
	448.6	10.3	504	876–1,356	1,380–1,860	8-car train, Toronto
Commuter rail train	85.0	10.5	1,100	200–1,200	1,300–2,300	Regular car, 10-car train

[a]In any transit vehicle the total passenger capacity can be increased by removing seats and by making more standing room available, and vice versa.
[b]Higher figures denote crush capacity; lower figures, schedule-design capacity.
[c]Presidents' Conference Committee Cars.

SOURCE: *Highway Capacity Manual,* Special Report 209, (Washington, DC: Transportation Research Board, 1985), p. 12–8.

10. *Commuter railroad car*: a standard railroad passenger car with high-density seating. It may be self-propelled by electric or diesel engines or designed for haulage by a locomotive.

11. *Downtown people mover*: a type of automated guideway transit vehicle operating on a loop or as a shuttle within a central business district.

12. *Ferry boat*: a passenger-carrying marine vehicle that provides frequent shuttle service on a published schedule over a fixed route between two or more points.

13. *Inclined plane*: a type of transit passenger vehicle operating over exclusive right-of-way on steep grades with unpowered vehicles propelled by moving cables attached to the vehicles and powered by engines or motors at a central location *not* on board the vehicle.

14. *Monorail*: a type of transit vehicle with a guideway formed by a single beam or rail from which an electrically powered transit vehicle or train of vehicles either straddles or is suspended.

Typical ranges for key dimensional and capacity characteristics of transit vehicles currently in significant use are indicated in Table 2–21. Lengths and widths shown are external body dimensions. Heights are from pavement or top of rail to roof. The lower end of the ranges of number of seats shown and the upper end of the ranges for standees generally apply to vehicles used on high-volume routes in large cities. The largest number of seats and the lowest number of standees occur on longer suburban routes, or where, for policy reasons, high levels of comfort are to be offered.

For rail vehicles, Table 2–21 also shows the maximum length of trains in common use, and the resulting total capacity per train.

Table 2–22 summarizes dimensional, capacity, and passenger accommodation characteristics for six representative contemporary North American urban transit vehicles. The standee capacities shown are those given by the manufacturers and do not necessarily represent policies of operators who purchase these vehicles.

The ranges of weights for different types of urban transit vehicles are shown in Table 2–23, which also lists the ranges of weight/design capacity ratios. The weight/design capacity ratio is one measure of efficiency, because propulsion energy depends more on vehicle weight than any other factor.

Traffic characteristics

This section discusses the key traffic characteristics that should be considered in the design and operation of the highway system, including traffic volumes, traffic volume growth, traffic composition, speed characteristics, speed trends, and spacing and headway characteristics.

Traffic volumes

Traffic volumes vary in both space and time. These variations are critical determinants of the way highway facilities are used, and they control many of the planning and design requirements for adequately serving traffic demand. The typical daily traffic volume on a highway is represented by the average daily traffic (ADT). When traffic volumes are counted over long periods of time, as is possible at permanent, continuous count stations, the annual average daily traffic volume (AADT) can be determined or estimated. The AADT is the total annual traffic volume divided by the number of days in the year.

Traffic demand varies by month of year, by day of week, by hour of the day, and by subhourly intervals within any given hour. Traffic volumes for hourly and subhourly intervals normally form the basis for design of highway facilities.

TABLE 2-22
Selected Characteristics for Some Contemporary North American Urban Transit Vehicles

Characteristics	Unit	Urban Transit Bus, Nonarticulated, General Motors, RTS II, 1978	Urban Transit Bus, Articulated, AM General/MAN, SG-220-18-2A, 1978	Light Rail Vehicle Articulated Boeing-Vertol, for Boston MTA, 1976
Length of vehicle body	ft	40.0	59.66	71.0
Maximum width of body	ft	8.5	8.5	8.85
Height, wheels to roof	ft	9.9	10.34	11.33
Seats		45[b]	73	52
Standees	Design/crush	23/47	37/72	167/230
Total passenger capacity	Design/crush	68/92	110/145	219/282
Maximum cars per train		—	—	3
Total train capacity	Design/crush	—	—	650/850
Width of seats	in	17.5	17.0	17.5[c]
Spacing between backs of transverse seats	in	27.4	27.0	30.0[c]
Area per seated passenger	in²	3.3	3.2	3.6[c]
Minimum aisle width	in	22.5	22.0	44.8
Clear doorway width per passenger lane	in	30.0 : 22.0	24.6	27.0
Doorways per side × lanes per door		1 × 2 : 1 × 1	2 × 2	3 × 2
Passenger load factor	Design/crush	1.5/2.0	1.5/2.0	4.2/5.4
Empty weight	lb × 10³	25.9	37.5	67.0
Weight per passenger (design/crush)	lb	380/280	340/260	305/240
Rated power of engines (motors)	hp	225	235	410

Characteristics	Rail Transit Car, Steel Wheels Pullman-Standard, R-46, 1975[a]	Rail Transit Car, Rubber-Tires Canadian-Vickers for Montreal, 1975[a]	Regional Rail Transit Car, Rohr Corp., San Francisco BART, 1972[a]
Length of vehicle body	75.0	56.5	71.0
Maximum width of body	10.0	8.3	10.5
Height, wheels to roof	12.1	12.0	10.5
Seats	73	40	72
Standees	142/277	118/169	60/156
Total passenger capacity	215/350	158/209	132/228
Maximum cars per train	8	9	10
Total train capacity	1690/2800	1420/1880	1320/2880
Width of seats	18.0[c]	17.5[c]	22.0
Spacing between backs of transverse seats	[d]	[d]	34.0
Area per seated passenger	3.3[c]	3.3[c]	5.2
Minimum aisle width	39.0	19.0	30.0
Clear doorway width per passenger lane	25.0	25.5	27.0
Doorways per side × lanes per door	4 × 2	4 × 2	2 × 2
Passenger load factor	3.0/4.8	4.0/5.2	1.8/3.2
Empty weight	87.1	55.3	58.7
Weight per passenger (design/crush)	405/250	325/265	445/255
Rated power of engines (motors)	460	400	560

[a]Data are average for the normal combination of "A" and "B" cars, or motorized cars and trailers, used to make up maximum length of trains.
[b]Maximum seating could be 47, but 2 seats were removed to provide more comfort for seated passengers.
[c]Estimated, based on seat plans.
[d]Transverse seats not arranged in files, or no transverse seats.

SOURCE: General Motors Corp.; AM General Corp.; Boeing-Vertol Co.; AMERICAN PUBLIC TRANSIT ASSOCIATES, *Roster of North American Rapid Transit Cars: 1945–1976* (Washington, DC: January 1977).

Seasonal and monthly variations. Seasonal fluctuations in traffic demand reflect the social and economic activity of the area being served by the highway. Figures 2–11 and 2–12 show monthly variation patterns observed in Illinois, and illustrate several significant characteristics:

○ Monthly variations are larger on rural routes than on urban routes.

○ Monthly variations are larger on rural routes serving primarily recreational traffic than on rural routes serving primarily business traffic.

○ Variations in traffic volume between days of the week are largest on recreation routes.

These observations lead to the conclusion that both commuter and business-oriented travel occur in more uniform patterns

TABLE 2–23
Ranges of Weight of Transit Vehicles

Transit Vehicle Type	Empty Weight (lb × 1,000)	Weight/Unit Design Capacity (lb/psgr)
Van	5–7.5	200–650
Minibus	7–17	200–700
Transit bus		
Single unit	14–26	175–340
Articulated	28–36	160–360
Double-deck	15–28	200–230
Streetcar		
Single unit	36–52	320–575
Articulated	45–110	250–600
Rail transit		
Steel wheel	35–90	200–500
Rubber tired	44–75	275–450
Commuter railroad car		
Regular	70–115	350–600
Double deck	100–125	400–750

SOURCE: *Lee Transit Compendium.* Vol. II (1975). Nos. 5, 6, 9; Vol. III (1976–1977). Nos. 5, 6, 9; various other databooks and specifications.

Figure 2–11. Examples of monthly traffic volume variations showing volume trends by route type on rural roads.

SOURCE: Data for Lake County, Illinois, in *Highway Capacity Manual,* Special Report 209 (Washington, DC: Transportation Research Board, 1985), p. 2–6.

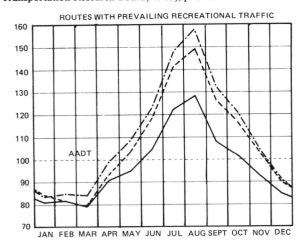

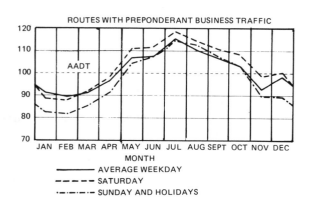

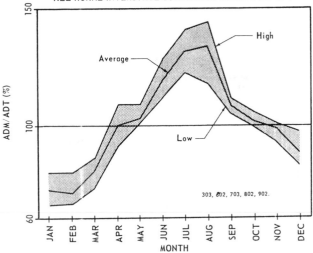

Figure 2–12. Ratio of traffic volume for average day of the month to AADT.

SOURCE: Continuous count data for Illinois from *Traffic Characteristics on Illinois Highways/1971* (Springfield: IL Department of Transportation, 1971).

than recreational travel, which is subject to the greatest variation among trip-purpose components of the traffic stream.

Daily variations. Volume variations by day of the week are also related to the type of highway on which observations are made. Figure 2–13 shows that weekend volumes are lower than weekday volumes for highways serving predominantly business travel, such as urban freeways. In contrast, peak traffic occurs on weekends on main rural and recreational access facilities. Further, the magnitude of daily variation is highest for recreational access routes and least for urban commuter routes.

Hourly variations. Typical hourly variations in traffic volume as related to highway type and day of the week are

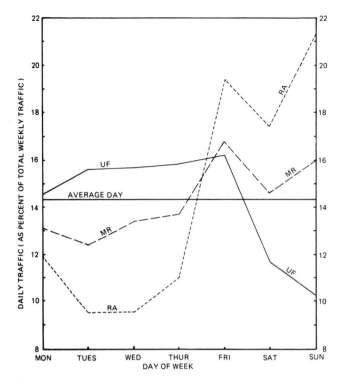

Legend:

MR, Four-lane rural freeway
RA, Two-lane rural recreational access route
UF, High volume six- and eight lane urban freeways

Figure 2–13. Examples of daily traffic variation by type of route.

SOURCE: 1980–1982 data for Minnesota in *Highway Capacity Manual*, Special Report 209 (Washington, DC: Transportation Research Board, 1985), p. 2–8.

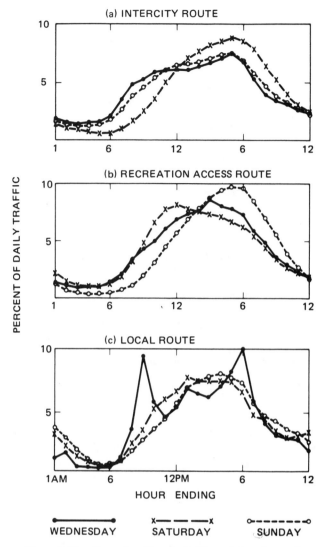

Figure 2–14. Examples of hourly traffic variations for rural routes.

SOURCE: Data for New York State in *Highway Capacity Manual*, Special Report 209 (Washington, DC: Transportation Research Board, 1985), p. 2–10.

illustrated in Figure 2–14. The typical morning and evening peak hours are evident for urban commuter routes on weekdays. The evening peak generally has somewhat higher volumes than does the morning peak. On weekends, urban roads show a peak that is lower and more spread out than on weekdays and that occurs in the early to mid-afternoon period.

Recreational routes also have single daily peaks. Saturday peaks on such routes tend to occur in the late morning or early afternoon (as travelers go *to* their recreational destination) and in the late afternoon or early evening on Sunday (as they return *from* their recreational destination).

On intercity routes serving a mix of traffic, late afternoon peaks are evident, and there is less difference between the patterns of variation for weekdays and weekends.

The repeatability of hourly variations is of great importance. The stability of peak-hour demands affects the viability of using available traffic data in design and operational analysis of highways. Figure 2–15 shows data obtained over a 77-day period in metropolitan Toronto. The shaded area indicates the range within which 95% of the observations are expected to fall. Although variations by hour of the day are typical for urban areas, the relatively narrow and parallel fluctuations among the 77 days illustrate the repeatability of the basic pattern. The observations depicted were obtained for one-way traffic only, as evidenced by the single

peak hour shown for either morning or afternoon. While the plots in Figures 2–14 and 2–15 are typical of hourly traffic variations, specific patterns do vary in response to local travel habits and environments, and these examples should not be used as a substitute for locally obtained data.

Relationship between ADT and peak hour volumes. Operational analyses of highways focus on the peak hour, because it represents the most critical period for traffic operations and has the highest highway capacity requirements. The peak hour volume, however, is not a constant from day to day, or from season to season.

In several studies of traffic volume characteristics, relationships have been developed between various hourly volumes and the average daily traffic (ADT), which is the total volume during a given time period (in whole days greater than one day and less than one year) divided by the number

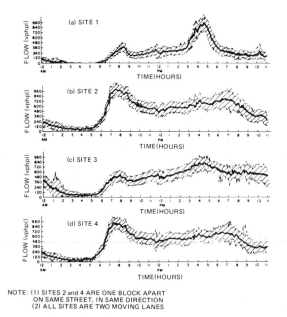

Figure 2-15. Example of repeatability of hourly traffic variations.

SOURCE: Data for four two-lane arterials in Toronto, Ontario, in *Highway Capacity Manual*, Special Report 209 (Washington, DC: Transportation Research Board, 1985), p. 2-11.

of days in that time period. A typical series of these relationships is represented by the plots in Figure 2–16, where both the percentage of annual average hourly volume and the percentage of annual average daily traffic are plotted in decreasing order of magnitude against hours of the year.

Although the plots approximately represent a family of similar curves, the functional characteristics of a given route dictate the size and rate of change of the ordinate in these plots for the highest hours of the year. Figure 2–16 illustrates that very large demands are present on recreational routes for only a few hours of the year, but on urban routes the volumes are more evenly distributed throughout the hours of the year.

Economic considerations in the planning and design of highways make it impractical to design for the highest expected hourly volumes. Instead, a design hourly volume (DHV) is typically selected from consideration of the specific relationship between the percentage of annual average daily traffic and the highest hours of the year given the functional class and operational characteristics of the route. Design hourly volumes should represent the traffic volumes and hourly variations forecast for a specified future year, known as the *design year*. In general, a pronounced break in curves such as those illustrated in Figure 2–16 has been thought to occur in the range from the 20th to the 50th highest hour. Recent studies have emphasized the difficulty in locating such a pronounced break, but the range from the 20th to the 50th highest hour remains a key element in determining highway design requirements.[30]

If an hourly value is selected to the left of the range discussed above (i.e., for some hour less than the 20th highest hour), a sizable increase in design requirements and highway construction costs may accommodate only a very few hours of the year. On the other hand, design for a volume to the

---

[30] N. Cameron, "Determination of Design Hourly Volume" (Calgary, Alberta: University of Calgary, May 1975).

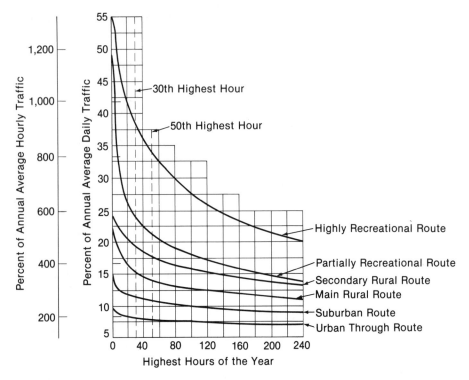

Figure 2-16. Distribution of hourly volumes on various types of routes found during the highest volume hours of the year.

SOURCE: T.M. MATSON, W.S. SMITH, AND F.W. HURD, *Traffic Engineering* (New York: McGraw-Hill Book Company, 1955). By permission of publisher.

right of the 50th highest hour produces relatively small decreases in the DHV. Many highway agencies have selected the 30th highest hourly volume (30 HV) as the appropriate DHV for most highways. In other words, the facility would be designed so that congestion would occur during a maximum of 30 hours during the design year. However, the 30th highest hour may not be appropriate for design of some recreational routes, as it only occurs on a few weekends.

The ratio of the DHV to the ADT for the design year is designated K. The standard practice for determining the DHV is to multiply the forecast value of ADT by a value of K appropriate for the functional class of the facility. Where values of K are based on the 30th highest hourly volume, the following characteristics have been noted:

o The value of K generally decreases as the ADT of the highway increases.
o The value of K decreases as development density increases.
o The highest values of K generally occur on recreational facilities, followed by rural, suburban, and urban facilities, in descending order.

The value of K for the 30th highest hourly volume for main rural highways generally averages approximately 15% and varies in the range from 12% to 18%. For urban facilities, the average value of K for the 30th highest hourly volume is about 11%, with a range from 7% to 18%. The ratio of any other highest hourly volume to the average daily traffic may also be selected as an estimate of K. The graph of hourly

volume as a percentage of the average daily traffic for the highest hours of the year in Figure 2–17 is an illustration of the range within which the value of this ratio is expected to fall 70% of the time for typical traffic conditions encountered on main rural highways.

Traffic volume growth

Traffic volumes on most highways grow every year except in areas affected by economic downturns or highway realignments that alter traffic patterns. Figure 2–18 illustrates the growth of traffic volumes and truck loadings on the rural Interstate highway system since 1970. The figure shows the cumulative percentage increase in total daily traffic and in total truck traffic as represented by equivalent 18,000-lb single-axle loads. Traffic volume growth rates for particular highways are highly site specific and must be based on forecasts from local data.

Traffic volume growth may be especially large in urban and suburban areas where new development of employment centers and residential subdivisions is occurring. The ITE *Trip Generation Handbook* provides guidance on the number of additional trips that are likely to be generated by specific types of development.[31]

Traffic composition

Another key characteristic of traffic volume is the composition of the traffic stream. Traffic composition is generally measured by percentage of trucks and buses at a given roadway location. Table 2–24 presents typical traffic volume composition for selected highway types based on data from the Federal Highway Administration (FHWA) Highway Performance Monitoring System.[32] The table shows that trucks constitute approximately 11% of traffic on rural highways and approximately 5% of traffic on urban highways. The percentage of trucks in the peak hour is generally smaller than these estimates. Buses generally constitute less than 1% of the traffic stream. Other vehicle types, including recreational vehicles such as motor homes and campers, have distinct operational characteristics that may be important at some locations, particularly on rural highways. The estimates in Table 2–24 are very general, and traffic composition must be evaluated on a site-specific basis for design purposes.

Speed characteristics

The quality of service provided by a highway facility is often associated with speed or its reciprocal—travel time. In fulfilling their travel desires, drivers select those speeds they consider appropriate for the conditions under which the trip is made. The variables that influence travel speeds can be classified as driver, vehicle, roadway, traffic, and

Figure 2–17. Relation between peak-hour volume and ADT for main rural highways.
SOURCE: *A Policy on Geometric Design of Highways and Streets,* 1990. American Association of State Highway and Transportation Officials, Washington, DC. Used by permission.

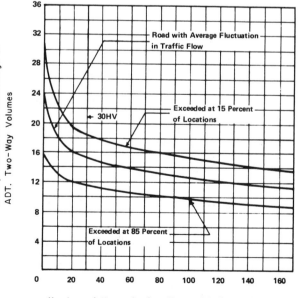

Number of Hours in One Year with Hourly Volume Greater than that Shown

[31] *Trip Generation Handbook* (Washington, DC: Institute of Transportation Engineers, 1989).

[32] *Highway Statistics 1987,* Report No. FHWA-PL-88-008 (Washington, DC: Federal Highway Administration, 1988).

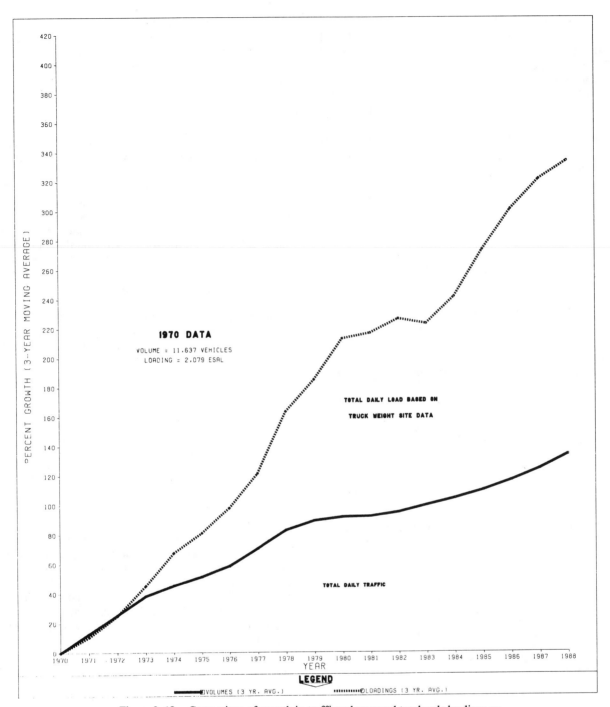

Figure 2–18. Comparison of growth in traffic volumes and truck axle loadings on rural interstate highways.
SOURCE: *Highway Statistics 1988,* Report No. FHWA-PL-88-088 (Washington, DC: Federal Highway Administration, 1988), p. 181.

environment. Vehicular speed is an important consideration in highway transportation because it has significant implications for the economics, safety, comfort, and convenience of highway travel for both motorists and the general public.

Speed fundamentals. Traffic speed is the magnitude of the velocity of vehicles in the traffic stream and is usually expressed in miles per hour (mph). Two distinct types of

average speed measures can be derived from the above definition to represent speed of the traffic stream. The first type of average speed is the time-mean speed, or spot speed, which is the arithmetic mean of a set of instantaneous vehicle speeds measured at one particular location on the highway. (In practice, spot speeds can be estimated reliably from the time required to travel very short distances, typically 100 ft or less.) Time-mean speed is calculated as the average

TABLE 2–24

Traffic Composition on U.S. Highways

	Percentage of Total Travel by:				
Highway Type	Passenger Cars[a]	Buses	Single-Unit Trucks	Combination Trucks	All Trucks
Rural Interstate	80.6	0.3	3.1	16.0	19.1
Other rural arterials	89.9	0.3	3.4	6.4	9.8
Other rural highways	92.7	0.5	3.8	3.0	6.8
ALL RURAL HIGHWAYS	88.9	0.4	3.5	7.2	10.7
Urban Interstate	91.5	0.2	2.3	6.0	8.3
Other urban highways	96.2	0.2	1.9	1.7	3.6
ALL URBAN HIGHWAYS	95.2	0.2	2.0	2.6	4.6
TOTAL RURAL AND URBAN	92.7	0.3	2.6	4.5	7.1

[a]Including light trucks and vans.

SOURCE: *Highway Statistics 1987,* Report No. FHWA-PL-88-008 (Washington, DC: Federal Highway Administration, 1988), p. 171.

of several spot-speed observations at a particular highway location, as follows:

$$\bar{u}_t = \frac{\sum_{i=1}^{n} u_i}{n} \qquad (2.9)$$

where:

$\bar{u}_t$ = average time-mean speed
u_i = spot speed of ith vehicle
n = number of vehicles whose spot speeds were observed

A second type of average speed is the space-mean speed, or travel speed, which is computed as a specified travel distance divided by the mean travel time of several trips over this highway section and is expressed by the following relationship:

$$\bar{u}_s = \frac{dn}{\sum_{i=1}^{n} t_i} \qquad (2.10)$$

where:

$\bar{u}_s$ = average space-mean speed
d = travel distance
n = number of trips whose travel times were observed
t_i = travel time for the ith trip

Time-mean speed is always greater than space-mean speed for a given sample of traffic flow, except for the special case in which all vehicles are traveling at exactly the same speed. In this special case, the time-mean and space-mean speeds are equal. The results of speed measurements summarized as time-mean and space-mean speeds are illustrated by a computational example in Table 2–25. An approximate relationship between time-mean and space-mean speeds is given by the following:[33]

$$\bar{u}_t = \bar{u}_s + \frac{\sigma_s^2}{\bar{u}_s} \qquad (2.11)$$

where:

$\bar{u}_t$ = average time-mean speed
$\bar{u}_s$ = average space-mean speed
σ_s^2 = variance of space-mean speeds

Space-mean speeds are directly related to the density of vehicles on the highway; time-mean speeds are related only to the number of vehicles passing a given point on the roadway.

Speed variations. The influences of travel variables on traffic speeds are conveniently classified as driver, vehicle, roadway, traffic, and environmental factors.

Driver factors. Although the effect of driver factors on speed has not been evaluated to a great extent, trip distance has the most significant effect on spot-speed characteristics, whereas the presence of passengers in the car and the sex of the driver influence driving speeds to a lesser extent. In general, individuals traveling long distances have newer cars and drive faster than do local travelers, and their speeds increase with trip length. Unaccompanied drivers tend to travel at higher speeds than drivers with passengers. Female

TABLE 2–25

Example of Computation of Time-Mean and Space-Mean Speeds

Vehicle	Time to Travel 176 ft (sec)	Velocity (mph)
1	1.9	63
2	2.1	57
3	2.1	57
4	1.8	67
5	2.3	52
6	2.0	60
7	2.2	55
8	2.0	60
9	1.7	71
10	1.9	63

Time-mean speed $= \dfrac{63 + 57 + 57 + 67 + 52 + 60 + 55 + 60 + 71 + 63}{10} = 60.5$ mph

Space-mean speed $= \dfrac{176\,(10)}{1.9 + 2.1 + 2.1 + 1.8 + 2.3 + 2.0 + 2.2 + 2.0 + 1.7 + 1.9} = 60.0$ mph

---

[33] J.G. Wardrop, "Some Theoretical Aspects of Road Traffic Research," Road Paper No. 36 in *Proceedings of Institute of Civil Engineers* (London: 1952).

TABLE 2–26

Comparison of Travel Speeds of Passenger Cars and Trucks

Speed Limit (mph)	Passenger Car Speed (mph)		Truck Speed (mph)	
	Mean	85th Percentile	Mean	85th Percentile
25	31.1	36.2	29.0	33.6
30	36.6	42.2	32.6	39.1
35	38.6	44.6	36.6	41.3
40	41.8	48.4	38.4	44.9
45	48.6	54.6	44.4	51.1
50	51.6	58.6	48.1	54.5
55	56.3	62.3	53.9	60.5

SOURCE: D.L. HARKEY and others, *Assessment of Current Speed Zoning Criteria,* Report No. FHWA-RD-89-161 (Washington, DC: Federal Highway Administration, June 1989).

drivers travel at about the same or at slightly lower average speeds than do male drivers. More men than women drive at excessively high speeds, and single persons tend to drive faster than married persons. Of course, these driver variables influence vehicular speeds to different degrees in different parts of the country.[34]

Vehicle factors. Among the various vehicle factors, vehicle type (passenger car, bus, single-unit truck, or combination truck) and vehicle age appear to have the predominant effect on vehicle speeds. Buses consistently travel at higher speeds than do passenger cars, and trucks consistently travel at lower speeds.[35] Table 2–26 presents the relative speeds of passenger cars and trucks from recent FHWA data. Passenger cars have wider speed ranges than do trucks or buses, whose drivers appear to choose much more consistent speeds. Average speeds of commercial vehicles decrease from light, single-unit trucks, to medium trucks, to heavy combination trucks. Within both the single-unit truck and combination truck

classifications, average spot speeds tend to decrease with increasing gross weight. For a given travel distance, the average speed decreases as vehicle age increases. Newer cars are generally driven faster than older cars because new cars go faster, ride more comfortably, travel more smoothly and quietly, and are generally in better mechanical condition.[36]

Roadway factors. Actual speeds adopted by drivers are greatly affected by various aspects of the roadway. Functional classification of the roadway, curvature, gradient, length of grade, number of lanes, and surface type have pronounced effects on vehicle speeds. Other factors that affect vehicle speeds include geographic location, sight distance, median type, lane position, lateral clearance, frequency of intersections, and type of development.

Traffic factors. Increases in traffic volume affect vehicle speeds, especially as highways approach congestion. Figure 2–19 illustrates the speed-volume relationship for freeways from the 1985 *Highway Capacity Manual.*[37] However, the effect of speed on traffic volume may be much smaller than previously supposed. The curves in Figure 2–19 are much flatter in the lower volume range than the comparable curves in the 1965 *Highway Capacity Manual.*[38] Furthermore, recent research on multilane rural and suburban highways indicates that traffic volume may have virtually no influence on traffic speeds at volumes below 1,400 passenger cars per hour per lane.[39]

Environmental factors. Environmental variables such as time and weather have important effects on vehicle speeds.

[34] J.C. Oppenlander, "Variables Influencing Spot-Speed Characteristics: Review of Literature." In *Special Report 89* (Washington, DC: Highway Research Board, 1966).

[35] Ibid.

[36] Ibid.

[37] *Highway Capacity Manual,* Special Report 209 (Washington, DC: Highway Research Board, 1985).

[38] *Highway Capacity Manual,* Special Report 87 (Washington, DC: Highway Research Board, 1965).

[39] W.R. Reilly and others, *Capacity and Level of Service Procedures for Multilane Rural and Suburban Highways,* Final Report of National Cooperative Highway Research Program Project 3-33, Highway Research Board, unpublished, 1989.

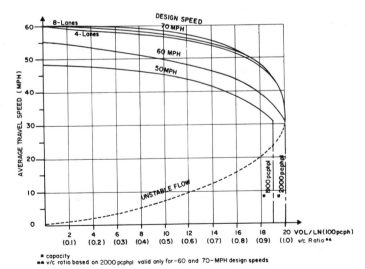

Figure 2–19. Speed-flow relationships for freeways under ideal conditions.

SOURCE: *Highway Capacity Manual,* Special Report 209 (Washington, DC: Transportation Research Board, 1985), p. 3–5.

Average traffic speeds tend to be highest in the fall and winter, intermediate in the spring, and lowest in the summer. Research results concerning daily and hourly fluctuations of traffic speeds are inconsistent. However, average spot speeds in the daytime are about 1 mph higher in urban areas and from 2 to 8 mph higher in rural areas than the corresponding speeds at night.

Reductions in average spot speed attributed to weather conditions range from 7% to 23% for poor visibility, from 4% to 38% for unfavorable road surface, and from 10% to 38% for both poor visibility and an unfavorable road surface.[40]

Speed trends. Traffic speeds have generally increased over the long term, but have been affected by changes in national speed limit policies in the last 15 years. From 1942 through 1973, average speeds on main rural highways increased at the approximate rate of 1 mph per year. Figure 2–20 illustrates more recent speed trends for rural Interstate highways. Speeds dropped sharply in 1974 with the introduction of the national 55-mph speed limit. Since 1974, average speeds on rural Interstate highways have increased by approximately 2 mph. As shown in Figure 2–20, approximately 80% of vehicles on rural Interstate highways exceed the 55-mph speed limit. Beginning in 1986, state highway agencies were permitted to increase speed limits to 65 mph on rural

Interstate highways, and today over 35 states have done so. The data for 1986 through 1988 in Figure 2–20 do not include any highways where the speed limit was raised to 65 mph.

The increase in speed limit to 65 mph on rural Interstate highways increased average and 85th percentile vehicle speeds by about 2 mph, as shown in Table 2–27. Over the same time period, speeds on rural Interstate highways increased by about 1.3 mph in states that did not raise the speed limit. This could represent a carryover effect of traffic from adjoining states with 65 mph speed limits traveling into these states. Speed increases on other classes of roads were comparable in states that raised the speed limit to 65 mph and states that did not.

Spacing and headway characteristics

The longitudinal distribution of vehicles in a traffic stream is represented by the pattern in spacing between common points of successive vehicles. This characteristic of traffic flow is measured as either some unit of time per vehicle or some unit of space per vehicle. A knowledge of spacing characteristics has application for estimating traffic delays and available gaps for vehicular or pedestrian crossings, for studying merging maneuvers between two streams of vehicles, for predicting vehicle arrivals at a point of interest, or for timing traffic signal systems.

Spacing fundamentals. Headway, which is inversely proportional to traffic volume, is the time interval between

---

[40] Oppenlander, "Variables Influencing Spot-Speed Characteristics."

Figure 2–20. Speed trends on rural Interstate highways, 1965–1989.
SOURCE: *Highway Statistics 1988,* Report No. FHWA-PL-88-008 (Washington, DC: Federal Highway Administration, 1988), p. 179.

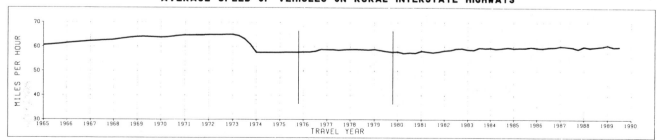

TABLE 2–27

Effects on Vehicle Speeds of the Speed Limit Increase to 65 mph on Rural Interstate Highways

Functional Classification	Number of States Reporting	Average Speed (mph)			85th Percentile Speed (mph)		
		Before[a]	After[b]	Change	Before[a]	After[b]	Change
States that raised rural Interstate speed limits to 65 mph[c]							
Rural Interstate[d]	19	60.64	62.60	+1.96	66.64	68.85	+2.21
Urban Interstate[e]	26	57.50	58.12	+0.62	64.13	64.82	+0.69
State primary[e]	28	56.39	56.44	+0.05	62.92	62.98	+0.06
States that did not raise rural Interstate speed limits							
Rural Interstate[e]	8[f]	58.60	59.92	+1.32	65.94	67.29	+1.35
Urban Interstate[e]	8	57.34	58.07	+0.73	65.31	66.13	+0.82
State primary[e]	8	55.22	57.23	+2.01	62.58	62.78	+0.20

[a]Before period: October 1986—March 1987.
[b]After period: October 1987—March 1988.
[c]Includes only states that raised their rural Interstate speed limit before September 30, 1987.
[d]Speed limit raised from 55 mph to 65 mph.
[e]Speed limit unchanged at 55 mph.
[f]Includes seven states plus Puerto Rico.

SOURCE: AMERICAN ASSOCIATION OF STATE HIGHWAY AND TRANSPORTATION OFFICIALS, *Effect of Raising the Speed Limit to 65 mph*, Washington, DC, July 20, 1989.

the passage of successive vehicles going by a fixed point on the roadway. The unit of time selected for headway measurements is usually seconds. Any point on the vehicle could be used in timing of headways, as long as the same point is used consistently for all vehicles. However, it is most common to measure headways from the front of one vehicle to the front of the next.

The longitudinal arrangement of traffic may also be represented by the spacing or distance interval between successive vehicles. Spacings are typically expressed in feet and are an inverse measure of traffic density. The relationship between spacing and headway depends on the average speed of the traffic stream that is being observed. Average headway and average spacing are related by Equation (2.12):

$$\overline{d} = c\,\overline{h}\,\overline{u} \qquad (2.12)$$

where:

$\overline{d}$ = average spacing (ft/veh)
$\overline{h}$ = average headway (sec/veh)
$\overline{u}$ = average speed (ft/sec)
c = constant (= 1.467)

Vehicle platooning and, thus, the headway distribution provide the key to understanding traffic service on two-lane highways. The 1985 *Highway Capacity Manual* defines traffic service on two-lane highways in terms of the *percent time delay,* which is the percent of total travel time that vehicles are involuntarily delayed in platoons.[41] The percent time delay can be approximated by the percentage of vehicles observed to travel with headways of 5 sec or less. This concept is developed further in Chapter 5, "Operational Aspects of Highway Capacity."

The headway distribution in conflicting traffic streams also controls the operation of turning and crossing maneuvers at intersections. The critical gap for a turning or crossing maneuver is defined as the median headway between successive vehicles in the traffic flow on a major street that

is accepted by drivers on a side street that must cross or merge with the major street flow.

Observed values. At any given flow rate in an individual lane, the average headway is the reciprocal of the flow rate. Thus, at a flow of 1,200 vehicles per hour per lane, the average headway is:

$$\frac{3{,}600 \text{ sec/hr}}{1{,}200 \text{ veh/hr}} = 3 \text{ sec/veh}$$

Vehicles do not, however, travel at constant headways. Vehicles tend to travel in groups, or platoons, with varying headways between successive vehicles. An example of the distribution of headways observed on the Long Island Expressway is shown in Figure 2–21. Lane 3 of Figure 2–21 has the most uniform headway distribution, as evidenced by the range of values and the high frequency of the modal

Figure 2–21. Example of time-headway distribution by lane for an urban freeway.
SOURCE: Data for Long Island Expressway in *Highway Capacity Manual,* Special Report 209 (Washington, DC: Transportation Research Board, 1985), p. 2–25.

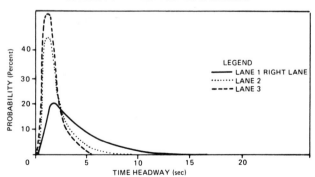

LONG ISLAND EXPRESSWAY, LEVEL OF SERVICE C

LEGEND
——— LANE 1 RIGHT LANE
·········· LANE 2
– – – LANE 3

PROBABILITY (Percent)

TIME HEADWAY (sec)

---

[41]*Highway Capacity Manual,* 1985.

value—the peak of the distribution curve. The distribution in Lane 2 is similar to that in Lane 3, with slightly greater scatter. Lane 1 shows a much different pattern—the range of headways is larger and the peak of the distribution curve is much lower. This difference reflects the lower flow rate usually observed in Lane 1 and the differing desires of Lane 1 drivers.

Very few headways in Figure 2–21 are less than 1.0 sec. A vehicle traveling 60 mph with a 0.5-sec headway would have a corresponding spacing of 44 ft. This spacing provides only 25 to 30 ft between vehicles, which is extremely difficult for drivers to maintain and allows little margin for driver error.

Table 2–28 shows the critical gaps for turning and crossing maneuvers used in the unsignalized intersection procedure in the 1985 *Highway Capacity Manual*.[42] Several sources in the literature report critical gaps for passenger cars that are 1 to 2 sec higher than the *Highway Capacity Manual* values.[43-45] Recent research has also found critical gap values for trucks to be 3 to 8 sec higher than the *Highway Capacity Manual* values.[46]

Poisson distribution. Under certain frequently encountered traffic conditions, the number of vehicles arriving at a point in an interval of time follows a random, or *Poisson, distribution*. The following requirements define required conditions for randomness of vehicle arrivals:

o Each vehicle is positioned by its driver independently of other vehicles.
o The number of vehicles passing a point in a given length of time is independent of the number that passes this point in any other length of time.

A counting distribution that satisfies the above requirements is the Poisson distribution, which is described by the following equation:

$$P(n \mid \overline{q}t) = \frac{e^{-\overline{q}t}(\overline{q}t)^n}{n!} \qquad (2.13)$$

where:

$P(n \mid \overline{q}t)$ = probability of the arrival of n vehicles at a point during the time interval t when the average volume is q vehicles per unit of time

n = number of vehicle arrivals

$\overline{q}$ = average traffic volume (vehicles per unit of time)

t = time interval (units of time)

e = base of natural logarithms = 2.718

$n!$ = n factorial = $n(n - 1)(n - 2) \ldots (2)(1)$

---

[42] Ibid.

[43] P. Solberg and J. C. Oppenlander, "Lag and Gap Acceptance at Stop-Controlled Intersections," *Highway Research Record 118* (Washington, DC: Highway Research Board, 1966).

[44] A.E. Radwan, *Development and Use of a Computer Simulation Model for the Evaluation of Design and Control Alternatives for Intersections of Minor Roads with Multilane Rural Highways: Field Study and Model Validation,* Report No. FHWA-IN-79-9 (Washington, DC: Federal Highway Administration, July 1979).

[45] K. Fitzpatrick, *Sight Distance Procedures for Stop-Controlled Intersections,* Doctoral Thesis, Pennsylvania State University, 1989.

[46] Harwood and others, *Truck Characteristics.*

TABLE 2–28

Critical Gaps in Traffic for Vehicles Entering or Crossing a Major Street

	Basic Critical Gap for Passenger Cars (sec)			
	Average Running Speed, Major Road			
	30 mph		55 mph	
	Number of Lanes on Major Road			
Vehicle Maneuver and Type of Control	2	4	2	4
RT from minor road				
Stop	5.5	5.5	6.5	6.5
Yield	5.0	5.0	5.5	5.5
LT from major road	5.0	5.5	5.5	6.0
Cross major road				
Stop	6.0	6.5	7.5	8.0
Yield	5.5	6.0	6.5	7.0
LT from minor road				
Stop	6.5	7.0	8.0	8.5
Yield	6.0	6.5	7.0	7.5

| Adjustments and Modifications to Critical Gap (sec) | |
Condition	Adjustment
RT from minor street: curb radius > 50 ft or turn angle < 60[a]	−0.5
RT from minor street: acceleration lane provided	−1.0
All movements: population ≥ 250,000	−0.5
Restricted sight distance[a]	up to +1.0

Maximum total decrease in critical gap = 1.0 sec.
Maximum critical gap = 8.5 sec.
For values of average running speed between 30 and 55 mph, interpolate.
[a]This adjustment is made for the specific movement impacted by restricted sight distance.

SOURCE: *Highway Capacity Manual,* Special Report 209 (Washington, DC: Transportation Research Board, 1985), p. 10–7.

If the spacing between two successive vehicles is considered as the occurrence of no arrival, then the headway between these vehicles must be greater than or equal to this time interval during which no other vehicles arrive. When the value of zero is substituted for the number of arrivals in the probability expression given above, the following equation results:

$$P(h >= t) = e^{-\overline{q}t} \qquad (2.14)$$

Comparisons of theoretical values computed for the Poisson distribution with observed headways evidence a close agreement for travel on four-lane rural highways. This goodness of fit is demonstrated in Figure 2–22 for a traffic volume in one direction of 500 veh/hr. However, the Poisson distribution is less effective in describing the actual headway patterns for two-lane highways in rural areas. This lack of agreement is explained by two factors that are characteristic of travel on two-lane roads: (1) although the Poisson distribution is continuous to a time value of zero, this condition is impossible on a two-lane highway because each headway must contain the time for a vehicle to travel at least its own length and (2) the lack of passing opportunities on two-lane roads, particularly as traffic volumes increase, causes platooning or bunching of traffic behind the slower-moving vehicles. These two situations explain the lack of fit

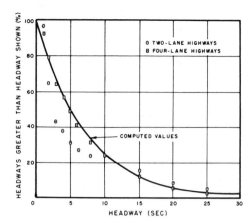

Figure 2–22. Computed and observed headways for typical two-lane and four-lane highways carrying 500 veh/hr in one direction.
SOURCE: *Highway Capacity Manual,* Special Report 87 (Washington, DC: Highway Research Board, 1965), p. 51.

in Figure 2–22 for computed and observed headways on two-lane rural highways.

Other probabilistic distributions have been proposed and validated for various traffic flow conditions. These relationships are addressed in several textbooks on traffic flow theory.

REFERENCES FOR FURTHER READING

A Manual on User Benefit Analysis of Highway and Bus Transit Improvements 1977 (Washington, DC: American Association of State Highway and Transportation Officials, 1978).

A Policy on Geometric Design of Highways and Streets (Washington, DC: American Association of State Highway and Transportation Officials, 1990).

CHING, P.Y., AND F.D. ROONEY, *Truck Speeds on Grades in California* (Sacramento: California Department of Transportation, June 1979).

CLAFFEY, P., *Running Costs of Motor Vehicles as Affected by Road Design and Traffic.* National Cooperative Highway Research Program, Report 111 (Washington, DC: Highway Research Board, 1971).

GLAUZ, W.D., "Future Changes to the Vehicle Fleet," in *Relationship Between Safety and Key Highway Features: A Synthesis of Prior Research,* State of the Art Report 6 (Washington, DC: Transportation Research Board, 1987).

HARKEY, D.L., AND OTHERS, *Assessment of Current Speed Zoning Criteria,* Report No. FHWA-RD-89-161 (Washington, DC: Federal Highway Administration, June 1989).

HARWOOD, D.W., AND OTHERS, *Truck Characteristics for Use in Highway Design and Operation,* Report Nos. FHWA-RD-89-226 and -227 (Washington, DC: Federal Highway Administration, December 1989).

Highway Capacity Manual, Special Report 209 (Washington, DC: Transportation Research Board, 1985).

Highway Statistics 1987, Report No. FHWA-PL-88-008 (Washington, DC: Federal Highway Administration, 1988). Published annually.

MATSON, T.M., W.S. SMITH, AND F.W. HURD, *Traffic Engineering* (New York: McGraw-Hill Book Company, 1955).

Parking Dimensions of 1990 Model Passenger Cars (Detroit: Motor Vehicle Manufacturers Association, 1990). Published annually.

ST. JOHN, A.D., AND D.R. KOBETT, *Grade Effects on Traffic Flow Stability and Capacity,* National Cooperative Highway Research Program Report 185 (Washington, DC: Transportation Research Board, 1978).

WINFREY, R., *Economic Analysis for Highways* (Scranton, PA: International Textbook, 1969).

Your Driving Costs (Falls Church, VA: American Automobile Association, 1989). Published annually.

3

TRAFFIC STUDIES

CHRIS D. KINZEL, P.E., *President*

TJKM Transportation Consultants

Introduction

This chapter contains descriptions of basic traffic engineering studies. These studies are designed to gather facts on traffic or parking conditions. Methods of data collection and analyses are presented, but more detailed descriptions of studies, equipment, personnel requirements, and forms are contained in the *Manual of Traffic Engineering Studies.*[1]

Studies described in this chapter must be adapted to a specific application so that the information gathered is timely, is as accurate and unbiased as possible, and the costs involved are consistent with the magnitude of the problem, available funds and personnel as well as the needed accuracy.

While the types of basic traffic studies have not changed appreciably over the years, the modern equipment available for studies has, in some cases, streamlined the methodologies and personnel requirements for such studies. This chapter updates the procedures for such studies affected by new technologies.

Inventories

An inventory is an accounting, tabulation, listing, or graphic display of information describing existing conditions. It must be readily available and periodically updated. Some inventories, such as traffic and parking regulations, parking facilities, and transit routes, may require frequent updating,

whereas others, such as street widths, will be changed less frequently.

Types of inventories

Administrative, environmental, and geometrics. Inventories of public streets and highways that will be used in traffic studies generally include existing rights-of-way; roadway, shoulder, and sidewalk widths; intersection and lane configurations; location of bike lanes; location of raised medians and landscaping; functional classification; adjacent land use and railroad crossings.

Traffic generators. Inventories of traffic generators such as office complexes, shopping malls, residential developments, schools, etc., may include driveway traffic counts, vehicle classification, number of employees, the gross building area, the gross leasable area, the net rentable area, the number of occupied rooms or dwelling units, the population, and the acreage of development.

Regulations and standards. Inventories that could be useful to traffic engineers include all traffic ordinances and regulations, intersection level-of-service standards, minimum geometric standards, and parking supply requirements.

Traffic control devices. Inventories of traffic control devices generally include all traffic signs, signals, flashing beacons, curb markings, striping and pavement markings. Such inventories provide information on the devices including the location, condition, and maintenance requirements.

The traffic signs, signals, flashing beacons and pavement markings should be inventoried on a specific location basis.

---

[1] Paul C. Box and Joseph C. Oppenlander, *Manual of Traffic Engineering Studies,* 4th edition (Arlington, VA: Institute of Transportation Engineers, 1976), 181. Currently being updated for publication in 1991.

The curb markings and pavement striping controls can be inventoried on a segment basis.

Transit facilities. An inventory of transit routes and schedules should include the location and length of stops and bus layover spaces. The location of any off-street terminal should also be noted. Any change-of-mode facilities, including transit parking and passenger pick-up/drop-off areas should be inventoried as well. Information on the characteristics of the transit vehicle mix may be included.

Parking facilities. Inventories of parking facilities typically are conducted for either a central business district of a city or a commercial shopping center. All available on-street and off-street spaces in the study area should be recorded. If applicable, the existing parking supply should be categorized by parking time limits or parking meter control. Special zones for truck loading, passenger loading, or taxi stands should also be noted.

Storage of inventories

Generally, inventories will require taking measurements and notes in the field as needed. Aerial photography is useful in the inventory of public streets and parking facilities. Photo logging, using either transparency film or video, may facilitate certain types of inventories. Photos form an excellent record of conditions at the time of the inventory.

Many types of inventories can be portrayed effectively on jurisdiction maps or specialized maps. Examples include:

Functional street classification
Intersection controls
One-way streets
Weight limits and truck routes
Overhead structure clearance
Transit routes and stops
Speed limits
Snow emergency routes
Daily traffic volumes
Land use and zoning
Locations of traffic signal systems and interconnection
Curb parking regulations
Location of parking meters
Off-street parking facilities
Street light systems
Roadway widths
Right-of-way widths

The use of computer files will facilitate maintenance and accessibility of most inventories and data files. Manual card files constitute an alternate means of storing data.

Separate files for each intersection and each different street and highway may be useful for storage of such information as special traffic studies, condition and collision diagrams, citizen complaints and responses, special reports, details of physical improvement changes, and dates of major control device changes.

Updating inventory records

Updating and maintenance of inventories is important. Work-order control systems, specified entry and revision routines, and other management devices are essential to this process. If the inventory data are not maintained as changes occur in the field, then the inventory becomes rapidly outdated. Periodic re-surveys to update or validate inventory records, usually at 3- to 5-year intervals, may also be needed.

Volume studies

Traffic volume counts are dynamic traffic studies. They include a tabulation, generally by time intervals such as 15 min, 30 min, 1 hour, 1 day, or 1 week, of the number of vehicles passing a specific point. Depending upon the collection method, data may be subclassified into vehicle types, occupancy, direction, or turning movement.

Methods and equipment

Manual counts. The manual count is the most basic volume collection procedure. It can include data on travel directions and turning movements at intersections or driveways, subclassifications of vehicle types or sizes, and number of pedestrian crossings. Manual counts may be taken for any desired time interval, with a common length of 1 hour with 15-min tabulation periods. Many of the studies suggested by the *Manual of Uniform Traffic Control Devices* (MUTCD) require the use of manually obtained volume counts. In addition, the checker may note other useful information such as queue lengths, obedience to control devices, unusual events contributing to delay, and use of unstriped areas for right turns. Traffic-counting personnel, after brief training and experience, may be able to identify deficiencies in signal timing, control-device placement, or intersection operational problems.

Most volume studies that involve manual counts are taken in peak periods that include the heaviest hours of morning and evening traffic. The counting period may depend upon the method used to obtain data and the planned use of the data. Counts are normally not conducted during special conditions such as holidays, sporting events, transit strikes, special sales, unusual weather, or temporary street closures, unless the count is for the specific purpose of studying the effects of these special conditions. Manual turning movement counts can be made for the following periods:

1. 12 hours; 7:00 A.M.–7:00 P.M.
2. 8 hours; 7:00 A.M.–11:00 A.M., 2:00 P.M.–6:00 P.M.
3. 4 hours; 7:00 A.M.–9:00 A.M., 4:00 P.M.–6:00 P.M.

For intersection counts, the number of people needed at a given location depends upon the total volume, the complexity of the area to be counted, and the degree to which secondary notations such as pedestrian volumes, queue lengths, or obedience to control devices are to be made. Generally, one person would be sufficient at a "T" intersection whereas two people would be needed for a four-way intersection. Personnel normally can count 3 to 4 hours without a break.

```
INTERSECTION:   2   I-680 SB Off  and Stoneridge Dr Pleasanton
COUNT DATE: 02/09/90     TIME: 7:00-9:00 AM
```

END TIME	SB RGHT	SB THRU	SB LEFT	WB RGHT	WB THRU	WB LEFT	NB RGHT	NB THRU	NB LEFT	EB RGHT	EB THRU	EB LEFT	LINE TOTALS
7:15	22	0	100	27	43	0	0	0	0	16	74	0	282
7:30	53	0	249	46	229	0	0	0	0	35	164	0	776
7:45	101	0	460	71	339	0	0	0	0	48	269	0	1288
8:00	153	0	665	101	589	0	0	0	0	60	445	0	2013
8:15	224	0	825	132	668	0	0	0	0	71	569	0	2489
8:30	284	0	962	158	806	0	0	0	0	86	673	0	2969
8:45	322	0	1040	186	885	0	0	0	0	106	753	0	3292
9:00	388	0	1125	208	995	0	0	0	0	131	847	0	3694

*********** CUMULATIVE COUNTS *************** CUMULATIVE COUNTS ************

END TIME	SB RGHT	SB THRU	SB LEFT	WB RGHT	WB THRU	WB LEFT	NB RGHT	NB THRU	NB LEFT	EB RGHT	EB THRU	EB LEFT	LINE TOTALS
7:15	22	0	100	27	43	0	0	0	0	16	74	0	282
7:30	31	0	149	19	186	0	0	0	0	19	90	0	494
7:45	48	0	211	25	110	0	0	0	0	13	105	0	512
8:00	52	0	205	30	250	0	0	0	0	12	176	0	725
8:15	71	0	160	31	79	0	0	0	0	11	124	0	476
8:30	60	0	137	26	138	0	0	0	0	15	104	0	480
8:45	38	0	78	28	79	0	0	0	0	20	80	0	323
9:00	66	0	85	22	110	0	0	0	0	25	94	0	402

************** PERIOD COUNTS ******************** PERIOD COUNTS ************

BEGIN TIME	SB RGHT	SB THRU	SB LEFT	WB RGHT	WB THRU	WB LEFT	NB RGHT	NB THRU	NB LEFT	EB RGHT	EB THRU	EB LEFT	LINE TOTALS
7:00	153	0	665	101	589	0	0	0	0	60	445	0	2013
7:15	202	0	725	105	625	0	0	0	0	55	495	0	2207
7:30	231	0	713	112	577	0	0	0	0	51	509	0	2193
7:45	221	0	580	115	546	0	0	0	0	58	484	0	2004
8:00	235	0	460	107	406	0	0	0	0	71	402	0	1681

*************** HOUR COUNTS ********************** HOUR COUNTS ************

Figure 3–1. Turning movement counts.
SOURCE: TJKM.

```
PEAK HOUR PERIOD:    7:15 TO    8:15   VOL=   2207
PEAK 15 MIN PERIOD:  7:45 TO    8:00   VOL=    725
```

In most cases, mechanical tally counters are used to record several types of data. The cumulative traffic counts for each specific movement are recorded every 15 min. A computer spreadsheet program used to reduce the period counts and to determine the peak-hour totals is shown in Figure 3–1.

Electronic count boards are available that will automatically store each individual turning movement count at the desired time interval. A photograph of an electronic count board is shown in Figure 3–2. After the count is completed, the counts may be downloaded directly from the board into a computer. The board manufacturer generally supplies a program designed to transfer the field data to a summary printout (see Figure 3–1).

Automatic counts. When counts are needed for longer time periods, automatic counters may be used. Portable devices are used for short-term periodic counts. They are battery operated and employ magnetic loop detectors or road tubes with air-impulse switches to actuate the counting mechanism. Inductive loop detectors work on the principle that the passage of a motor vehicle causes a disturbance in an electrical field or a change in the induction of the loop. The change in potential is amplified and an impulse sent to the counter. When the counter is used in conjunction with printed tape, it is called a *recording counter,* and volumes are printed on tape at selected intervals such as 15 min or 1 hour. The punched tape counter records the volume in binary code on special tape. This can be analyzed manually or computer processed with the aid of a translating device.

Electronic road-count machines work with detector loops or road tubes and air impulse switches plus an electronic

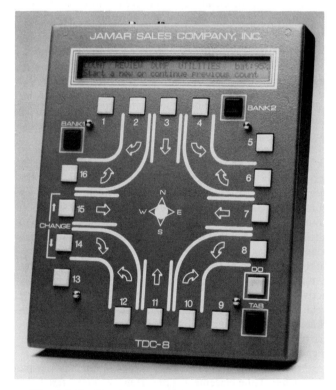

Figure 3–2. Turning movement counter.

memory. After the completion of the count, the data are downloaded directly from the machine into a computer. The manufacturer generally supplies a program designed to transfer the field data to a summary printout.

Short-period manual counts can be used in conjunction with automatic counters to estimate turning movements at an intersection, to determine vehicle classifications, or to make the automatic counts more usable.

A new technology traffic counter has been developed to be installed in the traffic lane with no need for rubber road tubes. The counter is made of a plastic material, is approximately 6 in × 12 in long × 1 in high, and is placed in the middle of each travel lane. The counter is fastened to the pavement using power nail fasteners and a power-driven tool. The counter contains an electronic sensor and a miniature computer that can determine the speed, volume, and classification of all vehicles. Because vehicles are classified by length, the user must set the parameters for vehicle classes. Data are stored in the internal microcomputer and can then be transferred to a portable microprocessor by telemetry. The portable microprocessor can then be interfaced with a personal computer for analysis.

Permanent-type counting stations are established at key locations along major highways. The stations can provide traffic data on a continuous basis. These counts are useful in comparing hourly and daily volumes and in developing reliable monthly and seasonal variation factors. Trend data in annual volumes can also be produced. These stations generally employ inductive loop detectors.

These permanent-type count stations can be used in monitoring traffic conditions along specific corridors or around a community. An on-line computerized traffic monitoring system is in the city of Pleasanton, California. The system is designed automatically to count turning movements at selected intersections, process and store the count data, perform capacity analyses, and produce intersection levels-of-service reports. The prime objective of establishing such a system is to provide the city with the ability to monitor traffic conditions on a regular basis and to record trends in travel patterns.[2]

Errors encountered in data gathering. Road tubes can be subject to counting error. Vehicles having more than two axles introduce an overcount error. The error increases with the proportion of multiaxle trucks. It can be compensated by computing correction factors from manual classification counts. However, the unadjusted count may be preferred to account for the fact that multiaxle vehicles occupy more space than do two axle vehicles. Such an unadjusted count is sometimes referred to as the *equivalent passenger car count*. Undercounting error can occur when vehicles simultaneously activate the road tube in adjacent lanes or when slow-moving vehicles fail to cause an activation. The counter can sometimes be rendered inactive by a vehicle parked with wheels on the tube. Also, the tube is subject to damage by sliding wheels, high-speed vehicles, vandalism, or pavement imperfections. A road tube cannot be used on a gravel road or during a period of snowfall. Tube placement is important because a fixed object such as a tree or pole is needed to anchor the recorder, and the tube must be placed normal to the paths of vehicles to avoid overcounting.

---

[2] Y. John Sun, "The Establishment of the Pleasanton Traffic Monitoring System," *ITE Journal* (November 1987), 41–46.

Other data gathering. Photographic equipment at elevated positions is sometimes used for counting. In the past, this generally consisted of time-lapse photography taken at speeds of 10 to 120 frames per min. Because of the availability and low cost of video equipment, video taping is becoming more practical to use in traffic counts. This procedure requires more equipment than manual counts but has the advantages of requiring less personnel in the field and providing a permanent record. However, additional labor is usually required in the office to summarize the data from the video or photographic media.

The photographic method provides an opportunity to obtain added field data from the film such as time delays, lengths of queues, etc., as long as field reference points are marked in the field and camera positions are carefully selected.

Spot counts

Manual spot counts typically are conducted at intersections or major generator driveways. The time intervals at which spot counts are desirable depends on many local factors including the importance of the location from a level of service standpoint and the amount of growth in the area. Generally, critical intersections in a growing community should be counted on a recurring basis and desirably once per year. The counts are used to check for needs of change in control devices, appropriateness of signal timing, need for improvements such as added turn lanes, or special signal phasing. Current counts are essential if the highway section is being studied for significant physical revision.

Spot counts may include classification of vehicle types. Typically, the "passenger vehicle" category is deemed to include also motorcycles, station wagons, and all two-axle, four-tire vehicles. The "truck" category includes all vehicles having more than four tires, any passenger vehicle pulling a trailer, and so on. Such simplified classification is normally adequate for operational studies, including those for design and capacity. In general, a minimum number of vehicle classifications should be established and checkers should be instructed that the total physical count is more important than an occasional misclassification.

More detailed information on vehicle classification is needed for *pavement design* purposes. Here the classification of trucks must be in at least two groups, with the single-unit type separate from tractor-trailer combinations. A standard method includes separate tabulations of two-, three-, four-, five-, and six-axle trucks.

Spot counts can also involve a summary of pedestrian activity at an intersection. During peak hour periods, the total number of pedestrian crossings in each crosswalk should be recorded. This information can be used to refine the signal timing for the pedestrian-crossing phase.

Estimated volume data may be obtained from manual short counts. Traffic at a given location is counted for 5 min out of each of several hours of several days of the week. The observers make a circuit of as many as six locations, utilizing 5 min for travel between locations. The resulting total at each spot is then multiplied by 12 to give the estimated total hourly flow. This procedure has been shown to yield estimates to

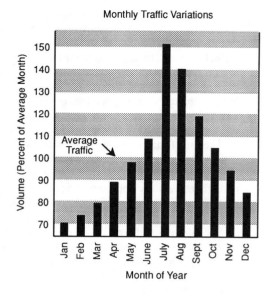

Monthly Traffic Variations

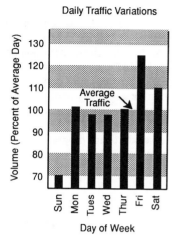

Daily Traffic Variations

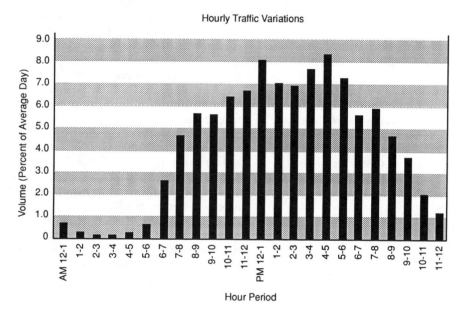

Hourly Traffic Variations

Figure 3–3. Traffic variations.

within several percent of the true total.[3] Temporary automatic (machine)-type counts are taken at spot locations to give estimates of daily traffic along sections of street or highway. It is usually desirable to apply day-of-week and seasonal adjustment factors.

Passenger car occupancy studies are made to obtain information on passenger-miles of travel (for planning purposes) and car pooling (for traffic system management purposes). This involves counting the number of persons in each vehicle as it passes the spot location and recording the vehicle in the appropriate passenger category.

Spot-count data are normally tabulated by 15-min intervals with hourly summations. For manual counts, the

tabulation normally involves each of the turning movements. A graphic presentation is typically prepared for the A.M. and P.M. peak hour data.

Hourly variation graphs are often prepared showing the percent of daily traffic as taken from automatic recording counts. An example of monthly, daily, and hourly variation in volume is given in Figure 3–3.

Areawide studies

It is desirable to obtain estimates periodically of the annual traffic volume on the major traffic street and highway network of cities, counties, and states. Such information is used to establish patterns of use and of growth.

Extensive counts for all links of a large area are of course impractical. Short counts, be they 1 hour or 1 day long, may

[3] Box and Oppenlander, *Manual of Traffic Engineering Studies,* pp. 29–35.

be considered as "samples." These can be expanded and adjusted for hourly, daily, and seasonal variation by use of a continuous master station. The ratio of the short-term data to a long-term data base at a continuous master station can be calculated and used to expand the short-term data at the specific site in the vicinity as long as there is some assurance that the traffic patterns have similar distribution patterns.

In the same way that 1- or 2-hour counts may be converted to average all-day counts, a 1-day count may be converted to an average weekly count or a monthly or annual count. Similarly, an average winter count may be converted to an average annual count.

Comprehensive areawide count programs may be established by the use of master count stations, control stations, key count stations, and local counts. For additional information, refer to the *Transportation Planning Handbook.*[4]

Speed studies

In this section, spot-speed studies and speed-study maneuvers are described. Information on study techniques and analysis is also presented.

Spot-speed studies

This study is used to measure speed of a vehicle at a specific location. It consists of observations of the individual speeds at which vehicles are passing a point on a roadway segment. These observations are used to estimate the speed characteristics of the entire traffic stream at the location in question, under the conditions prevailing at the time of the study. The speed characteristics have many applications, including:

Establishing speed zones

Timing traffic signals

Checking complaints on speeding

Checking for enforcement needs

Checking the need for posting advisory safe speed

Analyzing accident experience

Making before-and-after studies to evaluate a change in conditions

Checking speed trends by periodic studies at the same point

Because of the widespread availability of radar, many speed checks are conducted with such electronic equipment. Radar operates on the principle that a radio wave reflected from a moving target undergoes a frequency change proportional to the speed of the target. Individual radar-unit operating instructions should provide factors for calibration, optimum distance of survey, and optimum angle of survey. Only one person is required for the fieldwork. The operator should be familiar with the shortcomings of radar and the errors that can cause incorrect or false readings such as antenna positioning, beam reflection, and radio interference.

---

[4] *Transportation Planning Handbook,* Chapter 2, Washington, DC: Institute of Transportation Engineers, 1991.

It should be recognized that drivers are wary of radar use for enforcement purposes and, accordingly, may rapidly adjust their speed characteristics. A sudden change in speed reflects radar detection and potential speed-data bias. If this is a problem, then time-distance speed measurement procedure should be used. Other methods of measuring speeds include the use of loop detectors or road tubes connected to a roadside electronic unit; or hand computers used to measure times between two designated points on the pavement.

The use of automated speed monitoring equipment makes it possible to measure the speeds of every vehicle for a 24-hour period or longer. With radar, electronic stopwatches, and other manually operated devices it is impossible to measure the speed of every vehicle under many traffic conditions, and some type of sampling procedure must be used.

Speed checks are made in each section of a roadway where various conditions are relatively uniform. The specific location selected should be representative of the road section being surveyed and normally should have no traffic signal, stop sign, major intersection, or other factors nearby that influence the motorist's speeds. Also, the surveyor should make sure that there are no enforcement activities occurring upstream from the speed survey site. It is important that the surveyor and equipment not affect traffic speeds. For this reason, an unmarked vehicle in a concealed location is recommended, with the radar speed meter located as inconspicuously as possible. Normally, traffic is surveyed in both directions but with separate directional tabulations.

In order to get free-flow speeds, speed measurements normally will be taken during off-peak hours on weekdays such as 9:00 A.M. to 11:30 A.M., 1:30 P.M. to 4:00 P.M., or 7:00 P.M. to 10:00 P.M. On streets with low volumes, speed surveys may be done during peak hours or on more than one day to obtain the necessary sample size. However, if the study is based on citizen complaints, then the time period should reflect the nature of the complaint. Neither the weather nor other nontypical conditions should influence prevailing speeds. Observed conditions include the location of the spot speed survey, the direction of travel of vehicles surveyed, the date and day of the week, time of the survey, weather conditions, and any observations and notes that would help to interpret data. The existing posted speed limit is also noted. The minimum sample size for a speed check is generally 25 to 30 vehicles. Such a small sample should be used only for extremely low volume conditions. Normally, 50 to 100 or more vehicles are checked. For detailed information on selection of sample size, refer to the section "Engineering Study Techniques" at the conclusion of this chapter.

With low volumes it is usually possible to check nearly all vehicular speeds. At higher volumes, a sampling procedure is needed. Recommendations are:

1. Observe every nth vehicle (second, third, etc.). (Do *not* always observe the first vehicle in a platoon because the following vehicles may be restricted to travel at the speed of the lead vehicle.)
2. Select trucks for speed observation in proportion to their presence in the traffic stream.
3. Avoid sampling a large proportion of high-speed vehicles.
4. Measure only vehicles with a 4-sec or greater headway.

Analysis and presentation of data

In analyzing spot data, several characteristics can be developed. Some values are computed directly from the data and others are most easily determined from a graphic presentation. The first step in tabulation is to establish a frequency distribution. This involves selection of a group or class size. If too few (or too many) groups are selected, detail is lost in the data reduction. In general, the appropriate number of classes ranges from 8 to 20. After the field data have been collected, the range in measurements is determined by subtracting the lowest from the highest values. This range is divided by 8 and by 20 to estimate, respectively, the maximum and minimum class sizes that are reasonable for the observed data. A convenient class size is then selected within the limits of the minimum and maximum values.

After the class size has been determined, the class limits are selected to define completely the actual sample values within each class. These limits are written to the same precision as the original data. The midvalue or class mark is the middle point of the class. Also, it is important to select realistic classes for automatic speed classification equipment.

A typical frequency table is illustrated in Table 3–1. (The following discussion and illustration, although given in mph, are equally valid in km/hr.)

After the class limits have been recorded in the frequency table, each field observation is placed in its appropriate class. Adding the number of entries in each class gives the frequency of occurrences for each size classification in the speed check. The resulting table of occurrence in the various classes is defined as a *frequency distribution*. The sum of the occurrences in the several classes is equal to the sample size.

TABLE 3–1

Typical Frequency Table

Frequency Distribution for Spot Speed Data					
Class Boundaries	Class Midvalues (μ_1)	Class Frequencies[a] f_1	Relative Frequencies	Cumulative Frequencies Number	Cumulative Frequencies Relative
27.5					
	28.5	0	0.000		
29.5					
	30.5	1	0.005	0	0.000
31.5					
	32.5	2	0.011	1	0.005
33.5					
	34.5	14	0.075	3	0.016
35.5					
	36.5	7	0.038	17	0.092
37.5					
	38.5	20	0.108	24	0.129
39.5					
	40.5	38	0.204	44	0.237
41.5					
	42.5	29	0.156	82	0.441
43.5					
	44.5	35	0.188	111	0.597
45.5					
	46.5	15	0.081	146	0.785
47.5					
	48.5	12	0.065	161	0.866
49.5					
	50.5	9	0.048	173	0.930
51.5					
	52.5	4	0.022	182	0.979
53.5					
	54.5	0	0.000	186	1.000
55.5					
Total		186	1.000		

[a] As recorded in the samples.

SOURCE: Paul C. Box and Joseph C. Oppenlander, *Manual of Traffic Engineering Studies,* 4th ed. (Arlington, VA: Institute of Transportation Engineers, 1976), Table A-1.

Summary Calculations for Classed Spot Speed Data							
Class Boundaries	Class Midvalues μ_1	Class Frequencies f_1	$f_1\mu_1$	$f_1\mu_1^2$	$\mu_1 - \bar{x}$	$(\mu_1 - \bar{x})^2$	$f_1(\mu_1 - \bar{x})^2$
27.5							
	28.5	0	0	0	− 13.8	—	—
29.5							
	30.5	1	31	930	− 11.8	139	139
31.5							
	32.5	2	65	2,113	− 9.8	96	192
33.5							
	34.5	14	483	16,664	− 7.8	61	854
35.5							
	36.5	7	256	9,326	− 5.8	34	238
37.5							
	38.5	20	770	29,645	− 3.8	14	280
39.5							
	40.5	38	1,539	62,330	− 1.8	3	114
41.5							
	42.5	29	1,233	52,381	+ 0.2	0	0
43.5							
	44.5	35	1,558	69,309	+ 2.2	5	175
45.5							
	46.5	15	698	32,434	+ 4.2	18	270
47.5							
	48.5	12	582	28,227	+ 6.2	38	456
49.5							
	50.5	9	455	22,952	+ 8.2	67	603
51.5							
	52.5	4	210	11,025	+ 10.2	104	416
53.5							
	54.5	0	0	0	+ 12.2	—	—
55.5							
Total		186	7,877	337,335			3737

SOURCE: Modified from Paul C. Box and Joseph C. Oppenlander, *Manual of Traffic Engineering Studies,* 4th ed. (Arlington, VA: Institute of Transportation Engineers, 1976), Table A-2

Examination of Table 3–1 will show that the 85th percentile speed (that speed at or below which 85% of the vehicles are moving) is about 47.5 mph. The cumulative relative frequencies given in the last column reach $0.866 = 86.6\%$ with the 161st vehicle, and the upper class speed limit that includes the 161st is 47.5 mph.

Several measures of speed are commonly utilized in traffic engineering analysis. These include the time mean speed ("average"), the space mean speed, the median speed, the 85th percentile, and the 10-mph pace.

The *time mean* speed is obtained by dividing the sum of all the speeds in the sample by the number of vehicles in the sample. The general expression for the mean of unclassed data is represented by the equation

$$\overline{X} = \frac{\Sigma X_i}{N} \tag{3.1}$$

where: $\overline{X}$ = arithmetical mean
ΣX_i = sum of all observed spot speeds
N = number of observations

If the measurements have been placed into classes, such as in Table 3–1, the computation of time mean speed is as shown as follows:

$$\overline{x} = \frac{\Sigma(f_i\mu_i)}{\Sigma f_i} \tag{3.2}$$

where: $\overline{x}$ = arithmetical mean
$\Sigma(f_i\mu_i)$ = sum of produces of frequency and class midvalue of all classes
Σf_i = sum of frequencies of all classes

$$\overline{x} = \frac{\Sigma(f_i\mu_i)}{\Sigma f_i} = \frac{7,877}{186} = 42.3 \text{ mph}$$

The time mean speed is distinct from the *space mean* speed. The latter is obtained by dividing the sum of all travel times into the sum of all travel distances in the sample. Its primary use is to determine travel time accurately, and in freeway metering calculations.

The *median speed* (P50) is the speed below which one-half the vehicles in the sample travel and above which the other half travel. The median is a useful measure, because it is less affected by extreme values than is the time mean speed. In the sample, it lies in the class bounded by 41.5 mph and 43.5 mph.

The *pace* is the speed range within defined limits, usually 10 mph, which contains the largest number of observations. It is the range from 37.5 through 47.5 mph (6th through 10th classes) in this sample.

An important measure of variability is the *standard deviation,* which is the positive square root of the variance. The variance is the sum of the squares of the deviations of the observations from the mean, divided by the total number of observations less one. Therefore, the expression for

standard deviation is written in the following form for unclassed data:

$$s = \sqrt{\frac{(X_i - \overline{X})^2}{N-1}} \tag{3.3}$$

where: s = standard deviation
$\overline{X}$ = arithmetic mean
X_i = ith observation
N = number of observations

(using Equation 3.3 and Table 3–1 data)

$$s^2 = \frac{3,737}{185} = 20.2 \text{ and } S = \sqrt{20.2} = 4.5 \text{ mph}$$

The equation for standard deviation is applicable when the information is classed:

$$s = \sqrt{\frac{\Sigma(f_i\mu_i^2) - \frac{\Sigma(f_i\mu_i)^2}{\Sigma f_i}}{(\Sigma f_i) - 1}} \tag{3.4}$$

(using Equation 3.4 and Table 3–1 data)

$$s^2 = \frac{337,335 - \frac{(7877)^2}{186}}{185} = 20.3$$

$$s = \sqrt{20.3} = 4.5 \text{ mph}$$

The standard deviation reflects the dispersion of the observations around the mean. Most hand-held calculators permit calculating standard deviations very easily.

The skewness of the distribution can also be determined, although most people do not use this characteristic. If the data exhibit approximates a normal distribution, multiples of the standard deviation represent limits on either side of the mean within which will appear various percentages of the total observations in a selected sample. Under these conditions, 1, 2, and 3 standard deviations about the mean will contain respectively, 68.3, 95.5, and 99.7% of the observations.

The *cumulative frequency* diagram is prepared by plotting a graph with the class boundary of each class as the abscissa and the corresponding cumulative frequency of the class as the ordinate. The higher-class boundary is matched with the corresponding cumulative frequency when the frequency summation is from the small to the large values of the study variable. However, the lower-class boundary is selected for this matching if the cumulative frequency is summed from the large to the small observations. A smoothed S-shaped curve is used to connect the plotted points between two extreme class boundaries having cumulative frequencies of 0% and 100%. Such a diagram is shown in Figure 3–4, using the data in Table 3–1. This curve is characteristic of the cumulative frequency for a normal distribution and is a useful illustration for depicting mean and 85th percentile speeds.

Speed studies on curves

Studies may be conducted to determine the maximum comfortable speed on a horizontal curve for use in determining

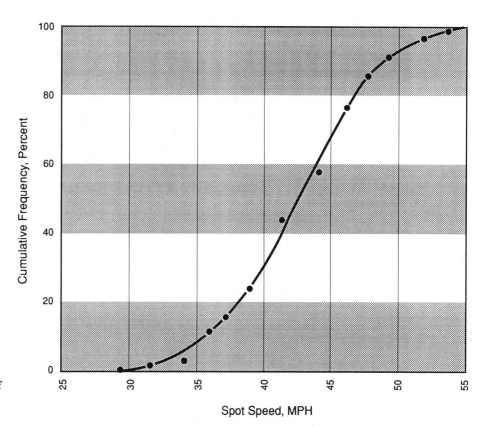

Figure 3–4. Cumulative Frequency of Spot Speeds.
SOURCE: TJKM.

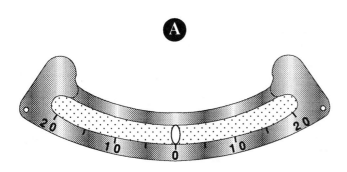

ZERO POSITION

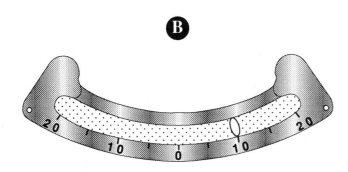

INDICATES 10° LEFT BANK

Figure 3–5. Ball-bank indicator.

whether there is need to communicate this information to the approaching motorists with advisory speed signing. Several trial runs are made through the curve in a test vehicle equipped with a ball-bank indicator. The ball-bank reading is a measure of the amount of lateral force (side friction) on the vehicle. Readings of 14° for speeds below 20 mph, of 12° for speeds between 20 and 35 mph, and of 10° for speeds above 35 mph are the usually accepted limits beyond which riding discomfort will be excessive and loss of vehicle control may occur. The reading at which loss of vehicle control by unprepared drivers can occur is a complex variable dependent on type and condition of pavement surface, inflation and condition of tires, condition of steering mechanism, and driver skill.

An illustration of a ball-bank indicator is shown in Figure 3–5. Two people are required to run the trial, namely a reader and a driver. Three runs should be made at the same speed that duplicates a ball-bank reading. Vehicles with no special suspension and standard automobile tires (which are properly inflated) should be used for the trials. The record for the runs should include the time, date, weather, location, existing speed posting, recorders' names, readings for each run, and general observations. It is advantageous to use a vehicle with cruise control so a constant speed can be maintained by the vehicle while negotiating the curve.

Travel time and delay studies

A *travel-time study* measures the time required to traverse a route. If information is also obtained on the location, duration, and cause of delays, it is called a *speed and delay study*.

Studies limited only to moving time are called *running-time studies.*

Travel-time data are needed to evaluate existing traffic conditions in transportation planning and for determining benefit for benefit-cost ratio calculations in transportation economic studies. When secured on an annual or other regular interval, trends in traffic conditions are identified. Travel time, or average speed, is used as the basis for determining levels of service on arterial streets and in weaving sections.[5]

For congested streets, speed and delay data are used in traffic control selection, evaluation, and revision. This information may also indicate locations where additional studies are needed to determine proper improvements. The data may be used in before-and-after studies to determine the effectiveness of changes in signal timing, circulation systems, parking prohibitions, or turn prohibitions. Delay may be considered to be made up of fixed delay caused by geometric design and traffic control, plus operational delay caused by interference or congestion.

Route studies

Several techniques can be used to obtain travel-time information,[6] as follows:

Test vehicle method. A series of "runs" are made through the section to obtain representative travel times. Two driving strategies may be used. In the "floating car" technique, the driver attempts to approximate the median speed by passing as many vehicles as pass the driver. The second driving strategy is the "average speed" technique, in which the driver travels at a speed that, in his or her opinion, is

representative of the speed of the traffic at every point and time (i.e., go with the flow). Tests of the latter method have shown excellent correlation with actual average travel times.[7]

Data are recorded by an observer in the vehicle or by the driver using a laptop computer or voice recorder. The older method is the use of an observer with two stopwatches: one to record time as control points are passed and the other to record stop delays. The observer starts the first stopwatch at the beginning of the test run and the time runs continuously, recording the cumulative lapsed time at successive control points along the route. Control points are predetermined to optimize the number of events to be recorded during the test run. The time, location, and cause of each delay is recorded by code on forms, computer, or by voice recorder. Table 3–2 shows a sample presentation of data collected from the test-vehicle method.

It is possible for a driver alone to obtain the desired information by using a laptop computer, thus eliminating the need for an observer. Laptop computers with internal clock functions preclude the need for stopwatches, and predetermined control points can be programmed into the computer. The internal clock and distance functions result in fully automatic recording of travel time at any location within the test run. Thus, the driver need only hit a computer key at delay and check points along the test run. A voice recorder allows the driver to record cause of delay. Simple computer programs have been developed that quickly calculate results.[8] In addition, the computer tests statistical significance on-site to optimize the number of test runs.

The sample size is based on the specific need for the information. The following suggested ranges of permitted errors

---

[5] Roger P. Roess and William R. McShane, *An Executive Overview of the 1985 Highway Capacity Manual* (Polytechnic University Transportation Training and Research Center for the Institute of Transportation Engineers, 1986), 5.

[6] Box and Oppenlander, *Manual of Traffic Engineering Studies,* pp. 93–105.

[7] William Walker, "Speed and Travel Time Measurement in Urban Areas," *Traffic Speed and Volume Measurements,* Highway Research Board Bulletin 156 (Washington, DC: Highway Research Board, 1957), 27–44. See also Felix J. Rimberg, "Urban Travel Time Measurement by Taxicab Speed Studies," *Traffic Engineering,* 31(9) (June 1961), 43–44.

[8] Dr. Alex Skabardonis, Institute of Transportation Studies, University of California at Berkeley.

TABLE 3–2

Speed and Delay Information

Cross Streets	Distance (ft)	Segment Speed (mph)	Travel Time	Stop Time	Speed Limit	Ideal Travel Time	Segment Delay[a]	Net Speed (mph)
Street Name: Stoneridge Dr.		Date: 08/10/88		Time: 13:53		Weather: clear		
Foothill Rd	0	—	0:00	0:00	—	0:00	0:00	
Springdale Dr	1,622	31.6	0:35	0:06	40	0:28	0:07	27.0
Stoneridge Mall Rd	603	34.3	0:12	0:12	40	0:10	0:02	17.1
Johnson Dr	2,485	43.4	0:39	0:00	45	0:38	0:01	43.4
Denker	1,715	40.3	0:29	0:08	45	0:26	0:03	31.6
Hopyard Rd	1,566	34.4	0:31	0:54	45	0:24	0:07	12.6
Chabot	681	35.7	0:13	0:00	35	0:13	0:00	35.7
Willow	868	34.8	0:17	0:16	35	0:17	0:00	17.9
Hacienda Dr	1,039	33.7	0:21	0:00	35	0:20	0:01	33.7
Gibraltar	991	30.7	0:22	0:00	35	0:19	0:03	30.7
W Las Positas	2,362	36.6	0:44	0:37	35	0:46	0:02	19.9
Totals	13,932	36.1	4:23	2:13		4:01	0:22	24.0

[a]Segment delay is the difference between observed travel time and calculated ideal travel time.

SOURCE: W. G. van Gelder, "City of Pleasanton, California travel time studies."

in the estimate of the mean travel speed are related to the survey purpose:[9]

- o Transportation planning and highway needs studies: ±3.0 to ±5.0 mph
- o Traffic operation, trend analysis, and economic evaluations: ±2.0 to ±4.0 mph
- o Before-and-after studies: ±1.0 to ±3.0 mph

If the required sample size is greater for the range found than the number of vehicles sampled, additional "runs" must be performed under similar traffic and environmental conditions. The use of a laptop computer programmed to make this calculation allows a quick determination of adequacy of sample size without leaving the study area.

License plate method. In the license plate method, travel-time information can be obtained by stationing one or more observers at each entrance and exit of the study section to record the time and license number (last four digits or characters) of each vehicle as it passes the observation point. The plates are matched later and the travel time for each vehicle (the difference between the two recorded times) is determined.

Equipment used in this technique consists of synchronized stopwatches and recording forms or voice recorders

---

[9] Box and Oppenlander, *Manual of Traffic Engineering Studies,* Table 7-1.

with or without audible time signals. Large sets of these data can be analyzed economically by use of computers; hand reduction is quite tedious. To determine adequate sample size, refer to Equation (3.10).

Photography method. Direct observation and timing of vehicles through an area is feasible if the observer can see both the entrance and exit points. In these situations, data can be recorded economically for longer periods of time using videotaping equipment with internal clock functions.

Interview method. An interview technique is useful when a large amount of information is needed with little expense for field observations. Employees of private companies or public agencies are asked to record their travel time to and from work on a particular day.

Presenting results

Isochronal charts. Travel-time information may be presented as the mean total time to travel a given distance or the mean overall speed maintained over that distance. In a study of travel from one point in all directions, *isochronal charts* are often drawn. These show the distance from a common origin (often the central business district) that can be reached in a given time. A typical chart is shown in Figure 3–6. Congested areas are evident wherever the isochrons are close together. Free-flowing freeways and major streets are evident where the lines become long "fingers" leading away from the common origin.

Figure 3–6. Isochronal chart.

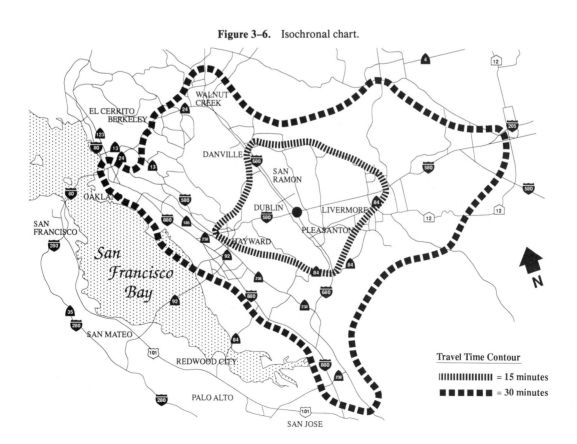

Vehicle delay rates. Another method of presenting travel-time data is to compute the vehicle delay rate. This consists of computing the mean travel time in minutes per mile and comparing this value with a standard adapted for the type of street, typically the posted speed limit. The difference between the two values (min/mi) is the *delay rate.* The delay rate multiplied by the volume gives a vehicle delay rate in vehicle-minutes per mile. These vehicle delay rates may be plotted on a map by using color codes or by using flow map techniques in which the width of the line represents the magnitude of the vehicle delay rate.

Travel-time data are also presented in a time-space diagram format. Typically, time is plotted vertically and distance plotted along the horizontal axis. Figure 3–7 presents a typical time-space diagram showing intersection delay.

Data gathered in speed and delay studies are subjected to a wide variety of analyses. Overall speed, running speed,

average delay, intersection delay, midblock delay, and duration and frequency of each type (cause) of delay by time and location are considered significant. Figure 3–8 presents a speed profile using speed and delay data. Average travel speed is the measure of effectiveness most often used for weaving sections and arterial streets.[10] Percent time delay is a common measure of effectiveness used to evaluate two-lane rural highways.[11]

Intersection delay studies

An intersection delay study will provide data on the amount of delay on one or more approaches. A travel-time or speed

---

[10] Roess and McShane, "An Executive Overview of the 1985 Highway Capacity Manual," p. 5.

[11] Ibid.

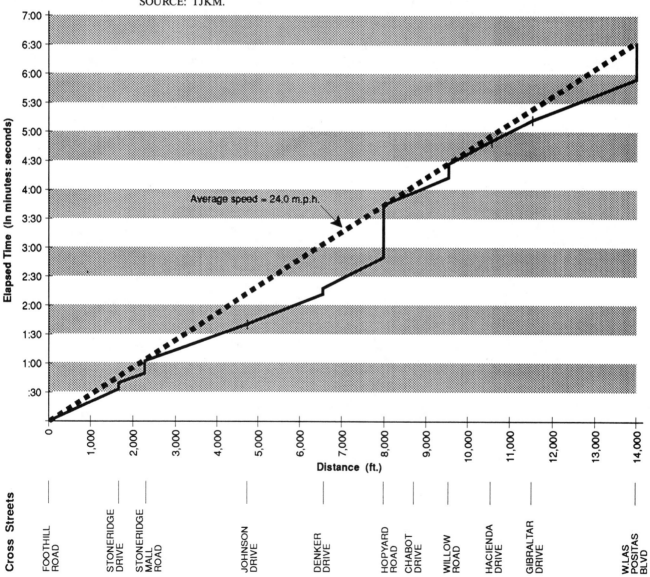

Figure 3–7. Travel time survey results: Foothill Road to West Positas Boulevard. SOURCE: TJKM.

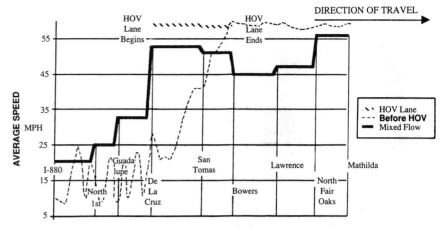

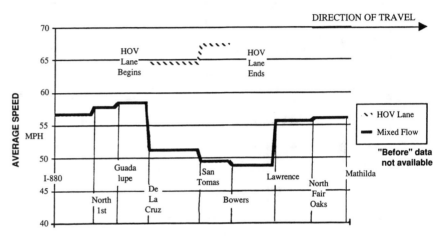

Figure 3–8. Speed profile.
SOURCE: "Before" data published in Santa Clara measure a strategic plan. "After" data collected by TJKM.

PM NB Peak Period Speed Profile

and delay study may indicate that certain intersections have undue amounts of delay and need more intensive study.

Average stopped-time delay is the measure of effectiveness for signalized intersections.[12] This study is useful in operational analyses of intersections, typically categorized as geometrics, signal control, and traffic. Geometric factors include the number and use of lanes, lane widths, grades, and other similar factors. Control variables involve the type, phasing, and timing of traffic signals. Traffic factors include vehicle composition percentages, parking and pedestrian activity, and vehicular flows by movement. The values that are typically computed from an intersection delay study include average individual or total delay by lane group, approach, or the entire intersection. Average delay per stopped vehicle and percentage of vehicles delayed can also be measured. Approach delay can be estimated from stopped-time delay measurements. Methods of estimating these values are presented in the 1985 *Highway Capacity Manual.*[13]

There are a variety of methods for field measurements of intersection travel time and delay. The travel-time study, when adapted to intersection delay, measures the travel time

from a point in advance of the intersection to a point in or beyond the intersection on one or more approaches. The direct methods used to obtain the data are as follows:

o Test car operated between key points
o License plates and times recorded at key points
o Time-lapse photographs or video taken from a vantage point to permit the timing of each vehicle shown on the film or videotape
o Observers stationed at a vantage point tracing individual vehicles through an intersection and recording times at critical locations

Some of these methods require extensive personnel or time for the collection and analysis of the data.

The most common method to estimate intersection stopped-time delay is based on direct observation of the number of stopped vehicles at frequent intervals, using the form shown in Table 3–3.[14] To avoid systematic error at signal-controlled intersections the sampling interval should not be an even subdivision of the length of the signal cycle. The length of the observed approach area should extend from a

[12] Ibid.

[13] *Highway Capacity Manual,* Special Report 209 (Transportation Research Board, 1985), Chapter 9, pp. 1–84.

[14] *Highway Capacity Manual,* p. 71.

TABLE 3–3
Vehicular Gap Size Study Format

Location: MAIN ST. & FIRST ST.

Date: 4/30/90   Time: 5:00 – 5:10   City: Pleasanton

PM				INTERSECTION DELAY WORKSHEET				
		(seconds)		Number of Stopped Vehicles				
MIN / SEC	+0	+20	+40					
5:00	2	4	2					
5:01	3	5	0					
5:02	6	3	5					
5:03	4	5	3					
5:04	2	2	4					
5:05	4	4	6					
5:06	5	2	1					
5:07	1	3	2					
5:08	4	4	3					
5:09	2	6	2					
Totals	33	38	28					

$I = 20$ secs. $V_s = $ 99   Delay $= \dfrac{\Sigma V_s \times I}{V} = $ 40 sec.

Volume, $V = $ 50

SOURCE: TJKM TRANSPORTATION CONSULTANTS.

point beyond the end of any expected queue to the far side of the intersection. The method is based on the flow-density-speed relationship.

The method requires one observer with a stopwatch for each approach studied. At each interval, the number of stopped vehicles is counted for each lane group and recorded in the appropriate time slot. During the entire period of the study, a volume count shall be maintained, counting vehicles as they cross the stop line of the lane group under study. The total volume count is recorded in the appropriate box on the form.

Average intersection delay is computed by multiplying the total number of stopped vehicles by the sampling interval (sec) and dividing the sum by the total volume of traffic entering the intersection during the study period. This can be expressed as:

$$\text{Delay} = \frac{(\Sigma V_s I)}{V} \qquad (3.5)$$

where: D = average delay per vehicle (sec)
ΣV_s = sum of stopped vehicle counts
I = interval between counts (sec)
V = total volume observed (number of vehicles)

Ten minutes of data for such a delay study are shown in Table 3–3. In this example, the total volume in the 10-min period was 50 vehicles (V). There were 99 vehicles (V_s) recorded during observations made at 20-sec intervals (I). The average stopped delay per vehicle was 40 sec per vehicle.

Equipment that facilitates the collection of vehicle delay data include a laptop computer or video camera. The use of videotape, however, often requires a substantial amount of time to reduce the data.[15]

Queue length study. In the queue length study, the instantaneous number of vehicles in a standing or slowly moving queue is counted in the field or from photographs. When the count is made at a signalized intersection, it should be made at the start of the green interval for each approach being studied, and also made at the end of the yellow interval. A notation should be made if the queue is completely discharged during the movement phase. At unsignalized intersections, the counts are usually made at equal intervals of time such as 30 sec or 1 min. This study is an important source of data for the determination of delay. It can also be used to measure the effectiveness of a change in traffic signal control. Queue lengths should be identified in any volume study at the end of a count period to differentiate flow from demand.

Queue length studies have an important application in determining the appropriate locations for secondary street intersections or major driveways when they will be in proximity to a signalized intersection. The study is performed by establishing potential access locations at varying distances from the stop line of the controlled intersection and observing the extent to which queue lengths reach or pass these locations. It is often possible to select existing features, such as fire hydrants, sign posts, poles, or other reference points. The study should be conducted during the peak hour of heaviest loading. The expected proportion of the hour during which entry or exit to each potential location will be blocked by backup of standing traffic is computed from the data. Separate stopwatches may be used for each of the reference points and operated continuously during periods when traffic backup blocks each location for a direct measure of the total blockage during the study period. The blockage time for each location is divided by the total length of study to produce the percent of time blocked.

Other traffic stream studies

Although volume and speed studies are the basic tools of the traffic engineer, several other characteristics of traffic streams are measurable and may be used directly for some purposes or that can be used to estimate volume, speed, or both. This section introduces the measurement of density, gaps, and vehicle spacing.

Highway operations studies

Density studies. Density refers to the number of vehicles per lane within a specific section of roadway. For capacity computations, it is often referred to as the number of passenger car equivalents per lane per mile of roadway. Density is a measure of effectiveness for basic freeway segments and multilane highways.[16]

Density can be obtained by observation, photography, or estimation using data from pavement detectors. Density can also be computed from input/output studies or from studies providing joint information on volumes and speed using the flow-density-speed relationship. In observational studies, frequent samples are made of counting the number of vehicles located on a road section. The average is taken as an unbiased estimate of the mean density during the sampling period. In photographic studies, the number of vehicles can be obtained by directly counting the vehicles in a section of roadway in each photographic frame. Pavement detectors can be used to determine occupancy for individual lanes, as described in the succeeding section, which can be translated into average density.

Occupancy. Occupancy is defined as the percent of time that a vehicle occupies a specific section of roadway. Detectors that identify the presence of a vehicle at the detector location provide an occupancy measure. Occupancy can be used to evaluate requests for nearby driveway access by identifying gaps between vehicles and queues during peak periods. Occupancy can also be used to compute average density, after accounting for vehicle length and effective detector length.

---

[15] R. David Henry, "Laptop Computers Measure Intersection Performance," *ITE Journal,* 57 (June 1987), 39–42.

[16] Roess and McShane, "An Executive Overview of the 1985 Highway Capacity Manual," p. 5.

$$\text{Occupancy} = \frac{100 \times \frac{(\text{average vehicle length} + \text{detector width})(\text{hourly flow rate})}{(\text{average operating speed})}}{(5{,}280 \text{ fpm}/3{,}600 \text{ sph})(3{,}600)}$$

where: fpm = ft per mile
      sph = sec per hour

For example, a single lane with a flow of 1,000 vehicles per hour with average vehicle lengths of 19 ft operating at 30 mph over a detector width a 6-ft effective detection length would register an occupancy of (in percentage):

$$\text{Occupancy} = \frac{(19+6)(1{,}000)(100)}{(30)(5{,}280/3{,}600)(3{,}600)} = 16\%$$

Conversely, if only the occupancy percent and the average vehicle and detection lengths are known, average density can be computed as:

$$\frac{\text{Average}}{\text{Density}} = \frac{(0.16)(5{,}280 \text{ ft/mile})}{(19+6)} = 33.8 \text{ vehicles per mile}$$

Gap studies. Time headways (measured between the front ends of successive vehicles) or gaps between vehicles (measured from the rear of a vehicle to the front of the following vehicle) are important to traffic engineers. The gap between vehicles is a critical variable in determining the level of service at unsignalized intersections.[17]

The frequency of adequate gaps is also used in warrants for adult protection at school crossings and for traffic controls at other pedestrian crossings. Gaps are also important in determining merging capacity at ramp junctions and in warrants for multiway stop-sign control.

Peak hour gap availability may also be used to check for possible control needs at a proposed new intersection or high-volume driveway along an existing roadway. Typically, gaps of about 6 to 9 sec are needed to allow the critical entry into the traffic stream of a major street. In this application, gaps occurring simultaneously across both directions of travel may be counted in groups, such as:

Gap Size (sec)	Factor
1–5	0
6–9	1
10–13	2
14–16	3
17–19	4
20–22	5
over 22	6

The *gap availability parameter* is calculated by multiplying the number of entry vehicles by the factor number listed. The summation of the product of the factors multiplied by the number in each group is called the *total effective gaps*. If the volume of traffic projected to enter from the cross street is about one-half the number of gaps available, no traffic control is likely to be needed. In practice, more vehicles can exit than the number of available gaps, because when several

vehicles are standing in a queue awaiting exit, the second and third drivers will accept shorter gaps. Thus, a 14- to 16-sec gap usually can accommodate three waiting cars.[18]

Ordinarily, side-street right turns are not critical, because available gaps for right-turning vehicles are needed in only one direction of flow, and for a given volume such gaps are far more frequent than the simultaneous available gaps in both directions. A few gap counters that can be connected to road tubes across a roadway have been commercially available. For manual use, an excellent timing device is a laptop computer that is actuated by a detector or human observer. Cumulative records of the successive times of arrival of vehicles are made and gaps subsequently computed. Vehicles can easily be classified using a preprogrammed laptop computer.

Stopwatches may be used to measure individual gaps. With data divided into the seven groupings, one person with a stopwatch can tabulate gaps and count directional flows having a combined volume of up to 1,500 vehicles per hour. This assumes no requirement for vehicular classification. A sample form is presented in Table 3–4.

Time-lapse photography and video systems can also provide records from which gaps may be measured. The data-reduction process from videotape can be facilitated with computer software that operates at real time. Gaps are keyed in and the software produces computations and data summaries.[19]

Conflict studies

This technique relates projected accident hazard to the frequency of observed intersection vehicular conflicts of various types. The existence of a conflict is inferred as a potential accident situation from an easily identifiable driver response. These responses include the application of the brake pedal (illuminated brake light) and/or sudden, forced lane changes. The causes of the action at intersections are categorized as relating to accident patterns for left-turn conflicts, cross-traffic conflicts, and/or rear-end conflicts. A secondary conflict occurs when a third vehicle must take evasive action caused by an original conflict maneuver. The secondary conflicts are recorded separately. The number of conflicts and the volume of traffic are counted. Relationships developed from experience are used to evaluate the accident potential of the intersection.[20]

Intersection studies

This section addresses specialized intersection studies including capacity (queue discharge), signal timing, queue studies, left-turn storage requirements, and multiway stop control.

[17] *Highway Capacity Manual*, p. 2.

[18] Louis Neudorff, *Candidate Signal Warrants from Gap Data*, Report Number FHWA/RD-82/152, 1983.

[19] Michael Kyte and Joseph Marek, "Collecting Traffic Data at All-Way Stop-Controlled Intersections," *ITE Journal*, 59 (April 1989), 33–36.

[20] M.R. Parker, Jr., and C.V. Zegeer, "Traffic Conflict Techniques for Safety and Operations," *Engineer Guide and Observers Manual*, Report Numbers FHWA-IP-88-026 and FHWA-IP-88-027, Federal Highway Administration, January 1989.

TABLE 3–4
Intersection Delay Worksheet

SURVEY DATE _____   LOCATION _____   CROSSWALK ACROSS _____

END OF SURVEY (To Nearest Minute) _____	NUMBER OF LANES "N" _____
START OF SURVEY (To Nearest Minute) _____	ROADWAY WIDTH "W" _____
TOTAL SURVEY TIME (Minute) _____	ADEQUATE GAP TIME "G" _____

GAP SIZE (Seconds)	NUMBER OF GAPS		MULTIPLY BY GAP SIZE	COMPUTATIONS
	TALLY	TOTAL		
8				
9				
10				
11				
12				
13				
14				
15				
16				
17				
18				
19				
20				
21				
22				
23				
24				
25				
26				
27				
28				
29				
30				
31				
32				
33				
34				
35				
36				
37				
38				
39				
40				
41				
42				
43				
	TOTALS			

t = Total time of all gaps equal or greater than G

Queue studies

Two types of queue studies can be made at signalized intersections: saturation flow rate (queue discharge) and queue length.

Saturation flow rate studies. Saturation flow rate is the flow in vehicles per hour that could be accommodated by the lane group assuming that the green phase is available for a full hour and there is a continuous queue of vehicles waiting to enter the intersection. Saturation flow rates are used in capacity analyses of signalized intersections and in signal timing.

In the saturation flow rate, or queue discharge study, several green phases are observed. The sample size should be determined based on Equation (3.10). A stopwatch is started when the green phase begins and stopped when the last vehicle in the delayed queue is discharged or when the phase ends. Times are recorded when the rear axle of the fourth, tenth, and last vehicles cross a reference point such as the stop line. The period defined as saturation flow begins when the rear

axle of the fourth vehicle in queue crosses the reference point and ends when the rear axle of the last queued vehicle at the beginning of the green crosses the same point. The fourth vehicle is used at the start of the measurement since the first three vehicles include "start-up delays." The effects of start-up delay are not included in a saturation flow rate, but are typically accounted for elsewhere in an intersection capacity analysis.

In the example shown in Figure 3–9 for Cycle 1, the time for the fourth vehicle was observed as 10.2 sec, and the time for the 14th vehicle was 30.7 sec; the average saturation headway per vehicle would be:

$$\frac{(30.7 - 10.2)}{(14 - 4)} = 20.5/10 = 2.05 \text{ sec/vehicle}$$

and the saturation flow rate would be:

$$\frac{3,600 \text{ sec/hr}}{2.05 \text{ sec/vehicle}} = 1,756 \text{ vehicles per hour of green per lane}$$

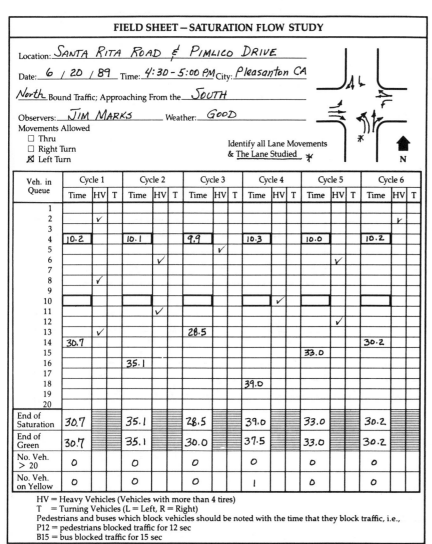

FIELD SHEET – SATURATION FLOW STUDY

Location: SANTA RITA ROAD & PIMLICO DRIVE

Date: 6 / 20 / 89 Time: 4:30 – 5:00 PM City: Pleasanton CA

North Bound Traffic; Approaching From the SOUTH

Observers: JIM MARKS   Weather: GOOD

Movements Allowed
- ☐ Thru
- ☐ Right Turn
- ☒ Left Turn

Identify all Lane Movements & The Lane Studied ✳

Veh. in Queue	Cycle 1 Time	HV	T	Cycle 2 Time	HV	T	Cycle 3 Time	HV	T	Cycle 4 Time	HV	T	Cycle 5 Time	HV	T	Cycle 6 Time	HV	T
1																		
2		✓															✓	
3																		
4	10.2			10.1			9.9			10.3			10.0			10.2		
5								✓										
6					✓									✓				
7																		
8		✓																
9																		
10									✓									
11					✓													
12														✓				
13		✓					28.5											
14	30.7															30.2		
15													33.0					
16				35.1														
17																		
18										39.0								
19																		
20																		
End of Saturation	30.7			35.1			28.5			39.0			33.0			30.2		
End of Green	30.7			35.1			30.0			37.5			33.0			30.2		
No. Veh. > 20	0			0			0			0			0			0		
No. Veh. on Yellow	0			0			0			1			0			0		

HV = Heavy Vehicles (Vehicles with more than 4 tires)
T = Turning Vehicles (L = Left, R = Right)
Pedestrians and buses which block vehicles should be noted with the time that they block traffic, i.e.,
P12 = pedestrians blocked traffic for 12 sec
B15 = bus blocked traffic for 15 sec

Grade 0   Area Type non-CBD

Figure 3–9. Field sheet: saturation flow study.
SOURCE: TJKM.

Saturation flow is described in more detail in the 1985 *Highway Capacity Manual*.[21]

Peak-hour factor determination

A peak-hour factor is used to analyze traffic conditions for periods less than 1 hour. Depending on the type of land uses in the area, peaking can be pronounced for short periods within the peak hour. The most common time period analyzed, other than the peak hour, is the peak 15 min.

Operational studies of the peak-hour factor can be obtained by counting volumes on each approach separately for each 15-min period of the peak hour (machine counts are very suitable for this) and calculated as follows:

$$\text{peak-hour factor} = \frac{\text{peak-hour volume}}{4 \times (\text{peak 15-min volume})}$$

Progression factor

A signal progression factor, referred to as *arrival type*, approximates the quality of signal progression on an intersection approach.[22] This is a critical traffic characteristic in the determination of level of service at signalized intersections.

Progression factors are best observed in the field, but could be estimated by examining time-space diagrams for the arterial under study. The following ratio is used:

$$R_p = PVG/PTG \qquad (3.6)$$

where:

R_p = platoon ratio
PVG = percentage of all vehicles in the movement arriving during the green phase
PTG = percentage of the cycle that is green for the movement; $PTG = (G/C) \times 100$.

Progression factors are used in level-of-service calculations for signalized intersections and arterial streets. This is a general categorization that attempts to quantify with some approximation the quality of progression on the intersection approach.

Sight-distance studies

Vertical curves. Sight-distance measurements are frequently used in operational studies, such as the setting of no-passing zones. Driver eye heights of 3.5 ft and obstacle heights of 4.25 ft are typically used. For continuous measurement of sight distance, a pair of vehicles, equipped with communication and distance-measuring devices, can be employed. The rear vehicle can keep the front vehicle barely in sight and command cumulative distance measures at selected check points. The recorded difference is the sight distance. Lines of sight may also be measured on foot using tapes or measuring wheels.

These studies are used to mark no-passing zones on two-lane highways by noting when sight distance is less than a specified distance to allow passing maneuvers. For this type of study the "following" vehicle maintains a specified distance behind the "lead" vehicle and marks locations where the lead vehicle cannot be seen.

Horizontal curves. Sight distance at horizontal curves is a function of roadway geometrics and roadside obstructions such as embankments and median barriers. AASHTO's *Policy on Geometric Design of Highways and Streets*[23] provides sight-distance measurement procedures.

Intersections. At low-volume intersections, the relative need for control, determination of a safe approach speed, and the selection between yield or stop signs often are dependent on intersection sight distance. AASHTO's *Policy on Geometric Design of Highways and Streets* provides sight-distance requirements for various intersection control based on desirable design parameters. Minimum intersection sight-distance requirements can be determined by field measurements and the time-distance relationship of vehicles using the intersection.

Multiway stop control studies

Multiway stop control is a useful and appropriate type of intersection control for certain unsignalized intersections. Under multiway stop control, all vehicles are stopped, with vehicles intended to depart in a counterclockwise rotation regime under the basic rules of the road, wherein the "vehicle on the right" has the right-of-way. Multiway stop control can be a low-cost solution at uncontrolled or two-way stop or yield intersections where either poor level of service or accident history has been experienced.

Field measurements of level of service at multiway stop-controlled intersections is performed by tracking progress of vehicles from the end of queue through an intersection. Four observers with stopwatches per approach track vehicles through the queue. Two are positioned at the end of the queue and two more observers are positioned at the stop line. Average delay per vehicle, which is an indicator of level of service, is determined in this manner.

A video camera reduces the personnel requirements from four to one person per approach. The camera should be positioned approximately 150 ft from the intersection with a clear view of queue formation and dissipation on the study approach and all turning movements at the intersection. The data reduction process from the videotape can be facilitated with computer software that transcribes data at the same speed as the videotape with vehicle starts and stops keyed in by an operator. This procedure is particularly useful when data from several locations have been collected. Vehicle delay is determined by calculating the sum of all differences between the departure and arrival times at the queue and comparing with typical flow rates.

[21] *Highway Capacity Manual*, pp. 73–74.
[22] Ibid., pp. 7–8.

[23] *A Policy on Geometric Design of Highways and Streets,* American Association of State Highway and Transportation Officials, 1990.

Pedestrian studies

Pedestrian studies consist of several different types including those related to traffic signal timing to allow for pedestrian walking time, clearance intervals, pedestrian grouping and street-crossing behavior, acceptance of gaps between approaching vehicles to cross the street, and others.

Pedestrian factors

Walking speed is an important factor for pedestrian studies. The key use of walking speed is in determining the minimum time for pedestrians crossing roadways. Walking speed is dependent upon the following factors:

o age of pedestrian
o sex of pedestrian
o volume of pedestrians
o steepness of grade
o distance to oncoming vehicle
o speed of oncoming vehicle
o width of crossing area
o width of sidewalk, walkway, or crosswalk
o type of group of pedestrians

Walking speeds are usually in the range of 2.2 to 4.0 ft/sec. The *Manual on Uniform Traffic Control Devices*[24] assumes a less conservative 4.0 ft/sec as the usual walking speed. Crossing speed studies are normally important at wide intersections with a large number of pedestrians and in locations where there are a great number of elderly pedestrians. See Chapter 1 for additional information.

Grouping of pedestrians

A pedestrian group-crossing study is conducted to identify pedestrian group size. This can be done by one or more observers tallying group sizes of pedestrians crossing the street. Pedestrian arrival rate may allow all group sizes to be tallied, or if the location is too busy, a sample may be taken to assure that the sample is random. Every "Nth" group size should be recorded where the size of N depends upon how busy the location is, and how many observers are assigned. The Nth group could be every second or third or fourth group . . . etc., depending upon a comfortable frequency for the observer to count and to record without confusion. Each group can be tallied and plotted to identify the distribution group sizes. The 85th percentile can be identified and used as the factor N to calculate the adequate gap size. An "adequate gap" is defined below.

Determining adequate gaps

Adequate gap is defined as the amount of time (in seconds) for one or more pedestrians to observe the traffic situation while standing at a safe location on one side of the roadway and then to cross the roadway to a point of safety on the opposite side. This time includes the initial reaction time and walking time for the pedestrian as well as consideration of group size. The formula for computing the adequate gap time for a pedestrian-crossing location is:

$$G = W/S + R + (N - 1)2 \qquad (3.7)$$

where:

$G =$ adequate gap time in seconds
$W =$ width in feet of the roadway to be crossed
$S =$ 15th percentile of the actual pedestrian walking speeds in ft/sec, or an assumed walking speed such as 3.5 ft/sec
$R =$ assumed to be 3 sec, the time which experience has shown for the typical pedestrian to look both ways, make a decision, and begin to walk across the roadway
$(N - 1)2 =$ the pedestrian clearance interval (additional seconds of time required for large groups of pedestrians to clear the roadway). For this calculation, it is assumed that pedestrians cross the roadway in groups or rows of five. Therefore, N is the 85th percentile group size divided by 5, and 1 represents the first row and 2 the time interval in seconds between rows.

The formula for adequate gap time provides an estimate of the required crossing time. Experience has shown that this formula is conservative and may be used for design requirements.

An adequate gap study can be based on a 30-min field survey consisting of six 5-min samples.[25] The average number of gaps per 5-min period can be calculated by Equation (3.8):

$$Gn = \frac{(G_t/P_t \times 6)}{(\text{pedestrian crossing time})} \times 6 \text{ periods} \qquad (3.8)$$

where:

$G_t =$ total usable gap time in seconds per 30 min
$P_t =$ pedestrian crossing time

The total usable gap time in seconds per 30-min period is divided by the pedestrian crossing time. The pedestrian crossing time is the street width (curb to curb) divided by the walking speed. The following is a sample calculation:

Usable gaps measured in each of six 5-min periods are observed to be 30 sec, 34 sec, 17 sec, 37 sec, 50 sec, and 72 sec for a total of 240 sec.

The street width is measured as 70 ft. The crossing time is this distance divided by the assumed walking time of 3.5 ft/sec so the crossing time is 20 sec. Six periods multiplied by the 20 sec yields 120 sec: 70/3.5 = 20 and 6 × 20 = 120.

The usable gap total of 240 sec divided by the pedestrian crossing time of 120 sec for six periods yields two usable gaps per 5-min period: 240/120 = 2.

---

[24] *Manual on Uniform Traffic Control Devices*, Federal Highway Administration, U.S. Department of Transportation, 1988.

[25] "School Area Pedestrian Safety," *Traffic Manual* (Sacramento: California Department of Transportation, 1987, updated periodically), 10–13.

Gap studies are appropriate for school area studies. Other studies near schools include safe routes to school as well as inventory and evaluation of factors affecting pedestrian safety. Refer to Chapter 1 for additional information.

Public transit studies

The purpose of public transit studies is to evaluate public transportation service provided within a given study area. The evaluation includes estimating the use of and determining the need for public transit. This is accomplished on either a regional or a local level. Transit studies at the local level tend to emphasize issues such as the location of transit stops, service frequency, and the effect of one-way streets on transit routing. This section describes three categories of transit studies: public transit service inventories, operations studies, and planning studies. A number of transit studies are described in detail in *Measuring Transit Service*[26] and *Urban Mass Transportation Travel Surveys.*[27] Forms are given in the *Manual of Traffic Engineering Studies.*[28] The *Transportation Planning Handbook*[29] contains more detailed information on public transit studies in Chapters 2 and 10.

Public transit service inventories

An inventory of public transit service supplies is essential background information for other transit studies and evaluations of a specific service. An inventory of the physical characteristics includes location of stops, transfer points, layout of routes, etc. An operations inventory collects information on service type, service frequency and hours of service, travel time between various points on the system, a summary of the rolling stock (showing its capacity, age, and condition), passenger volumes, and schedule of fares.

Operations studies

Three types of studies are used to determine transit operation characteristics: load checks, boarding and alighting checks, and speed and delay studies.

Transit load checks.   Transit load checks are made by observers stationed at one or more points along the routes. Generally, one check station is located at the maximum load point, where the number of passengers carried is known to be the greatest. If two or more points along one route are studied simultaneously, the travel time of each transit vehicle between these points can be measured. Each observer records vehicle identification, time of arrival and/or departure, number of

persons on board the vehicle when arriving, number of persons alighting, and number of persons boarding.

Boarding and alighting checks.   Boarding and alighting checks are conducted by observers traveling on transit vehicles. Each observer records the number of persons boarding and alighting at each stop, the number of persons on board between stops, the time the vehicle passes predetermined check points on the route, and, as required, the method of fare payment (cash, tickets, transfers). On lightly traveled transit routes, the observer may record each passenger's boarding and alighting stop to provide trip length information or a postcard survey can be distributed.

Speed and delay studies.   Transit speed and delay studies parallel similar studies for the entire vehicle stream described earlier in the chapter. Data are obtained by observers riding on the transit vehicles at various hours of the day. The time that each vehicle passes a check station and the cause and duration of delays are recorded. In addition to the types of delays found in general traffic speed and delay studies, the transit study also highlights boarding and alighting delays at bus stops and vehicle schedule delays.

Transit operations studies provide data on passenger volumes, transit vehicle trip characteristics, vehicle occupancy, travel times, and adherence to schedules. The studies are of value when they lead to improvement in transit service and scheduling. Traffic engineers use this information to evaluate bus stop location, turn prohibitions, and one-way street systems. Data are also useful in the investigation of street operating plans that favor transit vehicles, such as signal timing and designation of exclusive bus lanes.

Prototype on-board electronic sensing and communication equipment has been developed that makes it possible to conduct these studies continuously as a part of transit operations management. Automated data collection systems are being investigated by transit agencies for gathering ridership, fare revenue, and schedule adherence statistics. These systems are explained in more detail in *Modular Approach to On-Board Automatic Data Collection Systems.*[30]

Parking studies

Introduction

Parking studies are designed to identify inadequacies in the supply of parking, or to determine existing demand in order to plan for future parking. The existing demand may be in terms of actual vehicles parked at a specific site (or in a given vicinity) or may be translated into a parking rate. The study size may vary from an individual private lot to a city-wide study incorporating private and public lots as well as on-street parking.

Detailed descriptions of several techniques to conduct parking studies are described in several sources: *Conducting*

[26] National Committee on Urban Transportation, *Measuring Transit Service* (Procedure Manual 4A) (Chicago: Public Administration Service, 1958).

[27] Urban Transportation Systems Associates, *Urban Mass Transportation Travel Surveys* (Washington, DC: Federal Highway Administration, August 1972).

[28] Box and Oppenlander, *Manual of Traffic Engineering Studies,* pp. 153–170.

[29] *Transportation Planning Handbook* (Washington, DC: Institute of Transportation Engineers, 1991).

[30] National Cooperative Transit Research and Development Program, *Modular Approach to On-Board Automatic Data Collection Systems,* Report 9 (Washington, DC: Transportation Research Board, National Research Council, December 1984).

a Limited Parking Study and *Conducting a Comprehensive Parking Study*[31] and in *Parking Principles.*[32] In addition, the *Manual of Traffic Engineering Studies*[33] contains details of various study techniques including the forms for field work and data compilation. Other sources describe advanced techniques for conducting parking studies which involve the use of gravity models with the aid of microcomputers: "Development of Computerized Analysis of Alternative Parking Management Policies"[34] for analyzing the supply and management of parking facilities and "Estimating Downtown Parking Demands: A Land Use Approach"[35] for estimating and allocating parking demand in a downtown area.

The first step in conducting parking studies is to delineate the study area. If the study purpose is to determine the demand for a specific site, care should be taken to include vehicles parked off-site that are attributable to the site. In central business district (CBD) areas, the limits of the parking study should include areas within a reasonable walking distance, say, 500–1,500 ft. The second step is generally an inventory of the current parking supply. The remainder of the study depends on the requirements of the particular study.

Parking study requirements

Inventory of parking supply. This step is required in virtually all parking studies. The parking demand characteristics are more meaningful when presented in context with the supply (and vice versa). The value of the parking supply is important information in parking generation rate studies, even though the supply value does not directly figure into the calculations for parking generation rates. If the demand is measured to be within 90% to 100% of the actual supply, the true demand may not have been measured; there may be a latent demand that is unmet due to the lack of an adequate parking supply.

Curb spaces are inventoried by observers who traverse all streets (and alleys) and measure or estimate all categories of parking: unrestricted spaces, truck or passenger loading zones, time-limited parking zones, and prohibited parking zones. Each off-street facility is classified as to whether it is public unrestricted, public with charges or other restrictions, private (e.g., employees only), or commercial. When conducting a study of a large site with numerous categories of parking, these categories should be noted (e.g., a hospital with parking for doctors, employees,

visitors, volunteers, and outpatients). The capacity and parking fees (if any) of each off-street lot by classification are determined.

Inventory data are tabulated and may be presented graphically. Curb space inventories can be presented graphically by superimposing the curb parking restrictions on maps of the area (see Figure 3–10).

Current parking characteristics. This information is typically required when the purpose of the study is more than simply to determine the maximum demand. For example, the time of day the maximum demand occurs is important at sites such as hospitals or mixed-use commercial developments. This is because different users (e.g., outpatients vs. visitors or office workers vs. theatergoers) may be able to share parking because peak parking demands occur at different times of day. (For more information on the concept of shared parking and how it works, see *Shared Parking* by the Urban Land Institute.)[36] Also, the average duration that an individual space is occupied is important in determining whether time-limited on-street parking is appropriate and what the time limit should be.

Current and future demands. Current demand is useful both in determining a given parking generation rate and in determining what areas of a particular site or section of the central business district have inadequate or surplus parking. Future demand is projected from the current demand when a major impact to the parking demand may occur, either increasing the demand by constructing new facilities or buildings, or decreasing the demand by installing or improving new transit facilities.

Current and future revenue forecasts. This information is important when parking revenues, whether current, future or both, are expected to pay for new parking facilities.

Types of studies

Occupancy studies. These are designed to determine the number of parking spaces occupied at various times of day so as to determine the peak demand, the location of the peak demand, and surplus parking, if any. The data are obtained by conducting field observations.

Curb space occupancy is measured by counting each block face at regular intervals. The number of spaces occupied including commercial vehicles in loading zones and any vehicles illegally parked or double parked should be counted. Observers generally walk, but if the study area is large enough they may drive. The interval between successive observations of each facility will depend on the needs of the study. If the peak demand for a particular use or lot is all that is required and the area has an obvious and well-known peak period, then two or three observations within this peak period may be all that is required. (See special parking studies.) In other studies, counts at 1-hour intervals generally are sufficient throughout the business day. In small communities, two to five counts may adequately span the business day. One or two

[31] National Committee on Urban Transportation, *Conducting a Limited Parking Study* and *Conducting a Comprehensive Parking Study* (Chicago: Public Administration Service, 1958).

[32] *Parking Principles,* Special Report 125 (Washington, DC: Highway Research Board, 1971), 75–88.

[33] Box and Oppenlander, *Manual of Traffic Engineering Studies.*

[34] A.G.R. Bullen, "Development of Computerized Analysis of Alternative Parking Management Policies," *Transportation Research Record No. 845,* pp. 31–37 (Transportation Research Board, National Research Council, 1982).

[35] Herbert S. Levinson and Charles O. Pratt, "Estimating Downtown Parking Demands: A Land Use Approach," *Transportation Research Record No. 957,* pp. 63–66 (Transportation Research Board, National Research Council, 1984).

[36] *Shared Parking,* the Urban Land Institute, Washington, DC, 1983.

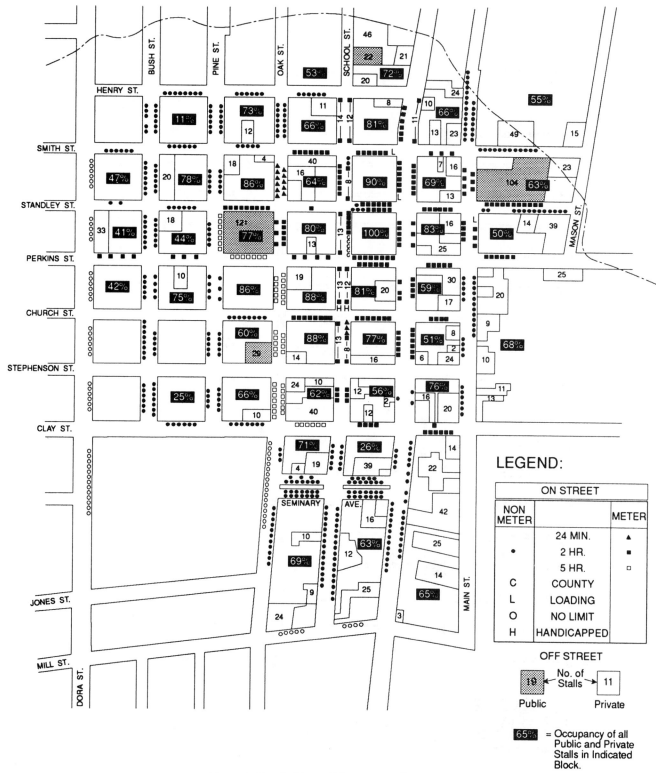

Figure 3–10. Parking stall inventory and occupancy.

Saturday or evening checks may also be required, depending on the needs of the study.

Off-street occupancy can also be measured using field observations at regular intervals. Another method is to use observers or recording traffic counters to conduct a cordon count of a single facility. If the off-street facility is very large or if it is of interest to compare the occupancy of different sections, visual counts may still be required. Another technique would be the use of aerial photography. Occupancy and duration were obtained satisfactorily in a study that identified cost

EDEN HOSPITAL 6/10/87

LOT	SPACES	9:00	9:30	10:00	10:30	11:00	11:30	NOON	12:30	1:00	1:30	2:00	2:30	3:00	3:30	4:00	4:30	5:00	MAXIMUM
1 - AUXILIARY/EMPLY	27	20	26	27	27	27	27	24	25	20	20	20	21	20	19	10	8	8	27
2A- VISITORS	70	51	60	67	68	70	70	68	65	69	70	67	69	70	66	62	60	57	70
2B- DOCTORS	26	24	27	27	27	21	21	23	24	24	23	21	21	19	17	21	20	18	27
3 - EMPLOYEES	49	49	49	49	49	49	49	49	47	47	49	49	47	48	39	17	13	16	49
4A- CARDIAC/DIETARY	22	22	21	22	19	19	19	18	16	16	14	17	18	20	20	19	16	16	22
4B- CONSTRUCTION	18	18	18	18	18	18	18	16	18	18	18	18	18	18	9	1	0	0	18
5 - EMPLOYEES	28	15	22	27	26	25	25	26	26	26	26	26	28	28	21	8	5	3	28
6A- SPECIAL PATIENT	35	26	31	31	31	28	28	23	22	30	30	34	33	34	31	22	23	24	34
6B- EMPLOYEES	14	11	12	12	12	10	10	12	13	9	8	11	11	13	10	13	12	12	13
7 - VISITORS/PATIENT	109	63	91	87	106	98	87	89	91	95	89	88	97	103	95	98	89	81	106
8 - EMPLOYEES	144	142	142	142	143	143	143	139	134	136	140	141	137	136	113	81	61	43	143
9 - EMPL -3 PM SHIFT	55	0	0	0	0	0	0	0	0	0	0	0	0	46	45	45	45	45	46
10- ON STREET PKG		14	19	25	20	15	15	12	10	7	7	8	8	8	8	8	6	4	25
TOTAL	597	455	518	534	546	523	512	499	491	497	494	500	508	563	493	405	358	327	563
% of PEAK	106.0%	80.8%	92.0%	94.8%	97.0%	92.9%	90.9%	88.6%	87.2%	88.3%	87.7%	88.8%	90.2%	100.0%	87.6%	71.9%	63.6%	58.1%	100.0%
% of CAPACITY	100.0%	76.2%	86.8%	89.4%	91.5%	87.6%	85.8%	83.6%	82.2%	83.2%	82.7%	83.8%	85.1%	94.3%	82.6%	67.8%	60.0%	54.8%	94.3%

EDEN HOSPITAL 6/11/87

LOT	SPACES	9:00	9:30	10:00	10:30	11:00	11:30	NOON	12:30	1:00	1:30	2:00	2:30	3:00	3:30	4:00	4:30	5:00	MAXIMUM
1 - AUXILIARY/EMPLY	27	23	25	25	25	25	24	24	26	26	22	19	19	20	19	8	7	8	26
2A- VISITORS	70	60	62	62	62	70	66	63	66	68	70	68	70	69	68	64	60	59	70
2B- DOCTORS	26	25	27	27	27	20	21	23	24	26	24	24	22	23	22	20	19	18	27
3 - EMPLOYEES	49	49	49	49	49	49	49	49	49	48	49	49	49	49	47	17	12	16	49
4A- CARDIAC/DIETARY	22	8	13	13	13	13	14	18	19	19	19	20	20	20	20	17	16	15	20
4B- CONSTRUCTION	18	18	18	18	18	18	18	16	18	18	18	18	18	18	8	1	0	0	18
5 - EMPLOYEES	28	26	27	27	27	27	27	25	26	25	26	27	28	28	21	10	6	3	28
6A- SPECIAL PATIENT	35	28	30	31	32	30	30	29	18	30	30	30	32	32	30	21	22	23	32
6B- EMPLOYEES	14	11	12	12	12	14	12	11	13	12	13	13	12	14	10	12	11	11	14
7 - VISITORS/PATIENT	109	78	86	90	95	98	93	85	83	91	100	106	107	107	97	96	90	77	107
8 - EMPLOYEES	144	128	130	133	134	137	133	128	123	124	125	129	129	120	104	82	60	39	137
9 - EMPL -3 PM SHIFT	55	0	0	0	0	0	0	0	0	0	4	10	12	46	46	44	46	46	46
10- ON STREET PKG		7	7	9	9	9	9	9	6	6	6	8	7	6	4	4	4	4	9
TOTAL	597	461	486	496	503	510	496	480	471	493	506	521	525	552	496	396	353	319	552
% of PEAK	108.2%	83.5%	88.0%	89.9%	91.1%	92.4%	89.9%	87.0%	85.3%	89.3%	91.7%	94.4%	95.1%	100.0%	89.9%	71.7%	63.9%	57.8%	100.0%
% of CAPACITY	100.0%	77.2%	81.4%	83.1%	84.3%	85.4%	83.1%	80.4%	78.9%	82.6%	84.8%	87.3%	87.9%	92.5%	83.1%	66.3%	59.1%	53.4%	92.5%

EDEN HOSPITAL 6/16/87

LOT	SPACES	9:00	9:30	10:00	10:30	11:00	11:30	NOON	12:30	1:00	1:30	2:00	2:30	3:00	3:30	4:00	4:30	5:00	7:00	7:30	MAXIMUM
1 - AUXILIARY/EMPLY	27	21	25	27	27	27	27	27	27	25	22	22	22	20	19	17	16	17	13	14	27
2A- VISITORS	70	54	62	70	70	68	68	70	67	70	69	69	69	68	67	60	60	65	68	66	70
2B- DOCTORS	26	23	27	27	27	25	26	23	23	23	22	21	21	20	18	20	21	17	15	14	27
3 - EMPLOYEES	49	48	49	49	49	49	49	49	49	49	49	49	48	43	35	25	19	17	25	25	49
4A- CARDIAC/DIETARY	22	8	12	18	15	15	15	18	14	18	17	17	16	14	20	21	22	22	19	19	22
4B- CONSTRUCTION	18	18	18	18	18	18	18	16	15	18	18	18	18	18	9	4	1	0	0	0	18
5 - EMPLOYEES	28	18	25	27	26	27	27	26	25	28	28	27	27	16	10	8	6	3	4	2	28
6A- SPECIAL PATIENT	35	27	29	35	33	29	26	22	27	30	30	32	35	30	32	26	26	26	24	21	35
6B- EMPLOYEES	14	10	11	14	12	9	10	14	14	11	13	13	12	12	11	13	12	11	10	10	14
7 - VISITORS/PATIENT	109	79	91	105	107	103	103	100	100	104	107	108	109	109	109	108	89	75	78	67	109
8 - EMPLOYEES	144	129	137	143	144	144	144	144	141	144	143	142	135	122	105	96	67	38	21	17	144
9 - EMPL -3 PM SHIFT	55	2	2	2	3	10	11	11	10	10	13	16	20	36	44	49	50	46	41	41	50
10- ON STREET PKG		4	7	9	10	11	11	11	8	12	12	12	8	8	7	7	4	6	5	3	12
TOTAL	597	441	495	544	541	535	535	531	520	542	543	546	540	516	486	454	393	343	323	299	546
% of PEAK	109.3%	80.8%	90.7%	99.6%	99.1%	98.0%	98.0%	97.3%	95.2%	99.3%	99.5%	100.0%	98.9%	94.5%	89.0%	83.2%	72.0%	62.8%	59.2%	54.8%	100.0%
% of CAPACITY	100.0%	73.9%	82.9%	91.1%	90.6%	89.6%	89.6%	88.9%	87.1%	90.8%	91.0%	91.5%	90.5%	86.4%	81.4%	76.0%	65.8%	57.5%	54.1%	50.1%	91.5%

Figure 3–11. Parking study results.

savings of up to 70% using colored aerial photographs in the CBD.[37] The aerial photographs could prove useful for other planning studies. If parking tickets are issued indicating the arrival as well as the departure time, they may be used to compile occupancy data.

Parking space occupancy data are compiled and summarized in a tabular format and can also be depicted graphically. Spreadsheet programs are extremely useful in summarizing the data. Some data processing programs may also have application for limited analysis.[38] The number of vehicles is summarized by hour and the peak hour of parking accumulation is identified. For curb space studies, comparisons can be made between legal versus illegally parked vehicles, and loading vehicles. If the observations were made over several days, the data can be averaged or the one-time maximum demand can be determined for each off-street lot or lot section that is restricted to different users. If certain spaces must be reserved for different users, the peak demand of the facility can be determined by summing the peak demand for each section of parking lot that is designated for the different users. An example spreadsheet compiling the results of a parking study over three days is depicted in Figure 3–11.

Duration and turnover studies. These are conducted to determine the length of time vehicles are parked in a given space (duration) and to determine the rate of space usage in the facility (called turnover). Time limits at curbs and off-street public facilities as well as the geometric and operational design are influenced by such information.

Duration and turnover studies can be conducted in a number of ways. Field observations of *curb spaces* involve noting the license plate numbers of vehicles in the study lot at frequent and regular intervals and the time of the observation (see Figure 3–12 for a sample field worksheet). The *Manual of Traffic Engineering Studies*[39] describes license plate studies in more detail. Care should be taken when recording just three characters of the license plate because in some geographical areas the same characters may be disproportionately represented. The interval between checks should be much less than the average parking length in order to observe the actual parking time lengths of short-term parkers. Because this is unknown until the study is completed, rational estimates must be made. One rule of thumb is to take checks at headways (starting intervals) of one-third to one-half the posted time limit. Thus, 5-min headways are used for 15-min limits, 20- to 30-min headways for 1-hour limits, 2-hour headways for unlimited parking, etc.

There is a technique based on the exponential distribution of the length of time parked that makes it possible to estimate the percent of parkers not seen in such a study.[40]

In some extremely high turnover areas, it may be necessary to observe the curb space continuously. The observer

[37] Thomas A. Syrakis and John R. Platt, "Aerial Photography Parking Study Techniques," *Parking,* Highway Research Record 267 (Washington, DC: Highway Research Board, 1969), 15–28.

[38] Peat, Marwick, Mitchell & Company, *A Guide to Parking System Analysis* (Washington, DC: Federal Highway Administration, 1972).

[39] Box and Oppenlander, *Manual of Traffic Engineering Studies.*

[40] D.E. Cleveland, "Accuracy of the Periodic Check Parking Study," *Traffic Engineering,* 33 (September 1963), 14–17.

LICENSE PLATE CHECK FIELD DATA SHEET

Street _____ Between _____ And _____

City _____ Date _____ Recorder _____ Side of Street _____

Codes: 000 last three digits of license number; ✓ for repeat number from prior circuit; ___ for empty space.

Space and Regulation	Begin Survey Times											

Figure 3–12. Parking duration field sheet.

identifies the cars by the spaces they occupy and some easily recognized characteristic of the vehicle (color, make, etc.). Ten to 20 spaces per observer are the practical limit for this study, which also may include notations as to the destinations walked to by some drivers.

Commercial off-street parking durations are most easily obtained by analyzing parking tickets showing times of arrival and departure. Durations can also be measured by continuous observance of a part or all of the facility or by recording license plate numbers at periodic intervals as described above. Turnover can be obtained by counting entering or leaving vehicles, provided that the beginning and ending accumulations are also shown.

In periodic-check parking studies, the time parked is estimated by multiplying the number of times a vehicle is seen by the time interval between checks. Because of the missed parkers, the average duration calculated from these data is higher than actual. Aerial photography, as previously mentioned, may also be useful in conducting duration studies. Parking duration and turnover are generally summarized in tables, by type of facility (curb-by different time limits, off-street-by retail lot, employee lot, etc.).

Parking turnover is calculated by dividing the number of different vehicles parked in a particular facility by the number of spaces. A turnover value must be related to a specific study period, such as 8, 10, or 12 hours, so that the unit is vehicles/space for k hours.

Truck loading studies. These usually involve the study of whether existing or new loading zones are appropriate at the curbside or on off-street locations. Indications that existing loading zones are inadequate are the consistent presence of double-parked commercial vehicles. Traffic enforcement personnel should be consulted for their input on the need for additional and/or use of existing loading zones. Additional data needed to conduct truck loading studies are the daily needs of the individual merchants affected by loading zone changes, such as the number and times of the deliveries and the size of the trucks involved. The impact on the public parking supply should also be considered.

In general, a truck loading zone can be used by all properties on a block. The time limits restricting the loading zone use to commercial vehicle use are typically business hours on weekdays. However, these hours vary by region and can vary from city to city.

Special parking studies. These are used to corroborate or develop zoning regulations and are often conducted to ensure that the zoning regulations for off-street parking supply are based upon realistic needs.[41] To develop zoning regulations, spot checks of parking demand should be made at peak times for typical land uses, such as:

- ○ residential—single family
- ○ apartments—studio or one bedroom
- ○ apartments—two or three bedrooms
- ○ restaurants
- ○ office buildings

- ○ shopping centers—neighborhood size
- ○ shopping centers—community size
- ○ shopping centers—regional size
- ○ banks
- ○ industrial plants
- ○ employee parking in residential zones

In making such checks, it is important to consider seasonal and day-of-week variations as well as hourly fluctuations. Residential uses peak between 1:00 A.M. and 5:00 A.M. Restaurants peak at noon (drive-ins or in the CBD) or early evening (sitdown or outlying areas). Offices have the heaviest demand Monday morning at 10:00 A.M. while shopping centers typically peak Saturday late morning through early afternoon. Banks peak Friday evenings or Saturday mornings near each month's end and often in the middle of the month as well.

Other special studies involve gathering information on *potential* sites for off-street parking development (dimensions, assessed value, and current use, if any). The modeling technique described in Bullen's article can be used to determine the impact of a new off-street facility.[42] If a city has a parking agency or a department responsible for curb and off-street public facilities, data are often gathered on gross revenues plus maintenance, operating, and collection costs. Financial statistics to show trends in net revenues are useful.

Parking demand and generation studies. These studies can be conducted using previously described field survey techniques for occupancy surveys. This information can also be determined using interview techniques or postcard surveys.

Interview surveys are used to determine the distances that drivers walk from the parking space to the destination, the trip purposes, and the origins or next destinations of the vehicular trip. The data are used to calculate the parking demand of an area, on a block-by-block basis. Occupancies, durations, and turnovers can also be obtained at the same time.

In an interview study, personal interviews are conducted at representative samples of curb and off-street parking facilities. Because of personnel limitations, the area may be subdivided and one portion studied at a time. Work is often limited to the peak hour of parking demand. Each interviewer is assigned from 12 to 15 curb spaces and attempts to interview each driver parking in these spaces. At off-street facilities, interviewers are often placed at entrances and exits.

The interviewer obtains the following information and records it on a form:

1. Place where the driver plans to go while the vehicle is parked
2. Purpose of the trip (e.g., social, work, etc.)
3. Origin of the trip that has brought the driver to the parking space (nearest intersection and city)

[41] *Parking Principles*, p. 32.

[42] Bullen, "Development of Computerized Analysis of Alternative Parking Management Policies."

The interviewer records the times of arrival and departure and may also note the location of the parking space and the number of vehicle occupants.

If the purpose of the study is to check the parking-generation characteristics of a particular building, the interviews may be conducted at entrances to the generator. Ideally, a parking-generation study is conducted at an isolated facility with parking restricted solely to the occupants of the building(s). In this case, standard occupancy survey techniques will suffice. For interview parking-generation studies, the following information is obtained:

1. Mode of travel (car, bus, taxi, bicycle, etc.)
2. If by car, whether a driver or passenger

3. If a driver:
   a. where parked
   b. trip purpose
   c. trip origin

This information is then translated into a parking rate by taking the sample results and applying them to the number of employees (if an office building) or average number of customers (if a commercial building) or other appropriate parameter. Employee parking characteristics will also need to be considered if a retail building is being surveyed; these may or may not have been determined by the interview survey.

A *postcard survey* requests the same information as an interview. Prepaid returned addressed postcards are placed

Figure 3–13. Sample mailback parking survey.

PARKING SURVEY

Dear Motorist:
Please help improve parking in this community by
answering these questions and mailing this postpaid card today.
Please do not mark in gray boxes!

I parked my car at this location today because (check one):

1 - I am employed or have my business downtown
2 - I am on a shopping trip (primary reason to be here)
3 - I am on a personal business trip (bank, doctor, etc...)
4 - Other reason (Please Specify) _____
_____

After parking my car here, I walked to (Please give address or name of building
visited): _____
_____

I drove to this parking space from: (Place where trip started): (Nearest street
intersection) _____
(City) _____

The length of time I parked here today was about:
_____ (hours) and _____ (minutes)

Comments: _____
_____
_____

THANK YOU FOR YOUR COOPERATION

under the windshields of vehicles parked in the study area. In off-street facilities, attendants may distribute the cards to their customers. Each postcard is given an identifying number to indicate the location of the parking space where it was issued. Distribution is usually made during the peak hour, as rapidly as possible. Figure 3–13 shows a sample card. Experience has shown that about one-third of the postcards are returned. The value of this type of survey is limited by the fact that response biases may cause expansions of the data to be unreliable.

An alternative technique can be used for *employees* at a specific building(s) or plant(s). Here, cards are handed out and collected by office managers, shift supervisors, and so on. Figure 3–14 shows an example of such a card.

Note that the interview cards do not ask for home address, but rather the nearest street intersection. This simplifies plotting the trip origin *and* increases the likelihood of returns since many people object to giving out their home address. Information on the city of origin is also usually needed.

Destinations of parkers within the study area are often shown graphically on maps. The trip purposes and distances walked are generally tabulated. Origins and destinations of vehicle trips to or from the study area are plotted on maps or on desire-line charts. Attempts have been made to use general origin-destination (O-D) data.[43] Unfortunately, the zone sizes are often too large. Parking demand estimates should use individual blocks.

Observance of control devices

One measure of the acceptance and understanding of traffic control regulations and devices is the degree to which motorists obey them. Lack of obedience to devices and regulations could be an indication of improper application, education, or enforcement. Studies to determine observance of devices and regulations may lead to recommended changes or to increased education and enforcement activities.

These studies are oriented toward motorist observance of stop signs, traffic signals, turn restrictions, parking regulations, and speed zones. Studies are done to evaluate traffic control device effectiveness, to measure before-and-after changes in compliance, to determine locations for selective enforcement, and to aid in educational efforts.

Compliance studies should be done so that motorists and pedestrians are not aware that they are being observed. The vehicles or equipment of the observer should not be conspicuous.

Sample size

The size of the sample necessary to produce reliable results is determined by the following formula:[44]

$$n = pqt_\alpha^2 / \varepsilon^2 \qquad (3.9)$$

where:

n = minimum sample size
p = proportion of drivers or pedestrians that observe the regulation
q = proportion of drivers or pedestrians that do not observe the regulation
t_α = constant corresponding to the desired confidence level
ε = permitted error in the proportion estimate of observance

The sum of p and q is always equal to 1.0. The constant t_α depends on the desired level of confidence. It can be found in Table 3–7. For example, for a 95% confidence level and for large samples, the value is 1.96, often rounded to 2.00.

Permitted error ε in the proportion estimate of observance is a measure of the precision required and is usually in the range of 1% to 10% (plus and minus).

The solution to the sample size equation determines the number of observations necessary to achieve the desired statistical precision. As an example, with a 15% violation rate, a 10% permitted error, and a confidence level of 95.5%:

Figure 3–14. Example of a handout card.

A traffic survey is being conducted in this area. Please answer the following questions and return the card to the office today <u>before</u> you leave.

1. Please check how you arrived:
   A. Drive automobile ____
   B. Rode with someone else ____
   C. Came by public transit ____
   D. Other means ____

IF YOU DROVE AN AUTOMOBILE, PLEASE ANSWER THE FOLLOWING QUESTIONS ALSO

2. Where do you live:
   A. City _____
   B. Nearest street intersection to your home
   _____
   _____
   _____

3. About what time did you arrive and when do you expect to depart:
   A. Arrive _____ Leave _____

THANK YOU FOR YOUR ASSISTANCE!

[43] Lawrence L. Schulman and Robert W. Stout, "A Parking Study through the Use of Origin-Destination Data," *Parking Analyses,* Highway Research Record 317 (Washington, DC: Highway Research Board, 1970), 14–29.

[44] Box and Oppenlander, *Manual of Traffic Engineering Studies,* p. 181.

$$n = 0.85(0.15)(2)^2/(0.10)^2 = 51$$

In this example, a minimum of 51 observations is required.

Observance of regulations

Speed limits. Compliance with speed zones is determined with spot speed studies. (Spot speed studies are described earlier in this chapter.) Usually it is desired to know the percent of motorists traveling at higher than the legal speed limit. Before-and-after spot speed studies can be made to determine the effectiveness of a change in speed limits or education and enforcement efforts. In the case of speed limits, the study to determine motorist compliance is essentially the same as the original spot speed study used as a guide in establishing the limit. Generally, speed characteristics reviewed are average speed, 85th percentile speed, top of pace speed, and percent of vehicles exceeding various speeds.

Stop signs. Traffic engineers frequently must deal with public pressure to install stop signs that do not meet established warrants or guidelines. It is sometimes useful to have information on how motorists obey existing signs at similar locations to those proposed. Compliance with stop signs is usually not measured as a simple yes/no event. One compliance measure is to make a qualitative field observation to determine the "type of stop" made by motorists. This could include *full or complete stop, rolling stop under 10 mph,* or *rolling stop over 10 mph.* Other studies are made with a radar device in which the minimum speed of vehicles approaching a stop sign is recorded. The results would be arrayed into speed groupings such as 0 to 5 mph, 6 to 10 mph, 11 to 15 mph, 16+ mph. Radar observations are most effective where there is a low compliance rate, where rolling stops at speeds greater than 5 mph are frequent, or both. In most cases, it is desirable to also record the condition of through traffic flow (conflict or no conflict).

Traffic signals. At signalized intersections, it is desired to know how drivers respond to clearance and red phases and right-turn-on-red conflicts. Observers could record the proportion of approaching vehicles entering the intersection not in compliance with state or local regulations. It may also be of interest to record the response of pedestrians to vehicular and pedestrian signal indications or to prohibited pedestrian movements.

Other traffic regulations. Many traffic regulations require simple and measurable responses by motorists and pedestrians. These include turn prohibitions and the use of pedestrian crosswalks at uncontrolled locations. Related studies would note the response or yielding of motorists to pedestrians at marked or unmarked crosswalks at unsignalized locations. In these cases, the proportion of motorists or pedestrians disobeying the regulations is computed and used for analysis and comparison purposes.

Parking regulations. Standard parking violations include overtime parking, parking in prohibited or restricted areas, double parking, and improper use of commercial loading or passenger loading areas. The level of detail used is determined by the specific purpose of the study. Most of the techniques utilized are similar to those described in the section on parking studies contained in this chapter.

Traffic impact studies

Introduction

Traffic impact studies are studies that project, describe, and suggest ways of off-setting the traffic effects of development of new activities within a geographic area. Traffic is broadly defined as circulation of people and goods by all surface transportation modes—automobiles, transit, trucks, pedestrians, and bicycle—in the vicinity of the proposed development project.

Often a traffic impact study is conducted within the context of a larger environmental impact study. Such studies are mandated by state and federal law for all projects conducted under the auspices of a public jurisdiction that may have significant deleterious impacts upon the natural and human environment. In these instances, the traffic impact study will be incorporated into a multidisciplinary research effort generally known as an environmental impact statement (EIS) or report (EIR).

Taken as a whole, an EIR/EIS has two primary purposes: determining and disclosing all significant environmental impacts of a proposed project; and, identifying "mitigation measures" that could reduce or otherwise compensate for such environmental disruptions. In many urban jurisdictions, traffic impacts and mitigations are routinely among the most visible and controversial aspect of environmental studies.

Need for a study

The Institute of Transportation Engineers (ITE) has attempted to establish certain minimum trip generation thresholds at which a traffic impact study should be conducted.[45] Nevertheless, it appears likely that the perceived need for both traffic impact studies and environmental impact studies will continue to vary considerably from locality to locality. Even jurisdictions that have established guidelines have found the need to revise such guidelines from time to time. Furthermore, most state environmental legislation categorically exempts certain projects from having to perform environmental impact analyses, regardless of the size of such projects.

The purpose of a traffic impact study is to assist planners in both the public and private arenas in making major land use and other development decisions with regard to circulation issues. The study quantifies the changes in traffic levels and translates these changes into transportation impacts in the vicinity of the project. The traffic impact study should then identify on-site and off-site transportation system

---

[45] *Traffic Access and Impact Studies for Site Development,* Proposed Recommended Practice (Washington, DC: Institute of Transportation Engineers, 1988).

improvements needed to accommodate the additional traffic associated with new development.

Study framework

Time period analyzed. The point in time at which traffic impacts of the project and other development are assessed is referred to as the "study period" or "horizon year." One appropriate study period is the year of completion of the project. Build-out of the project is the most typical study period for traffic impact studies.

While it may be desirable to perform a hypothetical "project-only" scenario in which only traffic generated by the project is added to existing traffic volumes, in reality it is rare for a single project to be the only potential source of new traffic in an area. Most jurisdictions consider it important to account for the traffic impacts of cumulative development in order to have a realistic traffic forecast.

Traffic impacts of development expected to be completed before or at approximately the same time as the project should therefore be examined in conjunction with the project's impacts. Some jurisdictions establish an "existing plus approved projects" scenario as the baseline scenario for assessing the relative impacts of a project. The underlying rationale behind such a procedure is that "there is no stopping" other development already approved by the jurisdiction, and that therefore the traffic impacts of approved development should be taken as a given.

The most common longer-term study period or "horizon year" is usually the forecast year of the areawide or regional traffic forecast of the community within which the development will take place. This forecast frequently corresponds to a period 20 to 25 years in the future or to full build-out of the community as currently envisioned by local general plans.

Study area. The study area should include all portions of the roadway network that the local jurisdictions and/or the transportation engineer performing the study suspect may be significantly affected by development of the project. In general, the analysis should include all critical segments of the surrounding transportation system where users are likely to perceive a change in the quality of transportation service. With respect to roadway intersections (which are typically the critical locations of a street network), significant impact has been operationally defined by some jurisdictions in terms of relative increases in capacity utilization, or average delay.

Project description. The project or projects under study should be characterized from a transportation standpoint. Two key elements of such a project description are:

o The type and intensity of the proposed land use (e.g., "250,000 square feet of office," or "700 multiple-family dwelling units").

o The location and number of access points (driveways). If these are new driveways, their relationship to the existing roadway network, including other roadways, should be described. This description should include distance from major intersections and other existing driveways.

Data needs. The following items of information are generally needed to perform an adequate traffic impact study:

1. Baseline traffic counts: peak-hour intersection turning movement counts and daily volumes for all streets and roadways in the study area.
2. Information on the number of lanes on streets in the study area.
3. Intersection geometry information (lane stripings and how lanes are utilized by motorists, regardless of striping).
4. Traffic signal phasing and timing information.
5. Accident records for study area intersections and key roadway segments, preferably dating back at least 3 years.
6. An inventory of transit service in the vicinity, including routes and frequency of service, as well as data on the degree of utilization (load factors).
7. Inventory of sidewalks and other pedestrian amenities.
8. Inventory of bikeways and other bicycle amenities.
9. Level-of-service standards.
10. Standard plans for roadway and other circulation facilities.

Traffic forecast assumptions. Forecasts of future traffic are developed based on reasoned assumptions regarding trip generation, trip distribution and assignment, and mode split. Each of these principal components of a traffic forecast is discussed briefly below.

Trip generation. Based on the type and intensity (or size) of the proposed land uses, trip generation should be estimated. This estimate may be made by multiplying trip rates for a given unit of development by the number of units of development contained in the proposed, or by use of regression equations that correlate trip activity with the scale as well as the type of land use. Whatever method is used, it is generally necessary to calculate the number of trips on a daily basis and in the peak hour or hours of the adjacent street network. In the peak hours, it is also important to estimate the inbound/outbound split of traffic.

Numerous sources and methods exist for estimating the amount of traffic that will be generated including conducting local surveys, consulting published comprehensive studies, and conducting an actual analysis of the project. The most widely recognized source of published data is *Trip Generation.*[46]

Trip distribution and assignment. Trip distribution refers to the geographic origin or destination of trips related to a project. Trip assignment refers to the addition of trips to a transportation network (streets, transit routes, pedestrian ways, etc.). Trip distribution is a general allocation of the traffic generated by the development under study; trip or traffic assignment is a specific routing of future traffic. These aspects of traffic impact studies may be relatively controversial, for the prediction of access

---

[46] *Trip Generation,* 4th edition (Washington, DC: Institute of Transportation Engineers, 1987).

patterns is relatively subjective. Several methods exist for estimating trip distribution and assignment.

Mode split. Mode split refers to the distribution of all *person* trips generated by development among the various transportation modes available—automobile (drive-alone or shared ride), transit, motorcycle, bicycle, and walking. In most suburban areas of the United States and Canada, the automobile is the dominant mode of access, but other modes should not be ignored. The importance of transit increases in more intensely developed areas; in some large central business districts, transit is the dominant mode during peak periods.

Impact analysis

Once baseline data have been gathered and the key assumptions of trip generation, distribution and assignment, and mode split have been determined, the analyst can project total traffic volumes for the project and other cumulative development scenarios. These projected volumes are then analyzed, using a variety of analytical techniques. The results of the analysis are the basis of study findings and recommendations. This section briefly describes the more important types of analytic techniques.

Roadway operations. Projected volumes at signalized study intersections should be analyzed using the same capacity analysis techniques employed to characterize baseline conditions. Unsignalized intersections should be analyzed using the same unsignalized analysis technique, unless signalization is warranted and planned. Traffic signal warrants, such as those contained in the *Manual of Uniform Traffic Control Devices,*[47] can be used to determine whether signalization is desirable. Such warrants consider overall volumes, delay to minor street traffic, accident history, and a variety of other factors. Impacts to service levels on key mainline roadway links should also be determined. Finally, the adequacy of storage space for turning vehicles at study intersections should also be analyzed. This analysis should consider signal phasing and overall signal cycle length, as well as vehicle volumes.

Site access and on-site circulation. In addition to off-site level of service impacts, the analysis should determine the adequacy of site driveways and the internal circulation scheme. Driveways should be designed considering the amount and type of traffic that will be using both the adjacent street and the driveway. The ITE guidelines for driveway design provide a comprehensive set of design principles for site access.[48]

Adequate access for service vehicles should be reviewed by determining the size and operating characteristics of service vehicles, particularly the turning radii of such vehicles. The use of truck turning templates on the site plan is an important technique for assessing the adequacy of access for service vehicles.

The number of on-site parking stalls (and off-site parking stalls if any are planned) should be compared with projected parking demand. Local surveys of similar land uses as well as national references such as the ITE *Parking Generation*[49] manual can be used to estimate parking demand. The Urban Land Institute's *Shared Parking*[50] suggests methodologies for assessing the degree to which parking demand may be reduced in mixed-use developments.

Impacts to transit facilities (routes, stops, shelters, etc.) should also be assessed. With high-density urban projects, impacts to pedestrian facilities should be determined using an appropriate methodology such as Pushkarev and Zupan's.[51] Bicycle impacts, such as removal of a bike lane, or increased vehicular conflict with bicyclists, should also be noted.

Safety. Locations having excessive accident rates should be given special attention, as should locations with high numbers of accidents. Measures capable of reducing traffic hazards at such locations should be identified if at all possible.

Neighborhood impacts. Various studies have found that the safety, neighborhood amenities, and overall livability of local or residential streets can be severely degraded by traffic volumes well before the physical capacities of such streets are reached. Consequently, if a project will add significant traffic to residential streets, the analyst should quantify such volumes. *Residential Street Design and Traffic Control*[52] (ITE) and Appleyard's *Livable Streets*[53] suggest guidelines for measuring the onus of such impacts, and also contain strategies and techniques for controlling the intrusion of nonlocal traffic onto neighborhood streets.

Mitigation measures

After the development impacts have been identified, the final phase of a traffic impact study is to determine what transportation system improvements and other mitigation requirements are required. The key questions raised in this phase of a traffic impact study are political as well as technical: What quality of transportation service is acceptable? What improvements will adequately mitigate projected impacts? What will these improvements cost? When will they be needed? Who should be responsible for implementing mitigation measures? Who will ensure that needed improvements are made?

Goals of mitigation. Generally, the object of the mitigation process is to maintain traffic service levels currently considered acceptable in the community. This may take the

---

[47] *Manual of Uniform Traffic Control Devices.*

[48] *Guidelines for Driveway Design and Location, A Recommended Practice* (Washington, DC: Institute of Transportation Engineers, 1985).

[49] *Parking Generation,* 2nd edition (Washington, DC: Institute of Transportation Engineers, 1987).

[50] *Shared Parking.*

[51] Boris Pushkarev with Jeffrey M. Zupan, *Urban Space for Pedestrians* (Cambridge, MA, and London, 1975).

[52] Wolfgang S. Homburger and others, *Residential Street Design and Traffic Control* (Englewood Cliffs, NJ: Prentice-Hall, 1989).

[53] David Appleyard, *Livable Streets* (Berkeley and Los Angeles: University of California Press, 1981).

form of level-of-service standards in local ordinances or other policy documents. Level-of-service standards mandate that intersections and other key links in the area's roadway system are maintained within the designated standards. The standards may or may not make exceptions for preexisting congestion problems at certain locations. Frequently a specific analysis methodology for determining service levels is specified.

In the absence of such a policy, conventional goals for peak-hour traffic conditions are: Level of Service C for smaller communities (area population under 25,000); Level of Service D for larger communities (area population greater than 25,000). In densely developed central business districts and other heavily trafficked areas of large metropolitan areas, Level of Service E conditions may be tolerated.

Mitigation strategies. Whatever the source of the mitigation goals, the analyst should identify a set of transportation system improvements (additional lanes, signal installation, changes to signal operations, etc.) capable of effecting an acceptable level of service. This should be done for all significantly impacted locations. Once such "concrete mitigations" are identified, the analyst should determine the feasibility of actually constructing the identified improvements. This is generally accomplished through field review and discussions with review agency personnel; this feasibility analysis should include estimating the right-of-way and construction cost of the improvements.

What is infeasible from a cost standpoint for a single project may prove feasible for a number of projects or the community as a whole. Thus it will often prove more effective to determine improvements for cumulative traffic, allowing each project to participate financially in proportion to its traffic contribution. To establish a funding base to mitigate cumulative traffic impacts, many local jurisdictions are adopting areawide improvement districts and assessing impact fees.

It may be desirable for local governments or larger developers to provide "front-end" construction funds for improvements so as to guarantee that they will be operational when needed—that is, before the projects which necessitate them are fully built and occupied. Traffic impact fees of individual developments are then used to reimburse the public or private provider of the front-end funding.

If any "concrete" mitigation measures are found to be infeasible, the intensity of the development can be reduced. "Operational" traffic mitigation measures should also be considered. Operational mitigation options include the entire spectrum of Transportation Demand Management (TDM) strategies and techniques. TDM is also referred to as Transportation Systems Management (TSM). Transportation Demand Management techniques and their potential for reducing site traffic are described in Table 3–5.

Reports

The traffic impact study report documents the foundations, analysis, and findings of the traffic study. It should do this comprehensively, yet its language should be as straightforward as possible. Technical terms should be kept to a minimum; when used, they should be clearly defined. Including both a one- or two-page summary at the beginning of the report and attaching technical appendices at the end of report will allow the dual needs of clarity and comprehensiveness to be served.

An outline of the elements of a comprehensive traffic impact report is shown in Table 3–6. This outline will vary somewhat from study to study. Environmental impact study reports frequently require a special format, and care must be taken that the project description (square footage, number of employees, etc.) matches the characterization of the project in other sections of the environmental document. In general, a traffic impact study report that follows the outline indicated in Table 3–6 could be reformatted into an environmental document relatively easily because it would contain all the required information and findings.

Whether the client and readers of the report are developers planning a site, or government planners reviewing a development proposal, the traffic impact study should be an accessible document that will allow decision-makers to assess fully a development proposal from the standpoint of circulation.

TABLE 3–5

Transportation Demand Management Techniques With Potential to Reduce Site Traffic Generation

Measure[a]	Likely Vehicle Trip Reduction[b]					
	Office	Retail	Industrial	Residential	Lodging	Event Centers
Substantial transit service to areas of trip origins[c]	T,P	T,PM	T,P	T,P	T,P	T,P
Carpool, vanpool	T,P	T,PM	T,P	T,P	—	T,P
Modified work schedules	P	—	P	P	—	—
Parking availability reduced below normal demand level or substantial increase in parking costs	T,P	?	T,P	T,P	T,P	T,P
Internal shuttle transportation when site is part of major development well served by shuttle	T,M	T,M	—	T,M	T,P	—
Transit subsidy (significant)	T,P	—	T,P	T,P	?	?
Quality pedestrian environment on-site (mixed-use developments only)—applies to each use	T,M	T,M	T,M	T,P,M	T,P,M	T,P,M

T = daily trips, P = peak-hour trips, PM = p.m. peak-hour trips, M = midday trips.
[a] Other measures may be applicable either separately or in combination with others. To be effective, each measure must be designed to generate usage of alternatives to single-occupant automobile.
[b] Significant likely impacts identified only.
[c] Requires convenient service to and from site and attractive boarding/waiting arrangements for significant impact.

SOURCE: *Traffic Access and Impact Studies for Site Development*, INSTITUTE OF TRANSPORTATION ENGINEERS, 1988.

TABLE 3–6
Sample Table of Contents—Site Traffic Access/Impact Study Report

I. Introduction and Summary
  A. Purpose of Report and Study Objectives
  B. Executive Summary
    1. Site location and study area
    2. Development description
    3. Principal findings
    4. Conclusions
    5. Recommendations
II. Proposed Development (Site and Nearby)
  Summary of Development
    1. Land use and intensity
    2. Location
    3. Site plan
    4. Zoning
    5. Phasing and timing
III. Area Conditions
  A. Study Area
    1. Area of influence
    2. Area of significant traffic impact (may also be part of Chapter IV)
  B. Study Area Land Use
    1. Existing land uses
    2. Existing zoning
    3. Anticipated future development
  C. Site Accessibility
    1. Area roadway system
      a. existing
      b. future
    2. Traffic volumes and conditions
    3. Transit service
    4. Existing relevant transportation system management programs
    5. Other as applicable
IV. Projected Traffic
  A. Site Traffic (each horizon year)
    1. Trip generation
    2. Trip distribution
    3. Modal split
    4. Trip assignment
  B. Through Traffic (each horizon year)

    1. Method of projection
    2. Non-site traffic for in study area
      a. Method of projections
      b. Trip generation
      c. Trip distribution
      d. Modal split
      e. Trip assignment
    3. Through traffic
    4. Estimated volumes
  C. Total Traffic (each horizon year)
V. Traffic Analysis
  A. Site Access
  B. Capacity and Level of Service
  C. Traffic Safety
  D. Traffic Signals
  E. Site Circulation and Parking
VI. Improvement Analysis
  A. Improvements to Accommodate Base Traffic
  B. Additional Improvements to Accommodate Site Traffic
  C. Alternative Improvements
  D. Status of Improvements Already Funded, Programmed, or Planned
  E. Evaluation
VII. Findings
  A. Site Accessibility
  B. Traffic Impacts
  C. Need for any Improvements
  D. Compliance with Applicable Local Codes
VIII. Recommendations
  A. Site Access/Circulation Plan
  B. Roadway Improvements
    1. On-site
    2. Off-site
    3. Phasing, if appropriate
  C. Transportation System Management Actions
    1. Off-site
    2. On-site operational
    3. On-site
  D. Other
IX. Conclusions

SOURCE: *Traffic Access and Impact Studies for Site Development,* Institute of Transportation Engineers, 1988.

Engineering study techniques

Statistical analyses in traffic engineering are used to determine the reliability of field data and to evaluate "before" and "after" conditions. This section describes how to choose a representative and accurate sample size. For a more complete description of statistical analyses and applications, refer to the *Manual of Traffic Engineering Studies*[54] and *Fundamentals of Traffic Engineering.*[55] The text for this section is taken from material found in *Fundamentals of Traffic Engineering.*

Some definitions

Population. The entire universe being studied.

Sample. Sample is that part of the population for which there are available data. The larger the sample, the more

likely it is to represent the characteristics of the population from which it was collected. However, large samples may be expensive or impossible to obtain. The trade-off between additional knowledge about the population obtained and the marginal cost of additional data collection and processing is the concern underlying the determination of sample size.

Degrees of freedom. Degrees of freedom describes the independence of the data points. The number of degrees of freedom is defined as the number of observations or class frequencies whose values may be assigned arbitrarily. In analyzing the distribution of highway funds to 10 counties, the average of nine allocations could be any value, but the tenth is fixed by the total funds that were available (nine degrees of freedom). Thus, the calculation of the sample mean of six independent values of travel time involves six degrees of freedom, but the calculation of the standard deviation requires the use of the mean, which allows five data points to have arbitrary values with the sixth fixed to produce the correct value of the mean. Therefore, one degree of freedom is lost, leaving five.

---

[54] Box and Oppenlander, *Manual of Traffic Engineering Studies.*

[55] Wolfgang S. Homburger and James Kell, *Fundamentals of Traffic Engineering,* 12th edition, Appendix, 1988.

Confidence. The value of the sample almost always differs from the true mean of the population. The goal is to determine the best estimate of the population mean as a number and an interval within which we are confident that the true value lies. The amount of confidence is expressed as a probability. To be "95% confident" in the sample means that there is a 95% probability that a sample collected from a population will yield a confidence interval that includes the true population mean. It is suggested that traffic engineers normally use a confidence level of 95% for studies, except for safety-related studies where a 99% confidence level should be applied.

Significance. Often we wish to know whether a subpopulation differs from its "parent" population or whether two populations differ from each other. Significance is defined in terms of the probability that the difference in sample parameters might have occurred by chance and that the populations are not different. Thus, "significant at the 5% level" means that the differences observed could have come from samples of the same population only 5% of the time. It is suggested that traffic engineers should consider results significant at the 5% level for studies except safety-related studies where a 1% significance should be applied.

Determining sample size

The selection of sample size for a particular type of study depends on the time and funds available for data collection and the feasibility of collecting data. Large samples are samples of such size that additional data points increase confidence slowly. Small samples are those whose accuracy, as expressed by confidence limits, increase substantially with each addition to the sample. The dividing point is at a sample size of 25 observations.

To determine the minimum sample size needed to obtain some selected level of accuracy (i.e., that size at which the standard error of the mean drops to a maximum acceptable value), the t-distribution is used.

$$\varepsilon = t_a \times \frac{s}{\sqrt{n}} \qquad (3.10)$$

where: ε = error of the mean at chosen confidence level
s = standard deviation of the sample
$t_a = (1 - \alpha)^{th}$ percentile of the t-distribution with $(n - 1)$ degrees of freedom
$\alpha = 1 - \dfrac{\text{percent of confidence level chosen}}{100}$
n = sample size

Note that $\dfrac{s}{\sqrt{n}}$ is the standard error of the mean.

In "large" samples, the values of t_a for any value of n approach those listed in Table 3–7 on the line showing ∞ degrees of freedom. Therefore, with a given value of ε and a first estimate of the value of s (to be confirmed from field data later), the equation can be solved for n.

Where data collection is expensive, small sample sizes (generally less than 25 observations) must be accepted. For example, in a travel time study by floating car the problem is

TABLE 3–7

Values of the *t* Distribution

Degrees of Freedom	Values of α		Degrees of Freedom	Values of α	
	0.05	0.01		0.05	0.01
1	12.706	63.657	16	2.120	2.921
2	4.303	9.925	17	2.110	2.898
3	3.182	5.841	18	2.101	2.878
4	2.776	4.604	19	2.093	2.861
5	2.571	4.032	20	2.086	2.845
6	2.447	3.707	21	2.080	2.831
7	2.365	3.499	22	2.074	2.819
8	2.306	3.355	23	2.069	2.807
9	2.262	3.250	24	2.064	2.797
10	2.228	3.169	25	2.060	2.787
11	2.201	3.106	30	2.042	2.750
12	2.179	3.055	40	2.021	2.704
13	2.160	3.012	60	2.000	2.660
14	2.145	2.977	120	1.980	2.617
15	2.131	2.947	∞	1.960	2.576

to determine what sample size should be requested to have 95% confidence that the mean value will be within 10% of the population mean if it is guessed that the mean travel time is $\overline{x}$ and the standard deviation of the travel time distribution is approximately 15 percent of $\overline{x}$. In this case the value of t_a is not constant because we have a small number of degrees of freedom. Equation (3.10) must then be solved by trial and error. In the example in Table 3–8, $\overline{x} = 30$ min ($s = 4.5$ min), $\varepsilon \leq 3.0$ min, and $\alpha = 0.05$.

Twelve runs should therefore be requested. Before completing the study, however, the standard deviation of the mean travel times must be calculated; if it is greater than the estimated 4.5 min, then recalculate ε; if this is higher than 3.0 min, the sample size must be further increased.

"Before" and "after" studies

"Before" and "after" studies measure the effect of a transportation improvement. It is appropriate to utilize such studies when it is reasonable to assume the "before" conditions would be repeated in the "after" period if the improvement had not been implemented. The study calculates the change in condition based on a change in number, rate, or percentage. The expression used is as follows:

$$\frac{(n \text{ in "after" period}) - (n \text{ in "before" period})}{n \text{ in "before" period}} \times 100$$

For example, in a study conducted to determine the before and after effects of a change in a municipality's level of traffic enforcement on accident rates, a "before" rate of

TABLE 3–8

Sample Size Example

n (Trial)	Degrees of Freedom	Value of t_a	Value of ε	Result
6	5	2.571	4.72	ε too high
8	7	2.365	3.76	Still too high
10	9	2.262	3.22	Still too high
12	11	2.201	2.86	OK

TABLE 3-9

Example Before and After Significance Calculation

Before		After		Before		After*	
1980	24	1983	25	1980	24	1983/4	16
1981	16	1984	8	1981	16	1984/5	9
1982	22	1985	13	1982	22	1985/6	11
$\bar{x}_1 =$	20.67	$\bar{x}_2 =$	15.33	$\bar{x}_1 =$	20.67	$\bar{x}_2 =$	12.00
$\sum(x_1 - \bar{x}_1)^2 = 34.67$		$\sum(x_2 - \bar{x}_2)^2 = 152.66$		$\sum(x_1 - \bar{x}_1)^2 = 34.67$		$\sum(x_2 - \bar{x}_2)^2 = 26.00$	

$$s = \sqrt{\frac{34.67 + 152.66}{4}} = 6.84$$

$$t = \frac{20.67 - 15.33}{6.84} \times \sqrt{\frac{3}{2}} = 0.96$$

$$s = \sqrt{\frac{34.67 + 26.00}{4}} = 3.89$$

$$t = \frac{20.67 - 12.00}{3.89} \times \sqrt{\frac{3}{2}} = 2.73$$

*Each analysis period runs from April 1 to March 31.

20.67 accidents was observed, then, after the level of enforcement was increased, an "after" rate of 12.00 was recorded. Using the expression above:

$$\frac{12.00 - 20.67}{20.67} \times 100 = 41.94\% \text{ reduction was observed}$$

To test the significance of differences of two means of small samples, the following formula is applied:

$$t = \frac{(\bar{x}_1 - \bar{x}_2)}{s} \sqrt{\left[\frac{n_1 n_2}{n_1 + n_2}\right]} \qquad (3.11)$$

where the term under the square root sign is the reciprocal of the sum of the reciprocals of the two sample sizes. The value of the combined standard deviation is:

$$s = \sqrt{\frac{\sum(x_1 - \bar{x}_1)^2 + \sum(x_2 - \bar{x}_2)^2}{n_1 + n_2 - 2}} \qquad (3.12)$$

Note that, since the mean of each of the two samples being compared has already been calculated, two degrees of freedom have been "lost" (denominator on the right side of the equation).

EXAMPLE

A municipality increased its level of traffic enforcement at the beginning of 1983. To determine whether this caused a significant change in traffic accidents, the data in the left half of Table 3-9 were obtained:

A glance at Table 3-7 shows that $t = 0.96$ is not significant for 4 degrees of freedom.

However, it may be that the first three months after the change (January–March 1983) represented a transition period. Data for January–March 1986 have become available, which can replace the transition period data. (Note that such substitution must be with data from the same months to avoid bias reflecting seasonal variations.) The new data and calculations are shown in the right half of the table. Note that $t = 2.73$ is almost significant at the 5% level.

REFERENCES FOR FURTHER READING

Box, Paul C., and Joseph C. Oppenlander, *Manual of Traffic Engineering Studies,* 4th edition (Arlington, VA: Institute of Transportation Engineers, 1976).

Transportation Planning Handbook (Washington, DC: Institute of Transportation Engineers, 1991).

Highway Capacity Manual, Special Report 209 (Transportation Research Board, 1985).

A Policy on Geometric Design of Highways and Streets (American Association of State Highway and Transportation Officials, 1990).

Traffic Access and Impact Studies for Site Development, Proposed Recommended Practice (Washington, DC: Institute of Transportation Engineers, 1988).

Trip Generation, 4th edition (Washington, DC: Institute of Transportation Engineers, 1987).

Parking Generation, 2nd edition (Washington, DC: Institute of Transportation Engineers, 1987).

Appleyard, David, *Livable Streets* (Berkeley and Los Angeles: University of California Press, 1981).

Homburger, Wolfgang S., and James Kell, *Fundamentals of Traffic Engineering,* 12th edition, 1988.

4

TRAFFIC ACCIDENTS AND HIGHWAY SAFETY

CHRIS WILSON, P.ENG., *Director General*

Road Safety and Motor Vehicle Regulation
Transport Canada

AND

TERRY M. BURTCH, P.ENG., *Director*

Traffic Safety Standards & Research
Transport Canada

Introduction

Collisions on highways, roads, and streets kill more than 500,000 people worldwide each year, and an additional 15 million are injured. The cost of American and Canadian accidents (more than 52,000 deaths and 4 million injuries) is estimated to exceed $60 billion annually. The per person cost of motor vehicle injuries is estimated at $9,062 per person injured, $352,042 per average fatal injury, $43,409 for a hospitalized person, and $1,570 for a person injured but not hospitalized.[1] These estimates include the damage to vehicles and cargo, hospital and medical expenses, police and emergency services, and damage to public property. Losses arising out of lost productivity are included; however, no amount of money could adequately account for the very real impact these deaths and injuries have on victims and their families and friends.

Studies, such as the *Tri-Level Study of the Causes of Traffic Accidents,*[2] have shown that human error is one of the main causes of highway accidents. Too often this fact is used to justify inaction on behalf of road authorities and vehicle manufacturers and it also leads to lack of public support for such measures. While it may be true that in 70% to 90% of all accidents, human error is recorded as a contributing factor on accident reports, road and vehicle improvements can greatly reduce the likelihood of human error or the consequences of the accident. For example, driver errors initiate most "single vehicle–lost control" accidents; however, the location of rigid roadside structures such as sign posts, utility poles, and culverts is a critical factor in determining the likelihood that off-roadway accidents result in collisions. Furthermore, the rigidity of the structure determines the severity of injuries received as a result of the original human error. Another example might be associated with seat belt use. Seat belts themselves are not normally a factor in the cause of a collision, but, when properly designed and used, they can greatly reduce the consequences of an accident.

As can be seen from the previous discussion, focusing on the reported causes of accidents can be misleading in searching for measures to reduce accidents and their consequences. Engineering solutions related to both the vehicle and the road offer greater safety opportunities than what simple accident statistics might indicate. The transportation engineering principles and practices described throughout this handbook, when applied correctly, will contribute to the reduction of accidents through the design, construction, and operation of safer streets and highways.

Accident characteristics

Data on injury and property-damage accidents are essential for accident analysis and in determining the total consequences of motor vehicle accidents; however, for the purposes of interjurisdictional comparisons, fatality data are used. Injury and accident numbers are defined and reported differently in various jurisdictions, and even fatalities are

---

[1] Dorothy P. Rice, Ellen J. MacKenzie, and associates, "Cost of Injury in the United States, A Report to Congress" (San Francisco: Institute for Health & Aging, University of California, and Injury Prevention Center, The John's Hopkins University, 1989), XXVII.

[2] J.R. Treat and others, "Tri-Level Study of the Causes of Traffic Accidents" (Institute for Research in Public Safety, School of Public and Environmental Affairs for the U.S. National Highway Traffic Safety Administration, March 31, 1977).

not reported in precisely the same manner in all countries. The standard definition of a traffic death recommended by the United Nations Organization, Geneva, is one that occurs within 30 days of the event; nevertheless, some countries still use different definitions such as death at the scene, death within 6 or 7 days, and even death within 12 months. Many countries do not have an effective means of updating fatality data when fatalities occur after the traffic event.

The total number of deaths attributable to traffic accidents should be viewed in relation to many other indicators of traffic, population, economic status, or safety in other aspects of society. Two broad categories of death rate are of general interest: traffic safety and public health.

Traffic safety is a measure of how the road system is performing; it is measured in terms of deaths per unit of travel, per registered vehicle or per unit of length of the road system. *Public health* measures, such as deaths per unit of population, allow for a comparison of traffic deaths against other causes of death such as cancer or work-related accidents. This can be particularly useful in assessing target groups for public health spending.

Gordon Trinca and colleagues produced an excellent table (Table 4–1) in their book *Reducing Traffic Injury: A Global Challenge*,[3] which shows death rates for personal safety (the measure of public health) and traffic safety and rates for motorization. Their book goes on to make other interesting comparisons of death rates by characteristics such as types of road user, age of population and drivers, sex, and per capita increase, while controlling for the degree of motorization.

The European Conference of Ministers of Transport (ECMT) publishes annual reports on road accidents in member countries.[4] Table 4–2, based on data from the 1987 report, shows the relative significance of different road users in traffic fatality statistics; for example, pedestrians represent 29.9% of the fatalities in Japan whereas the figure was only

---

[3] Gordon Trinca and others, "Reducing Traffic Injury: A Global Challenge." (Melbourne, Australia: Royal Australian College of Surgeons, 1988), 12.

[4] "Statistical Report on Road Accidents in 1987" (Paris: European Conference of Ministers of Transport, OECD Publications Service, 1989), 38, 39.

TABLE 4–1
Personal Safety, Traffic Safety and Motorisation

| | A HIGH NUMBER MEANS A LOW LEVEL OF SAFETY | | | |
| | Personal Safety | Traffic Safety | Motorisation | |
Country Listed in Descending Order of Motorisation	Deaths per 100,000 Population	Deaths per 10,000 Vehicles*	Vehicles* per 1,000 Population	Year of Data
1. USA	19.1	2.7	711	85
2. Canada	15.8	2.8	561	84/83
3. New Zealand	21.1	3.9	545	84
4. Australia	18.6	3.4	540	84
5. F.R. Germany	13.1	3.0	440	85/84
6. Kuwait	27.1	6.7	408	85/83
7. Japan	10.3	2.6	403	85
8. Sweden	10.0	2.5	397	84
9. Norway	10.7	2.7	397	84
10. Netherlands	11.3	3.2	355	84
11. Finland	10.7	3.2	340	84
12. Denmark	13.0	3.9	335	84
13. U.K.	10.3	3.2	322	84/83
14. Spain	16.4	6.9	239	80
15. Greece	21.1	12.0	176	84
16. Hungary	17.1	11.7	146	85/84
17. Singapore	11.4	8.3	138	85/84
18. S. Africa	30.5	24.8	123	84/83
19. Malaysia	23.9	21.5	111	84/83
20. Chile	13.3	17.9	74	83
21. Costa Rica	8.2	12.0	68	83
22. Jordan	14.9	26.1	57	85/84
23. Colombia	8.9	25.7	35	81
24. Turkey	11.8	44.3	27	84
25. R. Korea	18.3	97.4	19	85/83
26. Egypt	11.4	59.7	19	82
27. Philippines	4.4	24.4	18	80
28. Thailand	8.4	50.0	17	85/82
29. Papua New Guinea	9.7	63.8	15	81/80
30. Kenya	13.4	112.6	12	80
31. Pakistan	5.2	98.2	5	84/83
32. India	4.2	108.8	4	83
33. Ethiopia	2.5	168.5	1	83
	A HIGH NUMBER MEANS A LOW LEVEL OF SAFETY			

*Cars and commercial vehicles, but excludes motorcycles.

SOURCE: GORDON TRINCA and others, *Reducing Traffic Injury: A Global Challenge*, p. 12.

| 1987 | Pedestrians % | On Bicycles % | On Mopeds % | On Motorcycles % | In Cars | | Others[1] + Unrecorded % | Total Killed |
					Drivers %	Passengers %		
Austria	16.80	5.90	8.60	7.90	37.90	18.80	4.20	1424
Belgium	17.00	9.90	5.50	6.20	39.10	17.90	4.30	1922
Switzerland	22.70	6.20	7.90	15.90	30.70	12.70	4.00	952
Germany	21.20	9.20	2.60	11.00	35.90	17.40	2.70	7967
Denmark	20.20	12.50	4.70	6.30	35.40	15.60	5.30	698
Spain	18.40	1.80	6.60	5.50	31.50	26.60	9.60	7615
France	15.00	4.30	6.90	8.00	38.70	22.20	5.00	10742
United Kingdom	33.30	5.50	1.10	12.80	25.90	17.30	4.20	5339
Greece	24.40	1.40	8.10	11.50	18.50	17.20	18.80	1682
1986 Italy	16.70	6.10	8.80	10.50	31.50	19.00	7.40	7571
Ireland	30.80	7.80	→	15.20	22.10	17.40	6.50	461
Luxembourg	8.80	2.90	1.50	2.90	57.40	26.50	0.00	68
Norway	19.60	5.30	2.50	9.80	38.70	22.40	1.80	398
Netherlands	11.60	21.00	8.60	3.90	34.90	16.90	3.20	1485
Portugal*	28.10	4.90	27.00	2.50	15.20	13.50	8.70	2985
Sweden	18.30	7.40	3.00	7.50	41.30	19.60	2.90	787
Finland	23.40	14.80	4.30	3.80	27.70	20.10	5.90	581
1986 Turkey*	35.20				27.40	37.40	0.00	9510
Yugoslavia	30.70	7.20	2.70	3.50	22.00	19.50	14.40	4526
19 ECMT countries	23.00	5.30	5.60	7.10	31.00	22.20	5.80	66713
United States	14.50	2.00	0.20	8.40	35.90	18.20	20.70	46386
Canada	15.00	2.80	→	8.80	46.20	26.90	2.10	4280
1986 Australia	18.60	2.70	→	14.00	39.50	25.50	4.50	2888
Japan	29.90	9.80	9.40	16.30	25.20	10.50	0.40	12151
4 associated countries	17.60	3.60	–12.00–		34.70	17.60	15.00	65705

*Figures not converted to standard definition (death within 30 days).
[1]Others: in commercial vehicles, buses and coaches, cars, horse riders, etc.

SOURCE: ECMT, "Statistical Report on Road Accidents in 1987," pp. 38–39.

11.6% in the Netherlands; 21% of the fatalities occur to cyclists in the Netherlands compared to 2% in the United States. These differences are not surprising given the different character of road traffic in each country. Measures adopted to counter these road safety problems, or at least the priority given to different measures, are therefore related to the character of transportation in each country. The long-term trends in highway accident fatalities have, to a large extent, reflected economic growth and increases in the vehicle and driving population. In the United States, the trend of the fatality rate, measured in terms of deaths per unit of travel or vehicles licensed, has, over the long term, been downward (see Figures 4–1 and 4–2).

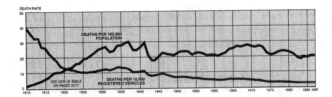

Figure 4–2. Trends in death rates.
SOURCE: NATIONAL SAFETY COUNCIL, *Accident Facts*, 1988 Edition, p. 49.

A number of theories have been advanced to explain the trend in fatalities. One theory is that as a country becomes more motorized, the road users become more "roadwise" and fatal accidents decrease. A second theory explains this trend by suggesting that there is a learning process which includes public demand for a safer transportation system. The safety community responds to this demand with safety measures that bring about a decrease in fatalities. A third theory is that discrete events such as economic recessions, safety laws, road-building programs, and so forth contribute to a long-term downward trend in fatalities. Each of these theories has weaknesses; however, the downward trend in fatalities in most motorized nations since 1980 is undeniable, and this trend has been experienced in a number of countries (see Table 4–3) with a variety of safety programs.

In North America, these encouraging trends continued even with a deteriorating road infrastructure, smaller

Figure 4–1. Travel, deaths, and death rates.
SOURCE: NATIONAL SAFETY COUNCIL, *Accident Facts*, 1988 Edition, p. 48.

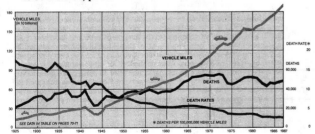

TABLE 4–3

Trends in Number of Deaths Within 30 Days

	1975	1980	1981	1982	1983	1984	1985	1986	1987	80/75[1] %	85/80[1] %	85/84[1] %	86/85[1] %	87/86[1] %	87/80[1] %	87/75[1] %
Austria	2390	1890	1839	1824	1905	1758	1477	1448	1424	−20.90	−21.90	−16.00	−2.00	−1.70	−24.70	−40.40
Belgium	2346	2396	2216	2064	2090	1893	1801	1951	1922	2.10	−24.80	−4.90	8.30	1.50	−19.80	−18.10
Switzerland	1243	1246	1165	1192	1159	1101	908	1034	952	0.20	−27.10	−17.50	13.90	−7.90	−23.60	−23.40
Germany	14,870	13,041	11,674	11,608	11,732	10,199	8,400	8,948	7,967	−12.30	−35.60	−17.60	6.50	−11.00	−38.90	−46.40
Denmark	827	690	662	658	669	665	772	723	698	−16.60	−11.90	16.10	−6.30	−3.50	1.20	−15.60
Spain	5,833	6,522	6,409	5,832	6,066	6,275	6,374	7,045	7,615	11.80	−2.30	−1.60	10.50	8.10	16.80	30.60
France	14,166	13,499	13,287	13,191	12,728	12,562	11,387	11,947	10,742	−4.70	−15.60	−9.40	4.90	−10.10	−20.40	−24.20
United Kingdom	6,679	6,239	6,069	6,150	5,618	5,788	5,342	5,618	5,339	−6.60	−14.40	−7.70	5.20	−5.00	−14.40	−20.10
Greece	1,187	1,372	1,516	1,744	1,776	1,908	1,908	1,625	1,682	15.60	39.10	0.00	−14.80	3.50	22.60	41.70
Italy	10,177	9,135	8,637	8,245	8,223	7,687	7,629	7,571	A	−10.20	−16.50	−0.80	−0.80			
Ireland	586	564	572	533	535	465	410	387	461	−3.80	−27.30	−11.80	−5.60	19.10	−18.30	−21.30
Luxembourg	124	98	100	75	85	70	79	79	68	−21.00	−19.40	−12.90	−0.00	−13.90	−30.60	−45.20
Norway	539	362	338	401	409	407	402	452	398	−32.80	11.00	1.20	12.40	−11.90	9.90	−26.20
Netherlands	2,321	1,997	1,807	1,710	1,756	1,615	1,438	1,529	1,485	−14.00	−28.00	−11.00	6.30	−2.90	−25.60	−36.00
Portugal	3,479	2,941	2,950	2,764	2,830	2,393	2,438	2,577	2,985	15.50	−17.10	1.90	5.70	15.80	1.50	−14.20
Sweden	1,172	848	784	758	779	801	808	844	787	−27.60	−4.70	0.90	4.50	−6.80	−7.20	−32.80
Finland	910	551	555	569	604	541	541	612	581	−39.50	1.80	0.00	13.10	−5.10	5.40	36.20
Turkey	6,663	4,839	5,004	5,893	6,761	7,420	7,229	9,510	A	−27.40	49.40	−2.60	31.60			
Yugoslavia	4,366	5,042	5,035	4,891	4,517	4,501	4,142	4,414	4,526	15.50	−17.90	−8.00	6.60	2.50	−10.20	3.70
ECMT countries	79,878	73,272	70,619	70,102	70,242	68,049	63,485	68,314	66,713	−8.30	−13.40	−6.70	7.60	−2.30	−9.00	−16.50
United States	44,525	51,091	49,301	43,945	42,589	44,257	43,825	46,087	46,386	14.70	−14.20	−1.00	5.20	0.60	−9.20	4.20
Canada	6,061	5,461	5,383	4,169	4,216	4,120	4,365	4,071	4,280	−9.90	−20.10	5.90	−6.70	5.10	−21.60	−29.40
Australia	3,694	3,272	3,321	3,252	2,755	2,821	2,942	2,888	2,771	−11.40	−10.10	4.30	−1.80	−4.10	−15.30	−25.00
Japan	14,206	11,752	11,874	12,377	12,919	12,041	12,039	12,112	12,151	−17.30	2.40	0.00	0.60	0.30	3.40	−14.50
Associated countries	68,486	71,576	69,879	63,743	62,479	63,239	63,171	65,158	65,588	4.50	−11.70	−0.10	3.10	0.70	−8.40	−4.20

[1]Percentage change: for example $1985/80 = \dfrac{(\text{Deaths in 1985} - \text{Deaths in 1980}) \times 100}{\text{Deaths in 1980}}$

A. Figures not available; figures from the previous year used to calculate 19-country total.

N.B. For all percentages in tables, a + sign is implicit where no sign is shown.

SOURCE: ECMT, "Statistical Report on Road Accidents in 1987," pp. 26–27.

passenger vehicles, and greater size differential between cars and trucks, all of which might have been expected to contribute to increased traffic fatalities. These factors may have been balanced by increased seat belt use and decreased driving while impaired by alcohol.

The National Safety Council publishes an annual report on accidents in America.[5] As shown in Table 4–4, traffic fatalities vary greatly from one part of the United States to another. In 1987, traffic death rates varied from a high of 4.1 deaths per 100 million vehicle miles in New Mexico to a low of 1.6 in Minnesota. U.S. forecasts for a number of factors that are expected to have an impact on travel and safety are shown in Table 4–5. As can be seen, most indicators point to increased driving with the potential for increased accidents and their consequences. The Federal Highway Administration also publishes a series of accident statistical reports titled "Fatal and Injury Accident Rates on Federal-Aid and Other Highway Systems" (1967–1981) and "Highway Safety Performance: Fatal and Injury Accident Rates on Public Roads in the United States" (1982–present). States provide the data on accidents, miles of travel, and highway miles by system for this report. These reports are useful sources of accident trend data by state and are the only source of trend data by highway system for those periods.

Road characteristics

A number of road, driver, and vehicle characteristics have been identified as having an important influence on traffic accidents and their consequences. In 1987, 43% of all fatal accidents in the United States occurred on rural roads and more than half (56.8%) occurred on straight, level roads (Table 4–6). Table 4–7 reveals that 80% (45.9% of 56.9%) of rural accidents take place away from junctions while only 65% of urban fatal accidents are at nonjunction locations.

In urban areas, road junctions without controls and with one-color traffic signals had about equal numbers of fatal accidents (Table 4–8). In rural areas, most intersection fatal accidents occurred at locations controlled by stop signs.

In 1987, 94% (34,026 out of 36,159 fatal accidents) occurred during dry weather and surface conditions (see Table 4–9).

Both the number and the proximity to the pavement of roadside fixed objects are directly associated with the frequency and severity of off-roadway accidents. From 1978 to 1987, nearly 470,000 people died in motor vehicle accidents in the United States. The first harmful event (FHE) in more

---

[5]National Safety Council, *Accident Facts,* 1988 edition (Chicago: National Safety Council, 1989).

TABLE 4-4

Motor-Vehicle Deaths by State, 1984–1987

State	Motor-Vehicle *Traffic* Deaths (Place of Accident)				*Total* Motor-Vehicle Deaths[a] (Place of Residence)			
	Number		Mileage Rate[b]		Number		Population Rate[b]	
	1987	1986	1987	1986	1985	1984	1985	1984
Total U.S.[c]	48,700	48,300	2.6	2.6	45,901	46,263	19.2	19.6
Alabama	1,116	1,086	3.2	3.2	1,013	1,035	25.2	25.5
Alaska	76	101	2.0	2.5	126	117	24.1	23.2
Arizona	933	1,002	3.9	4.4	869	875	27.2	28.5
Arkansas	638	603	3.5	3.4	583	572	24.7	24.4
California	5,500	5,222	2.4	2.4	5,291	5,324	20.1	20.6
Colorado	590	602	2.3	2.3	619	645	19.1	20.2
Connecticut	451	455	1.8	1.9	481	481	15.2	15.2
Delaware	147	138	2.5	2.4	119	132	19.1	18.6
Dist. of Col.	56	45	1.7	1.4	56	84	9.0	13.4
Florida	2,891	2,874	3.2	3.3	2,805	2,739	24.7	24.8
Georgia	1,604	1,542	2.7	2.7	1,440	1,448	24.1	24.8
Hawaii	138	117	1.9	1.7	129	136	12.3	13.1
Idaho	262	258	3.3	3.3	257	236	25.6	23.6
Illinois	1,685	1,617	2.2	2.2	1,752	1,726	15.2	15.0
Indiana	1,056	1,038	2.6	2.5	1,066	992	19.4	18.1
Iowa	491	441	2.3	2.2	523	483	18.2	16.6
Kansas	491	500	2.5	2.5	518	562	21.1	23.0
Kentucky	849	808	2.9	2.8	731	755	19.6	20.3
Louisiana	827	932	2.9	3.1	991	963	22.1	21.6
Maine	228	208	2.1	2.1	211	234	18.1	20.2
Maryland	830	790	2.3	2.2	766	666	17.4	15.3
Massachusetts	690	752	1.7	1.8	810	772	13.9	13.3
Michigan	1,622	1,632	2.2	2.3	1,684	1,634	18.5	18.0
Minnesota	530	572	1.6	1.7	676	676	16.1	16.2
Mississippi	757	766	4.0	4.0	688	703	26.3	27.1
Missouri	1,057	1,143	2.5	2.7	945	984	18.8	19.7
Montana	234	222	3.1	2.9	210	232	25.5	28.2
Nebraska	297	290	2.4	2.3	261	285	16.3	17.8
Nevada	262	233	3.0	2.9	214	211	22.8	23.0
New Hampshire	179	172	2.1	2.2	176	189	17.6	19.3
New Jersey	1,023	1,039	1.9	1.9	1,015	974	13.4	13.0
New Mexico	568	499	4.1	3.8	504	468	34.7	32.8
New York	2,327	2,114	2.4	2.2	2,176	2,191	12.3	12.3
North Carolina	1,598	1,645	3.0	3.1	1,518	1,489	24.2	24.1
North Dakota	101	100	1.7	1.8	103	122	15.0	17.8
Ohio	1,772	1,673	2.1	2.1	1,657	1,673	15.4	15.6
Oklahoma	608	711	1.9	2.3	775	887	23.4	26.8
Oregon	618	619	2.7	2.7	610	612	22.7	22.9
Pennsylvania	2,006	1,928	2.5	2.5	1,823	1,880	15.4	15.8
Rhode Island	100	114	1.8	2.1	112	109	11.6	11.3
South Carolina	1,087	1,059	4.0	3.7	924	946	27.7	28.6
South Dakota	134	134	2.2	2.1	147	158	20.8	22.4
Tennessee	1,248	1,230	3.2	3.1	1,137	1,160	23.9	24.5
Texas	3,261	3,568	2.3	2.4	3,799	4,005	23.2	24.9
Utah	297	313	2.4	2.6	342	309	20.8	19.0
Vermont	120	109	2.4	2.3	108	104	20.2	19.6
Virginia	1,022	1,118	1.9	2.2	998	1,007	17.5	17.9
Washington	790	714	2.1	2.0	797	835	18.1	19.2
West Virginia	469	437	3.6	3.3	443	436	22.9	22.3
Wisconsin	815	757	2.1	2.0	773	859	16.2	18.0
Wyoming	129	168	2.5	3.1	130	148	25.5	28.8

[a]Includes both traffic and nontraffic motor-vehicle deaths.
[b]The mileage death rate is deaths per 100,000,000 vehicle miles; the population death rate is deaths per 100,000 population. 1987 mileage death rates are National Safety Council estimates.
[c]Latest year available.

SOURCE: Motor-Vehicle *Traffic* Deaths from state traffic authorities; *Total* Motor-Vehicle Deaths from National Center for Health Statistics.

SOURCE: National Safety Council, *Accident Facts,* 1988 edition, p. 65.

TABLE 4-5

TABLE 4-5

Recent Trends in Demographic Factors and Their Effect on Future Vehicle Miles Travelled (VMT)

Factor	Recent Trends	Expected Effect on Travel
Population size	Increasing	Will cause VMT increase
Population location	Moving from northeast to south and southwest; moving from central city to suburbs	Will cause VMT increases in growing areas
Age distribution	Age group 35 to 49 will grow the most, primarily driving age	VMT will increase due to increase in driving age population
Household distribution	Household size declining	Smaller households have more VMT per person, VMT will increase
Size and type of workforce	Size will increase due to population increase, entry of women into workforce	People who work drive more than nonworkers, VMT will increase
Personal income	Will increase to the year 2000	VMT will increase due to increase in personal income
Industrial income	Expected to increase	Will cause VMT to increase, particularly in trucking
Vehicle fleet	Will increase in size due to population increase. Will become older	VMT will increase, but effect may be due to population and income changes, not vehicle fleet
Safety	Safety will continue to increase	Little or none
Alternate modes	Other modes will maintain their existing shares of freight and passenger travel	Little or none
Fuel availability	Cost of driving will become more apparent, fuel availability difficult to forecast	Increased fuel cost may limit VMT

SOURCE: U.S. HOUSE OF REPRESENTATIVES, PUBLIC WORKS AND TRANSPORTATION COMMITTEE, *Status of the Nation's Highways: Conditions and Performance,* June 1985.

TABLE 4-6

Fatal Accidents by Roadway Alignment and Roadway Profile, 1987

Profile	Alignment			Total
	Straight	Curve	Unknown	
Level	23,548	5,793	2	29,343
Grade	5,724	4,704	5	10,433
Hillcrest or Sag	745	382	1	1,128
Unknown	298	140	93	531
Total	30,315	11,019	101	41,435

SOURCE: NATIONAL HIGHWAY TRAFFIC SAFETY ADMINISTRATION, *Fatal Accident Reporting System,* 1987, p. 5.9.

than 130,000 of these fatalities (28%) was collision with fixed objects. Using 1987 data as an example, trees presented the greatest hazard, and were the first harmful event in 25.7% of fixed-object–accident fatalities. Utility poles had the second highest incidence (11.0%), followed by embankments (10.9%) and guardrail (10.3%). Table 4-10 lists the roadside objects struck most frequently in 1987 and the number of resulting fatalities.

Driver characteristics

The analysis of the causes of accidents and the potential to reduce accidents through traffic operations and design measures require an understanding of driver and vehicle characteristics. It is neither practical nor desirable in a handbook of this type to deal comprehensively with all the characteristics of drivers and vehicles that contribute to accidents and the development of solutions. However, the following sections will touch on a number of the more important ones. The sections in Chapter 1 dealing with driver and pedestrian characteristics probe the subject in a more complete manner.

The increase in the number of older drivers on the road has been identified as an area for concern. Table 4-11 shows the age distribution of drivers in both fatal accidents and all accidents during 1987. It may be noted that young drivers under 20 years old, and older drivers over age 74, are overrepresented in fatal accidents when compared to their representation in the driving population.

The design of roads and traffic control devices should take into consideration the elderly, who have a decreased ability to react quickly, move quickly (particularly as pedestrians), read signs, and understand complex traffic situations. Proposals have been made for extended pedestrian clearance intervals at locations frequented by older pedestrians; larger letters (and fewer messages) on highway signs; the location of traffic control devices to account for the poor peripheral vision and search capabilities of older drivers; and increased illumination at complex decision points to overcome deficiencies in vision that are more common in older drivers.

The use of alcohol by drivers has been identified as an important contributing factor in fatal accidents. Data reported in the *Fatal Accident Reporting System*[6] in the United States (Table 4-12) show that alcohol use by drivers involved in fatal accidents decreased between 1982 and 1987. The proportion of all drivers with blood alcohol concentration (BAC) of 0.10% or greater dropped from 30% to 25% during this period, and the portion of fatally injured

---

[6] National Highway Traffic Safety Administration, *Fatal Accident Reporting System, 1987* (Washington, DC: December, 1988) (DOT-HS-807-360).

TABLE 4-7

Fatal Accidents by Land Use, Junction Type, and Roadway Location, 1987

Land Use and Junction Type	On Roadway		Shoulder		Median		Roadside		Parking Lane		Other		Unknown		Total	
	Number	%	Number	%	Number	%	Number	%	Number	%	Number	%	Number	%	Number	%
RURAL																
Non-junction	8,884	35.4	998	55.3	469	43.9	3,986	61.4	7	8.5	4,677	68.2	14	26.4	19,035	45.9
Junction	3,238	12.9	34	1.9	11	1.0	171	2.6	2	2.4	222	3.2	0	0.0	3,678	8.9
Driveway, alley access	424	1.7	19	1.1	1	0.1	64	1.0	1	1.2	40	0.6	2	3.8	551	1.3
Rail grade crossing	271	1.1	0	0.0	0	0.0	2	0.0	0	0.0	6	0.1	0	0.0	279	0.7
Unknown	25	0.1	1	0.1	0	0.0	0	0.0	0	0.0	5	0.1	1	1.9	32	0.1
SUBTOTAL	12,842	51.2	1,052	58.4	481	45.0	4,223	65.0	10	12.1	4,950	72.2	17	32.1	23,575	56.9
URBAN																
Non-junction	6,813	27.2	659	36.5	511	47.8	1,879	29.0	70	85.4	1,610	23.5	29	54.7	11,571	27.9
Junction	4,922	19.6	86	4.8	75	7.0	351	5.4	2	2.4	252	3.7	2	3.8	5,690	13.7
Driveway, alley access	291	1.2	5	0.3	0	0.0	34	0.5	0	0.0	23	0.3	2	3.8	355	0.9
Rail grade crossing	172	0.7	0	0.0	0	0.0	0	0.0	0	0.0	4	0.1	0	0.0	176	0.4
Unknown	8	0.0	0	0.0	0	0.0	0	0.0	0	0.0	2	0.0	1	1.9	11	0.0
SUBTOTAL	12,206	48.7	750	41.6	586	54.8	2,264	34.9	72	87.8	1,891	27.6	34	64.2	17,803	42.9
UNKNOWN LAND USE	36	0.1	4	0.2	1	0.1	2	0.0	0	0.0	12	0.2	2	3.8	57	0.1
TOTAL	25,084	100.0	1,806	100.0	1,068	100.0	6,489	100.0	82	100.0	6,853	100.0	53	100.0	41,435	100.0

SOURCE: National Highway Traffic Safety Administration, *Fatal Accident Reporting System*, 1987, p. 5.13.

TABLE 4–8
Fatal Accidents at Road Junctions by Land Use and Intersection Traffic Controls

Traffic Control Device	Land Use							
	Urban		Rural		Unknown		Total	
	Number	%	Number	%	Number	%	Number	%
No Controls	2,043	35.9	1,315	35.8	12	54.5	3,370	35.9
One-Color Traffic Signal	2,032	35.7	267	7.3	3	13.6	2,302	24.5
Signal Flashing	108	1.9	137	3.7	0	0.0	245	2.6
Other Controls	101	1.8	85	2.3	0	0.0	186	2.0
Stop Sign	1,352	23.8	1,802	49.0	7	31.8	3,161	33.7
Yield Sign	34	0.6	67	1.8	0	0.0	101	1.1
Other	2	0.0	3	0.1	0	0.0	5	0.1
Unknown	18	0.3	2	0.1	0	0.0	20	0.2
Total	5,690	100.0	3,678	100.0	22	100.0	9,390	100.0

SOURCE: NATIONAL HIGHWAY TRAFFIC SAFETY ADMINISTRATION, *Fatal Accident Reporting System,* 1987, p. 5.14.

TABLE 4–9
Fatal Accidents by Roadway Surface and Atmospheric Conditions, 1987

Surface Condition	Weather					
	Normal	Rain	Sleet/ Snow	Fog & Other	Unknown	Total
Dry	34,026	16	6	356	13	34,417
Wet	1,536	3,756	142	175	16	5,625
Snow	163	21	376	6	0	566
Ice	302	38	226	21	0	587
Sand and other	78	3	0	17	1	99
Unknown	54	0	0	4	83	141
Total	36,159	3,834	750	579	113	41,435

SOURCE: NATIONAL HIGHWAY TRAFFIC SAFETY ADMINISTRATION, *Fatal Accident Reporting System,* 1987, p. 5.16.

TABLE 4–10
Distribution of Fatalities by First Harmful Event for Specific Fixed Objects, 1987

Fixed Object	Fatalities	Percent of Total Fatalities (46,385)	Percent of Fixed Object Fatalities (12,827)
Trees	3,299	7.1	25.7
Utility pole	1,406	3.0	11.0
Embankment	1,396	3.0	10.9
Guardrail	1,326	2.9	10.3
Curb or wall	861	1.9	6.7
Ditch	807	1.7	6.3
Culvert	586	1.3	4.6
Bridge	571	1.2	4.5
Sign or light support	538	1.2	4.2
Misc. pole or support	495	1.1	3.9
Fence	484	1.0	3.8
Concrete barrier	203	0.4	1.6
Building	108	0.2	0.8
Impact attenuator	18	0.0	0.1
Other fixed objects	729	1.6	5.7
Total	12,827	27.7	100.0

SOURCE: NATIONAL HIGHWAY TRAFFIC SAFETY ADMINISTRATION, *Fatal Accident Reporting System,* 1987.

TABLE 4–11
Age of Drivers: Total Number and Number in Accidents, 1987

Age Group	All Drivers		Drivers in Accidents					
			Fatal		All		Per No. of Drivers	
	Number	%	Number	%	Number	%	Fatal[a]	All[b]
Total	162,000,000	100.0	59,000	100.0	33,000,000	100.0	36	20
Under 20	14,100,000	8.7	8,200	13.9	5,200,000	15.8	58	37
20–24	16,900,000	10.4	10,300	17.5	5,800,000	17.6	61	34
25–34	39,800,000	24.6	16,300	27.6	9,100,000	27.6	41	23
35–44	33,000,000	20.5	9,500	16.1	5,400,000	16.4	29	16
45–54	23,400,000	14.5	5,600	9.5	2,900,000	8.8	24	12
55–64	18,500,000	11.4	4,000	6.8	2,200,000	6.6	22	12
65–74	12,500,000	7.7	2,900	4.9	1,500,000	4.5	23	12
75 and over	3,600,000	2.2	2,200	3.7	900,000	2.7	61	25

[a]Drivers in Fatal Accidents per 100,000 drivers in each age group.
[b]Drivers in All Accidents per 100 drivers in each age group.

SOURCE: NATIONAL SAFETY COUNCIL, *Accident Facts,* 1988 edition, p. 54.

TABLE 4-12

Blood Alcohol Concentration (BAC) for All Driver Fatalities and All Surviving Drivers Involved in Fatal Accidents

Fatally Injured Drivers

| | YEAR | | | | | | 1982–1987 Percent Change* |
	1982	1983	1984	1985	1986	1987	
0.00%	47	49	51	52	52	53	
0.01–0.09%	9	9	9	9	9	9	
≥0.10%	44	42	40	39	39	38	−14%
Total	24,690	24,138	25,589	25,337	26,630	26,831	

Surviving Drivers

| | YEAR | | | | | | 1982–1987 Percent Change* |
	1982	1983	1984	1985	1986	1987	
0.00%	72	74	75	77	77	77	
0.01–0.09%	9	8	8	7	8	8	
≥0.10%	19	18	17	16	15	15	−21%
Total	31,339	30,518	31,923	32,546	33,705	34,603	

Note: *1982–1987 percent change is difference in proportion of drivers with BAC ≥0.10%.

SOURCE: NATIONAL HIGHWAY TRAFFIC SAFETY ADMINISTRATION, *Fatal Accident Reporting System*, 1987, p. 2.5.

drivers with BAC of 0.10% or greater dropped from 44% in 1982 to 28% in 1987.

This decrease in the proportion of drivers intoxicated can be seen across most vehicle types in Table 4–13. Drivers of motorcycles in fatal crashes were most often impaired (38% in 1987) whereas the proportion of drivers of heavy trucks with BAC of 0.10% or greater was only 2%. Other driver-related characteristics that may affect the design of roads and traffic control devices are driver height (height of eye);

reaction time; perception time; hearing; visual acuity; and handicaps.

Enforcement and education play an important role in road-user safety. Public education has not been found to be an effective way of changing behavior; however, coupled with safety laws and enforcement, driver behavior with regard to seat belt use and driving after drinking has been modified, and accident and injury risk has been lowered accordingly. While public education may not affect behavior directly, it

TABLE 4-13

Percent of Drivers of Different Types of Vehicles Involved in Fatal Accidents and by Estimated Alcohol Level

Drivers of Passenger Cars Involved in Fatal Crashes

| BAC | YEAR | | | | | | 1982–1987 Percent Change** |
	1982	1983	1984	1985	1986	1987	
0.00%	60	61	63	66	66	66	
0.01–0.09%	9	9	9	8	8	9	
≥0.10%	31	30	28	26	26	25	−19%
Total	34,121	33,069	34,395	34,071	35,959	35,920	

Drivers of Vans Involved in Fatal Crashes

| BAC | YEAR | | | | | | 1982–1987 Percent Change** |
	1982	1983	1984	1985	1986	1987	
0.00%	66	68	70	74	74	74	
0.01–0.09%	8	7	8	6	6	7	
≥0.10%	26	25	22	20	20	19	−27%
Total	1,720	1,614	1,733	1,824	1,963	2,292	

Drivers of Special Vehicles Involved in Fatal Crashes

| BAC | YEAR | | | | | | 1982–1987 Percent Change** |
	1982	1983	1984	1985	1986	1987	
0.00%	58	60	60	64	64	64	
0.01–0.09%	8	8	9	8	8	9	
≥0.10%	34	32	31	28	28	27	−21%
Total	1,570	1,625	1,644	1,894	1,880	1,958	

TABLE 4–13 (Continued)

	Drivers of Motorcycles* Involved in Fatal Crashes						1982–1987 Percent Change**
	YEAR						
BAC	1982	1983	1984	1985	1986	1987	
0.00%	47	46	47	47	46	49	
0.01–0.09%	13	13	14	13	13	13	
≥0.10%	40	41	40	39	41	38	–7%
Total	4,490	4,288	4,650	4,598	4,558	3,848	

Note: *Does not include mopeds.

	Drivers of Light Trucks* Involved in Fatal Crashes						1982–1987 Percent Change**
	YEAR						
BAC	1982	1983	1984	1985	1986	1987	
0.00%	55	58	60	62	61	61	
0.01–0.09%	9	8	9	8	8	8	
≥0.10%	36	34	31	30	31	31	–14%
Total	8,308	8,209	8,952	9,126	9,767	10,302	

Note: *Under 10,000 pounds gross vehicle weight (gvw).

	Drivers of Medium Trucks* Involved in Fatal Crashes						1982–1987 Percent Change**
	YEAR						
BAC	1982	1983	1984	1985	1986	1987	
0.00%	91	89	90	91	92	94	
0.01–0.09%	4	3	3	3	3	2	
≥0.10%	5	8	7	6	5	4	–20%
Total	659	652	673	643	651	698	

Note: *Between 10,000 and 26,000 pounds gvw.

	Drivers of Heavy Trucks* Involved in Fatal Crashes						1982–1987 Percent Change**
	YEAR						
BAC	1982	1983	1984	1985	1986	1987	
0.00%	92	93	93	94	95	96	
0.01–0.09%	4	3	3	3	2	2	
≥0.10%	4	4	4	3	3	2	–50%
Total	3,923	4,138	4,383	4,448	4,364	4,311	

Note: *Greater than 26,000 pounds gvw.

Note: **1982–1987 percent change is difference in proportion of drivers with BAC ≥0.10%. Drivers of unknown vehicles type (2,105) were excluded from this table.

SOURCE: NATIONAL HIGHWAY TRAFFIC SAFETY ADMINISTRATION, *Fatal Accident Reporting System,* 1987, p. 2.8–9.

may raise awareness for safety issues and make it easier to obtain support for effective safety regulations and increased funding for safety countermeasures.

The use of safety belts is one of the most effective ways of reducing the severity of injuries. The installation of lap and shoulder belts in the front outboard seating position of all passenger cars has made this lifesaving device virtually universally available. Vehicle occupants have not voluntarily taken advantage of their availability. The use of safety belts was made mandatory in the State of Victoria, Australia, in 1970 and had been adopted in all eight Australian states by the end of 1973. Safety belt use in Australia is now over 90%.

By April 1990, 33 states and the District of Columbia in the United States and all 10 provinces in Canada had mandatory safety-belt-use laws in effect. Safety belt use in the United States was estimated to be 45% and 74% in Canada. In October 1989, seat belt use of 87.7% in the Canadian province of Saskatchewan, 85.2% in British Columbia, and 81.6% in Quebec was attained by a combination of vigorous enforcement and public education.

In November 1985, the Organization for Economic Cooperation and Development (OECD) sponsored a workshop, organized by the National Highway Traffic Safety Administration (NHTSA), on the effectiveness of safety-belt-use laws. James Hedlund of NHTSA and a member of the OECD working group that examined the issues and prepared background papers for the workshop prepared a paper titled "Casualty Reductions Resulting From Safety Belt Use Laws."[7] The usage and performance data available from 12

---

[7] *Effectiveness of Safety Belt Use Laws: A Multinational Examination* (Washington, DC: National Highway Traffic Safety Administration), 73–91.

TABLE 4–14
Belt Usage Changes and Casualty Reduction Performance of Belt Use Laws

Country or State	Usage Prelaw	Usage Postlaw	Years	Fatality Count	Fatality Perf.	Injury Count	Injury Perf.
Ireland	15%	45%	3	570	0%	4,900	0%
Victoria	15	48	4	2,670	40	71,000	42
Canada	24	50	11	34,000	37	1,700,000	20
New York	16	57	1	1,500	15	—	—
Denmark	19	67	2	640	13	15,000	27
Switzerland	37	76	2	1,000	35	30,000	35
Israel	10	80	4	220	41	930	27
Sweden	35	84	2	1,200	23	28,000	36
New Zealand	33	86	4	1,700	31	2,600*	43*
Norway	59	87	2	350	neg.	11,000	44
Germany	58	92	1	6,000	51	60,000	44
UK	40	94	4	7,700	32	106,000	38

Years: Total data collection period, pre- and post law.
Count: Approximate number of occupant casualties during the data collection period.
Perf.: Estimated belt law performance.
Injury: Defined differently in different countries.
*Driver only, 2 years of data.

SOURCE: National Highway Traffic Safety Administration, *Effectiveness of Safety Belt Use Laws: A Multinational Examination,* p. 89.

countries at that time are summarized in Table 4–14. The conclusions were as follows:

o belt usage laws reduce casualties
o belt law performance in reducing fatalities is quite variable, probably due to small numbers of observations and random variation
o belt law performance in reducing injuries increases as belt usage increases, consistent with the "selective recruitment" hypothesis
o fatality and injury reductions generally are consistent with a belt effectiveness of about 40% for 100% belt usage.

Vehicle characteristics

Vehicles with significantly different operating and design characteristics share the road; for example, bicycles share the same lanes with large trucks and buses. Table 4–15 lists vehicles by body style, not by vehicle use. During 1987, passenger cars comprised 75% of the registered vehicles but were involved in only 62% of fatal accidents. Trucks and motorcycles were both overrepresented in fatal accidents when compared to their representation in the registered vehicle population.

The energy crisis of the early 1970s prompted a move to smaller, more fuel-efficient cars. Although this trend has

TABLE 4–15
Types of Motor Vehicles Involved in Accidents, 1987

Type of Vehicle	In Fatal Accidents Number	In Fatal Accidents %	In All Accidents Number	In All Accidents %	Per Cent of Total Vehicle Registrations[a]	No. of Occupant Fatalities
All Types	**59,000**	**100.0**	**33,000,000**	**100.0**	**100.0**	[b]
Passenger cars	36,500	61.9	25,900,000	78.5	74.6	26,200
Trucks	15,900	27.0	6,000,000	18.2	22.2	7,100
Truck or Truck Tractor	10,100	17.1	4,800,000	14.6	21.4	[c]
Truck Tractor and Semi-trailer	4,600	7.8	800,000	2.4	} 0.8	[c]
Other Truck Combinations	1,200	2.1	400,000	1.2		[c]
Farm Tractors, Equipment	200	0.3	30,000	0.1	[d]	110
Taxicabs	400	0.7	100,000	0.3	0.1	90
Buses, Commercial	200	0.3	140,000	0.4	0.1	20
Buses, School	200	0.3	50,000	0.1	0.2	10
Motorcycles	3,900	6.6	360,000	1.1	} 2.8	3,900
Motor Scooters, Motor Bikes	300	0.5	20,000	0.1		300
Other[e]	1,400	2.4	400,000	1.2	[c]	970

[a]Percentage figures are based on numbers of vehicles and do not reflect miles traveled or place of travel, both of which affect accident experience.
[b]In addition to these occupant fatalities, there were 8,500 pedestrian, 1,400 pedalcyclist, and 100 other deaths.
[c]Data not available.
[d]Not included in total vehicle registrations; estimated number–5,700,000.
[e]Includes fire equipment, ambulances, special vehicles, other.

SOURCE: National Safety Council, *Accident Facts,* 1988 edition, p. 60.

Safety Technology Feature	Penetration	% Effectiveness per Accident Type					% Reduction in Fatalities
		Off-Road	Angle	Head-On	Rear End	Side Swipe	
Warning to Apply Brakes to Avoid Collision	50%	—	30	—	20	—	3.24
Warning of Vehicle in Blind Spot	50%	—	—	—	—	20	0.25
Driver Impairment Warning	50%	10	5	10	5	5	3.24
Nighttime Vision Enhancement	50%	5	5	—	5	—	1.5
Headway control, auto breaking	25%	—	20	—	60	—	1.5
Lane-Edge Tracking and Warning	25%	40	—	10	—	5	3.7
Robust Collision Prevention Systems	10%	70	70	70	70	70	5.5
Automated Highway	1%	90	90	90	90	90	0.01
Third Phase Total							18.9%

SOURCE: *Mobility 2000, Final Report of the Working Group on Operational Benefits.*

slowed in the 1980s, fuel efficiency to reduce automobile pollution and dependence on foreign oil supplies in the 1990s may rekindle interest in smaller cars. While some limited-access roads prohibit the use of very large or very small vehicles (parkways and freeways respectively), most of the nation's roads are accessible to all classes of vehicles. Road designs that separate very large and very small vehicles, such as truck lanes and bicycle paths, are desirable if volumes are large enough and if funds can be found to provide the facilities.

National governments throughout the world have established safety standards for motor vehicles, and the international harmonization of these standards has been encouraged and facilitated by the United Nation's Economic Commission for Europe. In North America, vehicle manufacturers and importers are responsible for certifying that each of their vehicles meets all federal safety standards. In all other countries, governments issue "type approval" to manufacturers and require that all of their production be essentially the same as the prototype approved.

A number of new innovations being introduced by manufacturers and/or considered for regulation by government agencies hold promise for improved vehicle safety:

Passive restraints—automatic seat belts and air bags to provide protection for those who do not use manual seat belts.

Daytime running lights—experience in Nordic countries has demonstrated that the use of lights in the daytime reduces multivehicle and vehicle/pedestrian accidents in both rural and urban areas; Canadian estimates suggest that relevant daytime accidents may be reduced by as much as 20% by this measure.

Antilock brakes—initially for trucks, but ultimately for all vehicles, will make controlled stops, particularly on slippery surfaces, more easily achieved.

Side impact protection—to reduce injuries and fatalities in side impact or angle collisions.

Heavy vehicle underride protection (or improved conspicuity)—to reduce the frequency and severity of collisions involving passenger cars running under the rear or side of slow-moving or stationary trucks, or being overrun by large trucks.

Intelligent Vehicle–Highway Systems (IVHS) - ("Smart Vehicles")—electronic control of some or all vehicle control features to reduce accidents and improve capacity. Electronic devices may be used to detect the presence of objects near the vehicle and to transmit important safety information to the drivers. IVHS is a joint public-private enterprise, and will require the parallel development of both highway infrastructure and the vehicle. The Benefits Group of Mobility 2000 prepared an estimate of the benefits that may accrue from the deployment of IVHS technology.[8] The estimates of benefits in Table 4–16 are based on data from the 1987 *Fatal Accident Reporting System* (FARS).

The net benefits in Table 4–16 are computed by multiplying the percent penetration of the hardware into the vehicle population by the percent effectiveness in reducing a given type of accident by the percentage of all fatal accidents that are of that type. For example, if a blind-spot warning system reaches a 10% penetration of the vehicle population and is thought to be 20% effective in preventing sideswipe accidents, which themselves constitute 2% of all fatal accidents, then we compute a reduction in the annual total of all fatalities as: $(0.10 \times 0.20 \times 0.02 = 0.0004$, or expressed as a percentage $= 0.04\%)$. Where more than one technology has payoff for a given accident type, the listed percentages are considered additive.

These estimates indicate that although IVHS offers the potential for very important accident and fatality reductions, the benefits for individual technologies are quite small and will be achieved only after major expenditures and commitments to research and development from all the active participants, private industry, state and local government, and university researchers.

Accident record systems

Appropriate traffic accident record systems are an essential part of any traffic safety program, and they are used by engineering, licensing, enforcement, medical, and insurance organizations, among others. Transportation engineers use

---

[8] *Final Report of the Working Group on Operational Benefits,* Mobility 2000 (Dallas, 1990).

data from these systems for a number of purposes, including identification and analysis of high-accident locations, evaluation of engineering measures intended to improve safety, and monitoring of safety trends. Accident record systems should include information from accident reports or files and from related inventory systems. The type and level of detail in accident record systems varies, but recommended data elements have been developed.[9]

Accident reports

The primary sources of accident data are local (city or county) and state or provincial police agencies. This information is usually recorded on accident report forms similar to that illustrated in Figure 4–3. Many agencies also receive information on minor accidents (noninjury producing), but no on-site investigations are performed in some jurisdictions.

Accident report forms are used to record some characteristics of the site of the accident, vehicle and driver actions, and information on other road users (bicyclists and pedestrians). They also include supporting narratives and diagrams. For transportation safety engineering analysis and assessment of specific sites, one of the most important parts of the report form is the sketch showing vehicle travel paths. Trained police officers often use sufficient detail in the sketch to indicate precollision movements of vehicles and other involved road users. The narrative portion of the report is equally important. In this section, officers may provide details on other, noninvolved vehicles whose actions may have contributed to the accident, or other information such as end-of-queue conditions (rear-end accident occurring in a traffic backup from a traffic control device). The narrative portion of the report is not usually contained in the electronic data systems although this information may be useful in analyzing a specific location. Engineering agencies should make provisions to retain copies of accident reports to assure that this information is available for future work.

The usefulness of accident reports depends on their accuracy and the inclusion of relevant details. Police officers benefit from an understanding of how this information is used by transportation engineers. In some jurisdictions, engineering and enforcement agency personnel meet routinely to discuss safety problems; such exchanges of information improve not only accident reporting but also the engineer's appreciation of enforcement findings.

A secondary source of accident data involves copies of the driver financial responsibility reports required in many states. Some states have established accident record bureaus to collect driver reports and some accident reports from various police agencies. Unfortunately, reports are not received for accidents involving property damage below a certain dollar value; this means that much useful accident information is missing. The quality of accident reporting and other relevant data as well as definition of what is reported should be reviewed carefully for potentially missing data. Data can be misinterpreted and the wrong conclusions reached if a portion of the data population is not reported.

Accident files

For traffic engineering purposes, copies of all accident report forms should be obtained from police agencies in the responsible jurisdiction. This is often best done by having a copy of the completed report form sent routinely by the police agency to the engineering office for storage and retrieval purposes.

All traffic engineering safety applications require the accident report information to be retrievable by location of the accident (intersection, road section, etc.). Because enforcement agencies often store information sequentially or by date, the engineering agency will usually set up a location-type file. Such files vary in sophistication from that of manually stored paper files in smaller jurisdictions to large computer-based systems. Several methods are available for establishing such a file.[10]

Inventory and data systems

Inventories are means of grouping accident and related information, in most cases by location such as intersections, city blocks, or highway sections. In general, there are two kinds of locations:

1. a spot, such as an intersection or short bridge, or
2. a road section, which might be as short as 50 ft or as long as 10 miles.

These inventories should also contain other site-specific information, such as adjacent land use; location of roadside fixtures; pavement type and condition; presence of medians, traffic control devices, and pavement markings; curb or shoulder type, etc., which are important when attempting to correlate accident causes with roadway design or operations, or when assessing the effects of safety measures (see section on "Evaluation" later in this chapter).

Computer systems

Accident data systems are usually built from accident report forms and involve storing data in standardized format. Information is normally interpreted, coded, keypunched, and subsequently stored on magnetic tapes or discs. Computer systems permit rapid tabulations of data, such as

- o periodic listings of accidents by specified location;
- o listings of "high-accident" locations (Table 4–17);
- o detailed tabulations of data for preparation of site-specific collision diagrams;
- o special summaries to relate accident frequency or rates to type of highway, geometric features, pavement type, or other expected causal factors.

Computerized accident record files should be designed to link up with, or at least be compatible with, other files such

---

[9] "Manual on the Classification of Motor Vehicle Traffic Accidents," American National Standards Institute, D16.1-1989.

[10] P. Box and J. Oppenlander, *Manual of Traffic Engineering Studies,* 4th edition (Washington, DC: Institute of Transportation Engineers, 1976). This manual is currently being updated for publication in 1991.

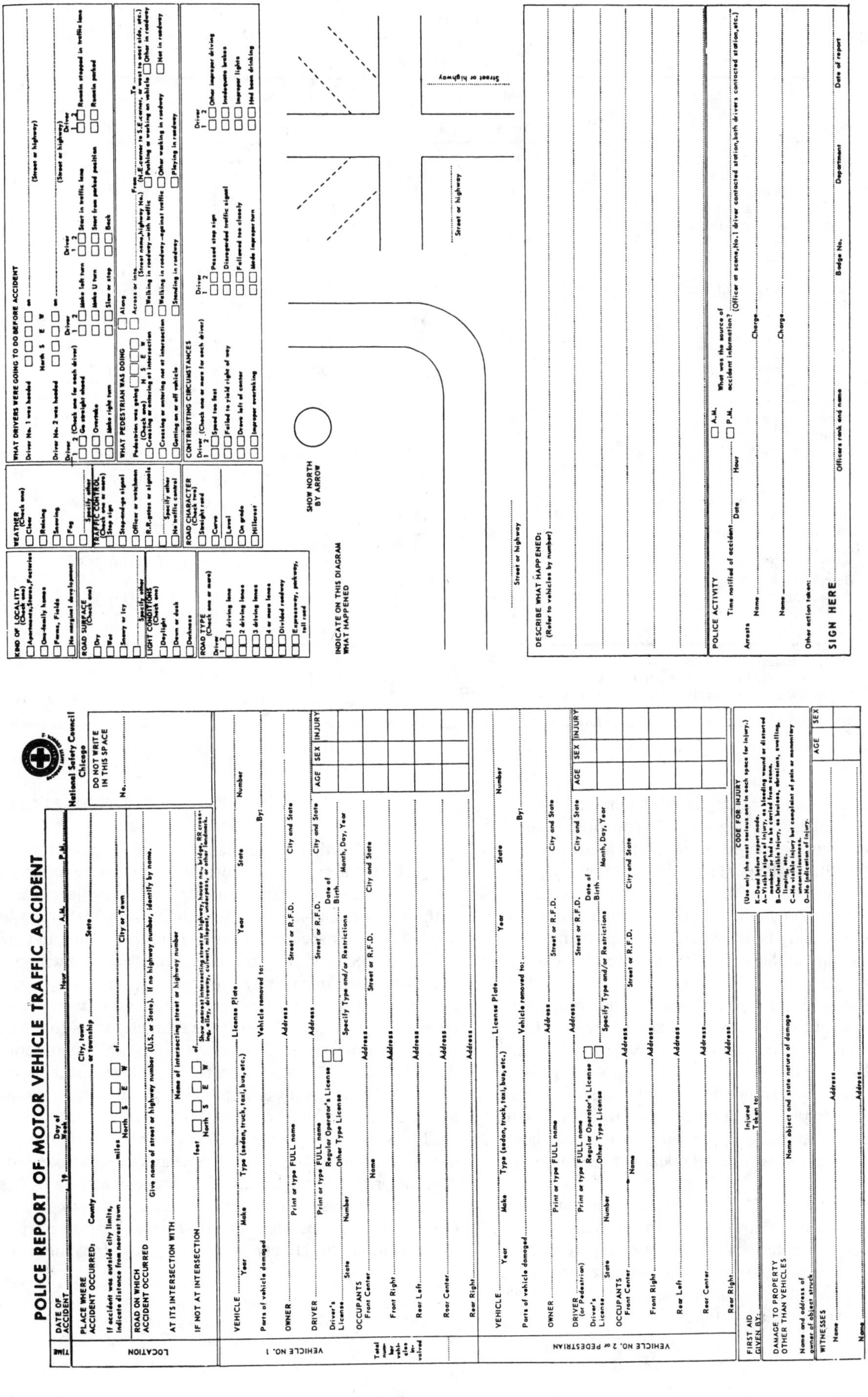

Figure 4-3. Sample of traffic accident report form.
SOURCE: ITE *Transportation and Traffic Engineering Handbook*, 2nd edition, 1982, p. 549.

TABLE 4-17

Example of Ranking
30 Highest Accident Intersections

Intersection	Total Accidents 3 Years	MEV[1] in 3 Years	Accident Rate per MEV	Rank[2] by Frequency	Rank[2] by Rate
Milwaukee/Touhy/Waukegan	193	134	1.4	1	10
Caldwell/Gross Pt./Touhy	122	94	1.3	2	(11)
Milwaukee/Oakton	111	45	2.5	3	4
Dempster/Milwaukee	108	54	2.0	4	(6)
Harlem/Howard/Milwaukee	85	53	1.6	5	8
Caldwell/Howard	61	22	2.7	6	2
Lehigh/Touhy	55	19	3.0	7	1
Harlem/Touhy	48	25	2.0	(8)	(6)
Milwaukee/Golf Mill South Driveway #9300	48	—	—	(8)	—
Golf/ Milwaukee	45	25	1.8	(9)	7
Ballard/Milwaukee	45	29	1.5	(9)	(9)
Oakton/Waukegan	40	30	1.3	10	(11)
Milwaukee/Main	38	14	2.6	11	3
Harts/Milwaukee	37	27	1.3	(12)	(11)
Gross Point/Howard/Lehigh	37	47	0.8	(12)	15
Milwaukee/Maryland	33	27	1.2	(13)	12
Dempster/Harlem	33	34	1.0	(13)	13
Dempster/Cumberland	26	11	2.3	14	5
Dempster/Ozark	24	18	1.3	15	(11)
Milwaukee/Golf Mill Center Driveway #9400	23	—	—	16	—
Milwaukee/Monroe	22	15	1.5	17	(9)
Milwaukee/Golf Mill North Driveway #9500	21	—	—	18	—
Dempster/Shermer	17	18	0.9	(19)	(14)
Milwaukee/Courtland	17	—	—	(19)	—
Harlem/Oakton	16	33	0.5	20	16
Howard/Waukegan	15	16	0.9	21	(14)
Dempster/Merrill	15	—	—	22	—
Harlem/Main	14	16	0.9	(23)	(14)
Central/Touhy	14	—	—	(23)	—
Caldwell/Oakton	13	9	1.5	24	(9)

[1] Million Entering Vehicles
[2] More than one location with same rank = (X)
Note: 40% of all accidents in city occurred at above intersections.

SOURCE: ITE, "Manual of Traffic Engineering Studies," 2nd edition, p. 57.

as traffic volumes, road location/classification, and type of traffic control. Linked or integrated systems offer several advantages to transportation professionals, primarily by enhancing the ability to analyze information from related files. This eliminates the extensive labor usually required to extract information from a number of files in order to create a special data set for analysis. In addition, linkage of different systems ensures comparable definitions are used in each system, improving consistency.

The increasing use and power of minicomputers and microcomputers permit engineers and analysts greater flexibility in data manipulation and analysis, particularly when accident and location inventory files can be linked together. This helps to overcome some of the limitations of central (state-level, for example) computer systems, which will not likely have location codes for local road systems, and whose standard reports are not sufficiently detailed for engineering analysis.

For statistical analysis, the level of detail from computer files is usually sufficient. However, for in-depth analysis, it is important that computer-based systems be supplemented with reasonable access to the detailed accident report forms where the narrative and diagram are essential, as well as information on the specific locations, including site inspections and interviews.

Accident analysis

Data applications. Accident data are required for many purposes by safety and transportation engineering professionals, including identification and ranking of high-accident locations, before-and-after studies of improvements, supporting expenditures for accident-reduction improvements, and identifying the need for other measures such as enforcement, parking restrictions, or roadway lighting. In general, however, the data are used to identify trends in accident occurrence and consequences (injury severity, for example) so as to develop solutions or measures that improve safety while maintaining mobility on our roads and streets. At the broadest level, data are compiled nationally or internationally to develop comparative measures of safety performance and characteristics of roadways, vehicles, or drivers that contribute either to accident causation or reduction.[11] These analyses can be used to develop priorities for funding and implementation of engineering measures to improve safety, such as design or operational changes. These data are also essential for the purposes of evaluating the effects of any changes implemented to

[11] "Framework for Consistent Traffic and Accident Statistical Data Bases," Scientific Expert Group T8, Road Research Programme, Organization for Economic Cooperation and Development, 1988 (IRRD 811327).

improve safety. Similar analyses are done at a more local level (state, province, or county level) to develop a better understanding of, for example, the safety effects of one form of traffic control versus another. Such studies contribute to our knowledge about which measures can be expected to improve safety in general.

The second general form of analysis is the study of specific sites or locations where a safety problem is deemed to exist. Accident data are used to identify the type of accident and effective countermeasures, and to monitor the effects of safety improvements. The analysis techniques in this case are quite different, depending less on development of statistical relationships and more on an understanding of the traffic and environmental conditions at each site that can be linked to accident causes.

High-accident locations. High-accident locations are those with the highest accident rate. This generally means the locations with the highest number of accidents when such locations are similar in design and function (intersections, two-lane roads, four-lane roads, etc.) and have similar traffic volumes (a measure of exposure to risk of an accident). For roadway sections, accidents per million vehicle–miles (MVM) are used to estimate accident rates. For intersections, the more usual rate is accidents per million entering vehicles (MEV).

Various techniques are available to identify such locations (Figure 4–4). Most of these techniques use statistical methods as a "screen" or "sieve" to retain only those locations with higher-than-expected numbers of accidents or accident rates (see section below, "Accident Rate Calculations").

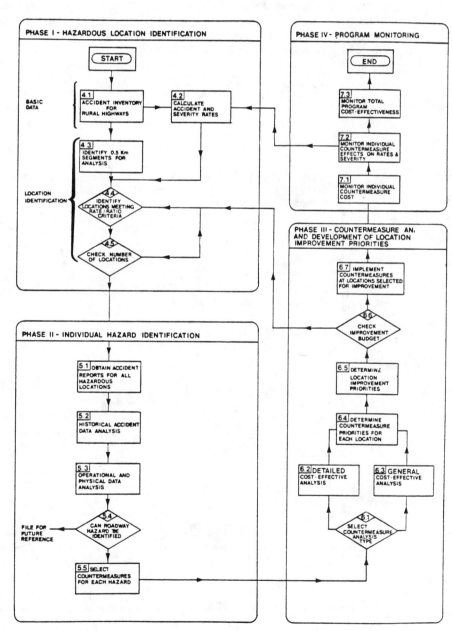

Figure 4–4. Flowchart for low-cost roadway safety improvements analysis.
SOURCE: TRANSPORT CANADA "Manual of Low-Cost Roadway Improvements for Rural Highways," 1981, p. 9.

These techniques generally group roadway types and then select locations that exceed either a certain number of accidents annually, a minimum accident rate, accident rates deemed significantly higher than average, or a combination of these.[12] Figure 4–5 illustrates one example of critical accident rates (those considered significantly worse than average) for different roadway types and traffic volumes in one jurisdiction. The calculation of critical accident rate depends on the average accident rate for the roadway locations being considered, the level of significance selected, and traffic volume, as illustrated in Figure 4–5.

For example, a collector roadway section with ribbon development (curve V) with an accident rate of 5.0 accidents per million vehicles entering and a sum of AADT's over the same period of 5,000 vehicles per day would exceed the "critical rate" (R_c) of 4.1 indicated in Figure 4–5. However, it is important to note that the selection criterion used can influence significantly how efficient the method is in correctly identifying locations as hazardous. Computerized analytical methods are available to improve the process of selection since the observed frequency of accidents or accident rate over a short period is not necessarily an accurate indicator of future accident occurrence.[13]

The spot map is another approach for identifying locations with concentrations of accidents, but this is most useful for special accident situations. These include pedestrian accidents (for example, whether in day or night conditions), parked-vehicle accidents, or accident occurrence by police patrol districts. Color-coded pins can be used to identify the different types of accidents. At high-frequency locations, "multiplier" pins may be required, where one pin represents 10 accidents. For traffic engineering purposes, the map should utilize a plan of the road system based on functional classification; freeways, major routes and collector streets should be separately coded from local streets.

Accident rate calculations. Accident rates are normally considered better measures of risk than accident frequencies alone, since they account for differences in traffic flow. The standard equation for calculating accident rates is

$$\text{rate} = \frac{\text{number of accidents}}{\text{exposure}}$$

[12] "Manual of Low Cost Roadway Safety Improvements for Rural Highways," prepared for Transport Canada by ADI Limited, Ottawa, Ontario, May 1981 (TP30375E/F).

[13] E. Hauer, and M. Montgomery, "Software to Aid in the Correct Interpretation of Accident Data," *ITE Journal* (February 1988).

Figure 4–5. Critical accident rates for various rural highway types.
SOURCE: Transport Canada "Manual of Low-Cost Roadway Improvements for Rural Highways," 1981, p. 17.

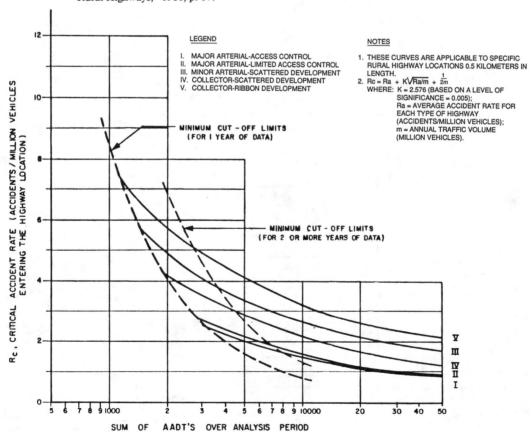

The rates can also be calculated using injuries, deaths, fatal or injury accidents.

Different rates are used for intersections and for roadway sections as follows:

$$\frac{\text{intersection}}{\text{accident rate}} = \frac{\text{annual number of accidents} \times 10^6}{\text{annual traffic entering}}$$

where rate = accidents per million vehicles entering the intersection annually (MEV)

$$\text{section rate} = \frac{\text{annual number of accidents} \times 10^6}{\text{annual vehicle–miles of travel}}$$

where rate = accidents per million vehicle–miles of travel (MVM)

where vehicle–miles = AADT (average annual daily traffic) × 365 days/year × section length in miles

Studies have shown that traffic accident rates sometimes increase with volume. For example, a six-lane freeway with 100,000 vehicles/day may have a higher accident rate than a four-lane freeway carrying 20,000 vehicles/day.

Rates can also be computed to adjust for the severity of accidents. In this case fatal accidents would typically receive more weight than injury accidents, which in turn receive more weight than property-damage accidents. This approach has been called an equivalent property damage only (EPDO) calculation, where the "weights" can be assigned arbitrarily or related to some other measure, such as the estimated costs of property damage, injury, and fatal accidents, as follows:

$$\frac{\text{Number}}{\text{(EPDO accident)}} = \text{PDO} + (\text{INJ.} \times \text{F1}) + (\text{FAT.} \times \text{F2})$$

Where PDO = number of property damage accidents
INJ = number of injury accidents
F1 = cost (injury accidents)/cost (PDO)
FAT = number of fatal accidents
F2 = cost (fatal accidents)/cost (PDO)

However, this type of analysis needs to be used with care because of the emphasis on fatal accidents. This emphasis, particularly where the number of accidents is not large, can result in attention being directed to locations that may not be the most appropriate for cost-effective improvements. It has been suggested that a more appropriate technique would be to combine the number of fatal and injury accidents and use a combined fatal-plus-injury average accident cost.[14] This approach can significantly alter assessments of cost-benefit and cost-effectiveness.

Various studies or assumptions also assign considerably different costs to each type of accident. For example, the technical advisory from the U.S. Federal Highway Administration[15] recommended the following costs based on a

"willingness-to-pay" concept: $1,700,000 per fatal accident, $14,000 per injury accident, and $3,000 per property damage accident (1986 figures). Using the formula above, this would result in the following weights: F1 = $14,000/$3,000 = 4.7; F2 = $1,700,000/$3,000 = 566.6. Most other estimates, using different assumptions, result in different costs (for example, Canadian authorities have estimated costs at $340,000 per fatal accident, $11,000 per injury accident, and $2,600 per property damage accident [1986 Canadian dollars]. In this case, the weighting factors would be significantly different (F1 = 4.2; F2 = 130). Even if the same assumptions are used, accident "costs" vary from country to country. Accordingly, this analysis should be restricted mainly to economic analyses of possible improvements.

Detailed analysis. When locations have been identified for further analysis, normal procedures include the following:[16, 17]

1. Obtaining all accident data and reports for a time period of at least two years;
2. Preparing a summary report of the accident data, including dates and times of accidents, weather conditions, road conditions, type of accident (sideswipe, angle, rear-end), type of vehicles involved, driver actions, and other information from report forms;
3. Preparing a collision diagram to identify patterns of accident occurrence that can assist the analyst in looking for engineering solutions (Figure 4–6);
4. Preparing a condition diagram or inventory sketch of the location, including physical features such as traffic control devices, pavement condition, utility poles and building lines, and others as appropriate;
5. Obtaining other data such as traffic speeds and volumes, vehicle classifications and signal timing (see also Chapter 3);
6. Visiting locations to observe and become familiar with specific site characteristics, traffic patterns, and other information not readily available from reports or inventories.

This process will assist the engineer in identifying and analyzing environmental, site, or traffic characteristics that may be contributing to traffic accidents.[18] When such causes are identified, the engineer or analyst can evaluate corrective measures that are known to be effective. Tables 4–18 and 4–19[19, 20] illustrate listings of engineering measures and their estimated effectiveness (see "Evaluation" below).

The final stage in the process is to evaluate the actual effectiveness of any measures after they have been put into

[14] "Motor Vehicle Accident Costs," Technical Advisory T7570.1 (Washington, DC: U.S. Department of Transportation, Federal Highway Administration, June 30, 1988).

[15] Ibid.

[16] Box and Oppenlander, *Manual of Traffic Engineering Studies.*

[17] ADI Ltd., "Manual of Low Cost Roadway Improvements."

[18] Goodell-Grivas, Inc., for the Federal Highway Administration, U.S. Department of Transportation, June 1981.

[19] "The 1989 Annual Report on Highway Safety Improvement Programs," Report of the Secretary of Transportation to the United States Congress, April 1989.

[20] V.S. Payne, "Accident Prevention and Reduction in Great Britain." Presentation to the Annual Conference, Institute of Transportation Engineers, Vancouver, Canada, 1988.

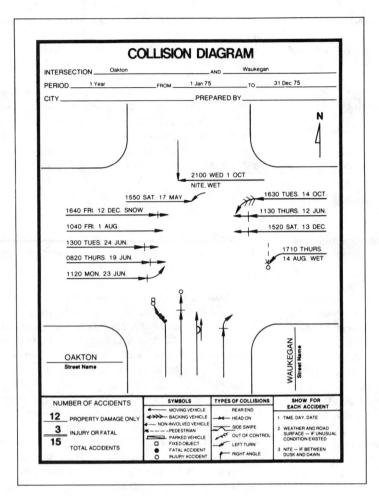

Figure 4-6. A collision diagram taken from an actual urban intersection.
SOURCE: ITE "Manual of Traffic Engineering Studies," 2nd edition, p. 62.

place. Evaluation techniques are well-documented elsewhere,[21-23] but a fundamental requirement is that any necessary data be identified and collected for both the period prior to and following implementation. The evaluation results, whether positive, neutral, or negative, should be documented to improve the knowledge base of the responsible agency and the profession as a whole.

Evaluation. Most evaluations of traffic engineering improvements for safety purposes are of the "before-and-after" type, and accordingly most of our estimates of the safety effects of improvements are based on this technique. Analysis appears reasonably straightforward, usually as follows:

$$\text{Effectiveness (\%)} = \frac{Nb - Na}{Nb} \times 100$$

where Nb = average accidents/time period before change
Na = average accidents/time period after change

[21] "Methods for Evaluating Road Safety Measures," Road Research Programme, Organization for Economic Cooperation and Development, 1981.

[22] "Highway Safety Evaluation, Procedural Guide," prepared for the U.S. Department of Transportation, Federal Highway Administration by Goodell-Grivas, Inc., March 1981.

[23] L.I. Griffin, "Three Procedures for Evaluating Highway Safety Improvement Programs," Presentation to the American Society of Civil Engineers, 1982 Annual Convention, New Orleans, October 1982.

However, it is necessary to establish whether or not the observed change in accident occurrence is truly significant. This is usually done with statistical tests using established procedures (see pp. 206–207, Appendix B, "Before and After Analysis" of the ITE *Manual of Traffic Engineering Studies*).[24] Figure 4–7 illustrates the relationship between number of accidents before any changes to the location, and the percentage reduction required to be considered significant. Other considerations are whether or not other factors may have had an influence on safety (new law or enforcement strategy, for example) or if there were changes in accident rates over time.[25]

This type of analysis is more reliable when the time periods before and after are at least of two years' duration, exposure did not change over time, and the number of accidents is large. However, it has been demonstrated that simple before-and-after analyses can produce biased estimates of effectiveness, due to a statistical effect known as "regression to the mean." In practice, this means that sites selected for corrective action usually have a higher number of accidents when selected, and often would have had a lower number of accidents in a later period even if no traffic engineering change had occurred (see Table 4–20).

[24] Box and Oppenlander, *Manual of Traffic Engineering Studies.*

[25] Griffin, "Three Procedures for Evaluating Highway Safety Improvement Programs."

TABLE 4-18

Evaluation of Safety Improvements by Construction Classification, 1974–1987

Type of Improvement and Construction Classification	Indexed Cost of Evaluated Improvements (millions)	Percent Reduction in Accident Rates After Improvements			Cost-Per-Accident Reduced (thousands)		Benefit/ Cost Ratio
		Fatal	Injury	Fatal + Injury	Fatal	Fatal + Injury	
INTERSECTION & TRAFFIC CONTROL	562.3	37	15	15	344.5	18.7	4.0
Channelization turning lanes	297.7	48	23	24	510.9	19.7	2.8
Sight distance improvements	7.8	44*	31	32	371.4	22.8	3.6
Traffic signs	19.6	34	3	4	59.3	15.7	20.9
Pavement markings and/or delineators	33.7	15	(1)*	(1)*	751.8	—	1.6
Illumination	13.2	45	8	9	122.6	16.8	10.3
Traffic signals upgraded	63.1	40	22	22	412.1	8.6	4.0
Traffic signals, new	127.3	49	21	21	344.5	10.5	5.1
STRUCTURES	307.7	50	28	29	752.2	92.1	1.7
Bridge widened or modified	79.8	49	22	23	1,077.9	103.8	1.2
Bridge replacement	156.9	72	47	49	1,201.6	159.3	1.1
New bridge construction	26.2	77*	40	43	1,637.8	223.2	0.8
Minor structure replacement/improved	39.0	36	20	21	277.1	39.8	4.5
Upgraded bridge rail	5.6	72*	41	45	189.5	39.4	6.5
ROADWAY AND ROADSIDE	1,971.3	31	13	13	722.2	54.0	1.8
Widened travel way	511.0	9*	7	7	4,041.3	174.2	0.4
Lanes added	212.9	(2)	13	13	—	66.8	0.1
Median strip to separate roadway	56.8	73	17	19	382.4	72.8	3.2
Shoulder widening or improvement	88.3	28	11	12	497.9	37.9	2.6
Roadway realignment	329.9	61	32	34	1,193.2	111.1	1.1
Skid resistant overlay	468.4	26	18	19	837.0	30.3	1.8
Pavement grooving	12.6	34*	15	15	377.3	14.5	3.8
Upgraded guardrail	149.7	42	8	9	151.7	31.5	8.1
Upgraded median barrier	7.4	45*	28	29	192.8	10.8	7.0
New median barrier	58.9	62	0*	3*	224.7	213.6	5.4
Impact attenuators	10.7	31*	36	36	390.0	8.0	4.0
Flatten side slopes/regrading	36.5	(25)*	9	8*	—	102.2	—
Bridge approach guardrail transition	4.8	61*	44	45	200.9	25.8	6.3
Obstacle removal	15.0	49	22	23	202.5	16.0	6.4
RAILROAD-HIGHWAY CROSSINGS	331.2	89	63	67	570.0	114.2	2.2
New flashing lights	55.0	91	74	77	551.4	103.5	2.2
New flashing lights and gates	138.1	92	84	86	591.8	115.3	2.1
New gates only	63.2	90	78	80	432.8	96.7	2.8

Note: Numbers in parentheses ( ) indicate increased accident rates.
*No significant change at the 95% confidence level.

SOURCE: Department of Transportation, "The 1989 Annual Report on Highway Safety Improvement Programs," 1989, p. 23.

TABLE 4-19

Accident Reductions for Low-Cost Treatments

Treatment	Accident Reduction (%)
Anti-skid (Calcined Rauxite)	50
Improved signing at junctions	46
Modifications to existing traffic signals	30
New traffic signals	46
New pelicans (crosswalks)	40
New zebras (crosswalks)	44
Superevelation	75
Lining and signing not at bends	32
Lining and signing at bends	55
Minor junction changes (not roundabouts)	52
Pedestrian refuges	19
Pedestrian barriers	43
New street lighting	24
Resurfacing	34
Alterations to roundabouts	20
Area studies	50

SOURCE: V. S. Payne, "Accident Prevention and Reduction in Great Britain," 1988, p. 14.

TABLE 4-20

Accident Count at 1,142 Intersections, 1974–1975

1 Number of Intersections	2 Number of Accidents per Intersection in 1974	3 Average Number of Accidents per Intersection in 1975
[n(x)]	[x]	[M(x)]
553	0	0.54
296	1	0.97
144	2	1.53
65	3	1.97
31	4	2.10
21	5	3.24
9	6	5.67
13	7	4.69
5	8	3.80
2	9	6.50

(2 intersections had 13 accidents, one had 16)

SOURCE: E. Hauer and others, "New Directions for Learning About Safety Effectiveness," 1986, p. 8.

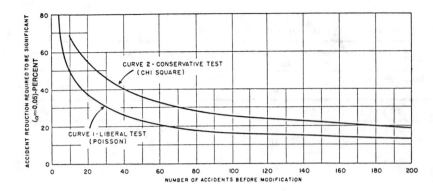

Figure 4–7. Curves of significance test for accident reduction.
SOURCE: ITE "Manual of Traffic Engineering Studies," 2nd edition, p. 207.

To account for this bias, analytical techniques have been developed and tested.[26, 27] These techniques provide an estimate of the "expected" number of accidents in an after period, rather than assuming that the accidents in the before period are an accurate predictor of what will happen in the future (see Table 4–21).

Other statistical analyses have been used to account for the regression-to-the-mean effect, as well as other statistical concerns with the simple before-and-after analysis. These include using a comparison group of similar sites where no improvements were made, and use of comparison groups with checks for comparability.[28]

International traffic safety organizations

Many international organizations have programs designed to address road safety problems. The United Nations has held conferences to develop conventions for traffic control devices and for rules of the road. Through its Economic Commission for Europe (ECE), it has established working groups to develop international standards for motor vehicles and other traffic safety programs.

---

[26] Hauer and Montgomery, "Software to Aid in the Correct Interpretation of Accident Data."

[27] "New Directions for Learning About Safety Effectiveness," report prepared for Transport Canada and the Federal Highway Administration by E. Hauer, J. Lovell, and B. Persaud, Ottawa, Ontario, January 1986 (7262E/F).

[28] Griffin, "Three Procedures for Evaluating Highway Safety Improvement Programs."

The Organization for Economic Cooperation and Development (OECD) has a Road Transport Research Program with a strong traffic safety component that brings experts in member countries together to prepare state-of-the-art reports on selected subjects and to identify future research requirements.

The World Health Organization (WHO) tries to focus attention on the traffic safety issue and to encourage developing countries to address the problem before it is too late. This is done through special conferences and seminars in various countries.

In the early 1970s, the North Atlantic Treaty Organization (NATO) brought the organizational and technical skills normally associated with military operations to bear on world road safety problems through a special study under the nonmilitary Committee on the Challenges of Modern Society. This project looked at a number of traffic safety issues, such as alcohol, drugs and driving, emergency medical services, and pedestrian safety, with the objective of developing action plans that could be adopted in member countries.

One of the important features of the NATO program was the application of goal setting to safety management. The project proposed the establishment of a national safety goal based on the potential to reduce deaths and injuries through an integrated program that embraced measures designed to improve roads, vehicles, and human performance on the road.

Even though all of these intergovernmental organizations have produced innovative findings, recommendations, and standards, their programs have not been well funded and

---

TABLE 4–21
Safety Effect by Accident Type in San Francisco

Accident Type	(1) Number of "Before" Acc.	(2) Number of "After" Acc.	(3) Expected No. "After" Acc.	(4) (1) – (2) / (1) Apparent % Reduction	(5) (3) – (2) / (3) Unbiased % Reduction
Right-Angle	129	16	93	88	83
Rear-End	10	16	4	−60	−300
Left-Turn	14	7	10	50	30
Pedestrian	6	2	6	67	67
Injury	48	9	35	81	74
Total	172	50	130	71	62

SOURCE: E. HAUER AND OTHERS, "New Directions for Learning About Safety Effectiveness," 1986, p. 21.

supported, and the recommendations have not been widely adopted. In addition to these large multidimensional organizations, there are many other international groups such as Prévention Routière Internationale (PRI), International Driver Behaviour Research Association (IDBRA), the Institute of Transportation Engineers (ITE), and the Permanent International Association of Road Congresses (PIARC) that bring together safety experts and that develop safety countermeasures and proposals.

These groups provide an essential service by setting international standards for vehicles, data, research, and program evaluation. One such service is the development of an accident data bank at BASt, the German Highway Research Institute. This data bank, developed with the support of the OECD and the European Economic Community (EEC), is designed to collect and store data that are of uniform quality to enable accurate comparisons of the road safety situation in different countries and the efficacy of unique safety countermeasures.

The success of this program and of other efforts to produce better standards and safety programs will only benefit those countries that are prepared to adopt the solutions developed and to contribute experience gained from research and program evaluation.

Safety management

National governments, at the highest level, should make a commitment to solving the road safety problem through leadership and by providing adequate resources. Countries such as France and Japan have demonstrated the power of placing the responsibility for developing a national road safety strategy in the office of the prime minister.

Research can be coordinated at the national level. It is perhaps easier to bring together, at the national level, academics and program managers from related disciplines such as medicine, engineering, human factors, economics, and statistics in a multidisciplinary approach to problem solving.

National governments should create and maintain national accident databanks and establish standards for accident reporting. These data can then be used to measure the performance of local safety agencies and to compare the effectiveness of safety programs.

Local responsibilities

Historically, local safety management was treated in one of two ways: Safety was declared to be "everybody's business" and no one group or individual was assigned responsibility; or an individual, usually with a very small staff, was identified as the safety expert with the responsibility to represent the organization at "safety events."

In recent years most jurisdictions have taken their safety responsibilities more seriously and have established clear lines of accountability for various aspects of highway safety. Highway or transportation agencies have created safety offices responsible for safety announcements and programs that do not normally include engineering activities.

The engineering aspects of safety are often still integrated with other activities such as maintenance, since many of the safety initiatives are implemented by maintenance staff; planning, since much of the data used by safety analysts are also used by planners; or the traffic operations office, because of the compatibility of professional skills. The disadvantage of integrating the safety responsibilities with others is the risk of losing independence and status when conflicts regarding funding and priorities arise.

Regardless of where the safety management responsibility lies, it is essential that all groups are involved in the safety program and that extensive consultations with nontransportation groups such as police departments, medical and para-medical organizations, advocacy groups, and special-interest groups (school boards, senior citizens, etc.) are carried out. While many of the safety solutions discussed in this chapter or those found in other chapters of the handbook are of an engineering nature, the safety problem is also a social and public health problem that affects everyone in a community, and there are other contributing solutions that are outside of the transportation-engineering spectrum.

Transportation safety professionals should bring together all the groups and organizations that can contribute expertise and support to address safety issues.

The responsibility for safety management should be readily identifiable. Having a group or individual clearly accountable for a jurisdiction's safety record ensures that the data needed to measure safety trends will be collected and that evaluations of the program and its components will be undertaken. It is not necessary to have all aspects of the safety program centralized; however, the person accountable for the safety record must have access to all program and performance information needed to demonstrate that public safety concerns have been addressed.

REFERENCES FOR FURTHER READING

Accident Facts, 1988 Edition (Chicago: National Safety Council, 1989).

ADI Ltd., "Manual of Low Cost Roadway Safety Improvements for Rural Highways," prepared for Transport Canada (TP3075E/F), Ottawa, Ontario, May 1981.

BOX, P., AND J. OPPENLANDER, *Manual of Traffic Engineering Studies*, 4th edition (Arlington, VA: Institute of Transportation Engineers, 1976).

Federal Highway Administration, *Synthesis of Safety Research Related to Traffic Control and Roadway Elements*, Vols. 1 & 2 (Washington, DC: United States Department of Transportation, 1982).

Economic Commission for Europe, "Statistics of Traffic Accidents in Europe, 1980, 1985, 1986, 1987, 1988" (New York: United Nations Organization, 1990).

Federal Highway Administration, Technical Advisory T7570.1, "Motor Vehicle Accident Costs" (Washington, DC: United States Department of Transportation, June 1988).

Goodell-Grivas, Inc., "Highway Safety Evaluation, Procedural Guide," prepared for the Federal Highway Administration, United States Department of Transportation, Washington, DC, March 1981.

GRIFFIN, L.I., "Three Procedures for Evaluating Highway Safety Improvement Programs," presentation to the American Society of Civil Engineers' Annual Convention. New Orleans, Louisiana, October 1982.

HAUER, E., J. LOVELL, AND B. PERSAUD, "New Directions for Learning About Safety Effectiveness," prepared for Transport Canada (TP7262E/F). Ottawa, Ontario, January 1986.

HAUER, E., AND M. MONTGOMERY, "Software to Aid in the Correct Interpretation of Accident Data," *ITE Journal* (February 1988).

HAUER, E., AND M. MONTGOMERY, "The Convincer: An Interactive Tutorial and Study Tool," *Traffic Engineering + Control,* London, Printerhall Ltd., May 1987.

HOMBURGER, W.S., L.E. KEEFER, AND W.R. McGRATH, *Transportation and Traffic Engineering Handbook,"* 2nd edition. Institute of Transportation Engineers. (Englewood Cliffs, NJ: Prentice-Hall, 1982).

Institute of Transportation Engineers, "Safety Programmes and Their Effect on Highway Standards" (Washington, DC: 1986).

Institute of Transportation Engineers, "The Correction of Hazards on Urban Streets" (Washington, DC: 1986).

Mobility 2000, IVHS Workshop, March 19–21, 1990 Proceedings. Dallas, June 1990.

National Highway Traffic Safety Administration, "Effectiveness of Safety Belt Use Laws: A Multinational Examination" (Washington, DC: United States Department of Transportation, February 1986).

National Highway Traffic Safety Administration, "Fatal Accident Reporting System 1987" (Washington, DC: United States Department of Transportation, December 1988).

PAYNE, V.S., "Accident Prevention and Reduction in Great Britain," presentation to the Institute of Transportation Engineers' Annual Conference, Vancouver, BC, 1988.

RICE, D.P., AND ELLEN J. MACKENZIE AND ASSOCIATES, "Cost of Injury in the United States, A Report to Congress" (Institute of Health and Aging, University of California and the Injury Prevention Center, The Johns Hopkins University. San Francisco, 1989).

Road Research Programme, "Framework for Consistent Traffic and Accident Statistical Data Bases" (Paris: Organization for Economic Cooperation and Development, 1988).

Road Research Programme, "Methods for Evaluating Road Safety Measures" (Paris: Organization for Economic Cooperation and Development, 1981).

"State's Model Motorist Data Base: Data Element Dictionary for Traffic Record Systems" ANSI D 20.1 (New York: American National Standards Institute, 1979).

"Statistical Report on Road Accidents in 1987" (Paris: European Conference of Ministers of Transport, OECD Publications Service, 1989).

STEWART, S.E., AND R.W. SANDERSON, "The Measurement of Risk on Canada's Roads and Highways," proceedings of the Third Symposium of Institute for Risk Research (Waterloo, Canada: University of Waterloo Press, 1984).

"The 1989 Annual Report on Highway Safety Improvement Programs," Report of the Secretary of Transportation to the United States Congress (Washington, DC: April 1989).

TREAT, J.R. AND OTHERS, "Tri-Level Study of the Causes of Traffic Accidents." Institute for Research in Public Safety, School of Public and Environmental Affairs for the National Highway Traffic Safety Administration (Washington, DC: March 31, 1977).

TRINCA, GORDON AND OTHERS, "Reducing Traffic Injury—A Global Challenge" (Melbourne: Royal Australian College of Surgeons, 1988).

United States House Public Works and Transportation Committee. "Status of the Nation's Highways: Conditions and Performance" (Washington, DC: June 1985).

5

OPERATIONAL ASPECTS OF HIGHWAY CAPACITY

WILLIAM R. REILLY, P.E., *Senior Vice President*

JHK & Associates

Principles of capacity

The *capacity* of a transportation facility is defined as the maximum number of vehicles (or pedestrians) that can reasonably be expected to use the facility in a given time period under prevailing roadway, traffic, and control conditions. The concept of *level of service,* first introduced in the 1965 *Highway Capacity Manual,*[1] comprises a set of defined operating conditions for each facility type. The maximum amount of traffic that can be accommodated while maintaining the defined operating conditions is termed the *service volume* for that level of service. These two concepts, capacity and level of service, are documented in the 1985 *Highway Capacity Manual*[2] of the Transportation Research Board. The 1985 *Manual* is considered to be the primary reference for American practice when considering the operations of highways, intersections, and other transportation facilities. The *Australian Road Capacity Guide*[3] and the *Canadian Capacity Guide*[4] are examples of research efforts and manuals that provide additional valuable information for capacity analyses.

There are three primary activities that traffic engineering professionals perform and that depend on capacity and level-of-service analyses. First, when new facilities are being planned or when existing facilities are to be expanded, the size in terms of width or number of lanes must be determined. Second, when existing facilities are to be upgraded using traffic operational improvements or when it is desired to maintain or achieve a given level of service without resorting to capital intensive projects, the operational characteristics of the facility must be studied within the framework of level of service. Finally, it is becoming more common for the traffic engineering profession to contribute to economic and environmental analyses when considering public or private improvements. The study of operating conditions and level of service provides base values for road-user costs, fuel consumption, air pollutant emissions, and noise.

The analysis procedures for capacity and level of service are continually being updated by the Transportation Research Board. This process began in 1986 following initial distribution of the 1985 *Manual* and is expected to continue through the 1990s. New updated chapters, errata listings, and enhancements to selected chapters are produced and are available through the Transportation Research Board in Washington, DC.

Capacity and level of service

The concepts of capacity and level of service are applied to a range of transportation facilities for both design and operational analyses.

Facility types. The facility types encountered by traffic engineering professionals cover freeways, two-lane and multilane highways, intersections, both signalized and

[1] *Highway Capacity Manual,* 1965, Special Report 87, Highway Research Board, Washington, DC, 1965.

[2] *Highway Capacity Manual,* 1985, Special Report 209, Transportation Research Board, Washington, DC, 1985.

[3] *Australian Road Capacity Guide—Provisional Introduction and Signalized Intersections,* Australian Road Research Board, Bulletin No. 4, 1968 (Replaced by ARR No. 123, 1981).

[4] *Canadian Capacity Guide for Signalized Intersections,* University of Alberta, 1984.

SOURCE: *Highway Capacity Manual*, 1985, Special Report 209 (Washington, DC: Transportation Research Board), p. 1–3.

unsignalized, urban arterials, bus transit, pedestrian facilities, and bikeways. Table 5–1 provides a summary of facility types included in the 1985 *Highway Capacity Manual*.

Capacity. The expression "capacity" depends on the units being observed (vehicles, passenger car equivalents, pedestrians), the time period, and the area of the facility being considered (lane, width in feet, area). Because each facility type covered by the 1985 *Highway Capacity Manual* has specific units for expressing capacity, Table 5–2 is presented as a summary. In terms of passenger car units, capacity under ideal conditions is characterized by 2,000 passenger cars per hour per lane (pcphpl) for uninterrupted flow along freeways and multilane highways. On two-lane rural highways, capacity ranges from 2,000 to 2,800 passenger cars per hour (pcph), total for both directions of flow, depending on the directional split of volume. At signalized

intersections, 1,800 passenger car units (pcu) can depart from the stop line for each hour of green time, on a per lane basis.

The 1985 *Highway Capacity Manual* lists some of the highest recorded traffic volumes by facility type. Although these values should not be considered typical, nor should they be considered as being achievable on all facilities of a given type, they do indicate the general upper limits of capacity. Note that in most cases the units are in terms of "vehicles" rather than passenger car units. Table 5–3 indicates the highest of the volumes listed in the 1985 *Manual*.

Levels of service. "Level of service" is defined as those operational conditions within a traffic stream as perceived by users of the traffic facility. The concept of level of service was originally defined as being a qualitative measure of operational conditions. Such a measure would ideally cover factors such as speed and travel time, delay, freedom to maneuver, traffic interruptions, comfort and convenience, and safety. In practice, levels of service have been and continue to be defined by one or two measures of effectiveness for each facility type. These measures relate much more to speed, delay, and density than they do to qualitative factors or to safety.

For each facility type, six levels of service, A through F, are defined. For uninterrupted flow along freeways and highways, Level of Service A represents free flow with individual vehicles being unaffected by the presence of other vehicles. Level of Service E represents operating conditions at capacity, while Level of Service F defines breakdown flow conditions. For facilities with interrupted flow, a measure of effectiveness is used to define level of service ranges. Table 5–4 lists the measures of effectiveness used for each facility type in the 1985 *Highway Capacity Manual*. The primary measure is listed, although secondary measures typically are used to complement the analysis of level of service.

TABLE 5–2

Capacity by Facility Type

Facility	Units of Traffic	Time Period[a]	Area	Capacity Units Expressed As	Capacity (Ideal Conditions)
Freeway-basic section	Passenger-car units	Hour	Lane	pcphpl[b]	2,000[c]
Freeway-weaving area	Passenger-car units	Hour	Lane	pcphpl	1,900
Freeway-ramp junction	Passenger-car units	Hour	Merge or diverge area	pcph	2,000
Freeway-one-lane ramp	Passenger-car units	Hour	Ramp roadway	pcph	1,700
Two-lane highway	Passenger-car units	Hour	Both lanes	pcph	2,800[d]
Multi-lane highway	Passenger-car units	Hour	Lane	pcphpl	2,000[c]
Signalized intersection	Passenger-car units	Hour of Green	Lane	pcphgpl	1,800[e]
Unsignalized intersection— stop or yield controlled	Passenger-car units	Hour	Lane or movement	pcph	1,000[f]
Unsignalized intersection— four-way stop	Vehicles	Hour	Entering lane	vphpl	450
Urban arterial	(Capacity usually measured and controlled by most restrictive signalized intersection)				
Transit-freeway bus lane	Buses	Hour	Lane	bphpl	1,330
Pedestrians-walkway	Pedestrians	Minute	Foot of effective width	p/min/ft	25
Bikeway	Bicycles	Hour	Lane (one-way)	bike/hr	2,150[g]

[a] Time periods of 1 hour are usually based on a peak 15-minute volume expanded to an "hourly rate of flow."
[b] Passenger cars per hour per lane.
[c] Research completed in 1989 (NCHRP Project 3-33) indicates that capacity may be 2,200 pcphpl for multilane highways.
[d] For 50–50 volume split by direction.
[e] Saturation flow rate, in passenger cars per hour of green per lane.
[f] Potential capacity with no conflicting volume and critical gap of 6.0 sec.
[g] Middle of reported range.

TABLE 5-3

Highest Reported Volumes

UNINTERRUPTED FLOW[a]			
Facility Type	Total Volume (vph)	Avg Vol Per Lane (vphpl)[2]	Location
Freeway—Urban Six-Lane	6,104	2,035	I-40, Nashville
	6,477	2,159	I-25, Denver
	6,786	2,262	CA17, San Jose
Multilane Suburban Four-Lane[c]	4,124	2,062	US101, Petaluma
Two-Lane Rural	2,450	—	Hwy 1, Banff
	2,364	—	CA84, Fremont

INTERRUPTED FLOW[a]			
Facility Type and Location	Avg Vol per Lane (vphpl)[b]	g/C Ratio	Avg Flow Rate Per Lane (vphgpl)[d]
Urban Arterial (Four-Lane)			
Antoine, Houston	1,155	0.65	1,777
US50, Cambridge, MD	1,582	0.60	2,637
Urban Arterial (Six-Lane)			
Almanden Expwy, San Jose	1,320	0.66	2,000
Ward Pkwy, Kansas City	1,159	0.61	1,900

[a]SOURCE: 1985 *Highway Capacity Manual,* unless noted.
[b]Vehicles per hour per lane.
[c]SOURCE: NCHRP Project 3-33, *Capacity and Level of Service of Multilane Rural and Suburban Highways.*
[d]Vehicles per hour of green per lane.

TABLE 5-4

Measures of Effectiveness for Level of Service

Type of Facility	Measure of Effectiveness
Freeways	
Basic freeway segments	Density (pc/mi/ln)
Weaving areas	Average travel speed (mph)
Ramp junctions	Flow rates (pcph)
Multilane Highways	Density (pc/mi/ln)
Two-Lane Highways	Percent time delay (%)
	Average travel speed (mph)
Signalized Intersections	Average individual stopped delay (sec/veh)
Unsignalized Intersections	Reserve capacity (pcph)
Arterials	Average travel speed (mph)
Transit	Load factor (pers/seat)
Pedestrians	Space (sq ft/ped)

SOURCE: *Highway Capacity Manual,* 1985, Special Report 209 (Washington, DC: Transportation Research Board), p. 1–4.

Service flow rates. When analyzing levels of service, the maximum rate of flow that can be accommodated within a level of service is typically calculated. The "service flow rate" represents the maximum hourly rate at which vehicles can traverse a roadway under prevailing conditions. Typically, the hourly flow rate is defined as four times the peak 15-min volume. When a service flow rate is given for a level of service, it is typically the maximum flow rate and thus it defines the boundary between its level of service and the next worst level of service.

Traffic flow

Three primary measures define the operational state of a stream of traffic. These three measures are speed, volume or rate of flow, and density.

Speed. The *speed* of a traffic stream is expressed in miles per hour (mph) or kilometers per hour (km/h). The most widely used measure of speed in traffic engineering analysis is *average travel speed.* It is computed by taking the length of highway or street segment and dividing it by the average travel time of vehicles passing through the segment. This measure can also be termed a "space mean speed." Another measure of speed is the *average running speed.* The distance traversed is divided by the time that the vehicle is in motion to obtain the average running speed. This measure excludes any time spent in the section when the vehicle is stopped. Average running speed is also considered a measure of space mean speed.

The measurement of individual vehicle speeds at a single point along a roadway results in a measure called "time mean speed." This type of measure is obtained by using a radar meter or a speed trap and is typically 1 to 3 mph higher than a space mean speed taken within the segment. For capacity and level-of-service analyses, the measures of space mean speed are the most appropriate.

Volume and rate of flow. Traffic *volume* is defined as the number of vehicles passing a given point or section of a roadway in a given time interval. Typically, volume data are collected on a 15-min or an hourly basis. If the analysis is to be based on a peak 15-min period, a conversion of the volume for 15 min must be made to achieve an hourly rate of flow so that the charts and tables in the 1985 *Highway Capacity Manual* can be used. For example, if the average per lane flow on a freeway during the peak 15 min is 450 vehicles, the hourly rate of flow will be 450 veh/0.25 hr, or 1,800 vehicles per hour (vph).

An example of the importance of traffic volume and of rate of flow is given by the following traffic counts taken over a 1-hour period.

Time Period (P.M.)	Volume (veh)	Rate of Flow (vph)
5:00–5:15	800	3,200
5:15–5:30	1,000	4,000
5:30–5:45	1,200	4,800
5:45–6:00	1,000	4,000
Total Hour	4,000	

The traffic count for the total hour is 4,000 vehicles; however, for each 15-min period, the rate of flow expressed on an hourly basis changes. It can be seen that the most critical 15-min period has a rate of flow of 4,800 vehicles per hour. If a facility is designed for the "average" rate, 4,000 vph would be used, but there would be a high probability of breakdown conditions occurring during the 5:30 to 5:45 P.M. period.

The peak hour factor (PHF) is a measure used to relate the maximum 15-min volume with the hourly volume. Peak hour factor is defined as

$$PHF = \frac{\text{Hourly Volume}}{4 \times \text{Peak 15-min volume}} \quad (5.1)$$

For the data listed above, the PHF would be calculated as 4,000/(4 × 1,200) or 0.833. Some calculations performed to assess operational characteristics begin with a 15-min period volume, which is converted to an hourly flow rate simply by multiplying by 4. Other analyses may begin with an hourly volume and a given PHF. To achieve the hourly flow rate representative of the peak 15-min period, the analyst simply divides the hourly volume by the PHF and uses the resulting flow rate in the capacity analysis. For the above example, 4,000 is divided by 0.833 (the PHF), giving 4,800 as the hourly flow rate.

Density. *Density* is defined as the number of vehicles occupying a given length of a lane or roadway, averaged over a given time period. Density is usually expressed in terms of vehicles per mile (vpm) and is considered a useful measure of effectiveness because it relates to the driver's ability to maneuver and to change lanes. The relationship among flow, speed, and density is

$$v = S \times D \quad (5.2)$$

where:

    v = rate of flow, in vph
    S = average travel speed, in mph
    D = density, in vpm

Typically, measurement of density is performed from a vantage point where a significant length of highway can be photographed or videotaped, and the number of vehicles in a given length can be counted. Calculation of a density value can be accomplished if both speed and volume are known.

Uninterrupted flow. For facilities such as freeways and multilane highways, uninterrupted flow conditions occur. Equation (5.2) indicates the relationship on such facilities among flow, speed, and density. As flow increases, speed tends to decrease, and density increases. At the point where capacity is reached, flow is at a maximum. If conditions begin to break down and flow becomes stop and go, speed

and flow are both reduced, while density continues to increase. The point at which such breakdown occurs on freeways and multilane highways is termed the *critical speed,* or *critical density,* or *point of capacity.* Figure 5–1 illustrates the relationships among flow, speed, and density.

Interrupted flow. Interrupted flow occurs on streets and highways with traffic signals, or stop or yield signs. For signalized intersections, traffic moves during green intervals, and stops during red. The proportion of the total cycle time available for green is termed the g/C ratio. If an approach to a traffic signal receives a 30-sec green interval during each 75-sec cycle, the g/C ratio is 30/75, or 0.40. This means that during 1 hour, the traffic movement in question can move 40% of the time.

Saturation flow rate at signalized intersections refers to the number of vehicles or passenger car units that can depart from the stop line in one lane if an entire hour of green were continuously available. Typically, for ideal conditions, the saturation flow rate is approximately 1,800 passenger cars per hour of green per lane.

Each time that the traffic signal goes red and then green, some of the efficiency of a continuously moving traffic stream is lost. At the beginning of the green interval, the first several vehicles moving away from the stop line do so more slowly than do following vehicles. The excess time that these first vehicles take is called "start-up lost time." At the end of each green interval, there is a change interval comprised of yellow, and sometimes an all-red interval. During at least a portion of this change interval, it is assumed that no vehicles will be entering the intersection. This period is considered as "clearance lost time." In practice, it has been observed that the amount of start-up lost time is from 2 to 3 sec and is incurred principally by the first three vehicles in each lane. At the end of the green interval, as the signal changes to yellow, it has been observed that drivers typically

Figure 5–1. Speed, density, and flow relationships.
SOURCE: *Highway Capacity Manual,* 1985, Special Report 209 (Washington, DC: Transportation Research Board), p. 1–7.

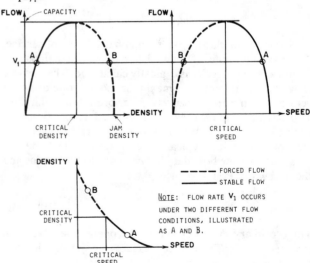

use approximately 2 sec of the yellow interval to continue entering the intersection.

In calculations of capacity and level of service for signalized intersections, the g/C ratio is taken to be the "effective" green time divided by the total cycle length of the signal. As a good approximation, the "effective" green time can be assumed to be equal to the actual green time. This is due to the "loss" at the beginning of a green interval being approximately equal to the "gain" at the end of the green when vehicles use the early portion of the yellow.

The key measure of level of service for signalized intersections is *delay*. The delay measure that is calculated is "stopped time delay," which is the time an individual vehicle spends stopped in a queue waiting to enter the intersection. The actual measure of effectiveness is based on the total stopped delay experienced by all vehicles in a lane, on an approach, or for all approaches combined divided by the total volume of vehicles using the lane, approach, or all approaches during a given time period. The measure is expressed as stopped delay per vehicle, in seconds. In applying the term *delay* and in using specific levels of delay as measures of effectiveness, the engineer is cautioned that the term is used in different ways for different technical applications. Great care must be taken when comparing "delay" values generated by various simulation models, air quality models, reported field measurements, and capacity and level-of-service techniques.

The other principal delay type can be termed "total delay" or "travel time delay." This delay is incurred by a vehicle approaching a signalized intersection, decelerating to a stop, moving slowly in a stopped queue, and eventually accelerating until it passes the stop line and enters the intersection. Travel time delay is defined as the difference between the time to pass straight through the intersection with no deceleration or stopping and the time to come to a stop and then accelerate and move into the intersection. Travel time delay on approaches to signals is typically 1.3 times as great as stopped delay.

Factors affecting capacity and level of service

The engineer involved with the operations of highways and transportation facilities often is able to effect changes in capacity and level of service by addressing those conditions that have an impact on capacity. The concepts described for capacity and level of service are dependent on a set of base "ideal conditions," which are then modified by actual operating conditions of a given facility. Some of these operating conditions can, in turn, be improved by the engineer.

Ideal conditions. The 1985 *Highway Capacity Manual* describes, for each facility type, the term "ideal conditions." Generally, when analyzing a given capacity problem, the analysis itself begins with the ideal conditions and then passes through a series of adjustments based on roadway, traffic, and control conditions.

For facilities with uninterrupted flow, ideal conditions include:

o Lane widths of 12 ft
o Lateral clearances of 6 ft or more

o All passenger cars in traffic stream
o Weekday or commuter drivers using facility
o Level terrain

For signalized intersections, ideal conditions include:

o Lane widths of 12 ft
o Level grade
o No curb parking
o All passenger cars in traffic stream
o No local transit buses stopping
o No turns being made from through lanes
o Intersection located away from CBD
o Green signal available 100% of time

Roadway conditions. Roadway factors include geometric conditions and design elements. In some cases, these factors are considered to have impacts on the capacity of a road, while in others, the factors may affect a measure of effectiveness such as speed, while not affecting the capacity or maximum flow rate that can be carried by the facility.

The width of lanes is considered to be an important factor, with wider lanes being able to carry more traffic or to carry traffic at a higher travel speed. The lateral clearance along the right edge of a roadway and along the median has the same effect, with greater clearances providing higher capacity or higher speed. The horizontal and vertical alignment of a roadway also can affect capacity. For uninterrupted flow facilities, the term "level terrain" implies that little or no impact occurs and that short grades of from 1% to 2% exist. Rolling terrain includes grades up to 4% and has a negative effect on speeds and capacity. Finally, mountainous terrain has steep grades or moderate grades over long distances, and the effect on speed and capacity is large, especially for heavy vehicles using the highway.

Traffic conditions. Whenever vehicles other than passenger cars (which include small trucks and vans) exist in the traffic stream, there is an effect on the number of vehicles that can be served. The conversion of each class of heavy vehicle to an equivalent number of passenger cars allows the analysis to be made on a consistent basis. There are three general categories of heavy vehicles: trucks, buses, and recreational vehicles (RVs). Other vehicle types can also be converted to an equivalent number of passenger cars, including motorcycles, and bicycles. Although each chapter in the 1985 *Highway Capacity Manual* uses "equivalency factors" that are specific to that facility type, the following lists provides a general set of average values.

Vehicle Type	Condition	Equivalent Passenger Cars
Truck (3 or more axles)	Level Terrain	1.5 to 2.0
Truck (3 or more axles)	Rolling Terrain	4.0
RV	Level Terrain	1.5
Bus	Level Terrain	1.5
Motorcycle	All	0.5
Bicycle	All	0.2

Lane distribution is a "traffic condition" on two-lane rural highways that has a major impact on the total capacities of such facilities. For example, with balanced flow under ideal conditions, a two-lane highway can carry 2,800 pcph. With a directional split of 70/30, the same highway would carry, at capacity, only 2,500 pcph. For signalized intersections, the distribution of volume by lane has an impact on both capacity and delay.

Control conditions. Traffic control has a major impact on signalized intersections, on unsignalized intersections, and on freeway ramps where ramp metering is used. Also, posted speed limits have been shown in recent research to have an impact on travel speeds independent of other factors that also impact capacity and level of service. Traffic regulations for curb parking, pedestrian movements, and restrictions on vehicle types using a given facility also impact capacity and level of service.

As an indication of the specific adjustments that are made to the base values under ideal conditions, Table 5–5 lists factors contained in the 1985 *Highway Capacity Manual,* by type of facility.

Factors beyond current procedures. The preceding sections have identified characteristics that have an effect on capacity and level of service. These variables are used in computations, and they are also variables that the engineer can address while developing operational improvements for transportation facilities. Other factors may or may not have an impact on capacity and level of service but, in fact, are not included in the analysis procedures. Such factors have either not been explicitly studied in earlier research, are considered to have a minimal impact on capacity, or are considered to be too complex or too expensive to study in the field. These factors are described in the following sections.

The principal roadway factor not included in the procedures of the 1985 *Highway Capacity Manual* is the state of repair and overall condition of the roadway surface. It is

TABLE 5–5

Adjustment Factors Used for Analyses

Facility	Factors		
	Roadway	Traffic	Control
Freeways—basic sections	Design speed Lane width Lateral clearance Grade Number of lanes	PHF Trucks Buses RVs Driver type	
Freeways—weaving	Same as basic sections, plus Configuration Length Number of lanes	Volume Ratio	
Freeways—ramp junctions	Adjacent ramp configuration Number of lanes	PHF Trucks Buses RVs	Metering rate
Freeways—ramp roadways	Design speed Lane width Number of ramp lanes	Heavy vehicles	
Two-lane highways	Design speed % no passing Lane width Shoulder width Grade	Directional Split PHF Trucks Buses RVs	
Multilane highways (1985 HCM)	Same as freeways, basic Sections, plus Development environment		
Signalized intersections	Lane width Area type Grade Number of lanes Type of lanes	PHF Heavy vehicles Right turns Left turns Pedestrian activity	Parking Bus stops Phasing Green time Cycle length Signal Progression
Unsignalized intersections	Grade Number of lanes Type of lanes Curb radius Sight distance Area population	PHF Heavy vehicles	STOP control YIELD control
Urban arterials	Function category Design category plus factors for signalized intersections	Free flow speed	
Transit	(LOS within vehicle depends on space per passenger)		
Pedestrians—walkways	Effective width	Peaking	
Bicycles—bike lanes	Number of lanes		

generally believed that poor surfaces containing irregularities or pot holes may reduce the capacity of uninterrupted flow facilities and may also reduce the saturation flow rate at signalized intersections. However, this factor does not enter into calculations described in this chapter.

The major traffic-related factor that is included in some chapters of the 1985 *Manual,* but not in others, is that of the type of driver using a given facility. This factor is included in freeway analyses, in multilane highway analyses (but current 1990 recommendations are that it be excluded when a new chapter is adopted), and in two-lane highway analyses. The factor is not included in any of the intersection-related analyses. The general concept is that persons who regularly use a facility may drive with shorter headways and thus higher capacities can be achieved. There is almost no documented evidence that this actually occurs, and it is somewhat controversial for use as an adjustment factor. It is currently used as a reduction of capacity, for areas of high tourism or areas where little commuter or repeat traffic occurs.

Factors relating to the ambient environment and to traffic control that are not explicitly included in capacity analyses are wet pavement or snow and ice conditions, darkness, fog, and the level of enforcement of speed limits or parking regulations. Because none of these factors is included in the calculations, it is assumed that all analyses are being performed for clear weather conditions during the daytime under reasonable levels of enforcement of traffic laws. There are indications that wet or icy pavements can reduce capacity by anywhere from 5% to 15%[5] The engineer should be aware of these factors and must use judgment if a design or an operational feature needs to be analyzed under any of these "nonstandard" environmental or enforcement conditions.

Application of procedures

When applying capacity and level-of-service analysis procedures, the level of detail required will determine the procedure to be used.

Levels of analysis. The 1985 *Highway Capacity Manual* lists three levels of analysis for each facility type. *Operational analysis* is the most detailed application and is typically used by traffic engineers to assess current conditions and to devise improvement programs that will increase capacity and improve level of service. *Design analysis* is similar to operational analysis in its level of detail but it tends to be used primarily to determine the sizing or number of lanes for a facility with a given set of traffic and roadway conditions and an objective level of service. Finally, *planning analysis* comprises very little detail and is used to obtain a preliminary estimate of facility sizing, usually for conditions 10 to 20 years in the future.

When studying a given highway or transportation facility, the engineer needs to select the level of analysis that will provide the appropriate level of accuracy and detail. There is a tendency to select the simplest of procedures (i.e., planning) even when data are available to perform a more accurate operational or design analysis. The analyst should be aware that all of the roadway, traffic, and control factors used in the more detailed analyses are implicitly included in a planning analysis but that the analyst has no control over the values. With the widespread use of both micro- and minicomputers, the more detailed procedures can be easily applied.

Level-of-service objectives. One of the cornerstones of all capacity and level-of-service work is the establishment of a "design level of service." Unless the analyst is simply making a determination of current or future level of service under a set of conditions, the establishment of a design level of service is one of the primary actions required of the jurisdiction or agency involved. Generally, political bodies and even technical agencies have only a modest level of understanding of the differences in traffic operation for different levels of service. One of the most useful techniques to overcome this lack of understanding is for the engineer to develop a series of photographs or videotapes that show the highway, the intersection, or the pedestrian facility operating under different levels of service. An early attempt to illustrate levels of service using videotape was made by the Maryland National Capital Park and Planning Commission.[6] The video presentation entitled "Capacity and Level of Service" has been used to show public officials, the public at large, and engineers and planners how traffic operations vary between Levels of Service A through F.

Table 5–6 provides a guideline on typical design levels of service that may be used by public agencies in the early 1990s. Although no standard national guideline has been set, it is apparent that in the past 10 to 15 years most jurisdictions have relaxed their guidelines (that is, have lowered the design level of service) in order to conserve scarce dollar resources. General observations of current practice indicate that smaller urban areas tend to demand slightly better levels of service than do large metropolitan areas. A study made for a southwestern city of 120,000 population showed that its network of arterial and collector streets would cost approximately 6% to 8% more if a peak-hour Level of Service C standard were used rather than Level of Service D. As a rule of thumb, in urban areas, the increase in cost to provide one level of service better might be on the order of 5% to 10%, exclusive of right-of-way costs.

TABLE 5–6
Suggested Design Levels of Service (LOS)[a]

Roadway Type	Small Areas[b]	Large Areas
Freeways	C	D
Rural Highways	B	C
Multilane Highways	C	D
Signalized Intersections	C	D[c]
Unsignalized Intersections	B	C
Urban Arterials	C	D[c]

[a]Levels shown are "acceptable." "Desirable" LOS can be considered as one level better than "acceptable." Guidelines suggested by AASHTO for "appropriate" design levels of service may differ from this table.
[b]"Small" areas are those with under 50,000 population.
[c]In very large cities, LOS E is sometimes considered "acceptable" for short periods during the peak period.

---

[5] *Highway Capacity Manual,* 1985.

[6] *Understanding Traffic Congestion,* Maryland-National Capital Park and Planning Commission, 1987.

A consideration in establishing a design level of service for an individual facility or for a network of roads is the time period considered as the "peak period." As indicated in Table 5–6, some agencies are making concessions to the peak period within the peak hour in that the level of service is allowed to creep into Level E for short periods of time. Most agencies currently use the peak 15-min within the peak hour as the design period. The use of hourly flow rate for the analysis requires conversion of the peak 15-min volume to an hourly rate (by multiplying by 4). If the short-term fluctuations and flow during a peak hour are simply allowed to occur and the analyst wishes only to consider the level of service averaged over the entire hour, the hourly volume is used directly. This implies that during the peak 15-min period, the level of service may in fact be worse than the design level of service.

Setting analysis volumes. Two basic factors are used by the engineer for capacity and level-of-service analysis. One is the *estimated* or *observed* capacity of the facility and the other is the *design* or *existing* volume. Depending on the availability of current or future volumes, the analyst must come to a point where the analysis volume is expressed as an hourly flow rate (except for pedestrian facility analysis, when 1-min and 15-min intervals are used).

Table 5–7 lists the typical sources of volumes that are the start point for establishing the volumes used within capacity and level-of-service computations. The engineer needs to be cognizant of the day or days of the week to be analyzed, the season of the year, and special events or unusual traffic conditions. Once a decision is made relative to what time period and what traffic conditions are going to be analyzed, an appropriate analysis volume can be computed.

TABLE 5–7
Volumes for Capacity Analysis

Source	To Convert to Peak-Hour Directional Volume (vph)	To Convert to Hourly Flow Rate
Average Daily Traffic (ADT)—2 way total	Multiply by K, D for peak hour[a]	Divide by PHF
Average Daily Traffic (ADT)—1 way	Multiply by K for peak hour	Divide by PHF
Peak Hour Volume (vph)—1 way	—	Divide by PHF
Peak 15 Minute Volume—1 way	—	Multiply by 4.0

[a] For two-lane highways, D is not applied. D is the directional split of traffic. K is the proportion of 24-hour traffic which occurs in the peak hour.

Use of field data

Throughout the 1985 *Highway Capacity Manual,* the analyst is encouraged to use locally derived values for key parameters. Because the 1985 *Manual* and other research studies are based on limited data bases that do not necessarily represent traffic and roadway conditions in all jurisdictions in North America, use of controlled and accurate local data is desirable. Some of the key parameters or data values for which the analyst might substitute locally derived values include:

- ○ Freeways—2,000 pcphpl as capacity for ideal conditions
- ○ Freeway weaving—Operating speeds for given levels of service
- ○ Freeway ramps—1,800 pcph as capacity of freeway-ramp junction
- ○ Multilane Highways—2,000 pcphpl as capacity for ideal conditions
- ○ Signalized Intersections—1,800 pcphgpl as saturation flow rate, ideal conditions
- ○ Signalized Intersections—1,400 vph as sum of critical volumes at capacity

For each of these factors, many agencies are finding that locally measured data are "less conservative." For example, sustained flows greater than 2,000 pcphpl for periods of 1 to 2 hours occur regularly on urban freeways in large metropolitan areas. Also, measured volumes at freeway ramp junctions often exceed 1,800 pcph. Saturation flow rates at signalized intersections, even under less than ideal conditions, can often exceed 1,800 pcphgpl. If the analyst has a sound local data base that indicates that any of these values or other key parameters used in the 1985 *Highway Capacity Manual* are not particularly applicable to local conditions, the local values should be used. Care must be taken to ensure that such local data are collected in a controlled manner and cover enough time periods and locations so that the data can be considered general in nature for the local community. One or two studies of 1 or 2 hours' duration should not be considered as yielding better or more applicable values than those found in the 1985 *Highway Capacity Manual.*

Calculations

Tools currently available to analyze capacity and level of service are of three general types: the 1985 *Highway Capacity Manual* comprising charts, tables, and analysis forms, computer programs for use on microcomputers, and nomographs for use in solving signalized intersection problems. Many engineers and analysts are now using microcomputer software as the principal tool for performing calculations. The entire 1985 *Highway Capacity Manual* has been placed into a software package that is available through the Federal Highway Administration (FHWA) and the McTrans Center at the University of Florida. Although this software package and other commercial packages provide effective means for calculating capacity, level of service, and associated measures of effectiveness, the engineer is cautioned to review results for reasonableness and consistency. Because the development of capacity software has generally been performed by persons other than those who developed the chapters of the 1985 *Highway Capacity Manual,* the translation from the manual to a software format has not always been completely accurate.

Nomograph solutions for signalized intersections are available through the Institute of Transportation Engineers.[7] Although nomographs can present a visual sense of concepts and calculations, they seldom are used because of the

---

[7] J. Leisch, *Capacity Analysis Techniques for Signalized Intersections,* (Washington, DC: Institute of Transportation Engineers, 1986).

greater convenience of software packages. Finally, the traditional procedures based on charts and tables, and using hard copy "fill-in" type forms, are still important both for a thorough understanding of a procedure and for situations where access to software and a microcomputer is limited.

Precision and accuracy

Precision of the results from capacity and level-of-service procedures relates to the amount of variation in a measure of effectiveness explained by the equations and concepts. For example, Chapter 9 of the 1985 *Highway Capacity Manual*, "Signalized Intersections," is based on research that had coefficient of determinations (r^2) in the range of 0.5 to 0.6. This means that the methods of Chapter 9 can explain only about 60% of the variation in field observed vehicle delay. The remaining 40% is due to factors not included in the method, to experimental error, and to unexplained variation in traffic conditions attributable perhaps to driver actions. Although operations on uninterrupted flow facilities may be somewhat easier to predict, no more than 60% to 80% of the variation in speed or delay as measured in the field can usually be modeled and explained by calculation procedures.

Because each of the factors used in capacity and level-of-service analyses is subject to a "plus or minus" accuracy, the results of such calculations should not be presented in a way that portrays them as absolute and precise values. For example, if a delay estimate at a signalized intersection is calculated at 25.4 secs, presentation of the results should be rounded to the nearest second and, in fact, the "true delay" might be in a range of 25 secs plus or minus 10%. In practice, the values calculated for measures of effectiveness might be assumed to have an accuracy range of plus or minus 5% to 10%. When future traffic volumes are used in the analysis, comment on the source of these volumes and the probability of the volumes actually occurring should be made when reporting the results.

Freeways

Freeways have become a critical element in both urban and rural highway networks. Flow on freeways is considered as "uninterrupted," and the relationship among speed, flow, and density is the basis for capacity and level-of-service analysis. Freeway operations, particularly in large metropolitan areas, have become an important part of traffic engineering and traffic enforcement work. The engineer is expected to develop operational improvements such as improved ramp geometry or restriping to achieve additional lanes, and is also becoming involved in surveillance and control programs and in incident-response activities (see Chapters 12, 13). These efforts are aimed at improving capacity and level of service for typical day-to-day traffic conditions, and also at minimizing the number of, and impacts of, breakdown or incident situations.

The analysis of freeway capacity is generally done for each component subsection: basic segments, weaving areas, and ramp junctions. For on-ramps, the influence of the merge maneuver is considered to be from 500 ft upstream to 2,500 ft downstream of the junction. For off-ramps, the influence of the diverge maneuver is considered to be from 2,500 ft upstream to 500 ft downstream of the junction. For weaving areas, impacts on mainline traffic are considered to occur along the length of the weaving area plus 500 ft upstream and 500 ft downstream of the ends of the weaving area. When performing an operational analysis of a freeway section or corridor, a sketch plan of the component subsections aids in selecting the appropriate analysis procedure.

Basic relationships

The basic relationships of freeway flow are based on assumptions of good pavement conditions, no traffic incidents, good weather, and no queuing or backups from downstream bottlenecks. The key traffic flow relationship for basic freeway segments is:

$$SF_i = MSF_i \times N \times f_w \times f_{HV} \times f_p \qquad (5.3)$$

where:

SF_i = service flow rate for LOS_i under prevailing roadway and traffic conditions for N lanes in one direction, in vph

MSF_i = maximum service flow rate per lane for LOS_i under ideal conditions, in pcphpl

N = number of lanes in one direction of the freeway

f_w = factor to adjust for the effects of restricted lane widths and/or lateral clearances

f_{HV} = factor to adjust for the effect of heavy vehicles (trucks, buses, and recreational vehicles) in the traffic stream

f_p = factor to adjust for the effect of driver population

Equation (5.3) allows for the calculation of the maximum hourly flow rate for a given level of service, in vehicles per hour, that can occur along a basic freeway segment under prevailing traffic and roadway conditions. Calculations for ramp junctions and weaving areas are based on other relationships described in subsequent sections.

Service volumes: basic sections

Level-of-service standards for basic freeway sections are based on density of the mainline traffic. Table 5–8 lists the maximum density for each level of service, and the maximum service flow rate per lane under ideal conditions. Note that Table 5–8 shows values for 70 mph, 60 mph, and 50 mph design speeds. Newer freeways and those that are currently being rehabilitated or constructed typically have design speeds of at least 70 mph. In some cases, older freeways or those with significant topographic constraints have lower design speeds. Figure 5–2 shows the speed-flow relationships under ideal conditions for basic freeway segments. Note that 2,000 pcphpl is considered as "capacity" for most freeways. Research of the late 1980s on multilane highways[8] has provided indication that the slope of the speed flow

---

[8] *Capacity and Level-of-Service Procedures for Multilane Rural and Suburban Highways*, NCHRP Project No. 3-33.

TABLE 5-8

Levels of Service for Basic Freeway Sections

LOS	Density (PC/MI/LN)	70 mph Design Speed			60 mph Design Speed			50 mph Design Speed		
		Speed[b] (mph)	v/c	MSF[a] (PCPHPL)	Speed[b] (mph)	v/c	MSF[a] (PCPHPL)	Speed[b] (mph)	v/c	MSF[a] (PCPHPL)
A	≤12	≥60	0.35	700	—	—	—	—	—	—
B	≤20	≥57	0.54	1,100	≥50	0.49	1,000	—	—	—
C	≤30	≥54	0.77	1,550	≥47	0.69	1,400	≥43	0.67	1,300
D	≤42	≥46	0.93	1,850	≥42	0.84	1,700	≥40	0.83	1,600
E	≤67	≥30	1.00	2,000	≥30	1.00	2,000	≥28	1.00	1,900
F	>67	<30	c	c	<30	c	c	<28	c	c

[a]Maximum service flow rate per lane under ideal conditions.
[b]Average travel speed.
[c]Highly variable, unstable.
Note: All values of MSF rounded to the nearest 50 pcph.

SOURCE: *Highway Capacity Manual,* 1985, Special Report 209, (Washington, DC: Transportation Research Board), p. 3–8.

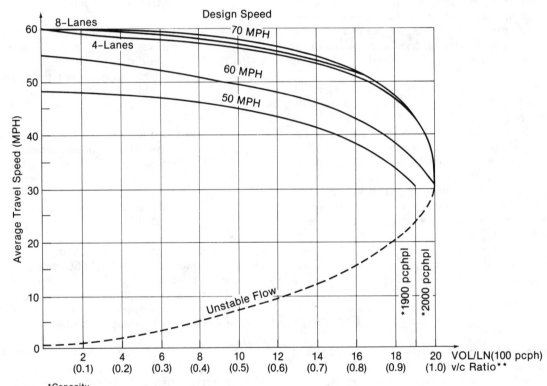

Figure 5–2. Speed flow relationships under ideal conditions (freeways).
SOURCE: Adapted from *Highway Capacity Manual,* 1985, Special Report 209 (Washington, DC: Transportation Research Board), p. 3–5.

curve may be much flatter than that shown in Figure 5–2. Until new research data taken on freeways are available, Figure 5–2 remains as the basic curve describing freeway traffic flow.

Figure 5–3 illustrates freeway conditions under Levels of Service A through F. Level of Service E is "capacity," with flow conditions producing maximum volumes.

Adjustment factors: basic sections

Three adjustment factors are used to reduce the maximum flow rate that can be achieved under conditions for a specific level of service. These factors relate to lane width and lateral clearance, heavy vehicles, and driver population. Each of the factors is represented in Equation (5.3).

Illustration 3-5. Level-of-service A.

Illustration 3-8. Level-of-service D.

Illustration 3-6. Level-of-service B.

Illustration 3-9. Level-of-service E.

Illustration 3-7. Level-of-service C.

Illustration 3-10. Level-of-service F.

Figure 5–3. Freeway conditions under LOS A through F.
SOURCE: *Highway Capacity Manual,* 1985, Special Report 209 (Washington, DC: Transportation Research Board), p. 3–9.

TABLE 5-9

Adjustment Factors for Restricted Lane Width/Lateral Clearance (Freeways)

Distance from Traveled Pavement[a] (ft)	Adjustment Factor f_w							
	Obstructions on One Side of the Roadway				Obstructions on Both Sides of the Roadway			
	Lane Width (ft)				Lane Width (ft)			
	12	11	10	9	12	11	10	9
	4-Lane Freeway (2 Lanes Each Direction)							
≥6	1.00	0.97	0.91	0.81	1.00	0.97	0.91	0.81
5	0.99	0.96	0.90	0.80	0.99	0.96	0.90	0.80
4	0.99	0.96	0.90	0.80	0.98	0.95	0.89	0.79
3	0.98	0.95	0.89	0.79	0.96	0.93	0.87	0.77
2	0.97	0.94	0.88	0.79	0.94	0.91	0.86	0.76
1	0.93	0.90	0.85	0.76	0.87	0.85	0.80	0.71
0	0.90	0.87	0.82	0.73	0.81	0.79	0.74	0.66
	6- or 8-Lane Freeway (3 or 4 Lanes Each Direction)							
≥6	1.00	0.96	0.89	0.78	1.00	0.96	0.89	0.78
5	0.99	0.95	0.88	0.77	0.99	0.95	0.88	0.77
4	0.99	0.95	0.88	0.77	0.98	0.94	0.87	0.77
3	0.98	0.94	0.87	0.76	0.97	0.93	0.86	0.76
2	0.97	0.93	0.87	0.76	0.96	0.92	0.85	0.75
1	0.95	0.92	0.86	0.75	0.93	0.89	0.83	0.72
0	0.94	0.91	0.85	0.74	0.91	0.87	0.81	0.70

[a]Certain types of obstructions, high-type median barriers in particular, do not cause any deleterious effect on traffic flow. Judgment should be exercised in applying these factors.

SOURCE: *Highway Capacity Manual,* 1985, Special Report 209 (Washington, DC: Transportation Research Board), p. 3-13.

For any freeway segment having lane widths less than 12 ft or objects closer than 6 ft from the edge of the travel lanes, the factor f_w is used in the calculation. Table 5-9 gives the adjustment factors for a range of conditions. Lane width is considered as the average width of the mainline lanes being analyzed. Also, when there are lateral obstructions along the right side of the travel lanes and along the median, the average width is used to enter the table-of-adjustment factors, entering on the right side of the table. Note that reduction in freeway lane width can reduce capacity by up to 22% for 9-ft lanes. Likewise, for conditions where there is no lateral clearance along the edges of the travel lanes, capacity can be reduced up to 19%, even when the freeway lanes are 12 ft wide. Results in the early 1990s from an FHWA research project[9] are expected to provide revised factors.

The adjustment factor for heavy vehicles can be applied in one of two ways. The freeway segment being analyzed can be considered as being a "general freeway segment," in which there are no significant specific grades. Such general segments are categorized as level terrain (short grades of up to 2%), rolling terrain (heavy vehicles reduce their speeds but do not operate at crawl speed), and mountainous terrain (heavy vehicles operate at crawl speed for significant distances). Table 5-10 contains the passenger car equivalents for each terrain type and for each type of heavy vehicle. The general

TABLE 5-10

Passenger Car Equivalents on General Freeway Segments

Factor	Type of Terrain		
	Level	Rolling	Mountainous
E_T for Trucks	1.7	4.0	8.0
E_B for Buses	1.5	3.0	5.0
E_R for RV's	1.6	3.0	4.0

SOURCE: *Highway Capacity Manual,* 1985, Special Report 209 (Washington, DC: Transportation Research Board), p. 3-13.

interpretation of this table is that for each heavy vehicle, an equivalent number of passenger cars would be required to have the same impact on traffic operations.

If a specific grade is to be analyzed, the equivalency tables are more detailed. Also, the 1985 *Highway Capacity Manual* provides separate equivalency tables for different types of heavy trucks. The most common application is for the analyst to assume that typical truck populations have weight/horsepower ratios of approximately 200 lb/hp. Another simplifying assumption that can be made combines all heavy vehicles into one category, followed by the use of the 200 lb/hp table for equivalencies. This assumption tends to overstate the equivalency of buses and RVs, but it is a rare case when these heavy vehicle types comprise any significant portion of all heavy vehicles. Table 5-11 is a composite of the equivalencies presented in the 1985 *Manual.* Factors for six- or eight-lane freeways are given for 200 lb/hp trucks and for RVs. Factors for buses are also listed. Equation (5.4) gives the

---

[9] *Capacity Adjustments for Lane Width and Lateral Clearance on Two-Lane, Two-Way Highways,* FHWA Research Project No. DTFH61-88-R-00063.

TABLE 5-11

Passenger-Car Equivalents for 6–8-Lane Freeways

Passenger-Car Equivalents for Typical Trucks (200 lb/hp)

Passenger-Car Equivalent, E_T

Grade (%) Percent Trucks	Length (mi)	4-Lane Freeways								6-8-Lane Freeways							
		2	4	5	6	8	10	15	20	2	4	5	6	8	10	15	20
<1	All	2	2	2	2	2	2	2	2	2	2	2	2	2	2	2	2
1	0-1/2	2	2	2	2	2	2	2	2	2	2	2	2	2	2	2	2
	1/2-1	3	3	3	3	3	3	3	3	3	3	3	3	3	3	3	3
	≥1	4	3	3	3	3	3	3	3	4	3	3	3	3	3	3	3
2	0-1/4	4	4	4	3	3	3	3	3	4	4	4	3	3	3	3	3
	1/4-1/2	5	4	4	3	3	3	3	3	5	4	4	3	3	3	3	3
	1/2-3/4	6	5	5	4	4	4	4	4	6	5	5	4	4	4	4	4
	3/4-1 1/2	7	6	6	5	4	4	4	4	7	5	5	5	4	4	4	4
	≥1 1/2	8	6	6	6	5	5	4	4	8	6	6	5	4	4	4	4
3	0-1/4	6	5	5	5	4	4	4	3	6	5	5	5	4	4	4	3
	1/4-1/2	8	6	6	6	5	5	5	4	7	6	6	6	5	5	5	4
	1/2-1	9	7	7	6	5	5	5	5	9	7	7	6	5	5	5	5
	1-1 1/2	9	7	7	7	6	6	5	5	9	7	7	6	5	5	5	5
	≥1 1/2	10	7	7	7	6	6	5	5	10	7	7	6	5	5	5	5
4	0-1/4	7	6	6	5	4	4	4	4	7	6	6	5	4	4	4	4
	1/4-1/2	10	7	7	6	5	5	5	5	9	7	7	6	5	5	5	5
	1/2-1	12	8	8	7	6	6	6	6	10	8	7	6	5	5	5	5
	≥1	13	9	9	9	8	8	7	7	11	9	9	8	7	6	6	6
5	0-1/4	8	6	6	6	5	5	5	5	8	6	6	6	5	5	5	5
	1/4-1/2	10	8	8	7	6	6	6	6	8	7	7	6	5	5	5	5
	1/2-1	12	11	11	10	8	8	8	8	12	10	9	8	7	7	7	7
	≥1	14	11	11	10	8	8	8	8	12	10	9	8	7	7	7	7
6	0-1/4	9	7	7	7	6	6	6	6	9	7	7	6	5	5	5	5
	1/4-1/2	13	9	9	8	7	7	7	7	11	8	8	7	6	6	6	6
	1/2-3/4	13	9	9	8	7	7	7	7	11	9	9	8	7	6	6	6
	≥3/4	17	12	12	11	9	9	9	9	13	10	10	9	8	8	8	8

Note: If a length of grade falls on a boundary condition, the equivalent for the longer grade category is used. For any grade steeper than the percentage shown, use the next higher grade category.

Passenger-Car Equivalents for Recreational Vehicles

Passenger-Car Equivalent, E_R

Grade (%) Percent RV's	Length (mi)	4-Lane Freeways								6–8-Lane Freeways							
		2	4	5	6	8	10	15	20	2	4	5	6	8	10	15	20
<2	All	2	2	2	2	2	2	2	2	2	2	2	2	2	2	2	2
3	0-1/2	3	2	2	2	2	2	2	2	2	2	2	2	2	2	2	2
	≥1/2	4	3	3	3	3	3	3	3	4	3	3	3	3	3	3	3
4	0-1/4	3	2	2	2	2	2	2	2	3	2	2	2	2	2	2	2
	1/4-3/4	4	3	3	3	3	3	3	3	4	3	3	3	3	3	3	3
	>3/4	5	4	4	4	3	3	3	3	4	4	4	4	3	3	3	3
5	0-1/4	4	3	3	3	3	3	3	3	4	3	3	3	2	2	2	2
	1/4-3/4	5	4	4	4	4	4	4	4	5	4	4	4	4	4	4	4
	≥3/4	6	5	4	4	4	4	4	4	5	5	4	4	4	4	4	4
6	0-1/4	5	4	4	4	3	3	3	3	5	4	4	3	3	3	3	3
	1/4-3/4	6	5	5	4	4	4	4	4	6	4	4	4	4	4	4	4
	≥3/4	7	6	6	6	5	5	5	5	6	5	5	5	4	4	4	4

Note: If a length of grade falls on a boundary condition, the equivalent from the longer grade category is used. For any grade steeper than the percent shown, use the next higher grade category.

Passenger-Car Equivalents for Buses

Grade (%)	Passenger-Car Equivalent, E_B
0-3	1.6
4[a]	1.6
5[a]	3.0
6[a]	5.5

SOURCE: *Highway Capacity Manual*, 1985, Special Report 209 (Washington, DC: Transportation Research Board), pp. 3-14, 3-16.

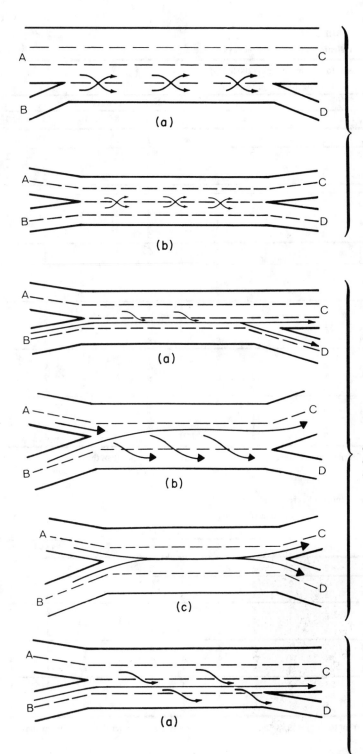

Type A weaving areas: (a) ramp-weave/one-sided weave, and (b) major weave with crown line.

Type B weaving areas: (a) major weave with lane balance at exit gore, (b) major weave with merging at entrance gore, and (c) major weave with merging at entrance gore and lane balance at exit gore.

Type C weaving areas: (a) major weave without lane balance or merging, and (b) two-sided weave.

Figure 5–4. Weaving area configurations.
SOURCE: *Highway Capactiy Manual,* 1985, Special Report 209 (Washington, DC: Transportation Research Board), pp. 4–3, 4–4.

calculation for combining the equivalency factors from the table into a single factor, f_{HV}, for use in Equation (5.3).

$$f_{HV} = 1/[1 + P_T(E_T - 1) + P_R(E_R - 1) + P_B(E_B - 1)] \quad (5.4)$$

where:

f_{HV} = the adjustment factor for the combined effect of trucks, recreational vehicles, and buses on the traffic stream

E_T, E_R, E_B = the passenger-car equivalents for trucks, recreational vehicles, and buses, respectively

P_T, P_R, P_B = the proportion of trucks, recreational vehicles, and buses, respectively, in the traffic stream

For downgrade conditions, equivalencies for level terrain are usually applied. For more severe downgrades (4% or greater and longer than 3,000 ft), the heavy vehicle equivalency is often taken as one-half of the corresponding upgrade equivalent. For composite grades the total rise in elevation is divided by the total length of section to develop an average grade. This average grade value is then used in the equivalency tables.

The third adjustment factor to adjust maximum service flow rates for freeway segments relates to driver population. The data upon which this factor is based are sparse, and the analyst should use this factor only when there is a strong indication that unfamiliar, weekend, or recreational drivers predominate in the traffic stream and drive in such a way as to reduce capacity of the freeway segment. The adjustment factors used are 1.0 for weekday or commuter type drivers and 0.75 to 0.90 for "unfamiliar" drivers. Because most analyses are focused on peak period conditions during weekdays when traffic is the heaviest, it is recommended that this factor not be used except for very special locations operating primarily with "unfamiliar drivers."

Calculations: basic freeway segments

A typical calculation would comprise the study of a freeway section with substandard conditions. If a six-lane freeway (three in each direction) has 11-ft wide lanes and obstructions along the median and along the edge 4 ft from the travel lanes, the adjustment factor f_w is 0.94. Assuming that this section lies in rolling terrain and has 6% trucks, 6% RVs, and no buses, the heavy vehicle factor f_{HV} is 0.77. If it is assumed that the predominance of drivers are commuters, the driver population factor f_p is 1.0. Given that this freeway section has a 70 mph design speed and three lanes in one direction, the service flow rate that can be accommodated at capacity is 4,340 vph. If the current hourly volume is 3,600 vph, current level of service would be estimated as LOS D (volume-capacity ratio, v/c, of 0.83 lies between 0.78 and 0.93, from Table 5–8).

Weaving areas

A weaving area on a freeway is defined as the crossing of two or more traffic streams traveling in the same general direction. As defined in the 1985 *Highway Capacity Manual* and shown in Figure 5–4, there are three types (A, B, and C) of freeway weaving areas. The length of a weaving section is considered to be measured from the merge area where the right edge of the freeway lane and the left edge of the merge lane are 2 ft apart to a point at the diverge where the two edges are 12 ft apart. The method for assessing weaving level of service is based on the calculation of the speed of weaving vehicles and the speed of nonweaving vehicles. The speed is considered as the average running speed of vehicles along the length of the weaving section. The two equations used to estimate speed have a common format, as follows:

$$S_w \text{ or } S_{nw} = 15 + \frac{50}{1 + a(1 + VR)^b (v/N)^c / L^d} \quad (5.5)$$

where:

a, b, c, d = constants

S_w = average running speed of weaving vehicles, in mph

S_{nw} = average running speed of nonweaving vehicles, in mph

VR = volume ratio

v = total flow rate in the weaving area, in pcph

N = total number of lanes in the weaving area

L = length of the weaving area, in ft

The constants used in Equation (5.5) are given in Table 5–12. Equation (5.5) is used with the "unconstrained" constants to achieve a preliminary result. Then, a check is made to determine if the unconstrained assumption was correct. This check is performed by using Table 5–13 and calculating the number of lanes required for unconstrained operation. If it is found that the conditions are constrained (i.e., that not enough space is being allocated to the weaving area) the analyst again uses Equation (5.5), but with the constrained constants. A final calculation is then made to estimate weaving and nonweaving speeds.

The level-of-service criteria for weaving sections on freeways are given in Table 5–14. Because the research and the data base for weaving sections were based entirely on

TABLE 5–12
Constants for Prediction of Weaving and Nonweaving Speeds

General Form: S_w or $S_{nw} = 15 + \dfrac{50}{1 + a(1 + VR)^b (v/N)^c / L^d}$

Type of Configuration	Constants for Weaving Speed, S_w				Constants for Nonweaving Speed, S_{nw}			
	a	b	c	d	a	b	c	d
Type A								
Unconstrained	0.226	2.2	1.00	0.90	0.020	4.0	1.30	1.00
Constrained	0.280	2.2	1.00	0.90	0.020	4.0	0.88	0.60
Type B								
Unconstrained	0.100	1.2	0.77	0.50	0.020	2.0	1.42	0.95
Constrained	0.160	1.2	0.77	0.50	0.015	2.0	1.30	0.90
Type C								
Unconstrained	0.100	1.8	0.80	0.50	0.015	1.8	1.10	0.50
Constrained	0.100	2.0	0.85	0.50	0.013	1.6	1.00	0.50

SOURCE: *Highway Capacity Manual*, 1985, Special Report 209 (Washington, DC: Transportation Research Board), p. 4–6.

TABLE 5–13

Criteria for Unconstrained vs. Constrained Operation

Type of Configuration	No. of Lanes Required for Unconstrained Operation, N_w	Max. No. of Weaving Lanes, N_w (max)
Type A	$2.19\, N\, VR^{0.571}\, L_H^{0.234}/S_w^{0.438}$	1.4
Type B	$N\{0.085 + 0.703\, VR + (234.8/L) - 0.018$ $(S_{nw} - S_w)\}$	3.5
Type C	$N\{0.761 - 0.011\, L_H - 0.005(S_{nw} - S_w) +$ $0.047\, VR\}$	3.0[a]

[a]For 2-sided weaving areas, *all* freeway lanes may be used as weaving lanes.
Note: When $N_w \le N_w$ (max), operation is unconstrained.
When $N_w > N_w$ (max), operation is constrained.

SOURCE: *Highway Capacity Manual,* 1985, Special Report 209 (Washington, DC: Transportation Research Board), p. 4–7.

TABLE 5–14

Level-of-Service Criteria for Weaving Sections

Level of Service	Minimum Average Weaving Speed, S_w (MPH)	Minimum Average Nonweaving Speed, S_{nw} (MPH)
A	55	60
B	50	54
C	45	48
D	40	42
E	35/30[a]	35/30[a]
F	<35/30[a]	<35/30[a]

[a]The 30-mph boundary is used for comparison to field-measured speeds.

SOURCE: *Highway Capacity Manual,* 1985, Special Report 209 (Washington, DC: Transportation Research Board), p. 4–9.

freeway facilities, applications of the equations and the level-of-service criteria to nonfreeway weaving conditions should be made with caution.

Calculations for freeway weaving areas are either "operational," in which speeds and levels of service are estimated, or "design," in which the length or number of lanes required in the weaving area are being determined. Equation (5.5) allows the analyst to do either. An example of an operational analysis is the determination of operating level of service for a weaving section 1,000 ft long with three mainline freeway lanes plus one auxiliary lane connecting an on-ramp with an off-ramp. The volumes are converted to hourly flow rates through use of a peak-hour factor and the application of the heavy vehicle, lane width and lateral obstruction, and driver population adjustment factors used for basic freeway segments. Assuming that these adjustments have been made, hourly flow rates, in pcph, are shown in Figure 5–5.

The critical volumes and ratios used in the speed equations are:

$$v_w = 600 + 300 = 900 \text{ pcph}$$
$$v = 900 + 4{,}000 + 100 = 5{,}000 \text{ pcph}$$
$$VR = 900/5000 = 0.18$$
$$R = 300/900 = 0.33$$

For the conditions of Figure 5–5, weaving speed is calculated as 42.6 mph and nonweaving speed as 50.4 mph. A check is made to see if the conditions imply unconstrained operations, and it is found that they do. Therefore, no recalculation

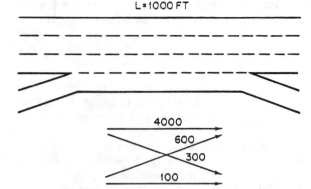

Figure 5–5. Weaving diagram.
SOURCE: *Highway Capacity Manual,* 1985, Special Report 209 (Washington, DC: Transportation Research Board), p. 4–13.

is needed. From Table 5–14, the estimated level of service for these conditions is LOS D for weaving and LOS C for nonweaving. As an indication of the sensitivity of speed and level of service to a change in the weaving section length, if all conditions remain the same but L is increased from 1,000 to 2,000 ft, estimated operational speeds are 49.0 mph for weaving and 56.8 for nonweaving. The levels of service for the 2,000 ft long section would be LOS C and LOS B for weaving and nonweaving vehicles, respectively.

The 1985 *Highway Capacity Manual* gives a set of "limitations" on the use of weaving equations. For example, a weaving length greater than 2,500 ft is considered to be beyond the limits of the data base used to establish the procedures. Although these limitations provide a set of guidelines, they do not imply that the weaving equations cannot be used or are invalid for conditions that go beyond the limitation threshold.

Procedures for analyzing capacity and level of service for weaving areas do not always produce results consistent with field conditions. Hence, the analyst is encouraged to perform field observations to verify the consistency of calculated values.

Ramp junctions

The point at which vehicles enter a freeway mainline from an on-ramp or the point at which mainline traffic diverges to an off-ramp are termed *freeway-ramp junctions.* The procedure for analyzing freeway-ramp junctions is dependent on the sum of the hourly flow rates, in pcph, of the freeway lane and the ramp lane. Table 5–15 gives the level-of-service criteria for merge and diverge junctions. Two important parts of the analysis are the estimation of Lane 1 (right-most freeway lane) volume and the estimation of the number of trucks that use Lane 1 upstream of the merge or diverge point. The calculation procedure is based on the computation of Lane 1 volume, the conversion of all volumes to passenger cars per hour using the peak-hour factor and the adjustment factors for basic freeway sections, and the computation of merge volume, or diverge volume, and of total freeway volume. Procedures are also available for left-side

TABLE 5–15
Level-of-Service Criteria and Ramp Flow Rates

Level of Service	Merge Flow Rate (PCPH)[a] v_m	Diverge Flow Rate (PCPH)[b] v_d	Level-of-Service Criteria for Checkpoint Flow Rates at Ramp-Freeway Terminals Freeway Flow Rates(PCPH)[c], v_f								
			70-mph Design Speed			60-mph Design Speed			50-mph Design Speed		
			4-Lane	6-Lane	8-Lane	4-Lane	6-Lane	8-Lane	4-Lane	6-Lane	8-Lane
A	≤ 600	≤ 650	≤1,400	≤2,100	≤2,800	d	d	d	d	d	d
B	≤1,000	≤1,050	≤2,200	≤3,300	≤4,400	≤2,000	≤3,000	≤4,000	d	d	d
C	≤1,450	≤1,500	≤3,100	≤4,650	≤6,200	≤2,800	≤4,200	≤5,600	≤2,600	≤3,900	≤5,200
D	≤1,750	≤1,800	≤3,700	≤5,550	≤7,400	≤3,400	≤5,100	≤6,800	≤3,200	≤4,800	≤6,400
E	≤2,000	≤2,000	≤4,000	≤6,000	≤8,000	≤4,000	≤6,000	≤8,000	≤3,800	≤5,700	≤7,600
F			WIDELY VARIABLE								

[a]Lane-1 flow rate plus ramp flow rate for one-lane, right-side on-ramps.
[b]Lane-1 flow rate immediately upstream of off-ramp for one-lane, right-side ramps.
[c]Total freeway flow rate in one direction upstream of off-ramp and/or downstream of on-ramp.
[d]Level of service not attainable due to design speed restrictions.

Approximate Service Flow Rates for Single-Lane Ramps[a] (pcph)

LOS	Ramp Design Speed (mph)				
	≤ 20	21–30	31–40	41–50	≥ 51
A	b	b	b	b	600
B	b	b	b	900	900
C	b	b	1,100	1,250	1,300
D	b	1,200	1,350	1,550	1,600
E	1,250	1,450	1,600	1,650	1,700
F		WIDELY VARIABLE			

[a]For two-lane ramps, multiply the values in the table by: 1.7 for ≤ 20 mph
1.8 for 21–30 mph
1.9 for 31–40 mph
2.0 for ≥ 41 mph

[b]Level of service not attainable due to restricted design speed.

SOURCE: *Highway Capacity Manual,* 1985, Special Report 209 (Washington, DC: Transportation Research Board), pp. 5–6, 5–15.

ramps. An NCHRP research project,[10] with final results expected in the early 1990s, will lead to a new updated set of data and new procedures that will replace current procedures for ramp junctions.

Another aspect of capacity at freeway ramps is the service-flow rate that can be achieved on the ramp itself. The 1985 *Highway Capacity Manual* contains guidelines, as shown in Table 5–15.

The use of ramp metering is becoming a commonly used operational tool. Ramp metering is typically applied along a series of on-ramps to protect the integrity of traffic conditions on the freeway mainline. By allowing a controlled number of vehicles to enter the freeway mainline at each on-ramp, speed and flow can be maintained just above the breakdown point at which conditions rapidly deteriorate. Observations in California indicate that once the freeway begins to break down, the per lane capacity is rapidly reduced from 2,000 pcphpl or greater to 1,400 or 1,500 pcphpl. This capacity loss of 25% to 30% for each lane creates an immediate bottleneck section that is reflected upstream. The use of ramp metering along critical sections of urban freeways can mitigate this problem. Typical practice sets metering rates at one vehicle every 4 to 6 sec (600 to 900 vph) for single-lane entries. For

double-lane entries, some 900 to 1,200 vph can be allowed to enter through a metered point. Not only does ramp metering provide the highway agency with a means to control the total loading along a freeway section, but it also provides a means to "break up" platoons or queues of vehicles, which, when entering the freeway mainline, can create disruptions in the right-most lane.

Figure 5–6 indicates the effects of a ramp metering operation. Demand on the ramp is greater than the service rate of the vehicles for about a 1-hour period. The longest queue on the ramp is about 30 vehicles, which would create a queue of some 600 to 700 ft. The longest delay incurred by a vehicle waiting on the ramp is about 5 min.

An important operational condition on many freeways is the capacity that can be achieved during maintenance or reconstruction activities. Because many freeways in the United States were constructed in the 1950s and 1960s, the 20- to 30-year design life is creating the need for extensive work-zone activities during the 1990s. Generally, the greater the number of lanes that are closed, the lower the capacity of the remaining lanes. Capacities of about 1,500 vphpl (equivalent to about 1,700 pcphpl) can be achieved under the best of circumstances for short-term work-zone activities. For construction sites set up on a long-term basis using portable concrete barriers, capacities approaching the capacity of normal freeway lanes have been observed. Table 5–16 illustrates

---

[10] *Ramp and Ramp Junction Capacity and Level of Service,* NCHRP Project 3-37.

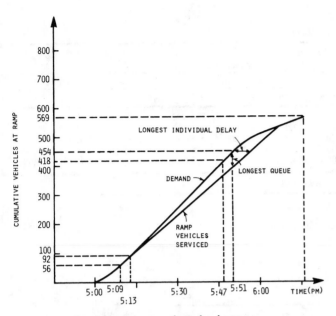

Plot of cumulative ramp demand and output.

Figure 5–6. Effects of ramp metering.
SOURCE: *Highway Capacity Manual,* 1985, Special Report 209 (Washington, DC: Transportation Research Board), p. 6–9.

observed capacities of short-term sites and long-term construction sites.

Another operational condition that is becoming more predominant on urban freeways is the use of high occupancy vehicle (HOV) lanes. HOV lanes are now being used for buses and for car pools. The general objective of such lanes is to allow some measure of priority movement for vehicles with greater numbers of persons. HOV lanes can be difficult to manage from an enforcement viewpoint and can also create difficult operations unless the HOV lane carries longer distance trips so that weaving across freeway lanes is minimized.

Operational improvements on freeways

Using the capacity and level-of-service procedures described in this section, the highway engineer can develop reasonable estimates of the impacts of various operational improvements. For example, conversion of the freeway

TABLE 5–16

Freeway Construction Zone Capacity

Measured Average Work-Zone Capacities

| Number of Lanes | | Number of Studies | Average Capacity | |
A Normal	B Open		(vph)	(vphpl)
3	1	7	1,170	1,170
2	1	8	1,340	1,340
5	2	8	2,740	1,370
4	2	4	2,960	1,480
3	2	9	2,980	1,490
4	3	4	4,560	1,520

Capacity of Long-Term Construction Sites with Portable Concrete Barriers

| Number of Lanes | | Number of Studies | Capacity Range (vphpl) | Average Capacity | |
Normal	Open			vph	vphpl
3	2	7	1,780–2,060	3,720	1,860
2	1	3	—	1,550	1,550

SOURCE: *Highway Capacity Manual,* 1985, Special Report 209 (Washington, DC: Transportation Research Board), pp. 6–11, 6–12.

shoulder to a moving traffic lane combined with a narrowing of all lanes will provide greater *total* capacity but will provide lower capacity *per lane* because of lane width and lateral clearance restrictions. Also, the lack of a shoulder for emergency situations and vehicle breakdowns may create significant safety and congestion problems, at least on an occasional basis. For weaving sections, the capacity procedures allow the engineer to determine the extent of improvement achieved through adding an extra lane or lengthening the weaving area.

Table 5–17 is a summary of design and operational improvements for freeways whose effects can be estimated through use of capacity and level-of-service procedures provided in the 1985 *Highway Capacity Manual.* Also, Table 5–17 includes an indication of the sensitivity of traffic flow conditions to the indicated improvement technique.

Multilane highways

Multilane highways exist in rural and suburban areas and cannot be classified as freeways because they typically lack full control of access and have some intersections at grade.

TABLE 5–17

Impacts of Freeway Operations Improvements

Action	Can Improve	Relative Improvement
1. Add basic lane	Capacity and LOS	25% to 50% (by adding one lane to four- or two-lane section, respectively)
2. Widen substandard lanes	Capacity and LOS	3% to 10% per foot of widening
3. Increase substandard lateral clearance	Capacity and LOS	1% to 7% per foot of increase
4. Prohibit heavy vehicles	Capacity and LOS	1% to 3% for each percent of HV in flow
5. Add truck climbing lane	Capacity and LOS	5% for each percent of HV in flow
6. Add lane to weaving section	LOS and Speed	+1 LOS, +5 mph for adding one lane
7. Lengthen weaving section	LOS and Speed	+1 LOS, +5 mph for doubling L
8. Convert left entry to right entry ramp	Capacity and LOS	Increase ramp capacity by 100 to 400 pcph
9. Add ramp metering	LOS	Reduces probability of freeway LOS F

Note: All "relative improvements" are for a given set of base conditions and will vary as conditions vary.

TABLE 5–18
Level-of-Service Criteria for Multilane Highways (1985 HCM)

LOS	Density (PC/MI/LN)	70 mph Design Speed				60 mph Design Speed				50 mph Design Speed			
		Speed[a] (mph)	v/c	MSF[b] (PCPHPL)		Speed[a] (mph)	v/c	MSF[b] (PCPHPL)		Speed[a] (mph)	v/c	MSF[b] (PCPHPL)	
A	≤ 12	≥ 57	0.36	700		≥ 50	0.33	650		—	—	—	
B	≤ 20	≥ 53	0.54	1,100		≥ 48	0.50	1,000		≥ 42	0.45	850	
C	≤ 30	≥ 50	0.71	1,400		≥ 44	0.65	1,300		≥ 39	0.60	1,150	
D	≤ 42	≥ 40	0.87	1,750		≥ 40	0.80	1,600		≥ 35	0.76	1,450	
E	≤ 67	≥ 30	1.00	2,000		≥ 30	1.00	2,000		≥ 28	1.00	1,900	
F	> 67	< 30	c	c		< 30	c	c		< 28	c	c	

[a] Average travel speed.
[b] Maximum rate of flow per lane under ideal conditions, rounded to the nearest 50 pcphpl.
[c] Highly variable.

SOURCE: *Highway Capacity Manual,* 1985, Special Report 209 (Washington, DC: Transportation Research Board), p. 7–7.

Generally, such highways can be considered as being between freeways and urban arterial streets. Their basic characteristics include signal spacings of 2 miles or greater, four or six lanes for through traffic, with or without median dividers, and some degree of side access in the form of driveways and cross streets. Although the 1985 *Highway Capacity Manual* contains calculation procedures for such highways, a major research project[11] undertaken in the late 1980s has provided a new data base and new recommended procedures for such facilities. This research, NCHRP Project 3-33, is currently under review by the Transportation Research Board and it is expected that the new procedures will be disseminated in 1991. A summary of the new procedures is found in the final section of this chapter.

Basic relationships

Calculation procedures found in the 1985 *Highway Capacity Manual* are very similar to those used for basic freeway sections. The basic equation has one additional factor, f_E, an adjustment for the development environment and type of multilane highway. Equation (5.6) is the basic equation for analyzing multilane highways.

$$SF_i = MSF_i \times N \times f_w \times f_{HV} \times f_E \times f_p \qquad (5.6)$$

The level of service criteria are based on density, and are given in Table 5–18. The adjustment factors for multilane highways are identical to those for freeway basic segments for heavy vehicles and for driver population. For lane width and lateral clearance, the multilane factors are almost identical to those for freeway segments, and the analyst can use the factors in Table 5–9. The only additional factor for multilane highways is the development environment adjustment, with values given in Table 5–19. The 1985 *Highway Capacity Manual* states that multilane highways with more than 10 driveways and/or intersections per mile on one side are classified as suburban, rather than rural.

TABLE 5–19
Development Environment Adjustment Factors

Type	Divided	Undivided
Rural	1.00	0.95
Suburban	0.90	0.80

SOURCE: *Highway Capacity Manual,* 1985, Special Report 209 (Washington, DC: Transportation Research Board), p. 7–13.

Operational improvements for multilane highways

In metropolitan areas, four- and six-lane suburban highways are becoming the focus for rehabilitation and traffic engineering improvements. Generally, these roadways are in areas of new development with rapidly increasing traffic volumes. Using the proposed new procedures, the engineer can assess the impacts of several traffic and roadway factors. At locations where intersecting streets require traffic signals, some jurisdictions are embarking on programs of grade separation. This action allows a better balance between capacity of the most restricted point on a multilane highway (the traffic signal) and the most unrestricted, the free flow portion covered by the analysis procedures described above. Table 5–20 provides a summary of actions that might be considered for improving operations on multilane highways. The table also indicates an estimate of the impact of each type of improvement.

Two-lane rural highways

Two-lane rural highways serve as the basic network of roads throughout the United States and Canada for connecting recreational areas, agricultural regions, rural towns, and cities. The principal difference between this type of facility and other highway types is that there is only one lane for use by traffic in each direction and that passing of slow-moving vehicles requires the use of a lane used by opposing traffic. Thus, traffic characteristics in each direction interact and both directions of travel must be analyzed as a unit. The 1985 *Highway Capacity Manual* contains two levels of analysis for two-lane highways: operational and system planning.

---

[11] *Capacity and Level-of-Service Procedures for Multilane Rural and Suburban Highways.*

TABLE 5-20
Operational Improvements for Multilane Highways

Action	Improves	Relative Improvement
1985 HCM PROCEDURE		
1. Add lane	Capacity and LOS	50% greater capacity, two levels of service better
2. Widen lanes	Capacity and LOS	3% to 15% per foot of widening
3. Improve substandard lateral clearance	Capacity and LOS	1% to 2% per foot of improvement
4. Install median	Capacity and LOS	5% to 12%
5. Install truck climbing lane	Capacity and LOS	5% for each percent of heavy vehicles
PROPOSED PROCEDURE		
6. Install median	Free Flow Speed	1.6 mph increase
7. Increase lane width	Free Flow Speed	1.9 to 4.7 mph increase per foot of widening
8. Improve substandard lateral clearance	Free-Flow Speed	0.2 to 0.4 mph increase per foot of improvement
9. Reduce No. of access points	Free Flow Speed	0.25 mph increase per access point
10. Install truck climbing lane	LOS	1 LOS better for each 5% of heavy vehicles

Note: All "relative improvements" are for a given set of base conditions and will vary widely as conditions vary.

Basic relationships

The analysis of two-lane highways is based on the two-way volume in pcph, the average travel speed of all vehicles, and the percent time delay. This last factor is defined as the average percent of time that all vehicles are delayed while traveling in platoons due to the inability to pass. As an approximation, any vehicle traveling at a time headway less than 5 sec is considered to be "delayed." Thus, if 25% of all vehicles in a section of highway have time headways of less than 5 sec, the percent time delay would be 25%. Figure 5-7 shows the basic flow relationships used to analyze two-lane highways. Table 5-21 gives the criteria for level of service for general segments on two-lane highways.

Figure 5-7. Flow relationships for two-lane highways (ideal conditions).
SOURCE: *Highway Capacity Manual,* 1985, Special Report 209 (Washington, DC: Transportation Research Board), p. 8-4.

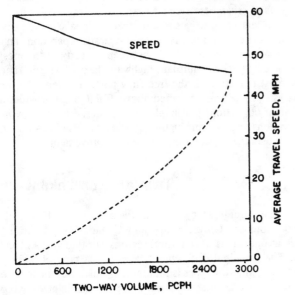

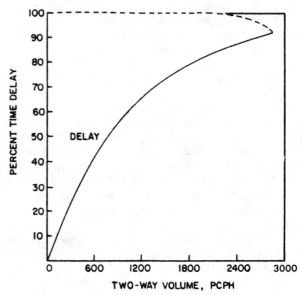

a. Relationship between average speed and flow on two-lane highways.

b. Relationship between percent time delay and flow on two-lane highways.

TABLE 5–21

TABLE 5–21
Level-of-Service Criteria for Two-Lane Highways

									v/c Ratio[a]												
		Level Terrain							Rolling Terrain							Mountainous Terrain						
	Percent Time	Avg[b]	Percent No Passing Zones						Avg[b]	Percent No Passing Zones						Avg[b]	Percent No Passing Zones					
LOS	Delay	Speed	0	20	40	60	80	100	Speed	0	20	40	60	80	100	Speed	0	20	40	60	80	100
A	≤ 30	≥58	0.15	0.12	0.09	0.07	0.05	0.04	≥57	0.15	0.10	0.07	0.05	0.04	0.03	≥56	0.14	0.09	0.07	0.04	0.02	0.01
B	≤ 45	≥55	0.27	0.24	0.21	0.19	0.17	0.16	≥54	0.26	0.23	0.19	0.17	0.15	0.13	≥54	0.25	0.20	0.16	0.13	0.12	0.10
C	≤ 60	≥52	0.43	0.39	0.36	0.34	0.33	0.32	≥51	0.42	0.39	0.35	0.32	0.30	0.28	≥49	0.39	0.33	0.28	0.23	0.20	0.16
D	≤ 75	≥50	0.64	0.62	0.60	0.59	0.58	0.57	≥49	0.62	0.57	0.52	0.48	0.46	0.43	≥45	0.58	0.50	0.45	0.40	0.37	0.33
E	> 75	≥45	1.00	1.00	1.00	1.00	1.00	1.00	≥40	0.97	0.94	0.92	0.91	0.90	0.90	≥35	0.91	0.87	0.84	0.82	0.80	0.78
F	100	<45	—	—	—	—	—	—	<40	—	—	—	—	—	—	<35	—	—	—	—	—	—

[a]Ratio of flow rate to an ideal capacity of 2,800 pcph in both directions.
[b]Average travel speed of all vehicles (in mph) for highways with design speed ≥ 60 mph; for highways with lower design speeds, reduce speed by 4 mph for each 10 mph reduction in design speed below 60 mph; assumes that speed is not restricted to lower values by regulation.

SOURCE: *Highway Capacity Manual,* 1985, Special Report 209 (Washington, DC: Transportation Research Board), p. 8–5.

The basic traffic flow relationship for level-of-service and capacity calculations for two-lane highways is

$$SF_i = 2,800 \times (v/c)_i \times f_d \times f_w \times f_{HV} \qquad (5.7)$$

where:

SF_i = total service flow rate in both directions for prevailing roadway and traffic conditions, for level of service i, in vph

$(v/c)_i$ = ratio of flow rate to ideal capacity for level of service i,

f_d = adjustment factor for directional distribution of traffic

f_w = adjustment factor for lane and shoulder width

f_{HV} = adjustment factor for the presence of heavy vehicles in the traffic stream

Adjustment factors

The analysis of two-lane highways can be performed for general-terrain segments or for specific grades. The analysis procedures are similar, with the general-terrain analysis being performed when grades are less than 3% or shorter than ½ mile. When analyzing a specific grade, an additional adjustment factor is included in the calculation. This factor adjusts for the operation effects of grades on passenger cars. The factor can be considered similar to the heavy vehicle factor in that the grade itself creates conditions for passenger cars that reduce their speeds and have a reducing effect on capacity. For this discussion, adjustment factors only for general-terrain segments will be identified. Table 5–22

includes factors for adjusting service flow rate, which are based on the directional distribution of traffic. Table 5–23 includes factors for the effect of lane width and shoulder width. Note than when conditions are near capacity (LOS E) narrow lanes and narrow shoulders have less of an effect on capacity than under more open conditions. Finally, Table 5–24 includes passenger car equivalents for each class of heavy vehicle by terrain type. The effect of heavy vehicles varies by the operating level of service.

A typical operational analysis of a general terrain segment would require calculations leading to an estimate of

TABLE 5–22
Adjustment Factors for Directional Distribution

Directional Distribution	Adjustment Factor, f_d
100/0	0.71
90/10	0.75
80/20	0.83
70/30	0.89
60/40	0.94
50/50	1.00

SOURCE: *Highway Capacity Manual,* 1985, Special Report 209 (Washington, DC: Transportation Research Board), p. 8–9.

TABLE 5–23
Adjustment Factors for Lane Width and Shoulder Width

| Usable[a] Shoulder Width (FT) | 12-ft Lanes | | 11-ft Lanes | | 10-ft Lanes | | 9-ft Lanes | |
	LOS A–D	LOS[b] E	LOS A–D	LOS[b] E	LOS A–D	LOS[b] E	LOS A–D	LOS[b] E
≥6	1.00	1.00	0.93	0.94	0.84	0.87	0.70	0.76
4	0.92	0.97	0.85	0.92	0.77	0.85	0.65	0.74
2	0.81	0.93	0.75	0.88	0.68	0.81	0.57	0.70
0	0.70	0.88	0.65	0.82	0.58	0.75	0.49	0.66

[a]Where shoulder width is different on each side of the roadway, use the average shoulder width.
[b]Factor applies for all speeds less than 45 mph.

SOURCE: *Highway Capacity Manual,* 1985, Special Report 209 (Washington, DC: Transportation Research Board), p. 8–9.

TABLE 5–24
Passenger Car Equivalents for Two-Lane Highways

| Vehicle Type | Level of Service | Type of Terrain | | |
		Level	Rolling	Mountainous
Trucks, E_r	A	2.0	4.0	7.0
	B and C	2.2	5.0	10.0
	D and E	2.0	5.0	12.0
RV's E_R	A	2.2	3.2	5.0
	B and C	2.5	3.9	5.2
	D and E	1.6	3.3	5.2
Buses, E_B	A	1.8	3.0	5.7
	B and C	2.0	3.4	6.0
	D and E	1.6	2.9	6.5

SOURCE: *Highway Capacity Manual,* 1985, Special Report 209 (Washington, DC: Transportation Research Board), p. 8–9.

TABLE 5–25

Operational Improvements for Two-Lane Highways

Action	Improves	Relative Improvement
1. Reduce "No Passing" zones	Capacity and LOS	2 to 10 pcph increase for each 1% of total length reduced
2. Increase lane width	Capacity and LOS	5% to 15% capacity increase per 1 ft increase
3. Improve shoulder width	Capacity and LOS	2% increase in capacity per 1 ft increase
4. Install truck-climbing lane	Capacity and LOS	1 LOS better for each 3% heavy vehicles

Note: All "relative improvements" are for a given set of base conditions and will vary widely as conditions vary.

current operating level of service. Consider a two-lane rural highway which carries 180 vph in the peak hour, with a 60-mph design speed, 11-foot lanes, and 2-foot shoulders, in mountainous terrain, with 80% of the 10-mile section having no passing zones. The directional split is 60/40, with 5% trucks, 10% recreational vehicles, and no buses.

First, the actual flow rate is computed as 180 vph/PHF. In this case, a PHF of 0.87 is taken from actual field data. The actual service flow rate is 207 vph. Using equation (5.7), the service flow rate for each level of service (A, B, C, D, E) is computed. The directional adjustment factor is 0.94 for all levels of service. The lane width/shoulder width factor is 0.75 for LOS A through D and 0.88 for LOS E. The factors for trucks and recreational vehicles are taken from Table 5–24 and the f_{HV} adjustment is computed for each level of service. Calculation of the service flow rate indicates 23, 127, 211, 371, and 941 vph for levels of service A through E, respectively. Comparison of the actual service flow rate of 207 vph with the calculated maximum flow rates shows that current conditions are operating just within Level of Service C. When the actual service flow rate exceeds 211 vph, conditions will move into Level of Service D.

Operational improvements for two-lane highways

In addition to factors that might be controlled by the highway agency, such as lane width, shoulder width, and to a lesser extent vehicle mix, there are a number of design and traffic treatments that are applied in practice. Among these are improvements in passing sight distance, paving shoulders, adding a third lane, constructing passing lanes, adding continuous two-way left-turn median lanes, and constructing climbing lanes. Better geometrics at intersections and turnouts for slow-moving traffic can also be used to improve operations. Table 5–25 provides a summary of operational techniques that can be used to improve level of service and capacity on two-lane highways. The table shows in a general way the impact a given improvement can have on two-lane highway operations.

Signalized intersections

Traffic signals provide a complex type of operation for traffic flow. Signals allocate green time between conflicting movements. This time allocation depends on signal timing and phasing and creates the need to stop each conflicting movement on a regular basis and then to allow it to flow. This stop-and-go operation creates not only lost time but also the need for change intervals between traffic phases. During the past several decades, several procedures have been developed for analyzing the capacity, level of service, and specific operating conditions at signalized intersections. Most of these procedures are based either on a series of equations and adjustments that lead to estimates of measures of effectiveness such as delay or queue length, or on a more simple calculation that provides a summation of traffic volume passing through the middle of the intersection. This latter type of procedure is generally termed "critical movement."

Regardless of the type of procedure used, the traffic, geometric, and control conditions at a signalized intersection all contribute to the operational efficiency. Thus, in the very simple "critical movement" type of analysis, a large number of these factors have simply been subsumed or have gone unrecognized. In the more complex and complete operational procedures that produce direct estimates of traffic measures, most or all of the measurable factors remain explicit in the analysis and can be varied and studied by the traffic engineer.

Critical movement analyses

This simple type of analysis provides a quick estimate of the ability of a signalized intersection to accommodate a total hourly volume of traffic. In the 1985 *Highway Capacity Manual,* this type of procedure is called the "planning procedure." For each phase of a traffic signal timed in a traditional manner, the analyst sums up the maximum volume per lane that conflicts on each of the two major streets. Thus, for the north-south street, the highest sum of a through-lane volume added to the opposing left-turn-lane volume is considered as the critical volume for north-south. The same summation is performed for the east-west street. The two critical sums are then added to obtain a critical volume for the intersection as a whole, in terms of vehicles per hour. The planning procedure does not contain a specific list of LOS A through F, but rather gives general ranges of operation defined by ranges of the critical volume. Table 5–26 lists the ranges used. Two other alternative critical movement procedures are briefly discussed at the end of this chapter.

TABLE 5–26

Capacity Criteria for Planning Analysis

Critical Volume for Intersection (vph)	Relationship to Probable Capacity
0 to 1,200	Under Capacity
1,201 to 1,400	Near Capacity
≥ 1,401	Over Capacity

SOURCE: *Highway Capacity Manual,* 1985, Special Report 209 (Washington, DC: Transportation Research Board), p. 9–21.

To illustrate the application of a critical movement procedure, Figure 5–8 is taken from the 1985 *Highway Capacity Manual.* Volumes used are mixed vehicles per hour and there is no adjustment for peak-hour factor. The critical sum of volumes is 1,307, while 1,400 is considered as the point where the capacity of the intersection is being reached.

When applying this procedure, local data and local values that indicate capacities consistently greater than 1,400 should be used to modify the threshold levels of Table 5–26. For example, in some cities the capacity threshold is taken to be 1,500 or even 1,600 vehicles per hour.

Operational analyses

When the engineer needs to consider the impacts of geometry, traffic, and control conditions at a signalized intersection, a detailed operational analysis is required. The procedure given in the 1985 *Highway Capacity Manual* is currently a widely used method for North American conditions. An illustration of this method is given as Figure 5–9. The analyst assembles all data on geometry, traffic, and signal timing, adjusts the volume for peak-hour factor and for distribution by lane, adjusts the saturation flow rate for less than ideal conditions, and then compares volume to saturation flow and capacity.

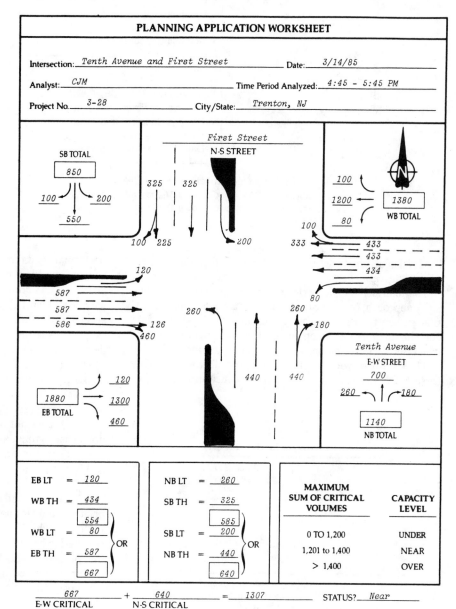

Figure 5–8. Planning analysis example— 1985 *Highway Capacity Manual.*
SOURCE: *Highway Capactiy Manual 1985,* Special Report 209 (Washington, DC: Transportation Research Board), p. 59.

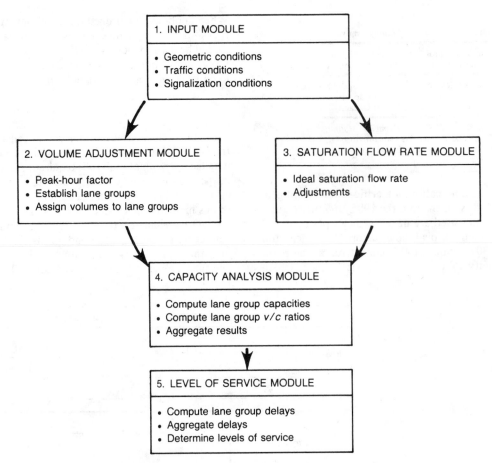

Figure 5–9. Operational analysis—1985 *Highway Capacity Manual.*
SOURCE: *Highway Capacity Manual,* 1985, Special Report 209 (Washington, DC: Transportation Research Board), p. 9–6.

The level of service is defined by criteria based on the average stopped delay per vehicle.

The analysis can be performed for a single lane, for an entire approach, or for the intersection as a whole. This ability to aggregate the results provides considerable flexibility for assessing the relative effectiveness of improvement measures at an intersection or among a group of intersections.

The level-of-service criteria for this procedure are given in Table 5–27. Because delay is a complex measure of traffic performance, its relationship to capacity and v/c ratio is not simple. The Level of Service E values are based on stopped delay and it is quite possible for an intersection approach to have very high delay (e.g., 55 sec) while operating with an acceptable v/c ratio well below 1.00. Conversely, it is possible that a lane or an intersection approach may have very low delay while operating with volumes that equal capacity. These varying conditions indicate that signal timing, signal phasing, and signal system platooning and synchronization have a significant impact on delay, regardless of the total volume being served. Thus, for signalized intersections, Level of Service F does not necessarily imply that more volume cannot be accommodated, but rather that the vehicles being served are suffering unacceptable levels of stopped delay.

Using the operational analysis procedure, the engineer is able to solve for the level of service, knowing details of geometry, traffic, and signal control. Also, solutions for appropriate signal timing to achieve a given level of service or solutions, which lead to an estimate of the number of lanes required to handle given traffic loads, may be obtained from this procedure. Some of the solution types require a trial-and-error procedure, which is not particularly difficult when using the microcomputer software that is available for calculations.

TABLE 5–27

Operational Analysis Levels of Service

Level of Service	Stopped Delay per Vehicle (sec)
A	≤5.0
B	5.1 to 15.0
C	15.1 to 25.0
D	25.1 to 40.0
E	40.1 to 60.0
F	>60.0

SOURCE: *Highway Capacity Manual,* 1985, Special Report 209 (Washington, DC: Transportation Research Board), p. 9–4.

Adjustment factors

For the analysis volumes, two adjustment factors can be used. First, when an hourly volume, in vehicles per hour, is known, it can be divided by the peak-hour factor to obtain an hourly flow rate based on the peak 15-min period. For most urban intersections, peak-hour factors are in the range of 0.85 to 0.95. The hourly flow rate can also be adjusted for the distribution of traffic among two or more lanes on the intersection approach. This "lane utilization factor" is 1.05 and 1.10 (times the average per lane volume) for two- and three-lane approaches, respectively. Generally, for volume conditions that approach capacity, the lane utilization factor is not applied in the analysis. When it is applied, the results can be considered as level-of-service estimates for the "worst" lane within a group of lanes.

The saturation flow rate for each lane or group of lanes is calculated by using Equation (5.8):

$$s = s_o N f_w f_{HV} f_g f_p f_{bb} f_a f_{RT} f_{LT} \qquad (5.8)$$

where:

s = saturation flow rate for lane group, in vphg
s_o = ideal saturation flow rate per lane, usually 1,800 pcphgpl
N = number of lanes in lane group
f_w = adjustment factor for lane width
f_{HV} = adjustment factor for heavy vehicles
f_g = adjustment factor for grade
f_p = adjustment factor for adjacent parking lane and activity
f_{bb} = adjustment factor for local buses stopping
f_a = adjustment factor for area type
f_{RT} = adjustment factor for right turns
f_{LT} = adjustment factor for left turns

The above factors comprise the adjustments to saturation flow (or capacity) that are explicitly considered in this procedure. These, along with the traffic signal timing and phasing factors, are the elements which a practicing engineer can focus on when developing improvements at signalized intersections. Other factors such as pavement condition, lack of pedestrian control, double parking, and poor delineation of lanes may, in fact, contribute to operational characteristics but must be considered through the use of engineering judgment since they are not part of the analytical procedure.

None of the above adjustment factors is explicitly considered in the critical movement procedures described earlier, although average values for the factors have been "built into" the critical movement "look-up" tables. If the analyst does not have data on the above factors or wishes to shorten the calculations, research[12] has shown that a good "average" saturation flow rate at many urban intersections is 1,600 vphgpl. This value takes into account typical adjustments applied to an ideal saturation flow of 1,800 pcphgpl. The operational procedure is based on data from around the United States at locations that have right-turn-on-red. Thus, no special

adjustment is needed when analyzing intersections operating with right-turn-on-red.

Table 5–28 lists the values for the adjustment factors to saturation flow. The table shows the factor for f_{RT} and f_{LT} for exclusive turn lanes with protected signal phasing. Other combinations of right-turn and left-turn lanes and phasings have specific values for these factors. Detailed tables for f_{RT} and f_{LT} are given in the 1985 *Highway Capacity Manual*.

Capacity and level of service

Once the adjustments have been made to volume and to saturation flow, the capacity of each lane or lane group being analyzed is computed using Equation (5.9):

$$c_i = s_i \times (g/C)_i \qquad (5.9)$$

where:

c_i = capacity of lane group i, in vph
s_i = saturation flow rate for lane group i, in vphg
$(g/C)_i$ = green/cycle ratio for lane group i

The v/c ratio for each lane group is computed by dividing the adjusted flow by the capacity computed by Equation (5.9). This v/c ratio is given the symbol X_i. When the sum of the X_i for critical lane groups exceeds 1.00, it is an indication that the geometrics and/or signal timing are physically inadequate to handle the traffic volume. In this case, the analyst can design improvement alternatives that will bring the sum below 1.0.

The final step in the calculation is the computation of delay and the estimate of level of service. Delay is computed using an equation with two parts. The first term of the equation accounts for uniform delay, while the second term accounts for incremental delay due to cycle failures when all of the vehicles do not clear on the first available green phase. Thus, when conditions are congested and volume is high, the second term of the equation takes on greater importance. Under normal volume conditions, the second term contributes very little to the estimate of total delay. Equation (5.10) is used to calculate average stopped delay per vehicle.

$$d = 0.38 \, C \, \frac{[1 - g/C]^2}{[1 - (g/C)(X)]} + 173 \, X^2 \, [(X - 1)$$
$$+ \sqrt{(X - 1)^2 + (16 \, X/c)} \,] \qquad (5.10)$$

where:

d = average stopped delay per vehicle, in sec/veh
C = cycle length, in sec
g/C = green ratio for the lane group
X = v/c ratio of the lane group
c = capacity of the lane group

As the final adjustment to the estimate of delay, d is multiplied by the progression adjustment factor, *PF*. This factor accounts for the impact that platoon arrivals and signal system operation can have on delay at a specific intersection. These adjustment factors were derived from field data taken in the United States in the early 1980s. Recent research results, described in the final section of this chapter,

---

[12] *Highway Capacity Manual*, 1985.

TABLE 5–28

Adjustment Factors for Saturation Flow

Adjustment Factor for Lane Width

Lane width, ft	8	9	10	11	12	13	14	15	≥ 16
Lane width factor, f_w	0.87	0.90	0.93	0.97	1.00	1.03	1.07	1.10	Use 2 Lanes

Adjustment Factor for Heavy Vehicles

Percent heavy vehicles, % HV	0	2	4	6	8	10	15	20	25	30
Heavy vehicle factor, f_{HV}	1.00	0.99	0.98	0.97	0.96	0.95	0.93	0.91	0.89	0.87

Adjustment Factor for Grade

	Downhill			Level	Uphill		
Grade, %	− 6	− 4	− 2	0	+ 2	+ 4	+ 6
Grade factor, f_g	1.03	1.02	1.01	1.00	0.99	0.98	0.97

Adjustment Factor for Parking, f_p

No. of Lanes in Lane Group	No. Pkg.	Number of Parking Maneuvers Per Hour, N_m				
		0	10	20	30	40
1	1.00	0.90	0.85	0.80	0.75	0.70
2	1.00	0.95	0.92	0.89	0.87	0.85
3	1.00	0.97	0.95	0.93	0.91	0.89

Adjustment Factor for Bus Blockage, f_{bb}

No. of Lanes in Lane Group	Number of Buses Stopping Per Hour, N_B				
	0	10	20	30	40
1	1.00	0.96	0.92	0.88	0.83
2	1.00	0.98	0.96	0.94	0.92
3	1.00	0.99	0.97	0.96	0.94

Adjustment Factor for Area Type

Type of Area	Factor f_a
CBD	0.90
All other areas	1.00

Adjustment Factor for Right Turns

Case	Type of Lane Group	Right-Turn Factor, f_{RT}
1	Exclusive RT lane; protected RT phasing	0.85

Adjustment Factor for Left Turns

Case	Type of Lane Group	Left-Turn Factor, f_{LT}
1	Exclusive LT lane; protected phasing	0.95

SOURCE: *Highway Capacity Manual*, 1985, Special Report 209 (Washington, DC: Transportation Research Board), pp. 9–12 to 9–15.

TABLE 5-29
Progression Adjustment Factors

Type of Signal	Lane Group Types	v/c Ratio, X	Arrival Type[a]				
			1	2	3	4	5
Pretimed	TH, RT	≤0.6	1.85	1.35	1.00	0.72	0.53
		0.8	1.50	1.22	1.00	0.82	0.67
		1.0	1.40	1.18	1.00	0.90	0.82
Actuated	TH, RT	≤0.6	1.54	1.08	0.85	0.62	0.40
		0.8	1.25	0.98	0.85	0.71	0.50
		1.0	1.16	0.94	0.85	0.78	0.61
Semiactuated	Main St. TH, RT[b]	≤0.6	1.85	1.35	1.00	0.72	0.42
		0.8	1.50	1.22	1.00	0.82	0.53
		1.0	1.40	1.18	1.00	0.90	0.65
Semiactuated	Side St. TH, RT[b]	≤0.6	1.48	1.18	1.00	0.86	0.70
		0.8	1.20	1.07	1.00	0.98	0.89
		1.0	1.12	1.04	1.00	1.00	1.00
	All LT[c]	all	1.00	1.00	1.00	1.00	1.00

[a]See lower part of table.

[b]Semiactuated signals are typically timed to give all extra green time to the main street. This effect should be taken into account in the allocation of green times.

[c]This category refers to exclusive LT lane groups with protected phasing only. When LT's are included in a lane group encompassing an entire approach, use factor for the overall lane group type. Where heavy LT's are intentionally coordinated, apply factors for the appropriate through movement.

SOURCE: *Highway Capacity Manual*, 1985, Special Report 209 (Washington, DC: Transportation Research Board), p. 9-20.

Arrival Type	Range of Platoon Ratio, R_p
1	0.00 to 0.50
2	0.51 to 0.85
3	0.86 to 1.15
4	1.16 to 1.50
5	≥1.51

SOURCE: *Highway Capacity Manual*, 1985, Special Report 209 (Washington, DC: Transportation Research Board), p. 9-8, Table 9-2.

may lead to revisions in these factors. Table 5-29 gives the factors as found in the 1985 *Highway Capacity Manual.*

Calculations

The 1985 *Highway Capacity Manual* provides worksheets that allow the analyst to proceed through each of the five modules in a logical manner. Calculations are more easily performed by using a software package representing the material in the *Manual*. Information on availability of the software can be obtained from the McTrans Center, University of Florida, Gainesville, Florida.

Improving intersection operations

Table 5-30 illustrates the sensitivity of stopped delay to a change in any one parameter affecting signalized intersection operations. A typical urban intersection was used as the "default" condition. Then, each factor in Table 5-30 was changed (holding other factors "constant") and the impact on each approach, lane group, and the intersection as a whole was determined. For example, compared with the base default case of 12-ft lane widths, the use of 11-ft lanes throughout the intersection increases total intersection delay from 39.8 sec per vehicle to 46.3 sec per vehicle. The overall intersection level of service changes from D to E.

Another comparison is made by changing the phase time on the east-west street by only 4 sec. The through traffic was given 4 sec less green time and the left-turn phase was given 4 additional sec. No change was made in signal timing for the north-south street. The change in total delay at the intersection for this relatively small change in signal timing is from 39.8 sec per vehicle to 44.0 sec per vehicle. The impact of heavy vehicles is shown in Table 5-30. If heavy vehicles on all approaches increase from 2% to 10% of total volume, average delay per vehicle increases from 39.8 to 47.6 secs.

Table 5-30 gives the sensitivity of stopped delay to changes in factors for an intersection that is operating under fairly heavy volume conditions. A similar comparison of sensitivity for light volume conditions is given in Table 5-31.

Unsignalized intersections

The analysis of intersections without traffic signals can be accomplished by applying a procedure based on the acceptance of gaps by traffic moving from a stop or yield sign on a side street, or by traffic making a left turn from the mainline. This procedure is based on experience from West Germany and Europe and has been adapted to conditions in the United States. The procedure can be used to study T-intersections and five-way intersections with stop or yield signs.

TABLE 5-30
Sensitivity of Stopped Delay to Various Factors (high volume conditions)

Delay (secs/veh)

Direction	Approach Lanes	Volume (vph)	Lane Group	Default Test[a]	Sensitivity Tests[b]								
					Lane Width (ft) 11	Arrival Type 2	Phase Time c	Grade E/W (+3%) N/S Level	PHF 0.80	Parking Maneuver/Hr 0	Heavy Vehicle (%) 10	Lane Dist. 1.00	Saturation Flow (pcphpl) 1,900
EB	4	120 (Left)	L	61.1	66.6	61.1	34.5	63.7	71.0	61.1	68.8	61.1	61.1
		1,100 (Through)	TR	25.4	27.5	28.1	45.4	26.4	29.9	22.5	23.9	22.1	23.0
		100 (Right)											
WB	4	135 (Left)	L	85.9	96.4	85.9	37.2	90.9	105.4	85.9	100.6	85.9	85.9
		1,000 (Through)	TR	21.5	22.2	24.8	27.7	21.8	22.8	20.3	20.9	20.1	20.5
		65 (Right)											
NB	4	130 (Left)	L	76.0	84.6	76.0	76.0	76.0	91.0	76.0	88.1	76.0	76.0
		1,200 (Through)	T	52.7	64.7	58.3	52.7	52.7	76.7	52.7	69.5	38.9	37.3
		100 (Right)	R	14.2	14.2	18.0	14.2	14.2	14.3	13.9	13.9	14.2	14.2
SB	4	95 (Left)	L	42.0	23.6	42.0	42.0	42.0	44.7	42.0	44.2	42.0	42.0
		1,200 (Through)	T	52.7	64.7	58.3	52.7	52.7	76.7	52.7	69.5	38.9	37.3
		120 (Right)	R	14.5	14.5	18.4	14.5	14.5	14.6	14.1	14.2	14.5	14.5
Overall delay				39.8	46.3	43.8	44.0	40.3	52.8	38.9	47.6	33.0	32.3
v/c				0.98	1.01	0.98	0.98	0.99	1.04	0.96	1.00	0.931	0.94
LOS				D	E	E	E	E	E	D	E	D	D
Percent Difference in Delay, Sensitivity Test Compared to Default Test				0	16	10	11	1	33	-2	20	-17	-19

TABLE 5-31
Sensitivity of Stopped Delay to Various Factors (low volume conditions)

Delay (secs/veh)

Direction	Approach Lanes	Volume (vph)	Lane Group	Default Test[a]	Sensitivity Tests[b]								
					Lane Width (ft) 11	Arrival Type 2	Phase Time c	Grade E/W (+3%) N/S Level	PHF 0.80	Parking Maneuver/Hr 0	Heavy Vehicle (%) 10	Lane Dist. 1.00	Saturation Flow (pcphpl) 1,900
EB	4	80 (Left)	L	37.2	37.9	37.2	31.0	37.6	38.4	37.2	38.2	37.2	36.2
		732 (Through)	TR	18.5	18.7	23.5	21.1	18.6	18.9	18.2	18.8	18.0	18.2
		66 (Right)											
WB	4	90 (Left)	L	40.1	41.3	40.1	31.6	40.6	42.1	40.1	41.8	40.1	38.3
		666 (Through)	TR	17.8	18.0	22.7	20.2	17.9	18.1	17.6	18.1	17.4	17.6
		44 (Right)											
NB	4	87 (Left)	L	39.1	40.1	39.1	39.1	39.1	40.8	39.1	40.6	39.1	37.6
		800 (Through)	T	18.4	18.8	21.2	18.4	18.4	19.2	18.4	19.0	17.9	17.8
		66 (Right)	R	13.7	13.7	17.4	13.7	13.7	13.7	13.6	13.7	13.7	13.6
SB	4	64 (Left)	L	34.5	34.8	34.5	34.5	34.5	35.0	34.5	34.9	34.5	34.1
		800 (Through)	T	18.4	18.8	21.2	18.4	18.4	19.2	18.4	19.0	17.9	17.8
		80 (Right)	R	13.9	13.9	17.6	13.9	13.9	13.9	13.7	13.9	13.9	13.8
Overall delay				19.8	20.1	23.3	20.6	19.8	20.4	19.7	20.3	19.5	19.3
v/c				0.655	0.676	0.655	0.655	0.66	0.694	0.646	0.683	0.621	0.621
LOS				C	C	C	C	C	C	C	C	C	C
Percent Difference in Delay, Sensitivity Test Compared to Default Test				0	2	18	4	0	3	-1	3	-2	-3

[a] For default test, the following parameter values were used: Lane Width—12 ft; Arrival Type—3; Grade—All Level; PHF—0.90; Parking Maneuver/Hour—20; Heavy Vehicle (%)—2; Lane Distribution—1.05; Saturation Flow (pcphpl)—1,800; Phase Time—E/W Approach, Left-Turn Phase Green Time 10 sec and Through-Phase Green Time 30 sec. N/S Approach, Left-Turn Phase Green Time 10 sec and Through-Phase Green Time 35 sec.

[b] For sensitivity tests, only one parameter was changed for each test. All other parameters were held equal to the default test values.

[c] On E/W approaches, the left-turn phase was given 4 additional sec, while the through-phase decreased by 4 sec.

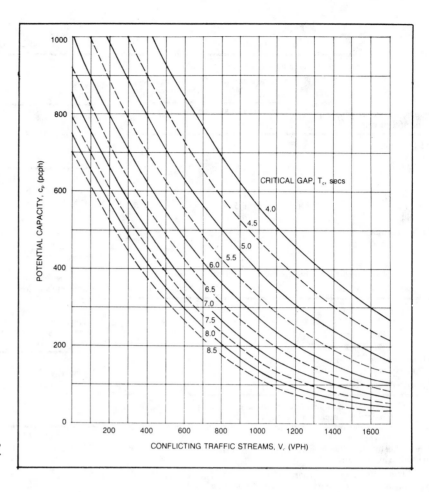

Figure 5–10. Potential capacity.
SOURCE: *Highway Capacity Manual,* 1985, Special Report 209 (Washington, DC: Transportation Research Board), p. 10–7.

The procedure does not apply to intersections with multiway stop control. The key elements of the procedure are the critical gap, in seconds, which is defined as the gap which 50% of all drivers will accept. Table 5–32 lists critical gaps for various movements. For cities with greater than 250,000 population, drivers are willing to accept slightly shorter gaps when moving away from a stop or yield sign. Also, when good geometry exists, some reduction in critical gap occurs. Figure 5–10 shows the potential capacity of a stop- or yield-controlled movement, in passenger cars per hour, given the critical gap and the volume of the conflicting traffic streams.

The level-of-service criteria are based on the reserve or unused capacity of the lane, in pcph. The unused capacity is simply the potential capacity less the volume using the lane. Table 5–33 gives the criteria for unsignalized intersections. Note that the reference to delay is not associated with the "stopped delay" used for signalized intersections. The two procedures were developed from different concepts using different data bases, and no direct correlation has been established between levels of service at traffic signals and at unsignalized locations. Local studies and data should be developed if the analyst requires such a correlation.

Multiway stop control

Although calculations of capacity at four-way stop-controlled intersections are not identified in the 1985 *Highway Capacity Manual,* Table 5–34 lists approximate service volumes at Level of Service C that can be accommodated in 1 hour. If the values in this table are multiplied by a factor of about 1.5, an approximate estimate of the maximum hourly volume (Level of Service E conditions), in vehicles per hour, can be made.

New research on multiway stop control is described at the end of this chapter.

Urban arterials

The capacity and level of service of urban and suburban arterials is based primarily on the operations of the traffic signals. This facility type is defined as having signalized intersection spacing of 2 miles or less, with relatively intense roadside development. The procedures for analyzing level of service are based on the average travel speed for all vehicles using the segment, section, or entire arterial under consideration. The average travel speed is computed by combining the running time along the arterial with the intersection approach delay. Note that intersection approach delay is considered to be 1.3 times the average stopped delay as computed by the procedure for signalized intersections.

Table 5–35 gives the level-of-service criteria for urban and suburban arterials. The arterial class is determined by the

TABLE 5-32
Critical-Gap Criteria

	Basic Critical Gap for Passenger Cars, Sec			
	Average Running Speed, Major Road			
	30 mph		55 mph	
	Number of Lanes on Major Road			
Vehicle Maneuver and Type of Control	2	4	2	4
RT from minor road				
Stop	5.5	5.5	6.5	6.5
Yield	5.0	5.0	5.5	5.5
LT from major road	5.0	5.5	5.5	6.0
Cross major road				
Stop	6.0	6.5	7.5	8.0
Yield	5.5	6.0	6.5	7.0
LT from minor road				
Stop	6.5	7.0	8.0	8.5
Yield	6.0	6.5	7.0	7.5

Adjustments and Modifications to Critical Gap, Sec	
Condition	Adjustment
RT from minor street: Curb radius > 50 ft or turn angle < 60[a]	−0.5
RT from minor street: Acceleration lane provided	−1.0
All movements: Population ≥ 250,000	−0.5
Restricted sight distance[a]	up to +1.0

NOTES: Maximum total decrease in critical gap = 1.0 sec.
　　　　Maximum critical gap = 8.5 sec.
　　　　For values of average running speed between 30 and 55 mph, interpolate.

[a]This adjustment is made for the specific movement impacted by restricted sight distance.

SOURCE: *Highway Capacity Manual*, 1985, Special Report 209 (Washington, DC: Transportation Research Board), p. 10-7.

TABLE 5-33
Level-of-Service Criteria for Unsignalized Intersections

Reserve Capacity (pcph)	Level of Service	Expected Delay to Minor Street Traffic
≥ 400	A	Little or no delay
300-399	B	Short traffic delays
200-299	C	Average traffic delays
100-199	D	Long traffic delays
0- 99	E	Very long traffic delays
[a]	F	[a]

[a]When demand volume exceeds the capacity of the lane, extreme delays will be encountered with queuing which may cause severe congestion affecting other traffic movements in the intersection. This condition usually warrants improvement to the intersection.

SOURCE: *Highway Capacity Manual*, 1985, Special Report 209 (Washington, DC: Transportation Research Board), p. 10-9.

TABLE 5-34
LOS C Volumes—Four-Way Stops

	LOS C Service Volume, vph		
	Number of Lanes		
Demand Split	2 by 2	2 by 4	4 by 4
50/50	1,200	1,800	2,200
55/45	1,140	1,720	2,070
60/40	1,080	1,660	1,970
65/35	1,010	1,630	1,880
70/30	960	1,610	1,820

SOURCE: *Highway Capacity Manual*, 1985, Special Report 209 (Washington, DC: Transportation Research Board), p. 10-14.

TABLE 5-35
Level-of-Service Criteria for Urban Arterials

Arterial Class	I	II	III
Range of free flow speeds (mph)	45 to 35	35 to 30	35 to 25
Typical free flow speed (mph)	40 mph	33 mph	27 mph
Level of Service	Average Travel Speed (mph)		
A	≥ 35	≥ 30	≥ 25
B	≥ 28	≥ 24	≥ 19
C	≥ 22	≥ 18	≥ 13
D	≥ 17	≥ 14	≥ 9
E	≥ 13	≥ 10	≥ 7
F	< 13	< 10	< 7

SOURCE: *Highway Capacity Manual*, 1985, Special Report 209 (Washington, DC: Transportation Research Board), p. 11-4.

TABLE 5-36
Arterial Classes

	Functional Category	
Design Category	Principal Arterial	Minor Arterial
Typical suburban design and control	I	II
Intermediate design	II	III
Typical urban design	III	III

SOURCE: *Highway Capacity Manual*, 1985, Special Report 209 (Washington, DC: Transportation Research Board), p. 11-8.

function and the design of a given highway. Table 5-36 lists the classes according to functional and design categories. Generally, Class I arterials have free-flow speeds of from 35 to 45 mph. Class II arterials have free-flow speeds of 30 to 35 mph, and Class III arterials typically have free-flow speeds from 25 to 35 mph. A flowchart of the procedure used to assess level of service is given in Figure 5-11. It can be seen that the procedure is relatively simple in that the analyst simply divides the distance by the sum of the running time and the intersection approach delays for a given section and computes the average travel speed. This computed speed is then compared with the level-of-service criteria.

The type of operational changes that can affect arterial level of service include any midblock project that allows running speeds to increase. Such improvements are readily identified on the arterial segment when the average travel speed is determined. The features that cause the greatest and most frequently occurring delays need to be analyzed for improvement. Such improvements might include wider lanes, control of driveway access points, installation of exclusive left-turn lanes, and control of interference from parking vehicles and pedestrians. Equally important for

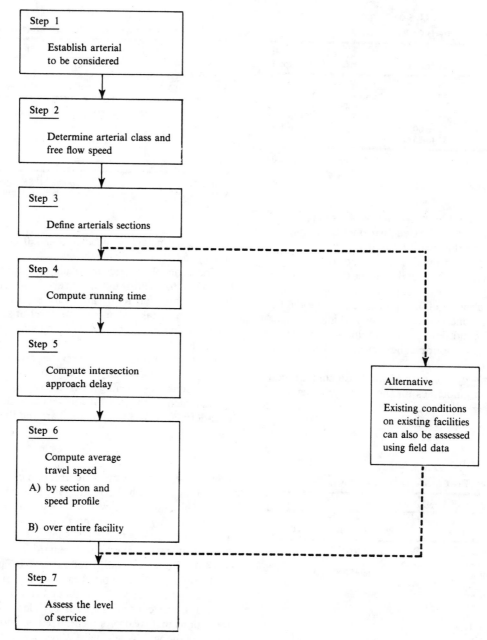

Figure 5–11. Arterial level-of-service procedure.
SOURCE: *Highway Capacity Manual,* 1985, Special Report 209 (Washington, DC: Transportation Research Board), p. 11–5.

efficient flow are operational improvements at the signalized intersections along an urban arterial.

Other capacity considerations

The 1985 *Highway Capacity Manual* has chapters covering capacity of transit systems, pedestrian facilities, and bicycle facilities. Although not applied as extensively as the procedures for intersections, freeways, and highways, the procedures for transit, pedestrians, and bike lanes provide useful guidelines in specific applications.

Transit capacity

There are two principal types of capacities for bus transit operation. First, the level of service inside the bus transit vehicle is an important consideration for scheduling drivers and equipment. Table 5–37 gives the criteria for level of service for buses. Note that the principal measure of effectiveness is the space per passenger, with anything less than about 4.3 square ft per passenger being considered as Level of Service F. A second type of capacity and level of service is that relating to exclusive or near exclusive bus lanes on arterial streets. Table 5–38 contains level-of-service criteria for bus lanes.

TABLE 5–37

Level-of-Service Criteria for Bus Transit Passengers

Peak-Hour Level of Service	Passengers	Approx. Sq Ft/Pass.	Pass./Seat (Approx.)
A	0 to 26	13.1 or more	0.00 to 0.50
B	27 to 40	13.0 to 8.5	0.51 to 0.75
C	41 to 53	8.4 to 6.4	0.76 to 1.00
D	54 to 66	6.3 to 5.2	1.01 to 1.25
E (Max. scheduled load)	67 to 80	5.1 to 4.3	1.26 to 1.50
F (Crush load)	81 to 85	<4.3	1.51 to 1.60

SOURCE: *Highway Capacity Manual,* 1985, Special Report 209 (Washington, DC: Transportation Research Board), p. 12–8.

TABLE 5–39

Level-of-Service Criteria for Pedestrian Walkways[a]

Level of Service	Space (Sq Ft/Ped)	Expected Flows and Speeds		
		Ave Speed, S (Ft/Min)	Flow Rate, v (Ped/Min/Ft)	Vol/Cap Ratio, v/c
A	≥ 130	≥ 260	≤ 2	≤ 0.08
B	≥ 40	≥ 250	≤ 7	≤ 0.28
C	≥ 24	≥ 240	≤ 10	≤ 0.40
D	≥ 15	≥ 225	≤ 15	≤ 0.60
E	≥ 6	≥ 150	≤ 25	≤ 1.00
F	< 6	< 150	.....Variable.....	

SOURCE: *Highway Capacity Manual,* 1985, Special Report 209 (Washington, DC: Transportation Research Board), p. 13–8.

If the criteria in Tables 5–37 and 5–38 are combined, level-of-service ranges can be established for the number of passengers per hour that can be carried on arterial street exclusive bus lanes. The maximum number of passengers per hour in a bus lane for Level of Service C conditions for both passengers inside the bus and for vehicle operations in the lane itself is about 4,000. If passengers inside the bus experience LOS E conditions and the bus vehicles are operating at LOS E conditions in the exclusive lane, about 10,000 passengers per hour can be carried in one exclusive bus lane.

The 1985 *Highway Capacity Manual* also contains considerable data and guidelines for the capacity of light rail vehicles, operating conditions and capacities of bus stops, and data on bus priority treatments.

TABLE 5–38

Level-of-Service Criteria for Exclusive Bus Lanes

Level of Service	Arterial Streets		
	Description	Buses/Lane/Hr	Midvalue
A	Free flow	25 or less	15
B	Stable flow, unconstrained	26 to 45	35
C	Stable flow, interference	46 to 75	60
D	Stable flow, some platooning	76 to 105	90
E	Unstable flow, queuing	106 to 135	120
F	Forced flow, poor operation	over 135	150

SOURCE: *Highway Capacity Manual,* 1985, Special Report 209 (Washington, DC: Transportation Research Board), p. 12–13.

Pedestrian capacity

The level of service for pedestrian flow depends on relationships among speed, density, and volume. As density of pedestrians increase, the space per pedestrian decreases as does the average speed. For pedestrian walkways, level-of-service criteria are given in Table 5–39. An important operational factor for pedestrian walkways is the *effective width.* The effective width is considered as the total width less the width preempted by street furniture, landscaping, commercial uses, or building protrusions. Preempted width due to traffic signs

and parking meters is considered as 2.0 to 2.5 ft. Benches preempt 5.0 ft, as do planter boxes.

The concept of space available to pedestrians can also be applied to queuing or waiting areas. Street corners, transit platforms, and lobby areas are examples of locations that can be analyzed using this level-of-service concept. The criteria for level of service for queuing areas is given in Figure 5–12.

Improvements for level of service of walkways are aimed principally at increasing the effective width by reducing obstructions. For queuing areas, the two principal means of improvement are increasing the area available and allowing well-defined paths for circulation around the queuing area.

Bikeway capacity

Bicycles have an impact on intersection capacity and on vehicular capacity when the bike lane is located on the street itself. The impact of bicycles in mixed traffic can be taken into account by using passenger car equivalents for bicycles. When lanes are less than 11 ft wide, a bicycle is considered equivalent to a passenger car. When the lane width is from 11 to 14 ft, bicycles are considered as 0.2 to 0.5 passenger cars.

The approximate reported capacity of a one-way bike lane or bike path, with a width of about 4 ft, is 2,000 bicycles per hour. Bikeway capacities are susceptible to the same type of reductions that roadways incur. That is, poor geometry and narrow lane widths will have a negative effect on bikeway capacity.

Other procedures and research findings

Freeways

During the early 1990s it is anticipated that considerable attention will be given to the measurement of traffic flow and speed on freeways. Many agencies have measured sustained flows in excess of 2,000 pcphpl with speeds in excess of 45 or 50 mph. These measured values, if confirmed by additional research and data, imply that the speed-flow curves for freeways in Chapter 3 of the 1985 *Highway Capacity Manual* may need to be revised. Recent research on

LEVEL OF SERVICE A

Average Pedestrian Area Occupancy: 13 sq ft / person or more
Average Inter-Person Spacing: 4 ft, or more
Description: Standing and free circulation through the queuing area is possible without disturbing others within the queue.

LEVEL OF SERVICE B

Average Pedestrian Area Occupancy: 10 to 13 sq ft / person
Average Inter-Person Spacing: 3.5 to 4.0 ft
Description: Standing and partially restricted circulation to avoid disturbing others within the queue is possible.

LEVEL OF SERVICE C

Average Pedestrian Area Occupancy: 7 to 10 sq ft / person
Average Inter-Person Spacing: 3.0 to 3.5 ft
Description: Standing and restricted circulation through the queuing area by disturbing others within the queue is possible; this density is within the range of personal comfort.

LEVEL OF SERVICE D

Average Pedestrian Area Occupancy: 3 to 7 sq ft / person
Average Inter-Person Spacing: 2 to 3 ft
Description: Standing without touching is possible; circulation is severely restricted within the queue and forward movement is only possible as a group; long term waiting at this density is discomforting.

LEVEL OF SERVICE E

Average Pedestrian Area Occupancy: 2 to 3 sq ft / person
Average Inter-Person Spacing: 2 ft or less
Description: Standing in physical contact with others is unavoidable; circulation within the queue is not possible; queuing at this density can only be sustained for a short period without serious discomfort.

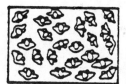

LEVEL OF SERVICE F

Average Pedestrian Area Occupancy: 2 sq ft / person or less
Average Inter-Person Spacing: Close contact with persons
Description: Virtually all persons within the queue are standing in direct physical contact with those surrounding them; this density is extremely discomforting; no movement is possible within the queue; the potential for panic exists in large crowds at this density.

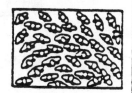

Figure 5–12. Level-of-service criteria for queuing areas.
SOURCE: *Highway Capacity Manual,* 1985, Special Report 209 (Washington, DC: Transportation Research Board), p. 13–12.

multilane highways[13] has led to a recommendation to revise speed-flow curves currently found in Chapter 7 of the 1985 *Highway Capacity Manual.* Coordination between Chapters 3 and 7 and the values within each chapter will be essential in the coming years.

The passenger-car equivalents used for heavy vehicles on freeways may also undergo revision in the early 1990's. Such revisions may be based on the multilane highway research[14] in which a new set of equivalency tables was developed. These new recommended values for multilane highways are given in the following section.

[13] *Capacity and Level-of-Service Procedures for Multilane Rural and Suburban Highways.*

[14] Ibid.

Multilane highways

The NCHRP Project 3-33[15] was undertaken in 1985 with the objective of establishing an extensive base of field data leading to a recommended new procedure for estimating level of service and capacity of multilane highways. In 1989, the draft recommendations were forwarded to NCHRP and to the Transportation Research Board (TRB) Committee on Highway Capacity and Quality of Service. Although some revisions in the draft material are probable, the basic concepts developed from the new data base are considered valid.

The key to the new recommended procedure is that a "free-flow speed" is established for the facility being analyzed. The free-flow speed is defined as the average travel speed of passenger cars operating under conditions of low to moderate traffic volume. The establishment of free-flow speed can be from field observations, from engineering estimates, or from estimates made from speed limit information or from 85th percentile speed measurements. The analyst is encouraged to collect field data to establish the free-flow speed. Table 5–40 gives the interrelationships between speed measures as established from data.

Once the free-flow speed of the facility has been established, the analyst uses a speed-volume relationship along with adjustment factors for lane width and lateral clearance, number of access points per mile on one side of the highway, heavy vehicles, and median type. The heavy-

15 Ibid.

vehicle adjustment factor is used to convert the analysis volume to passenger car units. The other three factors are used to adjust the travel speed on the facility being analyzed. The new procedure is a deviation from the 1985 procedure in which the adjustment factors were considered to impact capacity and service volume, not travel speed.

The shape of the speed-flow relationships derived from the new research is shown in Figure 5–13. The level-of-service criteria, still based on density, are given in Table 5–41. Proposed new passenger car equivalencies for heavy vehicles on specific grades tend to be lower than those found in the 1985 *Highway Capacity Manual,* and are given in Table 5–42. Table 5–43 contains the passenger car equivalents for general multilane highway segments, and Table 5–44 contains the proposed adjustment factors (adjustments to free-flow speed) relating to lane width and lateral clearance, type of median, and number of access points per mile along the right side of the multilane highway. This last factor is of particular

TABLE 5–40

Speed Relationships on Multilane Highways

Posted Speed Limit (mph)	Free-Flow Speed (mph)	85th Percentile Speed (mph)
40	48.6	51.0
45	53.1	58.0
50	54.2	58.0
55	59.5	62.0

SOURCE: NCHRP Project 3-33 (Field Data).

Figure 5–13. Proposed speed flow curves for multilane highways.
SOURCE: NCHRP Project 3-33.

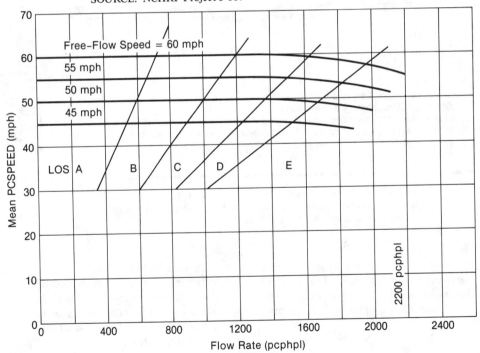

Note: These curves are based on data from NCHRP Project 3-33, Capacity and Level of Service Procedures for Multilane Rural and Suburban Highways, and have been accepted for use in future revisions of the Highway Capacity Manual.

TABLE 5–41

Proposed Level-of-Service Criteria for Multilane Highways

| | Free-Flow Speed | | | | | | | |
| | 60 MPH | | | | 55 MPH | | | |
Level of Service	Max Density (pc/mi/ln)	Min Speed (mph)	Max v/c	Max Flow Rate (pcphpl)	Max Density (pc/mi/ln)	Min Speed (mph)	Max v/c	Max Flow Rate (pcphpl)
A	12	60	0.33	720	12	55	0.30	650
B	20	60	0.55	1,200	20	55	0.50	1,100
C	27	59	0.75	1,630	28	54	0.69	1,500
D	34	57	0.89	1,950	34	53	0.83	1,820
E	40	55	1.00	2,200	44	50	1.00	2,200
F	40	55	*	*	44	50	*	*

*Highly variable, unstable.

| | Free-Flow Speed | | | | | | | |
| | 50 MPH | | | | 45 MPH | | | |
Level of Service	Max Density (pc/mi/ln)	Min Speed (mph)	Max v/c	Max Flow Rate (pcphpl)	Max Density (pc/mi/ln)	Min Speed (mph)	Max v/c	Max Flow Rate (pcphpl)
A	12	50	0.27	590	12	45	0.24	530
B	20	50	0.45	1,000	20	45	0.41	900
C	28	50	0.64	1,390	28	45	0.57	1,260
D	34	49	0.76	1,680	35	44	0.69	1,520
E	49	45	1.00	2,200	55	40	1.00	2,200
F	49	45	*	*	55	40	*	*

*Highly variable, unstable.

SOURCE: NCHRP Project 3-33.

TABLE 5–42

Proposed Passenger Car Equivalents for Multilane Highways

| | | E_T (4- or 6-lane highways) Percent Trucks and Buses | | | | | | | | |
Grade (%)	Length (mi)	2	4	5	6	8	10	15	20	25
<2	All	1.5	1.5	1.5	1.5	1.5	1.5	1.5	1.5	1.5
2	0–1/4	1.5	1.5	1.5	1.5	1.5	1.5	1.5	1.5	1.5
	1/4–1/2	1.5	1.5	1.5	1.5	1.5	1.5	1.5	1.5	1.5
	1/2–3/4	1.5	1.5	1.5	1.5	1.5	1.5	1.5	1.5	1.5
	3/4–1	2.5	2.0	2.0	2.0	1.5	1.5	1.5	1.5	1.5
	1–1 1/2	4.0	3.0	3.0	3.0	2.5	2.5	2.0	2.0	2.0
	>1 1/2	4.5	3.5	3.0	3.0	2.5	2.5	2.0	2.0	2.0
3	0–1/4	1.5	1.5	1.5	1.5	1.5	1.5	1.5	1.5	1.5
	1/4–1/2	3.0	2.5	2.5	2.0	2.0	2.0	2.0	1.5	1.5
	1/2–3/4	6.0	4.0	4.0	3.5	3.5	3.0	2.5	2.5	2.0
	3/4–1	7.5	5.5	5.0	4.5	4.0	4.0	3.5	3.0	3.0
	1–1 1/2	8.0	6.0	5.5	5.0	4.5	4.0	4.0	3.5	3.0
	>1 1/2	8.5	6.0	5.5	5.0	4.5	4.5	4.0	3.5	3.0
4	0–1/4	1.5	1.5	1.5	1.5	1.5	1.5	1.5	1.5	1.5
	1/4–1/2	5.5	4.0	4.0	3.5	3.0	3.0	3.0	2.5	2.5
	1/2–3/4	9.5	7.0	6.5	6.0	5.5	5.0	4.5	4.0	3.5
	3/4–1	10.5	8.0	7.0	6.5	6.0	5.5	5.0	4.5	4.0
	>1	11.0	8.0	7.5	7.0	6.0	6.0	5.0	5.0	4.5
5	0–1/4	2.0	2.0	1.5	1.5	1.5	1.5	1.5	1.5	1.5
	1/4–1/3	6.0	4.5	4.0	4.0	3.5	3.0	3.0	2.5	2.0
	1/3–1/2	9.0	7.0	6.0	6.0	5.5	5.0	4.5	4.0	3.5
	1/2–3/4	12.5	9.0	8.5	8.0	7.0	7.0	6.0	6.0	5.0
	3/4–1	13.0	9.5	9.0	8.0	7.5	7.0	6.5	6.0	5.5
	>1	13.0	9.5	9.0	8.0	7.5	7.0	6.5	6.0	5.5
6	0–1/4	4.5	3.5	3.0	3.0	3.0	2.5	2.5	2.0	2.0
	1/4–1/3	9.0	6.5	6.0	6.0	5.0	5.0	4.0	3.5	3.0
	1/3–1/2	12.5	9.5	8.5	8.0	7.0	6.5	6.0	6.0	5.5
	1/2–3/4	15.0	11.0	10.0	9.5	9.0	8.0	8.0	7.5	6.5
	3/4–1	15.0	11.0	10.0	9.5	9.0	8.5	8.0	7.5	6.5
	>1	15.0	11.0	10.0	9.5	9.0	8.5	8.0	7.5	6.5

Note: If a length of grade falls on a boundary condition, the equivalent from the longer grade category is used. For any grade steeper than the percent shown, use the next higher grade category.

SOURCE: NCHRP Project 3-33.

TABLE 5-43

TABLE 5-43

Proposed Passenger Car Equivalents on General Multilane Highway Segments

Factor	Type of Terrain		
	Level	Rolling	Mountainous
E_T for Trucks and Buses	1.5	3.0	6.0
E_R for RVs	1.2	2.0	4.0

SOURCE: NCHRP Project 3-33.

TABLE 5-44

Proposed Speed Adjustment Factors for Multilane Highways

ADJUSTMENT FOR MEDIAN TYPE

Median Type	Reduction in Free-Flow Speed (mph)
Undivided highways	1.6
Divided highways (including TWLTLs)	0.0

ADJUSTMENT FOR LANE WIDTH ON MULTILANE HIGHWAYS

Lane Width (ft)	Reduction in Free-Flow Speed (mph)
10	6.6
11	1.9
12	0.0

ADJUSTMENT FOR LATERAL CLEARANCE ON MULTILANE HIGHWAYS

Four-Lane Highways		Six-Lane Highways	
Total Lateral Clearance (ft)	Reduction in Free-Flow Speed (mph)	Total Lateral Clearance (ft)	Reduction in Free-Flow Speed (mph)
12	0.0	12	0.0
10	0.4	10	0.4
8	0.9	8	0.9
6	1.3	6	1.3
4	1.8	4	1.7
2	3.6	2	2.8
0	5.4	0	3.9

ACCESS-POINT ADJUSTMENTS

Access Points Per Mile	Reduction in Free-Flow Speed (mph)
0	0.0
10	2.5
20	5.0
30	7.5
40 or more	10.0

SOURCE: NCHRP Project 3-33.

importance for traffic engineers performing operational analyses. Table 5-44 indicates that for every 10 access points per mile along the right side of the roadway, average travel speed will decrease by about 2.5 mph. The adjustment factor for median type indicates that a physical divider or median provides a slightly better travel speed and level of service than does an undivided highway.

As an example of the calculations performed using the new procedure, consider an analysis of a four-lane highway with a speed limit of 45 mph. The highway is undivided, is in level terrain, carries 10% trucks with no buses or RVs, and

carries a peak-hour volume in one direction of 2,400 vph. Given that the lane width is 12 ft and that the lateral clearance on the right is 10 ft to the nearest obstruction, there is no reduction in travel speed due to these factors. The undivided median type produces a reduction in free-flow speed of 1.6 mph. There are 20 access points per mile on the right side of the roadway, giving an additional reduction in free-flow speed of 5.0 mph. The total reduction in free-flow speed from a highway with ideal conditions is 6.6 mph. Thus, a free-flow speed of 46.5 mph (53.1 mph estimated by using Table 5-40 less 6.6 mph for nonideal conditions) is estimated for this highway under existing conditions. Conversion of the volume in one direction to passenger car units is accomplished by using the equivalence for general segments of 1.5 for trucks. The result is 2,520 pcph, giving 1,260 pcphpl as the hourly flow rate for analysis. Using Table 5-41, the estimated level of service for a free-flow speed of 46.5 mph is just Level of Service C (1,260 pcphpl is at the LOS C/LOS D boundary).

Signalized intersections

An example of a critical movement procedure is found in *Transportation Research Circular,* No. 212.[16] The analyst assigns volumes to each traffic lane, identifies the type of signal phasing, checks to determine whether unprotected left turns can be accommodated, and then sums the critical volumes on each of the two streets. This procedure has more detail than does the planning method of the 1985 *Highway Capacity Manual,* but it is still considered a simple critical movement analysis. The criteria for level of service depend on the sum of the critical volumes and on the number of signal phases. Table 5-45 lists the level-of-service ranges for this procedure.

Another critical movement procedure that is used by some practicing engineers is termed "intersection capacity utilization," or ICU. This procedure,[17] like the others, is based on a summation of critical volumes expressed in terms of the proportion of capacity (i.e., the v/c ratio) used by each critical movement at a traffic signal. Levels of service are defined by

TABLE 5-45

Level-of-Service Criteria: TRB Circular 212

	Planning Applications (in vph)		
	Maximum Sum of Critical Volumes		
Level of Service	Two Phase	Three Phase	Four or More Phases
A	900	855	825
B	1,050	1,000	965
C	1,200	1,140	1,100
D	1,350	1,275	1,225
E	1,500	1,425	1,375
F	.... not applicable		

SOURCE: *Transportation Research Circular 212,* (Washington, DC: Transportation Research Board, 1980), p. 11.

---

[16] *Interim Materials on Highway Capacity,* Transportation Research Circular 212, (Washington, DC: Transportation Research Board, 1980).

[17] J.F. Gould, "A Comparison of the 1985 HCM and the ICU Methodologies," *Westernite,* (March-April 1990).

ranges of this summation of v/c ratios. ICU is a simple procedure and is used principally to make quick judgments about the general operating characteristics of an intersection.

Research continues to be performed in the area of signalized intersection capacity and level of service. Recent examples of such research include the study of passenger car equivalencies for large trucks.[18] Results from this work indicate that smaller trucks may have an equivalency of 1.6 passenger cars, whereas large five-axle combination trucks may have equivalencies of 3.1 to 4.1 passenger cars, depending on the truck position in queue at the intersection.

Another important research effort[19] that may lead to revisions in Chapter 9 of the 1985 *Highway Capacity Manual* relates to levels of service in shared-permissive, left-turn lane groups. This research may also lead to a reconsideration of the base saturation flow rate at signalized intersections.

Unsignalized intersections

Research on multiway, stop-controlled intersections[20] may lead to new procedures for analyzing delay and the level of service at stop-controlled locations. This new recommendation provides a means for estimating level of service, based on delay per vehicle, such that results from analysis of a stop-controlled intersection might be compared with results from analysis of a signalized intersection.

REFERENCES FOR FURTHER READING

EMOTO, T.C. AND A.D. MAY, "Operational Evaluation of Passing Lanes in Level Terrain—Final Report," (Sacramento: California Department of Transportation, July 1988).

McSHANE, W.R., AND R.P. ROESS, *Traffic Engineering*, (Englewood Cliffs, NJ: Prentice Hall 1990), p. 660.

FRUIN, J.J., B.T. KETCHAM, AND P. HECHT, "Validation of the Time-Space Corner and Crosswalk Analysis Method," *Transportation Research Record,* 1168 (1988), 39–44.

LEVINSON, H.S., "Streets for People and Transit," *Transportation Quarterly,* 40, (October 1986), 503–520.

SENEVIRATNE, P.N., AND J.F. MORRALL, "Level of Service on Pedestrian Facilities," *Transportation Quarterly,* 39, (January 1985), 109–123.

SHARMA, S.C., "Driver Population Factor in New Highway Capacity Manual," *Journal of Transportation Engineering,* 113, (September 1987), 575–579.

"Highway Capacity, Traffic Characteristics, and Flow Theory," *Transportation Research Record,* 1005 (1985), 128.

COLLIER, C.A., "Analysis of the Cost-Effectiveness of Improving Urban at Grade Intersections—Final Report," (Washington, DC: Federal Highway Administration, March 1988).

CROMMELIN, R.W., "Employing Intersection Capacity Utilization Values to Estimate Overall Level of Service," *Traffic Engineering,* (July 1974).

"Highway Capacity, Traffic Characteristics, and Flow Theory," *Transportation Research Record,* 1112, (1987).

"Traffic Flow Theory, Characteristics, and Highway Capacity," *Transportation Research Record,* 1091 (1987).

---

[18] Cesar J. Molina, "Development of Passenger Car Equivalencies for Large Trucks at Signalized Intersections," *ITE Journal,* (November 1987).

[19] *Levels of Service in Shared-Permissive Left-Turn Lane Groups at Signalized Intersections,* Report No. FHWA-RD-89-228.

[20] Michael Kyte, *Estimating Capacity and Delay at an All-Way Stop-Controlled Intersection,* University of Idaho, 1989.

6

ROADWAY GEOMETRIC DESIGN

TIMOTHY R. NEUMAN, P.E., *Senior Transportation Engineer*

CH2M HILL

Introduction

The term *geometric design* refers to the three-dimensional features of a highway that relate, affect, or are directly related to its operational quality and safety. These features, visible to the driver, include the cross section (lanes and shoulders, roadside slopes, and clear areas), intersections, channelization, interchanges, and the horizontal and vertical alignment of the highway.

Geometric design translates operational experience and research into the actual highway. It thus reflects the human characteristics of drivers, as well as the physical and operational characteristics of vehicles. Geometric design also involves a knowledge of construction and maintenance costs and other considerations.

For many reasons, geometric design is and will always be dynamic. The age and abilities of the driving population as well as their desires and expectations are changing over time. So, too, are the vehicles on the highways. Societal values also change over time, with resulting effects on what is considered reasonable or good geometric design. Environmental concerns, energy conservation, and other issues have influenced geometric design.

Despite the dynamic character of geometric design, the basic objectives of good design will always be to produce a highway that provides safe and efficient transportation, which reflects the characteristics of drivers and vehicles that will use it, and that represents a reasonable trade-off in terms of its costs and other impacts.

The evolution of geometric design standards and criteria in the United States is summarized in Table 6-1. The American Association of State Highway and Transportation Officials (AASHTO) has emerged as the source of most of the

TABLE 6-1

Evolution of AASHTO (AASHO) Design Policies in the United States

A Policy on Highway Classification, September 16, 1938
A Policy on Highway Types (Geometric), February 13, 1940
A Policy on Sight Distance for Highways, February 17, 1940
A Policy on Criteria for Marking and Signing No-Passing Zones for Two- and Three-Lane Roads, February 17, 1940
A Policy on Intersections at Grade, October 7, 1940
A Policy on Rotary Intersections, September 26, 1941
A Policy on Grade Separations for Intersecting Highways, June 19, 1944
A Policy on Design Standards—Interstate, Primary and Secondary Systems
Policies on Geometric Highway Design, 1950
A Policy on Geometric Design of Rural Highways, 1954
A Policy on Arterial Highways in Urban Areas, 1957
A Policy on Geometric Design of Rural Highways, 1965
A Policy on Design of Urban Highways and Arterial Streets, 1973
A Policy on Geometric Design of Highways and Streets, 1984
A Policy on Geometric Design of Highways and Streets, 1990

design values and criteria used in geometric highway design. Although most states and other agencies have developed their own design standards, the design approach and design values shown in the AASHTO policies are accepted by consensus and form the basis for individual state design practices. The Federal Highway Administration (FHWA) has adopted the AASHTO policies for design and construction and major reconstruction of Federal-aid highways.

As Table 6-1 illustrates, beginning in 1940 AASHTO design policies have been issued at 5- to 10-year intervals. Policy revisions are prepared in response to the continuing operational and safety experience, direct and applied research, changes in vehicle fleet characteristics, and other factors. Much of the research that forms the basis for AASHTO policy updates is conducted under programs sponsored by the FHWA, and by AASHTO and the individual states through the National Cooperative Highway Research Program (NCHRP).

Design types of highways

The most recent edition of the AASHTO *Policy* (1990) explicitly recognizes the importance of highway functional classification.[1] This term refers to the different types or classes of highways that comprise a complete system. These different classes are characterized by the nature and types of trips that take place, lengths of trips, and general traffic volume conditions. A complete highway system includes fully controlled access freeways, principal or major arterials, minor arterials, collector roads and streets, and local roads. This functional classification holds true for both rural and urban highways.

The structure of design criteria and standards under a functional-classification-based system is important to understand. Design values and level of service vary primarily according to the function of the highway facility. An important but secondary factor is the design volume of traffic.

What follows is a brief description of the major highway functional types.

Freeways

Although freeways serve the same function as principal arterials, they are sufficiently different in their characteristics to warrant separate treatment as a highway type. Freeways are fully controlled access highways, with no at-grade intersections or driveway connections. The design of both rural and urban freeways is discussed in Chapter VIII of the 1990 AASHTO *Policy*.[2] Design values and criteria are intended to provide for the highest level of safety and availability of service of all the highway types. Design standards for freeways also apply to other fully controlled access facilities such as toll roads.

Arterials

Principal and minor arterials carry longer-distance major traffic flows between important activity centers. They form the backbone of a highway system and should be designed to provide for as high a level of service as is practical. Chapter VII of the 1990 AASHTO *Policy* describes design for both rural and urban arterials.[3] *Guidelines for Urban Major Street Design,*[4] published by ITE, also contains design criteria for urban arterials.

Collector roads and streets

Collector roads link local streets with the arterial street system. They "collect" traffic in local areas, serve locally as

through facilities, and also directly serve abutting land uses. In their appearance collectors often are indistinguishable from local streets, although the volume of traffic along them may be higher. Chapter VI of the 1990 AASHTO *Policy* discusses design of collector roads and streets in rural and urban areas.[5]

Local roads and streets

Local roads and streets serve shorter local trips. Their primary function is to provide access to abutting land uses. Traffic volumes and speeds are generally low on such facilities. Chapter V of the 1990 AASHTO *Policy* addresses design of local roads and streets.[6] An additional reference, *Recommended Guidelines for Subdivision Streets,* published by ITE, also discusses local road design.[7]

Design controls and criteria

A number of basic design controls and criteria govern the manner in which a highway is designed. These controls, discussed below, apply to all highway types.

Control of access

Control of access refers to the legal limitation or restriction of access from private properties to public highway rights-of-way. The quality of operation and safety of traffic flow is greatly affected by the type and manner of access control. This should be established in location and design of a highway, and is significantly related to the functional classification of the highway.

Fully controlled access facilities include freeways, toll roads, and other similar facilities. No access is provided or allowed except at carefully designated and designed locations. Fully controlled access highways have no at-grade intersections or driveways along them, nor should they have at-grade railroad crossings.

Primary arterials, collectors, and even local streets are generally designed with some degree of access control. High-volume, high-speed arterials should be designed to limit the locations of at-grade intersections and to avoid an unreasonable number of driveways. Partial access control is accomplished on such highways through driveway consolidation, frontage roads, and other means. Even lower-class facilities may be designed with access control in mind. Driveways should be kept away from intersections to avoid multiple conflicts.

Design speed

The *design speed* of a highway is one of the basic controls that defines its physical and operational characteristics. Design speed is the maximum safe speed that can be maintained over a specified section of highway when conditions

---

[1] American Association of State Highway and Transportation Officials, *A Policy on Geometric Design of Highways and Streets.* (Washington, DC, 1990).

[2] Ibid.

[3] Ibid.

[4] Institute of Transportation Engineers, *Guidelines for Urban Major Street Design.* (Washington, DC, 1984).

[5] *A Policy on Geometric Design of Highways and Streets.*

[6] Ibid.

[7] Institute of Transportation Engineers, *Recommended Guidelines for Subdivision Streets.* (Washington, DC, 1984).

TABLE 6–2

**Recommended Range of Design Speeds by
Location and Functional Highway Classification
(speeds in mph)**

Terrain	Rural	Urban
	Freeways	
Flat	70–80	70
Rolling	60–70	60–70
Mountainous	50–60	50–60
	Arterial Highways	
Flat	60–70	30–60
Rolling	40–60	30–50
Mountainous	30–50	30–50
	Collector and Local Roads	
Flat	30–50	30–40
Rolling	20–40	20–40
Mountainous	20–30	20–30

are so favorable that the design features of the highway govern. This definition strongly implies that a design speed should be selected based on driver expectations, the type of highway, and terrain and topography.

The selection of a design speed is among the most critical, early decisions in the highway planning and design process. The design of the basic geometric features—including horizontal and vertical alignment, sight distance and in many cases cross section—is sensitive to design speed. While a higher design speed will generally result in slightly higher construction costs and/or more difficulties in design, there are serious implications of selecting too low a design speed. Drivers may overdrive critical locations, thus creating safety problems. Correction of a design problem related to too low a design speed is usually costly and often impossible to accomplish.

The recommended range for design speeds for the basic functional highway types is summarized in Table 6–2. These design speeds should be used regardless of political or legal restraints such as the 55 mph national speed limit or municipal speed restrictions.

Traffic volume considerations

Basic design controls are the volume and distributional characteristics of the traffic stream. For design of new highways or major reconstruction of existing highways, the design volume is a forecast for a designated design year. This design year is usually 20 years hence, which represents the maximum time period over which land use activity and the resulting traffic volumes and patterns can be reasonably forecast. Designers should recognize, however, that rights-of-way, structures, major grading, and other capital improvements will serve traffic well beyond the 20-year forecast time period.

For low-volume roadways and local roads and streets, the average daily traffic (ADT) is referenced in selecting geometric design controls. In such cases, peak-hour traffic volume is so low as to not approach the capacity of the facility. Variations in design by ADT are reflections of the cost trade-offs and safety considerations associated with such minor roadways.

For all other highways, the traffic volume basis for design is the *design hour*. Selection of an appropriate design-hour volume represents a trade-off between designing capacity for a recurring pattern of traffic, and the costs of providing additional capacity. Design-hour volumes typically range from the 30th highest hourly volume occurring in a year (generally applicable to rural highways) to the 50th highest hourly volume (typically applicable to recreational routes) to a 200th highest or typical weekday peak-hour volume (generally used in urban areas). Chapter 2 provides additional information on traffic volume variations.

Peak-hour or 30th to 50th highest-hour volumes are forecast using existing volume characteristics of similar highways in the general area of the facility. The characteristics include K, the percentage of ADT occurring in the design hour; D, the percentage of two-way design-hour volume occurring in the peak direction; and T, the percentage of trucks and other heavy vehicles in the traffic stream in the design hour. The above characteristics are included in the "Design Designation" of a highway, discussed later in this chapter.

Level of service and capacity

Forecast design-hour traffic is used to "size" a highway, to check its operating characteristics as design progresses, and to control refinement of certain design details. *Sizing* refers to a determination of the required number of lanes and the need for special features such as auxiliary lanes, passing or climbing lanes, and intersection channelization requirements.

Procedures for translating design-hour traffic to actual design requirements are outlined in the 1985 Transportation Research Board Special Report 209, *Highway Capacity Manual* (hereafter the HCM).[8] This publication, based on years of research and operational experience, describes the relationships of traffic volume to operational quality for freeways, two-lane and multilane rural highways, and urban streets and intersections.

In sizing any highway, the designer has the opportunity to determine the level of service intended for design year traffic conditions. Levels of service are defined in the HCM, ranging from A (highest quality of flow) to F (breakdown, failure), with Level of Service E representing the capacity of the highway facility. Chapter 5 contains a more detailed description of the factors and considerations associated with highway capacity and level of service. Guidelines for selection of a design level of service are shown in Table 6–3 for the range of highway types.

As a final note regarding traffic volume, the designer should always remember that design-year traffic is a forecast or estimate. It reflects a set of assumptions, including land use and demographic forecasts which could change easily and alter design-year traffic. For these reasons, design of a highway facility should never be strictly tied to one traffic volume projection. Sufficient flexibility should be retained in the design to accommodate a volume and pattern of traffic different from that which was forecast.

Similarly, while peak-period traffic volumes generally control the sizing of a highway, designers should recognize

---

[8] National Research Council, *Highway Capacity Manual.* Transportation Research Board Special Report 209. (Washington, DC, 1985).

TABLE 6–3

Recommended Design Levels of Service by Location and Functional Highway Classification

Terrain	Rural	Urban
	Freeways	
Flat	B	C–D*
Rolling	B	C–D*
Mountainous	B–C	C–D*
	Arterial Highways	
Flat	B	C–D*
Rolling	B	C–D*
Mountainous	B–C	C–D*
	Collector and Local Roads	
Flat	C	C
Rolling	C–D	D
Mountainous	C–D	D

*Design for Level of Service E (capacity) may be required in extremely constrained cases.

that most of the total travel will occur in off-peak periods. Hence, designs should safely accommodate the different speeds and other conditions that will occur in off-peak periods.

Design vehicle

The design vehicle is a selected motor vehicle, with dimensions and/or operating characteristics of such a critical nature that they influence or control the design of one or more highway elements. Part of the design process is the determination of which design vehicle(s) is/are appropriate for a given condition.

Many "design vehicle" assumptions are imbedded in current standards and criteria. These include driver eye height, vehicle headlight and tail light height, braking ability, and acceleration capabilities of passenger cars as well as heavy

vehicles. As the characteristics of the vehicle fleet have evolved over the years, AASHTO criteria have been adjusted.

The 1990 AASHTO *Policy* describes a range of possible design vehicles, as shown in Table 6–4.[9] These vehicles, and their operating characteristics (see Figure 6–1), are used as design controls in determining the radius and width of traveled way in intersection areas and in designing interchanges. The choice of design vehicle is influenced by the system or classification of highway, and by the proportions of the various types and sizes of vehicles expected to use the facility. The single-unit (SU) vehicle is generally the smallest vehicle used in the design of public highways. On most highways, one of the design semitrailer combinations should be considered in design, particularly where turning roadways are bordered by curbs or islands. Even for occasional use by semitrailer combinations, the pavement should be made sufficiently wide to accommodate such vehicles. Special vehicle configurations larger than the WB-50 combination are permitted on the interstate system and along adjoining roadways to access service facilities and terminal buildings. These vehicles include "double-bottom" semitrailers (designated as WB-65 and WB-70) and, in some western states in the United States, triple-bottom semitrailers.

Other design vehicle controls include vehicle height, which affects minimum vertical clearances:

Passenger vehicles (parking garages)	8 ft
Buses (bus terminals and parkways)	12.5 ft
Trucks	14 ft

On some highway systems or on isolated highways, provision for extra high loads is made, utilizing structure clearances of 15 to 17 ft, rather than the normal maximum legal clearance.

---

[9] *A Policy on Geometric Design of Highways and Streets.*

TABLE 6–4

Design Vehicle Dimensions and Turning Properties (in ft)

Vehicle Designation	L	WB	A	B	wb_1	wb_2	w	u	Minimum Turn U**	Minimum Turn F_A	Minimum Turn R_T
Semitrailer large WB-50	55.0	50.0	3.0	2.0	18.0	30.0	8.5	8.5	25.2	1.3	45.0
Semitrailer medium or small WB-40	50.0	40.0	4.0	6.0	13.0	25.0	8.5	8.5	20.1	1.5	40.0
Bus large B-40	40.0	25.0	7.0	8.0	—	—	8.5	8.5	16.8	4.5	42.0
Single-unit truck or bus-medium SU-30	30.0	20.0	4.0	6.0	—	—	8.5	8.5	13.6	2.0	42.0
Pass car-large or delivery van P	19.0	11.0	3.0	5.0	—	—	7.0	6.0	8.7	2.0	24.0
Pass car* medium P_m	18.0	10.0	3.0	5.0	—	—	6.5	6.0	8.5	1.8	21.0
Semitrailer* extra large WB-60	66.0	60.0	3.0	3.0	18.0	40.0	8.5	8.5	35.2	1.3	45.0

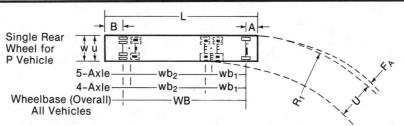

*Presently not part of AASHTO design vehicle designation
**Maximum track width for a 180 degree turn

SOURCE: *Turning Vehicle Templates Instruction Manual*, 1988, Jack E. Leisch & Associates.

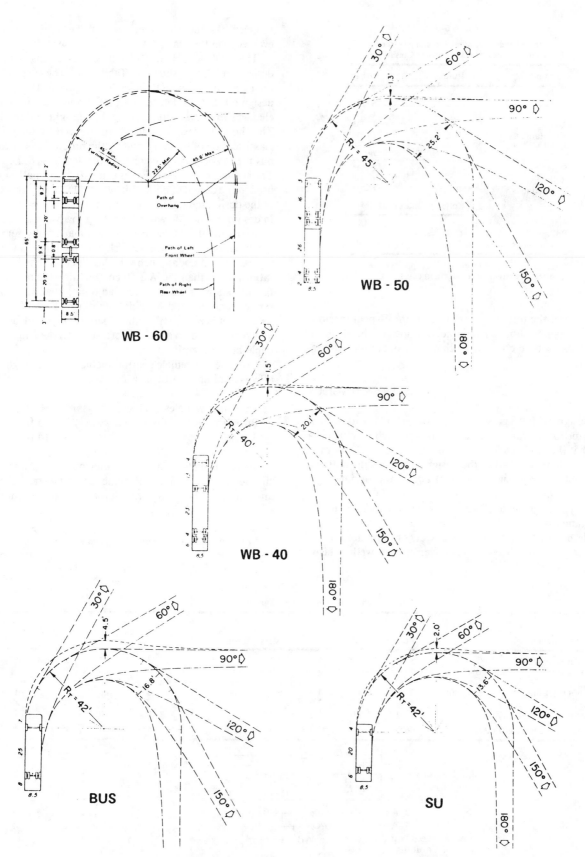

Figure 6–1. AASHTO design vehicles.
SOURCE: *Intersection Channelization Design Guide,* National Cooperative Highway Research Program Report 279, 1985.

Large-vehicle operating characteristics also control design of alignment. The weight to horsepower ratio (WT/HP) of a fully loaded truck affects its ability to accelerate or maintain speed on upgrades. A range of design vehicles, defined in terms of their WT/HP ratio, is identified by AASHTO.

Design driver

The concept of a *design driver,* while not explicitly noted within design criteria, is nonetheless apparent as a design control. Foremost within this concept is the assumption that the driver operating on the highway: (1) is alert and in control of physical and mental abilities (i.e., is not under the influence of drugs or alcohol nor overly fatigued); (2) has reasonable ability to see and perceive the roadway environment; and (3) has reasonable motor skills to enable steering, braking, and other necessary operations.

Perception-reaction times, tolerance to centripetal acceleration, gap acceptance, and speed perception are among the quantifiable human factors that influence design. Other less quantifiable but important human factors considerations include the capacity of a driver to absorb visual and other cues, make navigational decisions, and avoid dangerous situations. Designers should recognize human limitations and avoid geometrics that overload or severely tax the driver. These driver characteristics are presented in more detail in Chapter 1.

Safety considerations

Maximizing the safety of the highway should be the primary objective of design in all cases. Almost without exception, design standards and criteria have evolved with safety as a major consideration.

Designers should be aware of the safety implications of design alternatives. These implications differ greatly with the type of highway, volume conditions, and geometric characteristics. The following is a brief summary of important safety considerations related to design.

Access control. Full access control is the single greatest contributor to highway safety of any design control or feature. The construction and expansion of the interstate highway system has saved thousands of lives due primarily to adherence to full access control. Table 6–5 documents the difference in fatal and injury accident rates nationwide on interstate and other highways. Rates for both rural and urban highway types are from two to three times those for interstates.

On nonfreeway facilities, the frequency and location of intersections and driveways strongly affect safety experience. Rear-end, turning, and angle collisions result from speed differentials caused by turning drivers.

Cross section. The design quality of lanes, shoulders, and the roadside all contribute to safety. Figure 6–2 demonstrates the sensitivity of accident rates on two-lane highways to lane and shoulder widths. Note that relative accident rates are significantly higher for roadways with narrow (less than 11 ft) lanes and narrow (less than 4 ft) shoulders.

TABLE 6–5

Fatal and Injury Accident Rates on U.S. Highways in 1985
(accidents per 100 milion vehicle miles)

	Rural		Urban	
	FAR[a]	IAR[b]	FAR	IAR
Interstate	1.20	25.14	0.88	49.57
Federal aid primary	3.01	71.15	1.67	130.33
Federal aid urban				
Arterial	—	—	2.27	172.91
Collector	—	—	1.74	147.77
Federal aid secondary				
collector	3.66	104.87	—	—
Non-federal aid arterial	2.03	41.30	2.20	77.31
Non-federal aid collector	3.17	144.95	1.74	93.08

[a]Fatal Accident Rate.
[b]Nonfatal Injury Accident Rate.

SOURCE: *Fatal and Injury Accident Rates on Federal Aid and Other U.S. Highways,* U.S. Department of Transportation, 1985.

Alignment design. Accident rates on isolated horizontal curves are about three times the rates for tangents. Figure 6–2 also shows the sensitivity of accident rate to degree of curve for two-lane highways. Major contributors to high accident rates on curves are the roadside design features such as shoulder width, sideslope, and clear recovery zone width.

Sight distance. The relative safety of highway sections with limited sight distance varies significantly, depending on the presence of other features within the sight distance-related zone. Limited sight distance combined with intersections, driveways, narrow bridges, or sharp curves may pose particular problems.

Intersections. Intersections, as natural points of conflict, should be considered differently from other highway types. Traffic volume, type of traffic control and particularly protection for left-turning traffic are all factors affecting intersection safety (see Figure 6–2).

Traffic volume. Traffic volume plays a significant role in the types and rates of accidents on all highway types. On very low-volume highways, vehicle-vehicle conflicts are few, so single-vehicle accidents predominate. On high-volume roads, multivehicle accidents are more frequent. Figure 6–2 summarizes important traffic volume effects to be considered in design.

Selection of design standards

An early step in any highway design project is determination of major design controls that would apply, as well as design standards to be used. There are two different types of projects for which design standards may differ.

New construction and major reconstruction (4R). Design values and criteria described in the AASHTO policies are intended for use in projects involving new highways or where major reconstruction will occur. The latter may involve addition of lanes, alignment upgrading, or other major improvements.

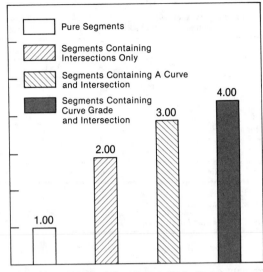

Relative Accident Rates (Ohio Data)

Relationship between Accidents and Combined Geometric Effects

Reference:

Accident Rates as Related to Design Elements of Rural Highways, National Cooperative Highway Research Program Report 49, 1968

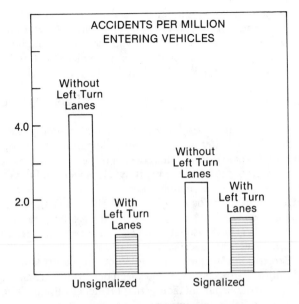

Relationship between Accidents at Intersections and Provision for Left–turn Lanes

Reference:

Evaluation of Left Turn Lanes as a Traffic Control Device, 1973, Ohio Department of Transportation, T.J. Foody and W. C. Richardson

Figure 6–2. Key geometric safety relationships.

In certain instances, adherence to one or more specific design values may present an undue cost or burden, or involve unacceptable environmental or societal impacts. *Design exceptions* to standards may be considered, but only following a rigorous evaluation of alternative designs. Design exceptions are often considered on major reconstruction projects, where the existing highway was designed to an outdated standard, and right-of-way availability is limited or costs to upgrade the highway to current standards would be prohibitive.

Resurfacing, restoration and rehabilitation (3R). Projects involving 3R work on existing highways are designed in a different manner from those implied by the AASHTO policies. There is no common or national set of values to which 3R design conforms; instead, different design approaches are employed in different parts of the country. Most states have adopted 3R design standards for their specific problems and conditions. A special study on 3R design[10] concluded that highway agencies should incorporate the following steps in the 3R design process:

1. Assess current physical and operational conditions, including accidents and traffic characteristics.
2. Incorporate intersection, roadside, and traffic control improvements that may enhance safety at the same time that pavement repairs and geometric improvements are considered.
3. Prepare a safety and design report that fully documents existing characteristics, applicable design standards, safety

problems, related design options, and rationale for any proposed design exceptions.
4. Review all safety and design reports with traffic and safety engineers prior to final approval and construction.

Design designation

Each design project, whether it is a new highway, 4R or 3R project, requires establishment of the major design controls prior to plan preparation. Design-study reports and contract plans should include a summary of the following information, used in developing the plan:

Design Year
Average Daily Traffic (current year)
Average Daily Traffic (design year)
Design Hourly Volume
Directional Distribution of Traffic
Trucks (percent of ADT and design hourly volume)
Design Speed
Control of Access
Design Level of Service

Other necessary information for geometric design includes:

Pedestrian volumes and locations of crossings
Type, location, and nature of parking (existing or additional required)
Design vehicle(s)

---

[10]National Research Council, *Designing Safer Roads—Practices for Resurfacing, Restoration and Rehabilitation.* Transportation Research Board Special Report 214. (Washington, DC, 1987).

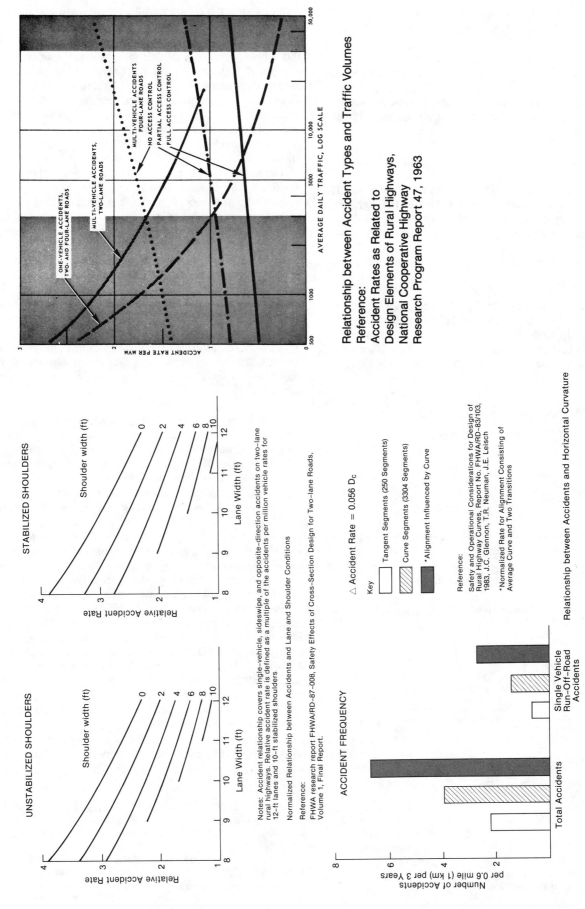

STABILIZED SHOULDERS

Shoulder width (ft)

0
2
4
6
8
10

Relative Accident Rate

Lane Width (ft)

UNSTABILIZED SHOULDERS

Shoulder width (ft)

0
2
4
6
8
10

Relative Accident Rate

Lane Width (ft)

Notes: Accident relationship covers single-vehicle, sideswipe, and opposite-direction accidents on two-lane rural highways. Relative accident rate is defined as a multiple of the accidents per million vehicle rates for 12-ft lanes and 10-ft stabilized shoulders

Normalized Relationship between Accidents and Lane and Shoulder Conditions

Reference:
FHWA research report FHWA/RD-87-008, Safety Effects of Cross-Section Design for Two-lane Roads, Volume 1, Final Report.

MULTI-VEHICLE ACCIDENTS, FOUR-LANE ROADS
NO ACCESS CONTROL
PARTIAL ACCESS CONTROL
FULL ACCESS CONTROL
ONE-VEHICLE ACCIDENTS, TWO- AND FOUR-LANE ROADS
MULTI-VEHICLE ACCIDENTS, TWO-LANE ROADS

ACCIDENT RATE PER MVM

AVERAGE DAILY TRAFFIC, LOG SCALE

Relationship between Accident Types and Traffic Volumes

Reference:
Accident Rates as Related to
Design Elements of Rural Highways,
National Cooperative Highway
Research Program Report 47, 1963

△ Accident Rate = 0.056 D_c

Key

☐ Tangent Segments (250 Segments)
▨ Curve Segments (3304 Segments)
▓ *Alignment Influenced by Curve

Reference:
Safety and Operational Considerations for Design of Rural Highway Curves, Report No. FHWA/RD-83/103, 1983, J.C. Glennon, T.R. Neuman, J.E. Leisch

*Normalized Rate for Alignment Consisting of Average Curve and Two Transitions

ACCIDENT FREQUENCY

Number of Accidents per 0.6 mile (1 km) per 3 Years

Single Vehicle Run-Off-Road Accidents

Total Accidents

Relationship between Accidents and Horizontal Curvature

Figure 6–2. Key geometric safety relationships (*cont.*).

Intersection turning-movement volumes

Type of traffic control required

Design elements

The basic design elements common to all highways include sight distance, horizontal and vertical alignment, and the cross section.

Sight distance

Sight distance is the length of highway visible to the driver. It results from the three-dimensional design of the highway, and is a primary design control for all highway types. Four types of sight distance must be considered in design—*stopping sight distance* (SSD), *passing sight distance* (PSD), *decision sight distance* (DSD), and *intersection sight distance* (ISD).

Stopping sight distance (SSD). Stopping sight distance (SSD) is the most basic form of sight distance. In concept, a highway should be designed to provide enough distance for the majority of drivers to stop safely to avoid collision with an object in the road. Figure 6–3 illustrates the AASHTO

Figure 6–3. AASHTO stopping sight distance model.

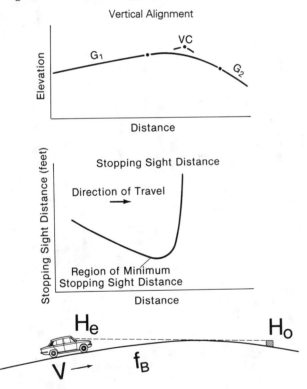

$$H_e = 3.5 \text{ feet}$$
$$H_0 = 0.5 \text{ feet}$$
$$V = \text{Design Speed (mph)}$$
$$f = \text{Coefficient of Braking Friction at Design Speed}$$

$$G = \text{Grade (percent)}$$
$$P = \text{Perception/reaction Time (2.50 sec.)}$$

$$SSD = 1.47\ PV + \frac{V^2}{30\ (f+G)}$$

operational model for SSD, which includes assumptions for driver behavior and visibility of an object.

Current AASHTO design policy recommends the use of this simple operational model, with the same basic assumptions for all highway types. A range of design values for SSD for a given design speed is shown in Table 6–6. This range is based on two assumptions—initial vehicle speed at a lower operating speed, and initial vehicle speed at full design speed prior to braking. The design value for coefficient of friction is intended to be conservative, representing a combination of relatively smooth tires on a wet pavement.

The limitations of the AASHTO SSD model should be noted and considered in design. It does not account for the longer braking distances of trucks. It assumes braking on tangent alignment, with no adjustments for additional needs on curves. It also assumes a relatively hard braking response by the driver, commensurate with collision avoidance.

Application of SSD requirements should reflect the more complex driving environment and variable safety conditions encountered on the highway. Designers should provide for SSD greater than the minimum requirements shown in Table 6–6 at locations where vehicle conflicts or hazardous conditions occur (e.g., on approaches to intersections, in advance of sharp curves, or prior to significant changes in the cross section). Plotting or calculating a SSD profile, also shown in Figure 6–3, is useful in identifying roadway segments with minimum SSD.

Passing sight distance (PSD). Passing sight distance (PSD) refers to that distance made available to drivers on two-lane highways to pass slower vehicles. The amount of PSD is a measure of both operational quality and safety. The AASHTO operational model for PSD considers it as a function of four distances: (1) distance during perception, reaction, and acceleration of the passing vehicle to encroachment on the opposing lane; (2) distance traveled by the passing vehicle in the opposing lane; (3) distance between the passing vehicle at the end of a pass and an oncoming opposing vehicle; and (4) distance traveled by an opposing vehicle for two-thirds of the time the passing vehicle occupies the left lane.

This model requires many assumptions about speeds of all vehicles involved, acceleration capabilities, and so forth. The resulting design controls for PSD are shown in Table 6–7. Both minimum PSD and minimum passing are shown in the table. (Note that PSD is measured from a 3.5-ft eye height to a 4.25-ft object height. PSD values are thus not directly comparable to SSD values shown in Table 6–6.) Recent research[11] has questioned the need for long passing zones implied by AASHTO. Table 6–7 also compares AASHTO PSD values with lengths recommended by the *Manual on Uniform Traffic Control Devices*[12] for marking passing zones; and with values suggested by Harwood and Glennon.[13] The latter figures are based on a different operational model of the passing maneuver.

[11] D.W. Harwood, and John C. Glennon, "Passing Sight Distance Design for Passenger Cars and Trucks," Transportation Research Board Record No. 1208 (1989).

[12] *Manual on Uniform Traffic Control Devices for Streets and Highways* (Washington, DC: U.S. Department of Transportation, 1988).

[13] Harwood and Glennon, "Passing Sight Distance Design."

TABLE 6–6
Stopping Sight Distance Design Requirements

Design Speed (mph)	Assumed Speed for Condition (mph)	Brake Reaction Time (sec)	Brake Reaction Distance (ft)	Coefficient of Friction f	Braking Distance on Level[a] (ft)	Stopping Sight Distance Computed[a] (ft)	Stopping Sight Distance Rounded for Design (ft)
20	20–20	2.5	73.3–73.3	0.40	33.3–33.3	106.7–106.7	125–125
25	24–25	2.5	88.0–91.7	0.38	50.5–54.8	138.5–146.5	150–150
30	28–30	2.5	102.7–110.0	0.35	74.7–85.7	177.3–195.7	200–200
35	32–35	2.5	117.3–128.3	0.34	100.4–120.1	217.7–248.4	225–250
40	36–40	2.5	132.0–146.7	0.32	135.0–166.7	267.0–313.3	275–325
45	40–45	2.5	146.7–165.0	0.31	172.0–217.7	318.7–382.7	325–400
50	44–50	2.5	161.3–183.3	0.30	215.1–277.8	376.4–461.1	400–475
55	48–55	2.5	176.0–201.7	0.30	256.0–336.1	432.0–537.8	450–550
60	52–60	2.5	190.7–220.0	0.29	310.8–413.8	501.5–633.8	525–650
65	55–65	2.5	201.7–238.3	0.29	347.7–485.6	549.4–724.0	550–725
70	58–70	2.5	212.7–256.7	0.28	400.5–583.3	613.1–840.0	625–850

[a]Different values for the same speed result from using unequal coefficients of friction.

SOURCE: *A Policy on Geometric Design of Highways and Streets,* 1990, American Association of State Highway and Transportation Officials.

TABLE 6–7

Passing Sight Distance Design Requirements

Design or Prevailing Speed (mph)	Passing Sight Distance (ft) as given by AASHTO Policy	Passing Sight Distance (ft) as given by MUTCD[a] Criteria	Passing Sight Distance (ft) for Passenger Car Passing Passenger Car	Passing Sight Distance (ft) for Passenger Car Passing Truck
20	800	—	325	350
30	1,100	500	525	575
40	1,500	600	700	800
50	1,800	800	875	1,025
60	2,100	1,000	1,025	1,250
70	2,500	1,200	1,200	1,450

[a]*Manual on Uniform Traffic Control Devices for Streets and Highways.*

SOURCE: D. W. HARWOOD and J. C. GLENNON, "Passing Sight Distance Design for Passenger Cars and Trucks," Transportation Research Board, Record No. 1208, 1989.

Decision sight distance (DSD). Decision sight distance (DSD) is that distance required for a driver to perceive an unexpected or complex situation, arrive at a decision regarding a course of action, and execute that decision in a reasonable manner. Such driver behavior produces considerably longer time and hence greater distance than that produced by stopping sight distance.

Table 6–8 summarizes DSD requirements for a range in design speeds and conditions.[14] While it is generally desirable to maximize sight distance wherever possible, DSD should be provided at or in advance of the following:

interchanges

intersections

lane drops

abrupt or major horizontal alignment changes

narrow bridges

toll facilities

Designers should avoid locating the above features where DSD is difficult or impossible to achieve.

---

[14] *A Policy on Geometric Design of Highways and Streets.*

TABLE 6–8

Decision Sight Distances

Design Speed (mph)	Decision Sight Distance for Avoidance Maneuver (ft) A	B	C	D	E
30	220	500	450	500	625
40	345	725	600	725	825
50	500	975	750	900	1,025
60	680	1,300	1,000	1,150	1,275
70	900	1,525	1,100	1,300	1,450

The following are typical avoidance maneuvers covered in the above table.
- Avoidance Maneuver A: Stop on rural road.
- Avoidance Maneuver B: Stop on urban road.
- Avoidance Maneuver C: Speed/path/direction change on rural road.
- Avoidance Maneuver D: Speed/path/direction change on suburban road.
- Avoidance Maneuver E: Speed/path/direction change on urban road.

SOURCE: *A Policy on Geometric Design of Highways and Streets* (Washington, DC: American Association of State Highway and Transportation Officials, 1990).

Intersection sight distance (ISD). At-grade intersections are inherent points of potential vehicle-vehicle conflict. A driver approaching an intersection should have an unobstructed view of sufficient length to permit control of the vehicle to avoid collision. The AASHTO guideline presents four cases for intersection control, each of which results in different intersection sight-distance requirements:

I. No control, with vehicles adjusting speeds to avoid collision
II. Yield control, with vehicles on the minor roadway yielding to the major roadway
III. Stop control on the minor roadway
IV. Signal control

Cases III and IV are the most common, with Case III representing the most critical conditions generally encountered. Within Case III are a range of possible operational assumptions regarding the stopped approach. Figure 6–4 shows sight distance for crossing the major roadway from the stop (Case III A). The distances shown on Figure 6–4 are a function of vehicle type and the width of the roadway being crossed. These values are measured from a 3.5-ft eye height to a 4.25-ft object height.

Horizontal alignment

The horizontal alignment of a highway is comprised of tangents, circular curves, and spiral transition curves. Circular curves are described as either curves of a constant radius, or as a degree of curve using the following mathematical relationship:

where:

$$D_c = 5729.6/R \qquad (6.1)$$

D_c = central angle of a curve subtended by a 100-ft length of arc
R = radius of curve (ft)

Design of horizontal alignment is based on the laws of mechanics as well as consideration of driver comfort. Superelevation is utilized on highway curves to counteract the centrifugal forces acting on a vehicle and driver when proceeding through the curve. The following relationship describes the interaction among curve design characteristics, vehicle speed, and superelevation:

where:

$$R = V^2/15(e + f) \qquad (6.2)$$

R = radius of curve (ft)
V = speed of vehicle (mph)
e = rate of superelevation (feet/ft of width)
f = coefficient of side friction between tire and road

Figure 6–4. Intersection sight distance.
SOURCE: *A Policy on Geometric Design of Highways and Streets,* 1990 American Association of State Highway and Transportation Officials, Washington, DC. Used by permission.

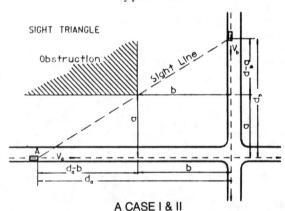

A CASE I & II
NO CONTROL OR YIELD CONTROL ON MINOR ROAD

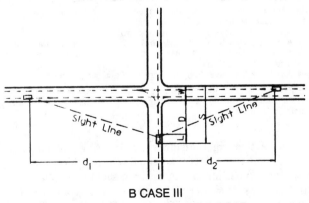

B CASE III
STOP CONTROL ON MINOR ROAD

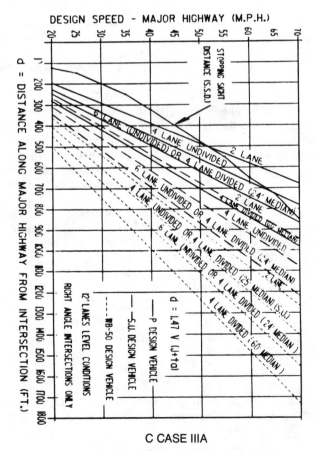

C CASE IIIA

Design controls for curves are determined by establishing limits on *e* and *f* based on practical design considerations.

Maximum superelevation. Maximum superelevation rates are established with consideration of operations at low speeds, and under snow and icing conditions. In North America, historical practice has been to design high-speed highways with maximum superelevation rates of 0.08 to 0.10 feet/ft. Under special conditions, rates of 0.12 feet/ft are used by some agencies. In urban areas, maximum superelevation rates of 0.04 or 0.06 feet/ft are common.

Side friction. Design values assumed for side friction are a function of both measured or observed coefficients of friction under various road conditions, as well as a consideration of driver comfort. AASHTO recommends design side friction factors as shown in Figure 6–5. These factors are intended to produce a margin of safety compared with actual conditions typically found on the highway system.

Maximum curvature. Evaluation of Equations (6.1) and (6.2) for ranges in *e* of 0.06 to 0.12, and *f* as given above in Figure 6–5, leads to maximum allowable degrees of curve for a given design speed, as shown in Table 6–9. Curves milder than the maximum allowable degree of curve for a given design speed are designed with superelevation less than the maximum value. The AASHTO *Policy* presents a series of design curves and tables that describe appropriate

superelevation rates for the full range of curvature allowable under a given design speed.

Design for horizontal curves as described above is based on an implied assumption that the vehicle tracks the curve as it is designed. It is acknowledged by AASHTO that this in fact is not possible if the curve is unspiraled. Research not only confirms that the AASHTO design assumption is not valid, but that the dynamics of driving on curves are quite different from the design assumptions.[15]

Typical and "critical" (i.e., aggressive) drivers track unspiraled curves in a manner that produces significantly greater friction demands on the tire/roadway interface than are intended by AASHTO design policy. As a result, the intended margin of safety in the AASHTO design policy is much less than anticipated. As a result of these findings, designers should strive to minimize the use of controlling or maximum curvature for a given design speed and use special transition curves for higher speed and sharper curve designs.

Curve length. Recent safety and operational research[16, 17] has also established the importance of considering length of curve in design. For an isolated location, designers should minimize the central angle of the curve, thereby avoiding too long or too sharp curvature. Central angles of less than 45° are considered desirable.

Horizontal SSD. One aspect of horizontal curve design is provision for enough stopping sight distance around the curve. Trees, buildings, retaining walls, and other objects may restrict the driver's view through the curve. Recommended design values for lateral offsets from the middle of an inside lane are shown in Figure 6–6. These design values should be considered as minimums, with greater values encouraged on high-speed highways because trucks have no eye height advantage over passenger cars for horizontal obstructions, and available pavement friction for braking is lower on curves than on tangents.

Spiral transition curves. Spiral curves are necessary geometric features of high-speed horizontal alignment. Spirals transition the roadway from tangent to circular curved alignment. Mathematically, the spiral is a curve of varying degree by length. Both simple spirals (connecting tangent and circular alignment) and combining spirals (between two different circular curves) are commonly used. Spirals are used for many reasons:

1. To fit the driver's natural path
2. To develop optimal superelevation
3. To facilitate pavement widening on curves
4. To enhance aesthetic design quality

Figure 6–5. Design friction factors for curves.
SOURCE: *A Policy on Geometric Design of Highways and Streets,* 1990, American Association of State Highway and Transportation Officials, Washington, DC. Used by permission.

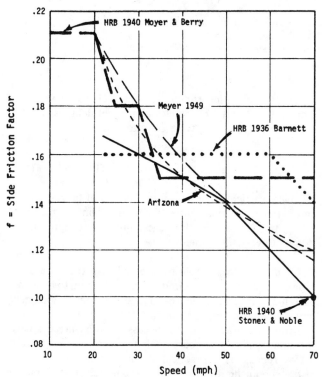

[15] J.C. Glennon, T.R. Neuman, and J.E.Leisch, *Safety and Operational Considerations for Design of Rural Highway Curves.* Federal Highway Administration Report No. FHWA/RD-86/035. (Washington, DC: Federal Highway Administration, 1985).

[16] Ibid.

[17] C.V. Zegeer, and others, *Cost-Effective Geometric Improvements for Safety Upgrading of Horizontal Curves.* Federal Highway Administration Report No. FHWA/RD. (Washington, DC: Federal Highway Administration, 1990).

TABLE 6–9
Maximum Degree of Curve and Minimum Radius Determined for Limiting Values of *e* and *f*

Design Speed (mph)	Maximum e	Maximum f	Total (e + f)	Maximum Degree of Curve	Rounded Maximum Degree of Curve	Maximum[a] Radius (ft)
20	.04	.17	.21	44.97	45.0	127
30	.04	.16	.20	19.04	19.0	302
40	.04	.15	.19	10.17	10.0	573
50	.04	.14	.18	6.17	6.0	955
55	.04	.13	.17	4.83	4.75	1,186
60	.04	.12	.16	3.81	3.75	1,528
20	.06	.17	.23	49.25	49.25	116
30	.06	.16	.22	20.94	21.0	273
40	.06	.15	.21	11.24	11.25	509
50	.06	.14	.20	6.85	6.75	849
55	.06	.13	.19	5.40	5.5	1,061
60	.06	.12	.18	4.28	4.25	1,348
65	.06	.11	.17	3.45	3.5	1,637
70	.06	.10	.16	2.80	2.75	2,083
20	.08	.17	.25	53.54	53.5	107
30	.08	.16	.24	22.84	22.75	252
40	.08	.15	.23	12.31	12.25	468
50	.08	.14	.22	7.54	7.5	764
55	.08	.13	.21	5.97	6.0	960
60	.08	.12	.20	4.76	4.75	1,206
65	.08	.11	.19	3.85	3.75	1,528
70	.08	.10	.18	3.15	3.0	1,910
20	.10	.17	.27	57.82	58.0	99
30	.10	.16	.26	24.75	24.75	231
40	.10	.15	.25	13.38	13.25	432
50	.10	.14	.24	8.22	8.25	694
55	.10	.13	.23	6.53	6.5	877
60	.10	.12	.22	5.23	5.25	1,091
65	.10	.11	.21	4.26	4.25	1,348
70	.10	.10	.20	3.50	3.5	1,637
20	.12	.17	.29	62.10	62.0	92
30	.12	.16	.28	26.65	26.75	214
40	.12	.15	.27	14.46	14.5	395
50	.12	.14	.26	8.91	9.0	637
55	.12	.13	.25	7.10	7.0	807
60	.12	.12	.24	5.71	5.75	996
65	.12	.11	.23	4.66	4.75	1,206
70	.12	.10	.22	3.85	3.75	1,528

Note: In recognition of safety considerations, use of e_{max} = 0.04 should be limited to urban conditions.
[a] Calculated using rounded maximum degree of curve.

SOURCE: *A Policy on Geometric Design of Highways and Streets,* 1990, American Association of Highway and Transportation Officials.

Research has also demonstrated that curves designed with spirals result in lower friction demands on the driver and vehicle, compared with nonspiraled horizontal curves, and that spiraled curves are safer than unspiraled curves.[18]

Unspiraled transitions. Where circular curves are designed without spiral transitions, special care is required in transitioning the pavement slope from the normal cross slope to the final superelevation in the curve. This is typically accomplished by developing superelevation partially on the tangent approach and partially on the curve. Practice among design agencies varies, with some designing for 50% of full superelevation at the point of curvature, and others developing 60% to 80% of full superelevation at the point of curvature. The length of highway needed to transition the pavement from a section with adverse crown removed to full superelevation is called *superelevation runoff.* The AASHTO *Policy* recommends 50 to 360 ft for superelevation runoff lengths, dependent on design speed and superelevation rate.

Vertical alignment

Highway vertical alignment is comprised of tangent grades and parabolic vertical curves. Geometric controls related to safety, operations, and drainage requirements apply to design of all vertical elements.

Grade. Grades should be selected with the terrain and type of highway in mind. Grades above 3% begin to influence passenger car speeds. Table 6–10 reports recommended maximum grades for open alignment.

Minimum grades are also necessary to ensure drainage. Minimums of 0.3% are commonly used. In very flat terrain, 0.15% represents a practical drainage minimum. Grades of 0% should be avoided. Such designs rely totally on the

---

[18] Glennon and others, *Safety and Operational Considerations for Design of Rural Highway Curves.*

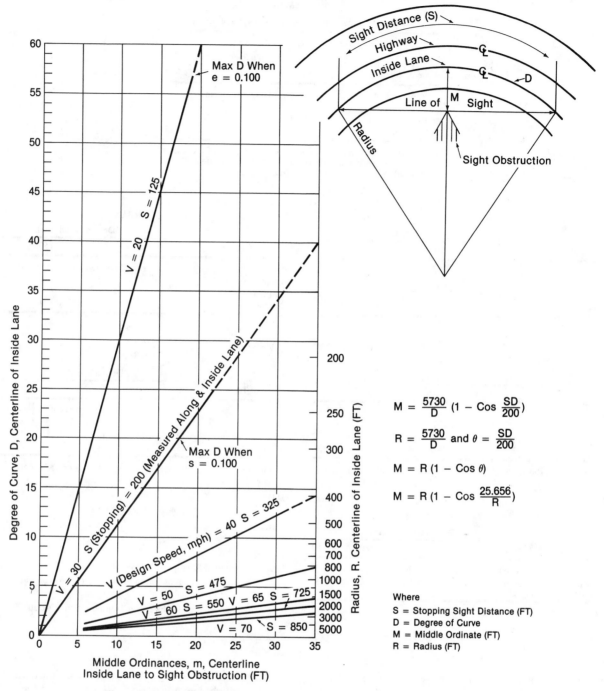

$$M = \frac{5730}{D} \left(1 - \text{Cos } \frac{SD}{200}\right)$$

$$R = \frac{5730}{D} \quad \text{and} \quad \theta = \frac{SD}{200}$$

$$M = R \left(1 - \text{Cos } \theta\right)$$

$$M = R \left(1 - \text{Cos } \frac{25.656}{R}\right)$$

Where

S = Stopping Sight Distance (FT)
D = Degree of Curve
M = Middle Ordinate (FT)
R = Radius (FT)

Figure 6–6. Horizontal stopping-sight distance requirements.
SOURCE: *A Policy on Geometric Design of Highways and Streets,* 1990, American Association of State Highway and Transportation Officials, Washington, DC. Used by permission.

normal cross slope for drainage. Transitioning of the cross slope for development of superelevation or warping of the pavement near an intersection could easily result in removal of any drainage slope.

The performance of trucks is greatly influenced by design of the vertical alignment. Very long and/or steep grades will influence the speeds of fully loaded trucks, affecting both the capacity and potential safety of the highway. Figure 6–7

depicts a basic design control, the combined grade, and length of grade. It is based upon a desirable maximum 10 mph speed reduction for a truck with a weight to horse-power ratio of 300.

Vertical curves. Vertical curves are designed to a parabolic form in the United States. Controls for the lengths of vertical curves are based on sight distance and driver comfort

TABLE 6–10

Maximum Grades for Design of Highways and Streets

RURAL AND URBAN FREEWAYS

Type of Terrain	Design Speed (mph)		
	50	60	70
	Grades (percent)		
Level	4	3	3
Rolling	5	4	4
Mountainous	6	6	5

Grades 1% steeper than the value shown may be used for extreme cases in urban areas where development precludes the use of flatter grades and for one-way downgrades except in mountainous terrain.

ARTERIALS

RURAL

Type of Terrain	Design Speed (mph)			
	40	50	60	70
	Grades (percent)			
Level	5	4	3	3
Rolling	6	5	4	4
Mountainous	8	7	6	5

URBAN

Type of Terrain	Design Speed (mph)			
	30	40	50	60
	Maximum Grade (percent)			
Level	8	7	6	5
Rolling	9	8	7	6
Mountainous	11	10	9	8

COLLECTORS

RURAL

Type of Terrain	Design Speed (mph)					
	20	30	40	50	60	70
	Grades (percent)					
Level	7	7	7	6	5	4
Rolling	10	9	8	7	6	5
Mountainous	12	10	10	9	8	6

URBAN

Type of Terrain	Design Speed (mph)					
	20	30	40	50	60	70
	Grades (percent)					
Level	9	9	9	7	6	5
Rolling	12	11	10	8	7	6
Mountainous	14	12	12	10	9	7

Maximum grades shown for rural and urban conditions of short lengths (less than 500 ft) and on low volume. Rural collectors may be 2% steeper.

LOCAL ROADS

RURAL

Type of Terrain	Design Speed (mph)				
	20	30	40	50	60
Level	–	7	7	6	5
Rolling	11	10	9	8	6
Mountainous	16	14	12	10	–

URBAN

Residential Streets—15% maximum

Streets in commercial or industrial areas—8% maximum

SOURCE: *A Policy on Geometric Design of Highways and Streets,* 1990, American Association of State Highway and Transportation Officials.

requirements. Lengths of vertical curves are expressed by the following relationship:

where:

$L = KA$

L = Length of vertical curve (ft)

A = Algebraic difference in grades

K = Constant   (6.3)

The constant K is based on the sight distance criterion used for design. For crest vertical curves this is usually a stopping sight distance criterion, although design may be based on passing sight distance. Figure 6–8 shows design lengths of crest vertical curves based on the AASHTO stopping sight distance model and assumed operation at design speed.

The K values for sag vertical curves are normally based on a headlight beam criterion. Where roadways are illuminated, or geometric constraints are severe, shorter vertical curves can be used. The limiting control in these cases is driver comfort. Table 6–11 summarizes K values for both crest and sag vertical curves.

Summary of alignment design

Assembling the horizontal and vertical elements of the highway into a complete alignment is not a simple or direct task. While the above controls establish guidelines and background, the designer must use care and good judgment in the application of these guidelines. Every effort should be made to provide balance in design, to avoid overloading the

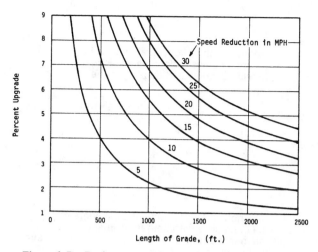

Figure 6-7. Design controls for combined length and grade.
SOURCE: *A Policy on Geometric Design of Highways and Streets,* 1990, American Association of State Highway and Transportation Officials, Washington, DC. Used by permission.

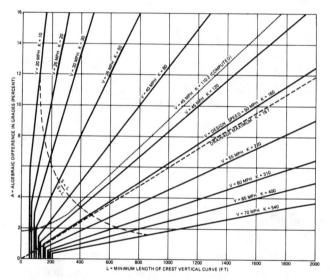

Figure 6-8. Crest vertical curve design requirements—for stopping sight distance and open road conditions (upper range).
SOURCE: *A Policy on Geometric Design of Highways and Streets,* 1990, American Association of State Highway and Transportation Officials, Washington, DC. Used by permission.

driver with difficult tasks, and to recognize and design around potential safety problems. The following is a summary of alignment design guidelines:

1. Select a design speed high enough to fit the terrain and driver's expectations. Drivers do not adjust their speeds relative to the type of highway as much as they do to the location, terrain, and adjacent land use.

2. Take special care in transitioning from a higher to a lower design speed. Avoid minimum vertical curves and/or controlling curvature at the beginning of the transition section. Limit differences in design speed between adjacent highway sections to 10 mph or less.

3. Provide more than minimum or controlling alignment in advance of potentially hazardous or conflict-inducing locations. These include at-grade intersections, changes in cross section, narrow bridges, railroad crossings, and so forth. Designers should not rely only on signing to warn or prepare drivers for a lower safe speed, but should adjust the highway geometry to provide a visual perspective of the impending condition.

4. Avoid horizontal or vertical alignment designed with short tangents between curvature in the same direction (referred to as "broken back").

5. Provide balance in the selection of alignment to match the terrain. Minimum alignment in flat or gently rolling terrain may appear out of place or awkward to the driver.

Coordination of horizontal and vertical alignment is essential to good design. Figure 6-9 summarizes recommended keys to achieving coordination.

Cross section

The third dimension of the highway—*cross section*—consists of the following elements: traveled way, shoulders, median, border area, and roadsides. Cross-section design varies significantly by type of highway and design speed. Figure 6-10 defines the design elements of the cross section for both rural and urban highways and streets.

Traveled way. The *traveled way* is that portion of the roadway reserved for traffic. It is generally comprised of two or more designated lanes. Widths of lanes and the resulting traveled way are a function of design speed, vehicle classification, and safety/operational considerations. Lane widths can range from 9 to 13 ft, but are usually 11 or 12 ft in width.

TABLE 6-11
Controls for Design of Vertical Curves: Values for *K*

| Design Speed (mph) | Crest Curves | | | Sag Curves | |
| | Stopping Sight Distance | | Passing Site Distance | Stopping Sight Distance | |
	Minimum	Desirable		Minimum	Desirable
20	10	10	210	20	20
30	30	30	400	40	40
40	60	80	730	60	70
50	110	160	1,050	90	110
60	190	310	1,430	120	160
70	290	540	2,030	150	220

Poor Design

Good Design

Highway Alignment

Coordination of horizontal and vertical geometry plays a significant role in providing safety and comfort to the motorist.

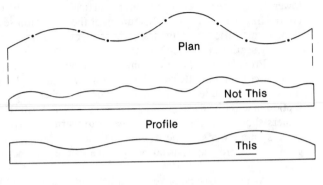

Plan

Not This

Profile

This

The horizontal and vertical alignments should be in balance with both horizontal and vertical elements of curves and tangents somewhat similar in length.

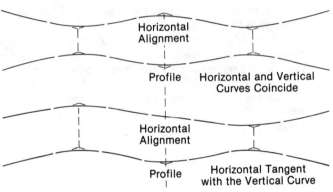

Horizontal Alignment

Profile Horizontal and Vertical Curves Coincide

Horizontal Alignment

Profile Horizontal Tangent with the Vertical Curve

Horizontal alignment elements should generally coincide with profile elements, both with respect to position and length.

Figure 6–9. Alignment design guidelines.
SOURCE: Adapted from "Notes on Fundamentals of Highway Planning and Geometric Design," Vol. 1, Jack E. Leisch & Associates.

On high-type, high-speed facilities such as freeways and primary arterials, lane widths of 12 ft are used almost exclusively. Ten- or 11-ft lanes may be used on lower-speed, lower-volume highways such as collectors and local roads. Table 6–12 summarizes lane width values generally used in design.

When less than 12-ft lanes are used, consideration should be given to widening the traveled way through horizontal curves. Additional width of 1 or 2 ft may be advised to account for the off-tracking characteristics of large trucks, and to provide for additional margin of safety through the curve.

Auxiliary lanes are special-purpose lanes adjacent to the continuous, through lanes of the highway. These can vary in width from 8 to 12 ft depending on the situation. On urban streets, auxiliary lanes are used for on-street parking. Special lanes for turning traffic at intersections are also provided. These are generally 10 to 12 ft in width, but may be 9 ft under low-speed, highly constrained conditions. Auxiliary lanes for passing slow vehicles are used on rural highways in rolling and mountainous terrain. Auxiliary lanes on freeways assist entering, exiting, and weaving traffic. These should be 12 ft in width.

All highways should be designed with a normal cross slope on tangent alignment to facilitate drainage. Cross-slope rates should be as low as possible to minimize effects on vehicle operations, yet be consistent with construction capabilities and structural stability. Cross-slope rates of 1.5 to 3% are generally used, as shown in Table 6–13.

Where more than two lanes require sloping in one direction, consideration should be given to increasing the slope on the outside lane(s) by 0.5%.

Shoulders. The shoulder is that part of the roadway immediately contiguous to the traveled way. Part or all of the shoulder may be paved or stabilized. Shoulders are extremely important features of all roadways. They serve a wide range of functions, including:

1. To aid in recovery of temporary loss of control, or to provide room to perform emergency evasive action;
2. To store a vehicle safely off the traveled way in emergency situations;

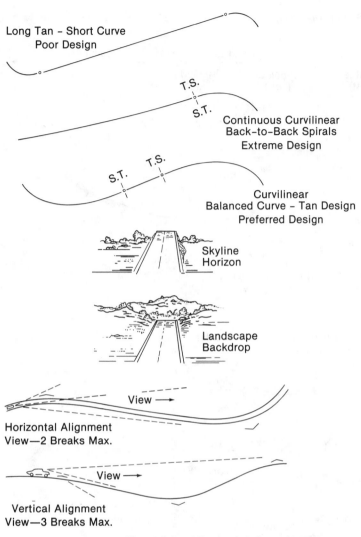

Long Tan – Short Curve
Poor Design

T.S.
S.T.
Continuous Curvilinear
Back–to–Back Spirals
Extreme Design

S.T.   T.S.
Curvilinear
Balanced Curve – Tan Design
Preferred Design

Skyline
Horizon

Landscape
Backdrop

View →
Horizontal Alignment
View—2 Breaks Max.

View →
Vertical Alignment
View—3 Breaks Max.

The preferred horizontal design provides a nominal tangent length between reverse spiral curves. Back to back spirals are less desirable, as are sharp curves without spirals.

A road with a crest which disappears into the sky can be disconcerting to the driver. This can be corrected by adjusting the location of the crest or providing a horizontal curve in combination with the profile crest.

Good coordination of horizontal and vertical geometry can be achieved by having no more than two breaks in horizontal alignment or three breaks in vertical alignment within the driver's view.

Figure 6–9. Alignment design guidelines (*cont.*).
SOURCE: Adapted from "Notes on Fundamentals of Highway Planning and Geometric Design," Vol. 1, Jack E. Leisch & Associates.

3. To provide a safe means of accomplishing routine maintenance operations;
4. To serve as a temporary traveled way during reconstruction, major maintenance, or emergency operations;
5. To serve as a primary clear area free of obstructions;
6. To enable provision for sufficient horizontal sight distance;
7. To maximize traffic flow and thereby capacity; and
8. To provide structural support to the pavement and traveled way.

In rural highways where space is generally available, shoulders of full width (12 ft) are desirable. Full-width shoulders are highly desirable on high-type urban freeways and arterials in recognition of the frequency of accidents, breakdowns, and other events that use the shoulder. Space limitations, however, often preclude the availability of 12-ft shoulders, necessitating narrower widths of 6 to 10 ft.

Shoulder cross slopes should be designed to avoid draining onto the traveled way. A key consideration in shoulder slope design is the difference between the shoulder and the adjacent lane. On tangent grades or along the inside of a horizontal curve, the shoulder is sloped at the same rate as the traveled way, or 1% to 2% greater. Along the outside of a horizontal curve, the shoulder should not be sloped so great that it contributes to loss of control for vehicles using it in emergencies. Maximum adverse or negative shoulder slopes should not exceed 6%.[19]

Medians. The *median* is that part of a divided highway that separates opposing traveled ways. Medians serve a number of functions, most of which are related to safety. Specifically, medians

1. physically separate high-speed, opposing traffic, thereby minimizing the chances of serious, head-on collisions;
2. provide for a clear recovery area for inadvertent encroachments off the traveled way;

[19]J.C. Glennon, T.R. Neuman, R.R. McHenry, and B.G. McHenry, *HVOSM Studies of Cross-slope Breaks on Highway Curves.* Federal Highway Administration Report No. FHWA/RD-86/035. (Washington, DC: Federal Highway Administration, 1985).

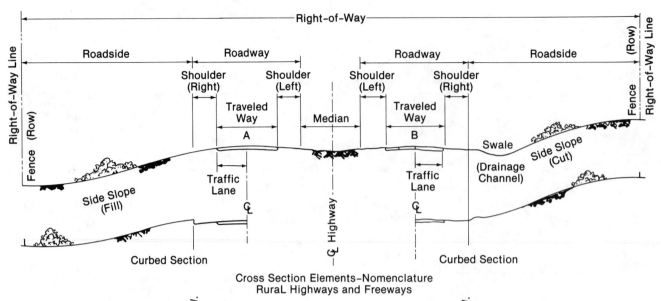

Cross Section Elements–Nomenclature
RuraL Highways and Freeways

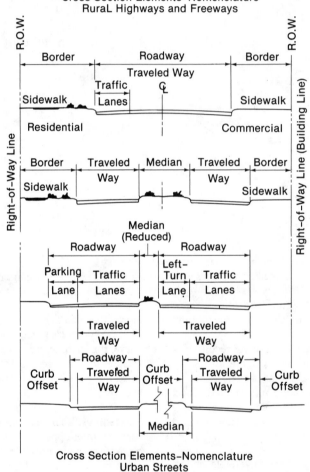

Cross Section Elements–Nomenclature
Urban Streets

Figure 6–10. Cross-section elements.
SOURCE: Adapted from "Notes on Fundamentals of Highway Planning and Geometric Design," Vol. 1, Jack E. Leisch & Associates.

TABLE 6–12

Lane Width Values for Design of Highways

FREEWAYS

RURAL	URBAN
12-ft lane widths used for all speed and volume conditions.[1]	12-ft lane widths generally used for all speed and volume conditions.

ARTERIALS

RURAL

Projected Design Traffic Volume	Width of Traveled Way(ft)[a]			
Design Speed (mph)	Current ADT Under 400	Current ADT 400 & Over	DHV 100–200	DHV Over 200
40	22	22	22	24
50	22	24	24	24
60	24	24	24	24
70	24	24	24	24

Width of Usable Shoulder (ft)[b] Each Side of Pavement				
All Speeds	4	6	6	8

[a]Width of traveled way may remain at 22 ft on reconstructed highways where alignment and safety records are satisfactory.
[b]Usable shoulders on arterials should be paved.

URBAN

Minimum	Desirable
10 ft—for low-speed, restricted areas with little truck traffic.	11 to 12 ft

COLLECTORS

RURAL

	Width(ft) for Design Traffic Volumes of:				
Design Speed (mph)	Current ADT Under 400	Current ADT 400 and Over	DHV 100–200	DHV 200–400	DHV Over 400
20	20	20	20	22	24
30	20	20	20	22	24
40	20	22	22	22	24
50	20	22	22	24	24
60	22	22	22	24	24
70	22	22	22	24	24

Width of Graded Shoulder—Each Side of Pavement					
All Speeds	2[a]	4	6	8	8

[a]Minimum width is 4 ft if roadside barrier is utilized.

URBAN

Minimum	Desirable
10 ft—for low-speed residential areas (not including width for parking)	11 to 12 ft (not including width for parking)

LOCAL ROADS

RURAL

	Width(ft) for Design Volumes					
Design Speed (mph)	Current ADT Less than 250	Current ADT 250–400	Current ADT Over 400	DHV 100–200	DHV 200–400	DHV 400 and Over
	Width of Traveled Way					
20	18	20	20	20	22	24
30	18	20	20	20	22	24
40	20	20	22	22	22	24
50	20	20	22	22	24	24
60	20	22	22	22	24	24
	Width of Graded Shoulder (Each Side)					
All Speeds	2	2	4	6	6	8

URBAN

Minimum	Desirable
9 ft—where attainable right-of-way is limited (not including width for parking)	11 ft

[1]Outside lane width of 13 ft may be used to enhance structural capability of pavement.

SOURCE: *A Policy on Geometric Design of Highways and Streets,* 1990, American Association of State Highway and Transportation Officials.

TABLE 6–13

Normal Pavement Cross Slope

Surface Type	Range in Cross-Slope Rate (percent)
High	1.5–2
Intermediate	1.5–3
Low	2–6

SOURCE: *A Policy on Geometric Design of Highways and Streets,* 1990, American Association of State Highway and Transportation Officials.

TABLE 6–14

Guide for Earth Slopes Design

Height of Cut or Fill (ft)	Earth Slope, Horizontal to Vertical, for Type of Terrain		
	Flat or Rolling	Moderately Steep	Steep
0–4	6:1	4:1	4:1
4–10	4:1	4:1	2:1[a]
10–15	4:1	2.50:1	1.75:1[a]
15–20	2:1[a]	2:1[a]	1.75:1[a]
Over 20	2:1[a]	2:1[a]	1.75:1[a]

[a]Slopes 2:1 or steeper should be subject to a soil stability analysis and should be reviewed for safety (See Traffic Barriers).

SOURCE: *A Policy on Geometric Design of Highways and Streets,* 1984, American Association of State Highway and Transportation Officials.

3. provide a means of safely storing stopped or decelerating left-turning vehicles out of the higher-speed through lanes;
4. provide a means of safely storing vehicles turning left out of a minor road or driveway as they await an available gap;
5. provide safe storage for pedestrians crossing the high-speed or wide, divided highway;
6. restrict or regulate left turns to and from adjacent businesses except at designated locations.

Median widths vary from nominal dimensions of 2 to 4 ft, to as wide as 100 ft in rural areas. Selection of an appropriate median width is based on available total width, economic concerns, an assessment of safety trade-offs, and future traveled way requirements. On rural freeways and expressways, median widths of 60 ft or more enable design without the need for barriers between opposing roadways. Variable-width medians are frequently used, with independent alignment developed for each direction of travel to provide for an aesthetic design or to minimize construction cost or both.

In urban areas, medians are generally 40 ft or less. On urban freeways, medians should desirably be wide enough to provide for full-width shoulders and a positive barrier. A dimension of 26 to 30 ft is considered optimal. On urban arterials, median widths of 16 to 18 ft provide for left-turn storage, with an additional 4 to 6 ft at intersections for channelization and for shadowing crossing-vehicle maneuvers.

Roadside design. Attention to good roadside design is among the most important safety considerations for a highway designer. Even the highest-quality alignment is subject to errant driving, resulting in roadside encroachments and possible accidents. The roadside should be designed to maximize the chances that a driver will be able to recover safely from a loss-of-control situation, or at least not suffer a serious injury or fatality should an accident occur.

Elements of the roadside include side slopes (also referred to as foreslopes), clear recovery areas, ditch design, and back slopes. For safety considerations, the roadside includes the shoulder and all other features outside the actual traveled way, with this total width considered the vehicle recovery zone.

Foreslopes of 6:1 or flatter are desirable for safe vehicular recovery. Slopes as steep as 4:1 operate reasonably well, and can be used where heights of cut or fill are moderate. Slopes steeper than 4:1 produce rollover and other serious accidents. Slopes of 2:1 or 3:1 may be necessary on high embankments. Drivers should be shielded from these by longitudinal barriers in many situations. AASHTO fore slope design values are summarized in Table 6–14.

A *clear recovery area* is that portion of the cross section outside the traveled way that is free of objects or hazards. Shoulders and fore slopes clear of trees, barriers, utility poles, etc., are all included in the clear recovery area.

On high-type facilities it is desirable to maintain at least a 30-ft clear recovery area. Clear-zone width requirements are based on average daily traffic and vehicle speeds, with extra width required on the outside of horizontal curves. In rolling or mountainous terrain or on difficult alignment, lesser dimensions may be required. Where slopes or objects are potential problems, guardrail or other barriers may be required. Designers should also consider providing greater clear recovery areas on the outside of horizontal curves, as drivers have a greater tendency to encroach on the roadside in these locations. The 1989 AASHTO *Roadside Design Guide*[20] provides guidance for design of clear recovery areas.

Barriers, signing and traffic control devices. Other features that should be considered in design include roadside barriers, signing for driver navigation, and traffic control devices.

Roadside barriers include guardrails, concrete barriers, and attenuation devices. The location and design of these must be carefully considered during design. The purpose of roadside barriers is to protect or shield the driver from an object or condition that, if struck, could have more severe consequences than collision with the barrier itself. Barriers are typically used in the following situations:

1. To separate opposing high-speed traffic flows;
2. To shield the roadway from steep side slopes, nontraversable ditches, or a series of culverts;
3. To shield the roadway from trees, rock outcroppings, or other natural features along the roadway;
4. To shield approaches to bridges;
5. To shield piers, columns, or other structural features; and
6. To restrict pedestrian movements.

There are many types of barriers in use. Each has a particular design application, defined by the length of need, traffic condition, and maintenance and construction cost implications. Roadside barrier design has evolved in recent years to respond to differing impact effects of very large trucks,

---

[20]American Association of State Highway and Transportation Officials, Roadside Design Guide. (Washington, DC, 1989).

and small cars, as well as typical passenger cars. Table 6–15 summarizes roadside barrier selection criteria.

A decision to use roadside barriers should be made with the understanding that the barrier itself is a continuous, longitudinal hazard. Procedures for assessing the trade-offs and for locating or selecting the appropriate barrier are shown in the 1989 AASHTO *Roadside Design Guide*.[21]

Figures 6–11 and 6–12 summarize two general guidelines for barrier design. The first shows barrier warrants for fill embankments, expressed as a function of the fore slope and height of fill. Note that slopes of 3:1 or steeper generally warrant a barrier. Figure 6–12 shows warrants for placement of a barrier in the median of a divided highway. Here, warrants are based on the median width and average daily traffic. A history of serious cross-median accidents would warrant barrier placement for those conditions where Figure 6–12 indicates barrier placement is optional.

An integral part of design is consideration of and designing for traffic control devices. These include warning and regulatory signs, pavement markings, traffic signals, navigational and other guide signs, and special devices such as flashing beacons, railroad crossing gates, and lane designation signing. Designers should recognize the need for these features. Traffic control devices should not, however, be relied upon to mitigate the effects of poor design features. The *Manual on Uniform Traffic Control Devices*[22] describes warrants, types of signs and markings, and design guidelines for traffic

---

[21] Ibid.

[22] *Manual on Uniform Traffic Control Devices for Streets and Highways* (Washington, DC: U.S. Department of Transportation, 1988).

TABLE 6–15
Selection Criteria for Roadside Barriers

Criteria	Comments
1. Performance capability	Barrier must be structurally able to contain and redirect design vehicle.
2. Deflection	Expected deflection of barrier should not exceed available room to deflect.
3. Site conditions	Slope approaching the barrier, and distance from traveled way, may preclude use of some barrier types.
4. Compatibility	Barrier must be compatible with planned end anchor and capable of transition to other barrier systems (such as bridge railing).
5. Cost	Standard barrier systems are relatively consistent in cost, but high-performance railings can cost significantly more.
6. Maintenance a. Routine	Few systems require a significant amount of routine maintenance.
b. Collision	Generally, flexible or semi-rigid systems require significantly more maintenance after a collision than rigid or high-performance railings.
c. Materials storage	The fewer different systems used, the fewer inventory items/storage space required.
d. Simplicity	Simpler designs, besides costing less, are more likely to be reconstructed properly by field personnel.
7. Aesthetics	Occasionally, barrier aesthetics is an important consideration in its selection.
8. Field experience	The performance and maintenance requirements of existing systems should be monitored to identify problems that could be lessened or eliminated by using a different barrier type.

SOURCE: *Barrier Guide,* 1989. American Association of State Highway and Transportation Officials.

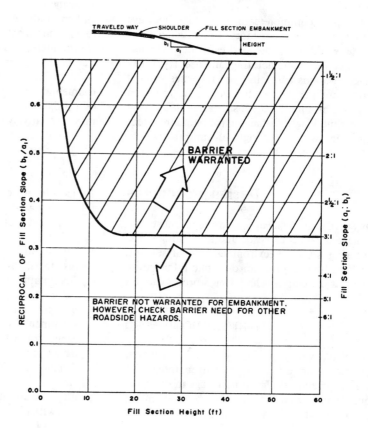

Figure 6–11. Barrier warrants for fill-section embankments.
SOURCE: *Roadside Design Guide,* 1989, American Association of State Highway and Transportation Officials.

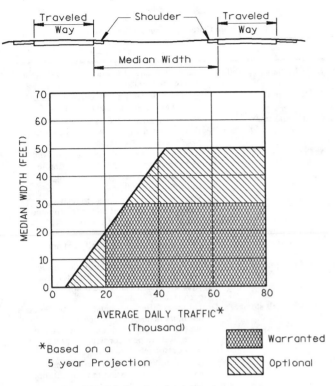

Figure 6–12. Median barrier warrants.
SOURCE: *Roadside Design Guide,* 1989, American Association of State Highway and Transportation Officials.

control. These are discussed in detail in Chapters 8 and 9 of the present text.

Noise barriers are a common feature on urban freeways through sensitive or narrow corridors. Constructed typically of concrete, wood, or landscaped earth berms, noise barriers are intended to shield abutting land uses (typically residential neighborhoods and sensitive properties such as schools, hospitals, and parks) from freeway noise. The need for noise barriers is an important consideration in location and design. Horizontal sight distance must be maintained, and sufficient space within the right-of-way provided to construct and maintain noise barriers.

Design procedures

Design of any significant highway improvement must consider a wide range of impacts, costs, and implications. Extensive engineering studies, environmental analyses and reports, and a coordinated program of public information and involvement in the planning and design process are all necessary.

Preliminary engineering

Highway project needs may be identified by emergencies, legislative direction, public requests, system maintenance, or local government requests. The initial stage in the process entails concept definition. This may be a brief statement or report describing the location, perceived need for improvement, and type of improvement proposed. This information is used to program and perform preliminary engineering studies.

Preliminary engineering studies have as their objective the identification of reasonable design alternatives and sufficient engineering to highlight significant differences among the alternatives. The following items are typically included in a preliminary engineering report:

1. General description of the proposed improvement, need for improvement, and relationship of the proposed highway improvement to the overall transportation system;
2. Description and discussion of alternative locations and designs;
3. Forecast design year;
4. Estimated total costs of construction and right-of-way;
5. Economic evaluation, including user costs;
6. Evaluation of consequences to the natural environment, including effects on land values, employment, community values and services (evaluation may be partly qualitative);
7. Evaluation of the sociological effect on the community, including relocation of residences and businesses, and accessibility to community services (evaluation may be mainly qualitative); and
8. A recommended plan for the proposed highway improvement.

Environmental process

The National Environmental Policy Act (NEPA) of 1969 directs all federal agencies systematically to assess the potential impact of a project on the human environment. Such assessment must include possible environmental impacts, and unavoidable adverse environmental effects, alternatives to a proposed action or project, local short-term uses of the environment, enhancement of long-term productivity, and irreversible and irretrievable commitments of resources.

Most highway projects involve impacts requiring environmental reviews. Depending on the size and importance of a project, the environmental process may be lengthy, or may be relatively brief. Three major types of environmental reviews are conducted for the range of highway types.

Type I—Full Environmental Impact Statement (EIS)

Projects for which it is known that significant environmental impacts will occur require preparation of an EIS. This includes, but is not limited to, projects that would:

1. have significant impact on natural, ecological or cultural resources, threatened or endangered species, wetlands, floodplains, prime or unique farmland, groundwater, natural resources, or fish and wildlife resources;
2. be highly controversial or considered unacceptable on environmental or legal grounds by a federal, state, or local agency, or by the public;
3. have significant residential, commercial, or agricultural displacement impacts;
4. cause a substantial disruption to an established community; disrupt orderly, planned development; be inconsistent with plans or goals that have been adopted by affected

communities, or adversely affect the economic vitality of an urban area;

5. have a significant impact on noise levels in noise-sensitive areas;
6. have a significant impact on air quality;
7. have a significant impact on water quality or a surface or subsurface public water supply system.

The full EIS process involves a Draft EIS, which is made available to the public and forwarded for review and comment to federal, state, and local agencies; a formal public hearing; and a Final EIS, which includes all review comments and input from the public hearing.

Type II—Environmental Assessment (EA) and Finding of No Significant Impact (FONSI)

An EA is prepared to provide sufficient evidence and environmental analysis to determine whether to prepare an EIS. An EA is prepared when the significance of the impacts cannot be clearly determined, or if some, but not all, of the EIS criteria can be met, or if the project is a large one.

Based on the review and findings of an EA and any public comments, a FONSI is prepared in the event the study concludes that the proposed action will not cause significant impacts. The FONSI is a conclusion to the EA and highlights data supporting the finding that no significant impacts will occur as a result of the action.

Type III—Categorical Exclusion (CE)

Actions on projects not producing significant effects on the environment are called Categorical Exclusions (CE). Examples include intersection improvements, pavement widening, or other minor improvements that do not require additional public rights-of-way.

Public hearings. Public hearings and information meetings are important parts of the highway design process. They are intended to inform the public of alternative plans, receive input or comment on the alternatives, and highlight trade-offs, local concerns, and specific project issues that may be controversial. Complex or large projects should be conducted with a series of public information meetings as work progresses. Public information programs including newsletters, toll-free "hotlines," press releases, and coordination with the local media have been successfully used on particularly controversial or sensitive highway projects. Public hearings are typically held near the end of the project, prior to decision making on a selected alternative.

Design plan completion

Following selection of an alternative, final construction plans are produced. This last step in the design process involves the detailing of many design features. During the final design, it is important to retain the essential operational and safety features that were included in preliminary plans.

Major highway projects involve complex trade-offs of traffic operations, environmental impacts, construction costs, and local public perceptions and needs. Addressing these needs in a systematic and comprehensive manner requires the expertise of many individuals. Design concept teams, formulated for major projects, may include the following specialized professionals:

Traffic engineers	Sociologists
Highway designers	Environmental specialists
Urban planners	Economists
Real estate specialists	Attorneys

Chapter 13 of the *Transportation Planning Handbook* contains more detail on the environmental process and its requirements for planning and design of highways.

Multiple-use development of rights-of-way

Conflicts between highway right-of-way requirements and availability or cost of right-of-way are common in dense, highly developed urban areas. In many parts of the country, such conflicts have been turned into joint opportunities to share the right-of-way between a highway and a development.

Joint use of right-of-way involves developing the air rights beneath an elevated highway, or over a depressed or at-grade highway. There is a wide range of types and intensities of joint use throughout the United States. These are described in *Highway Joint Development and Multiple Use*, a publication authored by the U.S. Department of Transportation.[23]

Joint use requires resolution of complex relationships between highway agencies that typically own the right-of-way, and private entities or local agencies who lease the air rights. Among the issues to resolve are the impacts of the development on the design, operation, and maintenance of the highway above or below the development. Granting of easements to utilities must also be considered. It should be recognized that, in most cases, once the development has been implemented, the ability to expand or revise the design of the highway is essentially lost. Therefore, it is important that design flexibility be considered in the highway and that there be assurance of the reasonableness of limiting future highway expansion before a major air rights project is implemented. Joint use of right-of-way must clearly be considered as early as possible in the highway planning stage.

Freeways

Freeways and other fully controlled access facilities are the highest form of highway. They are designed for high speeds, and with the highest operational quality and strictest design standards of any highway type.

The basic design controls for freeways are the same regardless of their location. These include the aforementioned criteria for horizontal and vertical alignment and cross-section features. Because of differences in traffic volumes, land use constraints, and traffic service needs, however, design of freeways in rural and urban areas differs.

---

[23] Highway Joint Development and Multiple Use. (Washington, DC: U.S. Department of Transportation, 1979).

Rural freeways

Rural freeways are typically constructed at ground level. Most rural freeways are four lanes (two each direction), although in high-volume intercity corridors or where very high truck volumes exist, six-lane rural freeways are necessary.

Alignment. The alignment design of a rural freeway should fit the natural terrain, produce uniform operation for all vehicle types at high speeds, and be aesthetically pleasing to the driver. Design speeds of 60 to 80 mph are appropriate for rural freeways. Vertical alignment should desirably produce truck speeds within 10 mph of full design speed on critical upgrades. Where this is not achievable, and truck volumes are significant, consideration should be given to providing auxiliary lanes on the upgrades to remove the slower trucks from the higher-speed lanes.

Cross section. Cross-sectional features of rural freeways should emphasize safety. Wide medians, full shoulders, and clear roadsides are important aspects of rural freeways. Typical cross sections are shown in Figure 6–13.

Median widths can vary from 50 or 60 ft to 100 ft or more. In rolling or mountainous terrain, variable-width medians are often used for environmental reasons or to minimize construction costs. Added safety features of wide medians are that head-on conflicts are essentially eliminated without the

Figure 6–13. Typical rural freeway cross sections.
SOURCE: *A Policy on Geometric Design of Highway and Streets,* 1990, American Association of State Highway and Transportation Officials, Washington, DC. Used by permission.

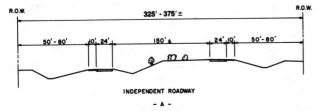

INDEPENDENT ROADWAY
– A –

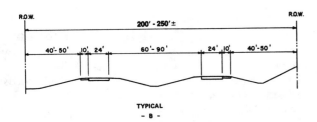

TYPICAL
– B –

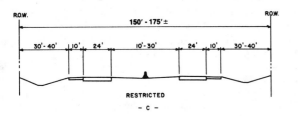

RESTRICTED
– C –

need to provide barriers between opposing traffic lanes, and headlight glare is minimized from opposing traffic streams.

Design of the roadside outside the traveled way is important. Full, 12-ft paved or stabilized shoulders should normally be provided. Shoulder widths should extend in full across all structures. In addition, the fore slope beyond the shoulder should provide for safe encroachments from errant vehicles at high speeds. Slopes of 6:1 or flatter for 30 ft from the traveled way are desirable; 4:1 slopes as minimums in difficult areas should be provided. Where excessive fill is necessary, a broken slope or "barn roof" design may be used.

Urban freeways

Urban freeways carry higher volumes in more constrained areas. They are generally a minimum of six lanes wide, and can be as great as 12 or 14 lanes in width, with special cross-sectional designs. To provide adequate lane requirements and minimize or avoid conflicts with surrounding land uses, urban and suburban freeways may be built on elevated alignment, in depressed alignment, or at-grade. Examples of all three are shown in Figure 6–14.

Urban freeways are typically designed for 50 to 70 mph design speeds, with 60 mph appropriate in many cases. Transitioning a freeway's design speed from 70 to 80 mph in rural areas to a 60 mph speed or lower should be done gradually to adjust the driver to the lower speed.

Cross section. Despite the constrained nature of urban areas and difficulty of construction, freeway cross-section design standards remain high. Full 12-ft lane widths are desirable on new or reconstructed urban freeways. Full 12-ft or 10-ft shoulders on the outside are considered mandatory because of the frequency of accidents, incidents, maintenance, and enforcement operations. Full-width shoulders on the inside are also recommended for wide freeways (i.e., six lanes or more).

In recent years, existing urban freeways in highly populated metropolitan areas have been rehabilitated with less than full-width lanes and/or shoulders. Elimination or narrowing of shoulders, and restriping of the traveled way to 11- or 11.5-ft lanes often can result in the addition of one more lane of capacity without the need to undertake costly corridor reconstruction. Care should be taken in considering such solutions to freeway capacity problems. The safety implications are unclear, and the loss of a useful shoulder width may produce undesirable operations because stalled or parked vehicles may have no place to be stored outside the traveled way.

Median widths of 26 to 30 ft are optimal. These provide room for full shoulders, barriers of the New Jersey-type design to separate opposing traffic, and additional space for bridge piers, light standards, and sign supports. Median widths of 40 ft or more are desirable if possible. These allow for barrier design to provide full horizontal sight distance, and room for long-range expansion in capacity.

Interchanges

Access to freeways occurs solely at interchanges. The location of freeway interchanges dictates the type, volume, and

AT-GRADE FREEWAYS

Normal Cross Section

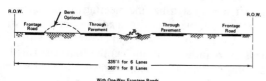

With One-Way Frontage Roads

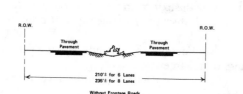

Without Frontage Roads

Restricted Cross Section

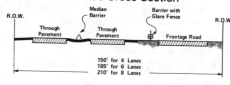

With 2-Way Frontage Road

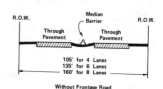

Without Frontage Road

ELEVATED FREEWAYS

Normal Embankment Cross Section

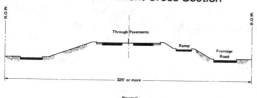

Normal

Restricted (with Retaining wall) Cross Section

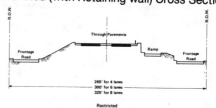

Restricted

DEPRESSED FREEWAYS

Normal Cross Section

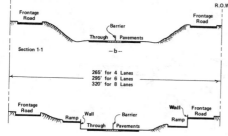

Restricted Cross Section

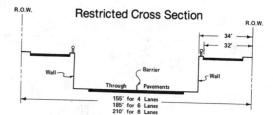

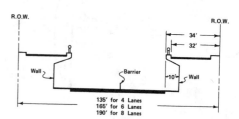

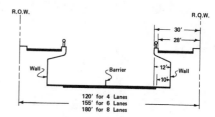

Figure 6–14. Typical urban freeway cross sections.

SOURCE: *A Policy on Geometric Design of Highway and Streets,* 1973, American Association of State Highway and Transportation Officials, Washington, DC. Used by permission.

pattern of traffic that will use the freeway. There are two basic types of interchanges—system and service. *System interchanges* are formed by the junction of two freeways or controlled-access facilities. *Service interchanges* occur between a freeway and a lower-class roadway, such as an arterial or collector. While clear differences in design requirements and constraints exist between rural and urban or system and service interchanges, the following are basic design principles that apply in all cases:

1. Avoid design of left-hand entrances and exits. Drivers expect ramps to be on the right. Research has shown that left-hand ramps produce accident rates over twice those of right-hand ramps.[24-26]

---

[24] J.I. Taylor and others, *Major Interchange Design, Operation, and Traffic Control.* Federal Highway Administration Final Report No. FHWA-RD-73-81. (Washington, DC: Federal Highway Administration, 1973).

[25] *The Suitability of Left-Hand Entrance and Exit Ramps for Freeways and Expressways.* Illinois Cooperative Highway Research Project Final Report IHR 61. Department of Civil Engineering, Northwestern University: Evanston, IL.

[26] R.A. Lundy, *The Effect of Ramp Type and Geometry on Accidents.* Highway Research Board Record No. 163. 1967.

2. Strive to achieve single-exit design, particularly at system interchanges. Single-exit designs are easier to sign, and less prone to navigational errors by unfamiliar drivers.

3. Place exits in advance of the crossroad. Visibility of the exit, ease of signing, and meeting driver expectations are all reasons for placing exits in advance of the crossroad.

4. Design the freeway and crossroad to allow for crossroads over (rather than under) the freeway. This has many advantages, including making gravity work with rather than against the driver for both exiting and entering the freeway; maximizing the exit ramp visibility to the driver; and minimizing costly disruption to freeway traffic during expansion of either facility.

5. Avoid designs that include weaving within the interchange on the freeway.

6. Provide for decision sight distance in advance of interchanges on the freeway.

7. Use auxiliary lanes, two-lane ramps, and special exit ramp designs to provide lane balance at all interchange ramp terminals. The principle of lane balance states that at exits, the sum of the lanes on the mainline and ramp should be one more than the number of lanes on the approach. Also, the number of lanes past the merge of an entrance ramp should be not less than one fewer than the sum of the lanes on the ramp plus the

Figure 6–15. Diamond interchange types.

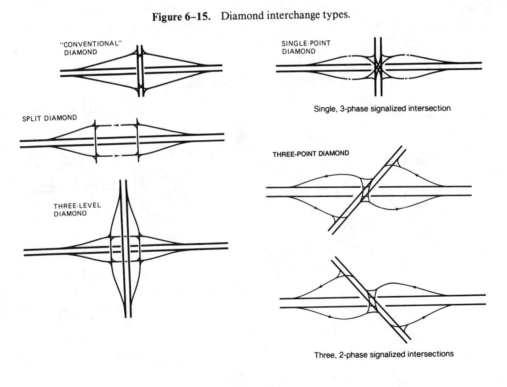

freeway prior to the merge. Lane balance minimizes lane changing.

8. Design interchanges to fit the principle of route continuity rather than forecast traffic patterns. This principle holds that the major marked route defines the through movement, with all entrance and exit ramps designed accordingly.

Service interchange types. There are many interchange types in existence today. Operational experience and research have demonstrated that only a few basic types of interchanges apply in almost every case.

The most common and simplest interchange form is the *diamond*. It is a complete interchange, with four one-way ramps and left turns made directly off the crossroad. There are various forms of the diamond to fit a range of geometric and physical constraints, as illustrated in Figure 6–15. Diamonds operate well under a wide range of volume conditions. In rural areas, the intersection of the ramps and the crossroad should be spaced far enough from the freeway structure to provide sight distance at the ramp terminal intersections. In urban areas, the intersection of the crossroads and the ramps can be as close as 300 to 400 ft apart if space limitations exist. Special traffic signal phasing schemes are required in this case. More conventional diamond ramp spacing of 600 to 800 ft between ramps is typical in urban areas.

Certain *partial cloverleaf* interchange forms provide greater capacity than do diamonds. In such typical applications, a high-volume left-turn movement is accommodated through a loop ramp. Partial cloverleafs with loops in advance of the crossroad, termed *Parclo A designs* (see Figure 6–16),

have distinct advantages over other forms. The left turns off the crossroad are replaced by loops, thereby (1) eliminating width required for one or more turning lanes across the structure; (2) resulting in high-capacity, two-phase traffic signal operation; (3) designing the lower speed loop movement as an exit off the lower speed arterial rather than the higher speed freeway; and (4) enabling single exit design in advance of the crossroad. *Parclo B designs* have loops beyond the crossroad. These offer some but not all of the advantages of Parclo A designs.

Other partial cloverleafs with loops in adjacent quadrants (Parclo AB) should be provided only where physical or special right-of-way constraints exist. These can produce weaving problems on one or more of the roadways, resulting in potential operational and/or safety problems. Full cloverleafs, with loops in all quadrants, have been proven to be inappropriate interchange forms for service interchanges.

Where very high arterial volumes occur and both through as well as ramp capacity demands are great, a three-level diamond can service traffic well. The first level is typically the freeway, with the second level reserved for intersections of ramp movements. This can occur at one intersection or four intersections, as shown in Figure 6–15. The third or upper level is for through arterial street traffic.

System interchanges. Interchanges between two controlled access facilities are referred to as *system interchanges*. These differ from service interchanges in that all ramp movements are designed to operate as free-flow diverges and merges.

Figure 6–16. Partial cloverleaf interchanges.
SOURCE: *Adaptability of Interchanges to Interstate Highways,* Jack E. Leisch, American Society of Civil Engineers.

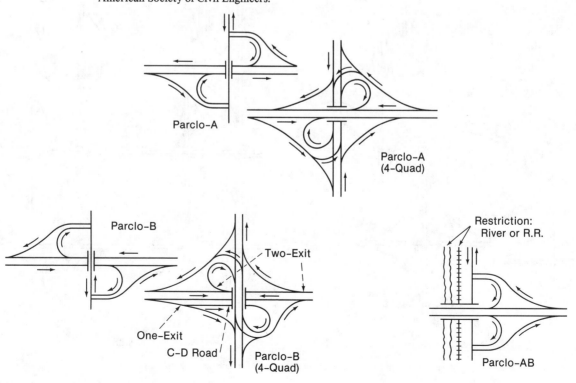

Although many system interchange forms are in place, only a few basic forms should be considered in new design or major reconstruction. These forms, shown in Figure 6–17, include the following highly desirable operational characteristics:

1. All exits are designed with single-exit design on the right.
2. There are no weaving sections within the interchange.

System interchanges can be designed with loops (referred to as semidirectional interchanges) or without loops (all directional). One interchange type formerly used extensively is the *full cloverleaf,* with loops in all four quadrants. This form has been shown to have limited applicability because of its poor safety history and operational problems associated with moderate to high volumes on the weaving sections. Cloverleaf interchange designs should not be used in urban areas. Their application is limited to rural areas where right-of-way is available and collector-distributor roads can be provided on both freeway facilities.

The Y interchange types are also shown in Figure 6–17. These incorporate the same features as the full, four-leg interchanges discussed above. The *trumpet* form is typically used on toll facilities where toll collection can be combined at one location.

Interchange ramp design.   Design features of a ramp include the exit ramp terminal, ramp proper, entrance ramp terminal, and ramp/crossroad intersection. Exit and entrance ramp terminals can be designed as either parallel lane forms, or taper forms.

Figure 6–18 illustrates typical exit ramp details for both forms. *Taper designs* usually utilize a tangent taper at a diverge angle of 3° to 6°, with the latter used on lower-speed urban freeways. At the point of physical diverge, also referred to as the *nose,* the ramp profile begins its independent alignment. Figure 6–19 shows typical entrance ramp design details. Taper designs use a 1° angle of convergence or 50:1 taper. *Parallel lane designs* should produce a comparable length of merge.

Figure 6–17.   System interchange types.

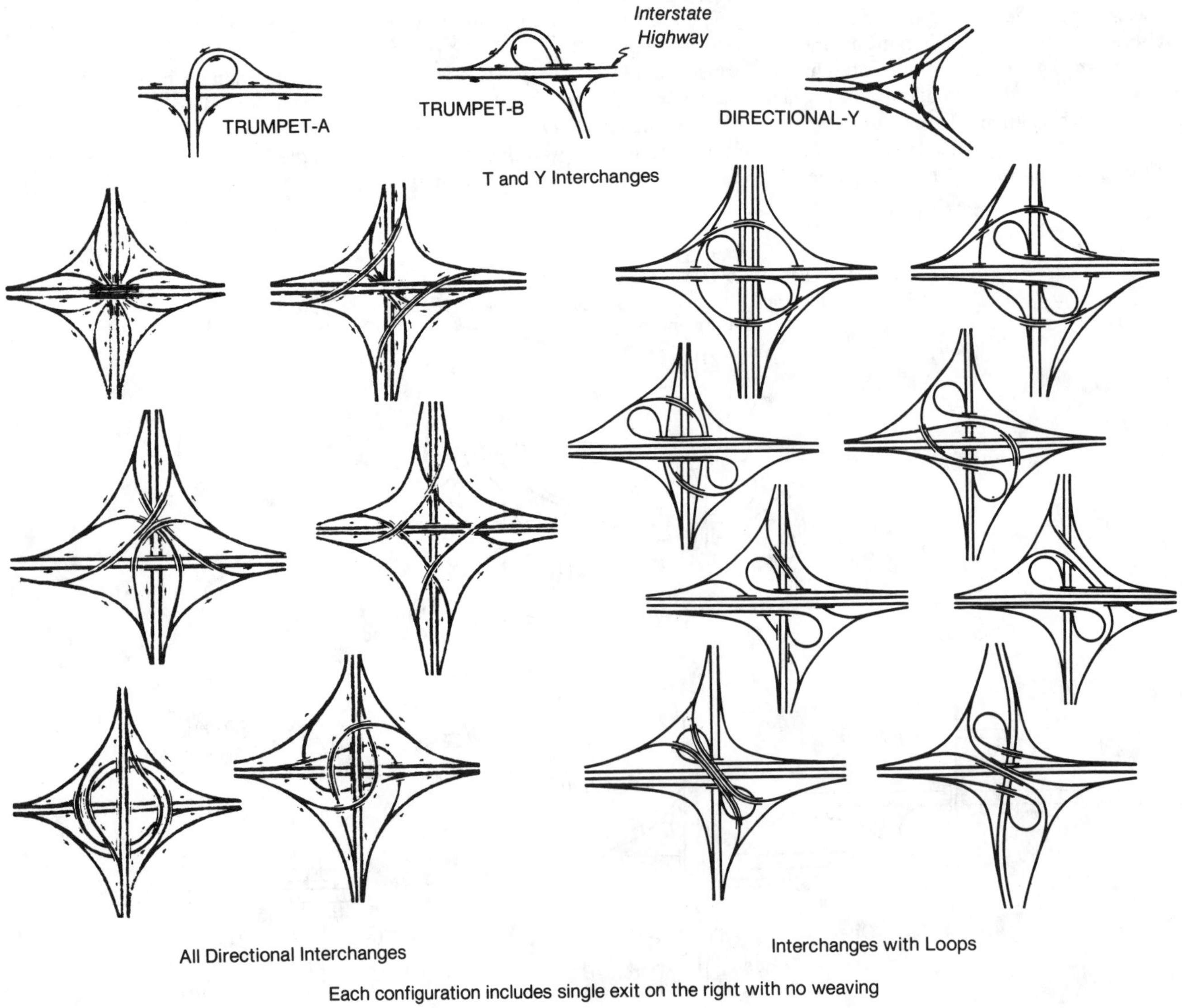

TRUMPET-A

TRUMPET-B

Interstate
Highway

DIRECTIONAL-Y

T and Y Interchanges

All Directional Interchanges

Interchanges with Loops

Each configuration includes single exit on the right with no weaving

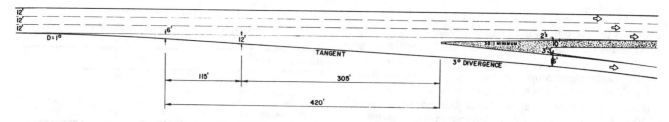

A. TAPERED DESIGN - TANGENT

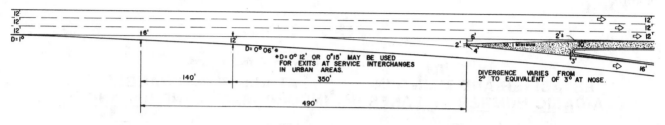

B. TAPERED DESIGN - CURVILINEAR

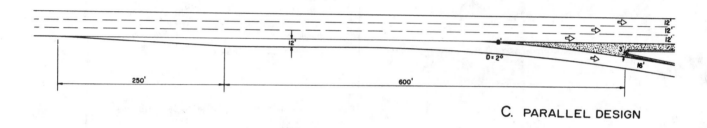

C. PARALLEL DESIGN

Figure 6–18. Typical exit ramp design.

A. TAPERED DESIGN - PREFERRED

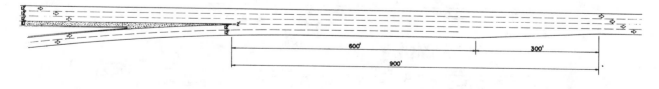

B. PARALLEL DESIGN

Figure 6–19. Typical entrance ramp design.

The *ramp proper* is the independent ramp alignment that connects the ramp terminal to another roadway. Design of this alignment should consider the needs of decelerating or accelerating vehicles. The design speed of controlling curvature on the ramp proper should not differ greatly from the roadway exited. Where a loop ramp or other low-speed alignment is necessary, a stem road or mild transition curve of 200 ft or more should be provided to transition the driver to the lower-speed alignment. Table 6–16 gives AASHTO-recommended guidelines for the design speed of ramps proper.

TABLE 6–16

Guide Values for Ramp Design Speed as Related to Highway Design Speed

Highway Design Speed (mph)	30	40	50	60	65	70
Ramp Design Speed (mph)						
Upper range (85%)	25	35	45	50	55	60
Middle range (70%)	20	30	35	45	45	50
Lower range (50%)	15	20	25	30	30	35

SOURCE: *A Policy on Geometric Design of Highways and Streets*, 1990, American Association of State Highway and Transportation Officials.

Figure 6–20. Operational criteria for design of urban freeways.

ESTABLISH AND MAINTAIN A BASIC NUMBER OF LANES

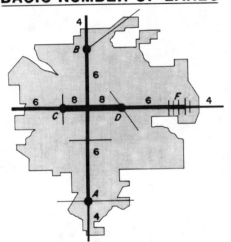

AT ALL INTERCHANGES INCORPORATE LANE BALANCE

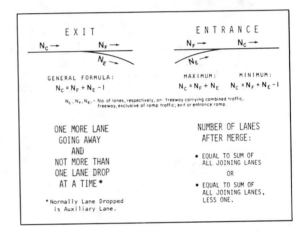

PROVIDE RIGHT-HAND EXITS ONLY AND USE AUXILIARY LANES TO MINIMIZE LANE CHANGING AND MAXIMIZE CAPACITY

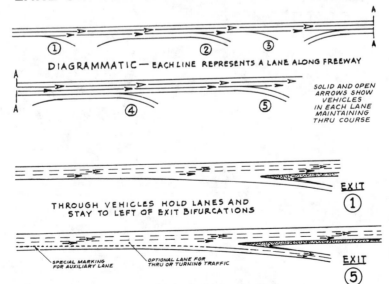

Urban freeway system design principles

Design of urban freeways must consider the operational needs along significant lengths of corridor. Problems at one location usually affect traffic flow well upstream. There are a number of important system considerations, which are summarized in Figure 6–20. These include (1) maintaining a basic number of continuous through lanes over a significant length of facility; (2) incorporating lane balance to minimize lane changing; (3) designing appropriate spacing between ramps to maximize capacity, enable signing, and optimize flow; (4) utilizing right-hand exits and auxiliary lanes to minimize lane changing; and (5) selecting appropriate interchange types. The above principles should be incorporated in the planning and preliminary engineering stages of a project. As design proceeds, it is vital that they be maintained.

Freeway system concepts

The volume, pattern, and type of traffic play a major role in the consideration of broad system concepts for a freeway. The term "system concept" refers to basic types of freeways, as illustrated in Figure 6–21. Most freeways are of a conventional form (Type I). Freeways with frontage

Figure 6–20. Operational criteria for design of urban freeways (*cont.*).

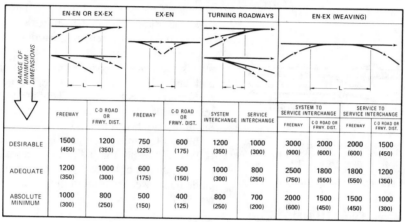

000 - DIMENSIONS IN FEET   (000) - DIMENSIONS IN METERS

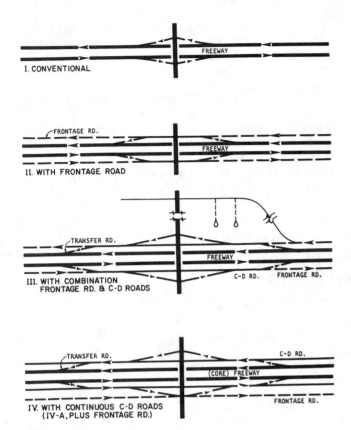

I. CONVENTIONAL

II. WITH FRONTAGE ROAD

III. WITH COMBINATION FRONTAGE RD. & C-D ROADS

IV. WITH CONTINUOUS C-D ROADS (IV-A, PLUS FRONTAGE RD.)

NOTE:
CONFIGURATIONS I, II & III MAY BE FURTHER EXPANDED TO INCLUDE ADDITIONAL ROADWAY WITHIN MEDIAN FOR REVERSIBLE TRAFFIC DURING PEAK HOURS OR FOR SPECIAL BUS LANES IN WHICH CASE THESE WOULD BE REFERRED TO AS CONFIGURATIONS I-A, II-A, & III-A

Figure 6–21. Linear freeway systems configurations.

roads (Type II) are also common. Types III and IV, freeways with frontage roads and collector-distributor roads, are typically found in corridors with very high traffic volumes. Collector-distributor roads are within the full access control of the freeway but physically separated from the higher-speed "express" or through lanes.

Figure 6–21 also notes that specialty-lane concepts are possible as designated variations of configurations I, II, or IV. These would include high occupancy vehicle (HOV) lanes, bus-only lanes, or reversible flow lanes.

The applicability of one or more system concepts is based on a number of factors:

- o Total design-year traffic demand
- o Directionality and peaking characteristics
- o Distribution of freeway trip lengths
- o Right-of-way and physical constraints
- o Local street system capacity

High occupancy vehicle (HOV) roadways. Growth in traffic has outstripped the ability of freeway corridors in many urban areas to serve the demand. A solution to this problem is to increase the person-carrying capacity of the corridor by increasing average vehicle occupancy. High occupancy vehicle (HOV) lanes, ramps, and other special facilities have been designed into existing urban freeway corridors during major reconstruction to achieve greater person-carrying capacity.

HOV lanes and facilities are reserved for buses, vans, and car pools with a designated minimum number of occupants, typically three or four. There are three basic types of HOV lanes, as illustrated in Figure 6–22. The highest type is a physically separated roadway, with separate ramp access. This type offers the greatest capacity, is most readily enforceable, but also is generally the most difficult to implement. Concurrent-flow HOV lanes typically involve designation of the inside or left lane of a freeway for HOVs. A 4-ft buffer with special pavement markings is provided to delineate lane use. Separate HOV-only ramps may or may not be provided off this lane. Concurrent-flow HOV lanes on the outside or right lane has been implemented in some locations. This concept poses inherent problems, in that all ramp conflicts occur on the lane designated for HOV use. Also, changes in the number of mainline lanes cannot be readily accomplished without disruption to the HOV lane continuity.

The third type, contra-flow operation, involves designation of a lane in the opposing direction of travel for HOVs. Because of the potential safety problems with this, contra-flow operation is limited to buses only (i.e., professional drivers) in its few current applications.

Other HOV features that may be included in design or reconstruction are separate priority ramps for HOVs, and park-and-ride lots adjacent to the freeway with special direct ramp access for buses. These are intended to promote transit usage, thereby increasing average vehicle occupancy. Planning and design for HOV must be within the context of an overall transportation plan for a corridor or urban area. HOV facilities have failed in some cases because they were not thought out or logically implemented. The following are guidelines for consideration in planning and design of freeway HOV facilities.[27]

For separate HOV facilities

1. The potential for HOV lanes on controlled-access facilities is limited to high-volume urban corridors. Because of the severe right-of-way and construction cost constraints associated with such corridors, consideration of sole-traveled-way HOVs should take place in the planning stage (prior to construction/rehabilitation of the corridor).

2. Projected corridor demand during peak period should indicate potentially serious capacity problems (Level of Service E throughout much of the route). A heavy peak-hour directional split of 65–35%, or 70–30% would be typical of such a corridor.

3. The corridor should be of sufficient length and should carry a high enough volume of traffic to provide the potential for large time savings to users of the HOV lanes. A minimum of 10 min is considered significant, with 15–20 min potential time savings considered desirable.

4. Transit bus demand should exist on the corridor. Such demand can be served by the HOV lanes, which, in turn, would promote additional transit usage.

5. HOV lanes on sole-traveled ways should normally be designed with a minimum of two lanes. (*Note:* In special cases one-lane sole traveled ways have been successfully implemented.) If possible, 12-ft lane width and 8–10 ft shoulders

---

[27] ITE Committee on HOV Guidelines.

CONCURRENT-FLOW HOV LANES

No Separation between HOV and General Use Lanes

| Full Shld. | 12' | 12' | 12' | 12' | 10' Shld. | 10' Shld. | 12' | 12' | 12' | 12' | Full Shld. |

Separation between HOV and General Use Lanes

| General Use Lanes | 10' Shld. | 12' | 10' Shld. | 10' Shld. | 12' | 10' Shld. | General Use Lanes |

Desirable Design

| General Use Lanes | 10'–12' Common Shld. | 12' | 4' Min. | 4' Min. | 12' | 10'–12' Common Shld. | General Use Lanes |

Common Shoulder Design

| General Use Lanes | 4' / 8' Buffer | 12' | 10' Shld. | 10' Shld. | 12' | 4' / 8' Buffer | General Use Lanes |

Minimum Separation

CONTRAFLOW HOV LANES

Flexible Posts

| 12' | 12' | 12' Buffer Lane Desirable | 12' 4' Min. | 4' Min. | 12' | 12' | 12' | 12' |

10' Shlds. Desirable

PHYSICALLY SEPARATED HOV LANES

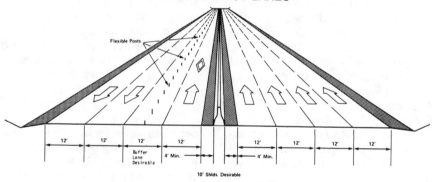

One-lane HOV Roadway

2'	12'	6'	(a)
2'	11'	2'	(b)
2'	11'	10'	(c)
4'	12'	10–12'	(d)

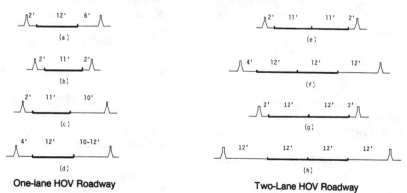

Two-Lane HOV Roadway

2'	11'	11'	2'	(e)
4'	12'	12'	12'	(f)
2'	12'	12'	2'	(g)
12'	12'	12'	12'	(h)

Figure 6–22. Freeway HOV concepts.
SOURCE: *Guide for the Design of High Occupancy Vehicle and Public Transfer Facilities,* 1983, American Association of State Highway and Transportation Officials.

should be provided in recognition of the requirements for large bus flows and the need to maintain a high level of uniform service.

6. Lane use vehicle occupancy requirements should be carefully considered if an acceptable level of service is to be provided for all users.
7. Access to the separate HOV lanes should preferably be restricted to special access ramps from grade-separated roadways.
8. In planning and design of sole-traveled-way HOV lanes, maximum operation flexibility should be preserved. Placement of special ramps, design of ramp terminals, etc., should enable conventional, reversible lanes or other configurations should the HOV concept become unworkable in practice.

For concurrent or contra-flow HOV lanes

1. The exclusive lane should be physically separated from the remaining freeway lanes. Physical separation provides additional safety and is the best deterrent to violators.
2. Physical modifications to the freeway that reduce lane widths must be approached with extreme caution. Lane width reductions to less than 11 ft should be avoided. Minimum shoulder widths should be maintained. Exclusive HOV lanes must be within a right-of-way wide enough to allow passing disabled vehicles. Lane widths for multiple lane exclusive HOV facilities should be at least 11 ft for mixed vehicular flow and at least 10 ft for bus-only operation.
3. Contra-flow priority lanes should be considered only when a considerable flow imbalance prevails during peak traffic periods and if the remaining off-peak direction traffic can flow at an acceptable level of service during these periods.
4. For safety reasons, contra-flow operations must be at all times clearly and unmistakably identified by opposing traffic. If a buffer zone to separate priority traffic from the general-use lanes cannot be provided, contra-flow lanes should be used by professional drivers only.
5. Concurrent-flow HOV lanes should be physically separated from the normal lanes by a buffer zone that will safely and effectively ensure the desired separation of vehicle classes.
6. Exclusive connections or bypass ramps should be considered where time savings to the HOV users are greater than the costs and when the ramp and through-lane capacities are not adversely affected.

Toll facilities

Toll facilities are fully controlled access roadways designed to the same high quality of design as freeways. Design speeds of 60 mph or more prevail. Cross-sectional features include full, 12-ft lane widths, 10- to 12-ft shoulders, 6:1 side slopes, and wide or protected medians.

The additional design considerations of toll collection affect the location and design of the facility. Two basic types of collection methods are used: mainline barriers at intermittent spacing, and toll collection at interchanges.

Where mainline barrier collection is used, the location of the barrier should be far enough from exit or entrance ramps to avoid weaving problems. Toll plazas should be located on tangent alignment with decision sight distance provided on the approach. Grades approaching the plaza should not exceed 0.5%. The design of the plaza itself should be based on the expected number of peak-period arrivals and the rate at which they can be processed. Sufficient toll collection lanes of great enough length should be provided to minimize the length of queuing at the plaza.

Where toll collection occurs at exit and/or entrance ramps, care should be taken in interchange ramp design. Plazas should be placed far enough upstream from the ramp diverge to provide for normal deceleration and braking to a queue.

Collector and arterial streets and highways

Most highway travel occurs on collector and arterial (i.e., nonfreeway) roads. A wide range of traffic volumes, traffic characteristics, speed, and adjacent land-use conditions can occur on such facilities. In general, collector and arterial highways should be designed to as high capacity as is feasible, and with consideration given to unfamiliar drivers.

Rural highways

Rural collectors and arterials are mostly two lanes wide, with some high-volume, nonfreeway "expressway" corridors designed to four-lane cross sections. Rural collectors and arterials hold the greatest potential for safety problems due to the combination of high-speed, high-volume operations and no or only partial access control.

Design speeds should be as high as practical given terrain and adjacent land use. Design-year level of service should be preferably B or C, which corresponds to operating speeds near design speed.

Cross section. Table 6–12 summarizes lane and shoulder width values for new design or reconstruction of collector and arterial highways. For all but very low volume roads, roadway widths of 22 to 24 ft are recommended, corresponding to 11- to 12-ft lane widths. For most highways, and all arterials with design hour volumes in excess of 400 vph, shoulder widths should be at least 8 to 10 ft.

Fore slopes of 3:1 are acceptable on lower-speed collector highways, but 4:1 slopes or flatter are preferred. For arterial highways, 4:1 slopes should be considered a minimum design, with 6:1 desirable. Longitudinal barriers may be necessary where steeper slopes are called for or space is limited.

In design of arterial highways in rural areas, consideration should be given to planning for expansion of capacity to a four-lane section. This may entail purchasing of additional right-of-way, and design of the initial two-lane section offset within the right-of-way to serve one direction of travel in the ultimate plan.

Four-lane arterials in rural areas should be designed with a median, preferably 30 ft or more, but at least wide enough to provide for intermittent left-turn lane deceleration and storage. Four-lane undivided rural arterials have been shown to produce high accident rates due to the lack of left-turn protection, head-on conflicts, and high speeds.

Intersections and driveways. Alignment design and location of intersections and driveways should recognize their significant influence on safety. Designers should provide more than minimum stopping sight distance along the highway in advance of intersections. Intersections should be relocated away from the crests of vertical curves, where sight distance is at a minimum.

Provision for left- and right-turn lanes off arterial and collector highways is important. These should be provided wherever feasible for all but very low volume highways. They are needed not for capacity, but rather to provide for safe deceleration off the higher speed through highway. Warrants and guidelines for design of channelized movements on rural highways are discussed in subsequent sections of this chapter.

Special features. Rural and arterial highways occasionally require special design features to aid their operation. These may include any of the following:

○ Three-lane sections for passing in rolling or mountainous terrain;
○ Auxiliary climbing lanes to increase capacity on a long upgrade and remove slow trucks from the higher speed passenger car traffic;
○ Truck escape ramps on downgrades to mitigate out-of-control trucks; and
○ Additional lanes and channelization at major intersections.

Urban highways

Urban collectors and arterials may carry substantial traffic volumes. A typical urban street network relies on major arterials spaced 1 mile apart, with supporting collectors and minor arterials. Urban collectors are typically two lanes wide, with provisions for parking, buses, and other needs. Arterials may be from two to as many as eight lanes wide, divided or undivided. In urban areas, there is typically no room for shoulders. Immediately outside the traveled way is a curb, used to control drainage and prevent errant vehicles from encroaching on adjacent property.

Cross section. Table 6–17 depicts minimum lane widths for urban streets. These are measured to the curb face. These dimensions imply a width of 32 to 36 ft for a two-lane collector with parking allowed on one side. Note that where bicycle traffic and/or truck traffic are substantial, greater widths are called for. Also note that higher-speed highways should be designed with 11-ft minimum and 12-ft desirable lane widths.

Urban streets are designed with curbs outside the traveled way. Curbs fulfill many functions, including control of drainage, delineation of the roadway, as a barrier against roadside encroachments (and therefore protection for pedestrians), and as a means of controlling access. There are two types of curbs—barrier and mountable, as shown in Figure 6–23. *Barrier curbs* are relatively high, steep-faced, and are

TABLE 6–17
Lane Widths for Urban Streets

Lane Type	Speed Under 40 mph[1]		Speed 40 mph and Over[1]	
	Minimum [2,3]	Desirable	Minimum[3]	Desirable
Curb parking lane only[4]	11	12	11	13
Curb travel lane[4]	11	12	11	13
Inside lane	10	12	11	12
Turn lane[5]	10	12	11	12

Note: 1 ft = 0.3m; 1 mph = 1.6 kph.
[1] The design speed for new facilities. For existing streets, use 85th percentile speed plus 5 mph but not less than the posted speed limit plus 5 mph.
[2] If medium or large trucks (includes buses) exceed 15% of Average Daily Traffic (ADT), use 11-ft width.
[3] On horizontal curves with radii of 500 ft and under, use 11-ft minimum for inside lanes and 12-ft minimum for curb lanes.
[4] If moderate to heavy bicycle traffic is expected in the street, a width of 15 ft is desirable.
[5] May be reduced by 1 ft under severely restricted conditions if not adjacent to curb on the right, and if few commercial vehicles are present. This does not apply to 2-way left-turn lanes.

SOURCE: *Guidelines for Urban Major Street Design,* Institute of Transportation Engineers, 1984.

used to prevent vehicle encroachments on the roadside. Heights of 6 to 9 in are typically used, which are effective at low to moderate speeds. *Mountable curbs* are used where crossing or encroachment of a vehicle is permitted or expected. Their design should not result in loss of vehicle control or undercarriage damage when struck.

The *border area* (see Figure 6–10) is the dimension between the edge of roadway and right-of-way line. It includes space for sidewalks, transit benches or shelters, traffic control devices, bikeways, traffic barriers, fire hydrants, utility access, and landscaping. Higher-volume, higher-class facilities such as major arterials should be designed with wide borders to accommodate all utilities and provide room for possible future expansion of the roadway.

Border widths of 10 ft are considered reasonable for accommodating typical needs such as a sidewalk, curb, drainage and other subsurface utilities, and utility poles. A minimum width of 7 ft for a border area is used by many cities in constrained locations.

The sum of all widths required for lanes, curbs, and border areas defines the right-of-way. Right-of-way acquisition or reservation should recognize the operational and safety needs of the highway as well as the border area. Table 6–18 summarizes recommended minimum and desirable right-of-way widths for urban arterials and collectors.

Alignment design. The design controls described earlier apply to urban collectors and arterials. Most agencies use maximum superelevation rates of 0.04 or 0.06 for urban streets. In many cases, lesser superelevation is necessary because of design constraints. Design speeds are generally 50 mph or less, with collector streets usually designed to a 30 mph or 35 mph design speed. The design of vertical alignment is often highly influenced by border conditions. An urban street must "fit" the adjacent properties which it serves. Driveways, sidewalks, and crossing streets will generally dictate edge of pavement and curb elevations. Further

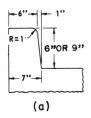

BARRIER CURBS

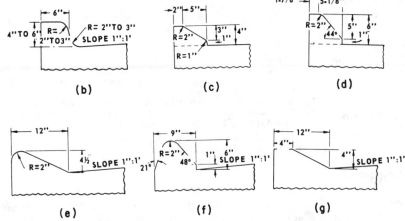

MOUNTABLE CURBS

Figure 6–23. Typical curbs.
SOURCE: *A Policy on Geometric Design of Highways and Streets,* 1990 American Association of State Highway and Transportation Officials, Washington DC. Used by permission.

TABLE 6–18

Right-of-Way Requirements for Urban Arterials and Collectors

	Number of Lanes	Minimum Right-of-Way (ft)	Desirable Right-of-Way (ft)
Undivided 2-directional	4 or 5	75	100
	6	86	100
	7	97	110
Divided 2-directional	4	78	100
	6	100	110

Note: 1 ft = 0.3m.

SOURCE: *Guidelines for Urban Major Street Design,* Institute of Transportation Engineers, 1984.

controls on street alignment design include drainage of both the street and the border area.

Intersections. Intersections generally control the capacity and operational quality of urban streets. Design of intersections is discussed separately in following sections of this chapter. Also, the reader is referred to the *Intersection Channelization Design Guide*[28] as a general reference.

Driveways. Driveways are necessary features of urban streets. They provide access to abutting property and are a service to the traveling public. Driveways also represent a potential safety and operational problem. They introduce additional points of conflict that can increase accidents and reduce the capacity of the street.

Two important aspects of driveways are their location and design characteristics. Both are discussed in an ITE publication, *Guidelines for Driveway Design and Location,*[29] which is summarized here. Driveway location and spacing are related to adjacent intersections and driveways to other properties. Keeping driveways away from intersections, even by small distances, helps to reduce conflicts. Figure 6–24 summarizes recommended basic driveway dimension guidelines. Different values for width of driveway and radius of return are shown for different property types. These values reflect that both higher volume and larger vehicles use commercial and industrial driveways.

Utilities. Placement of underground utilities is important in urban street design. Routine or emergency maintenance operations can significantly affect traffic flow, depending on the location of the utility relative to the traveled way. The following are guidelines summarized from the American Public Works Association *Manual of Improved Practice:*[30]

1. Sidewalk and median locations are preferable to through traffic lanes.

---

[28] T.R. Neuman, "Intersection Channelization Design Guide." National Cooperative Highway Research Project Report 279, (Washington, DC: Transportation Research Board, National Research Council, 1985).

[29] Institute of Transportation Engineers, *Guidelines for Driveway Design and Location.* (Washington, DC, 1975).

[30] *Manual of Improved Practice.* American Public Works Association.

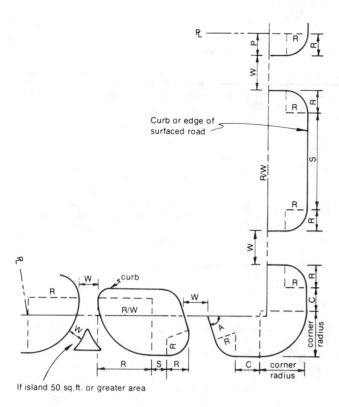

If island 50 sq.ft. or greater area

	Dimension Reference	Residential	Commercial	Industrial
Nominal Width[1]	W			
one-way		10	15	20
two-way		10	30	40
Right turn radius or flare[2]	R			
Minimum		5	15	20
Minimum spacing[3]				
From property line	P	0	0	−R
From street corner	C	5	10	10
Between driveways	S	3	3	10
Minimum Angle[4]	A	45°	45°	30°

1.0 ft. = 0.3m

[1]Residential driveway widths typically should not exceed about 24 feet (7 m). Commercial driveway widths may vary from about 24 feet for low volume activity (providing that 20 foot radii are used), to a maximum of 36 feet (11 m) for undivided design, higher volume activity. A 36 foot (11 m) driveway is usually marked with two exit lanes of 10 to 11 foot (3 m) width, with the balance used for a single, wide entry lane. Industrial driveway widths should not exceed 50 feet (17 m).

[2]On the side of a driveway exposed to entry or exit by right turning vehicles. The radii for major generator driveways such as shown in Figures 5 and 6 should be much higher than the values shown.

[3]Measured along the curb or edge of pavement from the roadway end of the curb radius or flare, except for conditions noted in Figure 7. For individual properties, a suggested limitation on the number of driveways is: 1 for 0-50 foot (0-15 m) frontage, 2 for 51-150 foot 16-50 m) frontage, 3 for 151-500 foot (51-150 m) frontage, and a 4 for over 500 (over 150 m) frontage.

[4]Minimum acute angle measured from edge of pavement, and generally based on one-way operation. For two-way driveways and in high pedestrian activity areas, the minimum angle should be 70 degrees.

Figure 6–24. Recommended basic driveway dimension guidelines.
SOURCE: *Guidelines for Driveway Design and Location,* Institute of Transportation Engineers, 1975.

2. Lanes normally used for curb parking are the next preferred location.
3. The outside travel lanes are next on the location priority list.
4. Centerline locations on two-way streets should be avoided so that both directions of flow are not affected.

5. Utilities with a history of high-frequency maintenance activities (gas and electrical) should be given first priority for occupying the median and parking lane areas.
6. Travel lane areas within intersections should be avoided where possible because these locations have maximum congestion potential.

Lighting. Major streets in urban areas, and some major intersections in rural areas, require lighting. Roadway lighting contributes to safe nighttime operation. In terms of design, the key consideration regarding lighting is placement of light poles. Chapter 10 discusses lighting design concerns and principles.

Special features. Urban arterial and collector highways experience special operational problems and conditions. On one hand, they are typically required to accommodate substantial through traffic and represent a major part of total network capacity. On the other hand, access to abutting properties is required, which creates midblock conflicts and degrades roadway capacity.

A number of design techniques have been used to optimize the conflicting needs of access and capacity. These include construction of raised medians to control, restrict, or consolidate access; driveway redesign; acceleration and deceleration lanes; and other features. Table 6–19 summarizes guidelines for consideration of these techniques. Along many arterials, accommodation of midblock left turns is a key requirement. Design solutions include continuous two-way left-turn lanes. The following are guidelines for their application:

Average daily traffic through volumes

10,000 to 20,000 vehicles per day for existing four-lane highways

5,000 to 12,000 vehicles per day for existing two-lane highways

Turning volumes

70 midblock left turns per 1,000 ft during peak hour
left-turn peak-hour volume of 20% or more of total volume

Minimum length

1,000 ft or two to three blocks is considered a minimum reasonable length

Adjacent land use

strip commercial or multiple unit residential

Other factors

closely spaced driveways
high midblock accident history

Table 6–20 summarizes recommended widths of continuous two-way left-turn lanes.

TABLE 6–19

Guidelines for Application of Access-Control Techniques

Access Control Technique	Highway Type	Speed (mph)	Average Daily Traffic	Number of Driveways Per Mile
Raised median to prohibit left turns	Multilane arterials	40+	10,000 vpd; peak-hour left turns 150+/mile	30 to 60
Raised channelization to limit access to right-in and right-out	Multilane divided highways	30 to 45	5,000 vpd; prohibited turns less than 100 vpd per mile	30+
Raised median with left-turn lanes	Multilane highways	30 to 45	10,000 vpd; peak-hour left turns 150+/mile	30+
Improved median opening geometry (tapers)	Multilane divided highways (4-ft medians as minimum)	30+	5,000 vpd	15+
Alternating left-turn lanes	All types with available width	35+	10,000 vpd; peak-hour left-turn demand at least 15% of through traffic	45+; 1,000 ft + between major intersections
Conversion of two-way driveway to two one-way driveways	All types	35+	10,000 vpd; peak-hour left turn of 40 vph	1 to 60
Conversion of two-way driveway to two two-way driveways with restricted access movements	Divided highways	35+	10,000 vpd; peak-hour left turns of 40 vph	1 to 60
Construction of local service road with limited, controlled access points	Primary divided arterials	40 to 55	20,000 vpd	60+
Physical barrier to prevent uncontrolled access to driveway	All types	All	10,000 vpd; driveway volume 500 vpd	45+ isolated locations
Widen narrow right lanes to assist right turns	Urban arterials	30+	5,000 vpd; right-turn driveway volume of 100+ vph per mile in peak hour	20+
Installation of right-turn deceleration lane	All	35+	10,000 vpd; driveway volume 1,000 vpd; right-turn volume of 40 vph in peak hour	Isolated locations
Continuous right-turn lanes	All	30+	15,000 vpd; right-turn volume per mile at least 20% of total	60+
Installation of right-turn acceleration lane	All	35+	10,000 vpd; right-turn egress of 75 vph in peak hour	Isolated locations

SOURCE: J.C GLENNON AND OTHERS, "Evaluation of Techniques for the Control of Direct Access to Arterial Highways," Report No. FHWA-RD-76-85, 1985.

TABLE 6–20

Lane Widths for Design of Continuous Two-Way Left-Turn Lanes

Prevailing Speed	Adjacent Land Use/ Vehicle Type	Appropriate Width of Lane
25–30 mph	Residential, business (passenger cars)	10 ft absolute minimum, 12 ft desirable
30–40 mph	Business (passenger cars, some trucks)	12 ft minimum 14 ft desirable
	Industrial (many large trucks)	14 ft to 16 ft
40–50 mph	Business	14 ft desirable

Local roadways

Local roads generally serve as access from land uses to the collector and arterial street system. A common attribute of local roads in both urban and rural areas is the low traffic volume along them. Speeds are also generally low.

Rural local roads

Cross-section design values for rural local roads are summarized in Figure 6–25. Low speeds and volumes can mean lane widths of 10 to 11 ft in many cases. Minimum usable shoulder widths of 4 ft are also provided. Alignment and stopping

Figure 6–25. Cross sections and right-of-way widths for 2-lane rural highways.
SOURCE: *A Policy on Geometric Design of Rural Highways,* 1965 American Association of State Highway and Transportation Officials, Washington DC. Used by permission.

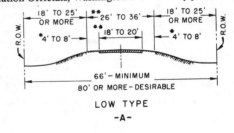

LOW TYPE
–A–

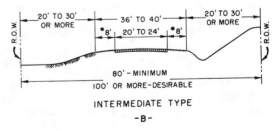

INTERMEDIATE TYPE
–B–

* Usable shoulder width
** For low volume roads with few trucks

sight distance should be provided for the design speed, which generally varies from 30 to 50 mph on such roads.

Urban local streets

Design of local streets in urban areas should emphasize not only the traveled way but also the land-use activities for which the road provides access. Sidewalks for pedestrians, border areas for utilities and clear view to driveways, and provision for cul-de-sacs and alleys are all important. Figure 6–26 shows dimensions for a typical urban street. Table 6–21 summarizes the many local street considerations and design values.

Intersection design

Intersections at-grade are unique elements of the highway. By definition, they represent points of potential vehicle conflict and are thus susceptible to accident potential. In urban areas, intersections are of such importance that they control the capacity of a street network.

Intersection design combines aspects of vehicle operational and driver performance characteristics. A prime consideration in intersection design is the type of traffic control

required. Table 6–22 summarizes the key human factors and vehicle characteristics that apply to intersection design.

Functional design principles

Two prime objectives of any intersection design are operational quality and safety. Stated differently, the design and traffic control scheme should optimize the operational quality of flow through the intersection; and the intersection should be designed to minimize accidents and their adverse consequences. These stated objectives are more precisely defined as follows:

1. Points of conflict should be minimized;
2. Conflict areas should be simplified;
3. Conflict frequency should be limited; and
4. Conflict severity should be minimized.

Designers have available at their disposal a range of intersection design elements that can be used to achieve the above functional objectives. Traffic islands separate conflicting movements. Street closures or realignment can be used to simplify an intersection. Turning lanes can remove slow or stopped vehicles from through lanes, thereby removing

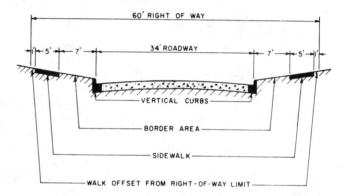

Figure 6–26. Typical urban local street cross section.
SOURCE: *Recommended Guidelines for Subdivision Streets,* Institute of Transportation Engineers, 1984.

TABLE 6–21
Local Street Design Guidelines

Terrain Classification → Development Density →	Level			Rolling			Hilly		
	Low	Medium	High	Low	Medium	High	Low	Medium	High
Right-of-way width (ft)	50	60	60	50	60	60	50	60	60
Pavement width (ft)	22–27	28–34	36	22–27	28–34	36	28	28–34	36
Type of curb (V = vertical face; R = roll-type; O = none)	O/R	V	V	V	V	V	V	V	V
Sidewalks and bicycle paths (ft)	0		4–6	0		4–6	0		4–6
Sidewalk distance from curb face (ft)	—	6	6	—	6	6	—	6	6
Minimum sight distance (ft)	←———— 200 ————→			←———— 150 ————→			←———— 110 ————→		
Maximum grade	←———— 4% ————→			←———— 8% ————→			←———— 15% ————→		
Maximum cul-de-sac length (ft)	1,000	700	700	1,000	700	700	1,000	700	700
Minimum cul-de-sac radius (right-of-way) (ft)	←———————————————————— 50 ————————————————————→								
Design speed (mph)	←———— 30 ————→			←———— 25 ————→			←———— 20 ————→		
Minimum centerline radius of curves (ft)	←———— 250 ————→			←———— 175 ————→			←———— 110 ————→		
Minimum tangent between reverse curves (ft)	←———————————————————— 50 ————————————————————→								

SOURCE: *Recommended Guidelines for Subdivision Streets,* Institute of Transportation Engineers, 1984.

TABLE 6-22

Human Factors and Vehicle Characteristics
Applicable to Design of Channelized Intersections

Human Factor	Design Values	Design Elements Affected
Perception/reaction time	2.0^a–4.0^b sec	Intersection sight distance
Driver height of eye	3.5 ft^b	Sight distance
Pedestrian walk times	3.0–4.5 fps^c	Pedestrian facilities

Vehicle Characteristics	Intersection Design Elements Affected
Physical Characteristics	
Length	• Length of storage lanes
Width	• Width of lanes
	• Width of turning roadways
Height	• Placement of overhead signals and signs
Operational Characteristics	
Wheelbase	• Nose placement
	• Corner radius
	• Width of turning roadway
Acceleration capability	• Acceleration tapers and lane lengths
Deceleration and braking capability	• Length of deceleration lanes and tapers
	• Stopping sight distance

^{a} *A Policy on Geometric Design of Highways and Streets* (1984).
^{b} Ibid.
^{c} *Traffic Engineering for Senior Citizen and Handicapped Pedestrians* (1983).

SOURCE: *Intersection Channelization Design Guide,* National Cooperative Highway Research Program Report 279.

potential conflicts. Islands can be used to store pedestrians safely off the traveled way.

Principles of channelization

The following nine principles of channelization apply to all intersection types:

1. *Undesirable or wrong-way movements should be discouraged or prohibited.* Channelization—traffic islands, raised medians and corner radii—should be used to restrict or prevent undesirable or wrong-way movements. Where such movements cannot be completely blocked, the channelization scheme should discourage their completion.

2. *Desirable vehicular paths should be clearly defined.* The design of an intersection—including its approach alignment, traffic islands, pavement markings, and geometry—should clearly define proper or desirable paths for vehicles. Exclusive turning lanes should be clearly delineated to encourage their use by turning drivers and discourage their use by drivers intending to proceed through the intersection. Traffic islands should not cause confusion about the proper direction of travel around them.

3. *Desirable or safe vehicle speeds should be encouraged.* Channelization should promote desirable vehicle speeds wherever possible. In some instances this means providing open alignment to facilitate high-speed, heavy-volume traffic movements. In other cases, channelization may be used to limit vehicle speeds in order to mitigate serious high-speed conflicts.

4. *Points of conflict should be separated where possible.* Separation of points of conflict eases the driving task.

Channelization techniques such as development of turning lanes, design of islands, and control of access points all serve to separate points of conflict. This enables the driver to perceive and react to conflicts in an orderly manner.

5. *Traffic streams should cross at right angles and merge at flat angles.* Crossing and merging of traffic streams should be accomplished to minimize both the probability of actual conflict or collision, and the severity of conflict. Channelization and alignment design should produce crossing vehicle streams at as close to right angles (90°) as is practical. Where vehicle streams merge, the alignment of the merging roadways should be accomplished at flat angles.

6. *High-priority traffic movements should be facilitated.* The operating characteristics and appearance of intersections should reflect and facilitate the intended high-priority traffic movements. Selection of high-priority movements can be based on relative traffic volumes, functional classification of the intersecting highways, or route designations.

7. *Desired traffic control scheme should be facilitated.* The channelization employed should facilitate and enhance the traffic control scheme selected for intersection operation. Location and design of exclusive lanes should be consistent with signalization or stop-control requirements. Location of traffic islands, medians, and curb returns should reflect consideration of the need to place signals and signs in locations visible to drivers.

8. *Decelerating, stopped, or slow vehicles should be removed from high-speed through-traffic streams.* Wherever possible, intersection design should produce separation between traffic streams with large speed differentials. Vehicles that must decelerate or stop because of traffic control or to complete a turn should be separated from through traffic proceeding at higher speeds. This practice facilitates safe completion of all movements by reducing rear-end conflicts.

9. *Provide safe refuge for pedestrians and other non-motor-vehicle users.* Channelization can shield or protect pedestrians, bicycles, and the handicapped within the intersection area. Proper use of channelization will minimize exposure of these vulnerable users to vehicle conflicts, without hindering vehicular movements.

Intersection design guidelines

Execution of the above design principles requires consideration of horizontal and vertical alignment, traffic control, vehicle acceleration, and deceleration characteristics. The following is a summary of intersection design guidelines.

Angle of intersection. Crossing roadways should intersect at 90° if possible, and desirably at no less than 75°. Where skew angles of 60° or less exist, geometric countermeasures such as reconstruction, or positive traffic control such as signalization, may be required.

Horizontal and vertical alignment. The alignment in advance of and through an intersection should promote driver awareness, operate well under frequent braking, and be easy to drive so the navigational task is not too difficult. More than minimum stopping sight distance should be provided on the intersection approach. Downgrades greater than −6% should be avoided on low-speed highways, and greater than

−3% on high-speed highways. Intersections should be designed on alignment no sharper than 3° of curve, and preferably on tangent.

Left-turn lane warrants and design. Left-turn lanes offer distinct operational and safety benefits at both signalized and unsignalized intersections. At signalized intersections, warrants are based on turning volumes, accident experience, and general capacity relationships. For safety reasons, separate left-turn lanes should be provided at high-speed rural signalized intersections regardless of volume or capacity consideration.

Elements of left-turn-lane design are shown in Figure 6–27. Design values for approach tapers, bay tapers, and lane

Figure 6–27. Design elements of left-turn bay channelization.
SOURCE: PAUL C. BOX AND ASSOCIATES, INC. (Skokie, IL).

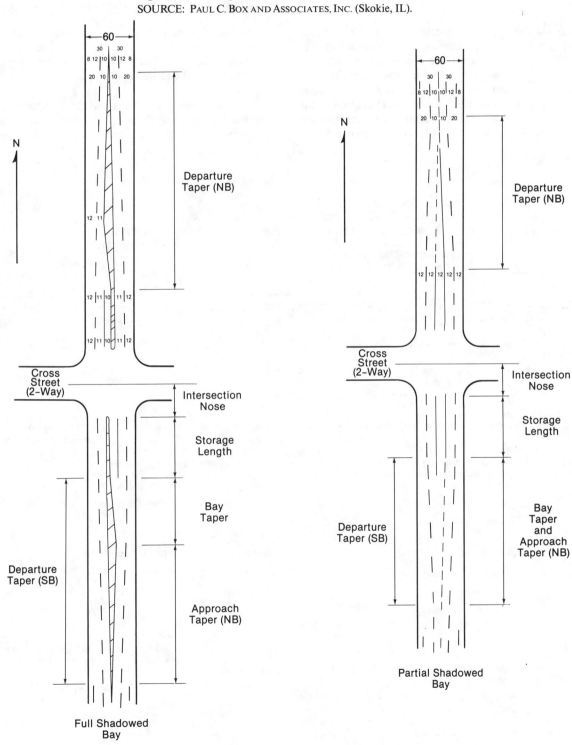

lengths are shown in Figure 6–28. These are based on deceleration in the lane, storage in the lane, or a combination of both. Note that storage requirements are a function of signal cycle lengths for signalized intersections.

Right-turn lane warrants and design. Factors to consider in providing for right-turn lanes include the volume of right turns, any history of right-turning rear-end accidents, speed of the highway, and land-use availability. Taper and lane-length design should be based on deceleration, storage, or both.

Corner radius design. The design of corner radii should be based on selection of a reasonable design vehicle (see Figure 6–29). There are cost and other trade-offs involved, with larger vehicles requiring more open intersections that increase cost and are more difficult to mark, signalize, and operate. Design for smaller vehicles can create operational problems should "oversized" vehicles have to use the intersection. Table 6–23 shows general guidelines to assist in selection of a design

vehicle. The actual radius or curb return design can be accomplished one of three ways. Simple circular radius designs are common on low-speed collector and local streets, and in downtown areas. Table 6–24 summarizes the operational characteristics of a range of typical corner radii. Alternative designs include three-centered, compound curves, or offset circular curves. These designs fit the paths of turning vehicles, thereby providing more efficient operations. Figure 6–30 illustrates such designs. Design of corner radii should also consider the needs of pedestrians. Larger radii or offset curb returns increase pedestrian crossing distances.

Widths of turning roadways. Turning-lane width guidelines are shown in Figure 6–31. The width of a turning roadway is based on the selected design vehicle, assumed operation, and the radius of the turn.

Traffic islands. *Traffic islands* are one tool of an intersection designer. Islands can be painted or raised, and they can serve a range of functions. Painted or "flush" channel-

Figure 6–28. Guidelines for design of left-turn lanes.
SOURCE: *Guidelines for Urban Major Street Design—A Recommended Practice*, Institute of Transportation Engineers, 1984, and *Intersection Channelization*, NCHRP Report 279, 1984, T.R. Neuman.

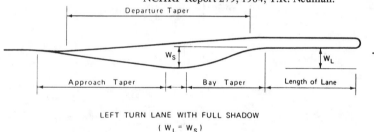

LEFT TURN LANE WITH FULL SHADOW
($W_L = W_S$)

T_a—Approach Taper Design

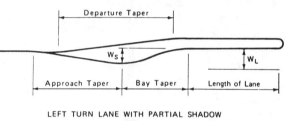

LEFT TURN LANE WITH PARTIAL SHADOW
($W_L > W_S$)

T_b—Bay Taper Design

Functional Basis: To provide a smooth lateral transition for all vehicles approaching the intersection
Form of Alinement: Tangent
Desirable Design: Provide a fully shadowed lane ($W_a > W_1$).
Recommended for high speed intersections and intersections in rural and open urban areas with no space constraints.

$$T_a = \frac{W_1 S^2}{60}$$
W_1 = Width of Lane (ft)
S = Speed (mph)

Typical Values for T_a*

S—Speed (mph)	W_1—Width of Lane (ft)		
	11	11.5	12
30	165	170	180
40	295	305	320
50	460	480	500
60	660	690	720

* Rounded to nearest 5 ft

Minimum Design: Provide a partially shadowed lane (see Figure 4.14) with $W_s < W_1$. Design as follows:

$$T_a = W_s S$$
W_s = Offset (ft)
S = Speed (mph)

As a minimum, a 10:1 taper ratio should be used.

Functional Basis: To direct left-turning vehicles into the turn lane
Form of Alinement: Tangent; or reverse curves with ⅓ of the total length comprised of a central tangent
Desirable Design: For fully shadowed left turn lane

$$T_b = \frac{W_1 S}{2.5}$$
W_1 = Width of Lane
S = Speed (mph)

Typical Values for T_b*

S—Speed (mph)	W_1—Width of Lane (ft)		
	11	11.5	12
30	130	140	145
40	175	185	190
50	220	230	240
60	265	275	290

* Rounded to nearet 5 ft

Minimum Design: Taper ratios of 8:1 can be used for tangent bay tapers. For constrained locations, ratios as low as 4:1 can be used with painted channelization.

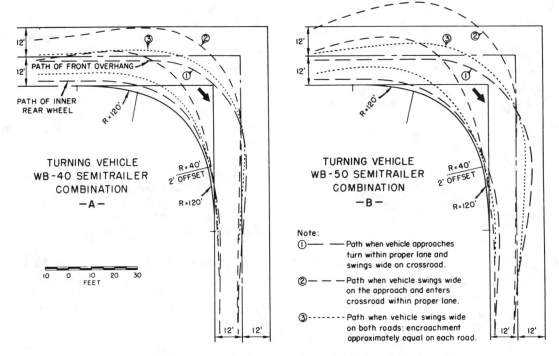

Figure 6–29. Minimum edge-of-pavement design for single-unit trucks and buses and necessary paths of larger vehicles.

SOURCE: *A Policy on Geometric Design of Highways and Streets,* 1965, American Association of State Highway and Transportation Officials, Washington, DC. Used by permission.

TABLE 6–23

Guidelines for Selection of Design Vehicle

Highway Type	Design Vehicle
Rural Highways	
Interstate/freeway ramp terminals	WB-50[a]
Primary arterials	WB-50[a]
Minor arterials	WB-50 or WB-40
Collectors	SU-30
Local streets	SU-30
Urban Streets	
Freeway ramp terminals	WB-50[a]
Primary arterials	WB-50 or WB-40
Minor arterials	WB-40 or B-40
Collectors	B-40 or SU-30
Residential/local streets	SU-30 or P

[a] Consideration of larger design vehicles, such as WB-65, and other "oversize" vehicles is important.

SOURCE: *Intersection Channelization Design Guide,* National Cooperative Highway Research Program Report 279, 1985.

TABLE 6–24

Operational Characteristics of Corner Radii

Corner Radius (ft)	Operational Characteristics[a]
<5	Not appropriate for even P design vehicles
10	Crawl speed turn for P vehicles
20–30	Low speed turn for P vehicles, crawl speed turn for SU vehicle with minor lane encroachment
40	Moderate speed turn for P vehicle, low speed turn for SU vehicle, crawl speed turn for WB-40 or WB-50 vehicle with minor encroachment
50	Moderate speed turns for all vehicles up to WB-50

[a] Assuming approach and departure occurs in curb lane.

SOURCE: *Intersection Channelization Design Guide,* National Cooperative Highway Research Program Report 279, 1985.

ization may be used on high-speed highways to delineate turning lanes, in constrained locations where room is not available for raised islands, or where snow removal is a major concern. Raised islands should be used where the primary function of the island is to shield pedestrians, locate traffic control devices, or prohibit undesirable traffic movements.

There are two basic types of islands—corner islands separating right turns, and median islands separating opposing traffic on an approach. Islands should be designed to "fit" natural vehicle paths. Sufficient offsets from the edge of the

island to traveled way should be provided to assist operation and minimize maintenance problems. Table 6–25 shows guidelines for island design, including offsets, lengths, widths, and corner radii.

Traffic control devices

It is particularly important to consider the effects of and requirements for traffic control devices in design of intersections. The *Manual on Uniform Traffic Control Devices* (MUTCD)[31] specifies desirable controls for location of signals, signs, and pavement markings. These should clearly be

---

[31] U.S. Department of Transportation, *A Manual on Uniform Traffic Control Devices.* (Washington, DC, 1988).

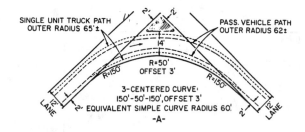

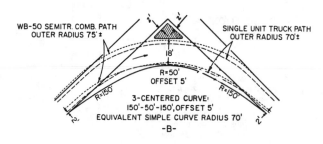

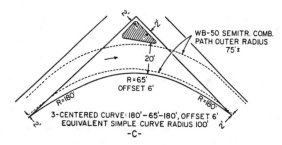

Figure 6–30. Designs for turning roadways with minimum corner island.
SOURCE: *A Policy on Geometric Design of Highways and Streets,* 1990 American Association of State Highway and Transportation Officials, Washington, DC. Used by permission.

considered as design proceeds. Chapters 8 and 9 of this text discuss details of traffic control.

One aspect of traffic control is that it can change. An intersection in an urban area may initially be stop controlled, but eventually converted to signal control as traffic volumes increase. The design should accommodate eventual conversion to signal control without the need for major reconstruction.

Bicycle facilities

The bicycle as a mode of transportation must be considered in design and operation of urban streets. Bicycle traffic may not constitute a significant portion of total travel, yet the presence of bicycles affects cross-section design, pedestrian facilities, and street operation. Concern for bicyclists is important given their vulnerability to injury, the presence of young bicyclists in the traffic stream, and the greater sensitivity of bicyclists to pavement surface irregularities.

There are two options in bicycle planning and design. The first is to do nothing and to let the existing street or network function with the bicycles. A second option is to provide for specially designed, marked, and signed bicycle facilities, as shown in Figure 6–32.

Cross section. Figure 6–33 summarizes recommended bikeway width values for bike lanes on existing streets. A minimum of 4 ft is shown, with greater widths on streets that have on-street parking.

Design speed. Design speeds of bikeways are generally 20 to 30 mph. The maneuverability and short braking distance of bikes means that for bikeways on existing streets, design speed is generally not an issue, as the design speed for motor vehicles is the controlling design parameter.

Alignment design. Design of bikeways should reflect the type of bicyclists who will use the facilities. Grades of 4% to 5% are considered reasonable maximums, although short grades of up to 15% may be used. Maximum curvature (minimum radius) is shown on Table 6–26 for design speeds of up to 40 mph.

Stopping sight distance. Table 6–27 summarizes recommended stopping sight distance design values for bikeways.

Grade separations. Conflicts between bikeways and other streets may be resolved with grade separations. Overpasses (bikeway over street) are easier to construct over an existing road, and they offer natural light and few security concerns. Underpasses are preferred in terms of bicycle operation. Design dimensions for grade separations are shown in Table 6–28.

Pedestrian facilities

Accommodating pedestrians is an important aspect of street planning and design in urban areas. The pedestrian's foremost need is safety, particularly in residential areas where children are prevalent. Comfort and convenience as well as capacity or level of service are also factors in designing for pedestrians.

Sidewalks

Sidewalks should be considered in design of new streets and in improvements of existing streets. Sidewalks may be included on one or both sides of a street or highway. Factors that should be considered in reserving space for sidewalks include type of abutting land use, amount of pedestrian activity, speed and volume of traffic on the street, and proximity to facilities such as schools, libraries, and parks. Design of sidewalks includes width, placement within the right-of-way, relationship to the traveled way, grades, and treatment at intersections.

Width considerations. Street type, adjacent land use, and frequency of use affect the width of sidewalks. Widths of

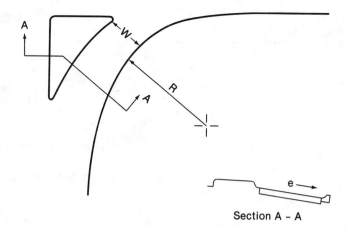

Section A – A

W—WIDTHS OF TURNING ROADWAYS

Radius on Inner Edge of Pavement, R (ft)	Case I One-Lane, One-Way Operation, No Provision for Passing a Stalled Vehicle				Case II One-Lane, One-Way Operation, With Provision for Passing a Stalled Vehicle by Another of the Same Type				Case III Two-Lane Operation, Either One- or Two-Way (Same Type Vehicle in Both Lanes)			
	Pavement Width (ft for Design Vehicle:											
	P	SU	WB-40	WB-50	P	SU	WB-40	WB-50	P	SU	WB-40	WB-50
50	13	18	23	26	20	29	36	44	26	35	42	50
75	13	17	19	22	19	27	31	36	25	33	37	42
100	13	16	18	21	19	25	29	34	25	31	35	40
150	12	16	17	19	18	24	27	29	24	30	33	35
200	12	16	16	17	18	23	25	27	24	29	31	33
300	12	15	16	17	18	22	24	25	24	28	30	31
400	12	15	16	16	17	22	23	24	23	28	29	30
500	12	15	15	16	17	22	23	24	23	28	29	30
Tangent	12	15	15	15	17	21	21	21	23	27	27	27

NOTE: P = passenger vehicles: SU = single-unit trucks; WB-40 = semitrailer combinations; WB-50 = semitrailer combinations.

R Radius (ft)	Degree of Curve	Range in Superelevation Rate—e for Intersection Curves with Design Speed (mph) of					
		15	20	25	30	35	40
50	—	.02–.10	—	—	—	—	—
90	63.6	.02–.07	.02–.10	—	—	—	—
150	38.2	.02–.05	.02–.08	.04–.10	—	—	—
230	24.8	.02–.04	.02–.06	.03–.08	.06–.10	—	—
310	18.5	.02–.03	.02–.04	.03–.06	.05–.09	.08–.10	—
430	13.3	.02–.03	.02–.03	.03–.05	.04–.07	.06–.09	.09–.10
600	9.6	.02	.02–.03	.02–.04	.03–.05	.05–.07	.07–.09
1,000	5.7	.02	.02–.03	.02–.03	.03–.04	.04–.05	.05–.06
1,500	3.8	.02	.02	.02	.02–.03	.03–.04	.04–.05
2,000	2.9	.02	.02	.02	.02	.02–.03	.03–.04
3,000	1.9	.02	.02	.02	.02	.02	.02–.03

NOTE: Preferably use superelevation rate in upper half or third of indicated range. In areas where snow or ice is frequent, use maximum rate of 0.06 or 0.08.

Figure 6–31. Design controls for turning roadways.

SOURCE: *A Policy on Geometric Design of Highways and Streets,* 1984 American Association of State Highway and Transportation Officials, Washington, DC. Used by permission.

TABLE 6–25
Guidelines for Island Design

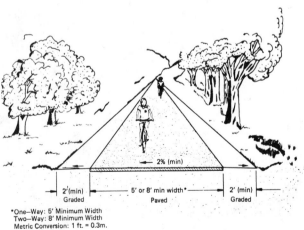

Recommended Offset Dimensions for Location of Traffic Islands

Offset in feet (see figure above)

O_a	O_b	O_c	O_d	O_e	O_f
2'–6'	1'–3'	2'–3'	2'–6'	2'–3'	0'–1'

Recommended End Radius Dimensions for Design of Traffic Islands

Radii in feet (see figure above)

R_1	R_2	R_3
2'–3'	2'–5'	1'–2'

Note: Offset values at the high end of the range are appropriate for high-speed roadways and large islands.

For roadways with shoulders, the island should be offset from the outside edge of shoulder.

SOURCE: *Intersection Channelization Design Guide,* National Cooperative Highway Research Program Report 279, 1985.

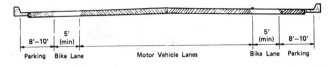

(a) CURBED STREET WITH PARKING

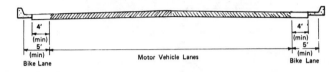

(b) CURBED STREET WITHOUT PARKING

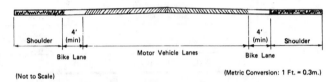

(c) STREET OR HIGHWAY WITHOUT CURB OR GUTTER

(Not to Scale)   (Metric Conversion: 1 Ft. = 0.3m.)

Figure 6–33. Typical bicycle lane cross sections.
SOURCE: *Guide for the Development of New Bicycle Facilities,* 1981, American Association of State Highway and Transportation Officials, Washington, DC. Used by permission.

TABLE 6–26
Design Radii for Paved Bicycle Paths

Design Speed—V (mph) (1 mph = 1.6 km/hr)	(e = 2%) Friction Factor—f	Design Radius—R (ft) (1 ft = 0.3 m)
20	0.27	95
25	0.25	155
30	0.22	250
35	0.19	390
40	0.17	565

SOURCE: *Guidelines for Development of New Bicycle Facilities,* 1981, American Association of State Highway and Transportation Officials.

4 to 6 ft are appropriate for most sidewalks, with 5 ft being a commonly used dimension. A 6-ft width should be considered in areas with frequent pedestrian travel such as near schools, parking facilities, and transportation terminals.

Location of sidewalks. Sidewalks are normally placed near the right-of-way line, with a nominal dimension of 1 ft reserved for fences, utility meters, etc. A desirable feature of sidewalk location, particularly important in residential areas, is provision for a setback from the curb line. Setbacks of at least 5 ft, and desirably 10 ft, are needed to:

1. provide a safety margin for children using the sidewalk;
2. minimize vehicle/pedestrian conflicts;
3. reduce splashing of pedestrians by passing vehicles;
4. provide space for utilities, traffic control devices, parking meters, etc; and

Figure 6–32. Bicycle path on separated right-of-way.
SOURCE: *Guide for the Development of New Bicycle Facilities,* 1981, American Association of State Highway and Transportation Officials, Washington, DC. Used by permission.

TABLE 6–27

TABLE 6–27

Design Stopping Sight Distances for Bicycles

Design Speed (mph)	Stopping Sight Distances (in ft) for Downhill Gradients of:			
	0%	5%	10%	15%
10	50	50	60	70
15	85	90	100	130
20	130	140	160	200
25	175	200	230	300

Note: Design values for stopping sight distances on bikeways can be developed in the same manner as on highways. The values shown were based on the following factors for wet pavement conditions.

Coefficient of skid resistance = 0.25
Perception-reaction time = 2.5 sec.
Eye height = 4.5 ft
Object height = 0 in
1 ft = 0.3 m; 1 mph = 1.6 kph.

SOURCE: *Bicycling in Tennessee, Planning and Design Manual,* Tennessee Department of Conservation and Transportation, 1975.

TABLE 6–28

Design Guidelines for Bikeway Grade Separations

	Minimum	Desirable
Vertical clearance (underpass) (feet)	8.2	10
Width (underpass or overpass) (feet)	8	14
Grade	15% maximum	10% desirable maximum

SOURCE: *Guidelines for Urban Major Street Design,* Institute of Transportation Engineers, 1984.

5. minimize the effect of pavement warping for driveway grades on the sidewalk itself.

Where setbacks cannot be provided because of limited right-of-way, 6- to 8-ft sidewalk widths are considered desirable.

Grades. Maximum grades for sidewalks are in the range of 5%, which is the limit that handicapped pedestrians can negotiate. Where greater grades result from the grade of the street, special design textures may be necessary. Where extended grades above 8% are required, handrails can be provided.

Treatment at intersections. Care should be taken to maintain a clear curb environment at intersections. Light poles, traffic signals, mail boxes, and other roadside furniture should be located away from the curb where the sidewalks meet at the intersection.

Crosswalks

The crosswalk is that portion of the traveled way designated for use by pedestrians crossing the street or roadway. Crosswalks are designated at intersections and at midblock locations, and they can be either marked or unmarked.

Crosswalks should be planned to maximize the safety of the crossing pedestrians. Wherever possible, right-angle crossings of the street should be used to minimize exposure to vehicles. At high-volume, high-speed intersections, crosswalks should be carefully located to avoid high-conflict quadrants or areas. Islands and medians should be used to store pedestrians outside the traveled way.

The *Manual on Uniform Traffic Control Devices*[32] contains guidelines on the marking of crosswalks. Crosswalk widths should be a minimum of 6 ft and as great as 8 ft. Crosswalks should be marked at all intersections on established routes to kindergarten or elementary schools. Special word and symbol markings can be used to supplement crosswalk markings.

Provisions for the handicapped

Design of sidewalks and crosswalks must include provision for handicapped pedestrians. The following guidelines address location and design of ramps for the handicapped:

1. Matching curb ramps should be provided at all intersection quadrants to provide maximum accessibility.
2. Utilities, drainage inlets, signs, and other fixed objects should not be located within the path defined by the curb ramp.
3. Curb ramps should only be constructed where sidewalks are provided.
4. Location of curb ramps relative to crosswalks and corner radii should follow the guidelines illustrated in Figure 6–34.

Grade separations

It is occasionally necessary to separate physically pedestrian movements from street traffic. Pedestrian overpasses or underpasses are used to accommodate crossing of high-volume and/or high-speed facilities in a safe manner. Locations where pedestrian grade separations are considered include those at schools along or near major arterial streets, over freeways, and near parks or other recreational land uses. Consideration of the type of separation to be used involves issues of cost, personal security, and physical feasibility.

Summary of geometric design: philosophy and approach

This chapter summarized and highlighted the basic elements of geometric design. There are clearly many considerations in highway design, including cost of construction, availability of right-of-way, character of adjacent land uses, compatibility with other infrastructure needs, and the use of appropriate design criteria and standards. The designer's task is a difficult one, requiring a balancing of many conflicting trade-offs.

It is essential that highway designers never lose sight of their primary objective, which is to provide for safe and efficient transportation. Application of design standards and criteria, and the inevitable design compromises that occur during design, must always take place within the context of maintaining safety and enhancing (or not degrading) operational efficiency.

---

[32] Ibid.

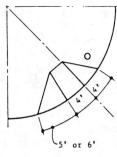

a. If the obstruction is located 0'-6' from the middle of the curb return, offset the ramp in the direction of the major pedestrian movement.

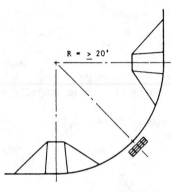

b. If a drop inlet is located 0'-6' from the middle of the curb return with a radius greater than or equal 20', parallel curb ramps should be installed. Parking should be restricted at least 10 ft. (20 ft. preferred) from the curb ramps.

If the curb radius is less than 20', the ramp should be offset in the direction of the major pedestrian movement as in part of this figure.

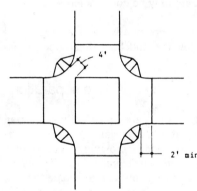

a. Middle of curb return (or diagonal) curb ramps.

b. Parallel curb ramps.

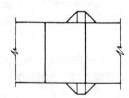

c. Parallel curb ramps located within crosswalks greater than or equal to 12 ft. in width.

Parallel curb ramps in a median. Medians may be made accessible by providing a break in the median or a crosswalk in front of the median.

For crosswalks or medians less than 12 ft. wide, center the ramp in the walk or median.

Parking should be restricted within 10 ft. (20 ft. preferred) of the curb ramp.

Figure 6–34. Guidelines for location and design of curb ramps for the handicapped.
SOURCE: B.H. Cottrell, Jr., "Guidelines for the Design and Placement of Curb Ramps," *Transportation Research Board Record* 923 (1983).

Indeed, the application of appropriate design standards is not an easy or direct task. A "cookbook" approach to design does not guarantee a safe or efficient roadway. As is demonstrated in both this chapter and throughout this text, many dynamic and complex relationships affect highway operations. Good design results from an understanding of how all elements of the highway—horizontal and vertical alignment, intersections and interchanges, and cross section—produce or contribute to safety and operational efficiency.

REFERENCES FOR FURTHER READING

A Policy on Geometric Design of Highways and Streets (Washington, DC: American Association of State Highway and Transportation Officials, 1990).

Guidelines for Urban Major Street Design (Washington, DC: Institute of Transportation Engineers, 1984).

Highway Capacity Manual (Washington, DC: Transportation Research Board Special Report 209, National Research Council, 1985).

Intersection Channelization Design Guide (Washington, DC: National Cooperative Highway Research Program Report 279, Transportation Research Board, National Research Council, 1985).

Manual on Uniform Traffic Control Devices for Streets and Highways (Washington, DC: U.S. Department of Transportation, Federal Highway Administration, 1988).

Recommended Guidelines for Subdivision Streets (Washington, DC: Institute of Transportation Engineers, 1984).

Relationship Between Safety and Key Highway Features. A Synthesis of Prior Research (Washington, DC: Transportation Research Board State of the Art Report 6, National Research Council, 1987).

Roadside Design Guide (Washington, DC: American Association of State Highway and Transportation Officials, 1989).

Synthesis of Safety Research Related to Traffic Control and Roadway Elements, Volumes I and II (Washington, DC: U.S. Department of Transportation, Federal Highway Administration, 1982).

7

PARKING AND TERMINALS

PAUL C. BOX, P.E., *President*

Paul C. Box & Associates, Inc.

Introduction

Cargo, conveyance, pathway, and terminal form the basic and integral elements in transportation. Whether the cargo is people or goods, and whatever the conveyance or pathway, a terminal facility must be present at the beginning and end points of each movement. When the terminal is inefficient, the resultant effects are delays and added cost. Impairment of other transportation facilities can result as well. Traffic engineers therefore must devote attention to the design and operation of terminals. This chapter deals with terminal facilities as a crucial element involved in the transportation equation, and it sets forth design and operational guidance to the professional. Automobile parking facilities of both surface and structure type are discussed, in addition to loading activities at curbside and off-street. Dimensions and elements of special-purpose vehicle parking also are included.

The planning elements of location, needs, zoning, programming, administration and financing of parking facilities are separately covered in the Parking Chapter of the Transportation Planning Handbook.

In the early 1950s the shortage of parking in major cities prompted the establishment of many parking authorities, commissions, and departments. Considerable research on needs and design was done, and many successful methods were developed to address parking problems. Although innovations have occured, primary parking solutions have remained relatively constant.

Parking operations

Types of facilities

A private motor car will usually be parked at or near the owner's residence for over half its life.[1] For convenience, safety, and access, this private parking should be as close to the residence as practical. For the single-family home, duplex, or row house, the best location is the building lot. Off-street parking for visitors, however, usually does not exist or is limited to one or two spaces in the driveway.

In larger residential developments, such as apartments, a common parking facility becomes mandatory—it serves residents, employees, and (desirably) visitors, and is usually integral with, or adjacent to, the basic development.

Individual commercial or industrial enterprises should have their own parking to serve employees, shoppers, and visitors. This private parking is also best located on the site.

In all three of these general cases, the parking is owner-supplied, is usually free, and is not available to the general public. It is a form of private, self-contained parking. Admittance is normally free-flowing, and the need for attracting parkers, collecting fees, or controlling revenue usually is not present.

Most of the so-called public parking facilities represent a more centralized and general-purpose use. Parkers usually have a choice of several destinations. Much of the time they also have a choice of alternate places to park. A facility open to the general public usually needs to attract parkers if the investment in its construction is to be justified. When

---

[1] *Parking Principles,* Special Report 125, Highway Research Board, 1971.

the facility is revenue-financed, the need is obvious. Even when the parking is free, justification is needed for the expenditure of benefit district assessment funds, parking meter revenue, or other public funds used to acquire land and to construct and operate the facility.

The design of a general-purpose parking facility must take into consideration the type of proposed operation—with attendants, by self-parking, or with a combination of the two. The most economical operation occurs where patrons park their own cars. In heavily used facilities where patrons pay for parking, it is sometimes feasible to utilize attendants to park the cars after the patron pulls into the lot. This also frequently occurs in older parking garages. Some facilities can operate on both systems, with certain areas reserved for self-parkers and other areas served by attendants.

The advantages of self-parking over attendant parking are considerable. The single rows and clear aisles permit faster and safer vehicle movement. Insurance costs are usually lower. Other advantages are that the owner may lock the vehicle, that the vehicle will not be handled by anyone but the owner, and delays to the owner will usually be less than under attendant operation. For self-parking, the unit space must be larger, and double parking or aisle parking cannot be allowed. This reduces total capacity and hence possible revenue. These factors apply, however, only when the facility is used to capacity. Most often the very substantial saving in labor cost achieved through self-parking more than offsets the difference.

A basic factor in design of a parking facility is the expected use by type of parker or type of generator being served. Parking duration can be either short-term or long-term, or a combination of both. Design dimensions are often larger in facilities for short-term parkers because of the high turnover rate and the need to provide easy access and circulation.

A facility may serve any and all types of parking generators, particularly in a downtown area. However, those that serve special events, such as sport stadiums, auditoriums, or other similar uses, require certain design considerations because of the parker characteristics. Persons attending special events generally arrive over a short time period, and nearly everyone will be leaving upon conclusion of the event. This places a severe strain on entrances, exits, and the internal circulation system. Adequate capacity must be provided.

Another operational element is whether a parking fee is to be charged. Fee-collection systems vary from self-deposit coin slots or gates (unattended lots), to monthly stickers and parking meters, to entry-ticket/cashier collection-on-exit. These are discussed under the design sections for parking lots and for garages later in the chapter.

Operational design elements

The design of a parking facility is very strongly influenced by its intended operation. The basic design elements and their associated operational features may be identified in successive steps as follows:

1. Vehicular access from the street system (entry driveway);
2. Search for a parking stall (circulation and/or access aisles);
3. Maneuver space to enter the stall (access aisles);
4. Sufficient stall size to accommodate the vehicle's length and width *plus* space to open car doors wide enough to enter and leave vehicle (stall dimensions);
5. Pedestrian access to and from the facility boundary (usually via the aisles) and vertically by stairs, escalators, or elevators in multilevel facilities;
6. Maneuver space to exit from the parking stall (access aisles);
7. Routing to leave the facility (access and circulation aisles);
8. Vehicular egress to the street system (exit driveway); and
9. Any revenue-control system (may involve elements of entry, exit, or both).

The simplest form of off-street parking is a single stall at a home. Assuming a straight driveway, steps 1 and 8 above use the same lane and curb cut, and step 9 does not apply. Steps 2 and 7 are rudimentary. Thus, a driveway serving a one-car parking stall or garage cannot be considered as representing a second parking space, if such parking *would block continuous access* to the basic stall. Step 6 usually involves backing out into the public street or alley, as part of steps 7 and 8. Herein lies the essential difference between low-volume parking and what generally should be practiced in facilities designed to handle more than a few cars. Except along alleys, the larger lots should have *all* parking and unparking maneuvers contained off-street. Frequent backing of cars across sidewalks and into public streets increases congestion and creates hazards.

For the larger facilities, and particularly garages, an operational concept necessarily precedes structural, architectural, and other design elements. The concept begins with the question, "What do we plan to serve?" From answers to this question, design features emerge such as user ease of access, security, vehicle circulation and walk patterns, signing, lighting, and equipment needs.

Design of off-street facilities

Elements of good design

In designing any off-street parking facility, the elements of customer service, convenience, and safety with minimum interference to street traffic flow must receive high priority. Drivers desire to park their vehicles as close to their destination as possible. The accessibility, ease of entering, circulating, parking, unparking, and exiting are important factors. Good dimensions and internal circulation are more important than a few additional spaces. Better sight distances, maneuverability, traffic flow, parking ease, and circulation are the results of a well-organized, adequately designed lot or garage.

Basic design principles

Site characteristics

Factors such as site dimensions, topography, and adjacent street profiles affect the design of off-street parking facilities. The relation of the site to the surrounding street system will affect the location of entry and exit points and the internal circulation pattern.

Access location

External factors such as traffic controls and volumes on adjacent streets must be considered—particularly in the location of driveways or garage ramps. It is desirable to avoid locating access or egress points where vehicles entering or leaving the site would conflict with large numbers of pedestrians. Similarly, street traffic volumes, turning restrictions, and one-way postings may limit points at which entrances and exits can logically be placed. It is important to investigate these factors at the beginning of design.

Driveways should be located to provide maximum storage space and distance from controlled intersections. Combined entry-exit points should preferably be located at midblock. At pay facilities, it is desirable to locate entry and exit points together so that attendants can monitor traffic both in and out from the same vantage point. However, along one-way streets, the "in" movement generally should not overlap the exit flow. Where entrances and exits are separated, the exit should preferably be placed in the downstream portion of the block, while the entrance should be placed as far upstream in the block face as practical.

Driveways

The basic, or nominal, design width for a two-way driveway serving a commercial land use is 30 ft with 15-ft radii.[2] With greater volumes (such as for a community shopping center), a 36-ft driveway may be appropriate, marked with two exit lanes (each 10 or 11 ft wide) and a single entry lane (14 to 16 ft wide) to accommodate the off-tracking path of an entering vehicle. For very high volumes, such as a regional shopping center or large office complex, twin entry and exit lanes, separated by a median 4 to 12 ft wide may be needed, with 20- to 30-ft radii.

Undivided driveways for trucks serving industrial developments should be up to 40 ft in width and have radii of at least 20 ft or more where practical.

One-way driveways may be narrower, such as 15 ft. Furthermore, the full-size radius is needed only on the side exposed to right-turn entry or exit.

Except in downtown areas, or other locations subject to high pedestrian conflict, the driveway design should expedite a rapid entry or exit flow. This reduces the conflict caused by speed differential with street traffic.

Auxiliary street lanes

Left-turn access from two-way streets represents the greatest accident potential at commercial driveways along major streets.[3] Such left-turn movements also create the most congestion. Depending upon the volume of turns into the driveway, and the volume of opposing flow, it is usually desirable to provide left-turn lanes (exclusive or as part of a continuous two-way left-turn lane), for service to moderate- or high-volume parking facilities.

Sometimes, right-turn lanes are also used (but these have far less value in terms of safety and reduction of congestion than is the case for left-turn lanes). The primary application

of right-turn, or deceleration, lanes is added capacity at signalized access points, or to reduce the speed differential effect where a high-volume driveway connects to a high-speed route.

Control of access

Normally, street access to a parking facility is not restricted. Where left-turn entry is to be prohibited, a barrier median is preferred. Without a barrier, NO LEFT TURN signs are an alternative. Left-turn exit is generally controlled by a channelizing island, plus NO LEFT TURN signs.

All exit movements must be controlled so as to assign right-of-way to street traffic. The basic traffic ordinance requiring drivers to stop before crossing a sidewalk or entering the street is often adequate. At higher-volume levels, Stop signs are sometimes used. At very high volumes, traffic signals may be appropriate.[4]

Access capacity

The number of vehicles that can enter or leave a parking facility, per lane, is related to the angle of approach (sharp turns have less capacity then straight-in runs), whether any control is used, the familiarity of the driver with the facility, the freedom of internal circulation (for entry), the amount of vehicular traffic on the streets (for exit), and the degree of conflict with pedestrians crossing the driveway. Considering only the factors of approach angle and driver familiarity, the findings from 71 studies in 24 states are given in Table 7–1.

For most parking facilities, far lower rates of flow are to be expected—especially when street and sidewalk conflicts are a factor. Looking solely at control methods and approach angle, a study in England found rates of 180 to 970 vehicles per hour as shown in Table 7–2.

More detailed capacity data related to type of control have been developed. A summary from four studies is given in Table 7–3.

In general, for a self-parking facility with no control, the capacity per lane ranges up to 800 vehicles per hour. One

TABLE 7–1
Vehicle Acceptance Rates of Large Parking Areas

Approach to Entrance	No. of Studies	Average Acceptance Rates Vehicles/hr/lane	
		Unfamiliar Entrance[a]	Familiar Entrance[b]
Straight approach (no turn movement)	20	850	1,100
90° right turn	15	750	1,000
90° left turn	24	830	900
Oblique angle, right	8	650	1,000
Oblique angle, left	4	720	[c]

[a] Includes racetracks, stadiums, and other facilities not frequently visited by the same individuals.
[b] Includes industrial plants, military bases, and other facilities where the same drivers enter daily.
[c] No data available.

SOURCE: A.A. CARTER, JR., "Vehicle Acceptance Rates of Parking Areas," *Public Roads*, Bureau of Public Roads (now Federal Highway Administration), October 1959.

---

[2] *Guidelines for Driveway Design and Location*, ITE Recommended Practice, 1987.

[3] Ibid.

[4] P.C. Box, "Signal Control of High Volume Driveways," *Municipal Signal Engineer* (November/December 1965).

TABLE 7–2
Parking Gate Capacities

	Vehicles/Hr
Entrance Lane	
Tight turn, automatic ticket issuance	350–450
Tight turn, no ticket	575–970
Straight approach, automatic ticket issuance	650–670
Exit Lane	
Pay variable fee	180–200
Pay fixed fee (per employee)	270

SOURCE: P.B. ELLSON; "Parking: Dynamic Capacities of Car Parks," *RRL Report LR 221*, Crowthorne, England: Road Research Laboratory, Ministry of Transport, 1969.

engineer has recommended a design value of 400 vehicles per hour,[5] while another has given this same figure for a lane *controlled* by an automatic ticket dispenser and gate.[6] Frantzeskakis has developed guidelines for considering capacities related to control methods, and also to street traffic (but not pedestrian sidewalk conflicts).[7]

If a facility is to be operated only with attendant parking, the capacity is, of course, directly related to the number of attendants. Ricker found rates of acceptance of 8 to 16 vehicles per hour per attendant, and delivery rates of 13 to 17, in studies of Central Business District garages.[8]

In summary, it can be seen that no one, all-encompassing value can be used to determine the lane capacity for access to or egress from a parking facility. This suggests a flexibility of design. When a two-way (entrance plus exit) driveway is used, the possibility of a three-lane design, with reversible center lane, should be considered.

Traffic circulation

The ideal movement into a parking facility is a left-hand turn from a one-way street. This places the driver position on the inside of the turn, which allows better visibility and more accurate judgment of the vehicle placement relative to curbs or other obstructions within the site.

Vehicle circulation on the site may be either two-way or one-way, depending on site dimensions and the angle of the parking stalls. Two-way circulation is generally allowed with 90° stalls, whereas one-way circulation is generally used with stall angles less than 90°. In any event, it is desirable to minimize traffic conflict points so as to reduce the accident and congestion potential.

Cross aisles are necessary in large facilities. Generally, not more than about 40 spaces should be provided without a cross aisle to reach exits or other parking spaces. In the ideal circulation pattern, an inbound driver is able to move in a continuous flow past all potentially available parking spaces. Upon unparking, the driver is able to reach the exit driveway while

TABLE 7–3
Parking Control Service Rates

		Vehicles Per Hour			
		Walker[a]	Crommelin[b]	West Germany[c]	UK[c]
Entrance and/or Exit					
Clear aisle no control	EA	800	800	700	772
	ST	379	—	—	—
Coded card	EA	400	400	360	—
	ST	257	—	—	—
Coin/token	EA	140	140	140	—
	ST	116	—	—	—
Fixed fee to cashier	EA	270	215	270	270
	ST	164	—	—	—
Fixed fee—no gate	EA	424	—	—	—
	ST	270	—	—	—
Entrance					
Ticket spitter— Automatic	EA	522	520	446	660
	ST	300	305	—	375
Ticket spitter— Push button	EA	480	—	400	—
	ST	257	—	—	—
Ticket spitter— Machine read	EA	400	—	400	—
	ST	232	—	—	—
Exit					
Variable fee to cashier	EA	144	150	200	150
	ST	120	—	—	—
Validated ticket	EA	300	340	—	—
	ST	212	—	—	—
Machine read tickets	EA	180	—	—	—
	ST	144	—	—	—
Machine read w/license plate check					
Front plate–manual	EA	110	—	—	—
Rear plate–camera	EA	80	—	—	—
Pay on foot					
Cashier		200	—	—	—
Machine		212	—	—	—
Exit token	EA	400	—	—	—

EA, easy approach; ST, sharp turn.
[a] Walker Research Study, 1985.
[b] Robert W. Crommelin, 1972.
[c] *Transportation Quarterly*, 36 (No.1), January 1982.

SOURCE: R.T. KLATT AND M.S. SMITH, "Access and Circulation Guidelines for Parking Facilities," *Compendium of Technical Papers*, ITE, 1987.

passing by only a minimum number of other stalls. In practice, these conflicting needs are best addressed by location and proper design of cross aisles in parking lots. In large parking structures, 'clearway' exit ramps (separate from parking aisles) are sometimes used to limit internal conflict during departure.

Parking dimensions: General

For the foreseeable future, a vehicle mix of widely varying sizes may be anticipated. There are three ways of handling layout: (1) design all spaces for Large-size vehicles (about 6 ft wide by 17 to 18 ft long); (2) design some spaces for Large vehicles and some for Small size (about 5 ft wide by 14 or 15 ft long); or (3) provide a composite layout with intermediate dimensions (too small for some Large cars, and too big for all Small cars). Subsequent sections of this chapter provide design dimensions for Large- and Small-size vehicles, followed by a discussion of composite layout.

[5] R.T. Hintersteiner, "Parking Control Guidelines for the Design of Parking Facility Portals," *ITE Journal* (January 1989).

[6] R.A. Weant, "Parking Garage Planning and Operation," Eno Foundation for Transportation, Inc., 1978.

[7] J.M. Frantzeskakis, "Traffic Flow Analysis for Dimensioning Entrances-Exits and Reservoir Space for Off-Street Parking," *ITE Journal* (May 1981).

[8] E.R. Ricker, "Traffic Design for Parking Garages," Eno Foundation for Highway Traffic Control, 1957.

In developing the design of any parking facility, it is customary to work with stalls, aisles, and combinations called "modules." A complete module is one access aisle servicing a row of parking on each side of the aisle (see Figure 7–1). In some cases, partial modules are used where the aisle only serves a single one-side row of parking. This arrangement is inefficient and should be avoided where possible.

The minimum practical stall width varies principally with turnover (frequency of stall use), experience of the parker, and the vehicle size. Commercial parking attendants can park Large cars in stalls less than 8 ft wide. With self-parking, stall widths that will accommodate most passenger cars and light trucks range between 8.3 ft and 8.8 ft, depending on anticipated parking activity. Site-specific circumstances will influence determinations of the most appropriate stall-width dimension. For example, a generous stall-width is suggested by conditions of high parking turnover, limited module width in which to develop the access aisle, or desire for a high level of user comfort and convenience. Where parking turnover is expected to be low, as for all-day employee parking, narrower stall widths are usually acceptable.

It is important to note that stall widths are measured *crosswise to the vehicle*. If the stall is placed at an angle of less than 90°, the width parallel to the aisle must be increased proportionally (see Figure 7–1).

The length of the stall should be appropriate to the overall length of most cars expected to use the space. The length refers to the effective *longitudinal dimension* of the stall (but not necessarily the length of the stall line marking). When rotated to angles of less than 90°, the stall depth perpendicular to the aisle increases up to 1 ft more, and then decreases.

Most parking aisles serve for both circulation and access to stalls. Exceptions concern intermediate crossover aisles, used to break up an otherwise excessive aisle length (generally more than 40 spaces long), or those at the ends of aisles. The access aisle width required to allow single-pass parking and unparking maneuvers varies principally with the angle of parking and secondarily with the stall width. It obviously is also related to the stall length. When dealing with large facilities, most parking designers work directly with the combinations of stall depth plus aisle width, or modules.

The total dimensions required for a parking module are produced by adding together the aisle width plus the stall depths (perpendicular to the aisle) on both sides. However, the *effective* stall depth depends on the boundary conditions of the module. If car bumpers contact a wall or fence on one or both sides, the maximum total module requirement is developed. If there is no boundary barrier of bumper height, but tires of parked cars contact wheel stops or curbing, the vehicle overhang must be considered. The curb must be set back if any bumper contact beyond the curb is critical. For 90° pull-in parking, the setback to the inner face (wheel side) of the curb should be about 2.5 ft. For back-in operation, a 4- to 4.5-ft setback of curbing is needed because of the greater rear overhang of some automobiles. Back-in parking is generally the best (a narrower access aisle width can be used) and potentially the safest (exit by pull-out rather than back-out, which greatly reduces the potential for striking other parked cars, moving vehicles, or pedestrians in the aisle). However, most drivers are unskilled in this maneuver and will resort to several pull-up-and-back efforts to park in a stall that could be readily entered by a single backing operation. While a significant number of commuter parkers (as at a rail station) may back into the parking stalls, aisle design for self-parkers is always based upon a pull-in type of entry.

The above setback dimensions are not adequate to furnish complete protection to any fences or decorative walls located on the perimeter. Unusual overhangs may be found, and it is also possible for tires to ride up on or over the blocks or curbing. When positive limitation is required, a bumper contact barrier such as a structural wall or highway guardrail should be used at the end of the stall. See sections dealing with boundary protection in parking lots and garage design later in this chapter.

For parking at angles of less than 90°, front-bumper overhangs beyond curbing are generally reduced with decreasing angle, and, for example, drop to about 2 ft at 45° angles.

Another type of module, the interlock, is possible at angles below 90°, as shown in Figure 7–1. There are two types of interlocks. The most common and preferable type is the bumper-to-bumper arrangement. The "nested" interlock can be used at 45°, and is produced by adjacent aisles having one-way movements in the same direction. This arrangement requires the bumper of one car to face the fender of another car. Wheel stops may be needed for each stall, and, even with their use, the probability of vehicular damage is much greater than for other parking arrangements.

Parking dimensions: Large cars

The long-term trend in American automobile design toward increased width has currently been reversed. However, more

Figure 7–1. Dimensional elements of parking layouts. SOURCE: Adapted from R.A. Weant, "Parking Garage Planning and Operation," Fig. 20, Eno Foundation for Transportation, Inc., 1978.

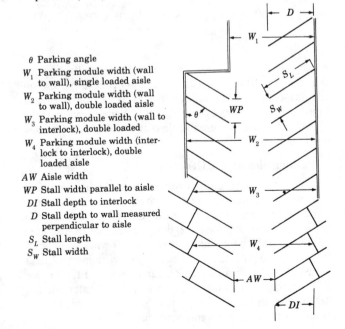

θ Parking angle
W_1 Parking module width (wall to wall), single loaded aisle
W_2 Parking module width (wall to wall), double loaded aisle
W_3 Parking module width (wall to interlock), double loaded
W_4 Parking module width (interlock to interlock), double loaded aisle
AW Aisle width
WP Stall width parallel to aisle
DI Stall depth to interlock
D Stall depth to wall measured perpendicular to aisle
S_L Stall length
S_W Stall width

efficient engines and increased use of lightweight materials may in the future allow some "regrowth" of vehicle size. The practical limits needed for door opening space between cars, and driver or passenger access to the vehicles, combine to produce an optimum stall width of about 8.5 ft for most applications today, unless vehicles are segregated by general size. Widths exceeding 9 ft are not recommended (except for handicapped stalls), because of inefficiency—wasted land and pavement area, unnecessary added maintenance (cleaning, lighting), decreased capacity for a given site, increased storm water runoff, and increased walking distances for users.

One approach to the range of stall-width needs is to consider a stall *classification* relating width to type of usage.[9] This might be roughly equated to the level of service concept, whereby parking delay and ease of access and egress vary with expected activity and type of user. Table 7–4 identifies four stall-width classes associated with typical turnover/user characteristics, for Large-size vehicles (6 ft wide by 17 to 18 ft long).

Table 7–5 lists design dimension guidelines for Large-size cars for typical parking angles, stall widths, and modules. In practice, a more rapid parking operation will be achieved if

[9] "Guidelines for Parking Facility Location and Design," ITE Committee 5D-8, May 1990.

the dimensions are increased. Slight reductions are also feasible as given in the Table notes.

Narrowed stall width in each class for parking angles of less than 90° is not desirable. A relation exists between stall width and aisle width, as shown in Table 7–5, but the stall-width needs are basically determined by door-opening clearances. Only at very flat angles of less than 35° can doors open ahead or behind the cars in adjacent stalls, and even then there can be little reduction in basic stall width.

TABLE 7–4

Stall-Width Classification

Class	Width (ft)[a]	Low	Medium	High	Typical Uses
			Typical Turnover		
A	9.00			X	Retail customers, banks, fast-food restaurants, other very high turnover establishments
B	8.75		X	X	Retail customers, visitors
C	8.50	X	X		Visitors, office employees, residential, airport, hospitals
D	8.25	X			Industrial, commuter, university

[a] For large-size vehicle, measured at right angles to stall.

TABLE 7–5

Large-Size Parking Layout Dimension Guidelines (see Figure 7–1 for description of elements)

1	2 S_w	3 WP	4 VP_w	5 VP_i	6 AW	7 W_2	8 W_4
						Modules	
Parking Class	Basic Stall Width (ft)	Stall Width Parallel to Aisle (ft)	Stall Depth to Wall (ft)	Stall Depth to Interlock (ft)	Aisle Width (ft)	Wall-to-Wall (ft)	Interlock to Interlock (ft)
	2-Way Aisle—90°						
A	9.00	9.00					
B	8.75	8.75	17.5	17.5	26.0	61.0	61.0
C	8.50	8.50					
D	8.25	8.25					
	2-Way Aisle—60°						
A	9.00	10.4					
B	8.75	10.1	18.0	16.5	26.0	62.0	59.0
C	8.50	9.8					
D	8.25	9.5					
	1-Way Aisle—75°						
A	9.00	9.3					
B	8.75	9.0	18.5	17.5	22.0	59.0	57.0
C	8.50	8.8					
D	8.25	8.5					
	1-Way Aisle—60°						
A	9.00	10.4					
B	8.75	10.1	18.0	16.5	18.0	54.0	51.0
C	8.50	9.8					
D	8.25	9.5					
	1-Way Aisle—45°						
A	9.00	12.7					
B	8.75	12.4	16.5	14.5	15.0	48.0	44.0
C	8.50	12.0					
D	8.25	11.7					

Notes: These dimensions are subject to slight reductions by local agencies under high cost conditions (such as garages) or slight increases in areas subject to special needs (such as extensive snowfall).
Column 1: See Table 7–4 for typical uses (A for high turnover; B and C for medium turnover, and C and D for low turnover).
Columns 5, 8: May also apply to boundary curb where bumper overhang is allowed.
Column 6: To vehicle corner.
Columns 6 to 8: Rounded to nearest foot.

SOURCE: "Guidelines for Parking Facility Location and Design," ITE Committee 5D-8, May 1990.

Parking dimensions: Small cars

Special dimensions for Small car parking have increased application in the United States. The percentage of such cars varies by year and also somewhat by geographical location. It is currently about 40% to 50%. The Small car design vehicle is about 5 ft wide by 14 to 15 ft long.

A suitable stall length for Small cars is 15 ft. Stall widths of 7.5 ft are appropriate for typical uses, and 8 ft for higher turnover conditions. Table 7–6 offers several layout dimensions for Small-size vehicles.

Composite parking dimensions

It would, of course, be highly desirable to utilize one "standardized" set of parking dimensions. Problems exist in separation of Small- and Large-size cars.[10] The smaller parking facilities probably do not warrant any consideration of separate dimensions for different size vehicles. In some developments, it will be impractical to provide separate size layouts.

An "ideal" design would utilize the values in Table 7–5. However, this will result in more area utilized per average car space than necessary. For this reason, composite dimensions have been proposed that represent a compromise between those necessary to adequately accommodate Large-size vehicles and those needed for the Small-car size. These rationalize dimensions based upon the expected proportions (weighted averages) of Large versus Small size.[11] While this concept unquestionably optimizes space-needed to space-provided, it also penalizes the larger-car parker.

[10] Ibid.

[11] Mary S. Smith, "The Level of Service Approach to Parking Design," *Parking*, (March/April 1987).

The net effects may be increased maneuvering to enter and leave the stall (hence possible added delay to other motorists using the access aisle), inability to open car doors a sufficient distance to enter or leave comfortably (increasing the potential for door-denting), and even complete denial of access by less experienced drivers to some stalls where adjacent vehicles are parked out-of-center.

Because of minimum car-door-opening clearances, composite size stalls should not reflect any narrower stall widths than the 8.25 ft minimum given in Table 7–5. However, the aisle width in Table 7–5 may be narrowed by 1 ft or so without experiencing a major increase in congestion and accessibility of parking stalls. This would particularly apply in structures with high construction cost.

Small car separation techniques

Greater use of smaller cars in North America and pressure to make parking more space-efficient have resulted in a variety of parking arrangements to take advantage of smaller car sizes. The objective of each layout is to maximize the number of stalls in a given area. These designs may involve the provision of some spaces designed for the exclusive use of Small cars, plus other spaces for Large cars. Several alternative layouts are possible. These are illustrated in Figure 7–2 and reviewed in detail in the "Guidelines for Parking Facility Location and Design," May 1990 draft by ITE Committee 5D-8.

Handicapped spaces

Handicapped (HC) parking spaces should be located close to elevators, ramps, walkways, and building entrances. They should also be located so that persons in wheelchairs can access the building. Curb cuts should slope up at a rate of no

TABLE 7–6
Small-Size Parking Layout Dimension Guidelines (see Figure 7–1 for description of elements)

1 Parking Class	2 S_w Basic Stall Width (ft)	3 WP Stall Width Parallel to Aisle (ft)	4 VP_w Stall Depth to Wall (ft)	5 VP_i Stall Depth to Interlock (ft)	6 AW Aisle Width (ft)	7 W_2 Modules Wall-to-Wall (ft)	8 W_4 Interlock to Interlock (ft)
2-Way aisle—90°							
A/B	8.0	8.0	15.0	15.0	21.0	51.0	51.0
C/D	7.5	7.5					
2-Way aisle—60°							
A/B	8.0	9.3	15.4	14.0	21.0	52.0	50.0
C/D	7.5	8.7					
1-Way aisle—75°							
A/B	8.0	8.3	16.0	15.1	17.0	49.0	47.0
C/D	7.5	7.8					
1-Way aisle—60°							
A/B	8.0	9.3	15.4	14.0	15.0	46.0	43.0
C/D	7.5	8.7					
1-Way aisle—45°							
A/B	8.0	11.3	14.2	12.3	13.0	42.0	38.0
C/D	7.5	10.6					

Column 1: See Table 7–4 for typical uses (Class A is for high turnover; class B and C for medium turnover; and C and D for low turnover).
Columns 5, 8: May also apply to boundary curb where bumper overhang is allowed.
Column 6: To vehicle corner.
Columns 6 to 8: Rounded to nearest foot.

SOURCE: "Guidelines for Parking Facility Location and Design," ITE Committee 5D-8, May 1990.

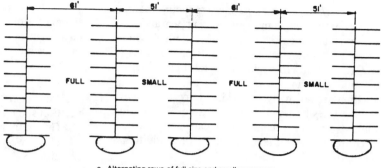

a. Alternating rows of full-size and small car spaces

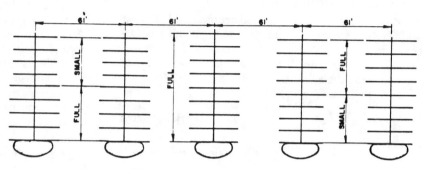

b. Small car and full-size spaces in same rows

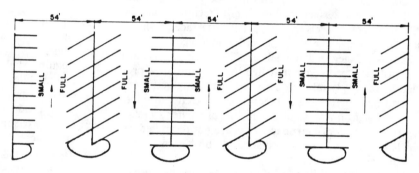

Figure 7–2. Parking layouts combining small and large car spaces.
SOURCE: "Guidelines for Parking Facility Location and Design," ITE Committee 5D-8, September 1989.

c. Cross-aisle separation of small car and full-size spaces

more than 1 ft in 8 ft (preferably not more than 1 ft in 12) and should be at least 4 ft wide in order to accommodate a wheelchair.

Curb cuts and ramp surfaces at right angles to the main walk have been found to be hazardous for pedestrians using the walk. Many people have been injured by slipping or stumbling on steep cross-slopes. To avoid this, cross-slopes (in the longitudinal direction of the walk) should not have a slope of more than 1 ft in 8. The areas of slope should be painted with a nonskid yellow coating.

A 12-ft wide area is the minimum to provide adequate space to park a vehicle and allow the driver or passenger to operate a wheelchair between parked vehicles. A narrower width of 10 ft is adequate where one side of the stall is open, such as at the end of a parking row when unobstructed adjacent pavement or walk area is available. By either pulling in, or backing in, the driver or passenger side of the vehicle can access toward the clear side, provided a 90° parking layout is used. Handicapped stalls should not be located on grades sloping more than 5% in one direction or 2% in the cross-direction.

Each HC parking space should be marked with a sign. Painting the wheelchair symbol on the pavement is another means that is used. Further details and suggested layouts are addressed in the ITE Committee 5D-8 report cited above.

State and local laws vary in both design and in arbitrary numbers of spaces required. However, a rational method of determining the number of spaces actually *needed* for typical land uses was developed by ITE Committee 5D-8 and published in an ITE journal article.[12]

Stall markings

Unless markings are used, parking will be difficult to enforce, and inefficient use of the area will result. Markings typically require some type of permanent surfacing. A paved surface has advantages in allowing proper drainage, reducing dust, facilitating snow removal and sweeping,

---

[12]"Handicapped Parking Supply," ITE Committee 5D-8, *ITE Journal* (September 1988).

providing an improved walking surface, reducing maintenance cost, and presenting a more pleasing appearance.

White is the standard color for marking stall lines in most communities, with yellow used only to delineate areas of NO PARKING. Yellow cross-hatching is most effective, for example, if other lines are white.

Striping lines are normally 4 in wide. Many owners have painted double lines between stalls and are of the opinion that this causes drivers to center their cars better within each stall. However, there is some question if the additional lines are worth the added cost.

Parking row sidewalks

Raised pedestrian sidewalks are sometimes used in large parking lots to separate rows of cars and to provide more favorable walking conditions. People walking to and from cars most often use the aisles, however, and the value of interior walkways is so debatable that these are seldom used today.

Parking lots

General elements and layout alternatives

Most of the design principles apply to all parking lots. While a 'lot' is a single surface facility, in hilly terrain it may be built on adjacent levels in a 'stepped' or terraced fashion, with short 'ramps' to connect the different levels.

Because of their lack of walls or cover, parking lots have no ventilation problems, and lighting is sometimes provided by relatively tall poles, thus affording high efficiencies and minimizing the number of poles. Generally, lots have clear sight lines and offer a feeling of greater security than in a more confined space. Lots are not restricted on vehicle heights and thus afford access to both commercial and emergency vehicles.

Most parking facilities are of the 'lot' variety, because of relatively low costs, coupled with ease and speed of design and construction. They are intrinsically more efficient per level than parking in structures because there are no requirements for ramps, stairs, elevators, or structural columns. It is often possible to use more generous dimensions for stall and aisle widths in lots, as compared with structures, because of lower construction costs.

Land cost generally is the determining factor when considering parking on the surface versus a structure. The trend to have most parking in lots, except in central areas, is expected to continue into the foreseeable future.

Generally, the layout of a parking lot seeks to strike a balance among maximizing capacity, maneuverability, and circulation. Rectangular sites afford the greatest opportunity to balance these factors. Arranging parking stalls along both sides of access aisles, with aisles parallel to the longer site dimension, provides greatest space efficiency. However, this is not always possible or desirable. The most appropriate layout depends on site-specific conditions. For two-way traffic, 90° parking is generally used. Often this is the most efficient layout if both lot size and shape are appropriate. Furthermore, the wide aisles are more inviting than the narrower ones used for space economy in flatter-angle layouts.

Much of the alleged difficulty with 90° parking has stemmed from inadequate aisle dimensions. Where proper measurements are used, a smooth and efficient operation can be achieved. The general advantages of 90° parking, as compared with lesser angles, are:

1. Most common and understandable;
2. Can sometimes be better fitted into buildings;
3. Generally most efficient if site is sufficiently large;
4. Uses two-way movement (can allow short, dead-end aisles);
5. Allows unparking in either direction. Thus, it can minimize travel distance and internal conflict;
6. Does not require any aisle directional signs or markings;
7. Wide aisles often provide room to pass vehicles stopped and waiting for an unparking vehicle;
8. Wide aisles increase separation for pedestrians walking in the aisle and between moving vehicles;
9. Wide aisles increase clearance from other traffic in the aisle, during unparking maneuvers;
10. Fewer total aisles (hence easier to locate a parked vehicle).

Several advantages and disadvantages of angle parking (usually 45° to 75°), are:

1. Easiest in which to park;
2. Can be adapted to almost any width of site by varying the angle;
3. Requires slightly deeper stalls but much narrower aisles and modules;
4. Drivers must unpark and proceed in original direction; hence producing greater out-of-way travel and conflict;
5. Unused triangles at end of parking aisles reduce overall efficiency;
6. To avoid long travel, additional cross aisles for one-way travel are required, which adds to gross area used per car parked;
7. Difficult to sign one-way aisles.

The relative efficiencies of various parking angles may be compared by the number of square feet required per car space (including the prorated area of the access aisle and entrances). Where the size and shape of the tract is appropriate, both the 90° and the 60° parking layouts tend to require the smallest area per car space. In typical lot layouts for Large-size vehicles, the average overall area required (including cross aisles and entrances) ranges between 310 and 330 square ft/car. A very flat angle layout is significantly less efficient than other angles.

Many conditions exist where one-way aisles are desirable. With angles of less than 90°, drivers can be restricted to certain directions; however, the angle should usually be no greater than 75° to avoid drivers going the wrong way. Adjacent aisles generally have opposite driving directions. Any multiple of modules can be used, depending on location of entrances and exits and the size of available land. However, at angles below 45° and with interlocking stalls, the module dimension may become too small to allow U-turn access between adjacent aisles by Large-size cars.

While the most efficient arrangement is usually found with aisles parallel to the long dimension of the site, parking aisles serving a specific generator should usually be oriented toward the building and/or pedestrian access points.

Car stacking units

The capacity of an attended parking lot can be increased by the use of small mechanical units of two- or even three-car height. These are sometimes used in parts of very-high-demand lots located on premium land. They can supplement the parking supply but are not appropriate for every stall; that is, they cannot double the surface capacity, for example.

Figure 7–3 illustrates a two-car unit.

Figure 7–3. Two-car stacking unit in New York City. SOURCE: HARDING STEEL, INC., Denver.

Access, fee collection and reservoir areas

Having entrances and exits at opposite ends of a parking lot tends to reduce conflicts between incoming and outgoing cars, but complicates revenue control. At entrances, care should be exercised to prevent backups onto the street. A well-designed parking lot can accept arriving cars as quickly as the street system delivers them. Aside from acceptance rate limitations caused by revenue-control measures, the principal causes of entry delay are sidewalk conflict with pedestrians, parking or unparking maneuvers just inside the entrance, and conflicting internal circulation, including vehicles waiting to exit.

Because driveway entrances to surface parking lots are at the same grade as the public sidewalks, it is generally impossible to avoid pedestrian conflict. The problem can sometimes be minimized by locating driveways on streets having lower pedestrian volumes and at points upstream from the heaviest pedestrian flow.

It is good practice to prohibit stalls so close to the entrance that unparking maneuvers would require backing onto the sidewalk. Depending on the size and turnover of the facility, several stalls near the entrance may best be kept out of active (high-turnover) use; however, this is more of an operational than a design element.

Where a fee is charged, collection can involve pre-pay, monthly stickers, or parking meters. Smaller lots may use coin-operated gates (sometimes with payment required on both entry and exit). Another method utilizes a coin slot box, keyed to each numbered parking stall, as shown in Figure 7–4. Other methods include coin-operated or currency-operated ticket dispensers.

Four elements have been identified as necessary to a successful unattended "honor" type operation.[13] These are:

1. Proper location.
2. Accurate and frequent lot checks or audits.
3. Accurate recording of violators.
4. Effective enforcement procedures.

If parking meters are used, care must be taken to locate the mounting posts where they are least subject to vehicular damage. When set back behind curbs, twin-meter posts are placed in projection of the alternate parking stall lines. When used inside lots, quad-meter mounts can often be attached to a post fixed in a raised (2-ft high) concrete base and located at the junction point of four adjacent stalls.

Parking lot operators that charge an hourly or daily fee often control revenue by issuing tickets at the entrance. If a ticket-dispensing machine is used, at least a two-space reservoir within the lot is usually needed. If tickets are manually dispensed from a booth, a larger inbound reservoir may be needed; however, this is highly dependent on the size (capacity) of the lot and its parker characteristics. Unfortunately, many smaller, high-turnover self-parking lots have cashier booths located adjacent to the public walk. The one-space reservoir in this design is often inadequate.

Where parking is done by attendants, a large inbound reservoir area may be needed. Such designs are seldom found, and congestion with backups into the street are common experiences.

The reservoir requirements for access in and out of lots serving major traffic generators are usually based on traffic signal control of the driveway intersection. This is discussed later in the Major Generator section.

End islands

At the ends of parking rows, drivers must circulate toward exit drives or continue searching for vacant stalls. The normal right-angle turn by a typical passenger car driver uses an inner radius of about 15 ft. During the turn, the vehicle sweeps a path some 10 ft wide at its extreme. The ends of parking rows in high-turnover lots can be treated with painted or curbed

---

[13] Bernard M. Meyer, "Unattended Parking Operations," *Parking,* (November/December 1986).

Figure 7–4. Advisory sign (above) and coin box (below) at a railroad commuter station, Glenview, Illinois.
SOURCE: PAUL C. BOX AND ASSOCIATES, INC.

islands. Design for 90° and 60° parking layouts are shown in Figure 7–5. These can be designed with either paint or curbing. A curbed island serves the following functions:

1. Limits parking encroachment into cross aisles;
2. Opens up sight distance at intersections of cross aisles with access aisles;
3. Provides a comfortable turning radius;
4. Provides a cart storage area at supermarkets;
5. Stores limited quantities of snow;
6. Protects directional signs and permits light pole or fire hydrant installations as needed;
7. Allows aesthetic plantings to avoid a "sea of paving" appearance;
8. Helps define the circulation and roadway system.

End islands should be considerably *shorter* than the effective stall depth so as to most closely delineate the swept path of a right-turning vehicle. If a curbed island is too large—that is, extends out to the end of stall—it will inhibit access. In 90° layout, it may block one vehicle from turning into a parking access aisle if another vehicle is awaiting exit.

In general, end islands are not needed in small lots of a few hundred cars; painted islands may be adequate for intermediate size lots of a thousand cars or so, with painted and/or curbed islands (sometimes alternating) desirable in large lots such as at regional shopping centers.

Boundary controls

There are four reasons for providing boundary controls:

1. to limit access to specific points;
2. to prevent encroachment of parked vehicles onto adjacent property, public sidewalks, streets or alleys;
3. to protect adjacent landscaping;
4. to prevent parked vehicles from rolling (forward or backward) into hazardous areas or down steep slopes.

Boundary controls can consist of 5- or 6-in-high curbing or continuously butted wheel stops, guardrails, closely spaced heavy posts, or structural walls. The type of control needed depends on the degree of protection required. A dropoff of a few feet at the edge of a lot would warrant more than a

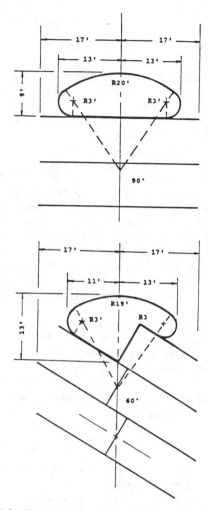

Figure 7–5. Examples of end islands for large car stalls.
SOURCE: "Guidelines for Parking Facility Location and Design," ITE Committee 5D-8, September 1989.

simple curbing, which can be mounted by a vehicle moving only a few miles per hour. Alternatively, a row of wheel stops or raised curbing, with a suitable setback of 2 or 3 ft, would normally be adequate to prevent parked vehicles from overhanging a public sidewalk or damaging landscaping.

Where parking directly abuts a building structural wall, no special protection may be needed. Decorative walls or low building windows may require wheel stops, posts, or even guardrails to protect the building more completely.

Wheel stops and speed bumps

In general, the ends of parking stalls within lots can be marked in a satisfactory fashion by only a paint line. Wheel stop blocks in the interior of a lot have disadvantages, for they may interfere with and present a hazard to people walking between cars, provide traps for blowing debris, and interfere with snow plowing in northern climates.

Parking adjacent to a building usually abuts a raised sidewalk—especially where entry doors are located. Such a walk should be about 3 or 4 ft wide to prevent bumpers from striking the building, and at least 6 or 7 ft wide if it is to be used by pedestrians entering doors into the building (such as for a row of small shops). When walks are constructed to separate parked vehicles from building faces, separate wheel stops are not needed and are undesirable.

Wheel stops are often used along the side boundaries of a lot, where large landscaped areas extend beyond the edge of pavement and an occasional override would present no significant hazard.

Wheel stops are sometimes used to form "islands" around fire hydrants, light poles, or trees, but a continuous curbing is more desirable. The wheel stops can be easily shifted and can represent a hazard to vehicles or pedestrians.

Speed bumps have been installed in some parking lots with an intent of slowing traffic speeds. A major study by the City of San Jose, California, found bumps on streets to be hazardous and recommended against their use.[14] Evidently, pedestrians can also trip and fall over such impediments. Use of speed bumps may be considered undesirable in parking lots.

Landscaping

Parking lots can be very effectively landscaped, and thus be aesthetically pleasing as compared to the interior of a structure. Trees and ground cover are excellent treatments to break up the pavement expanse and improve esthetics. Tree limbs below 7 ft should be avoided. Trees should be placed so as not to block car door openings or interfere with visibility of signs. The *type* of tree also is important. Varieties that drip sap or drop seeds can damage automobile finishes, and these should be avoided.

Care should be taken to use shrub, plant, and tree types that can withstand auto fumes and the concentrated heat arising from a large, paved surface. Landscaping can be an effective means in controlling pedestrian paths. Plantings of hedges can serve to funnel pedestrians into desired walk patterns within the site, but these should be kept low (24- to 30-in maximum height) for security purposes.

Sufficient setback must be provided for all plants so that the front and rear overhang of cars does not destroy them. Plantings adjacent to parking stalls, where persons enter or exit parked vehicles, should be limited to grass for a distance of at least 3 ft from the stall edge.

Extreme care should be exercised in locating shrubbery, low-growing trees, or other plants near driveways and internal crossovers so that *sight distances* are not restricted. This will require that the growth pattern of the plant be considered so that the small plant of today will not develop into a major sight restriction tomorrow. In general, the height of low plantings should not exceed 2 ft in any area where driver visibility is a factor.

Lighting

Adequate illumination of the parking area and access points is important. Ample mounting height and proper ratios of spacing to mounting height should be used to distribute acceptably

---

[14] C.D. Allen and L.B. Walsh, "A Bumpy Road Ahead?" *Traffic Engineering* (October 1975).

uniform amounts of light to the entire facility. Use of luminaires or spotlights of a sufficient intensity to illuminate all dark corners and areas between cars is impractical. The typical parking lot application is an average level of 1.0 to 2.0 horizontal footcandles (hfc), with the lowest point value being at least one-sixth this amount (about 0.15 to 0.35 hfc). Low-activity vehicle storage areas or access roads may need only 0.5 hfc on average, with the low point about one-fourth this. Poles range from 20 to 50 ft or more in height. If spotlights or floodlights are used, care should be exercised to minimize glare in the lot and on adjacent streets. Excessive spillover light on nearby residential property should be avoided. Refer to Chapter 10 for additional details on illumination design.

Pole locations should, of course, be coordinated with stall and aisle layouts. Where practical, poles should be kept at ends of parking rows. This allows flexibility for possible future changes in stall widths or layout. Where pole locations are required within parking rows, they should be at the junctions of adjacent stalls. This usually will not affect the parking capacity of the initial design layout.

Signs

In many cases, and especially in large lots, interior signs giving directions or aisle identification can be attached to lighting poles.

If a parking fee is collected, or if meters are employed, clearly marked and conveniently located signs indicating the conditions should be displayed at the entrance. These signs should be mounted sufficiently high (at least 5 ft to bottom) to allow drivers to see under them. If pedestrians will walk under the signs, a 7-ft minimum mounting height should be used.

Drainage

The lot pavement should have sufficient slope to drain properly. A minimum grade of 1% for asphalt and 0.4% for concrete should be provided, with up to 2% being desirable. Grades should not generally exceed 3% in directions longitudinal to parking stalls or 5% for cross-slopes or aisles. Catch basins or inlets are often placed down the interlock line (at the module limit line). This tends to reduce splashing of other vehicles or pedestrians.

Maintenance

The principal element of lot maintenance is cleaning. Markings and signs should be repainted as required for good visibility. Light fixtures should be cleaned at least annually, and lamps should be replaced prior to burnout. Shrubs, hedges, and trees should be kept trimmed to clear sight distance in the 2- to 7-ft-high zone above the pavement.

In certain climates, snow removal is an additional maintenance consideration. Snow cannot simply be pushed into the streets; it must be removed by truck, stored in nearby areas, or melted. Storing snow on the lot is doubly expensive in "pay" facilities, because the loss of revenue from spaces devoted to its storage must be added to the cost of moving the snow.

Garage design

Surface vs structure parking

Land cost is often the factor that determines economic viability of surface parking. As land costs increase, it may become economically justified to expand parking vertically in a structure, rather than expanding horizontally by the acquisition of additional land for surface parking.

Parking garage construction costs vary widely because of topography, architectural treatment, sophistication of revenue-control devices, foundations, and other factors. If land costs are not included, the cost of a parking structure is approximately four times the cost of an equivalent parking space in a surface lot. If the land cost *is* included, the figures can change dramatically. A survey of parking project costs by McCarthy Parking Structures (ca. 1988), found an average cost of garages to be $4,500 and for lots to be $1,000, per space.[15] The same source reported land costs of $15 per sq ft to represent the point to begin consideration of structures.

Many factors must be considered in making the decision whether to build surface or structured parking. In the case of private developers, these include service concept, demand characteristics, taxes, and financing. The primary considerations for all parking must be the walking distance of the patrons, maintenance, security, operation, and the availability of land and traffic access.

Site characteristics and access

The topography of a site may allow direct entry to more than one level of a garage. This will affect entry and exit locations as well as the interfloor travel system within the structure. The location of access points is more critical in garage design than in most surface lot designs, because of the increased number of spaces often available in structured parking. Street capacities, location of traffic controls, pedestrian sidewalk volumes, and other external factors must be analyzed to assure a design that is compatible with the surrounding street system.

When a structure is being designed, it may be possible to avoid all direct pedestrian conflict. This can be accomplished if the ramp is constructed within the sidewalk area, and the sidewalk is relocated inward (by first-floor building setback). If a building rises above the site, an arcade can be built over the sidewalk to afford pedestrians some protection from inclement weather.

Ramps may be constructed entirely within the street in those rare instances where surplus space is available. The ramp then curves under or over the sidewalk and into the parking facility. Regardless of whether or not the public sidewalk is set back, ramps may avoid pedestrian conflict and service parking floors above or below ground. Whenever garages are constructed adjacent to very high activity sidewalks, the sidewalk-setback ramped driveway design probably warrants consideration.

[15] Nichol Fran, "Protecting Paving Investments," *Shopping Center World* (June 1989).

Major use and operation

As in surface lots, the dominant type of use, whether by short-term or long-term parkers, or whether the facility will serve special events, will influence the design. Where vehicles enter and exit in short spans of time, consideration should be given to express ramps, particularly for exiting. The time required to empty the facility when filled to capacity is an important factor. This time should be kept to a minimum, with 30 to 45 min generally considered acceptable for most facilities. In some instances, exiting from parking facilities is constrained by external roadway conditions so that the desired time cannot be achieved.

Because revenue-control systems may affect both access and circulation, the intended operation whether as a free, or revenue, facility must be known initially. Of equal importance is whether the garage will involve self-parking, attendant parking, or a combination of these methods.

Because of labor, insurance, and liability costs, plus the difficulty of keeping low-cost workers for attendants, most parking is now operated as self-parking. When given the choice, parkers almost invariably prefer to park their own car for the following reasons.

1. Owners take better care of their cars than does the average attendant.
2. Self-parkers can lock their cars and take their keys with them.
3. Self-parkers avoid the long delay often associated with attendant retrieval of cars.

However, self-parking results in 10% to 15% fewer parking spaces for the same area as compared to attendant parking because of greater aisle and stall-width requirements.

Attendant parking may be justified as a convenience service and is popular at some restaurants and hotels. It may also be justified in the downtown areas of major cities, where rates and parking demand are very high. The additional capacity available by attendant parking and by double and triple stacking of cars may produce sufficient additional revenue to offset the higher cost of attendant parking. Labor costs for attendants may add as much as 40% to the parking cost.

Interfloor travel systems

The type of interfloor travel system may be either ramps or sloping floors, or various combinations. Only on a sloping site that permits direct access to each level are ramps unnecessary, but they still may be desirable for internal circulation. With sloping floor designs, the floors serve both as aisles and parking bays (see Figure 7–6). The sloping floor section is generally one or more parking modules in width. With other designs, the ramps are used exclusively for travel between floors. Combinations of the various ramps are possible. An express exit ramp may be incorporated into a sloping floor design. This ramp can be either a straight ramp along the side of the structure or a helical ramp.

An additional helical ramp variation is the double helix. This design consists of two independent interwoven ramps that drop two levels in one complete 360° turn. One ramp then serves the odd-numbered floors, while the other serves the even-numbered floors. This design is useful for tall structures, as it reduces the number of turns required in exiting from or entering the upper floors. It is desirable to keep the number of full turns to a maximum of five for self-parking facilities. With this criterion as a guide, 10 to 12 levels are possible with a double-helical ramp system. A single-helix ramp, however, should be limited to about six levels. For ramped or sloping floor designs, the number of levels should be limited to a maximum of six, because of the amount of turning required and the number of spaces a driver must pass. However, with express ramps (not illustrated for these designs), this may be increased.

Another human factor for consideration in determining the number of floors is the relative height of adjacent buildings. Many drivers develop a feeling of acrophobia in taller garages, particularly when they are driving at a level above the rooftops of adjacent buildings. Proper design of the parapet walls will reduce this effect by limiting the driver's view of surroundings. However, some drivers will experience claustrophobia when driving in enclosed, spiral ramps.

The layout of parking aisles and stalls for garages is similar to that used for surface lots. Stall and aisle dimensions generally remain the same, except that the aisles are sometimes narrowed slightly, as noted in Table 7–5.

Ramp grades, dimensions, and column placement affect the ease of circulation within the structure. Sloping floor grades should not exceed 4% to 5%. The parking angle on these sloping floors should be at least 60° to minimize the possibility of vehicles rolling out of the parking space and down the ramp. For 90° parking, sloping floor grades should not exceed 5%. Ramps without parking should be limited to about 10% to 12%, with grades up to 15% or 20% allowable in attendant parking structures. Driving ramps should be 14 to 18 ft wide with 12 ft sufficient for longer straight runs. A helical ramp should have a minimum outside radius of 32 ft, with a desirable radius of about 35 ft.

Self-parking ramp capacities normally range between 500 and 600 cars per hour per lane. Typical capacity used for design purposes is 400 cars per hour per lane. The exiting capacity is reduced to 150 to 200 cars per hour per lane when vehicles must stop at a cashier's booth on exiting. (See Tables 7–2 and 7–3).

Vertical clearance should be at least 7 ft, which normally results in floor-to-floor heights of about 10 ft.

Structural systems

There are normally four types of structural systems: (1) structural steel, (2) poured-in-place concrete, (3) precast concrete, and (4) post-tensioned concrete. Many factors must be considered in analyzing the various structural systems as they affect the relative economy and adaptability of the systems. A partial listing includes:

1. Building code requirements.
2. Maintenance.
3. Availability of materials and precast concrete fabricators.
4. Shipping distance and costs.

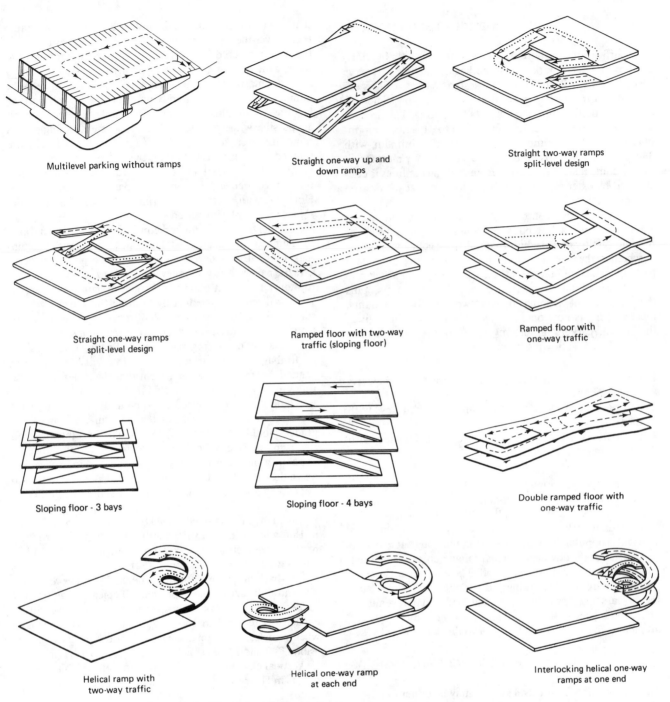

Figure 7–6. Illustration of ramp systems.

SOURCE: D. KLOSE; *Metropolitan Parking Structures,* New York: Praeger, 1965;
R.F. Roti, International Parking Design, "Square Foot Cost Averaging Principle for
Parking Structures," National Parking Association, Washington, DC.

5. Availability of contractors experienced in each structural system.
6. Environmental and atmospheric conditions.

Building codes may affect the structural system because of code restrictions on the type of structural system allowed. Some building codes may not allow exposed steel design or may require fireproofing of the external columns only.

Fireproofing of steel columns, beams, and girders can add greatly to the cost for this type of construction.

Certain materials may not be readily available in particular areas, and freight costs can be prohibitive. The lack of availability of good-quality precast concrete fabricators in some areas preclude this structural system. The post-tensioning of floor slabs of precast concrete structures requires contractors with personnel experienced in this work and on-site engineering supervision of the construction of post-tensioning elements and the process.

The relative maintenance costs of the various structural systems must be considered. Exposed steel (except so-called "weathering steel") requires periodic painting, and care must be taken to specify a sufficient number of shop-and-field coats of quality paint to retard deterioration of the steel. Atmospheric and environmental conditions can greatly influence the amount of maintenance required, particularly in heavy industrial areas, and near bodies of salt water. Weathering steels that develop a hard coat of rust do not require painting, but do cause difficult maintenance problems for a period of about 3 years until they develop their hard coating.

Precast structural systems can be advantageous where many structural members are duplicated throughout the structure. Once the forms are available, a great number of members can be cast quickly and economically. This advantage is lost when many different sizes of columns, beams, and girders are required, such as in a garage on an odd-shaped site.

Erection time at the site is also favorable for precast or steel structural systems. The construction of forms for poured-in-place concrete requires time not needed for the other systems. However, experience has shown that when time required to fabricate and ship the precast or steel members is considered, the total construction time from awarding of contract to occupancy of the building is practically the same for all structural systems.

It is not possible to say that one structural system is best or most economical in all locations and under all conditions. A value engineering analysis of each system should be made for each facility considering all the influencing factors, and only then should a decision be made on the structural system.

Short- vs long-span construction

Functional and operational considerations generally dictate that long- or clear-span construction be provided in most garages. There are certainly many advantages to this type of construction, particularly in a free-standing garage. However, when an additional structure (office, apartment, or other use) is planned above the parking garage, short-span construction may be advantageous. Long-span construction usually costs about 5% to 10% more than short-span, but this increase is offset by more parking spaces and better circulation. A comparison of short- and long-span construction is shown in Figure 7–7.

A list of advantages of long- or clear-span construction in free-standing garages includes the following:

1. Column-free floors.
2. Fewer columns and foundations.
3. More parking space.
4. Maximum operational efficiency and flexibility.
5. Faster and easier parking and unparking maneuvers.
6. Maximum flexibility to change stall sizes or parking angle.
7. Easy floor maintenance.
8. Unrestricted sight distance.
9. Fewer damaged cars.
10. Greater driver acceptance.

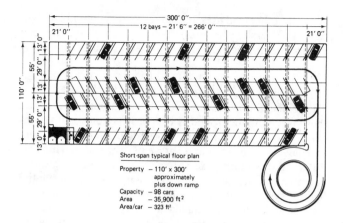

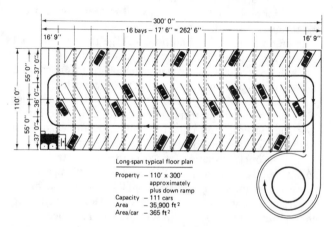

Short-span typical floor plan
Property — 110' x 300' approximately plus down ramp
Capacity — 98 cars
Area — 35,900 ft²
Area/car — 323 ft²

Long-span typical floor plan
Property — 110' x 300' approximately plus down ramp
Capacity — 111 cars
Area — 35,900 ft²
Area/car — 365 ft²

Figure 7–7. Comparison of short- and long-span construction.
SOURCE: R.C. RICH, RICHARD ASSOCIATES, "Methods of Construction and Construction Costs," National Parking Association, Washington, DC, 1966.

11. Minimum problem in locating ventilation and lighting equipment.

Disadvantages of long spans include:

1. Deeper floor construction.
2. Greater floor-to-floor heights.
3. Higher construction costs.

Single-purpose vs multiuse garages

A single-purpose garage is a structure for parking vehicles with little or no area devoted to other uses. A multiuse garage contains uses other than parking within the same structure, such as retail at sidewalk level, offices or residential above the parking levels, etc. In downtown urban areas, there is a trend toward multipurpose garages because of the need to utilize land more effectively, and where a single-purpose garage may be financially unfeasible. Pedestrian amenities may also be improved.

A garage can be included with many types of land uses—most often in conjunction with office, apartment, and retail developments. The parking can be provided either above or below the other land uses. In some cases (condominium high-rises or university campuses), roofs of parking structures can be used for athletic facilities such as tennis courts.

Underground garages

The principal difference between aboveground garages and those placed underground is the construction cost. An underground garage will usually cost 1.5 to 2 times more than an equivalent aboveground garage per car space. This cost differential is due to excavation, concrete retaining walls, potential continuous dewatering systems, a more substantial lighting system (with higher electrical operating costs due to lack of ambient light), added air circulation and exhaust systems, and the roof slab, which is not used for parking but often must support heavy landscaping.

A major problem with underground garages is water seepage through the walls or through the floor joints because of hydrostatic pressure. In addition, leakage through the roof may become a problem if the garage is located under a park.

The presence of a high water table may require increased size and depth of foundation to offset the effect of hydrostatic pressure. These factors generally tend to restrict the number of levels to three or four in an underground garage. Increasing construction costs and potential buildup of fumes and the problem of exhausting them generally prohibit more levels.

Stall layout, circulation, and operating characteristics of underground garages do not differ from aboveground garages. However, additional emphasis should be placed on directional signing and markings because of the lack of orientation underground. It is important to keep the driver and pedestrian oriented as to streets and nearby major land uses. Good lighting is especially important, as are other security measures.

Mechanical or elevator garages

In mechanical or elevator garages the vehicle does not move to the parking space under its own power, but rather is moved there mechanically. There are many types of mechanical garages in existence, but their use worldwide has been declining. This is primarily due to maintenance problems associated with the mechanical equipment. In the average mechanical garage, the equipment has tended to wear out in about 7 years—long before the garage has paid for itself. In addition, equipment failure can shut down the entire garage or a portion of it, creating bad customer relations when cars are stranded.

An additional problem with mechanical garages is their inability to accept surges of inbound or outbound traffic. Limited inbound capacity requires a large reservoir area to keep the cars off the streets. Low discharge capacity often creates customer irritation as people wait long periods of time for the return of their vehicles. Mechanical garages generally provide satisfactory service only when the parking demand is relatively uniform throughout the day with no large peak flows.

Lighting

Lighting within a garage is necessary to aid safety of movement and to discourage vandalism or acts of violence.

TABLE 7–7

Garage Lighting

Area	Footcandles	
	Day	Night
General parking and pedestrian areas	5	5
Ramps and corners	10	5
Entrance (first 50 ft)	50	5

SOURCE: Extracted from Fig. 14–27 in "Covered Parking Facilities," *IES Lighting Handbook, Application Volume*, Illuminating Engineering Society, 1987.

Table 7–7 presents recommendations for garage lighting by one national organization, but these values have not been verified by any rigorous research. A high light level is needed during the day at the entrance as a driver proceeds from bright sunshine into the garage.

Illumination of the access aisles and parking stalls is usually accomplished by luminaires mounted between the ceiling beams so that they do not affect the vertical clearances. This luminaire placement does, however, subject them to damage from high automobile radio antennas. Wire guards and/or polycarbonate covers are frequently used to minimize losses.

Safety and surveillance equipment

Any garage is a potential source of problems from loitering, vandalism, thefts, and crimes against persons. Garages can become havens for drunks and derelicts. Within a garage, most crimes occur in the elevators and stairwells. For this reason, these areas should be well lighted and should be capable of being closed off at night. Mirrors may be placed in stairwells or other places to improve visibility around corners or under steps. Glass-enclosed stairwells should be used in lieu of masonry whenever possible. Glass-enclosed stairwells may conflict with fire codes, but many cities have granted variances after considering the unique problems involved.

In garages that do not remain open 24 hours a day, there should be positive ways of closing off the garage, such as roll-down doors at entry-exit points. In addition, the garage should not have large openings around the building that would give access at other than designated points.

Sound and television monitoring systems are used as internal surveillance systems. The sound system monitors all sounds in the garage and in all elevators and stairwells. A speaker in the manager's office or cashier's booth allows the monitoring and identification of suspected trouble areas. Television monitoring uses cameras placed throughout the garage. This system is not widely used because of the necessity of having someone continually looking at monitors and because the light intensity in the garage is often insufficient to provide a good picture contrast. It is difficult to cover a large area with cameras, and cameras can be made inoperative by vandals. Security patrols are often used within garages at random times to discourage vandalism and acts of violence. Some authorities place emergency phones to permit patrons to call for assistance.

Fire protection

Many building codes require an inordinate amount of fire protection in garages. These requirements are based on old approaches to fire fighting that were incorporated into building codes many years ago. However, tests of auto fires have proven the futility of trying to ignite an adjacent auto that is parked beside a burning vehicle.

The combustible materials in motor vehicles consist of the gasoline, upholstery, and paint. The fire loading in a garage, or the number of Btu of combustible material per square foot of floor area, is extremely low compared to office buildings, apartments, and private houses. In underground garages, most building codes require sprinkler systems. Auto fires generally consist of smoldering upholstery, electrical fires under the hood, or the remote chance of a gasoline fire. Because the sprinkling of water or foam on the car is of little or no help, many cities now grant variances to eliminate sprinklers.

Other fire-detection equipment includes rise-of-heat indicators mounted in the ceiling of the garage. These indicators monitor temperature in an area and are usually designed to sound an alarm when the temperature in an area rises 15° in 1 min.

Placing large numbers of fire extinguishers in a garage has proven to be useless because of thefts. In lieu of small sprinklers, large extinguisher systems can be placed on dollies and situated at central locations in the garage. Too large to be easily stolen, they can readily be rolled to the source of any fire.

Boundary protection

The boundaries of each floor of an aboveground garage, and the roof deck, must be provided with an effective system to reasonably restrain cars that have overridden any curbing or wheel stops. Commonly used are parapet, structural walls at least 27 in high, or highway guardrail or bridge rail of similar height.[16, 17] Special structural calculations are needed relative to the restraint design and anchorage into the floor slab. Additionally, pedestrians must be protected against falls by rails 36 to 42 in above walk elevations along exposed edges of floors or ramps.[18]

Drainage and waterproofing

Many engineers consider water leakage through the floor slab to be the main unsolved long-term maintenance problem of parking structures. This leakage does not appear during the early life of the garage. The water, mixed with salts in colder climates, seeps down through hairline cracks and attacks the reinforcing bars. In the absence of adequate cathodic protection, these bars rust and expand, resulting in scaling of the concrete. Over time, this action can cause structural damage. Without proper maintenance, a garage can be at the point of structural failure within 10 to 15 years as a result of water leakage.

The best waterproofing system is one that gets the water off the floor and into the drains quickly. This is particularly necessary for the roof level. The riser system capacity should be designed adequately to accept a 10-year design rainfall. The floor of the garage should be sloped toward the drains with a minimum slope of 2%.

Some fire codes require flammable liquid separator drains.

There are numerous types of waterproofing systems. One common system consists of a mastic asphaltic material placed over the floor slab to a thickness of about 1/16 in. This material seeps and permeates down into the floor slab and forms an elastic, rubbery type of surface that seals small cracks. A wearing surface of 1/2- to 1-in of asphaltic concrete is then placed on top.

A second system consists of a plastic material that is usually sprayed onto the floor slab and seeps into the slab. For a wearing surface, a strong, tough rubberized plastic is then sprayed on the deck in several layers and built up to a thickness of 0.15 to 0.20 in. Flint chips or small gravel are often embedded in the wearing surface to aid traction.

A third waterproofing system is a membrane surface of sheet rubber that is placed over a concrete slab. A second slab is then poured over the membrane. However, some of these membranes begin to disintegrate after about 10 years. Also, it is nearly impossible to locate or repair a rupture in the membrane because of its position between the two slabs of concrete.

Other types of waterproofing compounds are generally spray-on plastic types; they have some waterproofing qualities but are of no value in preventing leakage when a hairline crack opens. Whatever sealant treatment is initially applied, experience with garages in the Northeast indicates that periodic resealing of concrete floor slabs is desirable.

Interior signing and marking

A parking facility is an extension of the street system. Directional and informational signs are generally needed and should conform as closely as possible to standard signs and markings used on streets. Many signs can be painted or attached to walls or columns.

The use of illuminated signs may be justified in some instances, such as when they must be viewed against a daylight background, or when aisle, ramp, or parking patterns are variable.[19] Well-placed, nonilluminated signs are often sufficient. Signs should direct motorists to parking spaces and exits, and they should inform the driver of conditions such as one-way aisles and ramps. In addition, signs should direct the pedestrian to exits, stairs, and elevators. Signs reading EXIT should not be used for *both* vehicular and pedestrian use, as this can be confusing. It is better to use terms such as STAIRWAY or ELEVATORS for pedestrians, where possible.

[16] "Guide for Selecting, Locating and Designing Traffic Barriers," American Association of State Highway and Transportation Officials, 1977.

[17] "Standard Specifications for Highway Bridges," American Association of State Highway and Transportation Officials, 1983.

[18] "American National Standard Safety Requirements for Floor and Wall Openings, Railing and Toeboards," American National Standards Institute, A12.1–1973.

[19] C.M. Bolden, "Signing and Graphics for Parking," paper presented at the 32nd Annual Convention, National Parking Association, New York City, May 1983.

It is difficult to determine exact sign location points from construction plans. It is better that the signing installation be postponed until the garage is nearly completed, and then to drive into the structure and determine locations from the vehicle. This gives the proper perspective with regard to sight lines, parked cars, and obstructions such as beams and columns.

Markings are used to define parking stalls, "no parking" areas (usually by yellow hatching), one-way aisles and turn (by arrows), and sometimes to help define the desired vehicle path outer limits. Raised curbing may be painted yellow—especially in areas likely to be contacted such as along ramps, at ticket dispensers, or at cashier booths. In order to increase the target value of yellow paint, it is preferable to use white for stall markings.

Revenue-control systems

In a large garage where fees are charged, a revenue-control system is a necessity. Such a system is designed to keep track of all entering and exiting vehicles and a record of tickets, monthly parkers or contract cards, cashiers on duty, and the amount of money handled. The system provides the information needed to check for revenue losses and thefts.

Six basic elements of a revenue-control system are needed if the record of parking activity and revenues is to be accurate.[20]

1. A correct count of all entering and exiting vehicles.
2. No one allowed to enter the parking area without taking a ticket or to exit without leaving a ticket or showing evidence of monthly or other contract arrangement.
3. All clocks operated accurately, synchronized, and designed so that cashiers cannot reset them.
4. Some type of validating device or cash register—not a time stamp—to record the outbound transactions, and tickets stamped with the appropriate information so that individual cashiers may be held responsible for their actions.
5. Ticket-issuing machines and entry lanes operated so that it is not possible for anyone to get more than one ticket from the machine, with future supplies of tickets kept in a safe place.
6. Parking gates or other devices used to prevent cars entering through an exit or leaving through an entrance lane.

In central cashiering, parkers return to a central location in the garage, pay their parking fee and receive a receipt, proceed to their car, and, on exit, hand the receipt to an attendant. Cedar Rapids, Iowa, opened the first garage with a totally automated payment system in 1978. The 800-car CBD structure is connected by walkway to a 300-room hotel and a 10,000-seat community center. The system uses two tickets, with a machine-coded one issued upon entry. This coded ticket is machine-read at a central auto-pay station, and an exit "pass" is issued. (A cashier may be needed to make change.) This ticket is read by an exit-gate reader that then opens the gate.

In exit cashiering, parkers return directly to their cars and drive to the exit, at which point they stop and pay their parking fee. In addition, under special conditions, such as at sporting events, the parker generally pays a flat fee on entering the facility and can exit without stopping.

Alternatives should be considered before selecting a cashiering method. The design and use of the facility can influence the choice. For example, in a garage serving a department store or theater, where parkers use a common path in returning, central cashiering may be a good choice. Exit time is reduced. With central cashiering, where a receipt only is picked up on exiting, vehicles can exit at a rate of 360 or more per hour per lane. With exit cashiering, the rate is about 180 vehicles per hour per lane. Central cashiering thus at least doubles exit lane capacity.

Design elements for large parking generators

Characteristics affecting parking design

The principal element of design importance for facilities serving major generators is the rate of arrival and the rate of departure—usually expressed as a percent of capacity in a given time period such as one-half to 1 hour. Shopping centers, downtown retail parking, and (to a degree) colleges, airports, and hospitals have both arrival and departure concentrations extending well beyond 1 hour. Other generators (offices and industrial plants) have very high peaking characteristics. Sports facilities have relatively high arrival rates and extremely high exit rates.

To accommodate the expected demands, the capacities of lanes must be estimated in order to calculate the number required. The type of revenue control (if any) also must be assumed. Knowledge of the parkers' characteristics is important—principally their walking distance limitations. Finally, the street capacities—both the area network and the direct access route—must be known or estimated. There obviously is no advantage in being able to discharge 10,000 cars in 30 mins, if the street system requires 1 hour or more to accommodate this loading.

Shopping centers

As of 1988, there were over 32,000 shopping centers in the United States, with a gross leasable area (GLA) of nearly 4 billion square feet.[21] About one-third of these exceeded 100,000 square ft and 1% exceeded 1 million square ft. Annual sales were over $641 billion.

Shopping centers reach their peak in the 4-week period prior to Christmas. The peak day is generally the Saturday before Christmas. Many shopping centers include uses other than retail, such as theaters, financial institutions, and office complexes. The peaks for those other uses do not occur at the same time as the peak for retail shopping. It has been found that up to 10% of the GLA can be added to a shopping center as office space without affecting the peak parking demand,

[20] J.M. Hunnicutt, "Safeguarding Your Parking Revenues," 1970 Annual Conference, American Association of Airport Executives, Las Vegas, May 1971.

[21] "1988 Census of Shopping Centers," as reported by *Shopping Center World* (February 1989).

and that centers of over 100,000 square foot GLA can accommodate 450 to 750 theater seats without providing additional parking.[22]

The presence of ancillary uses may be a factor in shopping center parking layout, access, and circulation. Parking for additional theater seating can sometimes be located in areas not suitable for retail customer use, such as behind buildings or on a little used side of the shopping center (provided doors are available connecting to the theater).

General principles of shopping center access are given in the ITE Committee 5-DD report.[23] A number of factors can influence the internal circulation and parking layout.[24] These include:

1. Size of the development and the magnitude and peaking characteristics of the generated traffic volumes.
2. Size and shape of the buildings proposed for development on the site.
3. Size and shape of the site.
4. Amount of road frontage available for access.
5. Distribution of site traffic at the site-access points.
6. Location, number, and design features of the proposed access points.
7. Local zoning constraints:
   a. Building setbacks.
   b. Landscaping requirements.
   c. Parking ratios and recognition of shared parking.
   d. Parking stall design standards, including recognition of car-size trends.

In the larger shopping centers, the incoming shopper should be guided onto an intermediate circulation system interconnecting all parking areas on the site. The intermediate road should *not* be adjacent to the building, but should be 300 to 500 ft distant, depending on the size of the development and configuration of the site. This circulation road not only provides intrasite circulation but also defines the primary parking area serving the development. Continuity of this roadway is essential to provide for the distribution of traffic entering from the external street system to the various parking areas. This design enables site traffic to enter from the adjacent street system at the first available opportunity. Vehicles are then able to circulate within the site rather than on the external streets where they can reduce capacity and increase hazards.

The most critical area of vehicle-pedestrian conflict usually occurs on the roadway adjacent to the storefronts. The primary traffic function of this road is to provide for passenger pickup and dropoff adjacent to store or mall entrances and for customer package pickup. This roadway also provides for emergency vehicle service and, where available, public transportation services. In most layouts, it also serves as part of the parking search circulation system and for exit when angle parking and one-way aisles are used.

Other elements of this system are connecting drives and aisles between the building circulation road and the major circulation road (intermediate ring road) that provide direct building access for service vehicles, and emergency vehicles, as well as passenger vehicles. Figure 7–8 illustrates the general principles of layout for a large regional shopping center. Of particular note is the need for substantial reservoir areas along the main access roads. For the inbound flow, these allow drivers to weave into the left- or right-turn lanes where they intersect the intermediate circulation road. The outbound reservoir area is needed for a similar weaving purpose relative to left or right turns into the major street, and also to minimize backup of exiting cars into the circulation road of the center.

Committee 5-2 (ITE) also developed suggested design guidelines for interior roads.[25] They are abstracted below:

1. Intermediate circulation road widths will vary between 24 and 48 ft, depending on the size of the development and whether parking or peripheral land-use activity occurs on both sides. In the smaller centers (less than 200,000 to 300,000 square ft), the intermediate ring road can be 24 to 30 ft wide. For intermediate-size centers (not over 600,000 square ft), a three-lane ring road, 33 to 36 ft wide, can be provided. The third lane typically becomes a left-turn lane at major access points and serves all the parking aisles when no peripheral development or parking is provided on the outside of the ring road.

2. The building circulation road has been as wide as 36 ft because of the circulation and pickup/dropoff (P/D) activity occurring in the vicinity of the building entrances. This width has encouraged parking and posed pedestrian safety and emergency vehicle problems. These roadways have been found to carry volumes as high as 3,000 vehicles per day and parking should not be permitted because of interference with pedestrian crossings.

In areas not located near building entrances and where no P/D activity occurs, the roadway width can be 26 to 28 ft. In areas where P/D occurs, two design treatments have been utilized: (a) a roadway width of 32 ft to accommodate two moving lanes of traffic and a P/D lane, and (b) a 26-ft roadway with a recessed 8-ft lane for P/D. In order to control parking along building fronts, it is recommended that the roadway be officially designated (by ordinance) as a fire lane and that NO PARKING, FIRE LANE signs be installed.

3. Delineation of both the intermediate and building circulation roads is essential for safe operation of large facilities. Curbed end islands, either landscaped or paved, as well as painted end islands are utilized. End islands (whether painted or curbed) are useful in several respects. An important one is to delineate the circulation road edge. Of equal importance is providing clearance (adequate sight distance) for vehicles desiring to exit parking aisles and enter the circulation road.

In order to be effective and cost-efficient, end-island designs should be the same regardless of the angle of intersection with the circulation roadway (see Figure 7–5). This allows a design template to be established for use in laying out the islands rather than having each one individually designed. Where islands are landscaped, their shape should

[22] "Parking Requirements for Shopping Centers," Urban Land Institute, 1982.

[23] "Guidelines for Planning and Designing Access Systems for Shopping Centers," *Traffic Engineering* (January/February 1975).

[24] N.S. Kenig, "Internal Circulation and Parking at Shopping Centers,"ITE Committee 5-2, Compendium of Technical Papers, ITE, 1983.

[25] Ibid.

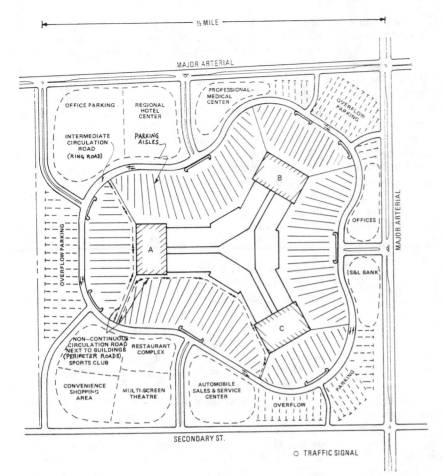

Figure 7-8. Example of regional shopping center layout.
SOURCE: Adapted from "Internal Circulation and Parking Design," Traffic Institute, Northwestern University, Stock 3859.

conform to a single design standard and the planting should be located and maintained so as not to affect sight distance. This typically requires open area in the zone from 2 ft to 7 ft above the road.

4. There are several basic types of parking areas in most regional shopping centers: prime, overflow, and employee. Different design standards can be applied to each.
   a. The *prime* parking area is usually located close to the building (within 200 to 300 ft) and in the immediate vicinity of a building entrance. It satisfies the peak parking demand on an average day and should be designed to a higher standard to accommodate the high turnover rate.
   b. *Overflow* areas, used mainly during seasonal peak shopping periods, should be designed to a lower standard than the prime areas. In many shopping centers, particularly the older ones, the intermediate circulation road separates the prime and overflow areas.
   c. *Employee* parking areas are usually located at the extreme fringe of the site or in areas nearer the building that are not readily associated with a major building entrance. Security often dictates that these areas be segregated from the major parking supply and be policed. Minimum parking dimensions should be used in the design of these low turnover spaces.

5. Parking layout includes considerations of:
   a. Orientation of parking aisles (perpendicular, parallel, or angled to the building)
   b. Traffic-flow patterns in the lot and means of control (one-way or two-way aisles, islands, desirable length of aisles without intermediate drive, and the effect of elevation changes caused by the larger centers with two levels that need to be served).
   c. As general design guidelines:
      o Parking aisles should be laid out so that pedestrian movements parallel the aisles and direct patrons to the building entrances.
      o Various parking areas or parking lots should be developed separately from one another and be limited by the intermediate circulation road, in the larger centers.
      o Generally, customer parking aisles should not exceed 300 to 500 ft in length without a break for circulation. These lengths may be exceeded when employee parking is added on the end and the circulation road represents the boundary for the shopping center. (Ideally, employee parking should be located outside the circulation road in order to maintain shorter aisle lengths.)
      o In those centers where the levels of parking slope to accommodate a two-level mall, the

primary concerns are the grades within the parking areas and the intersection treatment where the two levels meet the elevation of the intermediate circulation road.

6. The choice of whether to provide angle or 90° parking in a particular shopping center depends upon the shape and dimensions of the parking areas, the design and location of entrances and exits, the internal circulation plan, and the prevailing customer habits in the particular community. Either layout can serve a shopping center efficiently if appropriate design guidelines are observed. Generally, the method that provides the greatest total number of well-designed parking spaces within a given parking area is the most desirable.

7. The marking needs of shopping centers are similar to those for other parking lots; however, area *identification* signing is an added need in the larger centers. Signs are usually mounted on light poles with letters, numbers, and/or colors used to mark the specific parking areas and thus aid in locating vehicles.

It is highly desirable to consider transit bus service to all regional-size shopping centers. This service stop should be convenient to an entry point of the mall, but it should be designed to minimize conflict with individuals arriving by automobile. An ITE Informational report update gives useful design concepts and examples.[26]

Appropriate stall-width classes for shopping centers is A (9 ft) or B (8.75 ft) for prime areas, C (8.5 ft) for overflow, and C or D (8.25 ft) for employee areas (Table 7–4).

Office developments

The characteristics of office buildings, relative to parking design, are:

1. Moderately high arrival and departure rates (about 35% of daily total in each peak hour).
2. Relatively small amounts of visitor parking spaces needed (compared with the total supply).
3. A small demand for pick up or drop off of passengers.
4. A moderate demand for very short-term mail and package delivery and pickup.
5. Space for heavy delivery plus trash pickup.

Except for larger developments (over 200,000 square ft building area, or with access limited to a single major point), special reservoir areas are unlikely to be needed. The employee parking is usually oriented outward from the buildings, with suitable orientation for pedestrian access. Visitor parking is placed closest to the building—sometimes in a separate area. The pickup/drop-off area for passengers, mail and small packages, is usually informal for the smaller buildings (under 200,000 square ft), consisting of a curbed strip about 30 to 40 ft long in front of the main entrance. Larger buildings may have more elaborate loop driveways, but these often serve to increase walking distance for visitors and may not be very functional.

Heavy truck and trash loading should be *away* from the main entrance of most office buildings. Rear or side areas—sometimes screened from typical viewing angles—are used.

It is desirable that no reservations or stall assignments be made for any employee. Such reservations generate inefficiencies in that many stalls may be vacant at any one time. Stall assignments also may create enforcement problems.

Class B (8.75 ft) stalls (Table 7–4) are appropriate for visitors and Class C (8.5 ft) for general office employees.

Industrial plants

This type of development has higher peak-hour proportions of arriving and departing traffic, as compared with offices. Multiple-shift operation may require parking supply overlap (the second shift arriving before the first shift leaves). The proportionate number of visitor spaces tends to be less than for offices. Finally, the industrial use generates many truck movements, loading dock requirements, and usually trailer or truck storage areas. These may require security fences and gate control.

All industrial developments have an office force, in addition to the plant workers, and usually with a different schedule. Often, the office personnel parking area is separate from the plant worker area. A special close-in area (to the plant) may be reserved for high occupancy vehicles such as vans or buses and for motorcycles. Bicycle use is seldom significant, but if it is a factor the racks should be convenient to the plant-building entrances.

ITE Committee 6F-24 has prepared an information report.[27] Typical practices in parking layout are abstracted below:

1. The choice of providing angle or 90° parking depends on the shape and dimensions of the parking area, the design and location of entrances and exits, and the internal circulation plan. Either layout can serve efficiently if appropriate design principles are observed.

2. At large plants, blocks of parking by groups of 300 to 500 cars are preferable to larger aggregations. Pedestrian-vehicle conflicts can be reduced significantly with smaller parking areas; and assigned parking for different shifts and employment groups can be better controlled.

3. Visitor and special parking needs are usually located as close as possible to the office areas within the employment center. If the site is subject to tight security restrictions, then the visitor parking is typically placed close to the guard office or booth.

4. Care should be taken to ensure that proper routes to and from main pedestrian access locations are available for emergency vehicles. Fire routes should be designated around the periphery of buildings. All intersection corners should have sufficient radii for fire vehicles. These routes should be well signed, with parking prohibited as appropriate and preferably enforced under the authority of local municipal fire-route ordinances or laws.

---

[26] "The Location and Design of Bus Transfer Facilities," ITE, 1989 update (pending).

[27] "Employment Center Parking Facilities," ITE, 1988.

Pedestrian-vehicle conflicts are almost inevitable, but with minimum walking distances such conflicts can be reduced. Several guidelines can be followed to lessen the conflicts:

1. Parking lot assignments oriented to specific buildings.
2. Parking areas designed to focus on major walkways, which may be fenced or marked.
3. Parking aisles leading directly to the building. This will minimize inbound problems, because close-in spaces are taken first and late arrivals park farther away. For both inbound and outbound conditions, pedestrians can walk past parked cars rather than crossing aisles with cars arriving and departing. This works best with 90° parking.
4. Where pedestrians must cross service roads or access roads to reach parking areas, crosswalks should be clearly designated by pavement markings, signs, flashing lights, or even traffic signals, if warranted.
5. In some situations, the best means of separating pedestrians and vehicles is underpasses or overpasses at key points. Grade separation may be essential to prevent long delays and time losses, such as in freight handling areas, or to avoid exposure to hazardous plant operations. It also may be necessary where parking facilities and plant buildings are on opposite sides of major highways. If intersection capacity problems preclude the provision of a pedestrian phase in nearby traffic signals, grade-separated pedestrian crossings may be necessary. Pedestrian underpasses may present special security problems.

Unless an industrial center is under strict security measures, service vehicles usually use the same entrance and exit facilities as do employee vehicles. Where security control is required, separate truck gates may be provided with trucks checked by a guard. Controls may be required to eliminate possible conflict between trucks and other traffic, especially during periods of major employee shift arrivals and departures. Substantial volumes of truck traffic may require separate driveways or internal routes.

Within the site, commercial vehicles can be routed around the periphery to loading areas with no access allowed through employee parking lots. Turning radii should be adequate for the largest vehicle anticipated, including emergency vehicles.

Stall widths of Class C (8.5 ft) are appropriate for office personnel and visitors, with Class D (8.25 ft) used for plant workers (Table 7–4).

Airports

Because needs vary so widely, the parking layout for airports must be tailored to each site. General requirements are for separate areas for employees, airline passengers, car rentals, taxi and limousine holding, and the pickup/drop-off (P/D) next to the terminal. For the larger airports the P/D zones are often grade separated. Pedestrian/vehicle conflicts can be greatly reduced by pedestrian underpasses or overpasses connecting the parking areas with the terminal. Passageways, conveyors, and elevators should all be designed to handle people with hand luggage as well as baggage carts.

There also is need for a recirculation system, whereby drivers should be able to drop off passengers and/or luggage at the terminal frontage, and then reach the parking facility. Additionally, the converse unparking-to-terminal pickup movement must be provided.

Parking facilities at major airports tend to be very large. A 1985 summary of parking data for airports with over 2 million originating enplanements found on-site capacities of up to 25,000 spaces, with an average of 8,200 stalls.[28] In such large facilities, special consideration is needed for parked vehicle location. In addition to identification of garage levels, sections and aisles usually need separate letter and/or number coding.

The large numbers of stalls also create problems of access/egress capacity. Studies of two airport garages found outbound flows to peak at 22% to 24% of capacity, with similar demands during the inbound peak.[29]

Class B (8.75 ft) or C (8.5 ft) stall widths are appropriate for public parking, and C or D (8.25 ft) for employees (Table 7–4).

Rail station change-of-mode facilities

Parking facilities for commuter rail, rapid transit, and street car stations have proven extremely popular in many larger cities. The basic elements are P/D transfer between the train and buses, the taxi or private car P/D, bicycle and/or motorcycle racks, and general parking. It often is desirable to separate the general parking into Small and Large car sections. With two-car suburban families, the smaller car is usually used by the commuter, which creates a relatively high percent of Small cars.

Chapter 8 of *Parking Principles,* Special Report 125, summarized suggested priorities of location for several parking or loading elements relative to the boarding zone.[30] These are given below:

1. Bus loading-unloading.
2. Taxi loading-unloading (may intermix with buses or with cars).
3. Passenger car unloading (drop-off).
4. Passenger car loading (pickup).
5. Short-term parking.
6. Long-term parking.

Figure 7–9 shows how conflicts can be avoided by proper arrangement. Depending on the size and shape of the total site, the functions can be grouped in the manner shown, separated, or combined.

The facilities needed for the bus transfer operation include entry and exit driveway, sidewalk-level loading and unloading, and layover or holding area. The linear feet of loading space, the need for a bus bypass lane, and the capacity of the holding areas are related to the frequency of bus service. If taxi, airport limousine, or interstate bus connections are also included in the bus areas, additional space may be required.

Bus layover to maintain scheduled headways may be provided in two ways. The buses may wait at the loading curb

[28] Joseph P. McGee and William C. Arons, "Second National Parking Association Airport Survey," *Parking* (November/December 1986).

[29] Richard C. Rich, "Airport Parking Design," *Parking* (November/December 1985).

[30] *Parking Principles,* Special Report 125.

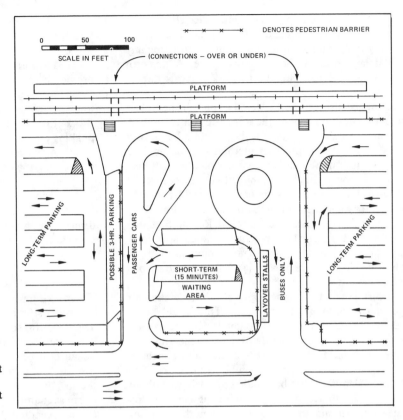

Figure 7–9. Example of a rapid transit change-of-mode terminal.
SOURCE: *Parking Principles,* Special Report 125, Highway Research Board, 1971.

itself, or they may be temporarily stored in an offset lane as shown in Figure 7–9. Curbing layover is limited to about two buses for the design shown, and a bypass lane would be essential. Late-arriving passengers should be able to board buses in the layover area without any vehicular conflict.

The principal loading areas should be sheltered and a covered walkway for the remaining distance to the train boarding area should be provided. The minimum shelter should be overhead as a protection against rain, with a 14-ft vertical clearance over the bus roadway area. In northern climates, three-sided or full enclosures are desirable for weather protection, with transparent walls to enhance security.

The automobile passenger operation contains three critical elements: (1) drop-off, (2) temporary standing while waiting to pick up an arriving passenger, and (3) the actual pickup. The drop-off is principally a morning activity, whereas the pickup typically occurs in the evening. An exception would be domestic-help arrival in the morning and departure in the evening.

Ideally, a car arriving for pickup purposes should be able to enter and exit from a holding area without having to pass by any drop-off activity. The vehicle should (during inclement weather) also be able to recirculate close to the train platform to make the pickup.

On-street or in-driveway pickup or drop-off is undesirable because it promotes congestion and hazard. The terminal area provided to handle this activity should be designed with a sufficiently high level of service to encourage use in a positive fashion. This can be done by (1) locating the train boarding area at a substantial distance from the nearest street, (2) placing the P/D lane close to the train area, (3) providing

adequate capacity by ample driveway widths, bypass lanes, and temporary standing areas, and (4) erecting barriers—fences and guardrails—to prevent pedestrian access from nondesignated loading areas.

Illinois studies found a pickup waiting area need for about 10% of lot capacity, or one space per 20 train passengers arriving during the peak hour. A modest over-design of any waiting area would appear to be far preferable to underdesign. Part or all of the spaces can readily be metered for intermediate term parking, and hence produce revenue during the day. They can be cleared for pickup use by imposition of a very short time limit (5 to 15 mins) regulation, commencing just prior to the evening commuter surge. This system has worked successfully for many years in Naperville, Illinois, for example.

To minimize walking distance, the parking lots should ideally radiate outward from the train boarding area. Aisles may be used by drivers walking to or from their cars, and special walks within the lots are usually unnecessary.

For lots of more than 200 spaces, at least one access point should be separate from the P/D area. For the very large lots (more than 500 spaces) it may be desirable to eliminate any access via the P/D area. In the latter case, train riders arriving as passengers in vehicles where the driver will park and also ride the train can be dropped off near the boarding area but from *within* the lot. Figure 7–9 shows such an inclement weather arrangement, which also allows recirculation to make an internal pickup.

If the parking facility is to be operated with a nominal fee, it will be similar to a commercial parking lot except that 24-hour access and egress should be provided. Fee collection at all hours can be handled in an excellent manner by free-in, pay-out coin gates. The gates can be locked open during the peak morning entry, thus eliminating any inbound delay or need for reservoir space. A detector should be provided to open the gate for arrivals during other hours.

A parking fee also can be collected as part of the transit fare collection or upon leaving the train fare control area. A token would be issued to operate the outbound gate of a free-entry, pay-exit system. This method has the obvious advantage of eliminating change-making problems. It also would assist in controlling a lot subject to use by nontransit riders such as local employees or apartment residents.

The need for some reduced-time-limit stalls separate from the bulk of all-day commuter use may exist. Reservations can be made for midday arrivals by posting parking restrictions, such as 3 hours, on selected spaces. These would normally be spaces located closest to the boarding area, to encourage off-peak use of the transit system.

Under certain conditions there may also be a demand for very long-term parking (several days). This would occur where the lot could be used by individuals desiring to park and then ride to an airport via the transit, connecting bus, limousine, or even a taxi. Such users should not necessarily be encouraged in preference to regular commuters. Any problem of this nature could be handled by setting a 24-hour time limit for the majority of spaces and leaving a few unregulated ones at maximum walking distance from the boarding area. Such spaces would be exposed to overnight parking, and security would be enhanced by a corner location near a public street.

Security also dictates an adequate lighting system for the entire parking facility plus periodic police patrol. Such a patrol, plus the possible regulation of time limits, illustrates the urgent need for coordination and cooperation between the local government agency and the change-of-mode terminal operator.

Class D (8.25 ft) parking-stall widths are appropriate for use at all commuter facilities because of the very low turnover (Table 7–4).

Freeway access park and ride

Parking lots of small to moderate size have been constructed in outlying areas along freeways approaching the central area. These are usually located in or near interchanges and provide opportunity for ride sharing (hence the name "park-and-pool" as they are called in Texas, for example). This concept originated in 1973 following a study done by the State of Utah. Typical "mini-lot" capacities of 50 to 100 spaces were proposed and many have been constructed.

Larger park-and-ride lots have been constructed with bus terminals. Express service may be limited to rush hours, but more frequent headways can be maintained—especially if the service is part of an established route. General design considerations abridged from a report are as follows:[31]

1. Minimize potential impacts on adjacent neighborhoods by providing links as direct as possible to the freeways.
2. Pave and light the lots, with shelter and telephone provided.
3. Minimize operating costs with self-parking and automatic fee collection (if a charge is made).
4. Provide adequate access capacity to accommodate peaking characteristics.
5. Avoid walking distances exceeding 1,500 to 2,000 ft, by locating the boarding point near the center of the lot.

More detail design criteria and layout examples are given in another report.[32] An exception concerns parking layout dimensions, which should use Tables 7–5 and 7–6 with Class D (8.25 ft) stall widths.

Sports and convention facilities

The parking needs at sports stadiums, convention centers, and other similar activities are dependent on many factors. The needs for such a facility in a CBD will be less than for a suburban location for the same size and type of generator because of better public transportation and the ability to use existing parking facilities to meet a portion of the demand. Early analyses should be made of the special-purpose facility to determine the number and type of events expected. Peak and average crowds should be estimated from which a design crowd can be determined. Crowds for different events exhibit different characteristics. For example, car occupancy for baseball games may average 2.5 persons per vehicle, whereas for football games it may be somewhat greater, at about 3.5 persons per vehicle. Also, studies have shown that the football fan will walk farther than will the baseball fan.

The time to empty a parking facility is especially critical for special-event or sports parking. This value is usually between 30 and 60 min, and efforts should be taken in design to minimize this time. One aid to reducing the time is to use pre-cashiering (collection of the parking fee on entry). The time then depends on the internal design of the parking facility, its ramps and exits, and the surrounding street system.

Separation of pedestrians and autos should be achieved whenever possible because of the large volume of pedestrians. This is especially true at vehicle exit driveways. Pedestrian ways, including overpasses or tunnels, should be considered. Some sports fans are willing to walk long distances (often to secure free parking at curbs or in well-removed vacant lots). This can increase the vehicle-pedestrian conflict at driveways.

ITE Committee 6A-5 developed an informational report that contains useful data on operations for special events.[33]

Because of low turnover, Class C (8.5 ft) or Class D (8.25 ft) parking-stall widths are suggested (Table 7–4).

---

[31] Raymond H. Ellis, John C. Bennett, and Paul R. Rassam, "Considerations in the Design of Fringe Parking Facilities," *Highway Research Record 474*, Highway Research Board, 1973.

[32] Douglas A. Allen, "Design Guidelines for Park and Ride Lots," *Compendium of Technical Papers*, ITE, 1979.

[33] "Traffic Considerations for Special Events," ITE, 1976.

Colleges and universities

The majority of campus parking occurs in lots because of the obvious economic constraints relative to garage construction. There are four general types of parkers: faculty and other staff, visitors, commuter students, and resident (dorm) students. Parking is usually free on smaller campuses, but charges are often levied at large institutions.

Relatively high rates of bicycle use occur—especially in warmer climates. Thus, provision for close-in bicycle and motorcycle parking is appropriate. For the commuter student arriving by car, very long walking distances are typical, as is use of Class D (8.25 ft) parking-stall width (Table 7–4). Parking for faculty, staff, and visitors should receive first priority for convenience, and utilize Class C stall widths (8.5 ft).

Hospitals

Hospital parking design must consider three-shift operation, separate areas for doctors, other hospital staff and nurses, visitors, and a small emergency parking area (police and limited private cars). Garage construction is typical to serve at least the doctors and visitors. It is desirable to provide pedestrian access routes that are not in conflict with vehicular movements.

A pickup/drop-off (P/D) entry system is needed, under shelter, and with full recirculation access to and from the visitor parking area.

Out-patient care is an expanding activity at hospitals. Depending upon site and external access characteristics, it may be desirable to consider a separate parking facility and P/D loop for the function.

Truck-loading needs are similar to those of office buildings—that is, avoidance of conflict with other activities and semi-concealment of trash.

Class C stall widths (8.5 ft) are appropriate for most hospital parking activities, except that Class D (8.25 ft) width is acceptable for general staff and nursing areas.

Residential

This type of parker spans the widest spectrum of the industry—from the single stall on a concrete ribbon driveway to underground garages with key- or card-controlled entry. The parking *ownership* generally differs from other land uses, in that the spaces are directly owned by the individual or are assigned as part of apartment rental or condominium ownership.

In earlier days (and consequently the current norm in older parts of most cities), parking was either at the curb or accessed via the rear alley—frequently in garages. The older garages are fairly consistent in their defects; inadequate setback from the alley for maneuver, doors too narrow, and structure too short. Today, many of these garages are used simply for other storage. Many have been removed to produce open parking pads. In some of these areas, new garages have subsequently been built (almost always with doors too narrow for efficient service).

For individual residential garage access from alleys, a minimum *setback* of 5 ft is suggested from all alleys in the typical 16- to 20-ft width range. The minimum garage depth should be 20 ft, although 22 to 24 ft should be provided whenever possible to accommodate the omnipresent miscellaneous storage needs of the average resident.

Garage door width should be 9 ft (single) and 17 or 18 ft (double) instead of the common 8- and 16-ft doors used.[34] This is especially critical where access requires a sharp turn. Interior *clear* widths of less than 11 ft for one-car garages, or 20 ft for two-car structures, should be avoided. Desirable dimensions are 12 and 22 ft, to allow side wall area for tool hanging, trash, and miscellaneous storage.

For front or side access garages or open parking pads, the minimum suggested driveway width is 10 ft. When serving a two-car garage, 20 ft is desirable. These widths provide pedestrian paths next to a parked vehicle in front of the garage.

A serious restriction of off-street parking use will occur unless access to each stall in *unobstructed.* Thus, a requirement of two spaces is met only by a two-car garage or parking pad about 20 ft wide, if serving a single-family home.

Parking for apartments, cluster housing, or multiple-tenant condominiums is commonly provided in lots, under open sheds, or in aboveground or underground structures. These should follow basic design guidelines for access, layout, and lighting. Security measures needed will vary with the location as well as the size of the facility. In parking lots, bushes or hedges large enough to conceal potential assailants should be avoided.

A typical stall class for multiple-space residential lots or garages is C (8.5 ft), although A or B may be more appropriate for condominiums (see Table 7–4).

Curb parking

Street purpose and typical curb uses

The primary function of the streets in any city is for the movement of vehicles. Parking on these streets (curb parking) must be considered a secondary use of street space, as should other uses, such as truck loading zones. When these secondary uses conflict with the movement of traffic, they should be removed to the degree practical.

Curb space along streets in any city has several categories of use. These include curb parking available to the general public (with or without limits on time of day or duration), truck loading zones, bus and taxi zones, passenger loading zones, and others. The amount of parking allowed at the curb varies with city size. In all cities, curb parking usually is restricted in certain sections to aid the movement of traffic. Many major routes and business district streets have total prohibition of parking, or at least rush-hour restrictions.

This section provides an overview of the congestion and accident problems generated by curb parking. It includes a discussion of the various restrictions that may be applied to curb parking to achieve better use of the limited street space. Parking meters, their application, and maintenance elements are also covered.

---

[34] "Recommended Guidelines for Subdivision Streets," ITE, 1984, (1991 update pending).

Disadvantages and problems

General factors

The curb parking problem involves parking-related accidents, traffic interference during the parking maneuver, and use of roadway area for parked vehicles. A single parked vehicle can create delays and hazards for hundreds of motorists. The elimination of curb parking usually improves the accident record. Because some curb parking is necessary in all communities, it should be regulated in such a way as to minimize congestion and hazard. The economic losses stemming from accidents and congestion due to curb parking are a strong incentive to pay for equivalent space in off-street parking facilities.

Congestion

There is no question that parking along a street significantly reduces the traffic-carrying ability of that street (see Figure 7-10). Traffic capacity is lost not only in the portion of street used for parking but also in the lane adjacent to the parking lane. Sometimes the entire width of a roadway is affected. The stopping, starting, and backing of vehicles during the parking maneuver physically restricts other traffic movements. The presence of vehicle passengers in the street, the opening of vehicle doors, or pedestrians walking out from between parked cars all tend to interfere with efficient vehicular movements.

There are many ways to lessen the adverse effects that parking has on the capacity of the street system. Foremost of these is the total prohibition of parking, stopping, standing, and loading along major streets. The function of a roadway system, especially the major streets, is to provide a travel way from one point to another and to provide access to abutting properties. A successful system is one that performs these functions efficiently and safely.

Congestion and accidents are measures of the failure of the system to operate well, and they occur as the result of a breakdown of the orderly and smooth flow of traffic. Many times this breakdown is caused by the side friction of curb parking. Even though a street with parking may be wide enough to carry the present traffic volume, greater capacity can be realized and a significant reduction in congestion accomplished through the use and vigorous enforcement of parking prohibitions.

Of particular note with reference to congestion are those areas adjacent to intersections. Where parking is permitted too close to intersections the result is blocked sight distances and poor visibility of vehicles and pedestrians. Also, vehicles parked close to intersections often block lanes that could be used by drivers to get around left-turning vehicles. At signalized intersections, parking spaces for vehicles should be even farther away from the intersection on both the approach and departure sides. This allows greater intersection capacity because left-turning vehicles may be more easily bypassed, waiting traffic can queue two or more abreast, and right-turn vehicles are accommodated.

Accident hazard

A clear relationship exists between accidents and vehicles parked along the curb. From a comprehensive review of accident data, it is safe to assume that curb parking is directly or indirectly responsible for at least one out of every five non-freeway accidents that occur in our cities each year.[35]

One study in a community of 65,000 population found that 43% of all local and collector-street accidents involved curb parking.[36] In this same city, annual frequencies of 14 parking accidents per mile were found on major streets, but only 1.8 parking accidents per mile on local and collector streets.[37]

A comprehensive study of curb-parking accidents in 10 cities and five states gathered street and accident data for over 170 miles of urban streets. This study related the magnitude and characteristics of urban street accidents to varying

---

[35] *Parking Principles,* Special Report 125.

[36] P.C. Box, "The Curb Parking Effect," *Public Safety Systems* (January/February 1968).

[37] P.C. Box, "Streets Should Not Be Used as Parking Lots," *Congressional Record* 112, 187, 21 November 1966.

Figure 7-10. Congestion by curb parking. A single parked car (in the foreground) forced this condition of dangerous lane changing and congestion. On two different occasions, the residents along this route voted (by postcard questionnaire) two-to-one for parking prohibitions. Yet two different political administrations refused to authorize the regulations.

SOURCE: Paul C. Box and Associates, Inc.

parking configurations, land uses, street widths, and street classifications.[38] Variables found to be associated with accident rates include functional classification of the street, utilization of parking, and abutting land use. Increases in parking utilization (the annual number of space hours occupied per mile) result in increases in accident rates for up to approximately 1.5 million annual space-hours per mile. For greater parking utilization rates, the accident rate was not found to increase. Parking prohibitions installed on major streets with parking utilization rates of about one million annual space-hours per mile or more could be expected to reduce *midblock* accident rates by up to 75%.

For all streets, an increasing accident rate was generally associated with changes from single-family residential to multifamily residential land uses. In addition, there is an increasing accident rate associated with changes from multifamily residential to office land uses and from office to retail uses.

Angle parking

Arranging parking at an angle to the curb provides more parking per unit of curb length than does parallel parking. The number increases as the angle increases, until at 90° almost 2.5 times as many spaces are available when compared with parallel parking. However, the greater angles increase the roadway space needed for maneuvering in and out of parking spaces. Furthermore, several studies have found angle parking to be more hazardous than parallel. Analysis of state highways in Nebraska concluded that, whenever practical, parking should not be allowed on the urban sections of these routes, and whenever parking cannot be restricted it should be of parallel rather than angle type.[39]

Some officials have thought that angle parking could be allowed on wide streets without the increased congestion and accidents that occur with its use on narrower streets. However, a 45° angle parking along both sides of an 80-ft-wide street will affect the *entire* street width. The angle-parked vehicles occupy nearly 15 ft of space, and their back-out operation directly affects an additional 12 to 15 ft. This can cause lane-change accidents in the second lane away from the angle parking. These problems may be readily noted by observing typical angle-parking activities.

The principal hazard in angle parking is the lack of adequate visibility for the driver during the back-out maneuver. Additional hazard results from the driver who stops suddenly upon seeing a vehicle ahead in the process of backing out. Because empty parking stalls are difficult to perceive with angle parking, motorists who are seeking a place to park must either proceed slowly (thus tying up traffic) in order to see the empty stall, or slow abruptly when they come upon an empty space.

There are, of course, conditions where one or more blocks can be closed to through traffic and converted into parking malls. The necessary factors are that (1) the street is not required for the traffic network, (2) through traffic can be effectively prohibited, (3) the parking is so urgently needed at the location that it is more important than block circulation, and (4) the traffic is slow moving and generally destined to adjacent business activities.

Warrants for parking prohibitions

Parking prohibitions can theoretically be warranted under three conditions—statutory, capacity effect, and hazard. A national consensus is available in the statutory warrants of the Model Traffic Ordinance.[40] The Basic Edition of the ordinance authorizes full-time prohibitions on both sides of roadways not exceeding 20 ft in width and on one side of those not over 30 ft wide.

Capacity studies have found that typical streets with parking have only two-thirds the capacity of those with curb parking prohibited. The *effect* of curb parking appears at volume levels that are only a fraction of typical capacities. This effect varies with the number of lanes and whether the location is midblock or at an intersection. On a four-lane street having parking on both sides (thus allowing only one moving lane in each direction), a single vehicle waiting to turn left at an intersection or driveway completely blocks the through traffic on its own side. However, if the street has at least two moving lanes for each direction, in addition to parking, a higher per-lane volume can flow effectively.

The criteria recommended by one source, as a warranting condition for parking prohibition, are given in Table 7–8.

A *local* street having a width of less than 16 ft requires prohibition of parking at all times on *both* sides, even with one-way or occasional two-way traffic. A width range of 17 to 24 ft generally requires full-time prohibition on one side. When the two-way movement is more frequent, widths of less than 26 ft should have parking prohibited on *both* sides, whereas widths of 27 to 31 ft may need one-side prohibitions. Obviously, the need for prohibitions is related to parking demand as well as vehicular volume.

Methods of prohibition

Problems of capacity during certain hours are often remedied through a part-time prohibition of parking along the

TABLE 7–8
Parking Prohibition Criteria for Major Streets

Type of Prohibition	Maximum Vehicles per Hour per Lane When Parking Allowed (one direction of flow)	
	1 lane	2 or more lanes
Midblock prohibition for entire street	400	600
Intersection prohibition up to 150 ft on approach and departure	300	500

SOURCE: P.C. Box, "Criteria for Regulation of On-Street Parking and Curb Loading Zones," Tenth Pan American Highway Congress, Montevideo, Uruguay, 1967, Table II.

[38] K.B. Humphreys, P.C. Box, J.D. Wheeler, and T.D. Sullivan, "Safety Considerations in the Use of On-Street Parking," *Transportation Research Record 722,* Transportation Research Board, 1979.

[39] P.T. McCoy, M. Ramanujam, and M. Moussavi, "Safety Comparison of Types of Parking on Urban Streets in Nebraska," paper proposed for presentation to the January 1990 Transportation Research Board meeting.

[40] "Uniform Vehicle Code and Model Traffic Ordinance," National Committee on Uniform Laws and Ordinances, 1968.

affected street. Peak-hour parking restrictions are in widespread use throughout the country. The theory is simply one of providing extra traffic lanes during the hours when traffic demands are heaviest. This type of control is prevalent in most large cities during the weekday peak commuter hours and times of special events such as sports contests. In this way, the street may serve a dual function—as a major route during the peak traffic load periods and as a facility providing more abutting property access during the lighter traffic periods.

In general, the need for restricting parking at the curb also dictates prohibiting the standing of vehicles—at least on major streets. This applies as well to loading because a single vehicle stopped for any purpose during the peak periods blocks a lane of traffic.

Types of restrictions

Control of curb parking is accomplished through the adoption of various parking regulations, implemented by conspicuous signing, and supported by enforcement. Several different regulations are currently in use:

1. *No Parking*—This regulation is used along those portions of roadway where occasional stopped vehicles will not impede the safe and efficient flow of vehicles. It may be used throughout the full 24-hour period on the major roads, or only during the peak commuter hours, times of special events, or like periods. The no-parking regulation permits the stopping of vehicles for the purpose of loading or unloading persons or goods, as such vehicles are not considered to be parked during a loading operation.

2. *No Standing*—This regulation allows a driver to stop for passenger pickup or dropoff, but ordinarily does *not* allow prolonged unloading of merchandise from trucks. It is used where the curb space must be kept clear practically all the time during the effective limit of the regulation. In most cities, the no-parking regulation, when properly enforced, is almost equally effective.

3. *No Stopping or Standing*—This regulation is used along those portions of roadway where the presence of vehicles stopped at the curb during any or all hours would constitute a critical impediment to the safe and expeditious flow of traffic. Such locations include those near fire houses, in tunnels and on bridges, at railroad tracks, or along the approaches to a signalized intersection where capacity problems are extremely critical. This regulation restricts the stopping of any vehicle (passenger car, truck, or bus) at the curb for *any* purpose during the times of the restriction, except in obedience to an officer or traffic control device. A high enforcement level of this regulation is very difficult to achieve and the regulation is not readily understood by many motorists.

4. *No Parking (Loading) Zone*—Various short-term no-parking regulations are necessary in urban areas to reserve space for truck loading and unloading, bus stops, passenger and taxi zones. These are discussed in the section dealing with special purpose zones.

Statutory prohibitions

Certain common restrictions that are set forth in the Uniform Vehicle Code do not require posting of no-parking signs.[41] Many cities are able to take advantage of the regulations by direct adoption of pertinent sections or by referencing provisions of state law. The more important standard regulations from Section 11-1003 of the Basic Edition of the code prohibits parking in the following locations (wording not verbatim):

1. On a sidewalk;
2. In front of a public or private driveway;
3. Within an intersection;
4. Within 15 ft of a fire hydrant;
5. On a crosswalk;
6. Within 20 ft of a crosswalk at an intersection;
7. Within 30 ft on the approach to any flashing beacon, stop sign, or traffic control signal located at the side of a roadway;
8. Between a safety zone and the adjacent curb or within 30 ft of points on the curb immediately opposite the ends of a safety zone, unless the traffic authority indicates a different length by signs or markings;
9. Within 50 ft of the nearest rail of a railroad crossing;
10. Within 20 ft of the driveway entrance to any fire station and on the side of a street opposite the entrance to any fire station within 75 ft of said entrance (when properly posted);
11. Alongside or opposite any street excavation or obstruction when stopping, standing, or parking would obstruct traffic;
12. On the roadway side of any vehicle stopped or parked at the edge or curb of a street (double parking); and
13. On any bridge or other elevated structure on a highway or within a highway tunnel.

The typical restrictions are outlined in driver license manuals. However, because these regulations are not necessarily known by all drivers, it is common practice for municipalities to install signs (especially in problem areas) for items 6 through 10 above.

Special-purpose zones

Loading zones

Curb zones are needed to provide space for the loading and unloading of commercial vehicles when alley and off-street loading areas, frequency of loading and unloading operations, and general curb-parking conditions might otherwise result in truck double-parking.

Cities frequently post on the LOADING ZONE signs the hours and days of the week when effective. This frees the zone for off-hour use by other vehicles. Freight zones typically allow any user (truck, car, etc.) to load or unload. These zones should not be restricted to the place of business they may abut, because the concept is one of a *public* loading zone. When cities levy fees for establishment and maintenance of the zone, friction can develop between the payer and nearby neighbors.

Zones should have sufficient length to allow parallel truck access. This will depend on the length of the trucks that will

---

[41] Ibid.

use the space and the location within the block face. Usual zone lengths are 30 to 60 ft. Extensions of existing no-parking areas (such as at driveways, fire hydrants, or intersections) are preferred because these areas can be used for easier maneuvering. It is usually not necessary to have the zone at the exact point of loading access.

Proper enforcement is necessary to prevent curb loading zones from becoming "private" parking for store owners, managers, or employees. Violations can often better be handled by parking tickets rather than zone removal. Removing the zone may simply create a double-parking problem with trucks.

The frequency of truck use to warrant a zone is a matter of local policy. An average of one loading per day is sometimes used as rule of thumb, and the use by several businesses should always be planned. The movement and delivery of urban goods is a problem in most business areas that significantly increases traffic congestion. Frequent double-parked trucks are an indication of insufficient curb loading space, and major control efforts may be needed.

Taxi zones

Taxi operations play a necessary part in the overall transportation and traffic patterns in our larger cities. Most cities allow taxis to utilize exclusive curb areas at strategic points, although they require them to comply otherwise with general stopping and standing regulations.

Zones are usually limited by the number of cabs. A two- or three-cab zone is typical in a suburban area or at a business district hotel. Use is sometimes restricted to one taxi company, but this may not be desirable for several obvious reasons.

The change from "call box" to radio-control of cabs has greatly reduced the need for stands at busy intersections. Although taxi owners may strongly prefer such locations, it is good practice to locate the stands well back (150 to 200 ft) from signalized intersections.

When removal of curb parking meters is necessary for installation or enlargement of a taxi zone, some municipalities have charged the companies for the lost meter revenue (gross revenue less maintenance and collection charges).

Length needed for each stall is about 20 ft, plus 5 ft of added maneuver access at each end of the row, if not otherwise clearly accessible.

Bus zones

At bus stops, sufficient curb space must be provided for the buses to enter and leave the traffic stream, as well as to stand and load or unload passengers. The required length of parking prohibition ranges from 50 to 145 ft, depending on bus size, number of buses at one time, and location of stop (nearside of intersection, far-side, or midblock). The appropriate location for bus stops depends on the intersection geometrics, the traffic volumes, turning movements, bus-route turns, and other factors. Midblock locations have certain advantages.[42]

It is common practice also to allow use of bus zones by passenger cars to load and unload passengers, but *not* to stand in the zone. Truck standing, even when engaged in loading or unloading merchandise, is ordinarily not allowed, since they usually would occupy the zone for too long a period of time.

Passenger zones

Zones for picking up and dropping off passengers by private vehicles and taxis may be required at many places within a city. Theaters, hotels, stadiums, and schools represent the more typical locations where provisions are made. Such zones are like bus zones in that they do not allow general parking or even standing. The success of these zones is dependent on rapid use of as small a space as possible by many vehicles.

A single stall will suffice for most passenger zones if it is properly designed. A vehicle should be able to pull into and out of the zone without any backing up. This requires 50 ft for most drivers, if parking is allowed at both ends of the zone. At certain high-demand locations such as hotels and change-of-mode terminals, two or more stalls may be required. A length of about 25 ft should be provided for each added stall.

When not required for passenger use, the zones can be released for general parking, but the opportunity for this is rare.

Public agency

Many cities have posted regulations such as parking for police, sheriff, or public officials only. The legality of such signs may be questionable, and the public reaction may be adverse. A preferable place for the spaces is in parking lots, where the reserved parking is less obvious than at the curb.

Time limits

The primary reason for imposing time-limit restrictions on curb parking spaces is to provide an efficient use through turnover. Although compensating for a lack of larger parking facilities, high turnover can only be effective where parker needs in the area are short in time and when proper *enforcement* exists. A maximum use is thus made of each individual space. Time-limit restrictions placed along curb parking areas are also useful in discouraging employee, commuter, and other long-time parkers from usurping space in retail and business areas. Further, very short time limitations can effectively provide the required rapid turnover at busy (short business time) places such as banks, post offices, and passenger or merchandise loading areas.

Time-limit restrictions are conveyed to the parker by signs. ONE HOUR PARKING 7 AM TO 6 PM and 15 MINUTE PARKING are examples of those used. Signing alone requires that sufficient enforcement be available so that the parking durations of vehicles can be checked. Further, enforcement must be by notation of vehicle registration license, marking of a tire, or other means of identification on a periodic basis.

Often preferred in addition to signing is the use of parking meters. Meters give a definite measurement of time and an instant reading of remaining time or violation. They also can promote short-term parking by charging a relatively high fee and may quickly pay for themselves, including maintenance and coin-collection costs.

Methods are under development to semi-automate ticket-writing for time limit (or other parking restriction) violations. Portable devices offer the potential to transmit license

---

[42] "Location and Design of Bus Transfer Facilities," ITE.

numbers to a central facility, which could allow immediate identification of vehicles wanted for unpaid tickets, theft, etc.

Parking meters

The parking meter as a mechanical time-measuring device generally indicating the available time remaining for a parked vehicle was developed in 1935. Some meters do not indicate the available time remaining, while others indicate the time over-parked. In proper application, they can greatly simplify the problem of enforcing parking regulations and encourage parking turnover. A 1985/86 study of meter performance in Ann Arbor, Michigan, found that violations per parked vehicle exceeded 50%, while less than 6% were ticketed.[43] Despite this, the study concluded that most meters efficiently allocated the premium short-term curb parking.

Types and installations

Two general types of parking meters are used: the manual and the automatic. The manual type requires the parker to insert a coin and turn a handle, which winds the clock and actuates the meter for a time period determined by the coin inserted and the duration the meter allows. In the automatic parking meter, a coin is inserted and the time automatically registers for that coin. However, the clock mechanism of the automatic meter must be wound periodically by maintenance personnel. In practical use, the two meters are interchangeable, with the same time limits and choice of coins. Suggested standard specifications for manual meters have been published and are available.[44]

Parking meters may be installed at either curb or off-street locations. For curb locations, the meters are mounted on a pipe generally placed about 18 in back from the curb and about 2 ft from the front edge of the parking stall. In some instances, two meter heads are mounted atop a single post. This can be done effectively in curb locations with "paired" parking where one post (with two meter heads) serves the parking stalls immediately ahead and behind the meters (see Figure 7–11), or in off-street facilities where two parking spaces face each other across an island.

Vending machines are in use, which dispense tickets showing expiration times. These are then placed on the parked vehicle dashboards. While this system is used in the United States for certain municipal lots, it is reportedly also used in Europe for curb parking time-limit control, in lieu of individual meters at each stall. The advantages are less clutter, lower maintenance and collection costs; while the principal disadvantage is lack of convenience—the parker must walk to the nearest vending machine.

Collection security

In major cities, the number of parking meters installed in on-street and off-street locations numbers in the thousands. The amount of money involved in the parking meter program is also substantial. For this reason, the security of parking meter funds is important. This involves the coins in the meter before it is collected and also from the time it

[43] A. Adiv and W. Wang, "On-Street Parking Meter Behavior," *Transportation Quarterly,* Eno Foundation, July 1987.

[44] "Parking Meter Specifications," *Technical Notes,* ITE, October 1980.

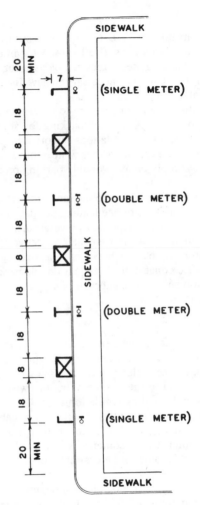

Figure 7–11. Example of paired parking meter layout.
SOURCE: *Parking Principles,* Special Report 125, Fig. 9.1, Highway Research Board, 1971.

is taken from the parking meter until it is deposited in the bank.

External security requires a parking meter with a good lock and a key that is difficult to duplicate. As no key is immune from duplication, no large municipal meter system should have all meters operated with the same lock-and-key combination. The lock should be designed so that it can be quickly and easily changed in the field to a different key combination whenever desired. This should be done particularly when a parking meter is stolen or a key disappears.

In parking meter revenue security, the coin-collection system is critical. The system should be designed so that coins go directly from the parking meter into the collection device without the collector having access to them. Several meter collection systems are available that provide a high degree of security. One uses a meter coin box that has a special top that can be inserted into a locked collection cart. The cart and the meter container have matching connections that release the money directly from the meter coin box into the collection cart. A similar system consists of a closed collection cart that connects by a flexible hose to a fitting on the meter, which releases the coins from the box directly into the cart. A third system has a long vacuum hose on a collection truck that connects directly to the parking meter collection box. A fourth system involves the use of two coin containers. The container in the meter containing the coins is replaced by a collector with a duplicate empty container. The locked containers that are removed with the coins are then carried to the collection point for emptying and counting.

New developments

Several innovative parking meters have been invented. In the spring of 1989, experimental installations were planned in 12 cities of one such unit. This meter is reported to operate on solar power, with 3-day storage capacity in full darkness.

Layout dimensions

Three types of stalls must be considered in dimensioning curb parking: end, interior, and "paired" parking stalls (see Figure 7-11). The end stall (because a vehicle can either be driven directly into or out of it) need only be long enough to accommodate a parked vehicle. A length of 18 ft is sufficient and often used today. Interior stalls must allow room for maneuvering, and a length of 21 to 22 ft is commonly used.

"Paired" parking has stall layouts so that two vehicles are parked bumper to bumper and the pairs of stalls are separated by maneuver areas. Stall lengths of 18 ft are used, with a well-defined marked maneuver area of 8 ft. The markings must be well-maintained.

The parking stalls should be defined by white lines extending perpendicular from the curb for 7 ft. The end stall line is generally marked with an L, while interior lines have a T shape.

A common mistake in layout is to crowd driveways and intersections too closely. In general, no stall should begin closer than 20 ft from the nearest sidewalk edge of *any* cross street. If the cross street is a major route, or the intersection control is a signal or four-way stop, the distance should be not less than 50 ft (100 to 150 ft is usually needed in such cases). These dimensions apply to both approaching and departing sides of the intersection.

Driveways should be cleared by a distance at least equal to the proper radius. This should be 15 ft from the point the driveway crosses the back edge of the sidewalk for most cases and no closer than 5 ft to the beginning and ending of the radius, if more than a 10-ft radius exists.

Truck facilities

Access and circulation

Driveways

In general, trucks use the same entrances to most sites as do employee vehicles and other traffic. The entrances and exits must be designed to accommodate the largest expected truck.[45] Additional vehicle tracking and off-tracking information is given in Chapter 6. If parking is allowed at the curb on the approach street, the vehicle path will be moved farther from the curb and result in a decreased entrance width and flare length. Adjustment of the property line lo-

cation will also change the entrance dimensions. Ease of turning into the site may be accomplished by use of "Y" or angle approaches. This may be particularly useful for access to and from a one-way street.

The minimum width of driveway required at gates is generally recommended at 16 ft for one-way operation, 28 ft for two-way operation, and 34 ft where pedestrian traffic is involved. If inbound trucks are stopped at the gate, it will be necessary to recess the gates so that sufficient storage space will be available for one truck, and preferably two, without backup into the access street.

Service roads

Service roads within the property should be at least 24 ft wide for two-way operation. Wherever practical, truck traffic should circulate counterclockwise, as the left turn is easier with large commercial vehicles because the driver's position is on the left side of the vehicle. Also, this places the truck in the most favorable position for backing into the dock. Parking should be prohibited where it may conflict with truck circulation or maneuvering.

A waiting or holding area for trucks is required next to the docks to accommodate trucks waiting for a dock space. The size of this area should be sufficient to provide space for the maximum number of trucks expected on the site, less the number of dock spaces provided.

Loading dock design

Type of expected vehicle

The type and size of truck is evidently the most critical factor in dock design. For suburban developments, the type of land use gives an indication of truck sizes requiring accommodation. In a CBD, the average truck size is likely to be smaller because of more constricted access. Table 7-9 gives the results of a Dallas study.[46]

Design dimensions

There are five major elements to consider in the design of a loading dock—all related to the size of truck.[47] Several of these are illustrated in Figure 7-12.

TABLE 7-9

Distribution of Delivery Vehicle Types, Dallas CBD

Vehicle Type	Percentage of Total Shipments Carried	Cumulative Percentage
Passenger car	18	18
Pickup truck	10	28
Van	27	55
Single-unit truck	40	95
Tractor-trailer truck	3	98
Other	2	100

SOURCE: D. CHRISTIANSEN; "Off-Street Truck Loading Facilities in Downtown Areas: Requirements and Design," *Transportation Research Record 668,* Transportation Research Board, 1978.

---

[45] *Guidelines for Driveway Design and Location,* ITE Recommended Practice.

[46] D. Christiansen, "Off-Street Truck Loading Facilities in Downtown Areas: Requirements and Design," *Transportation Research Record 668,* Transportation Research Board, 1978.

[47] C-J Chang, "Determination of Off-Street Truck Loading Space Requirements in Downtown Areas," *Compendium of Technical Papers,* ITE, 1985.

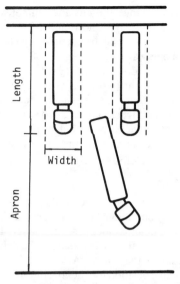

Figure 7–12. Physical design layout of loading spaces.
SOURCE: C-J Chang; "Determination of Off-Street Truck Loading Space Requirements in Downtown Areas," Compendium of Technical Papers, ITE, 1985.

Table 7–10 gives suggested dimensions for truck docks to serve the three general sizes of vehicles in use today. The vertical clearances used should be provided *throughout* the access route. If several different truck sizes are expected (this is usually the case), then different dock lengths may be provided, based upon the frequency of use by truck size. The width of loading space should consider needed clearances between trucks. Some are side-loading. In other cases, smaller delivery vehicles may, in peak periods, double-park behind trucks at the dock. Space for hand-cart movement between trucks is thus desirable.

Because vehicle design is not standardized, no single dock height can satisfactorily accommodate all vehicles. One approach is to provide several different dock heights, basing the design on the expected distribution of delivery vehicles. Another approach is to provide one continuous dock height to serve all vehicles, recognizing possible need to provide some type of adjustable dock-height equipment.

TABLE 7–10

Suggested Minimal Design Criteria for Off-Street Loading Spaces

Design Criterion	Size of Vehicle to Be Accommodated (ft)		
	Automobile, Pickup, Panel	Single-Unit Truck	Tractor-Trailer Truck
Vertical clearance	—[a]	13	14
Length	25	35	55
Width	11	12	12
Dock height	2–2.5	3–4.2	4–4.4

[a] Generally not a controlling design feature.

SOURCE: D. Christiansen, "Off-Street Truck Loading Facilities in Downtown Areas: Requirements and Design," *Transportation Research Record 668,* Transportation Research Board, 1978.

The total maneuver area (apron plus length of truck) required in front of docks depends on the overall length of trucks, the turning radii, direction of traffic circulation, and the width of berths. A maneuver length from the edge of the loading dock of not less than twice the overall length of the longest vehicle using the facility has been recommended.[48] Another recommendation is for a maneuver space of 105 ft with counterclockwise circulation and 165 ft with clockwise circulation to avoid blind right-hand backing maneuvers.[49]

Types of docks

Most docks built today place the face of the dock flush with the outside wall of the building. The flush-type dock can offer a covered, heated, closed dock operation without enclosing the trucks.

Enclosed docks completely enclose the truck when parked at the dock. The totally enclosed dock includes the maneuver area; trucks pull into the area, back into the dock space, and generally leave by another door. A second type of enclosed dock does not enclose the maneuver space, and trucks back through a door and straight into the loading dock. Care must be taken to eliminate the accumulation of vehicle exhausts in the enclosed area. However, enclosed docks have the advantage of protection of goods, control over pilferage, ability to erect crane systems or other overhead systems for loading and unloading of open trucks, and protection from weather. Modern industrial construction has often eliminated basements and dock-level buildings. Depressed approaches to docks can be used to create the elevation differential needed for the loading–unloading operation. These grades on the approach should not exceed 10%.

With a depressed approach, the top of the truck may contact the building wall before the truck bed contacts the bumpers if the dock is flush with the wall. To avoid this, the approach grade can be lessened, the building wall recessed, or the dock face extended. Another solution is the use of a level section 15 to 25 ft long immediately adjacent to the dock before beginning the approach grade.

Where sufficient maneuver space is not available for trucks to use docks parallel to the building wall, a sawtooth arrangement may be used. The number of berths that can be accommodated in a given dock length will be less with angular or sawtooth than with trucks at 90°. Berths at angles of 30° or less use twice as much dock space per berth as do those at 90°. Truck circulation must be one-way for backing into the berth. Wherever possible, circulation should be counterclockwise to avoid blind backing maneuvers.

Special-purpose vehicles

Buses

Typical school bus lengths range from about 20 ft (20 to 24 passengers) to about 35 ft (55 to 66 passengers). Their width

[48] "Parking Facilities for Industrial Plants," ITE, 1969.

[49] "Modern Dock Design," Kelly Company, Inc., Milwaukee, WI, 1968.

is 8 ft. Guidelines by AASHTO use a bus design vehicle 40 ft long (which fits the typical 47-passenger tour bus dimension).[50] This results in an outer off-trucking radius of 47 ft. Use of a similar dimension for school bus access is suggested.

Parking stall length should reflect the bus size to be stored. A stall width of 11 or 12 ft is appropriate.

While legal bus heights of 13.5 ft apply in most states, several allow 14 ft. These far exceed the height of most tour or transit buses, which is about 10 ft.

Light trucks

Full size, standard pickup trucks are about 18 ft long, with turning radii of about 24 ft. The widths are about 6.6 ft. Parking stall dimensions of 9 by 18 ft are suggested for design purposes. Full-size pickup trucks are readily parked in Class A stalls (see Table 7–4). They are frequently found in shopping centers—particularly in the West.

Recreational vehicles and vans

The AASHTO Design Motor Home measures 8 by 30 ft, has a 20-ft wheelbase, and an outer off-tracking radius of about 44 ft.[51] These dimensions greatly exceed those of the more common sport vans, which generally will fit within a Class A stall.

While some motor homes will be parked in public lots, such as at shopping centers, special areas set aside for their use are generally not warranted. Exceptions include roadside rest areas, amusement parks, and other vacation-oriented developments.

A survey reported in Rolla, Missouri, included both low and high turnover lots.[52] Pickups and vans totaled 24% in the low-turnover and 19% in the high-turnover lots. Large trucks, recreational vehicles, and motorcycles together totaled less than 1% in both types of lots.

Motorcycles, scooters, and mopeds

Typical motorcycles in use today have wheelbases of about 4.5 to 5 ft, with overall lengths of about 7.5 ft. While extended fork and heavy touring cycles may reach slightly over 8 ft, the use of an 8-ft stall length will accommodate the vast majority of motorcycles. Handlebar widths typically do not exceed 34 in, and cycles can readily be parked in 5-ft-wide stalls. A 10-ft-aisle width is ample.

Motor scooter wheelbases typically range from about 3.5 to 4 ft (although one model exceeds 5 ft). A 6-ft stall length will accommodate most scooters, as will 4-ft-wide stalls with 6-ft access aisles.

Mopeds and motorized bikes are roughly similar in size (about 6 ft overall lengths and 18 to 27-in width). A stall width of 3 to 4 ft is needed, with a length of 6 ft and an access aisle about 6 ft wide.

Lockers are available for secured and weather-protected moped parking. One two-unit lock is 4-ft wide and slightly over 6 ft long; however, it requires each vehicle to be backed in (on opposite sides), thus requiring access aisles on both sides of the locker. Another device is a vertical stand and heavy chain, arranged to allow recessing of the padlock so that it is not accessible to bolt-cutters.

Bicycles

Bicycles typically measure 5.5 to 6 ft in length, with handle bars spanning about 17 to 27 in. Parking racks, posts, and lockers of all types have been manufactured. For basic layout, a 2-ft-wide by 6-ft-long stall is appropriate, served by a 5-ft-wide aisle.[53]

Bicycles are quite vulnerable to theft; most owner-supplied locks, cables, or chains can be cut with bolt-cutters in a few seconds. Bicycle accessories are easily vandalized. The primary problem in bicycle parking is *security*. Using this measure, three classes of bicycle parking devices have been established:

Class I   Lockers or controlled-access areas where bicycles may be stored, and protected from theft, weather, and vandalism.
Class II   Devices that lock the bicycle frame and wheels, secured from theft of the unit. The individual may have to provide a padlock.
Class III Bicycle racks or fixed objects to which a bicycle may be secured by the individual's own locking device.

Conventional (Class III) devices do little to secure the frame and both wheels. They do not match the refined technology of the bicycle itself. State-of-the-art storage systems range from securing both wheels and the frame (Class II) to enclosing the bicycle completely within a locker (Class I). Controlled-access areas that provide protection from weather and vandalism also constitute a high-security, Class I facility. Strong, U-shaped locks supplied by the user are another effective means of securing bicycles, whether used in conjunction with a conventional rack or by attaching them to informal stationary objects. These locks are not as vulnerable to bolt-cutters as are conventional cables or chains.

The traditional bicycle "pipe" rack should not be considered as a Type III device for several reasons. These racks, by the nature of their design, may bend wheel rims and damage gearing mechanisms of the multispeed bicycles should the bikes fall over while chained to them.

The location of bicycle parking equipment on a site is as important a factor in determining security as the technology itself. An isolated location can lead to tampering or give the thief a chance to defeat the locks undetected. Parking should be near the building entrance and visually surveyed where possible. Combinations of equipment and location can enhance the security of bicycle parking. For example,

[50] "A Policy on Geometric Design of Highways and Streets," American Association of State Highway and Transportation Officials, 1984.

[51] Ibid.

[52] Charles E. Dare, "Consideration of Special Purpose Vehicles in Parking Lot Design," *ITE Journal* (May 1985).

[53] "The Denver Bicycle Parking Study," prepared for City of Denver by the Mountain Bicyclists' Association, November 1979.

Class II equipment located near an attendant in a parking garage is equivalent to Class I bicycle lockers. Problems relating to the use of bike parking may arise if bicyclists are unaware that parking is available or if they cannot find it at a given site. Publicity of location and simplicity of operation are important.

Time of use is another factor. Personal business and shopping visits have a lower security requirement, compared to all-day commuter bike parking or student parking that has nighttime use.

If bicycle parking is located in an automobile parking lot, physical barriers are needed to separate the parking areas. Even at very slow speeds a car can do extensive damage to a bicycle that is firmly attached to a fixed parking device.

REFERENCES FOR FURTHER READING

CHRISTIANSEN, D., "Off-Street Truck Loading Facilities in Downtown Areas: Requirements and Design," *Transportation Research Record,* 668, Transportation Research Board, 1978.

DARE, CHARLES E., "Consideration of Special Purpose Vehicles in Parking Lot Design," *ITE Journal* (May 1985).

Guidelines for Driveway Design and Location, ITE Recommended Practice, 1987.

"Guidelines for Parking Facility Location and Design," ITE Committee 5D-8, May 1990.

"Handicapped Parking Supply," ITE Committee 5D-8, *ITE Journal* (September 1988).

HINTERSTEINER, R.T., "Parking Control Guidelines for the Design of Parking Facility Portals," *ITE Journal* (January 1989).

KENIG, N.S., "Internal Circulation and Parking at Shopping Centers," ITE Committee 5-2, Compendium of Technical Papers, ITE, 1983.

Parking Principles, Special Report 125, Highway Research Board, 1971.

"Parking Requirements for Shopping Centers," Urban Land Institute, 1982.

"The Location and Design of Bus Transfer Facilities," ITE, 1989 update (pending).

WEANT, R.A., "Parking Garage Planning and Operation," Eno Foundation for Transportation, Inc., 1978.

WEANT, R.A. AND LEVINSON, H.S., "Parking," Eno Foundation for Transportation, Inc., 1990.

8

TRAFFIC SIGNS AND MARKINGS

ARCHIE C. BURNHAM, JR., *Engineering Consultant*

Introduction

Traffic engineers must use much more than the basic information contained in the *Manual on Uniform Traffic Control Devices for Streets and Highways.*[1] That manual (hereafter simply called the *Uniform Manual* or MUTCD) contains excellent material dedicated principally to descriptions of design requirements for the traffic control devices and the warranting conditions for their use. This chapter provides further information based on experience and practice. It does not pretend to be a substitute for the principles contained in the *Uniform Manual.* It does, however, outline basic concerns for the reader and prescribes ideas by recommendation and references to a wealth of information under specific categories of traffic control devices.

Traffic signs and markings are intricate traffic control devices that by simple display are used to regulate, warn, and guide traffic on all streets and highways. This chapter will address general information on both the need and use of the devices and will provide supplementary information to that found in the MUTCD for streets and highways. The *Uniform Manual* contains the warrants, standards, and design requirements for official signs and markings. The chapter is divided into sections that provide additional information and guideline for the following:

○ Sign Design and Other Factors
○ Pavement Markings
○ Miscellaneous Traffic Control Devices
○ Sign and Marking Warrants
○ Materials, Maintenance, Inventory, and Storage
○ Shop Operations

Background

International

In the interest of uniformity throughout the United States, all signs and markings should follow the accepted standards of the *Uniform Manual.* Some states have issued their own sign manuals, but most follow closely the *Uniform Manual* with additional applications and clarification specific to that given state. Other states have merely adopted the *Uniform Manual* by reference in its entirety. Local authorities should follow the standards adopted by their respective state for the design and application of traffic control devices (these standards are usually mandated by that state's law) for use on all public roadways. Practicing traffic engineers in other countries should follow their appropriate manual on traffic control devices such as those for the United Nations, Pan American countries, Canada, Australia, or European nations.[2-4]

---

[1] *Manual on Uniform Traffic Control Devices for Streets and Highways,* U.S. Department of Transportation, Federal Highway Administration, U.S. Government Printing Office, Washington, DC, 1978.

[2] S. Kirchner, "Traffic Signs and Markings in the German Democratic Republic," *Traffic Engineering and Control,* 12(6), incorporating International Road Safety Traffic Review (London) (October 1970), 316–318.

[3] *Manual of Uniform Traffic Control Devices for Canada,* Road and Transportation Association of Canada, Ottawa, Ontario, 1966.

[4] *Australian Standard Rules for the Design, Location, Erection and Use of Road Traffic Signs and Signals.* The Standards Association of Australia, Sydney, 1960.

The principal means of regulating, warning, or guiding traffic on streets and highways is through traffic control devices such as signs and markings. Because of the increasing competition for the driver's attention found within the highway environment, the need for well-designed and adequately maintained devices becomes paramount. Driver attention is directly proportional to the density of traffic, vehicle speed, complexity of maneuvering areas, and the relative operating environment in both urban and rural situations. Traffic signs and markings are most effective when they satisfy five fundamentals:

1. Fulfill a need
2. Command attention
3. Convey a clear and simple meaning
4. Command respect of the road users
5. Give adequate time for proper response.

The traffic engineer is usually the one designated to assure that the devices are effective and needed. Therefore, the following five basic requirements detailed in the *Uniform Manual* are to help the traffic engineer make the proper decisions so that traffic control devices are effective, understood, and satisfy the five fundamental purposes. These factors are:

1. *Design:* The combination of physical features such as size, color, and shape should command attention and convey a simple meaning. Consider the *Uniform Manual* criteria as governing guidelines with some flexibility for specific signs.

2. *Placement:* Devices should be installed so that they are within the cone of vision of the user to command attention and give adequate time for proper response. Vertical and lateral locations must be adjusted to fit the horizontal and vertical properties of the roadway.

3. *Operation:* The devices should be applied in a uniform and consistent manner so that they fulfill a need, command respect, and provide adequate time for response. Keep displays simple and assure they command attention by at least informally measuring their success.

4. *Maintenance:* The devices should be properly maintained to ensure legibility and visibility. In order to command respect and hold attention while meeting the needs of the users, the devices should be covered or removed when they are no longer appropriate.

5. *Uniformity:* This factor concerns the application of the same or similar devices in a consistent fashion for like situations so that they will fulfill the need of the user and command respect.

Traffic control devices must supplement each other in terms of providing a meaningful message to motorists. Traffic engineers should periodically make routine inspections, both day and night, to remove obsolete, contradictory, or unnecessary signs or roadway markings and to minimize any possible confusion. This is especially important in construction or maintenance work zone areas where conditions frequently change.

Signs and markings in the United States, Canada, Europe, and other parts of the world are displayed with a remarkable degree of uniformity. There has been close coordination so that with further adopting of new manuals, the standardization should improve (see Figure 8–1). One of the inherent problems, however, is with the revision of manuals between publications. The *Uniform Manual* of 1971 was not updated until 1978. Many significant revisions were adopted and practiced in the interim 7 years, but these were not incorporated until 1978. Likewise, a 10-year span passed after publication of the 1978 manual until various revisions, modifications, editorial changes, and other emendations were fully incorporated in the 1988 manual.

Each country does continue to display some devices that are unique to its own environment. For example, the United States, Canada, Germany, and Australia are the sole users of yellow markings for centerlines. Likewise, other countries have adopted specific signs or markings that are unique only to their jurisdiction. It should be noted, however, that, on the whole, a high percentage of standardization does exist in the Western nations and efforts should be continued to promote international conformity.

In the United States, signing tends to become nonstandard in certain types of environments more than in others. For example, as airports become larger and more complex, the signing at the airport is becoming more of a problem. It is complicated by signing for a multitude of activities that are not commonly used by infrequent airline users. The task of integrating airport signing is usually done by those who are not experienced in sign design and functional applications for the road user. Thus, the traffic engineer should exercise special care to assure airport signing does not create a problem that further contributes to airport users' confusion. Airport planning, design, and signing are covered in more detail in Chapter 12, "Surface Transportation Interface Areas," of the *Transportation Planning Handbook.*[5] Similar problems also exist in private developments such as shopping centers or office parks.

Federal safety standards

In the highway safety act of 1966, several standard areas were adopted that address traffic signs and markings.[6] These have become guidelines that have been incorporated into the operation of virtually every state and major metropolitan location within the United States. A general understanding of how signs and markings impact traffic operations and safety can be appreciated through a quick review of these guidelines. A summary follows:

1. *Number 9, Identification and Surveillance of Accident Locations:* Recommends the development of continuing programs to identify accident locations and methods for their

---

[5] *Transportation Planning Handbook,* Institute of Transportation Engineers, Washington, DC, 1991.

[6] U.S. Dept. of Transportation, "Highway Safety Program Standards," U.S. Department of Transportation, Federal Highway Administration, National Highway Safety Bureau, Washington, DC, 1974.

U.S.A. 1961	U.S.A. 1971	Canada 1966	United Nations Conference on Road Traffic Vienna, Austria 1968

Background – Yellow Legend and Border – Black	Background – White Legend and Border – Red	Background – Yellow Legend and Border – Black	Background – White Border – Red

		No Standard	
Background – White Legend and Border – Black	Background and Legend – White Circle – Red		Background – Red Bar – White

Background – White Legend and Border – Black	Background – White Legend Border and Arrow – Black Circle and Slash – Red	Background – White Legend Border and Arrow – Black Circle – Green	Background – White Legend Border and Arrow – Black Circle and Slash – Red

			No Standard
Background – White Legend and Border – Black	Background – White Legend and Border – Black	Background – White Legend and Border – Black	

Figure 8–1. Comparison of "old" and "new" U.S. signs with Canadian signs (1966) and those detailed for the United Nations Vienna Conference on Road Traffic (1968).

SOURCE: *Transportation and Traffic Engineering Handbook,* ITE, 4th Edition, 1976, p. 737.

correction. For example, signing, such as milepost, is needed as a means of locating and documenting the accidents in the field.

2. *Number 10, Traffic Records:* Recommends the collection and maintenance of records, including data on drivers, vehicles, accidents, and highways. The data pertaining to traffic control devices, such as signs and markings, are specifically listed under highways to provide an inventory to operate and maintain the system.

3. *Number 13, Traffic Control Devices:* Recommends the implementation of traffic control device improvements that bear directly on reducing accidents.

Identification and surveillance is assisted by an identification system in the field. The milepost sign has commonly served that purpose. Other identifying signs are utilized to accommodate the recording of active locations and control points for use in accident surveillance. Likewise, traffic signs are an important part of the traffic control devices necessary in the record-keeping aspects of standard number 10. Standard 13 calls for inventory analysis and accountability of traffic control devices including signs and markings. These are common elements of a professional traffic engineering program that can be met with a high degree of success when properly incorporated. Hamilton County, Tennessee, instituted an effective inventory application by using aerial photos in combination with computerized records to maintain a successful traffic sign inventory.[7]

[7] Stephen E. Meyer, "Traffic Sign Inventory—One County's Approach," *ITE Journal* (December 1985), 41–42.

Limitations and effectiveness

Frequent field review will ensure that traffic signs and roadway markings maintain their effectiveness. Routine maintenance and monitoring programs must be established to assure that damaged and defaced signs are replaced and those that have become dirty are cleaned. The frequency of the field review depends on many variables. Weather conditions such as snow and ice are important aspects that shorten the life of highway markings. High traffic volumes and especially high concentration of truck traffic also help reduce device longevity, as does vandalism, vehicle collision, and changeable traffic conditions. Material research is continually improving products to withstand these problems, but nothing substitutes for consistent follow-up inspections.

One of the key measures for the effectiveness of signs and markings can be found through the analysis of traffic accidents. Review of patterns developed over a period of time can often identify the need for change in devices to clarify the traffic operations. Many times traffic efficiencies can be improved when problems with traffic accidents or delays are identified and corrected through the display of additional signs or markings. These end products result from careful traffic engineering studies made to isolate primary causes of the accident or delay condition.

Objective, quantitative evaluation and analysis of existing traffic control devices often produce constructive means for corrective action. These result from items as elemental as those discovered from turning-movement counts. Sometimes they include complex analysis from extended observation of driver or pedestrian habits. For example, low-clearance signs are used on underpasses when the vertical dimensions measured will restrict truck movements. However, the combination of sharp sag vertical curves and long truck units requires a reduction in clearance dimensions because of the truck unit bridging across the sag curve. Damage to underpass structure, the truck unit, and hesitant movements in truck operations can be avoided with proper signing of vertical limitations.

Wherever data are collected regarding a traffic engineering problem, a potential solution involving change in traffic control devices may result. This change can include additional devices, but it may also include relocated or repositioned devices, or cleaning, repair or upgrading of size, reflectivity, intensity or mounting, or materials of existing devices. Also of importance is the necessity to remove traffic control devices when they are no longer applicable, as in situations where they may overload the driver's informational system or cause confusion.

Changes in traffic patterns or operations usually require a major change in necessary signs and markings. But other changes—such as additional regulation, increased enforcement, traffic study, or driver education—are also necessary. Once the need for adjustment and display of a certain sign or pavement marking is made, it is often desirable that the traffic engineer also identify other similar locations where that same condition exists. When patterns are evident, an effective public relations program can be developed to advise affected motorists. The information must convey to the public that a simple procedure has been adopted to alleviate a specific operational problem. The identified site should be used for relaying specifics, but it never should include actual names of people or incidents that could be critical either to individuals or classes of motorist.

Legal requirements

Under authority granted by Congress in 1966, the U.S. Secretary of Transportation has decreed that traffic control devices on all streets and highways in each state shall be in substantial conformance with the *Uniform Manual.* The MUTCD contains standards issued and approved by the Federal Highway Administrator to encompass both the design and warrants for the devices.

However, the decision to use a particular device at a specific location should be made on the basis of an engineering study of the location. Thus, while the MUTCD provides standards for design and warrants for the application of traffic control devices, it is not a substitute for engineering judgment. Therefore, it is the intent that the provisions of the MUTCD be the standards for the installation of traffic control devices, but not legal requirements for installation. The decision to use a specific traffic control device is based on an engineering judgment from a study of the problem location. Once the device is selected, it must conform to the provisions of the *Uniform Manual* relative to design, placement, and application. It should be noted that the *Uniform Manual* provides some flexibility in the selection of devices so that deviations from MUTCD requirements are permitted to satisfy specific field conditions. However, it is advisable to document the field conditions that support the engineering decision when major deviations occur. This will greatly assist in helping to avoid future litigation problems.

Qualified engineers are needed to exercise the engineering judgment inherent in the selection of traffic control devices. They are also needed to locate and design the roads and streets that the devices complement. Jurisdictions with responsibility for traffic control that do not have qualified engineers on their staffs should seek assistance from the state highway department. Other help is often available from their county, a nearby large city, or a traffic consultant. Even a periodic review by qualified persons or special consultation on a specific problem can eliminate significant operational or legal problems.

Human factors

Every traffic engineer becomes familiar with drivers' reactions and responses to factors in the operational environment. The proper use of this information is vital to success in traffic engineering solutions. Research has identified many information needs confirming that drivers must coordinate multiple actions to meet those needs,[8] which are present throughout the driving task and can be given priority.

The highest priority goes to those needs associated with the two main tasks of tracking and speed control of a vehicle. Both of these are sensitive to proper use of traffic signs and markings. Others, however, include avoiding an obstacle and maintaining the most safe and efficient course of travel in the

---

[8] G.F. King, and H. Lunenfeld, "Development of Information Requirements and Transmission Techniques for highway Users," *National Cooperative Highway Research Program, Report 123,* Highway Research Board, Washington, DC, 1971.

traffic stream. These needs relate to the effective deployment of signs and markings. Even the lowest priority information needs, which deal with trip preparation and direction finding, are helped with clear, consistent, and necessary informational signs and markings.

The traffic engineer builds the information system for the motorist. This individual must understand priority of information needs in order to select and display appropriate signs and markings effectively for the driver. For example, when drivers occupy a road section that will keep them busy with speed control or obstacle avoidance, efforts should be made to relocate directional signing to other noncompetitive road section locations. Directional or information signing should be installed in areas where simple steering and speed control maneuvers are the expected norm. By separating these anticipated points of conflict, drivers will not be overloaded with complex events that can render the traffic signs and markings ineffective. An example of this is found on major multilane urban arterials where the name of the next principal cross street (informational sign) is prominently displayed on an overhead support or located well in advance of the intersection (midblock location). This allows the motorist sufficient lead time to position the vehicle properly for any turning movement rather than to induce a sudden slowdown, congestion, or worse, as search is made for the critical information need. This effectively eliminates confusion at a critical traffic point, namely the middle of the intersection.

Professionals knowledgeable about the human factor have applied their expertise and understanding of driver characteristics to improve traffic control device application. They have also used this knowledge to analyze the field conditions of the roadway and traffic control devices to determine improvements that should be made. The process of applying traffic control devices in a more effective display for the driver has been referred to as "positive guidance." Some of the initial work in the highway areas was used to diagnose operational problems and recommend improvements where potential hazards exist. The Positive Guidance Process recommends a systematic analysis of the roadway, driver actions, and items that cause driver difficulty in understanding the driving tasks or that impact driver reactions (see Figure 8–2). The schematic of the Positive Guidance Process

The Functions and Activities for Developing Positive Guidance Information

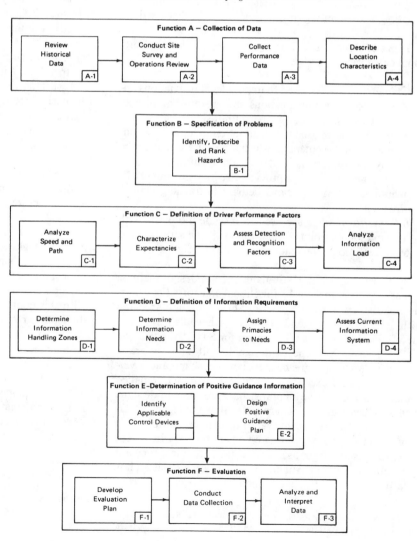

Figure 8–2. The Positive Guidance Process.
SOURCE: *A Users Guide to Positive Guidance,* FHWA, June 1977.

outlines the general considerations to be used in applying this analysis tool.

Human factors take on a magnified importance when mixed with the physical limitations of age in this changing society of transportation. Chapter 1 of this handbook addresses road user characteristics in more detail. The Transportation Research Board *Special Report 218* provides a good summary of the aging problem by noting that the driving population is changing. By the year 2020, it is predicted that 17% of the driving population will be 65 years of age or older, compared with only 5% in the same category 100 years earlier.[9] These older drivers have vision limitations and slower reaction times that mandate larger and/or improved placement of traffic control devices.

Traffic sign design and other factors

Traffic sign fundamentals

The following section will address the classification, types, and general specifications applicable to all signs. Some specific areas of concern regarding sign design, shape, color, symbol, message, and size, as well as reflectorization or illumination, and mounting configurations will be discussed.

Types of signs. Traffic signs are classified in four functional categories:

1. *Regulatory signs* are used to impose legal restrictions applicable to particular locations and are not enforceable without such signs, such as speed limit signs.
2. *Warning signs* are used to call attention to hazardous conditions, actual or potential, which otherwise would not be readily apparent, such as intersection warning signs.
3. *Guide signs* are used to provide directions to motorists, including route designations, detours, or traffic-generator directions.
4. *Informational signs* are used to provide motorists with information on points of interest, available services, and other geographic, recreational, or cultural sights.

The *Uniform Manual* stipulates certain design features that are inherent to signs by functional category. For example, regulatory signs are generally of a rectangular shape with red or black legend. However, they also are of a specific design such as the octagon for STOP, the triangle for YIELD, or X shape to designate rail-highway grade crossings. Warning signs have a diamond shape and a yellow background, except in construction or maintenance work zones, which have an orange background. Guide signs generally have a green background with white legend, as do informational signs, except that some can have a white or a brown background depending on where they are installed. Backgrounds and legends are reflectorized unless they are applicable only in daylight hours. As a practical matter, it is usually more cost-effective to mandate all backgrounds to

be reflectorized because of exchange, fabrication, warehousing, and other administrative processes.

General specifications. The fundamental qualities of a traffic sign are its ability to have an attention value, be legible, and obtain recognition. These terms are as follows:

1. *Attention value:* This characteristic of a sign demands attention by (a) *target value* where the quality makes a sign or group of signs stand out from the background or (b) the *priority value* where the quality makes it possible for a sign to be read first in preference to other existing signs. This is accomplished by sign shape, color, size, and placement.
2. *Legibility:* The characteristic of a sign that allows it to be read and understood because: (a) Pure legibility provides a distance in which a sign can be read in an unlimited time. In 1939, Forbes and Holmes[10] determined when a series "D" letter height is maintained, drivers with 20/20 vision will read and comprehend the legend at 50 ft for every 1 in of letter height. This remained the standard for 50 years, but recently Mace[11] found that by altering the letter height ratio to 1 in equals 30 ft, drivers of lesser visual acuity up to 20/40 could also read and understand sign legends at comparable positions in advance of the sign placement. Mace constructed a sign dictionary used in a model to determine Minimum Required Visibility Distance (MRVD). This standard is proposed as minimum distances for the sign to be read for proper driver reaction. (b) Glance legibility of a sign provides a distance in which a sign can be read at a glance (usually from 0.5 to 1.4 sec within a visual cone of approximately 5 ft by 100 ft). This vision cone is based on sign placement within the horizontal and vertical allowances set by the MUTCD. Sign placement outside drivers' cone of vision is often the explanation when drivers overrun a stop sign positioned on the shoulder behind a large radius intersection.
3. *Recognition:* This attribute of signs being quickly recognized and understood is due to the utilization of standard colors, shapes, and legends. The reading time per line of legend is a factor in establishing standards set in the MUTCD.

Other factors are noteworthy. For example, signs located over the highway are more likely to be seen before those located on either side of the highway. Additionally, the most important features involve brightness contrast of letters to the sign and the sign to the background. Improper balance with this contrast causes sign washout, making the legend difficult to read.

Sign design. Properties of a sign design include its shape, color, symbol, or message. These and other fundamental properties are reflected in the *Uniform Manual,* and adherence to them tends to provide the motorist consistency in sign information. This is helpful whether sign placement occurs in various jurisdictions of the United States or in other cooperating countries including Canada, Mexico, and

[9] "Transportation in an Aging Society," Transportation Research Board, Special Report 218, Washington, DC, October 1988.

[10] T.W. Forbes, and R.S. Holmes, "Legibility Distance of Highway Destination Signs in Relation to Letter Height, Width, and Reflectorization," *HRB Procedures,* 19 (1939), 321–326.

[11] Douglas J. Mace, "Sign Legibility and Conspicuity," *Transportation in an Aging Society, TRB Special Report 218,* pp. 270–293.

the European nations. The following is a brief outline of some specifics contained within those standards:

Shapes. There are nine standard sign shapes as illustrated in Figure 8–3.

1. Octagon: Exclusive for stop signs.
2. Equilateral triangle: Exclusive for yield signs in most countries.
3. Round: Exclusive for advance warning of a railroad crossing.
4. Pennant: Exclusive to warn of passing restrictions.
5. Diamond: Used to warn of existing or possible hazard.
6. Rectangle, long side vertical: Used for regulatory signs.
7. Rectangle, long side horizontal: Used for guide and informational signs.
8. Trapezoid: Used for recreational area signs.
9. Pentagon: Used for advance school and school crossing signs.

Figure 8–3. Standard shapes.
SOURCE: *Traffic Control Devices Handbook,* FHWA, 1983, pp. 2–7.

 The **diamond** shape is used for the majority of warning signs.

 The **rectangle,** with the longer dimension vertical, is used for the majority of the regulatory signs and some warning signs.

 The **rectangle,** with longer dimension horizontal, is used for the majority of guide signs and some warning signs.

 The **pentagon** with point up is used only for the School and School Crossing signs.

The **pennant** shape, with the longer dimension horizontal, is used only for the No-Passing Zone warning sign.

The **trapezoidal** shape is used for recreational guide signs only.

The **octagon** is used only for the STOP sign.

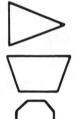  The **equilateral** triangle with one point down is used only for the YIELD sign. A triangle with the point up is used for warning signs in some European countries.

 The **round** shape is used for the Railroad Advance warning sign and for Civil Defense Evacuation route signing. The round shape is also used for prohibitory signs under the United Nations Signing Standards.

Colors. There are currently eight designated standard sign colors.

1. *Red:* Used as background color for STOP, DO NOT ENTER, WRONG WAY, and supplemental plates (4-Way STOP). Used as legend color for parking prohibition signs, route markers, and diagonal bars for prohibitory symbols. Also used for both border and message on YIELD signs.
2. *Black:* Used as background color for ONE WAY, WEIGH STATION, and NIGHT SPEED signs. Used as a legend color for white, yellow, and orange background signs.
3. *White:* Used as a background color for route markers, guide signs, and regulatory signs. Used for legend on brown, green, blue, black, and red signs.
4. *Orange:* Used as a background color for construction and maintenance signs.
5. *Yellow:* Used as a background color for all warning signs except for construction and maintenance signs.
6. *Brown:* Used as a background color for guide and informational signs related to recreational, scenic, or cultural interests.
7. *Green:* Used as a background color for milepost and guide signs except where brown or white are specified. Used as a legend color on white background for certain directional signs and permissive parking regulations.
8. *Blue:* Used as a background color for information and supplemental signs such as motorists services, rest areas, and evacuation route markers.
9. *Other:* Purple, light blue, coral, and strong yellow-green have been reserved for future use in designated sign categories.

The above designations are from the U.S. Standard Color Codes.[12] Although changes have not yet been made in chromaticity specification for highway colors, a recent study has indicated that yellow, green, and blue colors should be revised to improve legibility, contrast, and compliance with ASNI Z53.1 (1979) Color Codes.[13]

Standard symbols. There are five specific symbol shapes that have been standardized and adopted for signing applications.

1. *Circle:* Green if message is permissive. Red with diagonal bar if message is restrictive.
2. *Plus:* Used to indicate two intersecting roadways. It may be modified into a T or acute angle to represent various configurations for the number of legs of approach to the intersection.
3. *X:* Used exclusively for railroad crossings symbolizing the railroad "crossbuck" sign.
4. *Arrow:* Used with other symbols such as octagon or triangle to indicate STOP or YIELD AHEAD. Also used on

---

[12] National Institute of Standards, "Evaluation of Colors for Use on Traffic Control Devices," U.S. Department of Commerce, NISTIR 88-3894, Gaithersburg, MD.

[13] Belinda L. Collins, "Evaluation of Colors for Use on Traffic Control Devices," *Research Report,* Federal Highway Administration, McLean, VA.

guide signs to provide diagrammatic information or to be used as stand-alone in various construction, maintenance, and utility operations. The arrow is pointed upward to indicate ahead or this direction. It is pointed downward to designate this lane.

5. *Customized:* Symbols represent a message such as horse, truck on steep grade, pedestrians, etc.

In recent years, symbols have been used to improve the ability of traffic signs to communicate their messages. Several studies have concluded that symbolic signs are superior to alphabetic signs because they improve driver comprehension. However, less is known about the magnitude of this superiority in terms of legibility distance. Signs that seem most effective in symbol vs. alphabet include divided highway, school zone, narrow bridges, two-way traffic, signal ahead, yield ahead, and stop ahead. These have been compared in both sign and sign plus environment conditions to establish

their reliability. A symbol study by Paniati[14] found that the legibility distance for symbolic signs is about 2.8 times that of the equivalent alphabetic signs. Figure 8–4 illustrates the relative legibility distance of various symbols. The data also indicated a significant difference in legibility distance for older drivers and that bold symbols of simple design provide the best legibility distance for all age groups.[15]

Standard messages. Sign size and letter height have been designated in the *Uniform Manual*. The *Uniform Manual* categorically addresses all regulatory and warning signs by establishing criteria for letter series and spacing for highway signs. The U.S. Department of Transportation,

[14] J.F. Paniati, "Legibility and Comprehension of Traffic Sign Symbols," *Research Report*, Federal Highway Administration, McLean, VA.

[15] George Dale, Brant Williams, and Eugene Wilson, "Roadway Symbol Sign Evaluation," *ITE Journal* (January 1985), 29–32.

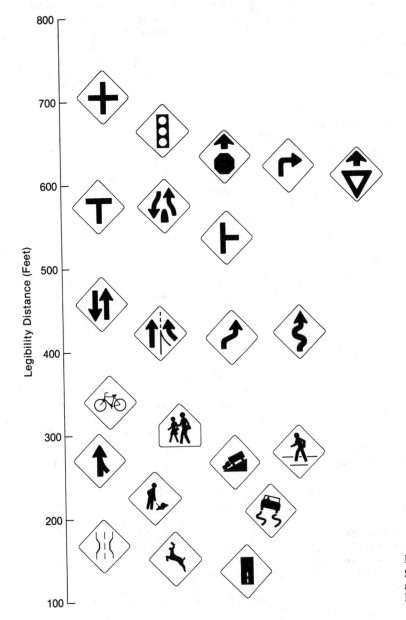

Figure 8–4. Symbol sign legibility distance. SOURCE: Jefferey F. Paniati, "Legibility and Comprehension of Traffic Sign Symbols," FHWA.

Federal Highway Administration (FHWA) publishes a manual of standard highway signs and pavement markings. This publication, titled "Standard Highway Signs," is available from the Government Printing Office, Washington, DC. The manual stipulates the various widths, heights, and strokes of letters in series from B through F.[16] As a general rule of thumb, in letter series E, the spacing requirements can be calculated by multiplying the number of letters in the message by the letter height.

Guide signs, however, usually have minimum legend sizes established to provide the motorist sufficient opportunity to read at normal approach speeds. In selecting the proper letter size, seven factors should be taken into account. These are:

1. Speed of approach vehicle.
2. Location of the sign.
3. Height, stroke, width and type of letters.
4. Illumination and reflectorization available.
5. Minimum required driver legibility distance.
6. Amount of sign copy.
7. Opportunity to use duplicative or repetitive signing.

Conventional highway signs do not normally require as large a message as is necessary for the freeway and expressway environment. When letter heights exceed 8 in, place names on guide signs should be developed in lowercase letters with initial letters capitalized. Commonplace names in uppercase and lowercase letters form a more recognizable word pattern for the traveling public. The initial uppercase letter is usually 1 1/3 times the loop height of the lowercase letters. Some jurisdictions provide increased emphasis by designating the initial uppercase letter with the minimum series E letter. Computer programs calculating horizontal letter spacing for Series B, C, D, E, and E-modified have been developed. One system developed by the North Carolina Dept. of Transportation is named SIGNSPAC. SIGNSPAC is on MS-DOS 2.1 and can be utilized with Lotus 1,2,3 and Procedure Sign Layout Sheets. Other computer programs are available to assist in the layout design of signs.[17]

Minimum size lettering on freeway guide signs should be 8 in. When minimum message heights are mandated, emphasis can be improved by increasing either all legend in uppercase letters or spreading the spacing between the letters. However, on interstate facilities, it is preferable for the names of all places, streets, and highways be designed in lowercase letters with initial caps. Major guide signs in advance of interchanges and in overhead display should be a minimum of 16-in caps, 12-in lowercase. These can be increased to 20-in caps with 16-in lowercase on heavy use, multilane controlled-access facilities. Specific recommendations for numerals and letters in accordance with interchange classification, type of sign, and the quantity of sign legend are specified in the *Uniform Manual.*

Guide signs often include diagrammatic symbols that show arrows that approximate the roadway geometrics and

are supplemented with legend in a clear and simple manner to provide glance-understandable message (see Figure 8–5). Guidelines for designing diagrammatic signs may be found in the FHWA publication on standard signs mentioned above.

The selection of lettering size for any sign must evaluate the needs of the user. Because of changes in automotive technology, the roadway system, and the population itself, the needs of the user are continually changing. By the year 2020, 17% of the population will be over 65 and have limited visual acuity. Based on physical attributes of the aged today, current use of letter height standards corresponds to the visual acuity of 20/25, which exceeds the visual ability of nearly 40% of the drivers between age 65 and 74. Thus, it behooves the traffic engineer in selecting letter height for a particular signing need to consider its effectiveness and the possible need for larger letter size than necessary for the average driver.

Standard sign sizes. The *Uniform Manual* establishes certain minimum and standard size for regulatory, warning, and for other signs such as route markers, logos or TODS (see discussion later in chapter). Using signs of larger size is dependent on the facility requirements for the signing need.

Figure 8–5. EXIT ONLY on left with diagrammatic (left-hand interchange lane drop).
SOURCE: *Manual on Uniform Traffic Control Devices,* 1978, p. 2F–25.

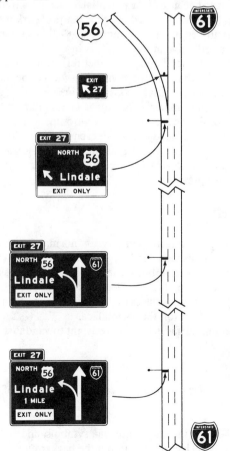

[16]"Standard Alphabets for Highway Signs and Pavement Markings," U.S. D.O.T., FHWA, Office of Traffic Operations, Washington, DC, 1977, pp. 5–41.

[17]R.L. Bleyl, and H.B. Boutwell, "A Computer Program for Guide Sign Design & Drafting," *Traffic Engineering* (Institute of Traffic Engineers, Washington, DC), 38 (March 1968), 22–26.

For example, high-speed roadways or those involving impaired visibility, multiple travel lanes, or located in heavy traffic areas all demand consideration for enlarging the sign and its message. For conventional highways, it should be noted that minimum legend size is currently under consideration for increases because of the aging driver, but for now the following sign sizes are standard in the *Uniform Manual*:

1. Stop: 30 in by 30 in
2. Yield: 36 in
3. Restricted Turn (Symbol): 24 in by 24 in
4. Do Not Enter: 30 in by 30 in
5. Parking signs: 12 in by 18 in
6. Regulatory signs: 24 in by 30 in
7. Guide signs: Minimum letter height of 4 in on low-speed urban and minor routes; 8 in or more on expressways and freeways or other limited-access roads.

Some jurisdictions on less important rural roads and low-speed conditions have reduced letter size in some categories, but efforts should be made to avoid this practice. Conversely, especially in limited-access environments, overhead legend should be greatly increased in size. The size is dependent on the types of mounting, exit, information, and message that must be communicated. Also, considerations for larger legend size may be necessary as driver population continues to age.

Reflectorization. Most signs require legibility at night because of the nature of their message. Even school signs are reflectorized because of the potential for school activities and meetings to be conducted during the hours of darkness. Thus any sign with regulatory, warning, or directional information must be as effective at night as it is in the day. This can be accomplished when the sign is either reflectorized or illuminated in order that nighttime visibility will not be impaired. Cost-effectiveness must be considered in selection of the type of reflectorizing material but other factors impact as well. Whether positioned in the road environment, mounted on the shoulder, or placed overhead, each placement provides a problem of angularity that must be satisfied. Regardless of the type of reflectorization used, the following general conditions must prevail:

General

o Sufficient retroreflectivity for normal approach distance to provide good target value and legibility.
o No glare that blots out the legibility of the message.
o At least 75% of retroreflectivity in rain or fog.
o Self-cleaning surfaces to discourage the heavy accumulation of dirt on the legend face.
o Durable surfaces that are resistant to vandalism.

Types

Sign reflectorization can be accomplished in any of the following ways:

o When the message, lettering, symbols, and border are totally reflectorized but not the background.

o When only the background is reflectorized and the message is opaque.
o Reflectorizing both the sign background and the message.

Special consideration should be taken when only the message is reflectorized. The lettering or symbol stroke must be sufficiently wide and the reflectorization must contain a high luminance so that adequate target value is achieved. Conversely, where both background and letters or symbols are reflectorized, there should be a brightness contrast so the message has a higher retroreflectivity value than the background material. Otherwise, a halo effect can occur from the bright background, partially obscuring the legend.

Methods

There are three methods of obtaining reflectivity. These are through reflector buttons, microprism sheeting, and spherical reflective sheetings. All three of these have the ability to reflect the incident light directly back toward its source.

o *Reflector buttons.* These are small reflector units set within a cutout form that depict a symbol or series of letters on the face of the sign. The prismatic reflectors are exposed on the back face of the button, which are encapsulated in a clear molding.
o *Microprism sheeting.* This is material containing many units of minute cornered cubed prisms per square inch that are embossed into a clear and durable sheet material. This reflective material is then cut in shapes to form specific symbols or letters or may be used to cover an entire background for selected signs.
o *Spherical reflective sheetings.* These materials are flexible adhesive-coated sheeting with various small glass spheres embedded in a plastic surface. Each glass sphere acts as a miniature reflector to direct the light back to the light source. See Figure 8–6.

Effectiveness measurement

The effectiveness of a reflective material is related to its quality and is measured by five specific parameters. These are:

o The amount of incident light redirected from the car head lamps back to the driver under normal conditions.
o The maximum entrance angle through which effective reflection is possible.
o Resistance to weather deterioration and vandalism.
o The ease of application and replacement.
o The cost per year of useful life.

Reflectorization angles are critical in measuring effectiveness of retroreflective material. Two specific angles that should be considered are the *observation* angle and *entrance* angle (see Figure 8–7).

Observation angle

The angle between the vehicle headlights and the observer's eye is called the *observation* angle. The driver's eye height

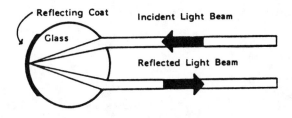

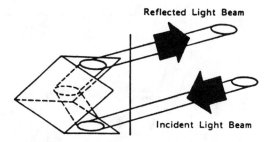

Figure 8–6B. Principles of cube-corner (prismatic) retroreflection.
SOURCE: *A Guide for Retroreflectivity of Roadway Signs for Adequate Visibility,* FHWA, 1987, p. 3.

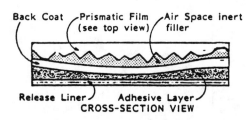

Figure 8–6A. Cross-section and top view of microprism sheeting.

above the vehicle headlights can range from 21 in for small cars to 64 in for large trucks. A wide observation angle is anything over 2°.

Entrance angle

The *entrance* angle is the angle between the vehicle path (headlights) and perpendicular to the face of the sign, which is either mounted in the ground or installed overhead. Note that the angle changes with the distance between the vehicle and the sign and is a function of sign location and vehicle placement. An entrance angle of 30° is considered wide for highway signing. Reflective materials must have a wide-angle response so that the light can be reflected even when the approaching headlights are substantially offset from being in front of the sign (see Figure 8–8). Standard reflective sheeting provides retroreflection for entrance angles up to 30° but wide-angle

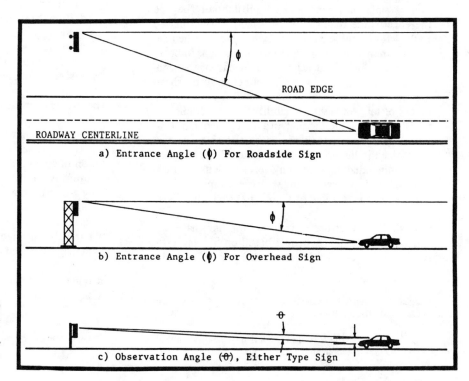

Figure 8–7. Illustration of entrance (∅) and observation angle (-θ-) under actual highway conditions.
SOURCE: *A Guide for Retroreflectivity of Roadway Signs for Adequate Visibility,* FHWA, 1987, p. 5.

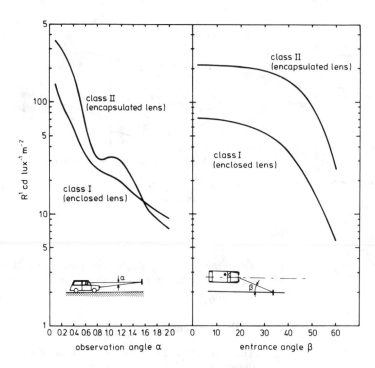

Figure 8–8(A) & (B). Typical performance data for white retroreflecting sheeting.
Note: The Relative Retroreflectivity will vary dependent on color and type of retroreflectivity material.
SOURCE: "Road Signs," International Commission on Illumination (CIE), Publ. 76, 1988, p. 58.

materials are necessary to give good reflectivity up to 40°. This is important for at least the following four reasons:

o Good visibility can be secured over a long range of approach distances including close viewing distance.
o Sign placements do not have to be so exact that their inaccuracies will nullify the effect of the message.
o Temporary damage to a supporting structure will not make the sign so out of position that it will not be legible.
o The sign will be functional in a multilane environment even when it has been located substantially beyond the edge of the roadway or be effective when placed to the left of a vehicle traveled path. Note that the relative luminance or retroreflectivity of the signs improve when the vehicle is at least 150 ft ahead of the sign. Reflectorized signs may not be effective in overhead locations unless there is at least 1,200 ft advance distances of tangent sight availability.[18] Some studies by CALTRANS have found that, in a lights-out demonstration project, overhead signs can be effective when reflectorized in lieu of being illuminated.[19] Illumination is recommended, however, on overhead signs for interstate, urban expressways, and wherever a rapid decision and response is required of a motorist (see Figure 8–9).

Recent studies on retroreflectivity of stop signs have recommended a replacement schedule based on SIA (specific intensity per unit area). These measurements of SIA can be

taken in the field in relation to various approach speeds.[20] Further studies currently underway at the FHWA and the TRB may outline more definitive retroreflectivity requirements to be considered for future adoption as standards.

Illumination. Illumination is considered a feasible application in traffic sign design whenever one or more of the following factors exist:

o The sign background is not reflectorized.
o The reflectorized sign is no longer effective.
o Additional target value is needed for a variable-message sign.
o Additional sign conspicuity is necessary for diagrammatic, lane control, or other overhead directional sign message techniques.

The two basic means of providing illumination are:

o Direct lighting where the light shines upon the surface of the sign itself so that it is illuminated from the direction of approach. It is necessary to aim the luminaire carefully to avoid hot spots and to provide uniform light distribution on the surface to be lighted.
o Indirect lighting where the light is produced by illuminating incandescent lamps or fluorescent tubes behind a translucent sign background upon which the sign message has been displayed.

In general, whether using direct or indirect lighting, care must be exercised to assure that the sign receives sufficient

[18] N. Bryan, D. Casner, R. Klotz and H. Knisley, "A Limited Evaluation of Reflective & Non-Reflective Background for Overhead Signs," PA D.O.T., *Research Report,* Harrisburg, PA, September 1978.

[19] CALTRANS, "Lights Out Overhead Guidesign Demonstration Project," *Report CA-TE-89-1,* Sacramento, CA.

[20] Juan M. Morales, "Retroreflective Requirements for Traffic Signs: A Stop Sign Case Study," *ITE Journal* (November 1987), 25–32.

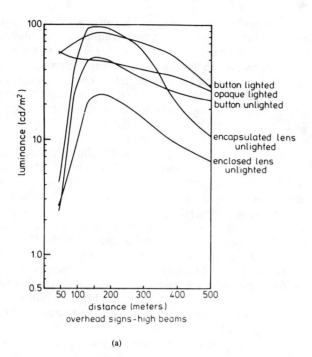

(a)

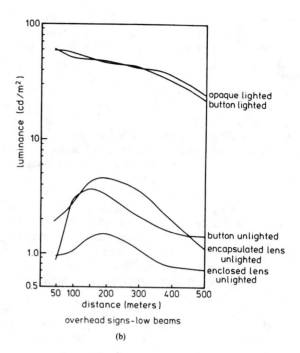

(b)

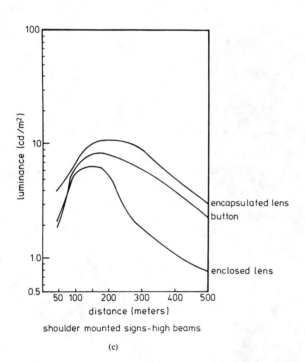

(c)

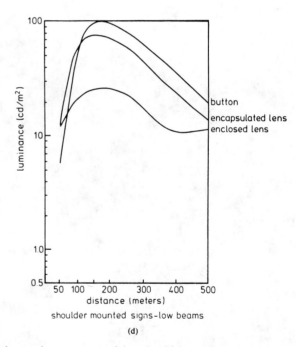

(d)

Figure 8–9. Typical sign luminance relative to sign type, material, and vehicle headlights.
SOURCE: "Road Signs," International Commission on Illumination (CIE), Publ. 76, 1988, pp. 94, 95.

illumination levels so that it is legible. Too much direct lighting can cause a sign message to wash out so that it is no longer legible. Indirect lighting must be uniformly distributed so that the message does not contain dark spots and become illegible.

Direct lighting can usually be installed so the lighting source is located above or below the message in such a way that it does not cast shadows on the face of the sign in daylight. It must still provide uniform distribution of light over the entire sign face. Special luminaires are devised for this purpose. This has been accomplished with incandescent, sodium, mercury vapor, and fluorescent units mounted in back of reflectors to assure uniform distribution of light over the entire sign face. Indirect lighting is generally accomplished by sectioning the traffic sign into panels and assuring that each panel is uniformly illuminated. An obvious disadvantage of this type sign is a conspicuous dark area due to light unit failure. This condition applies to external lighting as well, even though it may not be as noticeable in some cases.

Special signing. Signing is often employed in special situations to obtain a very specific effect. The three principal special signing situations are discussed below.

Active and passive signing. A highway sign is a passive device in that it displays a message constantly even though the message may not be as valid on every occasion that it is viewed. However, in a modified form of sign illumination, a passive sign can be placed in an active status with the attachment of a flashing light beacon. This will improve the target value and conspicuity for some critical-type signs. Red flashing lights have long been standardized with railroad crossings. Also, a red flashing beacon can be used in conjunction with a stop sign. Yellow flashing lights are used in connection with various types of warning signs. Some jurisdictions have had good success with the use of alternately flashing amber lights. These are displayed on school crossing signs and exercised actively only during the time of school crossings (see Figure 8–10). Two recent studies in Arizona, Phoenix, and Tucson indicated that the flashing school or pedestrian sign had no effect on vehicle speeds. Other common applications of this device include the use of "vehicle approaching when flashing" sign, and "too fast for curve when flashing" sign. There are other applications including factory or fire station entrances where there will be temporary problems.

The flashing beacon has further application in construction and maintenance signing areas when used in various

Figure 8–10. Phoenix flasher experiments
SOURCE: "Pedestrian Warning Flashers in an Urban Environment: Do They Help?" James W. Sparks and Michael J. Cynecké, *ITE Journal,* (January 1980), p. 32.

formats. Steady burning lights are often displayed to outline a particular traffic path through the construction area. Many times hazards can be clarified and improved with warning delineation by the incorporation of flasher units mounted on signs, barrels, barricades, or temporary restraining walls. Flashers are often used to draw attention to special conditions in construction areas, such as nonstandard overhead clearances, narrow lanes, shoulder drop-off, and other temporary conditions.

Signing is also used in an active sense when it is combined with other devices. Some of these are flags, flashers, or traffic control signals for identification, guidance, or informational purposes enhancing the traffic control system. For example, freeway ramp metering is dependent on signs, signals, and markings. The ITE has developed a recommended practice for ramp metering display that effectively combines the applications of signs and markings with signal display. The metering system can release one vehicle at a time or permit a second vehicle to proceed after the first is released on to the freeway. In either case, it is implemented with appropriate signs and markings to guide and channel the traffic. Additional guidelines for ramp metering controls are provided in Chapter 13.

Temporary signs. Most signs are used for conditions that exist all day, every day, and the signs are permanent in nature. However, one categorical exception is the use of temporary signs for construction or maintenance projects. Sometimes signs are needed only during periods where there is work activity on or near the roadway. Because of their temporary need, these signs can be applied on temporary mountings (see Figure 8–11).

The main use of a temporary sign in any functional class is to advise motorists of the condition and to guide them through or around the problem. This is true even though the operation may be temporary in nature. Most commonly in construction or maintenance work areas, the problems interfere with traffic flow on the road system as a result of work that is being conducted either within the right-of-way or adjacent to the right-of-way. The signs should be used by having the traffic control devices set up immediately before the beginning of the work and removed immediately when the work has been completed.

For those work activities that involve changing conditions, such as paving or mowing operations, the signs should be moved and revised with the progress of the work. For example, a paving operation over a 10-mile stretch of road that is completed in 1-mile increments daily should be signed in short increments. Signs are sufficient for the respective 1-mile increment under construction on any given day. This type of signing in the "active zone" is in addition to the standard advance signing advising of an "area" under construction that encompasses the entire 10-mile stretch.

Because of the special nature of application, the signs must always be placed where they will convey messages most effectively. They should not unduly restrict lateral clearance or sight distance. Temporary signs generally should be mounted on the right side of the roadway. Supplemental or double-indication signs may be located on the opposite side of the roadway, or left side, to emphasize the sign message or more readily attract the driver's attention.

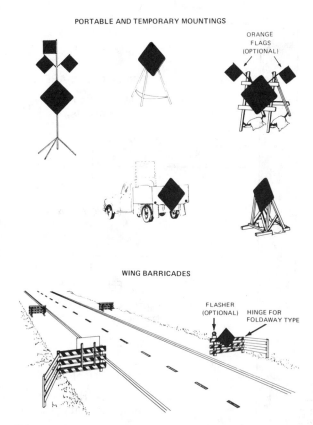

PORTABLE AND TEMPORARY MOUNTINGS

ORANGE FLAGS (OPTIONAL)

WING BARRICADES

FLASHER (OPTIONAL)   HINGE FOR FOLDAWAY TYPE

Figure 8–11. Methods of mounting signs other than on posts.

SOURCE: *Manual on Uniform Traffic Control Devices,* 1988, p. 6B–13.

The work zone for construction or maintenance activities is normally segregated into areas to facilitate traffic control, device application, and motorist compliance (see Figure 8–12). Area signing includes advance construction warning signs placed before the motorist reaches the work area. The advance distances are calculated in relation to speed. On rural highways, standard practice for heavy construction, restricted traffic movement, or detours call for advance notices at 1,500 ft, 1,000 ft, and 500 ft. Freeway signing often begins one-half mile before the work activity. Urban area applications usually are tied into a formula of 10 ft of advance distance for each mile per hour of approach speed. (NCHRP HR 20-5/20-03, "Work Zone Traffic Control and Safety on Urban and Suburban Streets," provides an excellent supplement for those interested specifically in signs and markings for urban arterials with work-zone activity.) This is followed by a transition area with channelization devices to direct traffic to the appropriate travel path followed by the activity area and termination area.

Stop, yield, and speed limit signs are usually the only regulatory signs used as temporary sign devices. However, if there are other specially authorized official restrictions, the sign should conform with standards exercised by the authority having responsible jurisdiction. Stop signs are commonly used on a temporary basis in transition to a signalized intersection. Texas examined many diamond

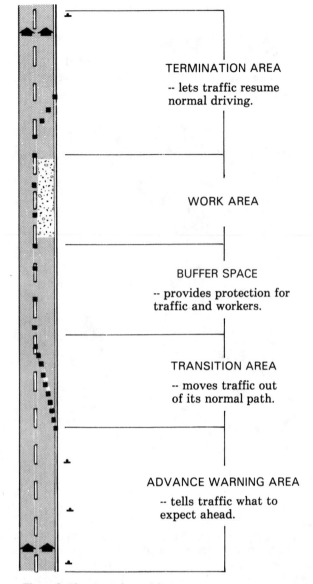

Figure 8–12. Areas in a work zone.
SOURCE: *Manual on Uniform Traffic Control Devices*, Part VI Rewrite, Graham-Migletz Enterprises, Inc., p. 10.

TERMINATION AREA

-- lets traffic resume normal driving.

WORK AREA

BUFFER SPACE

-- provides protection for traffic and workers.

TRANSITION AREA

-- moves traffic out of its normal path.

ADVANCE WARNING AREA

-- tells traffic what to expect ahead.

interchanges to discover the proper balance of when to remove stop signs in favor of signals.[21]

Advisory speeds are used in construction zone warning. These should be mounted on and in combination with other warning signs. They clarify that the advisory speed is applicable because of the special conditions during construction or maintenance activity. Advisory speeds are set by engineering studies that utilize a ballbank indicator, radar observations, or test runs by an experienced engineer. See the ITE *Manual on Traffic Engineering Studies* for additional details in determining advisory speeds.[22]

---

[21] Richard H. Oliver, "Relative Performance of Stop Sign versus Signal Control at Diamond Interchanges," *ITE Journal* (October 1987), 17–23.

[22] *Manual of Traffic Engineering Studies,* 4th ed. Institute of Transportation Engineers, Washington, DC, 1976. Currently, this publication is being updated with the 5th edition to be printed in 1991.

Temporary signs are utilized by governmental agencies, contractors, and utility companies. Many have developed separate manuals or handbooks on the proper control and use of traffic control devices and construction and maintenance applications. Some cities of reference are Saginaw, Michigan; Milwaukee, Wisconsin; and Overland Park, Kansas. On the project, special preconstruction meetings are required to review all phases of the traffic control plan. This is true especially for the phases that include a change in traffic signs and pavement markings. Individual questions arising on these matters should be referred to the proper state, provincial, or local government traffic engineer to clear any doubts before embarking on the work process.

Changeable message signs. These signs are used to inform drivers of regulations or instructions applicable only during certain periods of the day or under certain traffic conditions. They are most appropriately displayed in the changeable-message sign format. The need for and use of these signs have produced several types of electrical and electronic message boards that are often independently driven with a computer to select messages. The signs can be operated manually, by remote control, or by automatic controls that measure the conditions calling for special sign messages. Changeable-message applications are found in all of the functional-use classifications. Some examples of messages are:

1. Regulatory Signs:
   a. No Left Turn
   b. Speed Limit "_____"
   c. One-Way Traffic
2. Warning Signs:
   a. Accident Ahead
   b. Ice Ahead
3. Guide Signs:
   a. Route, Turn Right
   b. Stadium, Straight Ahead
4. Information Signs:
   a. Congestion Ahead
   b. Truck Scales Ahead

Wherever possible the changeable-message sign should conform to the same shape and color and be of the same dimensions as standard signs. Some of the larger electronic signs are not capable of meeting all of these conditions in their multiple use. However, at least one of the design attributes can be adapted for a particular sign so that it at least partially reflects the sign design standard. Other signs can be specifically configured to meet functional use requirements, such as:

1. Blank-out signs, which are signs that contain a fluorescent red or clear illumination behind cutout letters or symbols that are effective only when the grid is illuminated.
2. Signs that change message by a curtain drawn between a clear sign front and the electrical illumination provided behind the curtain or by means or a rotating drum.
3. Signs that change message by illumination of incandescent lights or fiber optics in appropriate patterns by selection of the light units that provide the desired letters, numerals, or symbols.
4. Signs that change their message by discs that are reflective white on one side and a background color on the reverse.

These discs rotate to form messages of alphabetical or numerical characters.

Sign location, mounting, and support. The effectiveness of any traffic control device is dependent on where and how it is displayed. The following sections point out good operating practice regarding these locations, mounting and support uses.

Location. The general rule is to locate signs on the driver side of the roadway. While this is usually followed, there are many instances when standardization of position cannot be attained. In those cases, alternate positions must be found where signs can be displayed for unobscured viewing by motorists. Considerations include locating the sign in order that it will not be splashed by mud, water, or other debris. The sign should not obscure another sign, not be hidden from view by other roadside objects, and not interfere with other roadside features such as driveways, hill crest, etc.

Mounting. Post-mounted signs should be placed at an angle of about 93° from the line of approaching traffic to minimize glare from their surface, and reduce mirror reflection without unduly reducing readability. The sign position should be selected with regard to road alignment of approaching traffic and not necessarily in relation to its predetermined angle from the edge of the roadway (see Figure 8-13). It is often desirable to tilt a reflectorized sign either forward or backward on the mounting so that the face of the sign will more squarely be directed to the approaching traffic (see Figure 8-14). Often, mounting heights will be increased to assure that the sign falls within the range of visibility for motorists driving uphill.

In general, only one sign should be mounted on each sign structure. This provides maximum emphasis for an individual sign that is decreased when it is required to compete

Figure 8-13. Orientation of roadside signs.
SOURCE: *A Guide for Retroreflectivity of Roadway Signs for Adequate Visibility,* FHWA, 1987.

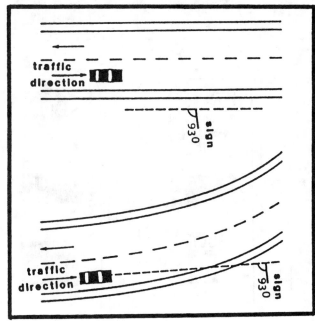

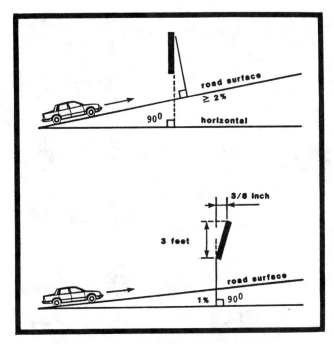

Figure 8-14. Orientation of overhead signs.
SOURCE: *A Guide for Retroreflectivity of Roadway Signs for Adequate Visibility,* FHWA, 1987.

among a group of signs. Notable exceptions to this would be the speed advisory sign, route marker assemblies, or any other plaque that should be attached to clarify the principal sign message. This practice of limiting signs to one per structure further enhances the usability of sign message should a sign be knocked down or temporarily damaged, in which case it does not hurt the existence of other signs in the same vicinity. Typical sign placement guidelines are illustrated in Figure 8-15.

The lateral clearance for regulatory and warning signs should be from 6 to 12 ft from the edge of the pavement or travelway in a rural area. Guide signs on conventional roads and streets should be placed laterally 6 to 12 ft from the roadway. The larger directional signs on high-volume highways should be placed to retain clearance of about 30 ft to minimize the chance of the sign being hit in an accident.

In urban areas, signs generally are mounted alongside the roadway in the space between the curb and sidewalk, behind the sidewalk, or above the sidewalk itself. The latter is preferable when the signs will not have to compete with advertising messages on adjacent buildings. Where practical, the signs should be placed as much as 10 ft from the edge of the nearest traffic lane but should be mounted at a height of at least 7 ft to be visible above other vehicles and not present a hazard to pedestrians.

Placement. Good judgment is required to assure that longitudinal placement of signs has been reasonably addressed. Obviously, regulatory signs must be placed at or near where the prohibition applies. Likewise, warning signs must be stationed in advance of the condition to which they call attention, and guide signs are posted as needed to keep motorists adequately informed or to give directions. There are four general guidelines for sign location:

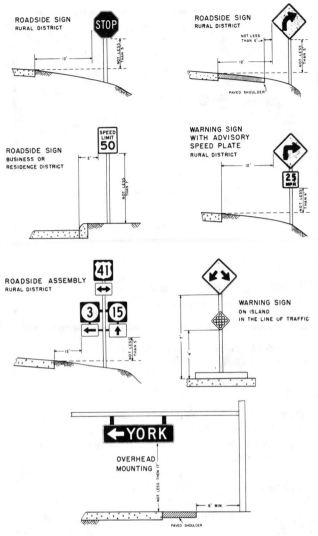

Figure 8–15. Height and lateral location of signs—typical installations.
SOURCE: *Manual on Uniform Traffic Control Devices,* 1988, p. 2A–14.

display guide sign information systematically so that motorists will become accustomed to the pattern and will be able to find the signs in uniform locations. Junction and advance turn arrows in rural areas are usually erected not less than 400 ft in advance of the intersection (see Figure 8–16). In urban or built-up areas, junction and advance turn signs should be located approximately midblock preceding the intersection but generally not more than 300 ft in advance. Major crossroads are often identified with larger street name signs located at midblock locations approaching the intersection and/or installed on traffic signal mast arms at the intersection.

Obviously, some judgment must be used in the selection of legend for use on guide signs. AASHTO has developed principles of priority regarding guide signs, which provide suggestions on the rank-order display of information. Depending on the type area and complexity of display, positive guidance

Figure 8–16. Typical route markings at rural intersections (for one direction of travel only).
SOURCE: *Manual on Uniform Traffic Control Devices,* 1988, p. 2D–16.

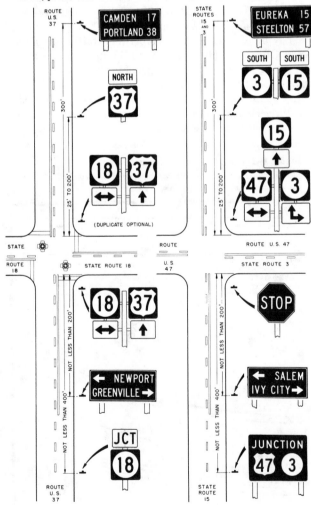

1. Stop and yield signs are placed within 50 ft of the intersecting roadway. In locations where this distance is not possible, supplementary pavement markings such as stop lines should be placed in the road at the intended point of compliance to supplement stop or yield signs. Traffic channelization can be employed to provide a small island for suitable location of the sign placement at wide throat and complicated intersections.

2. Warning signs must be placed somewhat in advance of the required point of compliance. Both urban and rural area locations are addressed in the warning sign placement table in the MUTCD. It should be noted that these placement distances are based on assumptions noted in the MUTCD table footnotes. Therefore, these distances may be adjusted owing to factors such as increased symbol sign legibility distances as noted above.

3. Guide signs are usually posted in advance of intersections and within the intersections itself. It is important to

principles should be applied. For example, in an area of multiple decision, the principle of "spreading" is appropriate to display to the motorist incremental information. Guide signs under the "spreading" principle are separated into individual installations rather than multiple signs on a single overhead sign structure. In areas where the sign message is vital, the principle of "redundancy" is often used to increase the probability the message is received by the motorist, or to confirm and reassure the motorist of the intended message. Redundancy requires that the message be duplicated on the left or repeated with an additional sign installation.

4. Destination signs in rural areas should be located no less than 200 ft or more than 300 ft in advance of the intersection. However, these distances may be increased to accommodate signs in advance of a special turn lane. In urban areas, somewhat shorter distances are permissible. Direction signs or route markers should be located at the intersection. Confirming route markers are placed 100 to 200 ft beyond the intersection in rural areas and 50 to 100 ft in urban areas to assure the motorist they are on the right route.

The selection of route numbers and control cities are both predetermined by AASHTO policy. Change or revision to the approved lists are considered by the respective AASHTO committees at each annual meeting. Lists of the current approved control cities and routes are available from the AASHTO office.

Interchange signs on expressways generally perform the function of route markers and destination signs. Thus they require greater advance warning distances than do signs on conventional highways. Where space allows, there should be a minimum of three sets of guide signs, not including signs placed as supplements to the regular sign. In all cases, the major guide sign should be placed in advance of the deceleration lane and spaced at least 800 ft from other signs.

In some situations there may be a need for two or more signs at approximately the same location. For these cases, it is necessary to establish a priority for the order of placements. Regulatory signs always take precedence over warning and guide signs. Guide signs are usually the least critical relative to location since there is more flexibility on placement particularly with some modification of the "action" portion of the guide sign legend. For either the regulatory or warning sign groups, these signs bearing the most important motorist regulation or warning should have priority on placement. An accepted order of priority is listed below:

1. Regulatory signs.
2. Warning signs.
3. Guide signs—Trail blazers, route markers, and destinations.
4. Emergency service signs—Hospital and telephone.
5. Motorist service signs—Fuel, food, lodging, and camping.
6. Public transportation signs.
7. Traffic generator signs—Museums, stadiums, historic sites, etc.
8. General information signs—Time zones, county lines, and city limits.

Additionally, signs should be placed so that they are compatible with each other. Speed zones should not be increased immediately in advance of major intersections, schools, or sharp curves. Also, the signing should provide a logical sequence of motorist messages to adjust speed, avoid potential hazards, and follow the desired route of travel.

Overhead supports. Highway signs should be mounted on overhead bridge structures if they are available. However, such structures are not normally found in the desired location, and the engineer must design a sign support that will handle both the deadload and windload of the completed sign and lighting system. These design features must further provide for the proper vertical and horizontal clearances to meet the minimum standards for the highway. Additionally, they must provide sufficient adjustment to compensate for acute angles and height clearances that may be in excess of that desired in a standard sign configuration. These are accomplished by changing the angles in aligning the signs so they are directly oriented to the motorists.

Overhead signs are often warranted in highly competitive driving environments. Whenever multilanes are involved or where there is major competition for drivers' attention, overhead signs become justified. Some of the warrants for overhead display include:

1. Traffic volume at or near capacity
2. Complex design
3. Three or more lanes in each direction
4. Restricted sight distance
5. Closely spaced interchanges or intersections
6. Multilane exits or turns
7. Large percentage of trucks
8. Background of street lighting
9. High-rise buildings
10. Consistency of sign message locations
11. Insufficient space for ground signs
12. Location near signal heads to make all controls visible.

Clearance. Overhead signs should provide a vertical clearance of no less than 17 ft for the entire width of the pavement and shoulders. When vertical clearances are less because of roadway structures, obviously the sign clearance can also be reduced. As a general rule, the sign should be at least 1 ft in excess of the minimum design clearance of the structure.

Signs erected at the side of the road in rural areas should be mounted with a clearance height of at least 5 ft measured from the bottom of the sign to the near edge of the pavement. Business, commercial, and residential districts—where parking or pedestrian movement is likely to occur—require the clearance to the bottom of the sign to be at least 7 ft. The height requirement for ground installations on expressways varies somewhat from those on conventional streets and highways. A minimum height of 7 ft above the near edge of the pavement to the bottom of the sign is the usual minimum dimension. If secondary signs are mounted below other signs, then the minimum dimension to the bottom of the major signs should be increased by at least 1 ft. Multilane facilities may require extension of the minimum length from 8 to 10 ft so the signs are visible over other vehicles. When a sign has been placed as much as 30 ft laterally from the edge of the roadway, the height to the bottom of the sign may be decreased to as little as 5 ft, especially where there are no competing roadside objects that interfere with or obstruct the view of the sign.

Supports. Signs can be correctly placed on existing supports that are used for other purposes such as traffic signal, street light, or even utility poles. This is not true, however, if utility crews are expected to climb the pole for any maintenance purpose. In those cases, the use of the pole for a secondary purpose of sign mounting is a false savings because of the possible hazard to utility crews while working on the pole. In any event, the utility company will usually object to this use of their poles.

Signposts and their foundations should be so constructed as to hold the sign in a proper and permanent condition that resists wind displacement or vandalism. On the other hand, the post must be constructed in such a way that it will yield or break away after being hit even by a small vehicle. Some typical small signpost mountings are shown in Figure 8–17, with breakaway post features illustrated.[23] The only exceptions would be if a breakaway support would cause increased hazard by allowing an overhead sign structure to fall onto the roadway. Another exception includes signs that are protected by curbing in urban areas or guardrails in rural areas, or if sign installations are a sufficient distance from the pavement edge to provide satisfactory recovery area for an errant vehicle.

In areas that do not have curbs or other sign protection, regulatory and warning signs are normally mounted on the side of the road on square or U-shaped posts that breakaway at pivot points or yield to any hit by a vehicle on impact.[24] A one post installation of 2 1/4 lb/ft is considered a breakaway post, but it will not support signs exceeding 36 in in dimension, or 10 sq ft in area. In those cases, a heavier steel pole (4 lb/ft) will support the signs up to 48 in wide and 16 sq ft in area at the standard mounting height. However, these signposts must be evaluated for their breakaway and yielding strength properties, and they are often fabricated with supplemental devices to allow them to break off at a splice point. A wooden post 4 in by 4 in will hold up to 15 sq ft of sign and a post 6 in by 6 in that has been predrilled for breakaway features can support a sign up to 50 sq ft. If signs are between 20 and 90 sq ft, mounting should be considered as 2- or 3-post installations, but proper breakaway features are needed. When breakaway features are necessary, they can be provided by drilling holes in the wood or providing specific design applications to aluminum or steel. Two designs are available:

1. *Texas design:* This breakaway features a slip joint at the base and a hinge joint below the sign to allow the post after impact to slip off the foundation and swing up and away from the vehicle.

2. *CBC design:* A load concentrated breakaway coupling is mounted at the base with a separate breakaway feature between the sign and the post. This allows the post to shear from the foundation on impact and swing away from the vehicle to avoid secondary collision.

Pavement markings

Kinds of markings

Pavement markings include all lines, longitudinal or transverse, as well as symbols and words that are applied to the pavement. Object markers, delineators, cones, or other roadway guidance devices used to delineate the proper path of travel for motorists are considered in this section even when they are not directly applied to the pavement. Pavement markings can be applied with paint, thermoplastic or preformed plastic marking, or reflective signing materials.

Function of markings

Pavement markings have a unique function in the proper control and regulation of both vehicular and pedestrian traffic. Mainly, markings channelize or guide the traffic into the proper positions on the roadway, but they also supplement other regulations and warnings displayed by signs and signals. Other functions of markings include barriers for opposing traffic, warning devices for restrictive sight distance or passing distance, and supplementary information for turning movements, school zones, and railroad crossings. As an aid to pedestrians, they channelize movement into safety crosswalks, and they provide an extension of the sidewalk superimposed across the highway. In general, pavement markings aid motorists but do not cause them to divert their attention from the roadway.

Temporary markings

Usually applications of markings are in areas where it would be desirable to have a permanent or long-life installation. An exception to this is in construction or maintenance areas where the need and use for temporary markings exist until the necessary roadway repairs or improvements have been made.

Temporary marking devices such as paint striping or other temporary traffic lines should be implemented by selecting the proper device to meet the needs for a given situation. A normal thickness of traffic paint would not be applied when the line must be removed in a few weeks, unless the line is applied on a temporary pavement. Likewise, plastic materials would not be applied with the normal-strength epoxy for temporary conditions.

Permanent traffic control devices are removed or covered when the message is inappropriate, and provisions must also be made so that the temporary installation of pavement markings can be removed when they are no longer needed. Otherwise, motorists can become confused by conflicting messages from the pavement markings.

Pavement marking design. The three fundamental types of pavement markings include the various types of

[23] AASHTO, "Standard Specification for Structural Supports for Highway Signs, Luminaires and Traffic Signals," American Association of State Highway & Transportation Officials, Washington, DC, 1975.

[24] N.J. Rowan, and R.M. Olson, "The Development of Safer Highway Sign Supports," *Traffic Engineering,* XXXVIII (1967) 46–50.

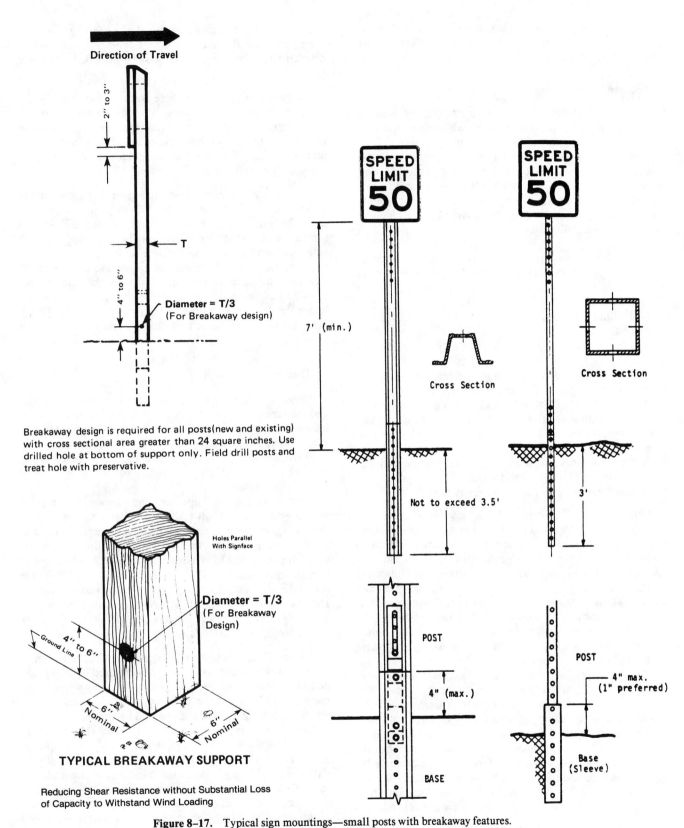

Figure 8–17. Typical sign mountings—small posts with breakaway features.
SOURCE: *Placement Guide for Traffic Control Devices,* U.S. Forest Service, 1981, p. 11.
SOURCE: *Guide of Roadside Improvements for Local Roads and Streets,* FHWA 1986, pp. 24, 26.

longitudinal lane markings, other transverse painted markings, and markers for objects and delineation. Each of these will be briefly discussed.

Longitudinal markings

General. Longitudinal lines are used to organize traffic into the proper lanes, advise motorists where passing is prohibited, and to supplement other warning devices. In the United States, the color of longitudinal markings can be white, yellow, or red. Black can also be used in combination with the three primary colors where the pavement does not provide sufficient contrast. Application and basic concepts of the standard colors have been established in the *Uniform Manual.* They are discussed below:

1. Yellow lines delineate the separation of traffic flow in opposite direction, and a vehicle will not normally be driven on the left side of the line except to pass another vehicle or turn.
2. White lines delineate the separation of traffic flow in the same direction.
3. Red markings delineate a roadway that shall not be entered or used by the viewer of that line.
4. Broken yellow and white lines are permissive in character. These are normally formed in segments and gaps in the ratio of 1:3 (a 10-ft stripe and a 30-ft gap).
5. Solid lines are restrictive in character, and a vehicle should not normally be driven across the solid line whether yellow or white in color.
6. Width of the lines indicate the degree of emphasis (with a 4- to 6-in line used as normal width).
7. Double lines indicate maximum restriction.

Center lines. These lines vary depending upon the width of the travelway and the type of area. A single broken yellow line used on the two-lane, two-way street indicates that passing is permitted from either direction. A solid yellow centerline on either side of the broken line indicates that passing is prohibited from that direction and immediately adjacent to the solid line. This is the type marking commonly used to separate a middle lane that has been designated for two-way left-turn movements. Double, solid yellow centerlines should be used in areas where passing is prohibited in both directions.

Double broken lines are used to designate a roadway when one or more traffic lanes are operated in a reversible pattern with either/or signs and signals. A single solid yellow line is used by some jurisdictions to indicate both the roadway centerline and partially restrictive passing requirements, although this application of pavement markings is not acceptable in conformance with the *Uniform Manual.* It should also be noted that a yellow no-passing barrier line is required for both horizontal and vertical sight distance restrictions in all cases where a broken yellow line is used.

Lane lines. These lines are normally broken white lines with the standard spacing of 10 ft painted and a gap of 30 ft. Some urban areas use a pattern of 5 ft painted and a gap of 15 ft. Solid lines can be used on approaches to intersections and in other areas where lane changes are discouraged. A dotted white line normally 2 ft in length with 4-ft gaps can

be provided to delineate extension of a line through intersections or across deceleration lane openings where special problems require that the motorist be provided additional guidance.

Pavement edge lines. These lines include solid white lines on each side of the travel roadway except that on divided highways and one-way roadways, the left edge of the direction of travel must be a yellow solid line. In some areas, and particularly with enforcement personnel, edge lines are commonly referred to as "fog" lines.

Channelizing lines. These lines are either extra wide or double lines and can be either white or yellow, depending upon the traffic patterns. If the lines provide traffic islands where travel in the same direction is permitted on both sides, the markings are white. If islands separate travel in opposite directions, the markings are yellow. In either case, any cross-hatching is the same color as the longitudinal lines. Often channelization provides the opportunity to enhance the delineation necessary to improve benefits for vehicle guidance and tracking tasks, especially for the older driver.

Transverse markings

General. Transverse markings are used to convey special messages to motorists either by themselves or in combination with signs. They include shoulder and curb markings, word and symbol markings, stop bars, crosswalk lines, railroad crossings, and parking space markings. All such markings are white, except they are yellow when they are part of the median markings.

Because of the low approach angle in which pavement markings are viewed, all transverse lines must be proportioned to give visibility equal to that of longitudinal lines and to avoid apparent distortion where longitudinal and transverse lines combine in symbols or lettering. Applications of these markings are provided in the *Uniform Manual* with the layout details illustrated in the FHWA publication "Standard Highway Signs." Details of pavement arrows are illustrated in Figure 8–18, which also depicts the necessary symbol elongation to compensate for the shallow viewing angle.

Crosswalks. These markings are a minimum of 6 in in width and at least 6 ft apart, and they are used both to guide pedestrians to the proper path and to serve as warnings to motorists of a pedestrian crossing location. A solid white boundary line up to 24 in wide should be used where vehicle speeds are over 35 mph or where crosswalks are not expected. Moreover, the area of the crosswalk may also be marked with white diagonal lines at a 45° angle or longitudinal lines at a 90° angle. Marking standards of crosswalks in some areas also provide for transverse lines on each side of the longitudinal line. Crosswalks greatly enhance the needs of elderly pedestrians by providing them a clearly delineated area that will keep them from wandering away from a direct path to the other side of the street particularly at complex intersections. The elderly do have difficulty in perpendicular perception, and some may rely on marked crosswalks to give them a false sense of security by assuming vehicles will yield to pedestrians.

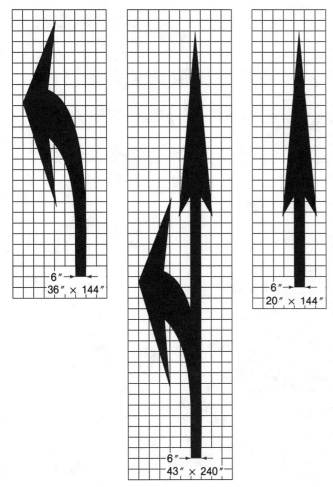

Figure 8–18. Arrows for pavement markings.
SOURCE: *Standard Highway Signs*, FHWA, 1979.

Stop bars. These lines are normally 12 to 24 in wide and should extend across all approach lanes. They should be placed at the desired stopping point and in no case more than 30 ft or less than 4 ft from the nearest edge of the intersecting roadway. Stoplines should ordinarily be placed 4 ft in advance of the nearest crosswalk line.

Railroad crossings. Pavement markings in advance of a railroad crossing consist of the letter X with the letters "RR," a no-passing centerline marking, and transverse lines on each side of the legend. They shall be placed on all paved approaches to railroad crossings where the roadway speed exceeds 40 mph and elongated to allow for the low angle in which they are to be viewed. A new pattern of pavement markings for railroad crossings has been approved and is illustrated in Figure 8–19. The "RR" has been relocated to the approach end of the symbol to reduce vehicle wear on the pavement markings.

Parking space markings. These markings encourage more orderly and efficient use of parking space where parking turnover is substantial. They are helpful in preventing encroachment on the fire hydrant zone. Parking space markings are useful to clarify where parking is prohibited, such as at bus stops, loading zones, approaches to the corner, and clearance space for islands.

Word and symbol markings. These markings are white and may be used for the purpose of guiding, warning, and regulating traffic. Any one application should be limited to not more than three lines of words and/or symbols. They can serve as regulatory (stop or right turn only); as warning (stop ahead or school zone); or as a guide for traffic (US 30 or SR 123). When used as a regulatory device, they must be supplemented by the appropriate signs. Symbols are preferable to words, but in all cases they must be elongated in the direction of the travel movement to provide the motorist the appropriate message to be viewed at a relatively flat angle. Large letters, symbols, and numerals should be used 8 ft or more in height. If the message consists of more than one word, it should read "up" with the first word being nearest the driver. Special applications are often used for repeat conditions where a lane is terminated or by creating a "trap lane."[25]

Curb markings. These markings can either be for roadway delineation or for parking regulation. When used as roadway delineation, the color should follow the guidelines for edgeline or channelization markings. When used as parking regulation, they should supplement standard signs and can be a special color as prescribed by local authorities. Commonly, both red and yellow have been used for parking restrictions, with blue used to delineate handicapped parking spaces.

Object markers and delineators

Object markers. A physical obstruction in or near a roadway constitutes a serious hazard to safe traffic movement and should be adequately marked. However, every effort should be made to remove the obstruction or to minimize its potential danger. In the United States, when obstructions are located within or adjacent to the roadway, they require marking. Markers should consist of an arrangement of one or more of the following designs (see Figure 8–20):

1. A Type 1 marker consists of nine yellow reflectors having a minimum dimension of approximately 3 in, mounted symmetrically on an 18-in diamond background either black or yellow in color; or an 18-in diamond, all yellow retroreflective panel.
2. A Type 2 marker consists of three yellow reflectors having a minimum dimension of approximately 3 in arranged either horizontally or vertically; or a 6 in by 12 in rectangular yellow reflector. Type 2 markers may be longer if conditions warrant.
3. A Type 3 marker consists of a vertical rectangular shape approximately 1 by 3 ft in size with alternating black and reflectorized yellow stripes sloping downward. The angle of slope should be 45° toward the side of the obstruction in which traffic is to pass. The minimum width of the yellow stripe should be 3 in. A better appearance can be achieved if the black stripes are wider than the yellow stripes.

---

[25] Gary Foxen, "The Signing and Marking of Trap Lanes," *ITE Journal* (June 1986), 48–49.

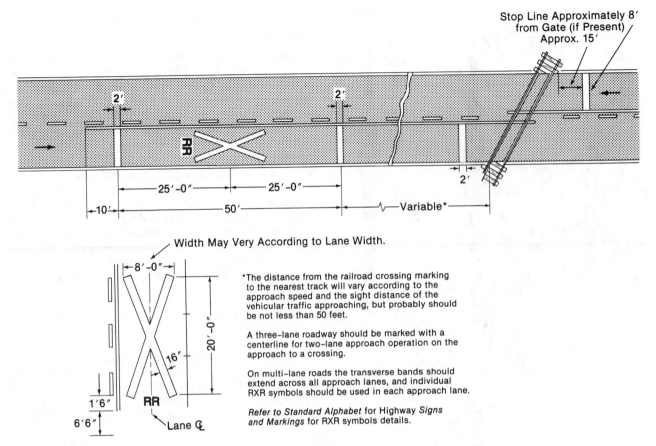

Figure 8–19. Revised railroad crossing pavement markings.
SOURCE: NATIONAL COMMITTEE ON UNIFORM TRAFFIC CONTROL DEVICES, 1990.

The maintenance procedures and schedules for these devices are quite similar to traffic signs. Any installation in roadway splash areas such as the median or the edge of the roadway should be washed at least twice a year. Because of their location, they are subject to vehicle damage and should be scheduled for frequent repair replacement.

Delineators. Road delineation markers with a minimum dimension of 3 in should be considered as guide markings rather than warning devices. Delineators may be used on long, continuous sections of highways or through short stretches where there are pavement width transitions or other changes in horizontal alignment, particularly where the alignment may be confusing.

Applications. Object markers and delineators must be applied in a very specific manner regarding the object on the road, adjacent to the road, or at the roadway end. These are described as follows:

1. Objects in the road should be marked with a Type 1 or Type 3 marker at a mounting height of at least 4 ft. In addition, large vertical surfaces can be painted with diagonal stripes 12 in or greater in width, similar in design to the Type 3 object marker. The alternating black and reflectorized yellow stripe should slope down at an angle of 45° toward the side of the obstruction on which traffic is to pass.

2. Objects adjacent to the roadway, which are close enough to constitute a hazard, should be marked with a Type 2 or Type 3 marker. The post should be installed so that the inside edge of the marker is to be in line with the inner edge of the obstruction at a mounting height of 4 ft.

3. The end of the roadway should be marked with a Type 1 marker, except nine red reflectors or an 18-in red panel can be used. The mounting height should be 4 ft above the roadway.

4. For delineation on through, two-lane, two-way roadways, single white reflective delineator units are placed on the right side of the roadway. Single yellow reflector units also may be placed on the left side of two-way roadways as additional guidance, particularly at sharp right-hand curves.

5. On expressway and freeway and other through roadway-type facilities, single white delineators should be placed continuously on the right side and on at least one side of interchange ramps. If used on the left side for additional guidance, delineators should be yellow to match the yellow edgeline.

6. Double delineators or vertical elongated yellow delineators should be spaced at 100-ft intervals along acceleration and deceleration lanes of freeway facilities.

7. Red delineators may be used on the reverse side of any delineator whenever it can be viewed by motorists traveling in the wrong direction. They can also be used to delineate the alignment of truck escape ramps.

Typical Type 1 Object Markers

18″x18″   18″x18″   18″x18″
Yellow Reflectors
Black Panel

Typical Type 2 Object Markers

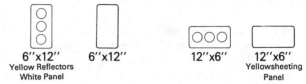

6″x12″   6″x12″   12″x6″   12″x6″
Yellow Reflectors   Yellowsheeting
White Panel   Panel

Typical Type 3 Object Markers

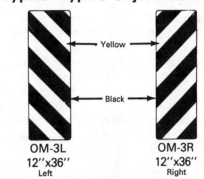

OM-3L   OM-3R
12″x36″   12″x36″
Left   Right

Typical End of Road Markers

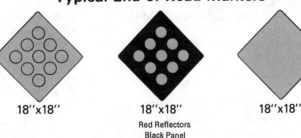

Figure 8–20. Types of object markers.
SOURCE: *Manual on Uniform Traffic Control Devices*, 1978, p. 3C–3.

18″x18″   18″x18″   18″x18″
Red Reflectors
Black Panel

8. Normally, delineators should be installed at a mounting height of 4 ft and spaced 200 to 528 ft apart. They must be adjusted to shorter spacings on approaches, horizontal curves, and crest vertical curves.

9. Curves or islands located on or near the line of traffic flow should be marked with white delineators if they are on the right or if traffic in the same direction passes on both sides of the island. If the curve or island separates traffic in opposing directions, then yellow delineators are used.

Special markings

General. Safe operation and needed capacity of many streets and highways depends on special applications of traffic markings. Several of the principal ones are as follows:

Lane reduction transitions. Pavement lane markings can be effectively used to supplement the standard signs that guide traffic when the pavement width is reduced to a fewer number of lanes.

Many variations are possible depending on which lanes must be offset or eliminated and on the amount of the offset. One or more lane lines must be discontinued and the remaining centerline and lane line must be connected in such a way as to safely merge traffic into the reduced number of lanes. This is especially important in the ending of a reversible lane section or roadway.

Lines marking pavement width transition should be a standard design for centerlane or barrier lines. Converging lines on roadways with a posted speed of 35 mph or greater should have a length of not less than that determined by the formula $L = SW$. On slower-speed streets and highways, the formula used is $L = WS^2/60$. In both equations, L equals the length in feet; S the off-peak 85th percentile speed in mph, and W the offset distance in feet. This provides a minimum transition slope of approximately 30:1 for slower speeds up to 50:1 for higher-speed highways. Table 8–1 shows the minimum taper length for several traffic lane widths and roadway speeds. The number of channelized devices such as traffic barrels or cones are also indicated for each transition.

Obstruction approach markings. Pavement markings are frequently used to supplement standard signs to guide traffic approaching a fixed obstruction within a paved roadway.

The obstruction may be in the center of the road in which case all traffic is usually directed to the right of it. These markings normally consist of a diagonal line extending from the center or laneline to a point 12 to 24 in to the right side of the approach end of the obstruction. The length of the diagonal markings can be determined by the same formula provided in the lane reduction transition. The diagonal lines should never be less than 200 ft in length in rural areas and 100 ft in length in urban areas. The use of obstruction approach markings and signs does not eliminate the need for adequate object markings on the obstruction itself (see Figure 8–21).

Colored pavement. The pavement may be colored to provide a significant contrast with adjacent paved areas to help control traffic. Colors should be limited as follows:

1. Red should not be used on approaches to a stop sign or yield sign, but may be used for curb restrictions.
2. Yellow should be used only for median-separated traffic flows in opposite directions.
3. White should be used only to provide contrast with other colors such as right-hand shoulders or for channelizing islands in two-way traffic.

The color application usually provides a stronger contrast in daylight operations since the colors tend to blend in with the adjoining pavement at night. However, street illumination can reduce this problem. Some of the special uses

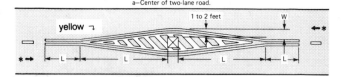

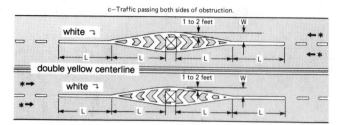

Figure 8–21. Typical approach markings for obstructions in the roadway.
SOURCE: *Manual on Uniform Traffic Control Devices,* 1978, p. 3B–19.

have been incorporated in airports and on private facilities where multilane traffic control is a concern so color-coded directional aids are provided on the pavement in conjunction with color-coded directional signing.

Channelization. Painted channelization can be used effectively to separate traffic safely, and it has the advantage of easy modification when warranted by driver behavior. If a more positive barrier is necessary, curbs and islands can be constructed at a later date. Paint channelization may well serve initially to establish the best arrangement before permanent construction is established. Some states have effectively used sandbags or flexible posts to lay out a temporary trial channelization that can be easily adjusted. Channelization is often a simple and cost-effective means of improving efficient use of a wide-paved roadway. This technique of channeling the traffic through the preferred spots of the intersection often improves safety and efficiency in traffic movement.

Raised pavement markers. Raised pavement markers form a semipermanent marking and improved visibility in nighttime, wet-weather conditions. The markers vary from 1/4 in to less than 1 in in height and can be used singularly as a reflective unit, in combination to supplement paint or thermoplastic lines, or in patterns to simulate and replace other pavement markings. (See Figure 8–22, which illustrates several types of raised markers.) The number and spacing of raised markers to simulate continuous lines and striping patterns are found in the FHWA *Traffic Control Device Handbook,* which is available from the U.S. Department of Transportation, Federal Highway Administration, Office of Traffic Operations, Washington, DC.

In snow-free areas, the raised pavement markers have received superior ratings compared to standard traffic paint in terms of durability, driver preference, night-wet visibility, and cost for a 10-year period.[26]

TABLE 8–1

**Length and Device Spacing
for Lane Closure and Channelization Tapers**

Speed Limit mph	Taper Length Lane Width (ft)			Number of Channelizing Devices for Taper[a]	Spacing of Devices Along Taper (in ft)
	10	11	12		
20	70	75	80	5	20
25	105	115	125	6	25
30	150	165	180	7	30
35	205	225	245	8	35
40	270	295	320	9	40
45	450	495	540	13	45
50	500	550	600	13	50
55	550	605	660	13	55

[a] Furnished for a 12′ lane. Divide spacing into taper length and add one.

SOURCE: *Traffic Control Devices Handbook,* FHWA, 1983, p. 6–10.

---

[26] John Dale, "Development of Formed In-Place Wet Reflective Markers," *NCHRP Report 85,* Washington, DC, 1970.

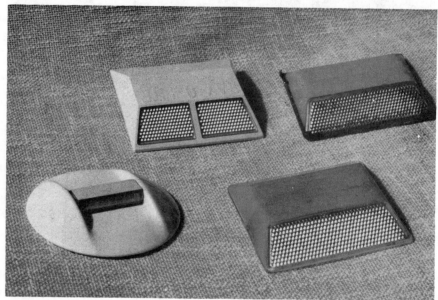

Figure 8–22. Non-reflective ceramic and reflective raised pavement markers.

SOURCE: *Traffic Control Devices Handbook,* FHWA, 1983, pp. 3–20, 3–21.

Almost all raised marker applications in the United States can be installed without specialized equipment. The individual units are applied to the pavement with epoxy cement in a standard pattern. Recessed units with a steel casing and two keels have been installed in snow-plow areas, but these units require special saws to cut grooves in the pavement.

A prototype machine was developed as part of a research project to clean the pavement, deposit epoxy resin, and secure ¼-in glass beads into the epoxy.[27] This machine installed one formed-in-place marker and then moved down the roadway to the location of the next unit. This type of marker has not been widely accepted.

In construction and maintenance areas, raised pavement markers become an effective marking technique for temporary and transition lines that change frequently. The increased retroreflectivity and improved visual qualities of raised pavement markers will emphasize the special roadway alignment variations. The marker can be easily removed when a new pattern is necessary because of construction work progress. Attempts to reduce lane width by adding raised pavement markers to control speed on residential streets have not been successful.[28]

---

[27] Bernard Chaiken, "Comparison of the Performance and Economy of Hot-Extruded Thermoplastic Highway Striping Material and Conventional Paint Striping," *Public Roads* XXXV-6 (1969), 135–156.

[28] Harry S. Lum, "The Use of Road Markings to Narrow Lanes for Controlling Speed in Residential Areas," *ITE Journal* (June 1984), 50–53.

Miscellaneous traffic control devices

General

Standards and specifications have been developed in the United States for additional traffic control devices not included in previous sections of this chapter. These additional devices are found in most sign and marking shops. They are used to guide traffic in and around work areas, to alert traffic to hazards ahead, and to provide a means of identifying specific locations on streets or highways. The devices are illustrated in Figure 8–23.

Barricades

These types of temporary devices are used to warn and alert drivers to hazards created by construction or maintenance activity in or near the travelway. Barricades also guide and direct drivers safely past the hazard. Barricades should be one of three kinds—Type 1, Type 2, or Type 3. Red and white striped barricades can be used to warn and alert drivers to the terminus of a road or ramp or when a lane is closed to traffic. Any permanent road closure requires the use of a red and white barricade. For nighttime use, they shall be reflectorized and/or equipped with lighting devices for maximum visibility.

Figure 8–23. Channelizing devices.
SOURCE: *Manual on Uniform Traffic Control Devices,* Part VI Rewrite, Graham-Migletz Enterprises, Inc., p. 62.

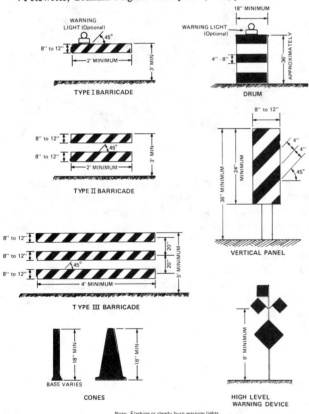

Traffic cones

Traffic cones are a portable, temporary device used for guiding drivers through or past an obstacle. Traffic cones and tubular markers of various configurations are available. They should be a minimum of 18 in high with a broadened base. They can be made of various materials to withstand impact without damage to themselves or the vehicles. Larger cones should be used where speeds are high or where more conspicuous guidance is needed. The predominant color of cones is orange. They should be kept clean and bright for maximum target value. For nighttime use, cones shall be reflectorized for maximum visibility.

Vertical panels

Vertical panels with a minimum dimension of 8 by 24 in can be used for traffic separation or shoulder delineation where space is at a minimum. They should be orange with white stripes and reflectorized in the same manner as barricades. The top of the panel should be mounted a minimum of 36 in above the road.

Drums

Drums with horizontal orange and white reflectorized stripes can be used to channelize or delineate traffic flow. When used, they should be approximately 36 in high and have at least two orange and two white stripes that are 4 to 8 in wide. Drums need not be cylindrical in shape but, rather, should follow the circular design with an 18-in minimum width dimension. Drums can be used singly or in groups to mark a specific hazard. When used in a roadway, they must be supplemented with appropriate advanced warning signs.

Barricade warning lights

Portable directional warning lights may be needed for mounting on drums or barricades to warn traffic of hazards ahead. The color of the light emitted should be yellow and they may be used in either a steady burn or a flashing mode. Barricade warning lights should meet the requirements of the ITE purchase specification for flashing or steady burning barricade warning lights. There are three kinds of warning lights:

1. Type A is a low-intensity flashing warning light most commonly mounted on a Type 1 or Type 2 barricade. These attempt to warn drivers that they are approaching a hazardous area and shall be visible on a clear night from a distance of 3,000 ft.

2. Type B is a high-intensity flashing warning light mounted on the advance warning signs to get drivers' attention and to alert motorists to conditions ahead. They shall be visible in direct sunlight from a distance of 1,000 ft. Extra attention to high-hazard locations within the construction area can be improved by mounting this Type B flasher on a Type 1 barricade in the vicinity of the hazard. These lights are effective in daylight and darkness and can operate 24 hours a day.

3. Type C is a steady-burn light intended to be used to delineate the edge of the travelway in construction areas. It is used where additional emphasis is needed to help guide drivers. Such lights shall be visible from a distance of 3,000 ft on a clear night.

It is recommended that when the warning lights depreciate below the above legibility distances they should be cleaned, relamped, new batteries installed, or the lighting unit replaced.

Rumble strips

Rumble strips are devices used to alert the driver of a change in conditions ahead. Strips can consist of sawed grooves in the pavement, a series of transverse grouped thermoplastic strips, or some other means of creating a tire rumble effect. Successful applications have been made in advance of sharp curves where motorists tend to approach too fast for existing conditions. Other applications are on approaches to stop signs, and in tapered areas where there is a reduction in pavement width. Some work has been completed on the development of effective patterns for these rumble strips.[29]

Portable changeable message signs

These include a flexible assortment of guidance, warning, and regulatory messages. They can be deployed on short notice at needed locations. These devices can display up to three lines of message and are especially effective in work zones. They should be located about one-half mile before the work site, with at least 850 ft legibility distance. The devices should conform to design, material, and placement criteria as provided in the existing ITE standard.[30]

Overspeed signs

Many effective uses of a combination warning sign and flasher signal have been reported with the "too fast for curve" signs and similar applications. Many factors have been isolated such as in the Michigan study to influence the effectiveness of reducing speeds in a deceptive curve condition.[31] However, the device offers the opportunity to combine the best features available from both the passive and active traffic control devices to provide increased effectiveness.

Advance warning arrow panels

These devices are helpful in both day and night conditions whenever a lane closure is necessary due to many factors.[32] These include slow-moving maintenance operations,

construction activity, or hazardous conditions that occur in high-density or high-speed conditions. Typical arrow panel specifications and operating modes are illustrated in Figure 8–24.

Sign and marking warrants

Traffic sign warrants

Sign effectiveness can be destroyed by overuse. Regulatory and warning signs are installed because of need and in some cases by warranting conditions. Guide signs, however, require more engineering judgment in order to determine how they can be used for the convenience and facilitation of traffic movement. Some general criteria can be established regarding the use of highway signs. The specific warrants for signs are contained in the MUTCD as general policy guidelines. When signing is displayed in accordance with these warrants, the usual result is to maximize efficiency and effectiveness. Some of the warrants will be discussed below.

Stop sign warrants. A stop sign may be warranted at an intersection where one or more of the following conditions exists:

1. Intersection of a less important road with a main road. An example could be the intersection of two subdivision streets, where application of the normal right-of-way rule is not unduly hazardous.
2. Intersection of a county road, city street, or a township road with a state highway.
3. Street entering a through arterial highway or street.
4. Unsignalized intersection in a signalized area.
5. Unsignalized intersection where a combination of high speed, restricted view, and accident records indicate a need for control by the stop sign.

The stop signs provide conflicting information at a traffic signalized intersection. Therefore, when an intersection has been signalized either in the normal or flashing operation, stop signs shall be removed. Additionally, stop signs are not intended for the sole purpose of controlling speed of the motorist and should not be utilized for that purpose. There are some indications that vehicle speed on urban arterial and collector roads are not influenced by stop-sign usage after traveling 300 ft from the sign location. Multiway (four-way or all-way) stop installations can be used as a safety measure at some locations. These are especially useful where the volumes on the intersecting roads are approximately equal and the following conditions have been established:

1. Where traffic signals are warranted, the multiway stop control is an interim measure that can be installed quickly while arrangements are being made for the signal.
2. When an accident problem as indicated by five or more reported accidents in a 12-month period are of a type susceptible to correction by a multiway stop condition and less restrictive controls have not been successful.

---

[29] W.R. Bellis, "Development of Effective Rumble Strip Pattern," *Traffic Engineering* (Institute of Traffic Engineers, Washington, DC), 39 (April 1969), 22–25.

[30] ITE Technical Committee 7S-3, "Portable Bulb-Type Changeable Message Signs for Highway Work Zones," *ITE Journal* (April 1988), 17–20.

[31] Richard W. Lyles, "Advisory and Regulatory Speed Signs for Curves: Effective or Overused," *ITE Journal* (August 1982), 20–22.

[32] ITE Technical Committee 4S-9, "Advance Warning Arrow Panels," *ITE Standard,* Washington, DC, 1984.

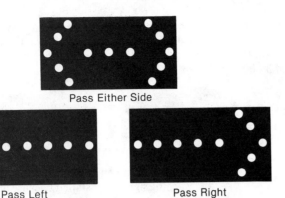

Pass Either Side

Pass Left   Pass Right

FLASHING ARROW PANELS-OPERATING MODES

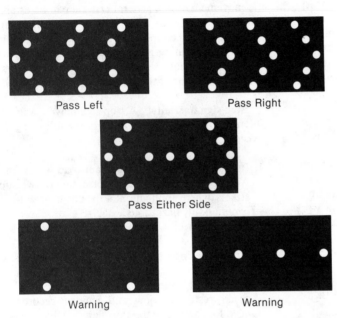

Pass Left   Pass Right

Pass Either Side

Warning   Warning

SEQUENTIAL ARROW PANELS-OPERATING MODES

Type	Minimum Size (Inches)	Minimum Number of Panel Lamps	Minimum Legible Distance (Miles)	Speed* (Mon)	Recommended Usage
A	24 by 48	12	1/2	20–40	Urban
B	30 by 60	13	3/4	40–55	Highway
C	48 by 96	15	1	55–65	Express/Freeway

Figure 8–24. Advance warning arrow panel specifications.
SOURCE: *Manual on Uniform Traffic Control Devices,* Part VI Rewrite, Graham-Migletz Enterprises, Inc., p. 104.

3. Minimum traffic volume
   a. where the total vehicle volume entering the intersection from all approaches averages at least 500 vehicles per hour for any 8 hours of an average day; and
   b. where the combined vehicular and pedestrian volume from minor streets must average at least 200 units per hour for the same 8 hours with an average delay to minor street traffic of at least 30 sec per vehicle during the maximum hour; but
   c. when the 85th percentile approach speed of the major street traffic exceeds 40 mph, the minimum volume warrants are 70% of the above requirement.

While multiway stops have recorded as much as 58% reduction in accident rate at some rural locations, there are specific cautions in this program use.[33] Some observers believe this type control has been unnecessarily proliferated at intersections in the last several decades. This is mainly due to the mistaken belief that a multiway stop control is a panacea for urban traffic problems.[34] Each potential location should

[33] "4-Way Stop Signs Cut Accident Rate 58% at Rural Intersections," *ITE Journal* (November 1984), 23–24.

[34] Everett C. Carter, and Himmat S. Chadda, "Multi-Way Stop Signs—Have We Gone Too Far?" *ITE Journal* (May 1983), 19–21.

be carefully studied with data collected and engineering judgment used to determine the safest and most efficient type of intersection control.

Yield sign warrants. These are established as follows:

1. On a minor road at the entrance to an intersection when it is necessary to assign right-of-way to the major road, but where a stop is not always necessary, and where the safe approach speed on the minor road exceeds 10 mph.
2. On the entrance ramp to an expressway where an adequate acceleration lane is not provided.
3. Within an intersection with a divided highway, where a stop sign is present at the entrance to the first roadway and further control is necessary at the entrance to the second roadway, and where the median width between the two roadways exceeds 30 ft.
4. Where there is a separate or channelized right-turn lane, without an adequate acceleration lane.
5. At any intersection where a special problem exists and where an engineering study indicates the problem to be susceptible to correction by the use of the yield sign.

Yield signs should not be used to control the major flow of traffic, approaches of more than one of the intersecting streets, or at intersections where there are stop signs on one or more approaches. The state of the art in the use of yield signs has been published in the ITE informational report "Yield Sign Usage and Applications."[35]

Lane-use control sign warrants. Criteria for installation of lane-use control signs at intersections are established as follows:

1. Lane-use control signs at intersections should be used whenever it is desired to require vehicles in certain lanes to turn, or to permit turns from an adjacent lane.
2. Lane-use controls permitting left (or right) turns from two (or more) lanes normally are warranted whenever the turning volume exceeds the capacity of one turning lane, and when all movements can be accommodated in the lanes available to them. Advance signs and pavement markings are desirable whenever lane control is used.

Warning signs. Where an engineering and traffic investigation indicates that attention must be called to an actual or potential hazardous condition, warning signs should be installed. Typical locations that may warrant such signs include:

1. Turns, curves, or intersections
2. Advance warning for stop signs, signals, or railroad crossings
3. Grades, depressions, or bumps
4. Narrow roads, bridges, or other limited clearance needs
5. Curves or areas where advisory speed is needed.

It should be recognized that the MUTCD is not an exhaustive listing of every type of warning sign applicable to a given situation. Other messages appropriate to a situation may be used on a diamond-shaped blank with the appropriate colored background of yellow or orange. The legend should be a clear indication of the potential hazard that may occur in or adjacent to the roadway. This is frequently true in situations involving construction, maintenance, or utility operation where more definitive information can be provided. For example, to supplement a work-zone sign, the message "Low Shoulder" is often appropriate. In other work activities adjacent to the rights of way, it is often appropriate to utilize the message "Trucks Entering or Leaving the Highway."

Other signs. There are various other signs that are miscellaneous in nature as authorized by the MUTCD. Most of these types of signs are used in support of local policies such as for low-volume roads, residential streets, school bus stop ahead, restricted sight distance at intersections, and others. However, several of these types are worthy of further discussion.

Logo and TODS. Since passage of the Federal Highway Beautification Act of 1965, some states have utilized specific service signing as an alternate, or to supplement billboard or advertising signs. These specific service signs provide motorist information regarding food, gas, lodging, and camping by display of a specific name or business symbol (logo) for the given service. Likewise, some states use standard signing to provide directions to significant traffic generators. These include recreational or historical attractions in the state, and they are identified with Tourist Oriented Directional Signs (TODS) for these facilities. Such signs are addressed in the *Uniform Manual,* and states are urged to develop compatible policies, guidelines, and administrative procedures for successful implementation of logo and TOD programs. The Transportation Research Board (TRB) has recently published a synthesis that provides helpful insight on this for both logo and TOD signs.[36]

Airport signs. Special attention has been given to the subject of airport signing by an Institute of Transportation Engineers Committee. A report from Committee 5-D-1 on airport roadway guide signs addresses the design process, the signing plan, and integration of the roadway signing into the airport design process.[37] Also, Chapter 12 of the *Transportation Planning Handbook* provides additional guidance on the integration of airport design and signing.

Driveway signs. A primary purpose of a public highway is to provide access to commercial establishments. However, there exists a bewildering conglomeration of reflectorized, nonreflectorized, lighted, and nonlighted signs with varying messages directing motorists in and out of many driveways to fast-food restaurants, banks, service stations, and other facilities. Some business activities provide access points

---

[35] ITE Technical Council 4 A-A, "Yield Sign Usage and Applications," *Transportation Engineering* (Institute of Transportation Engineers, Arlington, VA), 48 (October 1978), 37–43.

[36] NCHRP 162, National Academy of Sciences, Washington, DC, August 1990.

[37] ITE Committee, "Airport Roadway Guide Signs," *ITE Committee 5D-1,* March 1984.

with no signs at all. The result can be a very confusing traffic pattern jeopardizing the effective use of the facility. ITE Technical Council Committee 4A-3 addressed this subject to assist in providing improvement in the standardization of these signs.[38]

Wrong-way signs. Some divided roadway conditions present a confusing environment to the motorist. These include locations where wrong-way movement is possible such as with a combination on-and-off ramp, wide median intersections, and divided intersections with acute-angle geometry. In each case, specific improvement is possible with the application of a series of appropriate signs including, but not restricted to, the wrong-way sign.[39]

Milepost signs. Properly installed milepost markers can assist drivers in estimating their trip progress. They provide a means for identifying the location of emergency incidents and aid in highway maintenance and service. Mileposts may be erected on any section of a highway route. The zero mileage in any state usually begins at the south or west end where the route begins. The milepost marker is minimally a 6- by 9-in rectangular shape with 4-in white numerals, used on low-volume, low-speed rural roads. The standard size of 10 by 27 in or 10 by 36 in with 6-in reflectorized numerals on a green reflectorized background is normally applied to major roadways. Milepost signs should be mounted at minimum height and lateral placement in line with delineators. When a milepost system exists on the roadway, delineators can be keyed to the same system by spacing them at a 1/10 or 1/20 of a mile separation.

Chevron signs. This special-use warning sign is extremely effective when utilized on the one or two curves along a route that have a greater sharpness than all the others. However, they are used to supplement and not replace normal curve signing. The signs are spaced on the outside shoulder of the road in such a manner that the driver always has two of the signs in view while traversing the curve. They are also effective on loop ramps and on other horizontal curves that require a significant speed reduction.

Traffic marking warrants

Although pavement markings have definite limitations, their advantages include:

o Guide motorists without diverting attention from roadway.
o Has specific criteria for proper installation.
o Widely understood and accepted by motorists.

The *Uniform Manual* has specific guidelines for the use of markings in the United States. Some of these guidelines and studies include the following six areas:

Centerlines. It is desirable to maintain the centerline to separate the directions of traffic flowing in opposing directions. They should be used on paved highways when the following conditions are present:

1. A two-way road in rural districts is at least 16 ft in width and speeds are in excess of 35 mph.
2. The highways are in residential or business districts where there are significant traffic volumes.
3. An undivided highway has four or more lanes.
4. At other locations where an engineering study indicates a need.

Lane lines. These should be installed on multilane highways when an engineering study indicates the roadway will accommodate more lanes than when the road is not painted. In addition, lane lines should be installed:

1. On one-way streets where maximum efficiency and use of street width is desired.
2. At the approaches to important intersections and dangerous locations where the lines would better organize the traffic for roadway use.
3. On rural highways with an odd number of traffic lanes.

No passing zone. These markings are required on all roadways with centerline markings and shall be used at horizontal or vertical curves where the sight distance is less than the minimum necessary for safe passing at the prevailing speed of traffic. Conditions that justify such zones are set out in tables that correlate traffic speed vs. minimum sight distance.

Passing sight distance on a vertical curve is the distance in which an object 3.75 ft above the pavement can be seen from a point 3.75 ft above the pavement. Passing distance on a horizontal curve is measured in a similar manner. That is, measured when an object 3.75 ft above the pavement can be seen on a line tangent to an obstruction that cuts off the view on the inside of the curve.

Railroad crossing markings. These should be placed where rail highway grade crossing signals or automatic gates are operating and at all other crossings where the prevailing speed is 40 mph or greater.

Pavement edge markings. Should be placed on all interstate highways and on other roads where edge delineations are desirable to reduce driving on paved shoulders or refuge areas.

Crosswalks. These should be marked where studies show they could improve safety where there is substantial conflict between vehicles and pedestrian movements. They should also be installed at points of pedestrian concentration or where pedestrians cannot recognize the proper place to cross the roadway.

Sign materials, maintenance and inventory practices

Types of materials

Traffic signs are fabricated from a various assortment of materials including plywood, aluminum or steel sheeting,

---

[38] ITE Technical Council 4A-3, "Driveway Identification Signing and Marking," *ITE Journal* (September 1986), 13–15.

[39] John C. DuFresne, "Wrong Way Signing: A New Approach," *Technical Notes* (February 1982), 4–10.

fiberglass, and plastic. Each product has both strengths and weaknesses that must be evaluated for the specific use of the sign. For example, length of life, amount of maintenance, weight of the material, ease of handling, the accommodation of lighting, and the cost-effectiveness are general controls that lead one to a specific material.

Aluminum. Aluminum sign blanks should conform to the American Society for Testing and Materials (ASTM) specification B-209 alloy, 6061-T6 or 5154-H38. They are easily fabricated into any one of the standard shapes and sizes required for signing. A chemical conversion treatment provides coating conforming to ASTM-B449-67, Class 2. This produces a tight iridescent coating, and a satisfactory basis for application of reflective or nonreflective sheeting and paint. It also imparts increased corrosion resistance.

Steel. If commercial sheet steel is utilized, it must be coated or rust-proofed by one of the various chemical treatments recommended by the manufacturer. Use of 18-gauge steel is adequate for small signs up to 24 in for the longest dimension, whereas 16-gauge steel is recommended for any larger signs.

Plywood. High-density overlay plywood can be used for signs but is not readily available nor always a cost-effective material. High-density overlay plywood is available in thicknesses from 1/4 to 1 in, although thickness in the range 1/2 to 3/4 in is customarily used. Plywood sign backing has certain advantages over aluminum and steel because the need of additional back-bracing is minimized, and its tolerance to gunshot damage is higher.

Extruded panels. Signs used on structures over the roadway are usually larger and therefore appropriate for extruded or laminated aluminum panels. These can be bolted on the support structure with special clips manufactured for that purpose. Illinois has had success in using a louver sign that accommodates the visibility and legibility needs of the motorist while reducing the windload significantly so that lighter supports can be used or larger signs can be mounted on existing supports.[40]

Fiberglass or plastic. Sign blanks made with high-strength fiberglass or reinforced plastic have been used successfully for signs of all sizes. The material is often available at lower cost than aluminum. It can offer the advantage of specially colored backsides of the sign panel to be aesthetically pleasing in sensitive areas.

Sign face materials. The materials used for sign faces include paint, adhesive-coated plastic film, porcelain enamel, reflective sheeting, and reflective coatings of beads and a binder. Cutout letters, numerals, and symbols can be made of plastic reflective buttons in a porcelain enamel frame or of reflective sheeting. If signs are to be used exclusively in a daylight operation or if they are adequately

illuminated for nighttime visibility, consideration can be given for using paint, nonreflective plastic film, or porcelain materials in the legend. The following are minimum details that should be stipulated in any specifications or purchase orders for materials:

1. The colors are within the allowable tolerance in both daylight and night lighting conditions.
2. The material is weather resistant.
3. Minimum brightness levels are met for reflective signs.
4. There is adequate durability.
5. Materials are packaged and shipped so as to minimize damage.

State and federal agencies can provide guidance on specifications for sign materials. Usually these agencies have done extensive testing and/or research to determine the most cost-effective materials.

Maintenance of signs

An important element in effective traffic engineering is a maintenance management system for street and highway signs. Some of the pertinent considerations are listed below, but a detailed discussion can be found in NCHRP 20-5, 16-03.[41]

Inspections. To keep traffic control signs effective, adequate maintenance must be performed. Signs should be inspected during both daylight and darkness on a regular basis, several times per year, to assure that visibility and legibility have not become impaired.

The inspection process should identify those signs that require cleaning or replacement and to generate a work list for maintenance attention. Nontraffic engineering employees of the department may be recruited to participate in the sign inspection process. The use of simple forms can record damaged or obscured signs found at various locations. These employees can take notes on roads required for their travel during the regular course of their activity, and coupled with coordination to a master file, an entire system can be quickly covered. Obviously, signs located in heavy industrial areas will require a higher degree of maintenance. Inspections should assure that weeds, trees, shrubbery, or construction materials have not obscured the face of any highway sign.

Routine maintenance. This includes washing the sign with a good detergent or soap material as well as the removal of the sign in order that it might be cleaned with a steam generator, refurbished, or otherwise replaced in the cycle.

Illuminated signs demand greater attention. Because of their larger size, an occasional check is necessary for weather damage to either the sign or the mountings. It is quite common as a result of wind or lightning to have the mounting fixtures slip off-center. Additionally, lamp standards must be checked to eliminate hot spots that tend to wash out the effectiveness of the sign legend. Therefore, a regular cycle for lamp

---

[40] Illinois Dept. of Transportation, "Evaluation of a Louvered Panel for Use as a Freeway Background Sign," Division of Highways, Bureau of Traffic, Springfield, IL, 1971.

[41] NCHRP 20-5, 16-03, "Maintenance Management of Street and Highway Signs," Transportation Research Board, National Research Council, Washington, DC, April 1989.

inspection should be established so that the level of illumination can be maintained at a predetermined standard. Obviously, extreme temperatures of cold and hot will affect the lighting systems the most.

Special activities. Overhead signs and other freeway-mounted signs require special maintenance consideration since it is necessary to work over the traffic lanes. Problems can often be reduced by scheduling the work during the evening or light travel hours so that risks to motorists are reduced. Many of these signs require the removal of the legend so that the background can be washed and the legend replaced after refurbishment. Oftentimes the background panels can be overlaid before the new legend of reflective buttons or reflective sheeting copy is replaced.

Marking materials maintenance

General. Paint and glass beads have traditionally been the most common materials used for pavement and curb markings. However, there are newer materials now available that are more durable and sometimes more effective during inclement weather. While they may have a higher initial cost, the advantages of these newer products include lower maintenance, less interruption of traffic, better visibility, and increased legibility in winter.

Painted lines. Technological improvements also continue to be made in traffic-paint materials as well as beads, their gradations, and methods of application. The reflectorized line that has been installed at the minimum cost will serve 6 months or more on heavily traveled streets and highways. In most states, the highway department requires an application-wet-film thickness of traffic paint at 15 mil. The beads must be applied at a rate of approximately 6 lbs of drop-on beads per gallon of paint. The beads should have a standard refractive index of 1.5 or greater. Owing to large-quantity purchases, many states will prepare their own specification on paint and beads. Local jurisdictions often reference the state specification or obtain materials under a state contract.

One of the biggest improvements in paint has been in the development of rapid dry binders that can be applied with modified existing equipment on low-heat or with specially constructed high-heat equipment. Their main advantage is less disruption to traffic flow. As the costs of the materials are further reduced, there will be little difference in the applied cost. This is because the amount of equipment can be reduced and the need for cones or other devices during the drying period can be eliminated.

There are several methods used for the selection of traffic paint, but both the final selection and the purchase are generally based on price. The established specifications usually include the following:

1. A performance specification with a laboratory and service test procedure to be used to rate the submitted sample. Usually a committee composed of representatives of traffic engineering, materials testing, and purchasing evaluate the paint performance. Qualities evaluated may include general daylight appearance, color, film condition, bead retention,

and reflectance. Various rating methods have been used to evaluate these qualities.[42] Presently, both the FHWA and AASHTO are establishing regional traffic paint test and evaluation facilities. Their purpose is to select and to test materials for the states represented in the region. This reduces the cost of testing for an individual state and promotes uniformity among the states.

2. A formulation specification based upon rigid analysis of the ingredients and generally subject to defined laboratory test and occasional service test. It is impossible to determine all characteristics of a paint on the basis of a chemical analysis. Therefore, the specifications should include both lab and service tests for best results. Usually the buyer reserves the right to inspect the manufacturing facility. This type of specification should guarantee the buyer a paint well fitted to one's requirements. However, the buyer must have the means of determining the paint formula best suited for the area climate and road surface conditions. There are several serious disadvantages to this method: (a) The buyer will probably not be taking advantage of new developments in paint formulation unless they utilize a paint chemist whose own product development keeps pace with those of the paint manufacturer; (b) some paint manufacturers claim that this type of specification tends to stifle research and development; and (c) it may be impossible quantitatively to analyze all ingredients in a paint because of chemical changes during processing.

3. A specification that is by brand-name product or equal to the brand. This method is not generally advisable because it subjects the buyer to complaints of favoritism or discrimination. Many other factors must be considered to determine the equality of two products.

Both the maintenance procedure and schedule for traffic markings are generally dependent on the type of road, traffic volume, and the number of lanes. A schedule should be established that will provide for a usable line at all times. This scheduling can include many miles of streets and highways that require an application of paint only once a year. However, major arterials can require paint applications in the spring and fall or more often if weather permits. The maintenance procedure calls for the repainting of a new line directly over an earlier application. As needed, the old line should be cleaned when covered with dirt or pre-spotted by applying temporary marks on the highway when the previous lines have been worn away or eliminated by resurfacing. All resurfaced roads and new construction should be painted before they are opened to traffic.

Long-life lines. Hot- and cold-rolled thermoplastic markings have been successfully applied for a number of years in numerous countries throughout the world. The chief attribute of a thermoplastic marking is the serviceable life of the marking beyond that of paint.

Both hot and cold thermoplastic can be used for long-life applications or for crosswalks and legends. Experience in

---

[42] "A Model Performance Specification for the Purchase of Pavement Marking Paints," *Revised Standard, Transportation Engineering* (ITE, Arlington, VA), 46 (January 1976), 36–43.

Pennsylvania indicates that cold-rolled plastic is best used in areas with good ambient lighting and during a resurfacing contract. The material can be rolled into the surface during the final phases of the pavement surface composition.

The application of sprayed or extruded thermoplastic material with glass beads requires special equipment and is usually applied under a service purchase contract. Such contract should include specification for all materials, construction requirements, and warranty provisions.

Recently, new products such as the two-component thermosetting epoxy and polyester lane-striping materials have shown great promise. They outlast paint lines on the more heavily traveled streets and highways. In cold climates, it has been impossible to maintain a painted line through the winter months because of high traffic density, studded tires, snow plows, and the use of large quantities of abrasive materials. Thermoplastic, epoxy, and polyester lines appear to be one answer in these cases. Like painted lines, maintenance procedures and schedules call for reapplication of long-life markings of compatible material as the earlier line wears out or loses reflectorization. Thermoplastic markings are best applied prior to freezing weather. An excellent discussion on the qualities of existing marking materials is found in NCHRP 138.[43]

Temporary tape. Prefabricated tape markings with adhesive backing designed to conform closely to the texture of the roadway surface have found wide use for temporary markings. They are especially useful on projects where traffic must be rerouted during construction and for semipermanent markings such as pavement symbols, parking stalls, and parking lot markings.

In addition to providing the center and edge lines during a construction project, temporary lines are finding more and more use on new pavements. States are now requiring the contractor to install 1- to 4-ft strips of reflective tape on various spacings at the end of each day's paving operation. The FHWA has recently specified that a 4-ft stripe of not more than 50-ft spacing is required at the end of each day's paving, but these requirements are currently being reconsidered by the FHWA. These temporary markings are effective since they provide some guidance to the motorist until the permanent lines can be installed, which may be as long as a week or 10 days.

Glass beads. Glass spheres are used with many binders to provide night visibility where there is no continuous roadway lighting. These beads can be either premixed with the binder (i.e., paint) prior to application, dropped on by a gravity method, or applied by a combination of premix and drop-on. The beads are used commonly in paint or thermoplastic materials. The features of glass bead retroreflectivity are illustrated in Figure 8–25.

Specifications for glass beads are written primarily for the bulk purchase of beads for the drop-on application method. Specifications are written to cover the premix

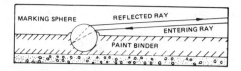

FOR FULL "RETRO-REFLECTIVITY" GLASS BEADS MUST HAVE 50 PERCENT OF THEIR SPHERE EXPOSED TO LIGHT SOURCE. BEAMS STRIKE THE FAR SIDE OF THE MIRRORED INTERIOR AND "BOUNCE" BACK TOWARD THE LIGHT SOURCE.

THE DIFFERENCE IN ENTRY AND EXIT ANGLES MEANS THE LIGHT BEAM WILL RETURN TO A POINT ABOVE THE LIGHT SOURCE TO THE DRIVER'S EYES. THE REFRACTING ANGLE ACTUALLY CORRECTS FOR THE 18" DIFFERENCE BETWEEN THE HEADLIGHTS AND THE DRIVER'S EYES.

RETRO–REFLECTIVITY

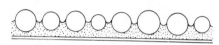

BISYMMETRIC BONDING BEADS

BISYMMETRIC BONDING BEADS ARE SPECIALLY COATED BEADS WHICH BOND AT THEIR EQUATOR FOR UNIFORM BEAD EXPOSURE, FIRM BONDING AND HIGH INITIAL REFLECTIVITY. THEY ARE GRADED FROM 40-80 MESH AND APPLIED AT 6 POUNDS/GALLON. SINCE THEY ARE A DROP-ON FLOTATION BEAD, A GREATER PERCENT OF BEADS ARE AVAILABLE FOR IMMEDIATE REFLECTORIZATION. HOWEVER, BECAUSE OF THEIR CONCENTRATION IN THE UPPER PART OF THE PAINT BINDER, THEY MAY PROVIDE A SOMEWHAT REDUCED REFLECTIVITY AS THE PAINT LINE IS REDUCED IN THICKNESS BY TRAFFIC WEAR.

BEAD SIZE RELATED TO DRY–FILM THICKNESS

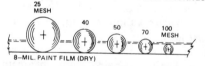

USING A PAINT BINDER OF 8-MIL DRY THICKNESS, THE 25 MESH BEAD IS TOO LARGE AND WILL QUICKLY WEAR OFF. THE 50 MESH WILL PROVIDE IMMEDIATE REFLECTIVITY AND WEAR LONGER, WHILE THE BURIED, 100 MESH BEAD HELPS MAINTAIN REFLECTORIZATION AS THE STRIPE WEARS DOWN. A MIXTURE OF BEAD SIZES WILL MAINTAIN FULL REFLECTORIZATION DURING THE PAINT BINDER'S LIFE.

Figure 8–25. Glass bead characteristics.
SOURCE: USDA FOREST SERVICE, *Placement Guide for Traffic Control Devices*, 1981, pp. 67, 69.

paint application method to a lesser degree. The essential requirements involved in a major glass bead specification are:

1. Gradation
2. Color and shape
3. Imperfections
4. Crushing strength
5. Index of refraction
6. Silicone content
7. Chemical stability
8. Reflectivity
9. Packaging

Marking equipment and operation

Paint marking machines. The varieties of paint machines range from small gravity-fed, manually operated equipment to large truck units designed for precision high-output striping work. Basic application problems make the

---

[43] NCHRP 138, "Pavement Markings: Materials and Application for Extended Service Life," Transportation Research Board, National Research Council, Washington, DC, June 1988.

smaller manually propelled equipment more efficient for painting word messages, stop lines, and other transverse markings. Supplemental equipment on the larger machines includes bead storage and dispensers to apply reflecting glass spheres. This equipment usually provides skipping devices to produce broken lines, paint pumps to load paint tanks, and various amenities to assist the operating crew.

Some machines are equipped with both inboard and outboard paint guns for maximum longitudinal marking flexibility. These machines can paint several lines at the same time and perform various changing patterns at low operating speeds. This is ideal for painting operations that are conducted on multilane highways as opposed to rural secondary roads. Extras found on the equipment include supplemental vats for holding up to 900 gal of paint, auxiliary paint guns and bead dispensers, paint heaters, paint pumps, dual steering, two individually steerable paint carriages for inboard and outboard operation, flashing arrow boards, and a complete communication system. The speed of operation varies considerably with the type of machine and the type of work. Under ideal conditions, approximately 15 to 30 machine miles of striping can be applied in a given 8-hour day.

On sections where there is no previous stripe available for guidance, a construction joint can be followed or a manual layout provided by placing spots of paint along the roadway. These are accomplished with a three-person crew, one flagging traffic while the other two make periodic measurements to premark a spot for guidance of the striping equipment. On tangents, control points should be not more than 600 ft apart. On curves, control points should be placed at closer intervals to assure the accurate location of the line. On multilane highways where many lines are parallel, offset attachments on many striping machines allow the painting of all parallel lines from only one layout line.

Continuing experiments by both operating agencies and manufactures of paint and paint equipment are underway to improve the application characteristics and durability. The common causes for premature paint failure include:

1. Insufficient cleaning of pavement
2. Over-thinning of paint
3. Damp or wet pavement on application
4. Application on windy days or when the temperature is below 40°
5. The presence of limestone or other alkaline materials that tend to break down the paint, causing it to be washed away
6. Insufficient paint film applied because of poor quality paint or improper mixture.

Glass beads can be entered separately into the paint with the aid of a special dispenser attached to a standard striping machine. Where the beads are premixed, however, the paint may be sprayed directly with conventional equipment.

Thermoplastic equipment. Long-life traffic lines can be installed by manually propelled thermoplastic machines for small applications such as crosswalks and stop lines or by large self-propelled machines with large tanks for heating reflectorized thermoplastic material. In either case, the machine must heat the material to approximately 400° F, with supplemental equipment such as bead storage and dispensers for spraying reflecting glass spheres onto the hot lines.

Skipping devices are a necessity for lane line and centerline applications if any volume of material is to be applied. If pavement markings are contracted to other firms, the contract should specify the following:

1. Material to be used
2. Amount of premixed glass beads
3. Amount of glass beads to be applied independently to the stripes
4. Width of line and the total length of the project
5. Minimum application rate
6. Warranty if desired

Temporary tape dispensers. Although temporary tape has been used for a number of years, it has generally been applied by hand. The procedure requires removing a backing material, placing the pressure-sensitive adhesive on the pavement, and rolling the tape. With increased emphasis placed on maintaining traffic lines through construction areas, new mechanical methods are now available to place lines quickly. Any person using temporary lines should review the problem with the tape manufacturer's representative.

Other markers. Object signs or markers can be mounted on the object and, if needed, placed on the approaches. Widespread use of concrete barriers has promoted a need for special delineation on the barrier. Drivers confronted with opposing traffic headlight glare are especially in need of assistance. The most favorable location for this delineation is on the barrier face, 26 in above the pavement.[44]

The best equipment for driving a post prior to mounting delineators or object markers is an air-operated or electrical post driver. Various models are available, from light post drivers for ground rods and stakes, to large-capacity, heavy-duty drivers for larger posts. A post extractor is very useful for pulling out damaged posts.

Inventory

Advantages. The inventory of official traffic control signs can be an important tool for the traffic engineer. In addition to helping establish control for the inventory process, maintenance, modernization, reduced tort liability, and management efficiency can benefit from an inventory program.

Such a system can be developed either by a manual survey or by review of a photo-logging process that can pinpoint deficiencies or damaged signs. This is usually a significant effort on the first attempt, but it can be more easily updated and kept current after it is once available.

Objectives. There are three objectives of a sign inventory:

1. To classify all traffic signs by condition, location, and size of sign and post by day, and reflectivity by night.

---

[44]Dwin U. Ugwoaba, "An Evaluation of Portable Concrete Barrier Delineators," *Technical Paper,* Transportation of Engineers, Seattle, WA, pp. 274–278.

2. To discover the conditions requiring change in design or size of signs, refurbishing, or reflectorization.
3. To establish existing and future sign needs and a plan to upgrade signs systematically to current standards.

Each governmental agency will want to develop a field-inventory procedure that will allow traffic control devices to be identified and tabulated with existing computer-processing equipment or inventorying processes. Local governments often want data broken down only by the nearest block.[45] Larger agencies, however, often require that locations be pinpointed with detailed systems like a milepost log.[46] To accommodate either system, the following dozen factors should be considered minimal needs in the process:

1. Block number or route number
2. Side of the roadway (north, south, etc.)
3. Direction the sign is facing
4. Sign code
5. Type post
6. Distance from the intersection or other reference point
7. Offset distance from the roadway
8. Installation date
9. Condition of sign
10. Other signs mounted on the same post
11. Condition of other signs
12. Visibility factor to the motorist.

Bulk storage. A bulk inventory system can be maintained by agencies using signs that have been completed or stored in a sign shop. This may be done in the form of inventory for raw materials such as sign blanks, post, etc., and/or in terms of finished sign product. Such an inventory is invaluable for a cost-effective operation that tends to backlog no more than a reasonable amount of product or finished signs to assure an efficient full-time operation with existing personnel.

Additionally, the inventory can be an aid to the monitoring of work completed by field crews or as the basis for budget forecasting. It also facilitates coordination of central headquarters operations with numerous field units. Many local jurisdictions like Columbus, Georgia, have benefited from a computerized material inventory system. Columbus developed an inventory of 1,370 items that could be grouped into five basic categories.[47]

Sign shop operations

General

The method of sign shop operations in a given agency is dependent on a cost-effectiveness analysis of local sign

fabrication vs. purchase of manufactured signs. Factors in the analysis include:

1. Number of signs to be used
2. Cost of labor
3. Time factor
4. Local policies on purchasing.

In addressing these issues, supplementary questions must be answered. Can exterior workers be used in the off-season as shop employees? Can emergency needs be filled? Are warehousing facilities available to stock needs in advance? Can economies be developed in the relation of bulk purchase vs. budget? Does the agency perform routine maintenance on major equipment needed for manufacturing?

Equipment needs

Sign production is greatly simplified if sign backings are purchased in proper blank sizes and ready for sheeting application. Then nothing further is required for painting or the application of sheeting. When additional processing is needed, a separate well-ventilated area should be provided, including a spray booth or a filtration system to purify the air of paint vapor.

Although some or most blanks are purchased to size, facilities are generally needed to fabricate other sign sizes. Large shears can be used to cut sheet steel or aluminum for original sign blanks and to salvage damaged signs by making smaller blanks. On the other hand, power saws are a necessary item in fabricating plywood signs.

Additional equipment depending on the size of the sign shop operation will include:

1. Equipment to paint or coat the new or used sign blanks
2. Machines to apply reflective sheeting, plastic film, or glass beads
3. Equipment to silk-screen and stencil messages
4. Ovens or infrared drying units and necessary drying racks
5. Storage racks, bins, and shelves.

In addition, some sign shops have facilities for stripping old finishes from reusable sign panels. Some stock special materials as countermeasures to discourage vandalism. These include property ID seals, vandal-resistant sign face material, agency decals for back of signs, tamperproof hardware and fasteners, and other items. Additional security is maintained by increasing the height of street-named signs and securing the post foundation to the ground. Toughnut, Vandalgard, Signfix, and Lexon are special-use products that are vandal-resistant sign mounting hardware.[48] See Figure 8–26.

In recent years, commercial companies have furnished the specialized service of refurbishing aluminum blanks for several states including Pennsylvania, Ohio, and West Virginia, at a cost of less than half the original cost of new aluminum sign blanks.

[45] H.A. Swanson, "An Urban Sign Inventory Procedure," *Traffic Engineering* (Institute of Traffic Engineers, Washington, DC), 40 (May 1970), 44–46.

[46] T.K. Datta and B.B. Madson, "Photologging: A Data Collection Method for Roadway Information Systems," *Transportation Engineering* (Institute of Transportation Engineers, Arlington, VA), 48 (April 1978), 19–25.

[47] "Computerized Materials Inventory: A Useful Tool," Institute of Transportation Engineers, pp. 6–8.

[48] Everett C. Carter and Himmat S. Chadda, "Sign Vandalism: Time for Action—Now," *ITE Journal* (August 1983), 16–19.

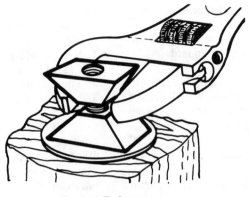

Typical Tufnut
(for 3/8″ Carriage Bolt)

VANDLGARD-NUT—INSTALLATION AND REMOVAL

INSTALLATION

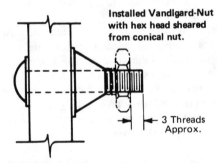

Installed Vandlgard-Nut
with hex head sheared
from conical nut.

← 3 Threads
Approx.

1. **Install Vandlgard Nut
by tightening hex until
it shears.**

REMOVAL

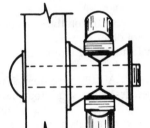

2. **Squeeze both nuts
firmly with vise-grips
and remove both nuts
together.**

Figure 8–26. Vandal-resistant sign fastener hardware.
SOURCE: USDA Forest Service, *Placement Guide for Traffic Control Devices,*
1981, pp. 13, 14.

Facilities for sign layout, letter and symbol design, and letter spacing will be needed as a close adjunct to the sign shop. Means for servicing illuminated signs must also be available.

Practices

Sign shop practices vary from state to state and from municipality to municipality as dictated by policy decisions. In some states, prison labor may cut letters and sign blanks from sheet metal, while in other states convicts may furnish the completed sign to the highway department. There are two common practices that evolve when highway or transportation departments manufacture their own signs. These are:

1. All signs are manufactured and supplied from one central, statewide sign shop.
2. Signs are manufactured at the district or regional level.

Obviously the selection of one of these processes calls for a different means of warehousing and cost-accountability than the other. Often an agency will have small routine signs manufactured on regional levels and reserve all of the large overhead or specialized signing for freeways to be manufactured or purchased by a central agency. In those cases, the central agency merely provides a backup for necessary replacement, overlay, repair, or temporary use.

Many cities over 50,000 population have sign shops that manufacture all or a part of their signs. The amount of work varies depending on the size of the municipality and its policies. Some buy fully fabricated signs (sign faces already applied to the sign blank) from commercial suppliers, while some buy completed faces and just apply them to the sign blanks. Others will apply sheeting to the sign blank and screen a message or else do all the signs that do not need a reflective background.

Governmental agencies should continue to review their policies of materials being used and their purchasing procedures. Guidance is available for local requirements from many sources. These include manufacturers of shop equipment, manufacturers of sign material, and from other agencies that already have sign shops in operation at the state, city, or county level.

REFERENCES FOR FURTHER READING

Manual on Uniform Traffic Control Devices for Streets and Highways, U.S. Department of Transportation, Federal Highway Administration, U.S. Government Printing Office, Washington, DC, 1988.

NCHRP 138, "Pavement Markings: Materials and Application for Extended Service Life," Transportation Research Board, National Research Council, Washington, DC, June 1988.

HR 20-5/20-03, National Academy of Sciences, Washington, DC, 1990.

National Institute of Standards, "Evaluation of Colors for Use on Traffic Control Devices," U.S. Department of Commerce, NISTIR 88-3894, Gaithersburg, MD.

TRB, "Transportation in an Aging Society," Transportation Research Board, Special Report 218, Washington, DC, October 1988.

Traffic Control Devices Handbook, U.S. Department of Transportation, Federal Highway Administration, U.S. Government Printing Office, Washington, DC, 1983.

9

TRAFFIC SIGNALS

ROY L. WILSHIRE, P.E., *Senior Vice President*

Kimley-Horn and Associates, Inc.

Introduction

No other device has such a daily impact on virtually every citizen as does the common, ever-present traffic signal. The trip to work is punctuated by stops at traffic signals, even on uncongested routes. School children obediently wait for the traffic signal to interrupt traffic so they can cross the busy thoroughfare. Drivers place their physical safety and that of their passengers confidently (sometimes too confidently) in the signal's ability to give them the right-of-way. A traffic signal's necessity is accepted by the public, and in fact demanded in some cases, as the solution of choice to assure safety and mobility.

In the scale of human history, traffic signals represent a rather recent development—with the world's first traffic signal using colored lights being installed in 1868 in London, England.[1] Since then, the development and application of traffic signals paralleled the development and use of the automobile, borrowing technology from railway signaling practice.

In recent years, the explosive development of computer and electronics technology has virtually eliminated all barriers to the implementation of highly flexible, sophisticated traffic control strategies. The capabilities of the tool no longer constrain its use. Instead, successful application is now limited by the practical ability of drivers to respond to variable, highly flexible control, and the jurisdictional issues associated with implementation.

Traffic signals are described as:

any power-operated traffic control device, other than a barricade warning light or steady burning electric lamp, by which traffic is warned or directed to take some specific action.[2]

This generalized definition is broad enough to include such things as intersection traffic control signals, lane-use control signals, and freeway entrance ramp meters. Even though signals have gained acceptance and use in freeway management systems, this chapter will deal primarily with the function and application of traffic control signals at street intersections, and their combination into interconnected systems.

Legal authority

For traffic control signals to achieve their operational and safety purpose, their message must be clearly understood and strictly obeyed. Of paramount importance is the requirement that signals be installed and operated in accord with proper legal authority so as to make compliance legally enforceable.

Because of the need for uniformity in the meaning of various signal indications, national standards have been developed. Chief among these are the *Manual on Uniform Traffic Control Devices for Streets and Highways (MUTCD),*[3] *Uniform Vehicle Code,*[4] and the *Model Traffic Ordinance.*[5] These

---

[1] *Manual on Uniform Traffic Control Devices for Streets and Highways,* U.S. Department of Transportation, Federal Highway Administration, Washington, DC, U.S. Government Printing Office, 1988.

[2] Ibid.

[3] Ibid.

[4] *Uniform Vehicle Code,* Washington, DC, National Committee on Uniform Traffic Laws and Ordinances, rev. 1968 and Suppl. 1, 1972, Secs. 11-201 through 11-206, 15-102, and 15-106.

[5] *Model Traffic Ordinance,* Washington, DC, National Committee on Uniform Traffic Laws and Ordinances, rev. 1968 and Suppl. 1, 1972, Secs. 4-1 through 4-9.

documents generally outline the legislation suitable for establishing the authority for traffic signal installations, the meaning of the signal indications, and the required obedience to these displays by motorists, bicyclists, and pedestrians. State and local jurisdictions, then, must individually adopt or enact these provisions in suitable legislation or ordinances to make them legally enforceable in that jurisdiction. Engineers should be familiar with the codes and ordinances applicable to their jurisdictions in order to assure compliance.

Benefits and drawbacks

Traffic signals that are appropriately justified, properly designed, and effectively operated can be expected to achieve one or more of the following:

- To effect orderly traffic movement through an appropriate assignment of right-of-way.
- To provide for the progressive flow of a platoon of traffic along a given route.
- To interrupt heavy traffic at intervals to allow pedestrians and cross-street traffic to cross or to enter the main street flow.
- To increase the traffic-handling ability of an intersection.
- To reduce frequency of occurrence of certain types of accidents.

In contrast, traffic signal installations, even though warranted by traffic and roadway conditions, can be poorly designed, ineffectively placed, improperly operated, or inadequately maintained. The following results from improper or unwarranted signal installations may occur:

- Increased accident frequency (especially of the rear-end type).
- Excessive delay for motorists and pedestrians.
- Disregard of signal indications.
- Use of less adequate routes in an attempt to avoid such signals.

Contrary to common belief, traffic signals do not always increase safety and reduce delay. Experience has indicated that, although the installation of signals may result in a decrease in the number (and severity) of right-angle collisions, signals will, in many instances, result in an increase in rear-end collisions. Further, the installation of signals may not only increase overall delay but may also reduce intersection capacity. Consequently, it is of utmost importance that the consideration of a signal installation and the selection of equipment be based on a thorough study of traffic and roadway conditions by an engineer experienced and trained in this field. This engineer should recognize that a signal should be installed only if the net effect expected to occur (balancing benefits vs. drawbacks) will improve the overall safety and/or operations of the intersection.

Determining need for traffic signal control

Signal warrants. In today's litigious society, it is obvious that the presence, absence, or improper operation of a traffic signal is a fertile field for legal claims and actions. This makes it even more evident that some system for establishing the need for a signal installation at a particular location is necessary. Such a system has been established, using a common denominator known as *signal warrants.*

The warrants for traffic signals are thoroughly described in the MUTCD. Because the MUTCD is considered to be a basic reference available to all traffic engineers, the warrants are not included here. It should be noted, however, that the 1988 version of the MUTCD includes significant changes to the warrants, specifically adding 4-hour volume, peak-hour delay, and peak-hour volume warrants. Individual states can also publish a version of the MUTCD, which should be referenced. These warrants should be considered as a guide rather than absolute criteria, and their use tempered with professional judgment based on experience and consideration of all related factors.

It should be noted that the MUTCD states that

> The satisfaction of a warrant or warrants is not in itself justification for a signal. . . . If these requirements are not met, a traffic signal should neither be put into operation nor continued in operation (if already installed).[6]

Stated differently, this means that the warrants represent the *lowest* threshold at which traffic signals *could, not should,* be installed. It should be recognized that even though traffic volumes may meet the minimum requirements for a traffic signal installation, the requirements for a signal should be thoroughly analyzed with the decision to install based on a demonstrated traffic need.

Clearly there is also a need periodically to examine existing traffic signal installations to make sure they are justified. For those that may no longer be needed, a practical procedure for signal removal is included later in this chapter.

Required studies. The decision to install a traffic signal should be based on a thorough investigation of physical and traffic flow conditions at the candidate site. This investigation will provide the data necessary for input into signal warrant analyses, and may also be used in the actual signal design process. Specific studies that are required to gather the necessary data include:

- *Traffic volume studies:* Intersection approach volume counts and pedestrian counts recorded at frequent intervals.
- *Approach travel speeds:* Spot speed studies on intersection approaches to determine the speed distribution of individual vehicles.
- *Physical condition diagrams:* An "as-built" scaled diagram showing physical conditions of the intersection and its approaches, including geometrics, channelization, grades, sight-distance restrictions, bus stops and routings, parking conditions, pavement markings, street lighting, driveways, location of nearby railroad crossings, distance to nearest signals, utility poles and fixtures, existing signals, and adjacent land use.

---

[6] *Manual on Uniform Traffic Control Devices.*

o *Accident history and collision diagrams:* Collision diagrams plot, by means of arrows and symbols, the paths of vehicles (or pedestrians) involved in all accidents at a specific location over some time period, usually a year. The symbols indicate types of collisions, vehicle types, time periods, severity classifications, lighting conditions, weather conditions, etc.
o *Gap studies:* Determining the size and number of gaps in the major road traffic can produce a measure of the equivalent number of adequate gaps (gap availability parameter) to serve side street traffic. Relating side street volumes and gap availability can show the average side street delay.
o *Delay studies:* Two methods are commonly used to measure the amount of intersection delay—the stopped time delay method and the travel time method.

For further information on study procedures, refer to Chapter 3, "Traffic Studies," and the *Manual of Traffic Engineering Studies.*[7]

Establishing priorities. Within agencies responsible for the traffic control function, it is not uncommon to find that there are a number of locations where the minimum warrants for traffic signals are met, but adequate resources are not available to permit immediate installation. In these situations, agencies must establish priorities as to which traffic signals are to be installed first.

How are such priorities set? In some agencies it is simply done on a chronological basis—that is, by date of warrant satisfaction. In other instances, the funding program and its restrictions may influence installation date. For example, state funds may exist for installations on designated routes, whereas the local city has no funds. Or, a city capital improvement program (CIP) may have been approved that provides funds for specific locations and prohibits shifting of funds to other locations.

Obviously a better approach would be to develop priorities on a defensible basis of need, perhaps taking into account the relative "degree" of warrant satisfaction and focusing on those locations with severe safety needs. Having detailed data available as described above will provide the facts on which meaningful priorities can be set. A priority list of signal improvements should not only include new installations but should also include upgrading of existing installations.

Signal control equipment

A functioning traffic signal installation at a typical intersection includes several components, including such fundamental elements as controller assemblies, signal heads, detectors, interconnecting wiring, and associated hardware. The specific equipment needed for each installation is generally determined through an engineering design

process that identifies the type of control required, the physical requirements of the specific location, and the need for uniformity and standardization.

The Institute of Transportation Engineers (ITE) has approved several standards for basic signal equipment. Among these are:

1. "Pedestrian Traffic Control Signal Indications," 1985
2. "Vehicle Traffic Control Signal Heads," 1985
3. "Traffic Signal Lamps," 1986
4. "Lane-Use Traffic Control Signal Heads," 1980
5. "Solid-State Pretimed Traffic Signal Controller Units," 1988
6. "Controller Cabinets," Proposed Equipment Standard, 1988

Other agencies have also produced equipment specifications that are appropriate for use in selecting signal equipment, including:

1. NEMA Standards Publication No. TS 1-1989, Traffic Control Systems, National Electrical Manufacturers Association, and subsequent revisions.
2. IMSA Wire and Cable Specifications, 1984.
3. James H. Kell and Iris J. Fullerton, *Manual of Traffic Signal Design,* Institute of Transportation Engineers, Prentice-Hall, 1982.
4. James M. Giblin and others, *Traffic Signal Installation and Maintenance Manual,* Institute of Transportation Engineers, Prentice-Hall, 1989.

In the following sections, each of the fundamental elements of signal equipment comprising a traffic signal installation is described.

Controller units

Definitions. In the discussion of traffic signal control equipment in this chapter, two phrases that are used extensively are defined as follows:

o *Controller Assembly*—The complete electrical mechanism mounted in a cabinet for controlling the operation of a traffic control signal.
o *Controller Unit*—That portion of a controller assembly devoted to the selection and timing of traffic signal displays.

Control types: An evolution. There are two basic types of control strategies implemented by controller units—*pretimed operation* and *actuated operation.* Early-day signal control was primarily of the pretimed version, where a fixed sequence of intervals of fixed duration was used to assign the right-of-way. In 1928,[8] an advance was made that permitted the controller unit to respond to actual vehicle demands, and the application of traffic-actuated control concepts began. Through the years, traffic-actuated control grew to include

[7]P.C. Box and J.C. Oppenlander, *Manual of Traffic Engineering Studies,* 4th ed., Institute of Transportation Engineers, Washington, DC, 1976. This manual is currently being updated, with publication due in 1991.

[8]Gordon M. Sessions, *Traffic Devices: Historical Aspects Thereof,* Institute of Transportation Engineers, 1971.

semi-actuated control, full-actuated control, volume-density control with "added initial, time waiting–gap reduction" features, and various other approaches. Today, virtually any control type can be implemented by a single controller unit, including some very sophisticated multi-phase, variable-phase sequencing, lead-lag left-turn alternative phasing combinations, even within coordinated system operation. (A discussion of phasing and timing is included later in this chapter.)

Pretimed operation. With basic pretimed control, a consistent sequence of signal indications, each of fixed duration, is regularly repeated. Although regularly repeated, this fixed timing can be varied on a time-of-day (or scheduled) basis, by operator command, or in response to variable-measured traffic patterns. This type of control is most appropriately used where traffic volumes and patterns occur on a stable, predictable basis. Other situations where pretimed control is appropriate include coordinated systems at locations where:

- ○ traffic progression is required on more than one phase, and
- ○ the needs of progressive movement are such that allowing the early return of the coordinated phase green, as often happens with actuated control, would result in unnecessary subsequent stops. For example, pretimed control is desirable at closely spaced intersections where complex phasing is used and precise interval timing is critical to efficient platooning of various traffic movements.[9]

Traffic-actuated operation. Actuated operation allows the controller unit to respond to the short-term variable nature of traffic flows as measured by traffic detectors and pedestrian push buttons. Complete phases with no demand can be skipped, and green intervals can be extended to match measured demand. Where this flexible timing is present on all phases, the controller unit is classified as *full-actuated.* Actuated controller units that have at least one phase that will not respond to detector or push-button activity are *semi-actuated.* It should be noted that full-actuated controller units in a system are generally operated in a semi-actuated mode when they are under system control.

Hardware evolution. If the progressive development of control concepts is considered to be *evolutionary,* then the development and change associated with traffic signal controller unit hardware can be called *revolutionary.* Early-day pretimed signal control was implemented by electromechanical controller units composed of synchronous motors, rotating dials, and cams that opened and closed electrical circuits. Vacuum-tube technology was subsequently applied to actuated controller units, followed by the use of solid-state, transistor, and, finally, digital electronic components. Today, microelectronic circuitry and computer technology with solid-state light switching dominate the new hardware available in the marketplace, making it possible for a more

universal, reliable controller unit to function as a pretimed, actuated, isolated or system (intersection or ramp) type controller unit.

Local operating agencies, however, do not typically operate in such a utopian situation. Their controller units have been acquired over many years from multiple vendors, embodying a full spectrum of hardware technology. The challenge is to manage the operation of this mix of hardware to provide interchangeability among different manufacturers' controller units while maintaining a high level of service to accommodate increasing demands.

NEMA and Model 170 approaches. Controller unit development has reached the stage where there are two basic design concepts, or approaches, in use—the NEMA standard and the Model 170. Figures 9–1 through 9–4 illustrate current versions of these controller units.

Under the NEMA standard approach, signal equipment manufacturers have voluntarily and cooperatively developed standards, under the auspices of the National Electrical Manufacturers Association (NEMA), to which their controller units must conform in order to provide compatibility and interchangeability among controller units. These standards describe physical, functional, interface, and environmental requirements for actuated controller units and auxiliary equipment. When provided in accordance with NEMA TS-1 standards, these units do not provide internal preemption, coordination, or communication capability; they permit only one vehicle detector per phase; and they provide a maximum of two-ring operation. NEMA-standard controller units are manufactured in various electronic arrays by several signal equipment vendors and are typically sold to users complete with software to satisfy the user's requirements. In practice, there are a number of NEMA+ controller units in use. These are NEMA standard units with additional connector(s) to permit internal and integral preemption, coordination, and communication capability. Unfortunately, there is no

Figure 9–1. NEMA-type controller unit. Courtesy, Traffic Control Technologies.

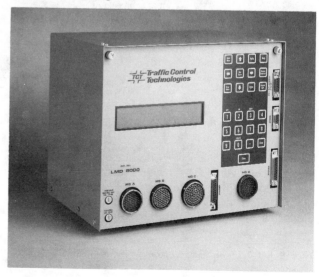

[9] Fred L. Orcutt, Jr., *Traffic Signal Applications Philosophy and Recommended Practice.* Published by Orcutt Associates, Fort Worth, TX, 1988.

Figure 9–2. NEMA-type controller unit. Courtesy, TRACONEX.

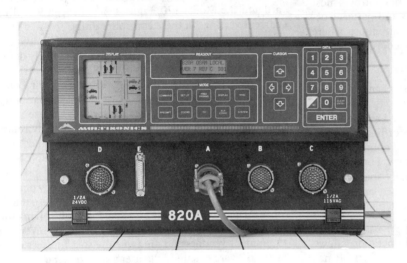

Figure 9–3. NEMA type controller unit. Courtesy, Winko-Matic Co./Multisonics Corp.

Figure 9–4. Model 170 controller unit. Courtesy, Signal Control Co.

standardization of connectors, so interchangeability, when permitted by logic voltages, requires adapter harnesses.

The Model 170 controller unit was described in a specification[10] jointly developed, with variations, by the states of California and New York. Rather than being a functional

specification, the Model 170 approach uses a hardware specification for standardization and provides a controller assembly as a package of standardized modules, complementary hardware and wiring harnesses, all housed within standard cabinets. Both California and New York (as well as other states) have continued to refine and update the specification, while preserving interchangeability, to reflect technological advancements and user needs. The controller unit includes a microprocessor that, with the proper software,

[10] "Type-170 Traffic Signal Controller System—Hardware Specifications." FHWA-IP-78-16, Federal Highway Administration, Washington, DC, 1978.

can perform in a variety of control applications. Software is typically supplied by the using agency or acquired from a third-party vendor. In practice, both approaches rely on microprocessors to implement very flexible control concepts. Prospective users should give careful consideration to both approach concepts and should recognize that standardization, however conceived or applied, is a means to the end of achieving safe, effective, and energy-efficient traffic control.[11]

Controller assembly

As defined earlier, the controller assembly includes the complete electrical mechanism configured to control the operation of a traffic control signal. It includes the controller unit and an array of auxiliary equipment, all mounted or housed in an appropriate enclosure called a cabinet.

Cabinet. The traffic signal control cabinet provides protection and support for the controller unit and auxiliary devices in an outdoor environment. An equipment standard for controller cabinets,[12] published by the Institute of Transportation Engineers (ITE), offers guidelines concerning materials, dimensions, configuration, mounting, internal wiring, and accessories.

Cabinet size is simply a matter of choosing an enclosure with sufficient space to physically house the various components (see Figure 9–5). Cabinet mounting is affected by size as well as the space available within the control area. For example, small pole-mounted or pedestal-mounted cabinets may be the preferred choice in dense downtown areas with cramped sidewalks, while base-mounted cabinets are used in outlying areas. Aesthetics play a role in cabinet selection and installation, and cabinet finish and color is a matter of individual preference and cost. However, cabinet finish and color may be of operational significance in hot climates where reflective characteristics and heat buildup are affected.

Model 170 cabinets and NEMA controller cabinets also tend to differ. Rack-mounted components are the norm for Model 170 cabinets, whereas shelves are provided in the NEMA approach on which are placed the various components, each of which is provided with its own enclosure. The design of cabinets should also consider the following practical factors:

- o *Access:* Cabinets contain expensive devices and potentially hazardous electrical currents. Access to cabinets should be restricted by appropriate locks to provide physical security for the equipment, and protection of the public from harmful electrical currents. Cabinets should be rainproof.
- o *Ventilation:* A ventilation system provides for the circulation of air to dissipate heat in order to keep the internal environment within tolerable levels. Filters should be provided to minimize dust, insects, and pollutants. Where necessary, the ventilation system should include a thermostatically controlled fan.
- o *Convenience receptacle:* Provision of an electrical outlet to accommodate test equipment is a thoughtful and practical consideration of the maintenance technician.
- o *Lighting fixture:* Signal control equipment can fail at night, and a light fixture within the cabinet can facilitate repairs/replacement. In cold climates, an incandescent fixture may also be used to provide heat to keep the cabinet's interior environment within a tolerable range.

Physical location of cabinet. In selecting a physical location for the controller assembly and cabinet, the following factors should be considered:

- o For safety reasons, the cabinet should be located as far away from the roadway as possible, taking care not to allow the cabinet to pose a vision obstruction for motorists.
- o Avoid exposing the cabinet to accidental damage caused by passing traffic.
- o The cabinet should not be placed where it would be hazardous to pedestrians, joggers, or bicyclists.
- o The cabinet location, when practical, should permit a person working on the controller unit to have a clear view of all intersection approaches.
- o The cabinet should be easily accessible and, if possible, near an area for parking repair vehicles.
- o The cabinet should not be located in a drainage ditch, in an area that could be under water, or where it can be exposed to direct saturation from landscape irrigation systems.

[11] R.L. Wilshire, and others, *Traffic Control System Handbook,* U.S. Department of Transportation, Federal Highway Administration, Washington, DC, FHWA-IP-85-11, 1985.

[12] "Proposed Equipment Standard—Controller Cabinets," ITE Technical Council Committee 7S-5, *ITE Journal* (July, 1988).

Figure 9–5. Traffic signal controller cabinets. Courtesy, Hennessy Products, Inc.

Additional equipment/features. Additional equipment includes separate units used with a controller unit to provide control functions and features. Some of these include:

1. *Flasher:* A mechanical or solid-state device used alternately to open and close signal circuits at a repetitive, frequent rate. Flashing operation is used when a controller unit malfunctions and the malfunction is detected by the conflict monitor; at night and during periods of low traffic; as an interim measure prior to full removal of a signal; or at new installations prior to automatic operation.
2. *Load switch:* The task of switching electrical power to signal heads and individual lamps is accomplished by solid-state load switches.
3. *Preemptor:* A unit (or software feature) to provide a special sequence of signal indications in response to an external command from a fire station, emergency vehicle, or other legitimate input.
4. *Auxiliary panel:* A panel assembly located behind an auxiliary door in the cabinet that permits the selection of flash/automatic (normal) operation, or lights off/on operation.
5. *Conflict monitor:* A device that externally monitors controller unit operation and signal indications. Upon detecting a conflict, absence of red, or improper operating voltage, the device automatically places the signal in flashing operation. An example of conflict monitor is shown in Figure 9–6.

Local controller coordination

Need for coordination. The coordination of signal operation between adjacent intersections offers an opportunity for significant benefits to motorists. On open highways, traffic flow is characterized as being random in that it is not normally influenced by upstream interruptions. Its arrival at a point is generally uniform throughout a selected time interval. In contrast, traffic flow on urban streets is generally less uniform because of interruptions, and it tends to

flow in pulsed groups of vehicles, or platoons. Signal coordination simply attempts to recognize this flow characteristic and coordinate signal operation to accommodate platoons with minimal stops. The success of signal coordination is influenced by the following factors:

o signal spacing
o prevailing speed of traffic
o signal timing (cycle length and split)
o magnitude of flow (volume)
o platoon dispersion
o midblock storage or contributions of traffic (e.g., parking garages).

When to coordinate. A simple tool for use in determining whether to try coordination is the *coupling index analysis.*[13] The coupling index is a dimensionless number computed as the ratio of two-way hourly volume between candidate signals to the distance, in feet, between the signals. Stated as an equation, the coupling index is:

$$I = V/L \tag{9.1}$$

where: I = coupling index
V = two-way volume, in vehicles/hr
L = distance between signals, in ft.

Use of the index has shown that a level of 0.5 or greater indicates that the signal should be coordinated during the time period examined.[14] This analysis also has use in defining the boundaries of subsystems within a controlled network of signals.

Obviously, the use of a mechanical "index" is not the only factor or approach to use in making the decision to coordinate signals. Safety and driver comfort (satisfaction) are

[13] H. Nathan Yagoda, Edward H. Principe, C. Edwin Vick, and Bruce Leonard, "Subdivision of Signal Systems into Control Areas," *Traffic Engineering,* 43, (September 1973), 42–45.

[14] Ibid.

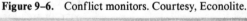

Figure 9–6. Conflict monitors. Courtesy, Econolite.

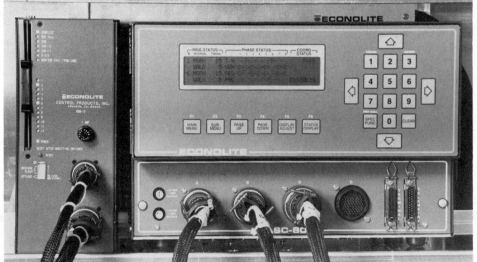

among the factors to be considered. Safety can be enhanced through progressive movement along thoroughfares where stops and delays are reduced. Driver comfort and satisfaction are influenced by the expectation of "system" operation where traffic moves smoothly with few stops, and trip times are generally repeatable along the same route. In fact, the layperson's view of good signal timing is where progression permits continuous movement with no random stops. Individual motorists understand and are able to observe route continuity and consistency but not optimized systemwide measures of effectiveness.

Methods of interconnection. Once a decision has been made to coordinate a group of intersections, there are two basic ways to interconnect the signals—direct means and indirect means. Direct methods employ a physical connection between controller assemblies; indirect methods rely on an air path or time-based approach. These are further described as follows.

Electrical cables (wires). The most widely used communication medium for traffic signal interconnection is the direct wire cable. Early systems employed multiconductor cables, typically seven-wire (AWG #14), that when energized singly or in various combinations of circuits would select appropriate dials, offsets, splits, and special functions.

Telephone-type cables. Sophisticated systems in use today employ digital data communication techniques rather than energized circuits, and the direct wire cables contain telephone-type twisted wire pairs. These wire pairs may be leased from a local communications company or installed and owned by the jurisdictional agency. In either event, the number of wire pairs can be minimized by using modems at both ends to encode and decode (multiplex) the data.

Coaxial cable. Coaxial cable (sometimes called "coax") has a large signal information capacity and has come into widespread use for cable-TV systems. Its "broadband" characteristics make it well suited for the transmission of video signals within narrow slices, or channels, of its bandwidth. Frequently, spare capacity is available on the cable-TV system's cable network to accommodate the traffic control function where one or two channels are sufficient. Shared use of the cable-TV system requires that the cable network be duplex—that is, capable of sending signals in both directions. Shared use of the cable-TV network should be carefully examined to assure an understanding of the potential problems of reliability, provision and control of maintenance services, tariff adjustments in a regulated environment, and other issues encountered by users of this approach. A user-owned coaxial cable network is also an attractive alternative, particularly in situations where closed-circuit TV (CCTV) is to be used, or where high-capacity communication trunk lines are needed—such as along a freeway for a freeway management system.

Fiber optic cable. Technology has advanced fiber optic cable to the point where it is analogous to a multicoaxial cable, but at a much smaller physical size and with enormous increases in capacity. Rather than transmitting electrical pulses, fiber optic cables accommodate multiplexed light pulses, through one or more fibers. For traffic signal applications, fiber optic cables are attractive for trunk communications because they are smaller and may be able to be placed in existing cramped ducts, and they are immune to electrical interference.

Radio. Radio represents an indirect method of interconnection that is not widely used, but that can be effective in certain situations. As generally used, it is a one-way form of data communication from a central transmitter to multiple receivers at intersection controller assemblies.

Time-based coordinators (TBC). One of the most popular advances in signal coordination has been the time-based coordinator. Working without cables, this device (often a microprocessor) functions as a very accurate clock to supervise a controller unit locally by the transmission of sync pulses and commands, much like a system master unit. By similarly equipping several adjacent controller assemblies, each operating from the same reference point, a high level of coordinated system operation can be achieved. The primary disadvantage is that two-way communication is not achieved.

Detectors

Types of detectors. Detectors in their transportation application have historically either recognized the presence of a moving or stopped vehicle, recognized passage of a moving vehicle by completing a circuit, or recognized changes in an electrical or magnetic field. Most detectors are comprised of three primary components: the sensor, the lead-in cable, and the interpreter/receiver (usually called an amplifier although it must do much more than simple amplification). When coupled with a sophisticated controller unit, detectors are used to derive volume, vehicle speed, lane occupancy, queue lengths, and to infer congestion, incidents, stops, and delays. The accuracy of these various derived parameters varies widely and needs to be understood before applying these data.

A number of detector types have been used, including the following:

1. Mechanical-force detectors
   o Pedestrian push buttons
   o Pressure (treadle) detectors
2. Energy-pattern change detectors
   o Magnetic
   o Magnetometer
   o Sonic
   o Radar
   o Inductive loops
   o Microwave
   o Infrared

Table 9–1 compares the functional capabilities of various types of detectors together with their advantages and disadvantages. Examples of shelf-mounted and rack-mounted detectors are shown in Figures 9–7 and 9–8.

Detector installation. Of all the factors affecting proper operation of detectors, perhaps the most important is proper installation. This is particularly true for those detectors with sensors located in the pavement. Proper

TABLE 9–1

Types of Detectors

Detector	Measuring Capability				Queue Length	Method of Operation	Advantages	Disadvantages
	Count	Presence	Speed	Occupancy				
Loop	Yes[a]	Yes	Yes	Yes	Yes	Vehicle passage cuts magnetic lines of flux that are generated around the loop, thereby increasing the inductance so that a change is detected and transmitted to an amplifying circuit	1. Size and shape of detection zone can be easily set by size of loop 2. Excellent presence detector 3. Capable of measuring all traffic parameters 4. Relatively easy to install 5. Relatively inexpensive to abandon loop and reuse amplifier at new location 6. Capable of detecting small vehicles 7. Under roadway location and not subject to damage except in poor pavement	1. Cost of installation may be excessive 2. Requires closing of traffic lane or lanes for short periods of time
Magnetic, nondirectional	Yes	No	No	No	No	Vehicle passage over wire coil embedded in roadway disturbs earth's lines of flux passing through coil and induces a voltage in the coil; voltage is amplified by high-gain amplifier to operate detector relay	1. Under roadway location and not subject to damage 2. Relative ease of replacement 3. Low maintenance	1. Nondirectional 2. Difficult to set detection zone 3. Subject to false calls where located near large DC lines 4. Cannot detect presence 5. Does necessitate closing of traffic lanes for installation
Magnetic, directional (two-coil version)	Yes	No	No	No	No	Same method of operation as nondirectional magnetic detector	1. Directional 2. Not affected by DC lines in vicinity 3. Well-defined detection zone 4. Low maintenance 5. Under roadway location and not subject to damage 6. Relative ease of replacement	1. Requires closing of traffic lane for installation 2. More expensive than nondirectional magnetic detector 3. Cannot detect presence
Magnetometer	Yes	Yes	No	Yes	Yes	Similar method of operation as nondirectional and directional magnetic detectors; makes use of small cylindrical sensing head that is placed below pavement surface; measures change in earth's magnetic field caused by vehicle	1. Relatively easy to install 2. Capable of measuring count or presence 3. Reliable 4. Not affected by DC lines in vicinity 5. Under roadway location and not subject to damage 6. Relative ease of relocation	1. Requires closing of traffic lane for installation 2. More expensive than nondirectional 3. May double count some vehicles due to magnetic material distribution 4. Poorly defined detection zone
Pressure	Yes	No	No	No	No	Weight of vehicle closure of metallic contacts to complete a circuit	1. Well-defined detection zone 2. Rugged construction 3. Reliable 4. Capable of detecting all moving vehicles, regardless of speed	1. Counts axles, which yields poor count accuracy 2. Does not measure presence 3. Installation may disrupt traffic for excessive period of time 4. Major resurfacing will require the use of a frame extension 5. Cannot be easily relocated

TABLE 9–1 (Continued)

Detector	Measuring Capability				Queue Length	Method of Operation	Advantages	Disadvantages
	Count	Presence	Speed	Occupancy				
Radar	Yes	No	Yes	No	No	Passage of vehicle reflects radar microwaves (Doppler principle) back to antenna to operate detector relay	1. Immune to electromagnetic interference 2. Does not necessitate closing of traffic lanes to install	1. Relatively expensive to purchase and install, particularly if existing poles not available for use 2. Requires FCC license to operate 3. Requires experienced personnel for installation and maintenance 4. Does not measure presence
Sonic, pulsed	Yes	Yes	Yes	Yes	Yes	Emits bursts of energy at a rate of approximately 20 times per sec; vehicle reduces wavelength resulting in the return signal arriving when receiver is open	1. Does not necessitate closing of traffic lanes to install 2. Does not require FCC license to operate 3. Can be used at locations with unstable pavement 4. Can classify vehicle by height	1. Same as 1 for radar detector 2. Somewhat inaccurate because of conical detection zone and wide variations in vehicle configurations and heights 3. Nondirectional 4. Sensitive to environmental conditions 5. Somewhat inaccurate under congested conditions
Sonic, continuous wave	Yes	No	Yes	No	No	Operates on Doppler principle, same as radar	1. Same as 1, 2, and 3 for pulsed sonic detector plus improved accuracy for speed measurement	1. Same as for 1, 4, 5 for pulsed sonic detector
Light emission photoelectric	Yes	Yes	Yes	Yes	Yes	Passage of vehicle between light emitter and photoelectric cell interrupts transmitted beam which operates a detector relay	1. Accurate for vehicle passage in a single lane 2. Most suitable for conditions of uniform light	1. Inaccurate for detection of more than one traffic lane
Infrared, interrupted beam	Yes	Yes	Yes	Yes	Yes	Same as photoelectric detector using infrared part of spectrum	Same as photoelectric	Same as photoelectric
Infrared, reflected beam	Yes	Yes	Yes	Yes	Yes	Overhead transmitter-receiver notes vehicle passage by change in reflectivity between vehicle and pavement	1. Most suitable for conditions of uniform light	1. Expensive 2. Sensitive to ambient light and color of pavement 3. Sensitive to weather conditions 4. Inaccurate because of reflective difference
High-intensity light (emergency vehicle detector)	Yes	Yes	No	No	NA	High-intensity light emitted from a device mounted on a priority vehicle is received by a light-sensitive detector indicating its presence	1. Provides a means to recognize selected vehicles	1. Expensive 2. Required equipment on vehicles

[a] Short loops (e.g., 6 ft) may be used to count. Long loops (e.g., 20 ft) do not count accurately because of the multiple occupancy on loop.

SOURCE: *Traffic Control Systems Handbook* ITE Publication No. LP-123, 1985.

Figure 9–7. Loop detector amplifiers. Courtesy, Indicator Controls Corp.

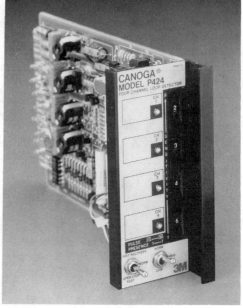

Figure 9–8. Rack-mounted detectors. Courtesy, 3M Corporation.

initial installation of these in-pavement elements is critical because

1. they are exposed to all extremes of weather, particularly moisture;
2. the pavement is subject to severe impact loading, shifting, and cracking;
3. there are typically so many detectors required that failure of even a small percentage creates a significant labor demand for timely repair; and
4. to accomplish repair, a traffic lane must generally be closed and workers exposed to accident risk. (Installation and maintenance techniques are thoroughly covered in the *Traffic Detector Handbook*[15] and the *Traffic Detector Field Manual*.)[16]

Signal heads and optical units

Definitions

1. *Signal face.* That part of a signal head provided for controlling traffic in a single direction.
2. *Signal head.* An arrangement of one or more signal lenses in signal faces that may be designated accordingly as one-way, two-way, etc.
3. *Signal section.* That part of a signal face containing an optical unit.
4. *Optical unit.* An assembly of lens, reflectors, lamp, and lamp socket.
5. *Signal indication.* The illumination of a traffic signal lens or combination of lenses at the same time.[17]

Number of signal faces and locations. A traffic signal's ability to communicate its message to a driver requires that it capture the driver's attention and be recognized in time for

[15] *Traffic Detector Handbook,* Institute of Transportation Engineers, Publication No. LP-124A, Washington, DC, 1991.

[16] *Traffic Detector Field Manual,* Institute of Transportation Engineers, Publication No. LP-125, Washington, DC, 1985.

[17] "Vehicle Traffic Control Signal Head," Standard of the Institute of Transportation Engineers, Publication No. ST-008B, Washington, DC, 1985.

the appropriate driver action. This requires the signal be of adequate brightness, proper location, and standardized configuration. For these reasons, standards have been developed concerning the number of signal faces, indication configurations, and placement of signal heads. The construction and arrangement of components of a typical signal face is illustrated in Figure 9–9.

The MUTCD specifies that, for through traffic, a minimum of two signal faces shall be provided and should be continuously visible to approaching traffic from a distance at least as far as the minimum visibility distance shown in Table 9–2. As noted, this distance varies with the 85th percentile speed. In cases where this minimum visibility cannot be provided, a suitable warning sign shall be erected, perhaps supplemented by a hazard beacon if the situation dictates. Further, required signal faces for through traffic on any one approach must be at least 8 ft apart. In addition, left-turn signal displays are also required where protected only and protected/permissive left-turn phasing is used.

Unless physically impractical because of the width of the intersecting street or other condition, at least one and preferably both of the signal faces must be located within the driver's cone of vision as described horizontally (and shown in Figure 9–10), being within 20° to the right and left of the center of the approach extended. Note that the center of the approach reflects inclusion of the left-turn lane when it is not separately controlled. At least one and preferably both of the required signal faces must be located within the 40 to 150 ft distance range beyond the stop line. Twelve-inch signals are required (1) for those faces within the 120 to 150 ft distance range (unless a supplemental near-side signal indication is used) and (2) for signal faces more than 150 ft from the stop line. In cases where the nearest signal faces are located beyond 150 ft, a supplemental near-side signal is required.

Vertically, signal heads mounted over a sidewalk or other nonstreet area must have at least 8 ft, but not more than 15 ft to the bottom of the signal face. Signals suspended over a roadway, on span wires or mast arm poles, must provide at least 15 ft but not more than 19 ft clearance.

Number and arrangement of lenses per signal face. Each vehicular signal face shall have at least three lenses, but not more than five. The lenses shall be red, yellow, or green and give a full circular or arrow-type of indication. The lenses in a signal face may be arranged in a vertical or horizontal straight line. Some frequently used arrangements of lenses in signal faces are shown in Figure 9–11. Vertical mounting of signal heads is the predominant method with horizontal mounting used to accommodate restrictive clearance situations.

Pedestrian signals. Pedestrian signal indications are intended for the exclusive purpose of controlling pedestrian traffic. They are required under the following conditions:

1. At signals installed under the Pedestrian Volume or School Crossing warrant.
2. When an exclusive pedestrian interval or phase is provided and all conflicting vehicular traffic is stopped.
3. When vehicular indications are not visible to pedestrians or cannot be conveniently viewed by pedestrians, or from

which pedestrians cannot readily and accurately deduce when they have the right-of-way.
4. At established school crossings at signalized intersections.

At complex, high-volume intersections, it may also be desirable to use pedestrian signal indications under the following conditions:

1. When pedestrian volume requires a pedestrian clearance interval to enhance pedestrian safety or minimize vehicle-pedestrian conflicts.
2. When pedestrians might be confused by split-phase, multi-phase indications of vehicular signals.
3. When pedestrians are permitted to cross only part of the street during a particular interval.

It is generally desirable to provide pedestrian signals at all new installations where there are pedestrians so as to provide them a clear and positive indication of pedestrian crossing intervals. Pedestrian signal faces are to be located at each end of a controlled crosswalk and mounted with the bottom of the housing not less than 7 ft nor more than 10 ft above the sidewalk. Either word message or symbols may be used as pedestrian signal indications. Portland orange DONT WALK or an upraised "hand" symbol, steadily illuminated, conveys the message that a pedestrian shall not enter the roadway in the direction of the indication. The flashing DONT WALK indication is used as a clearance interval during which pedestrians may complete their crossing, but not start to cross. The white WALK message or the "walking person" symbol means that a pedestrian may enter the roadway and cross in the direction of the indication. Even with a WALK indication, there may be possible conflicts with turning vehicles. However, a WALK indication may not be used when there is a conflicting arrow indication.

A previous practice of flashing the WALK indication to indicate a pedestrian-vehicle conflict was not readily understood by the public and has been deleted from the MUTCD as an appropriate indication.

Traffic signal lamps. Traffic signal lamps form a part of the signal optical unit consisting of light source, reflector, and lens. Lamps should be selected on the basis of the size of indications used, and they should conform to the latest published standards.[18] The MUTCD recommends the use of dimming devices for those locations where 12-in yellow signals with 150-W lamps are flashed at night.[19]

Hardware and mounting. Signal heads can be mounted within the driver's cone of vision in a number of ways—on poles or posts at the side of the roadway, suspended on span wires, or mounted on mast arms or signal bridges over the roadway. They can be in vertical or horizontal arrays as described earlier. A typical mounting arrangement is shown in Figure 9–12. In all cases, uniformity of display from a

---

[18] "Traffic Signal Lamps." Standard of the Institute of Transportation Engineers, Publication No. ST, Washington, DC, 1985.

[19] "Traffic Signal Dimming Devices," Technical Council Informational Report, *ITE Journal,* Institute of Transportation Engineers, Washington, DC, March 1982.

POLYCARBONATE VEHICLE SIGNAL REPLACEMENT PARTS

Ref. No.	Part Description
1	Housing, Green
	Housing, Yellow
	Housing, Black
2	Door Gasket, Neoprene
3	Door, Green
	Door, Yellow
	Door, Black
4	Hinge Pin (Quick-Click)
5	Lens Reflector Gasket
6	Connector (shown)
	Clamping washer
7	Carriage Bolt 1/4"-20 × 1 3/4"
8	Keps Nut 1/4-20
—	Polycarbonate Connector (not shown)
—	Signal Section Connector
—	Bolt 5/16"-18 × 2 1/2" Hex Head
—	Flat Washer 5/16"
—	Split Lockwasher
—	Hex Nut 5/16"-18
9	Terminal, Fast-Tab (Straight)
	Terminal, Fast-Tab (45°)
	Terminal, Fast-Tab (90°)
10	Terminal Screw (6-32 × 3/8")
	Terminal Screw (6-32 × 7/16")
11	Clip, Lens Gasket
12	10-32 × 3/8" Truss Head Screw
13	10-32 × 3/8" Fillister Head Screw
14	Visor, Tunnel
	Visor, Full Circle
15	Reflector Assembly
16	Bail, Reflector
17	Spring, Bail
18	Green Lens
	Yellow Lens
	Red Lens
	Green Arrow
	Yellow Arrow
	Red Arrow
19	Spiral Pin
20	Wingnut, Eyebolt Ass'y (Complete)

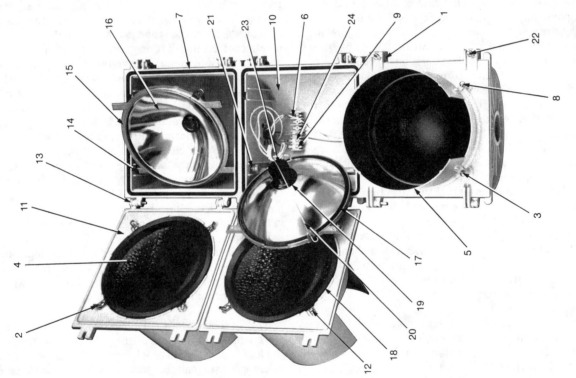

Figure 9–9. Traffic signal face. Courtesy, LFE Corporation, Traffic Control Division.

TABLE 9–2

Minimum Signal Visibility Distance for Varying Approach Speeds

85th Percentile Speed (mph)	Minimum Visibility Distance (ft)
20	175
25	215
30	270
35	325
40	390
45	460
50	540
55	625
60	715

SOURCE: *Manual on Uniform Traffic Control Devices for Streets and Highways,* U.S. Department of Transportation, Federal Highway Administration, 1988 edition, p. 4B–11.

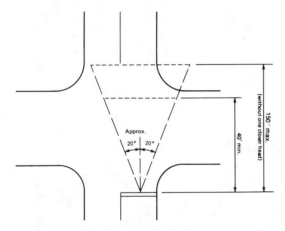

Figure 9–10. Desirable location of signal faces.
SOURCE: MUTCD, p. 4B–13.

driver's perspective should be a goal, and care should be taken to adhere to proper mounting and location layouts to enhance safety. Signal poles should not block crosswalks, should provide proper clearance from the curb, and should not be placed in locations where they would create a fixed-object hazard. Cabinets should not be placed in locations where a driver's visibility would be obstructed.

Because signal equipment is exposed to all extremes of weather, durable corrosion-resistant materials should be used for all mounting hardware. It should also be capable of withstanding expected wind loads and other local weather extremes.

Signal timing

Definitions

The discussion of signalized intersection control and related signal timing requires definitions of some basic terms. These definitions are provided below for convenient reference:

o *Cycle (cycle length).* A complete sequence of signal indications. In an actuated controller unit, a complete

Figure 9–11. Typical arrangements of lenses in signal faces.
SOURCE: MUTCD, p. 4B–9.

cycle, which includes all phases, is dependent on the presence of calls (actuations) on all phases.

o *Phase (signal phase).* The portion of a signal cycle allocated to any single combination of one or more traffic movements simultaneously receiving the right-of-way during one or more intervals.

o *Interval.* The part or parts of the signal cycle during which the signal indications do not change.

o *Phase sequence.* A predetermined order in which the phases of a cycle occur.

o *Split.* A percentage of a cycle length allocated to each of the various phases in a signal cycle.

o *Offset.* The time relationship (expressed in seconds or percent of cycle length) determined by the difference between a defined interval portion of the coordinated phase green and a system reference point.

Single intersection timing

Phasing. Phasing is the technique that alternately assigns right-of-way to conflicting traffic movements at signalized intersections. A traffic movement could be a single vehicular movement, a combination of vehicular traffic movements, a pedestrian movement, or a combination of vehicular and pedestrian movements. Precise nomenclature for defining the various signal phases and movements has been adopted and published by the National Electrical Manufacturers Association (NEMA). The assignment of NEMA traffic movement numbering is shown in Figure 9–13. Common

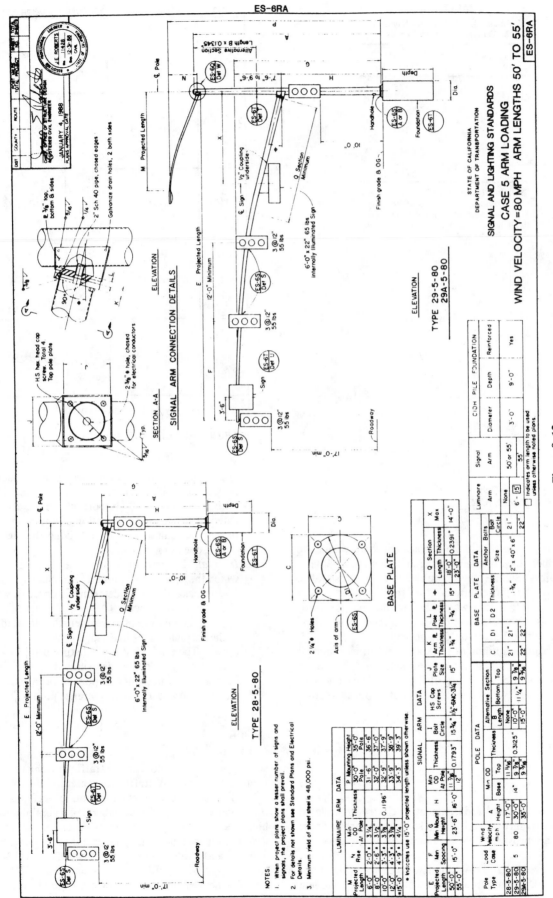

Figure 9–12.

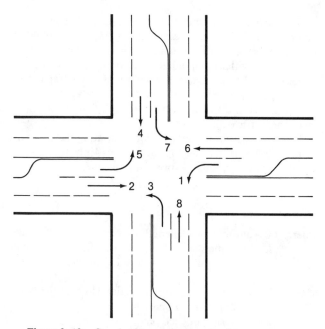

Figure 9–13. Standard movement nomenclature.

graphical techniques for representing a simple two-phase operation are shown in Figure 9–14, while a four-phased sequence is shown in Figure 9–15.

The concept of multiple-phase sequencing is illustrated in Figure 9–16 for two-, three-, and four-phase operation.

Figure 9–14. Single-ring 2-phase signal sequence.

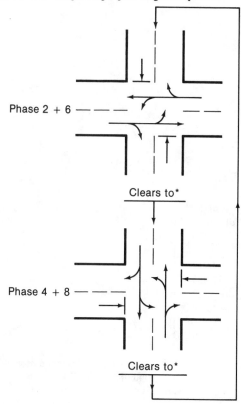

*Change and clearance intervals not shown

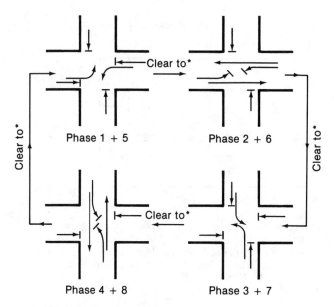

Figure 9–15. Single-ring 4-phase signal steady-demand sequence.

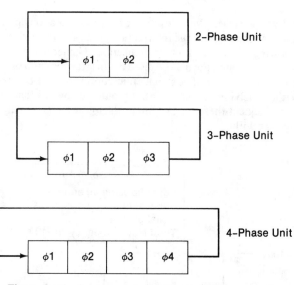

Figure 9–16. Sequence of phases—single-ring controller unit.

SOURCE: National Electrical Manufacturers Association (NEMA), *Traffic Control Systems* (Standards Publication-TS 1-1976). New York, 1976.

Note that the sequence of moving from phase to phase represents a loop, or "ring." Figure 9–17 expands this to illustrate two interlocking rings (hence "dual-ring" controller units), which are arranged to time in a preferred sequence and to allow concurrent timing of both rings, subject to the restraint imposed by the barrier, or compatibility line. At the barrier line, both rings cross simultaneously to select and time the phases on the other side.

In the "single entry" mode of operation in a dual-ring controller unit, a phase in one ring can be selected and

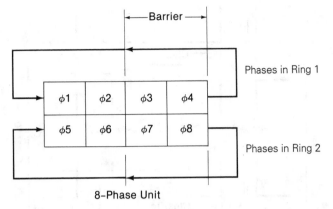

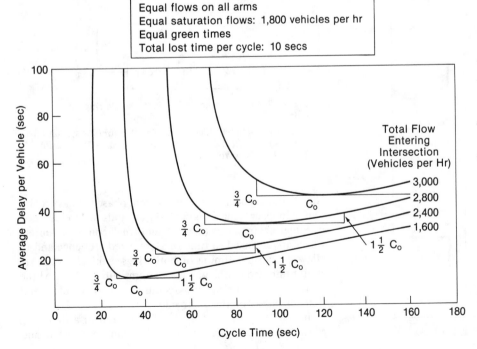

Figure 9–17. Sequence of phases—dual-ring controller unit.

SOURCE: National Electrical Manufacturers Association (NEMA), *Traffic Control Systems* (Standards Publication-TS 1-1976). New York, 1976.

timed alone if there is no demand for service on a nonconflicting phase of the parallel ring. The "dual entry" mode of operation, by contrast, requires the timing of a nonconflicting phase in the second ring.

Cycle lengths. Vehicles stopped at an intersection do not instantaneously enter the intersection at minimum headways following display of a green indication. Numerous studies have documented the delay caused by this start-up/stop operation created by a traffic signal's interruption of traffic flows. Without delving into specifics, it should be obvious that the more frequent these interruptions, the lower the anticipated capacity will be.

Just such a relationship has been developed as shown in Figure 9–18, where a nominal range of cycle lengths of 40 to 120 sec is suggested. The conclusion to be made is that longer cycles will accommodate more vehicles per hour, and cycles longer than necessary to accommodate the traffic volume present will produce higher average delays. The best rule is to use the shortest practical cycle length that will serve the traffic demand.

Webster[20] developed a model for computing the approximate cycle length that will minimize total intersection delay. His classic formula is expressed as follows:

$$C_o = \frac{1.5\,L + 5}{1.0 - Y_1 - Y_2 - \ldots - Y_n} \qquad (9.2)$$

where: C_o = optimum cycle length in sec
L = lost time per cycle, generally taken as the sum of the total yellow and red clearance per cycle, in sec
Y = volume divided by saturation flow for the critical approach in phase n
n = subscript for each phase.

Examples of the variation of delay with cycle length are shown in Figure 9–18 for multiple levels of flow. This figure further illustrates that the delay is never more than 10% to 20% above that given by the optimum cycle for cycles with 0.75 to 1.5 times the optimum cycle.[21]

---

[20] F. V. Webster, *Traffic Signal Settings,* Road Research Technical Paper No. 39. Road Research Laboratory, 1958, pp. 1–44.

[21] Ibid.

Figure 9–18. Effect on delay of variation of cycle length.

SOURCE: *Research on Road Traffic,* London, England: Her Majesty's Stationery Office, 1965, p. 306.

Vehicle signal change interval. A vehicle change interval is that period of time in a traffic signal cycle between conflicting green intervals. It is characterized by a yellow warning indication followed by a red clearance indication.

The duration of change and clearance intervals, as well as the appropriateness of red clearance intervals, is a topic where divergent views and strongly held positions are common. In 1985, the Institute of Transportation Engineers published a Proposed Recommended Practice[22] as the result of a technical committee's activity. This ITE Proposed Practice is currently being considered and discussed to arrive at alternatives that are acceptable as a Recommended Practice. This document describes a procedure that uses a formula based on a kinematic model of stopping behavior to determine the duration of the yellow indication. Next, the procedure calls for an engineering evaluation of the need for a red clearance interval and, if required, provides a second formula for its computation. The formula for computing the length of the yellow interval is:

$$y = t + \frac{v}{2a \pm 2Gg} \qquad (9.3)$$

where: y = length of yellow interval, to the nearest 0.1 sec
t = driver perception/reaction time, recommended as 1.0 sec
v = velocity of approaching vehicle, in ft/sec
a = deceleration rate, recommended as 10 ft/sec^2
G = acceleration due to gravity, 32 ft/sec^2
g = grade of approach, in percent divided by 100 (downhill is negative)

Depending on the policy of the local agency, the red clearance interval is determined by one of the following expressions:

$$r = \frac{W + L}{V}, \text{ or} \qquad (9.4a)$$

$$r = \frac{P}{V}, \text{ or} \qquad (9.4b)$$

$$r = \frac{P + L}{V} \qquad (9.4c)$$

where: r = length of red clearance interval, to the nearest 0.1 sec
W = width of intersection, in feet, measured from the near-side stop line to the far edge of the conflicting traffic lane along the actual vehicle path
P = width of intersection, in feet, measured from the near-side stop line to the far side of the farthest conflicting pedestrian crosswalk along the actual vehicle path
L = length of vehicle, recommended as 20 ft
V = speed of the vehicle through the intersection, in ft/sec

The recommended application of these formulae is to use (9.4a) where there is no pedestrian traffic, the longer of (9.4b) or (9.4c) where there is the probability of pedestrian crossings, and (9.4c) where there is significant pedestrian traffic or pedestrian signals protect the crosswalk.

Left-turn signal phasing. Left-turn signal phasing has experienced considerable change over the past few years. The power and flexibility of the microprocessor-based signal controller unit has made it possible to implement virtually any conceivable configuration of left-turn phasing, in some cases varying on a cycle-by-cycle basis in response to local demand and system progression needs.

The full range of phasing alternatives can be described in three basic categories:[23]

○ Unprotected left-turn phasing,
○ Protected-only left-turn phasing, and
○ Protected/permissive left-turn phasing.

Unprotected left-turn phasing. Unprotected left-turn phasing occurs when an exclusive phase is *not* provided for left-turning vehicles. Left turns are permitted to occur through gaps in the opposing traffic flow. Separate left-turn lanes may or may not be provided.

Protected-only left-turn phasing. When a separate interval is provided to accommodate a left turn without conflicting traffic, and left turns are prohibited during the rest of the cycle, protected-only left-turn phasing occurs.

Although the MUTCD provides no left-turn phasing warrants, the *Traffic Control Devices Handbook*[24] (Section 4C-1) offers suggested guidelines for separate left-turn phasing. These guidelines include:

○ *Volumes.* Consider further studies for separate left-turn phasing when the product of left-turning vehicles and opposing volumes during peak hours exceeds 100,000 on a four-lane street or 50,000 on a two-lane street, and the left-turn volume is greater than two vehicles per cycle.
○ *Delay.* Install separate left-turn phasing if a left-turn delay of 2.0 vehicle-hours or more occurs on a critical approach during a peak hour, and the left-turn volume is greater than two vehicles per cycle with an average delay per left-turning vehicle of at least 35 sec.
○ *Accident experience.* Install left-turn phasing if the critical number of left-turn accidents has occurred. For one approach, the critical number is four left-turn accidents in 1 year or six in 2 years. For both approaches, the critical number is six left-turn accidents in 1 year or 10 in 2 years.

Protected/permissive left-turn phasing. Protected/permissive left-turn phasing provides a protected phase (green arrow) during one interval and allows unprotected turns (on a

[22] "Determining Vehicle Change Intervals." Proposed Recommended Practice, *Institute of Transportation Engineers,* Washington, DC, ITE Publication No. RP-016, 1985.

[23] *Guidelines for Signalized Left-Turn Treatments,* Federal Highway Administration, Washington, DC, Implementation Package FHWA-IP-81-4, November 1981.

[24] *Traffic Control Devices Handbook,* U.S. Department of Transportation, Federal Highway Administration, Washington, DC, 1983.

circular green) to be made through gaps in the opposing traffic flow during another interval. One of the basic precepts of the protected/permissive technique is that the protected green arrow is displayed only when needed in a traffic demand condition. It is emphasized that the protected/permissive technique is an efficiency concept as opposed to an accident reduction concept, since it cannot provide the same degree of safety as an exclusive protected left-turn signal phase.

Lead/lag left-turn phasing. Leading and lagging left-turns are two alternatives for inclusion in protected left-turn phasing schemes. A basic set of the multiple combinations of these alternatives is shown in Figure 9–19. A leading left-turn sequence is one where the left-turn green arrow precedes the green interval for oncoming traffic. In contrast, the lagging left-turn sequence provides the protected left-turn green arrow following the green interval for oncoming through traffic. With protected intervals on opposing left-turn lanes, the dual lead, dual lag, and lead/lag sequences shown in Figure 9–19 can result.

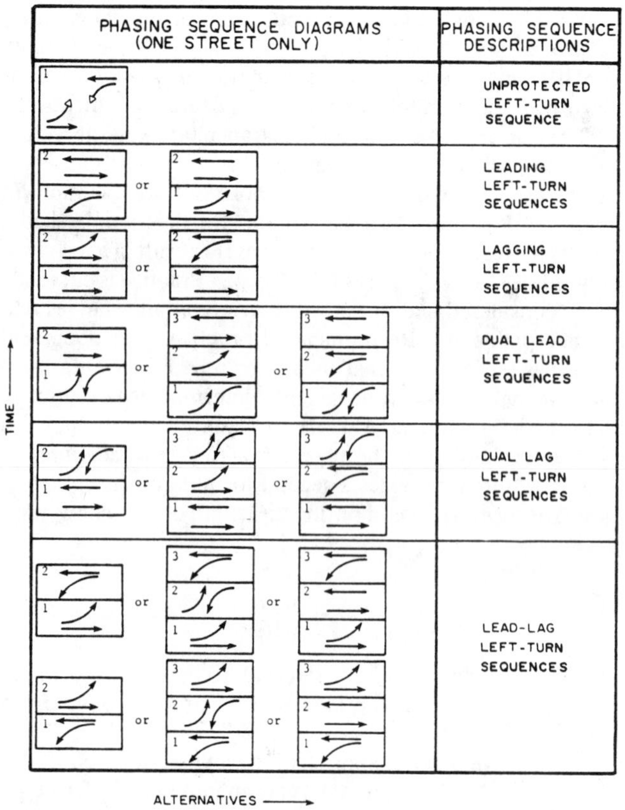

Timing for arterial routes

The basic approach to arterial street signal control is based on the concept of pulsed flow—that groups of vehicles (platoons) are released from a signal and travel in platoons to the next signal. Thus, it becomes desirable to establish a time relationship between the beginning of green at one intersection and the beginning of green at the downstream intersection such that vehicles that arrived during the red interval at the downstream intersection may receive a green before the next platoon arrives. This permits the continuous (progressive) flow of traffic along an arterial street and aids in reducing stops and delays.

This traffic control concept for arterial routes can be presented graphically by a time-space diagram, as shown in Figure 9–20. This figure introduces terms defined as follows:

○ *Through-band:* The space between a pair of parallel speed lines that delineates a progressive movement on a time-space chart.

Figure 9–19. Alternative left-turn signal phasing schemes. SOURCE: FHWA-IP-81-4.

○ *Band speed:* The slope of the through-band representing the progressive speed of traffic moving along the arterial.
○ *Bandwidth:* The width of the through-band in seconds indicating the period of time available for traffic to flow within the band.

The use of one-way or two-way operation is a major consideration in developing timing plans for an arterial street. If

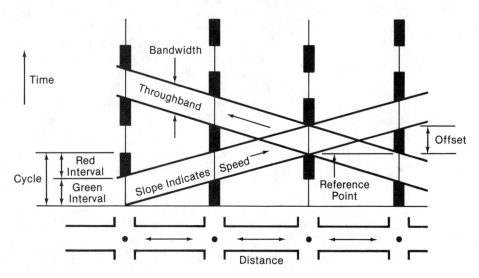

Figure 9–20. Time-space diagram.

the street is one-way, then full use of the through-band for progressive movement can be obtained. If the street is two-way, the problem of providing progression in both directions is much more difficult. Good two-way progression is dependent upon signal spacing, balanced directional flows, and intersection phasing, which allows large time periods for through-movement flows. These conditions rarely exist, and compromises must then be made in the bandwidth and progression speed achieved.

Since the offset relationship is obtained on the basis of time, it is necessary to utilize a common cycle or some multiple of a common cycle throughout the system. The cycle splits may vary at each intersection, but the point of beginning of the major street green must be constant.

Timing plan elements. To establish a control system for an arterial street (or open network), it is necessary to develop a timing plan for all of the signals in the system. Such a timing plan consists of the following elements:

o *Cycle length:* This must be the same (or some multiple) for all signals in a given area of the system. Cycle length is usually established by examining traffic flow and timing requirements at each intersection, then evaluating progression using various cycle lengths considered to be in an acceptable range, and, finally, selecting the timing plan that provides the most favorable flow.
o *Cycle splits:* A determination of the length of the various signal phases must be made for each individual intersection in the system. Phase lengths (splits) may vary from intersection to intersection.
o *Offset:* A determination of an offset value must be made for each intersection in the system. The offset is usually established with reference to one master intersection in the system.
o *Phase sequence:* A predetermined order in which the phases of a cycle occur under steady demand on all phases. The phase sequence in local controller units can be externally directed to change the order of left-turn service (e.g., dual lefts to lead/lag lefts) in order to provide more effective progression at various times of day or under various traffic conditions.

The bandwidth approach (as discussed earlier) for determining signal offsets to provide progression has been automated to utilize computer programs such as PASSER-II (and others as described later). Prior to the widespread use of computers for signal timing, the following types of timing plans were frequently employed:

o *Simultaneous:* Timing plan where a zero offset is used for several consecutive signals, along with a 50–50 cycle split. This system is utilized where extremely short block lengths occur and also where saturated flow conditions exist.
o *Single alternate:* Timing plan with half-cycle offsets that are alternated at each signal. This timing plan works well if the block lengths and/or cycle length permit the use of a half-cycle offset. For example, with a 60-sec cycle and a half-cycle offset, the offset is 30 sec.

If the block length is 900 ft, then a progression speed of 900/30 = 30 ft/sec is obtained.
o *Double alternate:* This is a variation of the single-alternate technique where two adjacent signals have the same offset and a half-cycle offset occurs every two signals. This system is often used for block lengths in the range of 450 ft. It should be noted, however, that the width of the through-band is reduced to one-half that of a single-alternate system. If the block lengths are shorter, a triple-alternate system might be utilized.

Traffic flow variations. In determining timing plans for arterial streets, one is faced with the problem of traffic flow variation. A given timing plan is developed for a given set of traffic conditions. When these traffic conditions undergo substantial change, the value of the timing plan is greatly reduced. Two basic types of traffic flow variations can occur as follows:

o *Traffic flow at individual intersections:* Traffic volumes can increase or decrease at one or more signal locations in the system. These changes can then alter the cycle-length requirements or the cycle split at the affected intersections.
o *Traffic flow direction:* The volume of traffic flow along the arterial can vary directionally on a two-way arterial. Using an inbound-outbound flow designation, three basic conditions can exist as follows:

1. Inbound flow is greater than outbound flow.
2. Inbound flow is approximately the same as outbound flow.
3. Outbound flow is greater than inbound flow.

For the first condition (usually in A.M. peak), one would want a timing plan to favor progressive movement in the inbound direction. For the second condition (off-peak flow), a timing plan that would favor inbound and outbound flow equally would be desirable. For the third condition (usually P.M. peak), a timing plan to favor outbound flow is desired.

Early control techniques usually attempted to provide at least three timing plans (A.M., off-peak, P.M.), which were selected on a time-of-day basis. Traffic-responsive control systems can now automatically adjust timing plans at shorter intervals based upon a measure of traffic flow on the arterial.

Timing plan development. In the analysis of procedures for determining timing plans for arterial streets, two basic categories of techniques can be identified as follows:

o *Manual techniques:* Manual calculations and/or graphical analysis to determine cycle length, splits, or offsets.
o *Off-line computer techniques:* Techniques that utilize computer software models to make necessary calculations to determine the timing plans. The term *off-line* indicates that the timing plans are determined and then implemented without direct interaction with the signal system or traffic flow conditions on the arterial.

While manual techniques form the basis for understanding how arterial timing plans can be developed, the process represents a trial-and-error juggling of multiple variables that can best be done using a computer. For a detailed discussion of the manual approach, the reader is referred to the *Traffic Control Systems Handbook* (p. 3.24).[25] Off-line computer techniques are most pertinent and are briefly discussed here.

Off-line computer techniques. Several computer models have been developed for signal coordination design. These models can be categorized into two broad categories. The first consists of models that maximize bandwidth. The second contains models that seek to minimize delay, stops, or other measures of disutility. The four example models discussed here are the following:

o Maximal Bandwidth
o Difference of Offsets
o PASSER II
o Arterial Analysis Package (AAP)

It is important to note that one cannot (or should not) blindly implement the computer-generated timing and offset settings. The engineer must carefully fine-tune the settings in the field based on observations of actual traffic flows.

Maximal bandwidth methods. The Maximal Bandwidth method optimizes signal offsets to produce maximal bandwidths along an arterial route given the cycle length, signal splits, signal spacing, and progression speeds, and subject to the following conditions:

1. If the platoons in both directions are equal, maximum equal bandwidths are provided for each direction of travel.
2. If the sum of the two bandwidths is greater than the sum of the two platoon lengths (in units of time), the individual bandwidths are made proportional (as far as possible) to the platoon lengths.
3. If the sum of the two bandwidths is less than the sum of the two platoon lengths, the larger platoon is first accommodated, if possible, and then as much bandwidth as can be arranged is given to the direction with the smaller platoon.

The input data required for this program are:

o Number of signals
o Cycle length range
o Average traffic volume in each direction
o Saturation flow headway
o Spacing between signals
o Minimum green durations for each movement
o Average speed in each direction between each pair of signals

Difference-of-offsets method. The British Transport and Road Research Laboratory developed a technique for optimizing offsets in a fixed-time signal timing plan for an arterial or network based on the minimization of vehicular delay. Given the traffic flows, the common cycle length, and the splits, the vehicular delay along a traffic link connecting a pair of signals depends on the departure and arrival patterns at the downstream signal and hence on the difference of offsets between the two signals.

From a knowledge of arrival and departure rates throughout the cycle at the downstream signal, the delay on the link can be computed for different values of the offset. For a two-way street section, the offset values are first weighted by the respective directional arrivals, and then combined to form an overall delay value.

In general, the Difference-of-Offsets method has perhaps the soundest logical base among the various offset-optimization techniques for arterial routes. This is because a quantitative traffic measure, namely delay, is directly considered and systematically minimized; however, the method does not necessarily minimize the number of stops or provide uninterrupted progression. In fact, to obtain minimum-delay offsets, traffic platoons have to be reformed at critical intersections in order that the green time will be fully utilized. Frequent reforming can be irritating to motorists.

PASSER-II model. The Progression Analysis and Signal System Evaluation Routine (PASSER) is an optimization model for progression along an arterial street considering various multiphase sequences. Further improvements in the processing algorithms and measures of effectiveness have been made by the Texas Transportation Institute, and the current version of the model is known as PASSER-II(87).[26]

The PASSER-II(87) model combines Brook's Interference Algorithm with Little's Optimized Unequal Bandwidth Equation, and extends them to multiphase arterial signal operations. The model inputs include turning movements, saturation capacity flow rates, distances between intersections, average link speeds, queue clearance intervals, permissible phasing sequencing, and minimum green times for each intersection. The program first determines the optimal demand-to-capacity ratios and uses them to determine the splits. Trial cycle lengths, phase, patterns, and offsets are varied to determine the optimal set of timings that maximize the progression bandwidth. Individual intersection capacity analyses, using methods paralleling the *Highway Capacity Manual*'s Signalized Intersection Analysis (including protected-only, permissive-only, and protected-permissive left-turns) are provided as part of the output.

PASSER-II(87) can handle up to 20 signalized intersections along a single arterial street, with up to four-phase sequences per intersection. It is written in FORTRAN IV for use on 16/32-bit computers and most microcomputers. The MS-DOS-based microcomputer version has a user interface program written in Pascal, to facilitate data input, editing, and output. The program is currently maintained by the Texas Department of Highways and Public Transportation.

Arterial analysis package (AAP). The Arterial Analysis Package is not so much a signal timing optimization program as it is a program that facilitates input to two of the more popular programs for arterial signal timing: PASSER II and TRANSYT-7F. The data base requirements for these two programs are similar, although the two programs use vastly

[25] Wilshire and others, *Traffic Control Systems Handbook.*

[26] A.S. Byrne and others, *Handbook of Computer Models for Traffic Operations Analysis.* Federal Highway Administration, U.S. Department of Transportation, Washington, DC, 1982.

different input file formats. The AAP allows a user to input one set of data and then to generate input for either timing program to take advantage of the features unique to both programs. The AAP is available in both 16/32-bit versions for mainframes and in a version for MS-DOS-based microcomputers.

Diamond interchange control

There are many variations of diamond interchanges: conventional full diamonds, half diamonds, split diamonds, and others. All varieties exist both with and without frontage roads. Typically, a diamond interchange signal is used to control the ramp terminal intersections with the cross street—in effect, creating two signalized intersections that are closely spaced (175 to 300 ft) on the cross street.

Several operational problems are possible with signalization of diamond interchanges. One problem occurs when there is a backup of cross-street through traffic from one of the ramp intersections through the other intersection. When this happens, traffic at the upstream ramp intersection may be partially or completely halted, thus reducing capacity of that intersection. When both ramps do this at the same time, the entire diamond interchange becomes locked-up and no vehicle can go forward. Another backup that can influence operation is when the left-turn pocket overflows and "spills back" into a through line. A third type of backup occurs when there is a long queue on the off-ramp of stopped or slowly moving vehicles that interferes with freeway main-lane traffic.

Initial signal control at full diamond interchanges was often provided by a three-phase signal sequence as shown in Figure 9–21. As traffic volumes continued to grow, operational experience indicated that improved signalization strategies were needed to increase throughput and reduce queue problems and associated delay.

The four-phase sequence represented a major advance in the signalization of diamond interchanges. One of its advantages was that left-turning vehicles are not stored in the center of the diamond. The four-phase, two-overlap sequence, shown in Figure 9–22, was developed by the Texas Transportation Institute in the late 1950s for use with pretimed and actuated controller units. This phasing sequence has been very popular, particularly under conditions of unbalanced heavy ramp traffic.

Diamond interchange control continues to advance in sophistication and flexibility, given the capability of microprocessor-based controller units. Caltrans, for example uses a Model 170 operating in a four-ring mode with either free or coordinated operation of the two signals selectable on the basis of either time-of-day or traffic data. Examples of two diamond interchange timing charts are shown in Figures 9–23 and 9–24 taken from the Caltrans Traffic Manual.

Timing for networks

The conventional method of network signal timing design is to provide preferential treatment for one or more arterials in the network. After favorable offsets are assigned to signals on

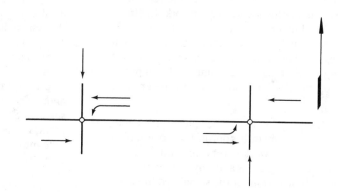

Figure 9–21. Three-phase pretimed diamond interchange phasing.

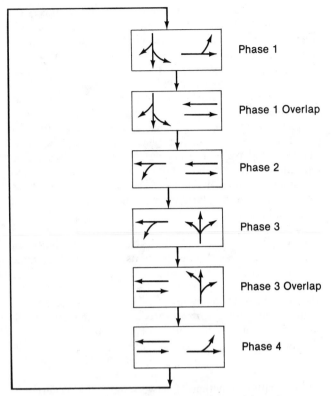

Figure 9–22. Four-phase pretimed diamond interchange phasing.

the preferred arterials or directions of travel, the remaining signals are adjusted to conform with the network. In effect, the network is reduced to a number of arterials (as shown in Figure 9–25) for easier analysis. The manual work involved in designating timings for a network is quite cumbersome and at times unmanageable. Fortunately, a number of computer-based optimization models for signal network design have been developed.

Manual techniques. The use of graphical trial-and-error techniques in a bandwidth approach to developing timing plans for a closed network is very difficult. A three-dimensional time-space diagram is actually needed if one is to consider flow in all directions at the signalized intersections. If the signal network is in a grid pattern, the common technique has been to utilize the basic concept of closed network timing and utilize timing plans of the simultaneous, signal alternate, or double-alternate types as previously described. For grid networks of one-way streets, the quarter-cycle offset plan was frequently used.

This timing plan works well with block lengths of approximately 450 ft. For example, with a cycle length of 60 sec, the progression speed would be 450/15 = 30 ft/sec.

It can be concluded that all of the timing plans that have been discussed are very dependent upon the geometrics of the grid network (specifically block length) and are very inflexible. With this basic manual timing approach, the only variables are the cycle length or type of plan (single alternate, double alternate, etc.).

Computer techniques. A number of algorithms and computer models have been developed with the objective of aiding a traffic engineer in the design of signal timing plans for cycle-based, pretimed control of traffic signal networks. These algorithms and programs were developed to function off-line (as opposed to real-time or on-line).

Computer techniques for off-line signal timing plan computation that are well documented and have received considerable testing and application include the following:

o SIGRID program
o TRANSYT-7F Model
o SIGOP-III Model

A relatively brief overview of these methods will be presented here in an effort to acquaint the reader with the basic state of the art in this area of traffic control. A considerable amount of documentation is available on most of these methods and will be referenced for the reader who desires to gain further knowledge of a given method.

SIGRID program

The SIGRID (Signal Grid) program[27] was developed by the Traffic Research Corporation for the Toronto traffic computer-control system. Given the cycle length, signal splits, link data, and a set of ideal or desirable offset differences for a signal network, the program calculates a set of optimum offset differences by minimizing the discrepancy between the two sets of values. In using the SIGRID program, the following points should be clearly understood:

1. The program minimizes only the differences between the ideal and actual offsets, and does not necessarily minimize the system delay.
2. The program only solves part of the problem involved in signal network optimization; it does not optimize the individual signal splits and link offsets. The program user has to predetermine the optimum splits and ideal offsets by another program or simply by experience. The desired offset, if calculated from speed and distance data by the program, is based on simplifying assumptions and does not take into account such factors as platoon dispersion and side-street traffic interruptions. The calculated offset is therefore not necessarily the best offset.
3. An inconsistency exists in the program in that the system delay function F is minimized and yet the calculated offsets corresponding to the lowest value of system average waiting times W are chosen as the optimum offsets.
4. The average waiting times (or delay-propensity factors) are computed based on oversimplified assumptions and do not necessarily reflect the true delay characteristics. The use of these values for system evaluation and for finding the "best" offsets is therefore questionable.
5. The program ignores the effects of changing cycle lengths on lost time and green utilization, and is therefore not suitable for making comparisons between different system cycle lengths.

---

[27] Traffic Research Corporation, "SIGRID Program: Notes and User's Manual," Metropolitan Toronto Roads and Traffic Department, Toronto, 1965–1973 (unpublished).

DIAMOND INTERCHANGE TIMING CHART
(Heavy Left Turn–200 vphpl or more)

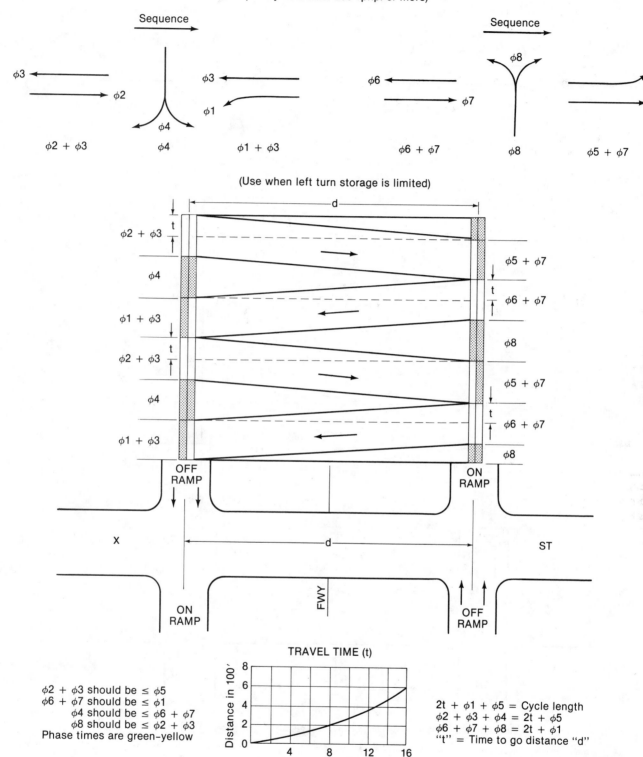

(Use when left turn storage is limited)

$\phi2 + \phi3$ should be $\leq \phi5$
$\phi6 + \phi7$ should be $\leq \phi1$
$\phi4$ should be $\leq \phi6 + \phi7$
$\phi8$ should be $\leq \phi2 + \phi3$
Phase times are green-yellow

TRAVEL TIME (t)

Average: 35 MPH Acceleration Time

$2t + \phi1 + \phi5$ = Cycle length
$\phi2 + \phi3 + \phi4 = 2t + \phi5$
$\phi6 + \phi7 + \phi8 = 2t + \phi1$
"t" = Time to go distance "d"

Note: These timing guidelines are ideal. Variations in timing may be necessary to provide proper splits to meet volume demands

Figure 9–23.

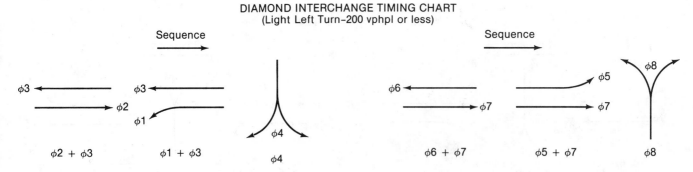

DIAMOND INTERCHANGE TIMING CHART
(Light Left Turn–200 vphpl or less)

(Use when left turn storage is limited)

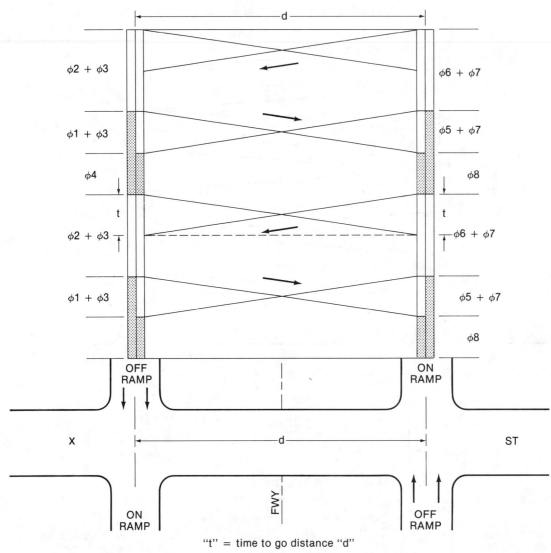

"t" = time to go distance "d"

Note: These timing guidelines are ideal. Variations in timing may be necessary to provide proper splits to meet volume demands
The green–yellow interval for phases 1,4,5 or 8 should equal time "t".

Figure 9–24.

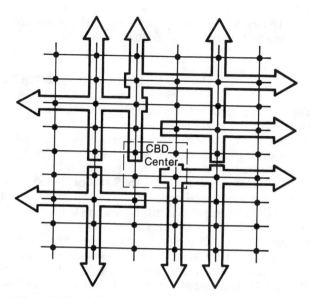

Figure 9–25. Preferential flow.

6. The platoon length (or green utilization) factor should be a link input instead of a system input, since green utilization varies considerably from signal to signal.

Despite its weaknesses, the SIGRID program represented a major breakthrough in the field of signal network optimization techniques when it was developed in 1964. Although it is less sophisticated than its more recent counterparts such as SIGOP, it is relatively simple to use. Another favorable aspect of SIGRID is that it involves very little computer process time and reasonable coding effort. It is simply a labor-saving device to manipulate signal offsets in a network so that they are as close as possible to the values demanded by the engineer. The program is flexible and extremely useful; with proper usage and sound engineering judgment it can produce meaningful results.

TRANSYT-7F model

The Traffic Network Study Tool (TRANSYT) is one of the most widely used models in the United States and in Europe for signal network timing design. It was developed in 1968 by Robertson[28] of the Transport and Road Research Laboratory (TRRL) in England, and since then, the TRRL has released several versions of this model. The version discussed here is TRANSYT-7F, where "7" denotes the seventh TRRL version of TRANSYT, and "F" symbolizes that this is the Federal Highway Administration's version of TRANSYT-7, which uses North American nomenclature on input and output.[29] The most current release of TRRL's TRANSYT is TRANSYT-9 and the most current release of the FHWA's version is TRANSYT-7F release 6.

---

[28] D.I. Robertson, *TRANSYT: A Traffic Network Study Tool.* Road Research Laboratory Report No. RL-253, Grothorne, Berkshire, England, 1969.

[29] C.E. Wallace and others, *TRANSYT-7F User's Manual.* U.S. Department of Transportation, Federal Highway Administration, Washington, DC, 1981.

TRANSYT-7F is used to optimize signal timing on coordinated arterials and grid networks. The mode of signal control considered by TRANSYT-7F is either pretimed, with 2 to 7 phases and fixed-phase sequence.

The structure of TRANSYT-7F consists of two main parts:

○ A macroscopic, deterministic traffic flow model used to compute the value of a specified performance index for a given signal network and a given set of signal timings. The performance index is a linear combination of measures of effectiveness (delays and stops) that are specified by the user.
○ A hill-climbing optimization procedure that makes changes to signal timings (splits and offsets) and determines whether or not the performance index is improved.

Input data for TRANSYT includes:

○ signal spacing
○ cycle length ranges
○ link speeds
○ lane configurations
○ minimum phase timings
○ phase sequencing
○ midblock volume inputs
○ saturation flow rates
○ left-turn treatment

TRANSYT-7F has a number of options that can be controlled by the user. These options include the following:

○ Buses can be modeled separately by including bus links. These can either be separate lanes or shared lanes.
○ Right-turn and left-turn delays caused by pedestrians can be reflected.
○ Overlap signal movements can be modelled.
○ Large networks can be subdivided into sections that can be handled by the program (i.e., 50 nodes and 250 links). The boundary nodes can be fixed from section to section so that their timings are not changed in the subsequent analysis. Another alternative is the expansion of program dimensional arrays to accommodate the larger networks.
○ Protected-only, protected-permissive, and permitted-only left turns can be modelled.
○ Unsignalized intersections controlled by stop signs on the cross streets as well as bottlenecks can be modelled.
○ Links can be prioritized to encourage development of a progression-oriented solution for arterial streets.
○ An estimate of network fuel consumption can be computer based on total travel, stops, and delay. The fuel consumption value includes fuel consumed at cruise, idle, and acceleration or deceleration. Fuel-consumption estimates are calculated for each link and then summed for the entire network or for individual routes.

The TRANSYT-7F model is written in FORTRAN IV for 16- or 32-bit computers and is available for MS-DOS-based microcomputers. Data input management programs, most notably EZ-TRANSYT and the T7FDIM program, exist to simplify the tedious data input process. A comprehensive user's manual was written to serve as an instructional guide for traffic engineers who desire to use the model.[30]

SIGOP-III model

SIGOP-III is an acronym for Traffic Signal Optimization Model, Version III. It was developed by KLD Associates, Inc., as an outgrowth and refinement of the original SIGOP model developed in the mid-1960s.[31] The similarities between TRANSYT-7F and SIGOP-III are: (1) both models are macroscopic signal timing and analysis tools, and (2) both models contain a traffic flow submodel and an optimization submodel that minimizes a user-specified "disutility" function. TRANSYT-7F considers delay and stops, whereas SIGOP-III considers delay, stops, and a term for queue "spillover."[32]

The basic inputs to SIGOP-III include flow rates, saturation flows (in terms of headways), minimum green times, yellow times, special phase times, and passenger car equivalent factors for trucks, buses, and turning vehicles.[33, 34] Outputs include: (1) data summary report, (2) signal timing report that contains offsets and splits for each phase, (3) performance analysis report that shows the value of the disutility function for each iteration of the model including the optimal value and detailed performance measures for each link and for the network as a total, and (4) user-specified time-space plots.

SIGOP-III is written in FORTRAN IV for use with IBM 360/370, CDC 6600, Amdahl 470, and MS-DOS-based microcomputer computer systems. Although limited to 80 signals and 230 links, the model documentation describes how it can be expanded to handle larger systems.

Like many signal-timing computer models, SIGOP-III has advantages and limitations. One of the major advantages of this model is the multiple cycle-length evaluation capability that can save the designer a considerable amount of time that would ordinarily be spent in preparing and running several jobs. The major limitations of SIGOP-III include the following:[35]

o Each signal cycle can accommodate a maximum of four phases; this does not adequately serve some users.
o The model does not consider bus links.
o Permissive and unprotected turns are not addressed explicitly by SIGOP-III. However, these conditions can be allowed for to some extent by restricting the capacities of such movements.

o The model does not explicitly consider nonsignalized intersections (stop-sign control, for example).
o The model lacks extensive field testing and evaluation.

UTCS control techniques

The development of efficient and effective real-time traffic control systems might be termed the ultimate objective of the traffic control engineer. The achievement of this objective is difficult, and research work in this area represents one of the leading edges of the traffic control field.

Most signal systems in the United States use the following approach originally developed for the Urban Traffic Control System (UTCS), in addition to other systems that utilize microprocessor-based controllers for developing on-line control strategies. The UTCS project was established by the Office of Research of the Federal Highway Administration in the early 1970s to provide for computer-supervised control of 200 signalized intersections in Washington, DC. An excellent overview of this project can be obtained from the publication *The Urban Traffic Control System in Washington, DC.*[36]

The UTCS research project on control strategies was divided into three generations of traffic control techniques, usually referred to as 1-GC, 2-GC, and 3-GC. First generation (1-GC) was referred to as a "table look-up" approach, where stored timing plans were selected in accordance with some criteria, e.g., time-of-day or traffic flow threshold. Second generation (2-GC) was characterized as an on-line timing plan development (or construction) concept, while third generation (3-GC) attempted to dynamically compute intersection timing without the constraint of system cycle length.

While 2-GC and 3-GC concepts did not emerge from research as tools for widespread implementation, the testing encouraged the development of techniques that have been referred to as 1 1/2-GEN. Characteristics of 1-GC and 1 1/2-GEN are described as follows:

First-generation control (1-GC). First-generation control uses prestored signal timing plans developed off-line and based on historical traffic data.[37] The plan controlling the traffic system can be selected on the basis of time-of-day, by direct operator selection, or by matching from the existing library a plan that is best suited to recently measured traffic conditions. The matching criterion is based on a network threshold value that incorporates traffic volumes and occupancies. The mode of plan selection is determined by the operator.

In the traffic-responsive mode, timing plans are usually updated once every 15 min. Smooth transition between different timing plans is provided by a transition routine that is part of 1-GC. This routine evaluates the magnitude of the changes, determines the time required for a smooth transition, and then controls signal-timing settings until the transition is complete. The same procedure of control is used for transition between computer control and standby.

[30] Ibid.

[31] Traffic Research Corporation, *SIGOP: Traffic Signal Optimization Program.* U.S. Bureau of Public Roads, Washington, DC, 1966.

[32] E.B. Lieberman and J.L. Woo, *SIGOP-III User's Manual.* U.S. Department of Transportation, Federal Highway Administration, Report No. FHWA-IP-82-A, Washington, DC, 1982.

[33] *Traffic Control Devices Handbook,* 1983.

[34] Traffic Research Corporation, *SIGOP: Traffic Signal Optimization Program.*

[35] Byrne and others, *Handbook of Computer Models.*

[36] Federal Highway Administration, *The Urban Traffic Control System in Washington, D.C.* U.S. Department of Transportation, Washington, DC (undated information brochure).

[37] J. MacGowan and I.J. Fullerton, "Development and Testing of Advanced Control Strategies in the Urban Traffic Control System" (three articles). *Public Roads,* 43 (Nos. 2, 3, 4) (1979–1980).

The pattern in effect is enhanced by a critical intersection control (CIC) feature, which is used to fine-tune the system at intersections that saturate frequently, by adjusting the allocation of green time (split) based on fluctuations in local traffic demand.

Certain intersections may be instrumented for bus priority. The decision to grant additional green time to buses is a function of passenger volumes, vehicular queues in and around the intersection, and the time of the arrival of a bus on the approach to the intersection. The program logic that makes this decision is contained within the bus priority system algorithm. The algorithm analyzes the actual intersection conditions in real-time and adjusts the signal cycle splits in response to actuated demands.

Timing plans for 1-GC can be calculated using one of the off-line computer software models described earlier; TRANSYT-generated plans were selected for testing in the UTCS research project in Washington, DC.

UTCS 1 ½-generation control

Users of first-generation control systems have found that stored timing plans tend to deteriorate in effectiveness over time due to changes in traffic flows. Fine-tuning to maintain their effectiveness requires extensive data collection and labor-intensive analysis. Because of a general lack of resources, timing plans are not updated as frequently as required to realize the full potential of 1-GC systems.

These difficulties could be overcome if the control system had the capability to compute signal timing plans for review and analysis by traffic engineering personnel. The term *1 ½-generation* has been used to describe systems having this capability. Systems in Overland Park, KS, and Winston-Salem, NC, are considered to be examples of systems that embody these concepts.

Signal maintenance

Effective signal maintenance begins with proper signal design and installation. It must not be simply left to the operator to "make it work." Selection of reliable and maintainable hardware, life-cycle costing, and standardization where possible are all factors that influence signal maintenance.

Definitions

1. *Malfunction.* Any event that impairs the operation without losing the display and sequencing of signal indications to approaching traffic. Malfunctions include timing failures, detector failures, loss of interconnected control, and other similar occurrences.
2. *Breakdown.* Any event that causes a loss of signal indication to any or all phases or traffic approaches. Breakdowns include controller unit failures, cable failures, loss of power, and the signal lamp burnout, which leaves no indication visible.
3. *Preventive maintenance.* Checks and procedures to be performed at regularly scheduled intervals for the upkeep of traffic signal equipment. It includes inspection, cleaning, replacement, and record keeping.

4. *Response maintenance.* The repair of failed traffic signal equipment and its restoration to safe, normal operation.[38]

Types of maintenance

Several categories of maintenance have been used through the years. Among them, Carlson[39] used the three categories of routine, preventive, and emergency maintenance. Parsonson adopted the three categories of preventive, response, and design modification in developing NCHRP Synthesis 114, *Management of Traffic Signal Maintenance.*[40] These classifications, originally developed for PennDOT,[41] are used here.

Preventive maintenance. The focus of preventive maintenance is on checking all equipment for proper operation and taking positive steps to repair or replace defective equipment. Proper attention to preventive maintenance can be expected to identify malfunctions, reduce breakdowns and associated response maintenance, significantly reduce liability exposure and road-user costs, and build confidence in and credibility for the operating agency. A basic preventive maintenance program could include the following elements:

- *Group relamping*—regular replacement of signal lamps before the end of their rated life. Done on an intersection basis within a geographical area, substantial technician travel time can be saved.
- *Signal lens cleaning*—done with relamping operation.
- *Pole and mast arm inspection*—to identify any cracks at joints or welds, inspect anchor bolts for tightness, spot paint rust areas or repaint exterior, check span wires, tethers, guys, and anchors.
- *Detector checks*—retune amplifier and check sensors for exposed wires or pavement deterioration.
- *Controller unit operation and timing*—to verify functional operation and use of appropriate timing.
- *Conflict monitor testing*—periodic replacement and bench testing.

A complete checklist for preventive maintenance items, along with suggested frequency of performance, is included in the *Traffic Signal Installation and Maintenance Manual.*[42]

Response maintenance. Response maintenance is the action taken in the event of a reported malfunction or breakdown. Response maintenance involves:

[38] James M. Giblin and others, *Traffic Signal Installation and Maintenance Manual,* Institute of Transportation Engineers (Englewood Cliffs, NJ: Prentice-Hall, 1989).

[39] W. Carlson, "How Do You Keep Your Traffic Signals Working?" *IMSA Signal Magazine* (July-August 1976), 10.

[40] Peter H. Parsonson, *Management of Traffic Signal Maintenance,* National Cooperative Highway Research Program, Synthesis of Highway Practice No. 114, Transportation Research Board, Washington, DC, December 1984.

[41] Edwards and Kelcey, Inc., *Maintenance of Traffic Signal Systems,* prepared in October 1981 for the Pennsylvania D.O.T. Revised by Penn D.O.T. for final printing in December 1982, p. 116.

[42] Giblin and others, *Traffic Signal Installation and Maintenance Manual.*

o *Notification:* Citizen complaints, police patrols, or roadway maintenance crews are usual sources of notification that there is a possible problem. Computer-based control systems with two-way communication ability also perform a level of self-evaluation and automatically report suspected failures, as do intersection controller units with telephone call-up capability.

o *Response:* From a liability perspective, a timely response is imperative. Such a response may require a field trip to the site to verify and determine the severity of the problem.

o *Repair:* Emergency repair would temporarily restore safe operation of the traffic signal, while final repair would return the installation to its design operating condition.

Design modification. Design modification is considered to include any change to the approved design and/or operation of a properly justified traffic signal. Such changes are typically made in response to a recurring operational problem. Examples of design modifications include the addition or removal of a phase or special function, or perhaps changes to the signal display. Conversion from unprotected left-turns to a protected/permissive left-turn operation would encompass both example areas.

Maintenance records

Adequate traffic signal maintenance records are essential. They have application in determining mean-time-between-failure (MTBF) rate for various hardware items, which helps in establishing preventive maintenance schedules. They are also useful in detecting failure trends and correcting recurrent problems. Their use in protecting the operating agency in the event of a lawsuit is evident.

Adequate maintenance records and documentation begin with a complete file of construction record drawings and maintenance/user manuals for each controlled location and primary hardware element. In today's age of the computer, the use of Computer-Aided Design and Drafting (CADD), bar codes, data-base management systems, and inventory software, the task of relating equipment location and maintenance activity becomes achievable. Maintenance service records can then be maintained by intersection (location), hardware item (serial number), and type of failure/repair. Signal timing records are also an important part of the record system for each signalized location and for system control.

Special signals

Lane-use control signals

Lane-use control signals are special overhead signals having indications to permit or prohibit the use of specific lanes. They are most commonly used for reversible-lane control to accommodate highly directional flows such as along radial thoroughfares to central business districts, approaches to major generators, near toll plazas, and in tunnel applications. They also have application in freeway management systems (FMS) where it is desired to:

1. Keep traffic out of certain lanes during certain times or flow conditions to facilitate merging operations.
2. Alert drivers to a lane that ends, such as near a freeway terminus.
3. Manage traffic flow on a freeway, bridge, or tunnel where a lane may be temporarily blocked by an accident, incident, or maintenance activity.

Lane-use control signals consist of a downward green arrow, red X, or yellow X indications. Their meaning, design, location, and operation are described in the MUTCD.

Railroad-highway grade-crossing signals

Of the approximate 222,000 highway-railroad grade crossings in the United States, about 63,000 (less than 30%) are protected by active warning devices, and more than 36,000 are grade separated. In 1987, all of these crossings experienced 5,859 accidents, 598 deaths, and 2,313 injuries. In addition, 531 accidents resulting in 116 injuries and 26 fatalities occurred at the 121,419 private crossings.[43]

Determination of need and selection of traffic control devices at a grade crossing is made by the public agency having jurisdictional responsibility. Subject to such determination and selection, the design, installation, and operation must be in conformance with the provisions of the MUTCD.

Traffic control devices (TCD's) for crossings are classified as *passive* devices and *active* devices. Passive rail-highway crossing TCD's include regulatory, warning, and guide signs, and supplemental pavement markings. Active TCD's include flashing light signals, automatic gates, bells, and advance warning devices.

Owing to the large number of significant variables to be considered, there is no single standard system of active traffic control devices universally applicable for grade crossings. Based on an engineering and traffic investigation, a determination is made whether any active traffic control system is required at a crossing and, if so, what type is appropriate. Before a new or modified grade-crossing traffic control system is installed, approval is required from the appropriate agency within a given state.

It is not uncommon for traffic signalized intersections to be close or adjacent to grade crossings. Where traffic signals are within 200 ft of the grade crossing and active crossing protection signals are provided, the two signals should be interconnected in order to provide an appropriate preemption condition while the crossing signals are in operation.

Flashing beacons

A flashing beacon is generally used at intersections where full traffic signals are not warranted. As typically used, a flashing yellow display is provided to the major roadway, and flashing red is provided to the minor street, supplementing its normal stop-sign control.

In addition to intersection use, flashing beacons also serve a useful purpose where the flashing yellow is used to help

[43] "Rail-Highway Crossings Study" Report of the Secretary of Transportation to the United States Congress, April 1989, Publication No. FHWA-SA-89-001.

warn motorists of a particularly hazardous condition. Such condition might be an unusually sharp curve or turn, an obstruction in the roadway, a narrow bridge or underpass, or intersection ahead hidden from view because of a vertical curve. Flashing beacons can also be used in advance of remote, isolated signalized intersections such as the approach to the first signal encountered when entering an urban area from a high-speed rural roadway.

Other special applications of flashing beacons include:

○ Approaches to metered ramps on freeway control systems.
○ School zones where reduced speed limits may be imposed during specific times of certain days.
○ Special warnings to specific drivers, such as might be detected by overheight vehicle detectors on approaches to low-clearance bridges.

Traffic signals at movable bridges

Warning gates and signals on the approaches to movable bridges (drawbridges) are intended to keep vehicles and pedestrians from entering an area where hazards exist owing to the operation of the bridge. Standard three-color traffic signal indications are generally used when movable bridge operation is quite frequent. In other cases, two red signal indications in vertical array separated by a STOP HERE ON RED sign may be used. The control mechanism for the signal is interconnected with the bridge control and timed so that a warning indication is shown in advance of the bridge opening to not only prevent entry, but to permit all traffic on the bridge to clear.

Ramp meter signals

The use of signals on entrance ramps to freeways is becoming widely accepted as a part of the active management of the demand capacity relationships on urban freeways. Two indications using two- or three-section signal heads are required on each ramp, and the green interval is typically short so as to permit the release of single vehicles for entry.

(*Note:* This chapter is not intended to cover this broad application of signals on freeways. For further information, the reader is directed to *Freeway Management Handbook.*)[44]

Signal control by emergency vehicles

Preemption of signalized intersections by emergency vehicles (usually ambulance and fire trucks) is done through the use of radio or light-beam transmitters on the vehicles to receivers on the signalized approach. Simple push buttons at the fire station or dispatcher activation of a predefined route timing may also be used.

When the receiver detects an emergency vehicle demand for service, special equipment in the controller assembly determines the current status of the green display on the emergency vehicle preempted approach. If green, the green

indication is simply extended to permit passage of the emergency vehicle. If not green, the controller unit is advanced in accordance with the timing constraints to display a green signal for the approaching emergency vehicle. Following vehicle passage, the controller unit returns the traffic signal to its normal operation. Traffic signals under preemptive control should be operated in a manner designed to keep traffic moving. Prolonged all-red or flashing signal sequences are to be avoided.

Evaluation techniques

In evaluating the performance of traffic signals and signal systems, a number of measures of effectiveness (MOE's) have been used, including:

○ Travel time
○ Speed
○ Delay
○ Queue length
○ Volume
○ Volume/capacity ratios
○ Fuel consumption
○ Accidents/conflicts

The overall performance of a signal system is primarily determined by the degree of vehicular mobility provided and the amount of traffic being serviced. The MOE's that measure these are discussed below.

Service rate (total travel)

Total travel, expressed in units of vehicle-miles, gives a measure of the amount of traffic being serviced. It is computed for each link by multiplying link volume by link length. For a network, it is expressed as:

$$TT = \sum_{i=1}^{N} V_i L_i \qquad (9.5)$$

where: TT = total travel
V_i = volume on link i
L_i = length of link i
N = number of links

Mobility (total travel time)

Total travel time, expressed in units of vehicle-hours, is the number of vehicles using a link during a given time period multiplied by the average travel time of the vehicles. For a network, it is expressed as:

$$TTT = \sum_{i=1}^{N} t_i V_i \qquad (9.6)$$

where: TTT = total travel time
t_i = average travel time along link i
V_i = volume on link i
N = number of links

It has been hypothesized that a linear relationship exists between system travel time and system service rate. Thus,

[44] R. Sumner and others, *Freeway Management Handbook,* Federal Highway Administration, Washington, DC, DOT-FH-11-9706 (4 vols., May 1983).

for a given network operating under a specific signal plan, a linear regression equation can be developed and used as a means of performance comparison between different signal plans.

From the driver's perspective, the ability to move at an acceptable speed with few stops is considered important. As a result, each user judges system performance from an individual perspective without benefit of knowing (or caring) about total system performance at optimum levels. Because of this, judgment must be exercised in implementing signal control and evaluating its performance solely on the basis of system-level MOE's without considering the "driver's viewpoint."

Signal removal

As discussed in the introduction to this chapter, traffic signals are not always the best solution for a particular situation. Further, conditions change, and traffic flow patterns that at one time warranted signals may have changed sufficiently to render an operating signal no longer necessary. Responsible operating agencies should periodically evaluate the effectiveness and necessity for traffic signals against current signal warrants and remove those that are no longer warranted.

Figure 9–26. Signal removal—preliminary screening.
SOURCE: FHWA-IP-80-12.

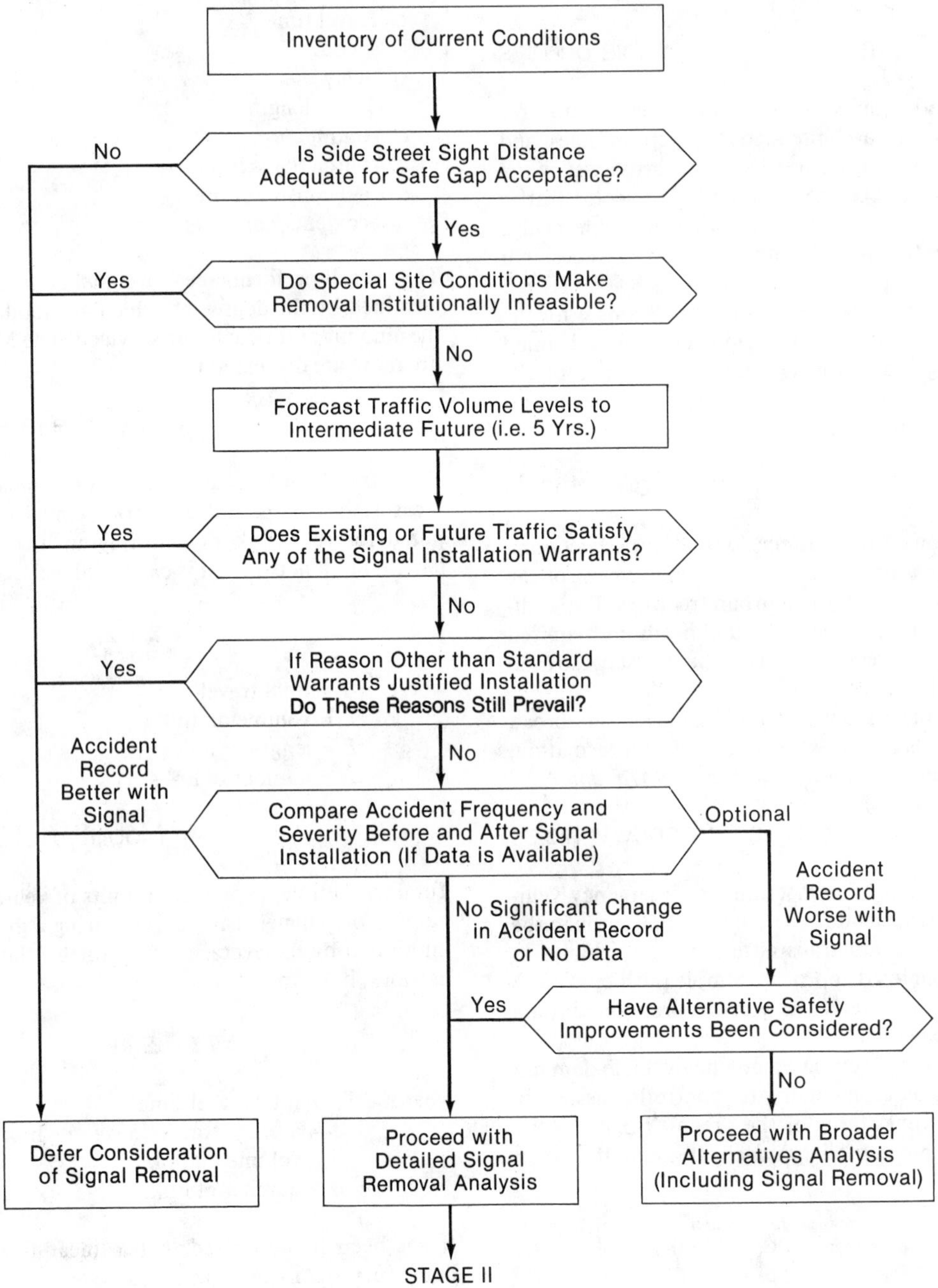

In 1980, the U.S. Department of Transportation, Federal Highway Administration, commissioned a study[45] to establish criteria for the removal of signals that are no longer needed. The study recommended the use of a two-stage process for making the signal removal decision, including:

○ *Preliminary screening*—a general process (shown in Figure 9–26) for application to a large group of signals that screens out those requiring further detailed analysis; and

○ *Detailed analysis*—a process of detailed investigations (shown in Figure 9–27) of technical and social impacts of signal removal.

Use of such a documented engineering study approach is valuable in satisfying potential liability concerns. It further is useful in convincing the political decision-makers and the general public that the signal removal decision was carefully assessed and expected to produce quantifiable and beneficial results.

Having made the removal decision, care should be taken in its implementation. The following procedure has been suggested:[46]

○ Advise motorists of the impending change by placing the signal into flashing operation—main street yellow and cross-street red. Supplemental information signs may be desirable.

○ Observe the operation at the intersection at the time the change to flashing operation is made to verify that it operates as planned. If it does, install stop signs on the cross streets.

○ Observe the intersection again after about 1 week to verify that it is operating as planned. If it is, remove the traffic signal. It is recommended that this removal include all aboveground equipment such as signal heads, span wires, poles, controller cabinet, etc., leaving all foundations and anchor bolts in place, properly protected, for about 6 months. This permits cost-effective reinstallation if required because of operational experience.

○ Monitor the intersection's accident experience and operation closely for at least 6 months.

It should be remembered that public reaction with regard to signals is typically to demand their installation in response to a perceived need. Rarely does the public demand the removal of a traffic signal. As a result, to be successful, signal removal must be "sold," both to the citizenry and at the political level.

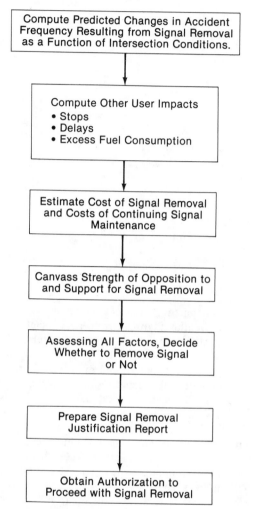

Figure 9–27. Signal removal—detailed analysis.
SOURCE: FHWA-IP-80-12.

[45] J.L. Kay, L.G. Neudorff, and F.A. Wagner, *Criteria for Removing Traffic Signals,* Technical Report (FHWA-RD-80-104) and User Guide (FHWA-1P-80-12), Alexandria, VA; JHK & Associates, Inc., and Wagner-McGee Associates, 1980.

[46] Orcutt, *Traffic Signal Applications Philosophy and Recommended practice.*

10

ROADWAY LIGHTING

RICHARD E. STARK, P.E., *President*

ERS Engineering

Purpose of roadway lighting

Roadway lighting encompasses all types of street, highway, pedestrian, and bikeway lighting. The lighting of streets dates back to 1558 in Paris when torchlights mounted on buildings provided nighttime illumination for pedestrians and carriages. Today, roadway lighting has become an integral part of the urban landscape.

The value of roadway lighting in various aspects of community enhancement has been attributed to crime prevention, business promotion, community pride, and traffic safety.

The transportation engineer is interested in all aspects of community development, especially in the area of traffic safety. Road lighting allows road users, both pedestrians and motorists, to proceed with greater safety, comfort, and convenience at night.

It is extremely important that motorists, whose operational requirements exceed that of pedestrians, have a clear visual scene of their surroundings. Motorists' visual requirements are related to their driving tasks. What drivers must see to operate their motor vehicles satisfactorily and how to provide for the visibility required is the challenge facing the traffic engineer.

Benefits of lighting

The benefits of roadway lighting are in several areas. Those that are the most significant and that are subject to analysis involve the relation of lighting to accidents. Numerous studies of the value of fixed road lighting have been made in North America and the rest of the world.

The draft document, "Value of Public Roadway Lighting," published by the Illuminating Engineering Society,[1] reviews the important studies of the lighting–accident relationship.

The findings indicate that the provision of adequate lighting—designed, installed, and maintained properly—can usually effect a significant reduction in nighttime accidents. Standards have been set nationally for lighting practice and when followed should provide adequate visibility. Vehicle headlights alone cannot completely provide for the nighttime motorists' visual driving needs.

Studies have shown that beneficial results from providing fixed lighting occur in both urban major streets and on freeways. Urban areas have reductions of 21% to 36% in nighttime accidents while reductions of up to 40% in freeway accidents have been predicted. Rural studies indicate improved safety as well.[2]

Worldwide studies by the International Commission on Illumination (the French abbreviation CIE is used)[3] have shown a 9% to 75% reduction in nighttime accidents after lighting or improved lighting has been installed.

Finally, the increased numbers of older drivers (over 65) whose visual capabilities are reduced make it important to provide better roadway visibility through fixed lighting.

---

[1] Illuminating Engineering Society, "Value of Public Road Lighting" (Draft 1989).

[2] R.H. Wortman, and M.E. Lipinski, "Interim Report: Development of Warrants for Rural At-Grade Intersection Illumination," Illinois Cooperative Highway Research Program, Ser. No. 135, University of Illinois at Urbana-Champaign, 1972, p. 37.

[3] International Commission on Illumination (CIE) 1988, "Road Lighting as an Accident Countermeasure," Report Number 8.2, Commission Internationale de L'Eclairage, Paris (in process).

Definition of terms

What follows are general definitions of several technical terms used herein and the explanation of some factors considered in lighting design. An understanding of these terms and factors is essential for proper design consideration.

Average initial illuminance. The average level of horizontal illuminance incident on the pavement area of a traveled way at the time the lighting system is installed when lamps are new and luminaires are clean; expressed in average footcandles (fc) for the pavement area.

Average initial luminance. The average level of luminance of the road surface at the time the lighting system is installed when the lamps are new and luminaires are clean; expressed in average candelas (cd)/sq ft.

Average maintained illuminance. The average level of horizontal illuminance incident on the roadway pavement when the output of the lamp and luminaire are diminished by the light loss factors; expressed in average footcandles for the pavement area.

Average maintained luminance. The average level of luminance when the output of the lamp and luminaire are diminished by the light loss factors; expressed in average candelas/ sq ft.

Candela. The unit of luminous intensity. Formerly the term "candle" was used (abbrev. cd).

Candela per square foot. The unit of luminance (photometric brightness) equal to the uniform luminance of a perfectly diffusing surface emitting or reflecting light at the rate of 1 lumen (lm)/sq ft, or the average luminance of any surface emitting or reflecting light at that rate. One candela/ sq ft equals 10.76 cd/sq meter (m).

Candela per square meter. The International System (the French abbreviation SI is used) unit of luminance (photometric brightness) equal to the uniform luminance of a perfectly diffusing surface emitting or reflecting light at the rate of 1 lm/sq m, or the average luminance of any surface emitting or reflecting light at that rate. One candela/sq m equals 0.0929 cd/sq ft.

Candlepower. Luminous intensity in a specified direction; expressed in candelas.

Complete interchange lighting. The lighting of the freeway through traffic lanes through the interchange, the traffic lanes of all ramps, the acceleration and deceleration lanes, all ramp terminals, and the crossroad between the outermost ramp terminals.

Equipment factor. A factor used in the illuminance or luminance calculations that compensates for light losses due to normal production tolerances of commercially available luminaires when compared with laboratory photometric test models. It is common practice to approximate these losses using a 5% to 10% loss factor.

$$\text{Equipment Factor (EF)} = 0.95 - 0.90.$$

Footcandle. The illuminance on a surface 1 sq ft in area on which there is uniformly distributed a light flux of 1 lm. One footcandle equals 10.76 lux (abbrev. lx).

Illuminance. The density of the luminous flux incident on a surface; it is the quotient of the luminous flux by the area of the surface when the latter is uniformly illuminated.

Light loss factors. The depreciation factors that are applied to the calculated initial average luminance or illuminance to determine the value of average maintained (depreciated) luminance or illuminance at a predetermined time in the operating cycle, usually just prior to relamping, and which reflect the decrease in effective light output of a lamp and luminaire during its life. It is made up of several variables, and judgment must be exercised in arriving at a suitable factor. The more important variables that should be considered when determining the light loss factors are:

1. Decrease of lamp lumen output with burning hours (LLD, Lamp Lumen Depreciation);
2. Reduction of some discharge lamp light output when operated in other than the vertical position;
3. Schedule of lamp replacement;
4. Frequency and effectiveness of luminaire cleaning (LDD, Luminaire Dirt Depreciation);
5. Equipment factor (EF); and
6. Operation of light sources at other than rated current or voltage.

Lumen. A unit of measure of the quantity of light (abbrev. lm). One lumen is the amount of light that falls on an area of 1 sq ft, every point of which is 1 ft from the source of 1 cd (candle). A light source of 1 cd emits a total of 12.57 lm.

Luminaire. A complete lighting unit consisting of a lamp or lamps together with the parts designed to distribute the light, to position and protect the lamps, and to connect the lamps to the power supply.

Luminaire cycle. The distance between two luminaires along one side of the roadway.
(*Note:* This may not be the same as luminaire spacing along the centerline, considering both sides of the road.)

Luminance. The luminous intensity of any surface in a given direction per unit of projected area of the surface as viewed from that direction.

Luminous efficacy of a source of light. The quotient of the luminous flux emitted by the total lamp power input. It is expressed in lumens per watt (lm/W).

Lux. The International System (SI) unit of illuminance. It is defined as the amount of light on a surface of 1 sq m, all points of which are 1 m from a uniform source of 1 cd. One lux equals 0.0929 footcandle (abbrev. fc).

Partial interchange lighting. Lighting that consists of a few luminaires located in the vicinity of some or all ramp terminals. The usual practice is to light those general areas where the exit and entrance ramps connect with the through traffic lanes of the freeway and generally light those areas where the ramps intersect the crossroad.

Uniformity of illuminance. The ratio of average footcandles of illuminance on the pavement area to the footcandles at the point of minimum illuminance on the pavement. It is commonly called the *uniformity ratio*. A uniformity ratio of 3:1 means that the average footcandle value on the pavement is three times the footcandle value at the point of least illuminance on the pavement.

Uniformity of luminance. The average level-to-minimum-point method uses the average luminance of the roadway design area between two adjacent luminaires, divided by the lowest value at any point in that area. The maximum-to-minimum-point method uses the maximum and minimum values between the same adjacent luminaires. The luminance uniformity (avg/min and max/min) considers the traveled portion of the roadway, except for divided highways having different designs on each side.

Veiling luminance. A luminance superimposed on the retinal image, which reduces its contrast. It is this veiling effect produced by bright sources or areas in the visual field that results in decreased visual performance and visibility.

Basis of seeing at night

Light

Light is that portion of the radiant energy spectrum capable of producing visual sensation. This sensation results from stimulation of the retina of the human eye by radiant energy of the proper wavelength being emitted or reflected in a sufficient quantity and within the visual field.

Contrast

The contrast of an object is determined by the luminance relationship between the object and its background. An object or target that has a luminance L_T is considered to be in negative contrast if the background luminance L_b is higher in value (silhouette-type discernment). When background luminance L_b is lower in value than the target luminance L_T, the target is in positive contrast (reverse silhouette discernment). The difference between the target and its background (L [$L = L_T - L_b$]) must be significant to perceive the target. Generally, bi-directional roadway lighting results in targets having both negative and positive contrast. Proper lighting geometry can improve contrast, thus revealing a greater percentage of objects.

Discernment by surface detail

Discernment by surface detail depends on a high order of direct illumination on the face of the object toward the driver.

The object is seen by variations in brightness or color over its own surface without general contrast with its background. Discernment by surface detail may be a principal method of seeing in heavy traffic when the complexity of the situation requires considerable visual detail.

Visual acuity

Contrast sensitivity, or the ability to distinguish luminance (brightness) differences, provides for the detection of objects, but the identification of most objects is accomplished by visual acuity. By definition, visual acuity is the ability of the eye to resolve small details. In driving, two kinds of visual acuity are of concern: static and dynamic visual acuity.

Static visual acuity occurs when both the driver and the object are stationary and is a function of background brightness, contrast, and time. With increasing illumination, visual acuity increases up to a background luminance of about 3 cd/sq ft, and then it remains constant despite further increases in illumination. Static visual acuity also increases with increasing contrast of the object. Optimal exposure time for a static visual acuity task is from 0.5 to 1.0 sec when other visual factors are held constant at some acceptable level.

When there is relative motion between the driver and an object, such as occurs in driving, the resolving ability of the eye is termed *dynamic visual acuity*. Dynamic visual acuity is more difficult than static visual acuity because eye movements are not generally capable of holding a steady image of the target on the retina. The image is blurred and, therefore, its contrast decreases. The conditions favorable for dynamic visual acuity are slow movement, long tracking time, and good illumination. These are rarely found in the nighttime driving environment except in sign reading, an important dynamic visual task.

Glare

Two types of glare affect driving performance: discomfort glare and disability glare.

Discomfort glare does not reduce the ability to see an object, but it produces a sensation of ocular discomfort. Research is being carried out to establish a criterion for evaluation of this effect on the driver.

Disability glare, also known as *veiling luminance* or stray light at the eye, alters the visual field in such a way as to reduce the brightness of a viewed object and its background, thereby making the driver's visual task more difficult. Veiling luminance can be easily calculated and as such has become a part of current standards. The American National Standards Institute (ANSI) practice limits the ratio of maximum veiling luminance to the average pavement luminance as produced by the luminaires.

Pavement reflectance

Pavement plays a very important part in the visibility of objects on the road. As explained earlier, objects are generally seen by contrast with the background, either positive or negative. The light the motorist sees, other than that directly from the luminaire or reflected from objects, is pavement luminance. Therefore, it is extremely important to know the type of pavement and pavement reflectance characteristics. The current standards for roadway lighting where luminance criteria are used—ANSI,[4] AASHTO,[5] CIE[6]—require the use of pavement reflectance. Horizontal illuminance levels in the ANSI and AASHTO standards also require the selection of a pavement type with regard to reflectance characteristics.

Current standards follow the pavement reflectance characteristics of the CIE[7] system. Table 10–1 describes the four pavement classifications. Classifications are based on lightness and specularity of the pavements. Studies have shown that commonly used pavements can be grouped into four classes having similar reflectance characteristics. Tables providing the reduced reflectance coefficients for each of the four pavement classifications are known as *r*-Tables.

The *r*-Tables for standard surface R1 (concrete) and R3 (asphalt) are those most commonly used in North America.

TABLE 10–1
Road Surface Classifications

Class	Description	Reflectance
R-1	Portland cement concrete road surface. Asphalt road surface with a minimum of 15% of the aggregates composed of artificial brightner (e.g., Synopal) aggregates (e.g., labradorite, quartzite).	Mostly Diffuse
R-2	Asphalt road surface with an aggregate composed of a minimum 60% gravel (size greater than 10 mm). and Asphalt road surface with 10 to 15% artificial brightener in the aggregate mix. (Not normally used in North America).	Mixed Diffuse and Specular
R-3	Asphalt road surface (regular and carpet seal) with dark aggregates (e.g., trap rock, blast furnace slag); rough texture after some months of use. (Typical Highways)	Slightly Specular
R-4	Asphalt road surface with very smooth texture.	Mostly Specular

SOURCE: ANSI/IES RP-8, 1983.

Time as related to seeing

The time available to the driver to perform the visual task, whether recognition of an object or perception of a traffic situation, is very important (see Chapter 1). This time decreases as vehicular speed and situation complexity increase. The time factor is extremely critical in low-illumination areas or in complex high-background brightness areas where the eye is in a continual state of adaptation.

---

[4] *"American National Standard Practice for Roadway Lighting"* (RP-8) (New York: Illuminating Engineering Society/American National Standards Institute, 1983).

[5] "An Informational Guide for Roadway Lighting" (Washington, DC: American Association of State Highway and Transportation Officials, 1984).

[6] "Publication 12-2, Recommendations for the Lighting of Roads for Motorized Traffic," Commission Internationale de L'Eclairage (CIE) (1977).

[7] "Calculation and Measurement of Luminance and Illuminance in Road Lighting," CIE 30 (Washington, DC: National Bureau of Standards, 1976).

Driver visual information needs

Visual information needs associated with the driving task can be organized in accordance with three driver-performance levels. These levels and the general information needs are as follows:

1. *Positional performance:* needs associated with routine steering and speed control. These needs are satisfied primarily through pavement markings, curb delineation, and delineation of road edges, lane divisions, and roadside features.

2. *Situational performance:* needs associated with required changes in speed, direction of travel, or position on the roadway as a result of a change in the geometric, traffic and/or environmental situation. These needs are as varied as the number and types of road and traffic situations encountered in driving.

3. *Navigational performance:* needs associated with selecting and following a route from an origin to a destination. These needs are satisfied primarily by formal information sources (signs, markers, etc.) and informal information sources (landmarks, etc.).

Visibility concept

The main criteria for acquiring necessary information required for driving safely at night are whether objects, lane lines, pavement areas, and other sign and signal devices are sufficiently visible to allow for guidance, necessary maneuvering, or stopping when required. These various areas or objects must be illuminated by headlights, fixed lighting, off-roadway sources, or be self-illuminated.

The visibility of an object depends on several factors: (1) the contrast of the object's surface luminance and the luminance of the background (pavement, sky, etc.); (2) the adaptation of the eye relative to that object; (3) the magnitude of the disability glare (veiling luminance) at the eye; (4) the transient adaptation (reduction of equivalent contrast due to readaptation from one luminous background to another); (5) the visual complexity of the background, and motion dynamics; (6) the color, size, and shape of the object, and (7) the age and visual characteristics of the motorist.

A number of similar visibility models have been proposed to incorporate the above conditions. Experiments have been carried out by the Roadway Lighting Committee of the Illuminating Engineering Society, and a proposed new standard will be based on the visibility model of Adrian.[8]

The visibility model can be expressed as follows:

$$V_L = \frac{\Delta L \text{ actual}}{\Delta L \text{ threshold}}$$

where ΔL threshold indicates the threshold at which a target of a certain angular size, positive or negative, becomes visible on a background of a certain luminance. ΔL is the actual difference in luminance between the target and its background. V_L has no units, but is a ratio to indicate how much a target is above threshold.

---

[8] W. Adrian, "Model to Calculate the Visibility of Targets," *Lighting Research and Technology,* 21 (4) (1989).

The new standard calculates ΔL threshold for a 23-year-old observer with fixation time of 0.2 sec. The observer is located 272.55 ft from the target looking down at the target from an eye height of 4.76 ft. The target used in this model is a 7-by-7-in flat upright surface with 20% reflectance. This is a relatively small target in comparison with other targets and roadway objects; hence the term Small Target Visibility (STV). The standard sets V_L levels for various roadway situations.

This standard using the concept of Small Target Visibility is being proposed as a criterion where safety is the main consideration in lighting design. This is especially important where the design involves higher speed roadways (over 35 mph).

Traffic criteria and warranting conditions

Roadway and area classifications

Most highway and street systems encompass several classes or types of roadway and walkways. At one extreme are high-speed, high-volume facilities carrying through traffic, with no attempt made to serve abutting property, pedestrians, or local trips. At the other extreme are local highways, streets, or roads that carry low volumes, at low speeds, with a primary function of land access rather than vehicular movement.

A comprehensive lighting program requires that the roads and streets be classified on the basis of intended function. The following classifications are those recommended by the Illuminating Engineering Society.

Classification definitions

1. *Freeway:* This is a divided major roadway with full control of access and with no crossings at grade. This definition applies to toll as well as nontoll roads.
   a. *Freeway A:* This designates roadways with greater visual complexity and high traffic volumes. Usually this type of freeway will be found in major metropolitan areas in or near the central core and will operate through much of the early evening hours of darkness at or near design capacity.
   b. *Freeway B:* This designates all other divided roadways with full control of access where lighting is needed.
2. *Expressway:* A divided major roadway for through traffic with partial control of access and generally with interchanges at major crossroads. Expressways for noncommercial traffic within parks and parklike areas are generally known as parkways.
3. *Major arterial:* That part of the roadway system serving as the principal network for through traffic flow. The routes connect areas of principal traffic generation and important rural highways entering the city.
4. *Collector:* The distributor and collector roadways servicing traffic between major and local roadways. These are roadways used mainly for traffic movements within residential, commercial, and industrial areas.

5. *Local:* Roadways used primarily for direct access to residential, commercial, industrial, or other abutting property. They do not include roadways carrying through traffic. Long local roadways will generally be divided into short sections by collector roadway systems.
6. *Alley:* Narrow public ways within a block, generally used for vehicular access to the rear of abutting properties.
7. *Sidewalk:* Paved or otherwise improved areas for pedestrian use, located within public street rights-of-way, which also contain roadways for vehicular traffic.
8. *Pedestrian walkway:* A public facility for pedestrian traffic not necessarily within the right-of-way of a vehicular traffic roadway. Included are skywalks (pedestrian overpasses), subwalks (pedestrian tunnels), walkways giving access to parks or block interiors, and midblock street crossings.
9. *Isolated interchange:* A grade-separated roadway crossing that is not part of a continuously lighted system, with one or more ramp connections with the crossroad.
10. *Isolated intersection:* The general area where two or more noncontinuously lighted roadways join or cross at the same level. This area includes the roadway and roadside facilities for traffic movement in that area. A special type is the channelized intersection in which traffic is directed into definite paths by islands with raised curbing.
11. *Bikeway:* Any road, street, path, or way that is specifically designated as being open to bicycle travel, regardless of whether such facilities are designed for the exclusive use of bicycles or are to be shared with other transportation modes.
    a. *Type A—Designated bicycle lane:* A portion of roadway or shoulder that has been designated for use by bicyclists. It is distinguished from the portion of the roadway for motor vehicle traffic by a paint stripe, curb, or other similar device.
    b. *Type B—Bicycle trail:* A separate trail or path from which motor vehicles are prohibited and which is for the exclusive use of bicyclists or the shared use of bicyclists and pedestrians. Where such a trail or path forms a part of a highway, it is separated from the roadways for motor vehicle traffic by an open space or barrier.

Area classifications (abutting land uses)

1. *Commercial:* A business area of a municipality where ordinarily there are many pedestrians during night hours. This definition applies to densely developed business areas outside, as well as within, the central section of a municipality. The area contains land use that attracts a relatively heavy volume of nighttime vehicular and/or pedestrian traffic on a frequent basis.
2. *Intermediate:* Those areas of a municipality often characterized by moderately heavy nighttime pedestrian activity such as in blocks having libraries, community recreation centers, large apartment buildings, industrial buildings, or neighborhood retail stores.
3. *Residential:* A residential development, or a mixture of residential and small commercial establishments,

characterized by few pedestrians at night. This definition includes areas with single-family homes, town houses, and/or small apartment buildings.

Certain land uses, such as office and industrial parks, may fit into any of the above classifications. The classification selected should be consistent with the expected night pedestrian activity.

Warrants for lighting

Warrants are factual evidence compiled for the purpose of justifying the installation of roadway lighting. Warrants should be based on conditions relating to the need for roadway lighting and the benefits that may be accrued therefrom. Factors such as traffic volume, speed, road use at night, night accident rate, road geometrics, and general night visibility are important considerations in determining the minimum conditions justifying lighting. Justification for lighting may also be based on economic effectiveness, such as reduction in personal injuries and property damage caused by accidents, improved operational efficiency, and other societal benefits.

Two principal sources of lighting warrants are available: "An Informational Guide for Roadway Lighting," published by AASHTO (see note 5) and "Warrants for Roadway Lighting," NCHRP Report No. 152.[9] Both will be discussed briefly; the lighting designer or administrator should have access to both documents.

The American Association of State Highway and Transportation Officials (AASHTO) warrants are based primarily on experience. They set forth a description of operational, geometric, and developmental conditions that must be matched or exceeded in order to justify the installation of roadway lighting. AASHTO warrants are developed for six principal categories of roadway lighting:

○ Freeways
○ Interchanges
○ Tunnels and underpasses
○ Roadway safety rest areas
○ Roadway sign lighting
○ Streets and highways

The AASHTO warrants place greatest emphasis on freeway-type facilities. For lighting on urban streets, NCHRP Report No. 152 ("Warrants for Roadway Lighting") provides a more detailed analysis. The warrants procedure outlined in that report embodies an analytical assessment on the effect of geometric, operational, and developmental conditions and night accident experience on driver visual information needs. Further, that approach is applied uniformly to all classes of major roadways, specifically:

○ Noncontrolled-access facilities
○ Intersections
○ Controlled-access facilities
○ Interchanges

Both warrant procedures are well documented in the *Roadway Lighting Handbook* published by the Federal Highway Administration.[10]

Light sources

The most important element of illumination equipment is the light source. It is the principal determinant of visual quality, illumination efficiency, energy conservation, and the economic aspects of the illumination system. Although numerous types of light sources have been used in roadway lighting, major emphasis is placed on those sources currently used in modern designs: mercury, metal halide, high-pressure sodium, low-pressure sodium, and fluorescent.

Gaseous-discharge lamps produce light by the excitation of gas or metal vapors in the arc tube of the lamp. When an electrical potential is applied across the electrodes of the arc tube, the gas is ionized and electrons flow through the arc tube. These electrons collide with the atoms of the gaseous medium, momentarily altering their structure. When the atoms return to their normal state, energy is released, resulting in the emission of light.

Light sources are normally compared on the basis of four major characteristics: (1) luminous efficacy (the number of lumens produced per watt of energy expended); (2) color rendition (color quality); (3) lamp life (number of operating hours); and (4) optical control. Specific data for the light sources indicated above, and others, are summarized in Table 10–2.

A brief discussion of the five major light sources is contained in the following paragraphs.

Mercury. The mercury lamp has been used for street lighting purposes for decades. It was developed during the 1930s and was extremely popular until the appearance of more efficient light sources. Its luminous efficacy is only fair, its color is fair for the clear lamp, and good if the lamp is phosphor-coated. The optical control is good for the clear lamp. The lamp life is exceptionally long and dependable. It has been used for practically all outdoor applications. However, it is being replaced rapidly by other sources in general roadway lighting applications because of its relatively low efficacy.

Metal halide. Metal halide lamps produce better color at higher efficacies than do mercury lamps. Their life, however, is somewhat shorter, and they are more sensitive to lamp orientation (horizontal or vertical) and vibrations. Excellent results have been obtained with these lamps in high-mast lighting. If exceptionally good color rendition is required, the metal halide lamp should be considered.

---

[9] "Warrants for Roadway Lighting," NCHRP Report 152 (Washington, DC: National Academy of Sciences, 1972).

[10] *Roadway Lighting Handbook,* 1978, U.S. Department of Transportation, Federal Highway Administration, RD&T Report Center, HNR-11, 6300 Georgetown Pike, McLean, VA 22102.

TABLE 10-2

Typical Area and Roadway Lighting Lamp Characteristics[a]

	Lumens Per Watt				
	(Including Ballast Losses[b])	Lamp Only	Lumens	Wattage Range	Rated Ave. Life (hrs.)[c]
Incandescent[d]	N/A	11–18	655–15300	58–860	1500–12000
Tungsten-Halogen	N/A	20–22	6000–33000	300–1500	2000
Fluorescent	58–69	70–73	4200–15500	60–212	10000–12000
Mercury-Clear	37–54	44–58	7700–57500	175–1000	24000+
Mercury-W/Phosp.	41–59	49–63	8500–63000	175–1000	24000+
Metal Halide	65–110	80–125	14000–125000	175–1500	7500–15000
High Pressure Sodium	60–130	83–140	5800–140000	70–1000	20000–24000
Low Pressure Sodium	78–150	131–183	4650–33000	35–180	18000

	% Maintenance Output at End of Life	Color Rendition	Optical Control	Cost	
				Initial (Lamp)	Operational (Power)
Incandescent[d]	82–86	Exc.	Excellent	Low	High
Tungsten-Halogen	93	Exc.	Exc. Vertical Poor Horiz.	Moder.	High
Fluorescent	68	Good	Poor	Moder.	Moder.
Mercury-Clear	62–82	Fair	Good	Moder.	Moder.
Mercury-W/Phosp.	50–73	Good	Fair	Moder.	Moder.
Metal Halide	58–74	Good	Good	High	Low
High Pressure Sodium	73	Fair	Good	High	Low
Low Pressure Sodium	100[e]	Poor	Poor	High	Low

[a] All figures show operating ranges typical for lamp sizes normally used in area and roadway applications.

[b] Ranges shown cover low wattage lamps with regulated type ballasts (worst condition) through high wattage lamps with reactor type ballasts (best condition).

[c] Rated average life is based on survival of at least 50% of a large group of lamps operated under specified test conditions at 10 or more burning hours per start.

[d] Larger sized incandescent lamps (up to 2,000 watts) for floodlighting applications are available. Depending on operating conditions, the luminious efficacy and life change considerably for these lamps from the typical values shown. Lamp schedules should be consulted for details.

[e] Low pressure sodium lamps maintain initial lumen rating throughout life, but lamp wattage increases. Considering this change in wattage, the luminous efficacy of these lamps (including ballast losses) at 18,000 hours is 67–117 lumens per watt.

SOURCE: *Roadway Lighting Handbook* (Washington, DC: U.S. Dept. of Transportation, 1978).

High-pressure sodium. This is the newest of the family of discharge-type lamps and provides excellent luminous efficacy, good lumen maintenance, long life, and fair color. The lamp represents a good economic compromise ideally suited for general roadway lighting applications.

Low-pressure sodium. The principal advantage of this light source is its exceptionally high luminous efficacy. Its disadvantages are its monochromatic color and its large size. The lamp is an excellent source where color and optical control are less important than the quantity of light produced per unit of electrical energy. It has been used successfully in roadway lighting and in tunnel lighting where high illumination levels are required.

Fluorescent. Fluorescent light sources have good efficacy (70 lm/W) and provide excellent color rendition, but the lamps are large and, therefore, are most effectively used where a large long-source size lends itself to the distribution pattern required. This lamp has been used most successfully in sign lighting, underpass, and tunnel applications. The fluorescent lamp light output is more sensitive to ambient temperature variations and has a shorter life than the high-pressure sodium or mercury lamps.

Ballasts

Most lamps used for roadway lighting are of the gas-discharge type: fluorescent, low-pressure sodium, mercury, metal halide, and high-pressure sodium. Each type requires an electrical component called a *ballast,* which serves to provide the proper voltage, waveform, and current to start and operate a gas-discharge lamp.

Five types of ballasts are in common use: reactor, autotransformer reactor, regulator, autotransformer regulator, and electronic. Of these, the reactor is the simplest and therefore the lowest priced. The disadvantage of the reactor type is that it will not tolerate an input voltage drop of more than 5% without seriously affecting lamp performance. Where the input voltage is not sufficient to start and operate the lamp, an autotransformer is used to raise the voltage. The normal reactor ballast has a low power factor; this can be corrected by adding the proper size capacitor to the circuit.

The regulator ballast has a circuit that will permit proper operation of a lamp over a range of input voltages: ±13% for mercury lamps and ±10% for high-pressure sodium. The autotransformer has a lower cost and slightly less ballast power loss than the normal regulator ballast. The electronic ballast will hold lamp output to 2% over a ±10% change in input voltage.

A high-pressure sodium lamp requires, in addition to the normal ballast components, a starting device. This device produces a pulse of at least 2,500 V with a duration of 1 μ sec across the lamp.

Most ballasts presently used in roadway lighting are of the integral type; that is, the ballast is installed within the luminaire housing. Integral ballasts have the following advantages over ballasts designed for pullbox or pole-base installation: low cost, low operating temperature, and accessibility for replacement of defective components.

Luminaire design and placement

Design and types

Webster's defines "luminaire" as "any body that gives light," but in street lighting the term *luminaire* describes the complete lighting assembly, less the support assembly. The modern-day luminaire, as shown in Figure 10–1, consists of a weatherproof housing enclosing the light source, a reflector, and in many cases the electrical ballast for discharge-type lamps. A refractor comprises the lower part of the enclosure and serves, with a reflector, to control the distribution of light on the roadway. The refractor is generally a molded glass element that provides prismatic control of light (Figure 10–1). A flat glass lens rather than a refractor is used where reduced glare is desired.

Luminaires are designed and identified primarily on the basis of the area of coverage (i.e., the width and length of the area to be lighted and the "allowable beam angle"). The higher beam angles permit greater spacing of luminaires for uniform coverage, but higher beam angles mean more glare and reduced effectiveness of the lighting system.

To standardize luminaires for manufacture and design purposes, the Illuminating Engineering Society has assigned type numbers (Types I through V) to luminaires that produce different lateral light distribution patterns used for various purposes. These pertain mainly to the street width and the location of the luminaire in relation to the roadway. A brief description with illustrative sketches of each luminaire type is given in Figure 10–2. A more detailed and technical description of luminaire types can be found in the *American National Standard Practice for Roadway Lighting* (see note 4).

Luminaires are also classified on the basis of vertical light distribution: the ability to spread light along the length of the roadway. Short, medium, and long distributions are established on the basis of the distance from the luminaire where the light beam of maximum candlepower strikes the roadway surface, which is defined as follows:

1. *Short distribution:* The maximum candlepower beam strikes the roadway between 1.0 and 2.25 mounting heights from the luminaire.
2. *Medium distribution:* The maximum candlepower beam strikes the roadway at some point between 2.25 and 3.75 mounting heights from the luminaire.
3. *Long distribution:* The maximum candlepower beam strikes the roadway at a point between 3.75 and 6.0 mounting heights from the luminaire.

On the basis of the vertical light distributions, theoretical maximum spacings of luminaires are such that the maximum candlepower beams from adjacent luminaires are joined on the roadway surface. With this assumption, the maximum luminaire spacings would be 4.5 mounting heights for a short distribution, 7.5 mounting heights for a medium distribution, and 12.0 mounting heights for a long distribution. These spacings will not, however, satisfy the design criteria outlined below.

In practice, the medium distribution is most widely used, and the luminaire spacing normally does not exceed 5 mounting heights. Short distributions are seldom used for reasons of economy, and long distributions are not used to any great extent because the high beam angle of maximum candlepower produces excessive glare.

It is important that the distribution of light flux emission above the beam of maximum candlepower be controlled. Light flux emission at the higher vertical angles generally contributes substantially to increased pavement brightness, but it also contributes greatly to increased disability and discomfort glare. To achieve balanced performance, it is

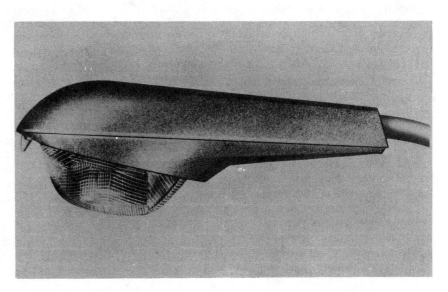

Figure 10–1. Typical roadway luminaire.
SOURCE: *Roadway Lighting Handbook* (Washington, DC: U.S. Dept. of Transportation, 1978).

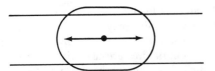

Type I—A luminaire designed for center mounting over streets up to 2.0 mounting heights in width.

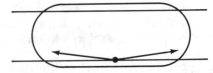

Type II—A luminaire designed for mounting over the curb line of street widths less than 1.5 mounting heights.

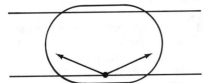

Type III—A luminaire designed for mounting over the curb line of street widths up to 2.0 mounting heights.

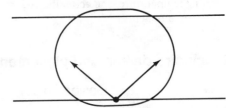

Type IV—A luminaire designed for mounting over the curb line of street widths greater than 2.0 mounting heights.

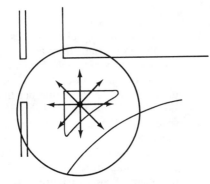

Type V—A luminaire designed to distribute light equally in all lateral directions.

Figure 10–2. Illustrations and descriptions of IES luminaire-type nomenclature. SOURCE: *American National Standard Practice for Roadway Lighting* (New York: Illuminating Engineering Society, 1977), p. 8.

necessary to control the light flux emission above the beam of maximum candlepower. The three categories of control are:

1. *Cutoff:* A luminaire light distribution is designated as cutoff when the candlepower per 1,000 lamp lm does not exceed 25 (2.5%) at an angle of 90° above nadir and 100 (10%) a vertical angle of 80° above nadir.
2. *Semi-cutoff:* A luminaire light distribution is designated as semi-cutoff when the candlepower per 1,000 lamp lm does not exceed 50 (5%) at an angle of 90° above nadir and 200 (20%) at a vertical angle of 80° above nadir.
3. *Non-cutoff:* The category when there is no candlepower limitation in the zone above maximum candlepower.

Placement

Luminaire placement is an integral part of the design of an effective lighting system. Luminaires are mounted at a given height above the roadway, depending on the lamp output, and at specific points along the roadway, depending on the character of the roadway to be lighted. For roadways not having medians, the luminaire is normally installed in a "house-side" location, which may be further described as a "one-side" system, a "staggered" system, or an "opposite" system. For streets having wide medians, and where barriers are to be installed, a "median lighting system" provides very effective lighting at less cost because of the saving in luminaire supports and electrical conductors (see Figure 10–3).

Mounting height is generally determined by the lamp output and the desired average illumination on the roadway and by the required uniformity of the distribution. Figure 10–4 indicates the relationship between maximum candlepower and mounting height based on disability glare calculations. In general, higher output luminaires are mounted higher.

Mounting heights of 65 ft or higher have been utilized in a special roadway lighting technique called *high mast lighting.* In high mast lighting, which is used mainly to light large areas such as interchanges, wide freeways, intersections, toll plazas, and parks, luminaires are arranged in combinations in order to provide a total system output of up to 1 million lm distributed over a large area. The justifications for high mast lighting pertain mainly to the removal of luminaire supports near the traffic area for safety reasons and to provide a panoramic view of the entire area.

Luminaires for high mast lighting include (1) the IES Type V luminaire, which provides a circular distribution; (2) the IES Type II or III luminaires, which provide an asymmetric distribution and can be used to cover various

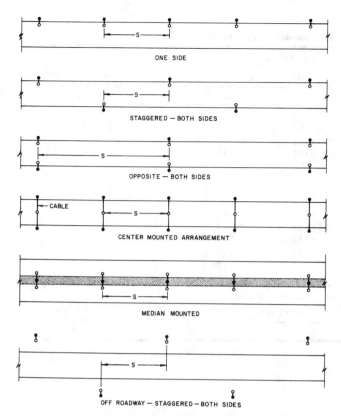

Figure 10–3. Typical mounting configurations (Luminance patterns repeat at spacing boundaries indicated)

SOURCE: *Roadway Lighting Handbook* (Washington, DC: U.S. Department of Transportation, 1983).

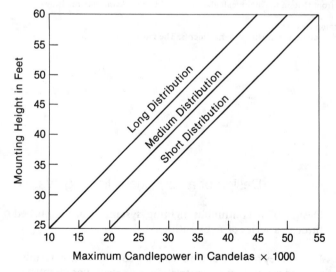

Figure 10–4. Minimum luminaire mounting heights based on current practice and DVB (Disability Veiling Brightness) calculations.

SOURCE: ANSI/IES RP-8, 1983.

roadway configurations; and (3) various other specialized luminaires usually combined to provide specific patterns.

Design of lighting systems

General design criteria

In 1983 the American National Standards Institute (ANSI) approved a new standard for roadway lighting. (see note 4.) This standard was different from past standards in that it introduced the concept of specifying luminance levels for roadway lighting. Illuminance (horizontal footcandles) levels were retained as a second method, but were now specified according to road surface reflectance. The standard was now more consistent with the international CIE (International Commission on Illumination) standard.[11] AASHTO's "Informational Guide for Roadway Lighting" is similar to the ANSI standard.

The ANSI standard also introduced the concept of restricting disability glare (veiling luminance). The maximum veiling luminance—that is, the maximum level the motorist encountered—would be compared with the average pavement luminance to produce a ratio of 0.3 to 1.0 or less for major roads and 0.4 to 1.0 or less for minor roads to obtain satisfactory control of disability glare.

Additionally, luminance ratios of maximum to minimum luminance and average luminance to minimum luminance are required. Table 10–3 indicates the recommended maintained levels of luminance and illuminance for the various road and area classifications. These levels represent the lowest in-service luminance or illuminance values for each classification.

Table 10–4 provides levels for walkways and bikeways. These levels are expressed in illuminance (horizontal footcandles) since pavement brightness is not the required design goal.

The current standard practice for roadway lighting allows the computation of lighting levels using horizontal illuminance or luminance, but recommends luminance as the preferred method.

The preference for luminance, however, does not apply to two special cases, pedestrian way lighting where both horizontal illuminance and vertical illuminance (for security purposes) are specified (see Table 10–4), and high mast lighting where uniformity of illuminance and surround lighting modify the visual task complexity.

Hand calculations made for the luminance method are very laborious and take several hours to get a few points. Microcomputers can, once the information has been entered, calculate luminance and other features of a complete lighting system in a few minutes. Although hand calculation of horizontal illuminance is a relatively easy task, it still requires graphical information from the manufacturer of the luminaire.

Most of the technical information is now supplied on floppy disks, and many illuminance programs are available,

---

[11] "Publication 12-2, Recommendations for the Lighting of Roads for Motorized Traffic."

(a) Maintained Luminance Values

Road and Area Classification		Average Luminance L_{avg} (cd/ft^2)	Luminance Uniformity		Veiling Luminance Ratio (maximum) L_v to L_{avg}
			L_{avg} to L_{min}	L_{max} to L_{min}	
Freeway Class A		0.06	3.5 to 1	6 to 1	0.3 to 1
Freeway Class B		0.04	3.5 to 1	6 to 1	0.3 to 1
Expressway	Commercial	0.09	3 to 1	5 to 1	
	Intermediate	0.07	3 to 1	5 to 1	0.3 to 1
	Residential	0.06	3.5 to 1	6 to 1	
Major	Commercial	0.11	3 to 1	5 to 1	
	Intermediate	0.08	3 to 1	5 to 1	0.3 to 1
	Residential	0.06	3.5 to 1	6 to 1	
Collector	Commercial	0.07	3 to 1	5 to 1	
	Intermediate	0.06	3.5 to 1	6 to 1	0.4 to 1
	Residential	0.04	4 to 1	8 to 1	
Local	Commercial	0.06	6 to 1	10 to 1	
	Intermediate	0.05	6 to 1	10 to 1	0.4 to 1
	Residential	0.03	6 to 1	10 to 1	

(b) Average Maintained Illuminance Values (E_{avg}) in Footcandles

Road and Area Classification		Pavement Classification			Illuminance Uniformity Ratio (E_{avg} to E_{min})
		R1	R2 and R3	R4	
Freeway Class A		0.6	0.8	0.7	3 to 1
Freeway Class B		0.4	0.6	0.5	
Expressway	Commercial	0.9	1.3	1.2	
	Intermediate	0.7	1.1	0.9	3 to 1
	Residential	0.6	0.8	0.7	
Major	Commcercial	1.1	1.6	1.4	
	Intermediate	0.8	1.2	1.0	3 to 1
	Residential	0.6	0.8	0.7	
Collector	Commercial	0.7	1.1	0.9	
	Intermediate	0.6	0.8	0.7	4 to 1
	Residential	0.4	0.6	0.5	
Local	Commercial	0.6	0.8	0.7	
	Intermediate	0.5	0.7	0.6	6 to 1
	Residential	0.3	0.4	0.4	

Notes:

L_v = veiling luminance

1. These tables do not apply to high mast interchange lighting systems, e.g., mounting heights over 65 ft.
2. The relationship between individual and respective luminance and illuminance values is derived from general conditions for dry pavement and straight road sections. This relationship does not apply to averages.
3. For divided highways, where the lighting on one roadway may differ from that on the other, calculations should be made on each roadway independently.
4. For freeways, the recommended values apply to both mainline and ramp roadways.
5. The recommended values shown are meaningful only when designed in conjunction with other elements. The most critical elements are:
   (a) Lighting system depreciation
   (b) Quality
   (c) Uniformity
   (d) Luminaire mounting height
   (e) Luminaire spacing
   (f) Luminaire selection
   (g) Traffic conflict area
   (h) Lighting termination

SOURCE: Adapted from: ANSI/IES RP-8, 1983.

some provided by luminaire manufacturers. Luminance programs are also available from various sources.[12]

The design of a lighting system is carried out on the basis of selecting various components of the system and making computations to verify their compliance with accepted criteria.

---

[12] Illuminating Engineering Society Computer Committee, "Available Lighting Computer Programs," *Lighting Design + Application*, (June 1986), 42–43.

Design of a continuous lighting system

Design of a continuous lighting system is accomplished through a series of steps outlined below:

1. A survey must be made of existing conditions to determine roadway and area classification, pavement type, number of lanes and lane width, basic traffic conditions both vehicular and pedestrian from a speed and accident standpoint, and extent and complexity of surround lighting.

TABLE 10–4

Recommended Average Maintained Illuminance Levels
for Pedestrian Ways[a]

Walkway and Bikeway Classification	Minimum Average Horizontal Levels (E_{avg})	Average Vertical Levels for Special Pedestrian Security (E_{avg})[b]
	ft cd	ft cd
Sidewalks (roadside) and Type A bikeways:		
Commercial areas	0.9	2.0
Intermediate areas	0.6	1.0
Residential areas	0.2	0.5
Walkways distant from roadways and Type B bikeways:		
Walkways, bikeways, and stairways	0.5	0.5
Pedestrian tunnels	4.0	5.0

[a]Crosswalks traversing roadways in the middle of long blocks and at street intersections should be provided with additional illumination.
[b]For pedestrian identification at a distance. Values are 5.9 ft above walkway.

SOURCE: Adapted from: ANSI/IES RP-8, 1983.

2. Based on the information from item 1, minimum levels of illuminance or luminance are selected from Table 10–3 or Table 10–4. Because these are minimum maintained values, other factors may require higher values such as television surveillance or unusual geometrics. These higher levels should be justified.

3. The selection of system components to be used in accomplishing the design is the next procedure. This process involves the location of luminaire supports, mounting height, size of lamp, maintenance factor, equipment factor, and luminaires to be used.

4. Luminaire photometric data must be acquired before any calculations can be made from luminaire data. A significant number of roadway luminaire manufacturers now furnish luminaire photometrics on floppy disks to be used with computer calculation programs. If hand or graphic calculations are to be made, then either a coefficient of utilization—isofootcandle data sheet—or a coefficient of luminance utilization—isoluminance data sheet—is required.

5. Light loss factors (LLF) must be determined. Lighting systems should be designed by considering certain light loss factors. These factors can be divided into various categories: maintenance factor (MF), field factor (FF), and equipment factor (EF). Maintenance factors are time-dependent.

   a. The maintenance factor consists of the lamp lumen depreciation (LLD), which is the loss of light output of the lamp over time, the luminaire dirt depreciation (LDD), and the system depreciation due to lamp burnouts. Lamp lumen depreciation can be obtained from manufacturers' data, such as the lamp lumen depreciation curve. This information will provide lamp lumen output versus burning hours. Luminaire dirt depreciation (LDD) can be estimated from the chart in Figure 10–5. This chart assumes no cleaning of luminaires.

   b. Field factors (FF) relate to the equipment supplied. Since most photometric testing is done at a specified lamp wattage, which is held constant using laboratory ballasts, a significant difference may exist between the photometric laboratory and field performance of the equipment, where voltage is relatively constant[13] and wattage varies considerably.

   c. Equipment is also affected by the ambient temperature of the area where the equipment is operated. Data on light output vs. temperature should be considered.

---

[13]I. Lewin, and L. Stafford, "Photometric and Field Performance of High Pressure Sodium Luminaires," *Journal of the Illuminating Engineering Society* (Summer 1987).

Figure 10–5. Chart for estimating roadway luminaire dirt depreciation factors for enclosed gasketed luminaires.

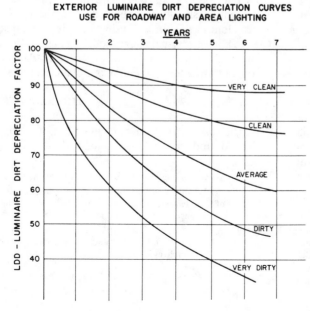

EXTERIOR LUMINAIRE DIRT DEPRECIATION CURVES
USE FOR ROADWAY AND AREA LIGHTING

Select appropriate dirt curve from kind of conditions described below for type luminaire to be used.

Areas—clean-pavement-grass. No open loose ground. Slow traffic. Little or no adhesive qualities in atmosphere. Most rural areas, residential roadways, slow traffic, no trucks.

Areas—as above except average car and truck traffic, downtown open areas, intermediate and freeways in open areas.

Areas—as above but slightly more exposure. Residential, intermediate, local minor roads, few trucks.

Areas—confined. Greater than average cars and trucks, expressway, freeways. Downtown, major adhesive dirt.

Ind/Comm. areas. Trucks, buses, adhesive dirt, confined areas, heavy traffic.

d. Variations in operating voltages (voltage drop factor, VD) must be considered in selecting ballast types, taking into account possible line voltage drops.

e. The lamp and ballast combination luminaire factor (LF) may produce a different light output from that indicated by the laboratory photometrics cited above.

All of the above factors must be considered in the design process as well as the geometric and environmental concerns. In general, the maintenance, field, and equipment factors will change the light output and luminance or illuminance levels.

For example, if a luminance level of 0.09 cd/sq ft represents the initial design value for new equipment, a maintenance factor (MF) with a group relamping system might be 0.73 (LLD) × 0.95 (LDD) = 0.6935 (MF).

The field factor (FF), assuming a high intensity discharge (HID) source not affected by temperature, would be:

Field factor (FF) = 0.982 (LF) × 0.92 VD factor = 0.90

Assuming an equipment factor (EF) of 0.95 due to manufacturer's tolerances, the initial light level of 0.09 cd/sq ft would then be multiplied by 0.6935 (MF) × 0.90 (FF) × 0.95 (EF) = 0.053 cd/sq ft. The maintained light level would be 0.053 cd/sq ft.

Computational procedures

Once the preliminary arrangement of poles, mounting height, and mast arms have been selected and the luminaire and light loss factors chosen, initial computations can begin.

It is recommended that a reliable computer program be used to calculate the luminance or illuminance levels, uniformity ratios, and veiling luminance. This will result in considerable time saving and greater accuracy. If hand calculations are the only option, then it is recommended that the computation procedures in either the *Roadway Lighting Handbook* and the "Addendum to Chapter Six" of the *Roadway Lighting Handbook*[14] (for luminance) published by the Federal Highway Administration, or the *American National Standard Practice for Roadway Lighting,* ANSI (RP-8), and the *Designing with the IES Roadway Lighting Practice* (RP-8),[15] published by the Illuminating Engineering Society, New York, be used.

Results of the first calculation should be compared with the selected levels, uniformity ratios, and the veiling luminance ratio required for the roadway and area classifications. Adjustments are then made with subsequent calculations. Several calculations are usually required before the proper relationship among lighting levels, uniformity ratios, and glare (veiling luminance) ratios are correct. Changes in luminaire spacing are most often adjusted to arrive at the proper design, but different luminaire types, mounting heights, mast

arm lengths, luminaire tilts, and even changes in the geometric lighting configurations may be necessary to accomplish the design goal. The relationship among these changes and luminance criteria are explained in detail in the "Addendum to Chapter Six" of the *Roadway Lighting Handbook*[16] (pp. 18, 19).

Partial lighting

Where partial lighting rather than full lighting is necessary (because of economic considerations), caution should be used when installing a few poles to light visually critical areas. The primary benefits of this type of lighting are to act as a beacon, giving approaching motorists advance warning of a conflict area, and to provide some illumination at the conflict area. Because these are areas of traffic conflict and decision making the probability of a car-pole collision is much greater than for a continuously lighted roadway.

Additionally, research[17] has shown that there are some visual problems introduced when the motorist, who is visually adapted to a lower relative brightness, is briefly exposed to higher levels and accompanying disability glare. Partial lighting should only be installed where simple straightforward geometrics are involved using as few luminaires as possible. Where road curvatures and other changes occur, the partial lighting should be extended beyond those critical points, and under these conditions complete lighting should be strongly considered.[18]

Both ANSI RP-8 (*American National Standard Practice for Roadway Lighting*) and the AASHTO guide[19] have examples of partial lighting installations. Lighting design should be accomplished by the luminance method with strong emphasis on the reduction of disability glare.

Isolated traffic conflict areas

Where complete lighting is provided at an interchange, intersection, or other traffic conflict area that is separated from a continuously lighted area by 20 sec or more of driving time, it is referred to as an *isolated traffic conflict area.*

Lighting should be extended from the traffic conflict area along each approach using a driving time distance of 2 sec to allow for visual adaptation. The lighting design for the extension should be as required in the roadway classification, but using the residential area classification.

Traffic conflict areas

Traffic conflict areas are intersections, weaving sections, areas of reduced or substandard geometrics, etc.; and

---

[14] *Roadway Lighting Handbook,* "Addendum to Chapter 6" (Designing the Lighting System Using Pavement Luminance), 1983, U.S. Department of Transportation, Federal Highway Administration, RD&T Report Center, HNR-11, 6300 Georgetown Pike, McLean, VA 22101.

[15] *Designing with the IES Roadway Lighting Practice* (New York: Illuminating Engineering Society).

[16] *Roadway Lighting Handbook,* "Addendum to Chapter 6."

[17] National Cooperative Highway Research Program Report 256, "Partial Lighting of Interchanges," 1982, Transportation Research Board, National Academy of Science, 2101 Constitution Avenue, N.W., Washington, DC 20418.

[18] R.S. Hostettler and others, "Trade-off Between Delineation and Lighting on Freeway Interchanges," Federal Highway Administration—R. D. 88-223, U.S. Department of Transportation, Federal Highway Administration, RD&T Report Center, HNR-11, 6300 Georgetown Pike, McLean, VA 22102.

[19] AASHTO, "Informational Guide for Roadway Lighting."

locations where there are increased potential for vehicular-to-vehicular, vehicular-to-pedestrian, or vehicular-to-fixed-object collisions.

Current standards require that the lighting level at intersections and weaving areas be the sum of the levels of the individual roadways, and where high volume driveways enter public streets, levels should be 50% higher than the average route value. Future standards will require higher visibility levels at all of the above locations.

High mast lighting

Lighting of large areas by means of a group of luminaires at mounting heights of 65 ft or more is generally referred to as *high mast lighting.* High mast lighting is mainly used on interchanges and wide roadway sections, but is being used also for continuous freeway sections. Additionally, high mast lighting at grade intersections is being employed more frequently.

The advantages of high mast lighting are several. The overall lighted area appears more like the daylight scene. The off-roadway areas, connecting ramps, landscaping, and other components of the road and its environs are illuminated, in contrast to the conventional illumination of the road surface only. Drivers are able to relate to the entire road complex.

Fewer poles located farther from the edge of the pavement (recommend a minimum of 50 ft unless protected) reduce the possibility of car–pole collisions and improve aesthetics.

Where high mast lighting is used on isolated interchanges, intersections, or at the beginning or end of continuously lighted roads, transition lighting is easily achieved because of the gradual reduction of light levels at higher mounting heights.

Continual growth of traffic and the aging of road systems have resulted in widening, reconstruction, and rehabilitation of roadways. In many cases, where high mast lighting had been installed, no changes in the lighting system were required since all poles were located at a substantial distance from the road edge. The lighting system also serves to illuminate the construction site.

The disadvantages of high mast lighting are the required extra energy to illuminate areas other than the road surface and the installation of a more complex lighting unit.

The current 1983 *American National Standard Practice for Roadway Lighting* (RP-8) treats high mast lighting separately by requiring that only illuminance rather than luminance levels be met (see Table 10–5). These values are somewhat lower than required for conventional lighting. Complete design procedures for high mast lighting are detailed in the ANSI Publication.

High mast lighting hardware. High mast lighting units are more complex in design, installation, and maintenance than are conventional lower mounted units. The simplest system uses luminaires fastened directly to a mast and requires that the pole be climbed to service the luminaires. The most complicated installations have automatic built-in lowering systems such that luminaires can be serviced at ground level. Where lowering systems are used, safety

TABLE 10–5

Recommended Maintained Illuminance Design Levels for High Mast Lighting[a,b]

Road Classification	Horizontal Illuminance (E_{avg}) in ft cd		
	Commercial Area	Intermediate Area	Residential Area
Freeways	0.6	0.6	0.6
Expressways	0.9	0.7	0.6
Major	1.1	0.8	0.6
Collector	0.7	0.6	0.6

[a]Recommended uniformity of illumination is 3 to 1 or better; average-to-minimum for all road classifications at the illuminance levels recommended above.
[b]These design values apply only to the travelled portions of the roadway. Interchange roadways are treated individually for purposes of uniformity and illuminance level analysis.

features should be incorporated in the design to avoid accidental falling of luminaire rings.

Pole and foundation design is more critical in high mast installations because of increased wind load on the poles and the large luminaire configurations.

Selection of safety features

While lighting provides improved visibility and reduction in night accidents it also requires the installation of luminaire supports or lighting poles. These poles can be a hazard if not properly located and made breakaway such that the errant motorist either avoids them or the resultant car–pole collision produces relatively little damage or injury.

The most important and first step in the design of a lighting system is pole location. Areas of possible collisions should be singled out and no poles located in those places. Poles located on the outside of roads should be set back from the traveled way as far as possible. Where possible, poles should be placed behind guardrails and on concrete barriers or walls.

Poles located near the roadside without protection should be equipped with breakaway devices. There are several different types of bases used, including, but not limited to, frangible transformer bases, slip bases, and breakaway couplings (see Figures 10–6, 10–7, and 10–8). Breakaway characteristics must comply with the *Standard Specifications for Structural Supports for Highway Signs, Luminaires, and Traffic Signals.*[20] The Federal Highway Administration has test facilities to evaluate breakaway features for compliance.

The question of whether or not to use a breakaway device in areas where poles are located close to pedestrian walks or where developed property is close to the roadway should be investigated. Studies should be made relating car–pole collision frequency vs. the frequency of the presence of pedestrians in the projected area of a possibly felled pole. This will help to determine whether or not breakaway devices would be hazardous to pedestrians should a collision occur.

[20] *Standard Specifications for Structural Supports for Highway Signs, Luminaires, and Traffic Signals,* 1985, Code LTS-1. AASHTO, 444 North Capital Street, N.W., Suite 225, Washington, DC 20001.

Figure 10–6. Frangible transformer base.

Figure 10–7. Frangible breakaway coupling.

Figure 10–8. The multidirectional slip base.

Tunnel and underpass lighting

Tunnels and underpasses require special consideration. Ordinarily, roadway lighting is used to illuminate a roadway in open space. Tunnels and underpasses are semiclosed spaces, or a covered section of roadway. There are height restrictions and special installation and maintenance requirements for tunnels and underpasses.

Underpasses. An underpass is defined as a portion of a roadway extending through and beneath some natural or fabricated structure, which, because of its limited length-to-height ratio, requires no supplementary daytime lighting. Length-to-height ratios of 10:1 or lower will not, under normal conditions, require daytime underpass lighting.

Underpass lighting is required at night on continuously lighted roadways while underpass lighting on unlighted roads may be warranted under special circumstances where frequent pedestrian traffic is present or there is unusual or critical roadway geometry.

Design values for underpass lighting should be equivalent to that of the adjacent roadway. Uniformity ratios must also be met that may result in somewhat higher levels of lighting.

Tunnels. Tunnels can be classified into two types—short tunnels and long tunnels. A tunnel is classified as short if the length of the tunnel is equal to or less than the wet pavement minimum stopping sight distance (as recommended by AASHTO)[21] for the tunnel approach roadway. A long tunnel would be one where length exceeds the AASHTO minimum wet pavement safe stopping sight distance.

Tunnels require daytime lighting with high levels in the entrance area to provide proper adaptation between the approach and surround brightness and the entrance area (threshold zone) of the tunnel. Tunnel length and approach can be separated into several sections: (1) the approach (access zone CIE)[22]; (2) portal; (3) threshold zone; (4) transition zone; (5) interior zone; and (6) exit zone (CIE).[23]

Short tunnels have one interior zone, the threshold zone. In some cases, short tunnels having less than the safe stopping sight distance with high ratios of width to height, and where visibility is provided by the high luminance of the exit portal, no supplemental daytime lighting may be required.

The Recommended Practice for Tunnel Lighting[24] lists luminance levels for the threshold zone. The transition zone is the area where the lighting level from the threshold zone to the interior zone is reduced in steps. The first step should be equal to or greater than one-quarter of the threshold zone luminance. The last step should be equal to or less than twice the interior zone luminance. Intermediate steps should be equal to or greater than one-third of the preceding zone. Daytime interior zone luminance should be not less than 0.5 cd/sq ft. Nighttime tunnel luminance should be not less than 0.23 cd/sq ft average for the entire tunnel.

Tunnel lighting design involves techniques somewhat different from roadway lighting. Therefore, it is recommended that other references be consulted for this application (see notes 5, 22, and 24).

Rural at-grade intersection illumination

Based on a comprehensive investigation of rural at-grade intersection illumination conducted at the University of Illinois, the following is recommended as the basis for rural at-grade illumination warrants.

[21] *Policy of Design Standards for Stopping Sight Distance,* 1971, Code PSD, American Association of State Highway and Transportation Officials (AASHTO), 444 North Capitol Street, N.W., Suite 225, Washington, DC 20001.

[22] *Guide for the Lighting of Road Tunnels and Underpasses,* Commission Internationale de L'Eclairage (CIE), Draft Publication No. 26, 1989.

[23] Ibid.

[24] *Recommended Practice for Tunnel Lighting* (RP-22) (New York: Illuminating Engineering Society, 1987).

Rural intersections should be considered for lighting if the average number of nighttime accidents (N) per year exceeds the average number of day accidents (D) per year divided by 3. All the accident data available since the date of the last modification to the intersection should be used when calculating these averages. If N is greater than D/3, the likely average benefit should be taken as N − D/3 accidents/year.

The likely benefits of lighting new or modified intersections should be estimated from previous experience. It is recommended that illumination be provided whenever an intersection is channelized. The estimated cost of lighting the intersections, which show a benefit using the above criteria, should be computed. The lighting program should then be based on the resulting list of intersections ranked in priority order by means of the benefit/cost ratio (expressed as annual reduction in accidents/annual cost).[25]

The recommended warrant is designed to give decision-makers the most information possible based on current knowledge. It is implicitly assumed that the highway improvement budget is limited, and thus interest is focused on maximizing the benefits of a limited budget. For this reason, reductions in number of accidents rather than accident rates are used. One important implication of this approach is that the distribution of funds for lighting improvements tends to be directed into the areas of high traffic volumes. Thus, if intersections are ranked on a statewide listing, the distribution of the budget would not be the same as one distributed by listing intersections on a district basis. The latter would spread improvements more uniformly throughout a state, but at a lower overall benefit/cost ratio.

Cost considerations

The most important factor in the design of a lighting system is providing for the visual needs of the motorists using that facility. If these needs are *not* met then the financial investment is unsatisfactory. Currently, the meeting of these needs is specified in various standards such as the ANSI-RP-8 and AASHTO guide.

Designs used in meeting these standards can vary and therefore the installation cost and subsequent maintenance will vary as well. Systems can be designed with many low-wattage luminaires mounted at low heights or fewer high-wattage luminaires at higher mounting heights. Equally important in the initial installation investment is the selection of the lamp, luminaire, and luminaire distribution.

The following items should be evaluated from the range of possibilities while still meeting the visual criteria:

1. Arrangement or placement of lighting units with respect to the roadway (median, one-side, staggered on opposite sides, opposite on two sides, and median plus side-mounting)
2. Mounting heights
3. Pole and base type
4. Lamp characteristics (type, initial lumens, lumen maintenance, wattage)
5. Luminaire type

6. Light distribution type
7. Type of distribution system
8. Burning hours per year
9. Type of ownership and maintenance.

Experience, especially from a maintenance standpoint, has provided insight into particular designs that are least costly to maintain and provide for safer roadway operations.

The provision of a full shoulder that allows service vehicles to be clear of through roadway lanes provides for safer maintenance. Installation of lighting units on concrete median barriers or between median barriers reduces the number of auto–pole collisions.

Some of the roadway features that affect lighting installation are as follows:

1. Median width
2. Presence or absence (or plans for such) of a median barrier and its type
3. Number of traffic lanes
4. Width of roadway shoulders
5. Overall roadway cross section
6. Number and type of exit and entrance ramps
7. Number and type of intersections and intersecting roadways and streets.

Sign lighting

Warrants

Sign lighting is an extremely important asset for the driver's visual information needs with respect to navigational ability. Quickly and accurately acquiring necessary directional information is essential to the night motorist. The *Manual on Uniform Traffic Control Devices* (MUTCD),[26] section 2E-6, requires that all overhead sign installations should normally be illuminated. The type of illumination chosen should provide effective and reasonable uniform illumination of the sign face and message.

Standard roadway signs are usually externally illuminated although internally illuminated and luminous-source message signs have been used.

The warrants for sign lighting contained in AASHTO's "Informational Guide for Roadway Lighting" are simply that a sign that is not adequately visible at night should be lighted. Signs that have been moved for roadside clearance reasons far from the edge of the road, or overhead signs where headlights may not provide adequate illuminance of the sign panel, should be lighted.

Area classification

The levels of sign luminance or illuminance are determined by the ambient luminance or brightness of the background against which the sign is viewed.

[25] Wortman and Lipinski, "Interim Report: Development of Warrants for Rural At-Grade Intersection Illumination."

[26] *Manual on Uniform Traffic Control Devices for Streets and Highways,* Superintendent of Documents, U.S. Government Printing Office, Washington, DC 20402.

1. *Low*—Rural areas where objects at night are visible only in bright moonlight. There is little or no other lighting.
2. *Medium*—May contain small areas of commercial lighting and/or roadway lighting.
3. *High*—Central business districts, high-level lighted roadways, brightly lighted commercial advertising signs, or highly illuminated parking facilities.

Table 10–6 gives the lighting levels for the three areas. Uniformity of maximum to minimum incident light on the whole sign face must not exceed 6:1, with 4:1 being desirable.

TABLE 10–6
Luminance and Illuminance for Sign Lighting

The Following May Be Used as a Guide for Lighting Levels: Ambient Luminance or Illuminance		
Low	Medium	High
Luminance[a] 2.2–4.5 cd/ft²	4.5–8.9 cd/ft²	8.9–17.8 cd/ft²
Illuminance 10–20 fc	20–40 fc	40–80 fc

[a]Maintained reflectance of 70% for white sign letters.
SOURCE: IES RP-19, 1983.

Sign color

Standardized sign colors have been established as described in the MUTCD. The lighting designer should be sure that the light source will adequately illuminate and preserve the colors on the sign. The *Recommended Practice for Roadway Sign Lighting, RP-19,* [27] contains light source color recommendations and other design parameters for sign lighting.

Operation and maintenance

Electric power considerations

All lighting installations need electrical power. This requires coordination with an electric utility to obtain the necessary voltage at the designed point of delivery. For small systems the operating voltage is 120 or 240 V. For larger systems fed from a single point of delivery, operating voltages of 480 V single-phase and 277-480 V three-phase can be used.

Luminaire/ballast units are available for operation from 120, 240, 277, and 480 V circuits. The building-type wire normally used has insulation rated at 600 V, making it suitable for any of these voltages.

The most popular means of switching roadway lighting on and off are light-sensitive switches or photoelectric controls. This type of control turns on the lighting when the daylight level drops to a certain value, and turns it off when the dawn light level reaches a certain value. A single control can be used to switch a number of lighting units in a system, or one can be plugged into a luminaire provided with a socket to switch that one luminaire.

Electric power is available both metered and unmetered or "flat rate." In the latter case, a monthly cost is charged for each lighting unit based on the lamp size. The selection of metered or unmetered service for a particular system requires an economic analysis that includes consideration of the system size and the rate structure of the electric utility.

Design of the electrical system

The electrical design of a lighting system usually falls outside the traffic engineer's responsibility. An excellent text on this subject is *Highway Lighting Engineering,* [28] by Anatanas Ketvirtis, which provides very detailed descriptions, plans, and specifications for all electrical design associated with highway lighting.

Energy conservation

The use of electrical energy for roadway lighting must be carefully employed. Appropriate lighting design can keep energy consumption to a minimum. Fortunately over the last few decades great strides have been made in the improvement of the electric lamp. The advent of the gaseous-discharge lamps has resulted in a fivefold to tenfold increase in lamp efficacy (lm/W) over the incandescent lamp. It is important that the most efficient light sources be used in new designs as well as replacing older and less efficient light sources in existing installations.

Careful analysis of lighting system geometry can also result in lower energy consumption. Pole-mounting heights and luminaire size and efficiency are important factors in energy-conservation designs.

Maintenance

Proper maintenance of the lighting system is essential. It must produce the type of results that were contemplated in the system design. Light output from lamps diminishes and dirt accumulates on the luminaire over a period of time. This requires a restoration process. Maintenance intervals and the extent of maintenance vary from a 6-month cleaning interval to once every 4 years, and at the extreme only when there is a lamp burnout. Lamp replacement varies from replacing individual lamps to group relamping on a 3- to 5-year basis. The average rated lamp life in excess of 24,000 hours for mercury and high-pressure sodium lamp fit into the above group relamping periods quite well.

Annual or biennial cleaning of luminaires will keep the luminaire in good condition. Different environments (see Figure 10–5) will require appropriate cleaning intervals. As luminaires are being designed that have better systems of sealing and filtering the optical enclosures, cleaning of the exterior lens only except when relamping may be sufficient to maintain light output.

[27] IES Roadway Lighting Committee, *Recommended Practice for Roadway Sign Lighting, RP-19* (New York: Illuminating Engineering Society, 1983).

[28] A. Ketvirtis, *Highway Lighting Engineering* (Toronto: Foundation of Canada Engineering Corporation Ltd., 1967).

Inventory and records

Accurate inventory of lighting equipment is essential to efficient system operation. Records of routine maintenance and repairs require that up-to-date inventory of field installations be kept. The advent of computerized data base programs permits instantaneous review of system performance. Programs have been developed that will provide current information on each luminaire such as when it was relamped, repaired, replaced, or cleaned. The maintainer can then analyze the incidence of various equipment needs and take appropriate action. This requires an accurate field reporting system. Field maintenance forces must record and transmit information as to the required maintenance at each field installation in order that the data be kept current.

REFERENCES FOR FURTHER READING

ADRIAN, W., "Model to Calculate the Visibility of Targets," *Lighting Research and Technology,* Volume 21, Number 4, 1989.

"American National Standard Practice for Roadway Lighting" RP-8 (New York: Illuminating Engineering Society/American National Standards Institute, 1983).

"An Informational Guide for Roadway Lighting" (Washington, DC American Association of State Highway and Transportation Officials, 1984).

"Calculation and Measurement of Luminance and Illuminance in Road Lighting," CIE 30, Washington, DC: National Bureau of Standards, 1976.

Designing with the IES Roadway Lighting Practice (New York: Illuminating Engineering Society).

Guide for the Lighting of Road Tunnels and Underpasses, Commission Internationale de L'Eclairage (C.I.E.), Draft Publication No. 26, 1989.

IES Roadway Lighting Committee, *"Recommended Practice for Roadway Sign Lighting RP-19,"* New York Illuminating Engineering Society, 1983.

Illuminating Engineering Society, "Value of Public Road Lighting," Draft (1989).

Illuminating Engineering Society Computer Committee, "Available Lighting Computer Programs," *Lighting Design + Application,* June 1986.

International Commission on Illumination (CIE) 1988, "Road Lighting as an Accident Countermeasure," Report Number 8.2, Commission Internationale de L'Eclairage, Paris, France (In process).

KETVIRTIS, A., *Highway Lighting Engineering* (Toronto: Foundation of Canada Engineering Corporation Ltd. 1967).

LEWIN, I., AND L. STAFFORD, "Photometric and Field Performance of High Pressure Sodium Luminaires," *Journal of the Illuminating Engineering Society,* Summer 1987.

National Cooperative Highway Research Program Report 256, "Partial Lighting of Interchanges," 1982, Transportation Research Board, National Academy of Science, 2101 Constitution Avenue, N.W., Washington, DC 20418.

"Publication 12-2, Recommendations for the Lighting of Roads for Motorized Traffic," Commission Internationale de L'Eclairage (CIE), 1977.

Recommended Practice for Tunnel Lighting (RP-22) (New York: Illuminating Engineering Society).

Roadway Lighting Handbook, 1978, U.S. Department of Transportation, Federal Highway Administration, RD&T Report Center, HNR-11 6300 Georgetown Pike, McLean, VA 22102-2296.

Roadway Lighting Handbook, Addendum to Chapter 6 (Designing the Lighting System Using Pavement Luminance), 1983, U.S. Department of Transportation, Federal Highway Administration, RD&T Report Center, HNR-11, 6300 Georgetown Pike, McLean, VA 22101-2296.

Standard Specifications for Structural Supports for Highway Signs, Luminaires, and Traffic Signals, 1985, Code LTS-1. AASHTO, 444 North Capital Street, N.W., Suite 225, Washington, DC 20001.

Warrants for Roadway Lighting, NCHRP Report 152 (Washington, DC National Academy of Sciences, 1972).

WORTMAN, R.H., AND M.E. LIPINSKI, "Interim Report-Development of Warrants for Rural At-Grade Intersection Illumination," Ill Cooperative Highway Research program Ser. No. 135, University of Illinois at Urbana-Champaign, 1972.

11

TRAFFIC REGULATIONS

H. Richard Mitchell, P.E., *Principal Associate*

TJKM Transportation Consultants

AND

Roy A. Parker, P.E., *Transportation Administrator*

City of San Leandro, California

This chapter addresses the purpose and scope of traffic regulations and includes discussion of one-way streets, reversible lanes and roadways, turn regulations, transit and carpool lanes, pedestrian-only streets, restrictions on types of vehicles, speed regulations, emergency regulations, residential street controls, and other regulations used to obtain maximum efficiency and safety on street transportation systems. Right-of-way regulation at intersections by use of signs or traffic signals is discussed in Chapters 8 and 9.

Uniformity of traffic regulations and legal responsibility for maintenance of traffic control devices are briefly discussed, since these are essential considerations for the proper functioning of the overall street and highway system. Legal authority for traffic regulations is discussed in Chapter 14.

Purpose and scope of traffic regulations

The purpose of traffic legislation is to provide for the necessary and reasonable regulation of street and highway traffic, and "to insure, as far as this can be done by law and its application, that traffic shall move smoothly, expeditiously, and safely; that no legitimate user of the highway, whether in a vehicle or on foot, shall be killed, injured or frustrated in such use by the improper behavior of others."[1]

In the United States, traffic regulations are embodied in laws and ordinances of local, state, and federal jurisdictions. Although traffic legislation has traditionally been a state

and local function in the United States, Congress has passed several major traffic laws, including the Interstate Commerce Act of 1935, the National Traffic and Motor Vehicle Safety Act of 1966, and the Highway Safety Act of 1966.

These acts included provisions for establishing safety regulations for interstate trucks and buses, motor vehicle and tire safety standards, and highway safety program standards and guidelines. Until 1966, the promulgation of laws concerned with street and highway traffic regulations in the United States was left largely to the states. Since 1966, there has been significant federal intervention, although it is expected that states will continue to exercise primary jurisdiction in the field. Local governments are empowered by the states to enforce laws within their boundaries and also to establish local traffic regulations, such as:

1. Regulation of stopping, standing, and parking of vehicles.
2. Regulation of traffic by police officers or traffic control devices in conformity with the state motor vehicle code.
3. Regulation of speed in conformity with the state code.
4. Designation of one-way streets, through streets, and truck routes.
5. Establishment of turn prohibitions and no-passing zones.
6. Control of access and removal of sight-distance obstructions.

Uniformity of traffic regulations

Over the years, an increasing emphasis has been placed on the need for interstate and international uniformity in traffic laws and regulations, as well as in traffic control devices that are used to convey these laws and regulations to the road user. This emphasis has resulted in the development of

---

[1] National Committee on Uniform Traffic Laws and Ordinances, *Uniform Vehicle Code and Model Traffic Ordinance*, Evanston, IL, 1987, Foreword, p. v.

uniform traffic codes and ordinances, which along with the *Manual on Uniform Traffic Control Devices*[2] (discussed in Chapters 8 and 9) provide a consistent basis for regulating roadway traffic flow.

The Uniform Vehicle Code. The Uniform Vehicle Code (UVC) is a model set of motor vehicle laws designed and advanced as a comprehensive guide or standard for state motor vehicle and traffic laws. The UVC is promulgated by the National Committee on Uniform Traffic Laws and Ordinances, composed of about 150 representatives from federal, state, and local governments, and from business, industry, and professional societies. Most states have incorporated the recommendations of the UVC into their vehicle regulations and "rules of the road."

Model Traffic Ordinance. The Model Traffic Ordinance (MTO) is a model set of ordinances prepared by the National Committee for guidance in formulating local traffic regulations. As an example of MTO use, in California, the League of California Cities has prepared a California Uniform Traffic Ordinance based on the MTO as a guide for both small and large cities in that state. Other states have adopted portions of the MTO by reference.

In the United States, the "rules of the road"—for the actual regulation and control of traffic—are almost exclusively within the sphere of the states and local jurisdictions, and they are generally formulated to work together with specific controls and devices. The rights and duties of pedestrians and bicyclists, as well as drivers of motor vehicles, are delineated in the various state vehicle codes.

The UVC and state codes establish minimum requirements for certain vehicle equipment, including head and tail lamps and directional signals, brakes, horns and warning devices, mufflers and emission control devices, mirrors, windshield wipers, and seat belts. The UVC calls for compulsory vehicle inspection at least annually to check on conformance with minimum equipment requirements.

In several states, the vehicle code provides for inspection of emission control devices and measurement of exhaust chemistry periodically for all vehicles registered and garaged in a "nonattainment area"—an air basin not attaining specified ambient air quality standards. Inspection of vehicle safety components by police officers is also authorized if there is probable cause to believe there is a violation. (The California *Vehicle Code* provides an example of such regulations.)[3] These restrictions have generally shown an improvement in air quality resulting from motor vehicles.

The federal Motor Vehicle Safety Act of 1966, the Air Quality Act of 1967, and later provisions also empower the U.S. government to specify vehicle safety standards and to control vehicle emissions. It is expected that the control of motor vehicle emissions will become more restrictive in the future.

---

[2] *Manual on Uniform Traffic Control Devices,* U.S. Department of Transportation, Washington, DC, 1989.

[3] California Department of Motor Vehicles, *Vehicle Code,* Sacramento, CA, 1986.

Legal responsibility of public agencies

Governmental and individual officials' responsibilities for maintaining public highway safety systems are more important than ever before. Responsibility for traffic safety and traffic systems management begins with every state and local agency employee who participates in the operation of highway systems and may have opportunities for identifying and reporting traffic safety problems. Whether the problem involves an overgrown shrub that obstructs a stop sign, an unsigned sharp curve, or a deteriorated pavement, the responsibility of the government employee remains the same. Proper reporting and corrective action should be initiated promptly.

Public agencies and their employees used to rely on *sovereign immunity* for protection against being named as defendants in civil lawsuits concerning the design, operation, and maintenance of street and highway systems. Particularly since the 1960s, the law has changed so that both the agency and the individual administrator or employee may be found legally liable for all or a portion of damages awarded for loss under certain circumstances. (Statutory protection in many states may prevent suits against individual employees, except in cases of negligence.)

Refer to Chapter 14 for more discussion of tort liability.

One-way streets

Most major streets and highways are originally designed for use by two-way traffic. The need for the adoption of one-way traffic regulations may arise from increased traffic usage, conflicts among vehicular flows and between pedestrians and vehicles, and the resulting congestion and accidents. Conversion to one-way street operation (often in conjunction with parking restrictions) may also be needed to provide additional capacity to serve new development.

In major activity centers, such as the central business district of a city with many high-traffic, closely spaced intersections, one-way regulations are frequently used because of traffic signal timing considerations and to improve street capacity. In the development of new activity centers such as shopping malls, sports arenas, and industrial parks, one-way regulations are sometimes included in original street and traffic plans.

Some minor street and alleys are also designated for one-way operation because of limited width or in order to prevent through traffic within a neighborhood.

One-way streets are generally operated in one of three ways:

1. A street on which traffic moves in one direction at all times.
2. A street that is normally one-way in a particular direction but at certain times is operated in the reverse direction to provide additional capacity in the predominant direction of flow.
3. A street that normally carries two-way traffic but which during peak traffic hours is operated as a one-way street. Such a street may be operated in one direction during the

morning peak hour and in the opposite direction during the evening peak hour, with two-way traffic during all other hours.

Advantages and disadvantages

One-way regulations are generally used to reduce congestion and to increase the capacity of a street network. One-way streets may also affect safety and the types of uses on adjacent land. An intersection of two one-way streets has substantially fewer potential conflicts than does an intersection with two two-way streets, as shown by Figure 11-1.

The following advantages may be expected in terms of capacity, safety, and operating conditions:

Effect on capacity. Traffic conflicts and delay at intersections are a principal cause of congestion and longer travel time on two-way urban streets. On one-way streets, turning movements are not delayed by opposing vehicular traffic, but they may be obstructed by heavy pedestrian volumes and thus encounter significant delay. With one-way streets, more complete use may be made of street pavements with unusual width. The capacity of a street may be increased by as much as 50% by use of one-way regulations (see Chapter 5).

The increased capacity afforded by one-way regulations may also make it possible to permit parking either part- or full-time on streets that, if operated as two-way streets, could not be used for parking. More efficient signal timing can also increase street capacity because of improved traffic progression between signalized intersections, as discussed in Chapter 9.

Effect on safety. One-way streets with traffic signal controls at major intersections are more likely to have gaps in traffic for safer crossing movements by pedestrians and vehicles at other cross streets and driveways along the route. In addition, drivers and pedestrians crossing one-way streets need be concerned with and wait for traffic from only one direction.

Numerous studies have shown that the conversion of two-way streets to one-way operation reduces total accidents on an order of 10% to 50%.[4] In some cases, specific kinds of accidents are reduced even more.

However, vehicles turning left out of one-way streets appear to hit pedestrians significantly more frequently than do all other turning vehicles, probably because of automobile roof support pillars blocking the view of the crosswalk, which

---

[4]J.A. Bruce, "One-Way Major Arterial Streets," *Improved Street Utilization Through Traffic Engineering,* Highway Research Board Special Report 93, Washington, DC, May 1967.

Figure 11-1. Intersection conflicts.
SOURCE: *Manual of Geometric Design Standards for Canadian Roads,* Roads and Transportation Association of Canada, Ottawa, 1986, p. D15.

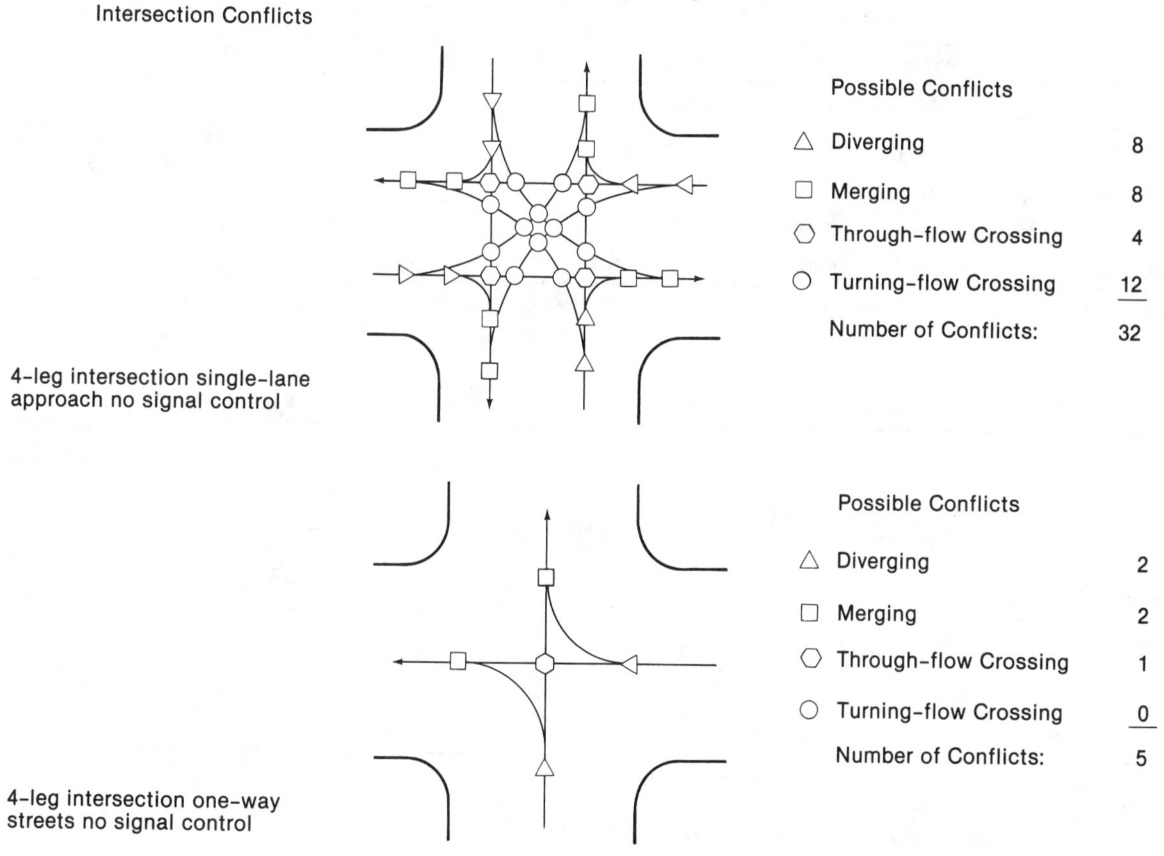

Intersection Conflicts

4-leg intersection single-lane approach no signal control

Possible Conflicts		
△ Diverging		8
□ Merging		8
◯ Through-flow Crossing		4
◯ Turning-flow Crossing		12
Number of Conflicts:		32

4-leg intersection one-way streets no signal control

Possible Conflicts		
△ Diverging		2
□ Merging		2
◯ Through-flow Crossing		1
◯ Turning-flow Crossing		0
Number of Conflicts:		5

is parallel to the original direction of travel.[5] Minor midblock collisions have been known to increase as a result of improper weaving by drivers to position themselves for an available parking space or to get in the proper lane for a turn. In addition, transition areas between one-way and two-way operations are frequently hazardous and require special traffic control treatment.

Effect on operating conditions. A primary reason for use of one-way streets is to improve traffic operations and reduce congestion. The degree of improvement in operating conditions, travel time, and safety depends, of course, upon the particular operating elements of the previous situation. Generally, travel times can be reduced from 10% to 50% and accidents by the same rate even with a slight increase in total traffic volumes.[6] (See Tables 11–1 to 11–3.)

Such general improvement in traffic operations must be balanced against the following disadvantages:

_____

[5] P.A. Habib and others, _Analysis of Pedestrian Crosswalk Safety on One-Way Street Networks,_ Report DOT-OS-70057, U.S. Department of Transportation, Washington, DC, September 1978.

[6] P.A. Mayer, "One-Way Streets," _Traffic Control and Roadway Elements—Their Relationship to Highway Safety,_ Highway Users Federation for Safety and Mobility, Washington, DC, 1971, Chapter 10.

1. Some motorists must travel extra distances to reach their destination. Overall, this extra distance will likely increase the amount of fuel used and the travel time.
2. Changes in travel patterns will eliminate turning movements at some intersections and increase them at others, possibly resulting in new control problems at different locations in the area.
3. Strangers may become confused with the one-way street pattern, especially if network geometry is irregular or the one-way pattern is not uniform. Additional directional signing, pavement markings, channelization, and signal indications may be required to handle unexpected travel routing.
4. Transit operations may be adversely affected if vehicles are forced to operate on two streets instead of one. Where a narrow strip of trip generators exists along one street, walking distances to the nearest bus stop for the desired travel direction may increase.
5. Emergency vehicles may need to take a more circuitous route to reach some destinations.

Effect on area economic conditions. In many cases, improved traffic movement and increased safety can produce broad economic benefits both to adjacent land users and to the general public. Nevertheless, when implementing a one-way street system, especially one involving commercial

TABLE 11–1

Change in Traffic Volume, Trip Time, and Number of Stops after Conversion to One-Way Operation, Fifth Avenue, New York City

Section	Average Daily Traffic Volume			Average Trip Time (min)			Average Number of Stops		
	Before	After	Change (%)	Before	After	Change (%)	Before	After	Change (%)
Washington Sq. to 23rd St. [0.8 mi (1.3 km)]	15,265	18,722	+ 23	4.7	2.4	− 49	3	1	− 67
23rd St. to 42nd St. [0.9 mi (1.45 km)]	21,725	23,591	+ 9	7.3	2.9	− 60	5	1	− 80
42nd St. to 57th St. [0.7 mi (1.1 km)]	26,130	29,965	+ 15	7.4	4.4	− 39	5	3	− 40
57th St. to 138th St. [4.1 mi (6.6 km)]	11,592	14,953	+ 29	22.4	16.4	− 28	14.8	7	− 53
Totals (averages)	(16,411)	(19,595)	(+ 19)	42.1	26.4	− 37	27.8	11	− 60

SOURCE: J. A. BRUCE, "One-Way Streets," _Improved Street Utilization through Traffic Engineering,_ Highway Research Board Special Report 93, May 1967.

TABLE 11–2

Accident Changes and Traffic Characteristics on One-Way Streets, London, England

Street	Percent Change in Traffic (Average Weekday)			Percent Change in Travel Time				Percent Change in Accidents	
	Mileage	Volume	Vehicle-Miles	Off Peak Each Direction		P.M. Peak Each Direction		Injury	Pedestrian
Tottenham Ct. Rd.*	5.1	+ 4	+ 8	− 49	− 34	− 43	− 14	− 21	− 33
Baker St.*	2.1	+ 2	+ 3	− 48	− 35	− 65	− 55	+ 4	− 3
Earls Ct. Rd.†	6.3	+ 10	+ 12	− 33	− 15	− 27	− 16	− 27	− 13
Kings X*	2.5	− 2	+ 18	− 28	0	− 27	+ 40	− 33	− 40
Bond St.†	1.3	+ 9	+ 14	− 26	− 38	− 15	− 38	0	0
Piccadilly*	1.3	− 4	0	− 19	− 12	− 5	− 12	− 14	− 32

*6 months before and after.
†3 months before and after.

SOURCE: J. T. DUFF, "Traffic Management," Conference on Engineering for _Traffic,_ 1963, p. 49.

TABLE 11-3

Accidents and One-Way Streets, New York City

Street and Length Made One-Way	Period	Number of Accidents					Total Accidents	Total Injured	Accident Rate*
		Angle	Rear End	Turning	Other	Pedestrian			
Madison Ave.,	Before	23	49	53	67	54	246	167	16.7
23rd St. to 135th	After	23	34	24	45	32	158	101	9.3
St. [5.7 mi (9.2 km)]	% change	0	−31	−49	−33	−41	−36	−40	−44
Fifth Ave.,	Before	40	65	68	84	63	326	190	20.4
Washington Sq.	After	38	53	52	73	45	261	156	13.7
to 38th St., [6.5 mi (10.5 km)]	% change	−5	−18	−23	−13	−29	−18	−18	−32
Both streets	Before	63	114	121	151	117	572	357	18.6
	After	61	87	76	118	77	419	257	11.6
	% change	−3	−24	−37	−22	−34	−27	−28	−38

*Accidents per million vehicle-miles.

SOURCE: J. A. BRUCE, "One-Way Major Arterial Streets," *Improved Street Utilization through Traffic Engineering*, Highway Research Board Special Report 93, May 1967.

streets, traffic engineers should expect objections from affected business owners, who may contend that one-way streets will adversely affect their trade.

Studies made in various parts of the United States have generally tended to disprove such claims. Moreover, where one-way systems have once been implemented, many business owners formerly opposed to the one-way street plan have become supporters.

Although the economic and environmental impact on converting to a one-way street system will undoubtedly vary from one place to another, a study by the Michigan Department of State Highways revealed that opposition tended to come from property owners immediately adjacent to one-way streets, with more support from others in the area. Despite fears of losses in business and property values, there was no indication of adverse economic impact on either business activity or residential property values.[7]

Trends in one-way street usage

The number and total mileage of one-way streets have increased significantly over the years. In 32 European towns, the total mileage of one-way streets increased from 225 to 575 km in a 10-year period after the end of World War II.[8] Figures are not readily available for the United States, but general observation suggests a similar trend. It may not be realistic to expect continued expansion of one-way street systems in large cities, but increased usage in many smaller and medium-sized cities has been noted.

Criteria for use of one-way streets

Legal background. Although the Model Traffic Ordinance[9] directs that the traffic engineer be authorized to

determine and designate one-way streets and alleys (Section 32-301), many cities and counties require the approval of the governing body. Following such approval, if needed, the traffic engineer arranges for the placement and maintenance of the necessary traffic control devices, giving public notice thereof. The *Manual on Uniform Traffic Control Devices* (MUTCD) specifies the design and location of such signs.

Traffic studies. An engineering evaluation is needed to determine the advisability of one-way operation in a given street network. Such a network may range in size from two parallel streets to all streets in an area. The evaluation should include:

1. Physical inventory of existing system to determine:
   a. Widths and adaptability to one-way operation.
   b. Termination points where needed traffic control devices can be effectively provided.
   c. Transit operational needs within the network.
   d. Existing traffic control devices.
   e. Parking needs and practices.
   f. Major street and driveway intersection locations.
   g. Heavy pedestrian crossings.
2. Traffic volume studies on each street involved, including:
   a. Hourly directional counts.
   b. Turning movement counts during peak hours at critical intersections with streets and major driveways.
   c. Counts on streets parallel to the one-way pair(s) being considered, to estimate the effects of possible traffic diversion.
3. Speed and delay studies in both peak and off-peak periods to provide data on overall travel times and the locations and causes of major delays.
4. Traffic signal studies to evaluate existing progression programs and to determine the improvement that might be gained from one-way operation.
5. Parking studies to determine the feasibility of curb parking prohibitions on one or both sides during all hours or only in peak periods as an alternative or supportive measure to one-way operation.

---

[7] *The Economic and Environmental Effects of One-Way Streets in Residential Areas,* Department of State Highways, Lansing, MI, 1969.

[8] E. Nielsen, "Experience from 10 Years' Fight against Traffic Congestion," 36th International Congress, International Union of Public Transport, Brussels, Belgium, 1965, p. 15.

[9] National Committee on Uniform Traffic Laws and Ordinances, *Uniform Vehicle Code and Model Traffic Ordinance.*

6. Comparative capacity analyses of various alternative forms of operations.
    a. Capacity restrictions in the existing system that might be alleviated.
    b. Directional capacity of the existing network.
    c. Directional capacity of the proposed network.
    d. Directional capacity with parking prohibitions on the existing and proposed systems.
    e. Directional capacity using unbalanced operation techniques (two-way streets with off-center movement to encourage traffic to use one street in one direction and the other in the opposite direction, with progressive signal timing favoring the direction having more lanes) or reversible lanes (see next section).
7. Estimates of added travel distance and increase in total travel time in the network.
8. Feasibility studies with respect to transit routing and location of stops.
9. Investigation of probable effect on movement of emergency vehicles.
10. Investigation of probable effect of one-way operation on businesses, passenger loading zones (hotels, theaters, etc., may be on the "wrong" side of street), parking facility entrances and exits, and other land-use or curb-use activities.
11. Analysis of frequency, severity, and types of accidents along the proposed one-way street, with estimates of possible changes.
12. Pedestrian studies to evaluate the possible effects of one-way operation.
13. Economic evaluation of the costs of various types of operation in relation to the overall benefits that are anticipated.[10]

Planning considerations. The amount of data to be collected and analyzed in planning for one-way traffic regulations will depend largely on the size and complexity of the one-way system under consideration. The following questions should be considered:

1. Is the layout of the street system such that one or more pairs of one-way streets can be implemented on a practical basis? In other words, will it be logical and make sense and be accepted by the public?
2. What effect would the proposed one-way street(s) have on transit operations and patronage?
3. Must parking be restricted in certain areas to provide the proper number of traffic lanes?
4. What changes need to be made in signs, markings, parking meters, traffic signal indications and detectors, and other traffic control devices?
5. What impact would one-way traffic have on freight delivery and truck routing?
6. Are there major traffic generators on the streets to be considered for one-way operation, and what, if any, effect would there be on such generators?

[10] W. S. Homburger and J. H. Kell, *Fundamentals of Traffic Engineering*, 12th edition (Berkeley: University of California, Institute of Transportation Studies, 1988), p. 25-2.

7. Are the geometric elements of the street sections proposed for one-way operation such that the transition to two-way traffic (or termination at an intersection) would not cause safety or congestion problems?

As a general rule, two-way streets should be made one-way only if:

1. It can be shown that a specific traffic problem will be alleviated and the overall efficiency of the transportation system will be improved.
2. One-way operation is more efficient, safe, and cost-effective than alternative solutions.
3. Parallel streets of adequate capacity, preferably not more than a block apart, are available or can be constructed.
4. Such streets provide adequate traffic service to the area traversed and carry traffic through and beyond the congested area.
5. Safe transition to two-way operation can be provided at the end points of the one-way sections.
6. Proper transit service can be maintained.
7. Such streets are consistent with the master street or highway plan and compatible with abutting land uses.
8. Thorough study shows that the overall advantages significantly outweigh any disadvantages.

Benefits of one-way traffic regulations

Increased capacity. One-way streets will often:

1. Reduce intersection delays caused by vehicle turning movement conflicts and pedestrian-vehicle conflicts.
2. Allow lane-width adjustments that increase the capacity of existing lanes or provide an additional lane.
3. Reduce travel time.
4. Permit improvements in public transit operations, such as routings without turnback loops (out on one street and return on a parallel street).
5. Permit turns from more than one lane and doing so at more intersections than would be possible with two-way operation. (Care must be taken that designated turning lanes are clearly marked and do not block needed through lanes.)
6. Redistribute traffic onto adjacent streets to relieve congestion.
7. Simplify traffic signal timing by:
    a. Permitting a wider range of offsets for progressive movement of traffic.
    b. Permitting offsets to achieve wider through bands.
    c. Reducing multiphase requirements by eliminating left-turn conflicts and/or making minor streets one-way away from complex intersections.

Increased safety. One-way streets are likely to:

1. Reduce vehicle-pedestrian and vehicle-vehicle conflicts at many intersections.
2. Prevent pedestrian entrapment between opposing traffic streams.
3. Improve drivers' fields of vision at some intersection approaches.

Improved economy and environmental protection. One-way streets may:

1. Provide additional capacity to satisfy traffic requirements for a substantial period of time without large capital expenditures for new street construction.
2. Permit stage development of a master plan.
3. Meet changing traffic patterns quickly and at a relatively low cost.
4. Facilitate the loading and unloading of commercial vehicles with minimal impact on traffic flows.
5. Preserve sidewalks, trees, and other valuable frontage assets that would otherwise be lost because of the widening of existing two-way streets.
6. Be used to prohibit traffic from entering a residential neighborhood by making short lengths of street one-way outbound from the neighborhood.
7. Provide for parking on one side of a street that would otherwise be too narrow to permit parking and adequate clearance or sight distance for safe operation.
8. Be part of a freeway, expressway, rotary, or other system utilizing ramps, frontage roads, or connecting streets that handle movements that are essentially unidirectional in nature.

Roadway requirements

Although one-way systems will differ in details, there are certain basic factors to consider in developing a network of one-way streets:

1. The capacity of the street(s) in one direction should approximately balance the capacity of the street(s) in the opposite direction. If capacities cannot be balanced, the street having the lower capacity must have adequate capacity for current traffic and, if possible, for some time into the future.
2. Preferably, the one-way pair should be adjacent streets (although systems are operating satisfactorily where there are intervening parallel streets).

Design of termini

Some street patterns readily lend themselves to good traffic operations at one-way system termini—as when two streets join in a "Y" pattern to become one. In a gridiron pattern, however, the one-way system usually ends at a typical four-way intersection. When the one-way system would normally terminate at a major cross arterial, it is usually desirable to extend the system one block beyond that point. This is particularly true of the one-way street carrying traffic toward the crossing arterial. Construction of diagonal connections to facilitate transition from two-way traffic to one-way traffic should be considered when one-way streets are part of an arterial system.

Reversible lanes and roadways

A reversible lane system is potentially one of the most efficient methods of increasing rush-hour capacity of existing streets under proper conditions. With relatively low capital cost, unused capacity may be assigned to the direction of heavier flow, with the result that all lanes are more fully utilized. The system is particularly useful on bridges and in tunnels, where the cost to provide additional capacity would be high and, perhaps, impossible.

Arterial routes that are normally operated as two-way streets, particularly those in urban areas, can experience much greater peak-hour traffic volumes in one direction than in the other. This condition can result in wasted street capacity. Reversible lanes can provide for better utilization of such streets.

Under a reversible lane system, one or more lanes are designated for movement one-way during part of the day and in the opposite direction during another part of the day. On a three-lane road, for example, the center lane might normally operate as a two-way left-turn lane, but during the peak hours operate as a one-way lane in the direction of greater flow. The purpose of a reversible system is to provide for the needed capacity through the use of an extra lane or lanes for the dominant direction of flow.

Three methods that have been used rather extensively are (1) to reverse the flow of an entire one-way street during peak-hour periods, (2) to make a two-way street operate as totally one-way during that period, and (3) to operate a multi-lane two-way street with an unbalanced number of lanes during peak periods—reversing one or more lanes to favor the predominant traffic flow in each peak. In Phoenix and Tucson, Arizona, a reversible lane is used during off-peak periods as a two-way left-turn lane, and during peak periods it is allocated to through-traffic movement in the peak direction.

The advantages and disadvantages[11] of reversible lane operation are summarized below:

Advantages

1. Extra capacity is provided in the direction and at the time needed. Both morning and evening peaks can be accommodated on the same street.
2. No "paired" street developed to arterial standards is needed, as in the case of one-way streets.
3. The existing system of arterials is utilized more efficiently (e.g., two parallel arterials of 6 lanes operated as two-way streets or as a one-way pair provide 6 lanes in the peak direction; the same arterials both operated with 4–2 lane imbalance provide 8 lanes in the peak direction).
4. Minor direction traffic does not have to shift to another street, as in the case of a reversible one-way street.

Disadvantages

1. Cost of installation (i.e., control devices) and/or operation (e.g., moving cones) of reversible lanes may be high.
2. Increased accidents may occur if the control methods used are not clear and positive.
3. Concentrated enforcement efforts may be needed to prevent violations of the lane-use regulations.

_____

[11] Ibid., p. 25-5.

4. Changeover problems before and after peak periods may be difficult to solve. Frequently, one or more lanes are removed from operation during the changeover.
5. Provision of adequate capacity for the minor direction may be difficult, especially on 4-lane streets; curb parking prohibitions may be necessary, adding to enforcement costs and inconvenience to the public.

Criteria for installation of reversible lanes

Although reversible lane operation is principally used on existing streets and roadways, it can also be designed into new streets, freeways and expressways, bridges, and tunnels. Applications to older limited-access facilities is difficult because most such roadways have fixed medians separating the two directions of traffic. By constructing special median-crossing locations and by properly using traffic control devices, however, even these facilities can be used in a reversible manner. In such cases, extreme care must be exercised to provide positive control of lane usage in order to maintain safe operation.

Specific criteria should be examined prior to installation of reversible lane or roadway segments:

Traffic studies. General traffic studies[12] required are the same as those described for one-way streets. Additional circumstances that favor the use of unbalanced operation are:

1. Lack of adequate adjacent streets rules out the consideration of one-way operation (e.g., a 6-lane arterial with all parallel streets being 2-lane residential streets).
2. Wide streets (five or more moving lanes) with ratios of major to minor flows exceeding 2 to 1 are especially conducive to this type of operation.
3. High proportion of commute-type traffic which desires to traverse the area without turns or stops is facilitated by this type of operation.
4. Terminal conditions are such as to permit the full utilization of the additional lanes. This is an essential requirement.

Evidence of congestion. If the level of service during certain periods decreases to a point where traffic demand is in excess of actual capacity, the use of reversible lanes should be considered.

Time of congestion. It should be determined that the periods during which congestion occurs are periodic and predictable. Traffic lanes can usually only be reversed at a fixed time each day.

Ratio of directional traffic volumes. Lane reversal requires that the additional capacity for the heavier direction be taken from the traffic moving in the opposite direction. Traffic counts by lane will determine whether or not the number of lanes in the counter direction can be reduced, how many lanes should be allocated to each direction, and when the reversal should begin and end. On major streets,

there should remain at least two lanes for traffic flowing in each direction.

Capacity at access points. There must be adequate capacity at the end points of the reversible lane system, with an easy transition of traffic between the normal and reversed-lane conditions. Installation of a reversible lane system with insufficient end-point capacity may simply aggravate or relocate a congestion or accident problem.

Lack of alternative improvements. Cost factors and right-of-way limitations preclude widening the existing roadway or providing a parallel facility on a separate right-of-way.

Methods of implementation

Once a reversible lane system is determined to be necessary and feasible, the method of designating lanes to be reversed and the direction of flow must be selected. Three general methods are used:

1. Special traffic signals suspended over each lane.
2. Permanent signs advising motorists of the changes in traffic regulations and the hours they are in effect.
3. Various physical barriers, such as traffic cones, signs on portable pedestals, and movable divisional medians.[13]

Control techniques

Positive means of controlling lane usage[14] is required on streets operated with reversible lanes. Various techniques have been used, including:

Curb-mounted signs. Curbside signs alone are the least form of control and are not adequate for most streets, since strangers as well as regular commuters may be in the traffic stream during the period of unbalanced operation indicated on the signs. Curb-mounted signs alone also do not comply with the *Manual on Uniform Traffic Control Devices* (MUTCD) but must be accompanied by overhead signs or lane-use control signals.

Overhead signs. Signs are mounted above the reversible lane(s) by mast arms or span wires, either in conjunction with lane-use signals or by themselves. The signing can be either changeable message or static signs indicating the specific lane restrictions.

Lane-use control signals. Lane-use control signals are a positive form of lane control. Standards for such signals are set forth in the MUTCD.[15] Prior to use, the existence of laws authorizing such signals and establishing legal duty of drivers should be verified.

Movable pedestals, tubes, or traffic cones. These channelizing devices, sometimes used with "Keep Right"

---

[12] *Ibid.*, pp. 25-5, 25-6.

[13] *Manual on Uniform Traffic Control Devices*, Secs. 4E-7–4E-12.

[14] Homburger and Kell, *Fundamentals of Traffic Engineering*, pp. 25-5, 25-6.

[15] *Manual on Uniform Traffic Control Devices*, Secs. 4E-8–4E-12.

messages, are placed along lane lines to separate directions of traffic. This method, although effective, is costly in terms of maintenance since the devices must be placed in position and moved or removed for each peak period, and the placement of these devices may present dangerous conditions for workers. However, they are especially effective for temporary situations.[16] More substantial dividers, such as movable guardrail "trains," hydraulically operated fins, or movable concrete (Jersey-type) barriers, are better suited to reversible operations on high-speed expressways and freeways. Some high-volume facilities have been built with reversible lanes incorporated into the median as shown in Figure 11–2.

Figure 11–2. Use of movable physical guard-rail barriers to affect reversible-lane-control on Interstate 70 in St. Louis, which is built with 8 traffic lanes operated as 3 lanes in each direction at all times, with the center two lanes reversible. The top photo indicates west-bound vehicles entering the reversible lanes by changing lanes through the opening created when the movable guard rail is in the open position. The bottom photo shows the movable guard rail in the closed position with eastbound vehicles moving out of the reversible lanes into the regular traffic lanes. The reversible-lane section is 5.5 miles in length, with five movable guard-rail trains for affecting reversible-lane operation.

SOURCE: Missouri State Highway Department, Traffic Engineering Division, Jefferson City, MO.

---

[16] Homburger and Kell, *Fundamentals of Traffic Engineering,* p. 25-6.

Turn regulations

Conflicts between turning vehicles and pedestrians and between turning vehicles and other vehicles approaching from the opposite direction can cause congestion delay and safety problems at intersections and driveway access points. Turn restrictions are a means of eliminating such conflicts and reducing congestion and accidents. In countries where traffic on two-way streets and highways moves on the right side of the centerline or median, left turns create vehicle-vehicle conflicts. Where traffic moves on the left side of the roadway, right turns are the cause of such conflicts.

Since turn regulations are a constraint on the freedom and flexibility of motorists to choose a specific traffic route, the advantages and disadvantages of such restrictions should be carefully evaluated before action is taken.

Factors to be considered

Turning vehicles conflict with pedestrian movements, must yield the right-of-way to oncoming traffic, and often block the movement of vehicles behind them unless separate turn lanes are provided. Turn regulations can be an inexpensive method of alleviating turning movement problems, particularly if signing alone is sufficient to indicate the regulations to drivers. However, other methods should not be overlooked: separate turn lanes, separate signal phases, all-way pedestrian walk signals, and elimination of conflicting crosswalks (by use of pedestrian barriers if needed).

A turn prohibition compels drivers to use an alternate route. This usually involves a longer travel distance, and, consequently, additional vehicle-miles of travel on the street system. The prohibition of a turn at one intersection may merely move the problem to one or more other locations. Hence, turn prohibitions should be established only after careful analysis of the new route(s) which drivers will have to use. The turn restrictions must be generally acceptable to motorists or heavy enforcement will be required to obtain reasonable compliance. If at all possible, a positive barrier, such as a channelizing island at an intersection, will provide the most obvious and most easily enforced form of turn restriction.

It is not always necessary to prohibit turning movements at all times in order to alleviate a congestion or accident problem resulting from conflicts produced by turning vehicles. Turn prohibitions may be in effect continuously or only at certain hours of the day. The latter method is generally preferable because it causes the least amount of restriction and inconvenience to the driver. However, observance may be better if restrictions are permanent, since drivers are more likely to get into the habit of using an alternate route.

Turning-movement problems should be identified by time of day and restrictions considered first for those hours when study data indicate that congestion or accidents are occurring and when a suitable alternative route is available. When part-time restrictions are used, the signs notifying motorists of the restrictions must be designed and placed so that the time of the restriction is clearly visible to approaching motorists. An "active" sign (e.g., one with a flashing beacon) is desirable for part-time restrictions to help alert drivers when the restrictions are in effect.

At intersections controlled by traffic signals, turns can be restricted to certain phases of the signal operation by use of separate signal displays and appropriate signs. This type of turn restriction is generally most effective when a separate lane is provided for turning vehicles. The signal phasing techniques can be used to eliminate the conflict between turning vehicles and pedestrians as well as between turning vehicles and opposing traffic. Some jurisdictions prohibit peak-hour turns altogether even though separate turn signals and turn lanes are present.

Types of turn-restriction devices

Regulatory signs. Regulatory signs are placed at the near right corner of intersections and, if left-turn prohibitions are involved, also at the far left corner. Signs may indicate that turns are prohibited or that they are allowed or required from designated lanes approaching an intersection. Overhead installations of signs, particularly at signalized intersections, may provide greater visibility, but they are supplemental to the corner installations. The hours during which the restriction applies should be shown, where applicable.

Changeable message signs. Changeable message signs are more expensive; however, the hours when restrictions are in force need not be shown on the sign, thus simplifying the message. The message of such signs is also usually more noticeable to drivers when the regulations are in effect.

Turn restrictions with signals. Turn prohibitions (or required turns from designated lanes) in effect at all times can be indicated by using traffic signal arrow indications instead of circular lenses for the *permitted* movements. However, if the authorized movements conflict with pedestrian streams crossing during the same phase, green arrows cannot be used. Turn restrictions associated with the operation of traffic signal phases are discussed in Chapter 9.

Pavement markings. Pavement markings may be used to supplement signs or signals. They are especially suitable for left-turn prohibitions on multilane roads, where regulatory signs may be hidden from drivers in inside lanes by other vehicles. However, it should be recognized that pavement markings are not a barrier to a turn and, therefore, extra signing may be necessary.

Channelization. Channelization can be constructed or placed in a manner that makes turning movements physically impossible. This control method is not suitable if the prohibition is needed only during certain hours of the day.

Lane-use control signs. Lane-use control signs may be used in advance of an intersection to supplement mandatory turn signs, signals, or pavement markings.

Criteria for installation of lane-use control signs at intersections are established as follows:

1. Lane-use control signs at intersections should be used whenever it is desired to require vehicles in certain lanes to turn, or to permit turns from an adjacent lane.

2. Lane-use controls permitting left (or right) turns from two (or more) lanes are normally warranted whenever the capacity of one turning lane is exceeded, and when all movements can be accommodated in the lanes assigned to them.[17]

Traffic and network considerations

1. The amount of congestion and delay caused by turning movements should be determined through traffic counts, speed and delay studies, and accident records.
2. The number of collisions involving vehicles making each turning movement should be evaluated with respect to the overall accident experience at the intersection.
3. Suitable alternative routes should be available prior to implementing a plan to restrict turns.
4. The possible impact of rerouted traffic on congestion and accidents should be estimated for intersections that would be required to accommodate the traffic diverted by the turning restriction.
5. Noise and other possible adverse environmental impacts caused by the rerouted traffic should also be considered.
6. The feasibility of alternative solutions, such as provision of separate storage lanes for the turning movements and, at signalized intersections, the use of special turn-movement phasing, should be explored.
7. The advantages gained by the use of separate signal phasing for left-turn movements must be weighed against the loss of capacity for through traffic caused by adding additional phases to the cycle and reducing the proportion of green time available. A commonly used criterion for provision of a separate left-turn signal phase is that the numerical product of the left-turn peak-hour volume multiplied by the opposing through traffic peak-hour volume must be at least 100,000. Part-time left-turn signal phasing or "protected-permissive" types of phasing should be considered in lieu of the more restrictive protected-only type, if local conditions permit.

Effect of turn restrictions on accidents

Because turn restrictions cause a change in travel routes, reliable data on the full impact of turn restrictions on accidents are difficult to obtain. Data compiled in San Francisco indicated that accidents at four intersections with turn restrictions were reduced by 38% to 52%. All of the intersections were high-volume intersections used by 30,000 to 55,000 motorists in an average day.[18]

The prohibition of left-turn movements at driveways between intersections is frequently accomplished by construction of a median divider. A study in Wichita, Kansas, reported that prohibition of turns between intersections by use of a median reduced accidents between intersections by amounts ranging from 43% to 69% during the first 3 years after the

[17] *Manual on Uniform Traffic Control Devices*, Secs. 2B-17, 2B-18.

[18] R. T. Shoaf, "Traffic Signs," *Improved Street Utilization through Traffic Engineering*, Highway Research Board Special Report 93, Washington, DC, May 1967.

median was installed. During the same time period, accidents at intersections where turns were not prohibited increased by amounts ranging from 12% to 38%. However, since accidents between intersections originally represented more than 60% of the total accidents on the street section affected by the construction, the median construction resulted in a net accident reduction.[19]

An alternative to turn restrictions is the designation of a separate lane for storage of vehicles waiting to make left turns. This traffic control technique can take the form of a "continuous two-way left-turn lane" that can be used by motorists proceeding in either direction. Left-turn storage lanes can also be established with pavement markings for one direction of traffic only on approaches to intersections where left turns create accident or congestion problems. Designation of such lanes may, however, require that parking be prohibited, and this could create a need for a study of curb parking supply and demand. The advantages of the turn lane must then be compared to the impact of parking restrictions in the determination of the best course of action.

Special problems to consider

U-turns. U-turns have an adverse effect on capacity and safety, especially if the street is too narrow to permit the maneuver to be completed easily. It is usually necessary to prohibit U-turns at all times at extremely busy intersections. This may be indicated by the posting of regulatory signs, though in some states the prohibition may be established by state law or local ordinance to cover entire areas. If signal cycles with protected-only left-turn phasing are used at signalized intersections, U-turns are permitted only during the left-turn phase.

Multiple turn lanes. Two or more lanes may be designated as turning lanes to the right or left. The lane on the outside of the turning movement may have optional turning (right or left turn permitted) but the other lanes must have compulsory provisions (right or left turn *only*) to prevent vehicle conflicts. For double-turn lanes, the control is usually indicated by pavement markings and ground-mounted signs (curb-mounted for double right-turn lanes and median-mounted for double left-turn lanes). Overhead signs are usually used at signals and where needed to assure good observance and effectiveness of the regulation. (If three or more lanes are designated as turning lanes, the use of overhead signs is almost always necessary.)

"Free" right turns. The right turn movement may be separated from the rest of the traffic by means of channelization and permitted to move under separate control from other traffic on the same approach. Typically, the right turn might be uncontrolled or controlled by a yield sign, whereas the other traffic on the same approach would be under signal or stop-sign control. If the "free" right turn is provided with a separate lane to turn into, or if volumes in the right lane on

the intersecting street are sufficiently light, this movement may move on essentially a continuous basis, subject only to intermittent vehicle or pedestrian conflicts.

At more heavily traveled locations, the right turn may be subject to substantial delay and conflict. At signalized intersections, a separate signal phase may be provided for the right-turn movement in order to control conflicts and enhance the efficiency of the movement. Except for very light volumes, the effectiveness of the "free" right turn is dependent upon a separate right-turn lane approaching the intersection; otherwise the right-turn vehicles may frequently be trapped in a line of vehicles waiting to proceed straight through the intersection.

This type of control is usually used at locations where a single right-turn lane is sufficient to accommodate right-turning volumes. It is generally not suitable for use with multiple right turns because of the problems associated with complex merging of two or more lanes with the through traffic on the intersecting street.

Railroad crossing signs and signals. Railroad crossing signs and signals may include turn prohibitions from streets paralleling and close to the railroad, either alone or in conjunction with adjacent street traffic signals, in order to prevent turning movements across the tracks during times when trains are approaching or using the crossing. Effectiveness is enhanced by use of special indications visible only at the time of need (i.e., signs that are legible only when railroad restrictions are activated), rather than reliance on permanent NO RIGHT TURN ON RED signs or red-arrow signal indications that may not convey immediately the special nature of the prohibition.

Turning movements on red-signal indications

Right turn on red. Right turn on red (RTOR) is now allowed in all areas of the United States, after a full stop during the red indication and after yielding to pedestrians and cross traffic, unless signs have been posted prohibiting this maneuver. (It should be noted that New York City is an exception to the RTOR rules—with RTOR permitted only when indicated by signs.) RTOR is a means of increasing capacity and reducing delay; however, there has been some concern about accident increases, especially in areas where RTOR was previously not permitted. Where special safety conditions warrant, the signs prohibiting RTOR should be posted. (This regulation should not be confused with the right turn permitted by a green arrow and not requiring a stop, which may be displayed simultaneously with a circular red for through traffic.)

Left turn on red. Left turn on red at the intersection of two one-way streets (out of a one-way street into a one-way street) is also permitted in many areas after the vehicle has made a full stop. The turning vehicle crosses no other vehicle paths, but the driver must look for conflicts with merging cross traffic and pedestrians. A few states have permitted left turn on red at intersections where one or both streets are two-way, provided that signs are installed permitting this

[19] R. Johnson and B. McKinley, unpublished paper submitted to Technical Award Committee, Missouri Valley Section, Institute of Transportation Engineers, 1971.

maneuver. The MUTCD and most jurisdictions preclude this latter practice, however. It is generally held that movements that involve an unexpected crossing of pathways of moving traffic should not be indicated during any green interval, except when the movement involves only slight hazard, serious traffic delays are materially reduced by permitting the conflicting movement, and drivers and pedestrians subjected to the unexpected conflict are effectively warned thereof.

Transit and carpool lanes

The use of traffic regulations that restrict certain lanes on public streets and highways to transit or high-occupancy vehicles (HOV's), such as carpools and vanpools, has been expanded as greater efforts are made to obtain maximum transportation capacity from existing facilities. A model law for the establishment of HOV lanes and facilities is available from the Federal Highway Administration (FHWA).

Preferential treatment of HOV's is a common method of Transportation Systems Management (TSM), directed toward achieving maximum efficiency—expressed in terms of *persons* moved rather than vehicle flow. In street and highway operations, this means giving priority or exclusive use of lanes to buses and other HOV's in the traffic stream.[20]

Exclusive-use lanes may either be "contra-flow," meaning that traffic moves against the flow of traffic on that side of the street centerline or freeway median, or "with flow," in the normal direction of traffic on that side of the centerline or median. Exclusive-use facilities include:

1. A lane that is reserved for exclusive use by transit or HOV's at all times. This type of restriction is most often used on limited-access facilities.
2. A lane that is limited to use by transit or HOV's only during peak traffic periods. This type of restriction is often used on city streets where there is a demand for curb parking or curbside freight loading during nonpeak traffic periods, or where the amount of transit and HOV traffic during nonpeak periods is insufficient to justify the restriction.
3. A freeway entrance ramp that can only be used by transit or HOV's, or which includes a bypass lane for HOV's around ramp entry controls.

The introduction of lane usage restrictions, while beneficial to transit or HOV's, limits the ability of other traffic to use the facility with such lane restrictions. Therefore, proposals for lane-use traffic regulations must be evaluated carefully in order to determine that such regulations will improve the total capacity of the facility to move people and goods in a safe and efficient manner.

Factors to be considered

An engineering analysis of the capacity of the facility at a desirable level of service and of the occupancy of existing vehicles should be made to determine the persons-per-hour capacity of the facility without lane-use restrictions. Observations of travel time and delay under existing conditions should be measured so that they can be compared to values that can be anticipated if an existing lane is restricted to use by transit and HOV's. (In most freeway projects, the analysis deals with whether an HOV lane should be *added* without changing the regulations for the existing lanes.) In evaluating travel time, the average travel time per person becomes an important factor.

In some areas, congestion has now reached such high levels in central cities and suburbs as to suggest consideration of the use of exclusive HOV lanes even if they do not result in an immediate increase in persons moved so as to encourage long-term growth in vehicle occupancy.

If an exclusive lane is on a street that also provides access to adjacent property, the impact of a restriction on passenger and freight loading and unloading must be evaluated, and careful attention must be given to driveway access. Lanes on city streets reserved for exclusive transit use are often along the curb, and loading and unloading of passengers by taxis and other automobiles can interfere with transit operations in the exclusive lane. The closing of entrance ramps to freeways to all but transit or HOV's can create additional traffic congestion on adjacent parallel facilities, and such disadvantages must be weighed against the advantages to the users of the facility with restricted entry.

On freeways, the length of special lanes for HOV's must be great enough so that weaving problems are not caused by vehicles entering and leaving the lanes. The purpose of HOV lanes is to provide a path for HOV's to get past all significant "bottlenecks" on a route segment. Little is gained unless the HOV lanes are long enough to carry vehicles beyond the last bottleneck point.

Transit and HOV lanes on freeways

Two types of exclusive freeway lanes have been used:

1. With-flow lanes operate in the same direction as the rest of the freeway roadway of which they are a part. These may be physically separated by a buffer lane from the unrestricted lanes, or be merely designated by signs and pavement markings.
2. Contra-flow lanes operate in the direction opposite to that of the roadway in which they are located. These are separated from the unrestricted lanes at least by cones or tubes spaced at frequent intervals, or preferably by a blocked-off buffer lane.

A study by the California Department of Transportation concludes that an exclusive bus lane on a freeway can efficiently accommodate 800 to 1,000 buses per hour. Short exclusive lane sections less than a half mile long can accommodate up to 1,200 buses per hour according to this study.[21] The potential passenger capacity of such a lane is, of course, dependent upon the seating capacity of the buses, the feasibility of standing loads, and other factors. Actual peak-hour

---

[20] Homburger and Kell, *Fundamentals of Traffic Engineering*, p. 25-8. HOV's include taxis, vans, autos, and other motor vehicles with 2, 3, 4 or more occupants. The definition of what constitutes a high-occupancy vehicle varies from project to project.

[21] California Department of Transportation, *California Ridesharing Facilities,* Sacramento, CA, January 1984.

volumes on an exclusive bus lane on I-495 at the Lincoln Tunnel in New York have been reported to be as high as 25,800 persons per hour.[22]

A 21-week experiment with an exclusive with-flow lane for buses and carpools on a 12-mile segment of the Santa Monica Freeway in Los Angeles produced mixed results. The lanes, previously open to all traffic, were reserved for HOV's from 6:30 to 9:30 A.M. and 3:00 to 7:00 P.M. Monday through Friday. The evaluation of the project indicated an increase in the number of carpools (vehicles carrying three or more persons) and an increase in transit usage. However, the lane restriction increased congestion and travel time in remaining freeway lanes, and the result was a 7% reduction in the use of these lanes during the 7-hour peak travel period. This, in turn, created additional traffic on adjacent city streets as motorists sought alternative routes to avoid the freeway congestion, with a resulting substantial increase in travel time. Another impact was a 30% increase in injury accidents on the adjacent city streets and a 100% increase in accidents on the freeway with the restricted lane.[23]

In fact, it has been found to be very difficult to convert a regular traffic lane moving in the peak travel direction into an HOV lane. At some locations, where exclusive contra-flow lanes were added to existing freeway facilities, more favorable results have been achieved (see Table 11-4). HOV lanes also appear to work best if they originate and terminate in areas of roadway that are free-flowing. This may be because free-flow conditions mean there is only a minor difference in speed between the HOV lanes and the regular traffic lanes. HOV lanes are also not needed in free-flowing segments of roadway, unless additional bottlenecks are to be encountered.

Exclusive bus lanes on city streets

Exclusive bus lanes have been used for many years in a number of large cities to improve transit operating characteristics

---

[22] *Priority Techniques for High Occupancy Vehicles—State of the Art Overview,* Transportation Systems Center, U.S. Department of Transportation, Washington, DC, November 1975.

[23] Q. Gillard, "Los Angeles Diamond Lanes Freeway Experiment," *Traffic Quarterly,* ENO Foundation for Transportation (Westport, CT, April 1978).

in congested areas and on arterial streets without access control. Transit vehicles, because of their need to make fairly frequent stops to pick up and discharge passengers, are generally unable to achieve travel times comparable to private automobiles. When the number of persons using transit on a street segment is equivalent to the passengers per lane in private automobiles, the reservation of a lane for transit vehicles can usually be justified. Such restrictions may also be justified at lower transit passenger levels in order to provide continuity of an HOV lane and thereby permit passengers in such vehicles to reach their destinations more quickly, and if the average travel time of all persons using the street segment can be improved.

Objections to transit only or transit and carpool lanes on surface streets are frequently offered by taxicab operators, delivery firms, and the owners or users of adjacent property who may be denied curbside loading privileges because of such restrictions.

Some of the forms these exclusive lanes take on city streets are discussed below.

With-flow curb lanes. One lane of an arterial street with curb side bus stops can accommodate over 100 buses per hour with preferential treatment. This meets the needs of most locations; only the most densely traveled corridors in the largest cities have a greater demand. It is relatively simple and inexpensive to implement reserved curb lanes for buses operating in the direction of normal traffic flow. Where volumes of buses are high, the operation is self-enforcing. When bus volumes are low, other vehicles will attempt to use the lane. If right turns cannot be prohibited, they are accommodated by allowing right-turn vehicles to use the lane for a limited distance in advance of their turn and by locating bus stops on the far side of intersections.

Contra-flow curb lanes. Operation of buses in the opposite direction to normal traffic flow on a one-way street has proven effective in a number of locations. Such a lane helps utilize unused capacity when peak flows are unbalanced. These lanes tend to be self-enforcing and usually provide higher operating speeds than normal flow lanes. There have been some reports of accident problems, especially of collisions of buses with pedestrians or vehicles

---

TABLE 11-4

Freeway Contraflow Bus Lane Projects

Location	New Jersey	Boston	New York	Marin Co. California
Route	I-495 (Lincoln Tunnel approach)	Southeast Expressway	I-495 (Long Island Expressway)	U.S. 101
Length (mi)	2.5	8	2	4
Year started	1971	1972	1972	1972
A.M./P.M.	A.M.	A.M.	A.M.	P.M.
Remaining traffic lanes	2	2	2	2
Buffer lane	No	No	No	Yes
Typical bus volumes	500	35	120	70
[per peak hour (peak period)]	(900)	(70)	(200)	(150)
Typical passenger volumes	21,000	1,400	6,000	3,000
[per peak hour (peak period)]	(35,000)	(3,000)	(10,000)	(6,000)

SOURCE: INSTITUTE OF TRANSPORTATION ENGINEERS, *Transportation and Traffic Engineering Handbook,* 2nd ed., © 1982, p. 813. Reprinted by permission of Prentice Hall, Inc., Englewood Cliffs, New Jersey.

entering the roadway and failing to look for buses coming from the contra-flow direction.

Contra-flow center lanes. Reserving interior lanes for transit use is quite complex and normally is suited only for express-run segments. On Kalanianaole Avenue in Honolulu, a contra-flow bus lane is provided on a two-way, four-lane, undivided section. Cones are used to separate the opposing flows. In another section where the arterial is a six-lane divided arterial, the preferential lane is a median contra-flow lane. Left turns are prohibited at most intersections during the period of preferential operation.

Median bus lanes. Bus lanes can be provided in or adjacent to medians on divided streets. Major problems in loading and unloading limit most such lanes to express runs. A notable exception is in New Orleans, where streetcar tracks in a wide median were replaced by bus lines; this still allows buses to load and unload with adequate safety. Another median lane operation is on the six-lane divided section of Kalanianaole Avenue in Honolulu. Still another type of median lane operation is the use of a center lane (which may be a two-way left-turn lane or a reversible lane) as an exclusive bus lane during certain periods. Such an operation was in place on N.W. 7th Street in Miami prior to the opening of exclusive lanes on I-95. Buses preempted signals along the route. Estimates indicate a 20% increase in speed because of the signal preemption and an additional 10% increase because of the exclusive bus lane. This represents a 6-min saving in travel time.

Bus-only streets. Streets reserved for "buses only" provide the most effective means of separating transit vehicles from other vehicular traffic. When implemented in conjunction with other improvements this can provide a very desirable focus for a major activity center as well as enhancing the transit element. Both Washington, DC, and Chicago have created short bus streets that act essentially as on-street terminals with several major routes radiating from the bus streets.[24]

Exclusive freeway access for high-occupancy vehicles

The technique of metering access to freeways as a means of improving operations and capacity has been used in many cities, beginning with manual methods in St. Louis in 1959, and with automatic means in Chicago in 1961. A more recent variation of that technique is to restrict certain freeway access ramps to transit or high-occupancy vehicles. Such restrictions, of course, force low-occupancy vehicles to use adjacent parallel facilities, which in most cases result in longer travel times for trips of comparable length. The State of California has numerous successful locations with HOV bypass lanes around signal-controlled access ramps.

Bicycle lanes and bikeways

In many of the world's urban areas, bicycles have long been a significant transportation mode—accounting for 15% or more of daily trip-making in some European and Asian cities with favorable terrain and climatic conditions.

Usage of bicycles for transportation in the United States has expanded in recent years, and increased development of bicycle lanes and bikeways has also occurred. As a result, new attention has been directed to standards and criteria for the planning and design of these facilities.

Planning and design of bikeways

Two publications deal with the planning and design of bicycle lanes and bikeways: *A Bikeway Criteria Digest*[25] and *Planning and Design Criteria for Bikeways in California.*[26] Both offer guidance for planning and locating such facilities, stressing that usage is generally either recreational or utilitarian and that bikes are usually accommodated through shared use of existing streets and highways. Bikeways do not necessarily have to be separate facilities; in fact, the enhancement of user convenience and safety sometimes may be served better by education, enforcement, and maintenance of existing facilities than by construction of new physical facilities.

Types of bikeways. Bikeways are classified into bicycle paths, lanes, and routes. Bicycle paths are completely separated from motor vehicle traffic and are contained within an independent right-of-way. Another option is the bicycle lane, which is established within the roadway directly adjacent to the outside motor vehicle lane or the shoulder. Bicycle lanes are designated by signs and pavement markings. Many existing bikeways are bicycle lanes, with the highway or street dictating the design, including alignment, grades, and drainage.

Bicycle lanes must be developed as one-way facilities. The use of bidirectional lanes (two-way operation on one side of the street) is not recommended because:

1. They require unconventional turns at intersections.
2. They present problems at the transition from one-way to two-way operation since some bicyclists have to weave across traffic to bike in the proper lane when the special bike lane begins or terminates.
3. They require that bicyclists travel in a direction opposite to the adjacent motor vehicle traffic lane.
4. They may result in accident problems, due to collisions of bicycles with pedestrians or vehicles entering the roadway and failing to look for bicycles coming from both directions.

The bicycle route option is a road designated by signs for bicycling, where the bicycle must share the road with motor

---

[24] Homburger and Kell, *Fundamentals of Traffic Engineering,* pp. 25-8, 25-9.

[25] *A Bikeway Criteria Digest,* U.S. Department of Transportation, Federal Highway Administration, Washington, DC, 1977.

[26] *Planning and Design Criteria for Bikeways in California,* U.S. Department of Transportation, Federal Highway Administration, Washington, DC, 1979.

vehicles. These signs are intended to alert motorists to the presence of bicyclists and to guide bicyclists to use streets that have been determined to be suitable for bicycle usage.

Use of the sidewalk as a bikeway is an option that has very limited application but may be developed in unusual circumstances by either letting the bicycle share the sidewalk with pedestrians or by designating a selected portion of the sidewalk for bicycles. In no case, however, should use of the roadway by bicyclists be prohibited.

Use of sidewalks by bicycles may present problems and is not generally advisable because of the potential conflict among bicyclists, pedestrians, and motorists. Among the factors contributing to this experience are:

1. Poor sight distances often prevail at driveways.
2. Poor visual relationships between bicyclist and motorist occur at intersections.
3. Bidirectional operation compounds sight distance and visual relationship problems.
4. Pedestrians of all ages, but especially small children and older persons, are uneasy when meeting bicyclists along the sidewalk, with resulting conflict and confusion.

Bikeway widths. The actual determination of bikeway pavement width depends on bicycle width, maneuvering allowance, clearance between oncoming and passing bicycles, and edge conditions.[27] The recommended design guidelines for bicycle facilities are provided in Chapter 6.

Intersection treatments. Accident statistics indicate that about two-thirds of bicycle/motor vehicle accidents occur at intersections. Typical bicycle/auto movements at intersections are shown in Figure 11–3.

Two movements are particularly dangerous to the bicyclist. The first is the conflict with a right-turning motorist. The second is the left-turning bicyclist in conflict with through vehicular traffic. Consideration should be given to the minimization of these conflicts. For example, providing a bike lane to the left of a right-turn-only lane channels bicyclists out of the path of the right-turning motorists, as illustrated by Figure 11–4.

"Pedestrian-only" streets

The closure of a street to all vehicular traffic on a permanent basis may be viewed, in one sense, as more than a traffic

[27] *Design and Construction Criteria for Bikeway Construction Projects,* Notice of Proposed Rulemaking, *Federal Register,* Vol. 45, No. 151, U.S. Department of Transportation, Federal Highway Administration, Washington, DC, August 4, 1980.

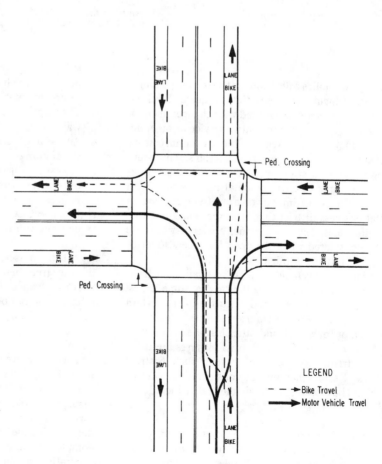

Figure 11–3. Typical bicycle/auto movements at intersection of multilane streets.

SOURCE: *Planning and Design Criteria for Bikeways in California,* State of California, Business and Transportation Agency, Department of Transportation, June 1978, p. 27.

LEGEND
- - - → Bike Travel
—— → Motor Vehicle Travel

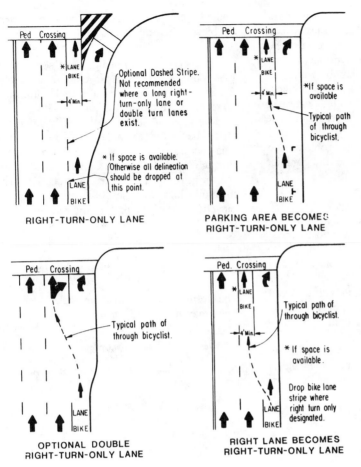

RIGHT-TURN-ONLY LANE

Optional Dashed Stripe. Not recommended where a long right-turn-only lane or double turn lanes exist.

* If space is available. (Otherwise all delineation should be dropped at this point.)

PARKING AREA BECOMES RIGHT-TURN-ONLY LANE

*If space is available

Typical path of through bicyclist.

OPTIONAL DOUBLE RIGHT-TURN-ONLY LANE

Typical path of through bicyclist.

RIGHT LANE BECOMES RIGHT-TURN-ONLY LANE

Typical path of through bicyclist.

* If space is available.

Drop bike lane stripe where right turn only designated.

Figure 11–4. Bike lanes approaching motorist right-turn-only lanes.

SOURCE: *California Highway Design Manual,* State of California, Business and Transportation Agency, Department of Transportation, January, 1987 p. 31.

regulation, since it implies a decision that a street is no longer needed to serve vehicular traffic in that particular location. Such closures are frequently associated with changes in circulation and parking (possibly with shuttle service or other access improvements), which make use of the street unnecessary for vehicles and more desirable for pedestrians. The issues surrounding permanent street closure are often more concerned with economic or environmental matters than with traffic matters. They sometimes involve an attempt to save an aging central business district that has experienced a sharp decline in retail sales. The traffic issues are similar to those that must be considered when the need for a new street or highway segment is being decided.

Closure of streets to vehicular traffic on a *part-time* basis is clearly a matter of traffic regulation. In some cities, streets in congested areas are closed to traffic during certain hours of each day. The French Quarter of New Orleans and the central area of Innsbruck and other European cities are examples. More common types of street closures are those that occur less frequently, for special events such as street fairs or parades. Factors that must be considered in deciding if a street should be closed to all vehicular traffic include the following:

1. The economic or environmental gains anticipated.
2. The ability of adjacent streets and intersections to accommodate traffic that would be diverted by the closure,

including the higher volumes of turning vehicles that will be created at certain locations.
3. The impact of the closure on transit routes and passenger walking distances.
4. Access to off-street parking facilities in the vicinity of the street being considered for closure.
5. Access to adjacent properties for loading and unloading of freight, refuse collection, utility and building maintenance, and so on.
6. Emergency vehicle (e.g., police, fire, and ambulance) access and routings.

All businesses, individuals, and agencies who regularly use the street being considered for closure must be notified well in advance of the proposed action. If the closures are intermittent but occur on a regular basis, consideration may need to be given to permanent improvements to increase the capacity of adjacent streets to accommodate the traffic diverted by the closure action.

An alternative to complete closure of a street is the limiting of traffic to certain types of vehicles, such as buses or taxis. Nicollet Street in Minneapolis, Chestnut Street in Philadelphia, and two streets in Portland, Oregon, are examples of this form of traffic regulation. The factors to be considered in determining the feasibility of such actions are the same as for complete closure, except that the impact on transit routes is generally favorable.

Restrictions on certain types of vehicles

Traffic regulations restricting movements of certain vehicle types and sizes may be of the following types:

1. Weight-limit restrictions at bridges and viaducts.
2. Weight-limit restrictions on pavements of a low design standard.
3. Truck prohibitions on parkways, boulevards, or other streets that serve primarily residential or recreational land uses.
4. Restrictions on truck loading and unloading in congested districts during peak traffic hours.

Procedures for determining the theoretical load capacity of bridges and viaducts have been developed by the American Association of State Highway and Transportation Officials (AASHTO).[28] Careful judgment must be used, however, in the use of information derived from theoretical formulas in the establishment of load limits. Consideration must be given to the availability of alternate routes, the condition of pavements and structures on those routes, the economic losses resulting from additional travel distances, and other factors.

Many jurisdictions prohibit all trucks on streets designated as "parkways" or "boulevards." Such restrictions are generally based on the concept that large, heavy vehicles create excessive noise and interfere with the pleasure of driving on facilities that have been designed with special emphasis of visual aspects of the roadway and adjacent lands. When such restrictions are adopted, provisions must be made for parallel arterial street facilities that can be used by all classes of vehicles.

Restrictions on truck loading and unloading during peak traffic hours are limited in use to large cities with significant congestion in the downtown core areas. When such restrictions are under consideration, studies should be made to determine:

1. The number of truck loading and unloading operations that take place at various hours of the day in the area under study.
2. The effect of such operations on street capacity, congestion, and accidents.
3. The operating hours of business firms and other facilities served by truck loading and unloading operations.
4. The impact of truck loading restrictions on the cost of operations of delivery firms.
5. Prohibition of overnight parking of trucks (or other overlength vehicles) on streets in residential neighborhoods.
6. Prohibition of parking of vehicles over a certain height within a specified distance of an intersection in order to maintain sight distance for moving traffic.
7. Route restrictions on hazardous cargo carriers.
8. Oversize vehicles or oversize loads.

Provision of off-street space for loading and unloading operations is a more desirable method of alleviating street traffic problems related to truck operations than special traffic regulations alone.

When trucks are prohibited in certain areas, "truck routes" can be designated to guide commercial vehicles to the best route around such restrictions. Trucks may be required to use such routes, with deviation allowed to reach their destination by the shortest possible route. The installation of signs identifying special routes for trucks should be done only after study to make certain that the routing is suitable for safe usage by large commercial vehicles.

Speed regulations

Proper use of speed regulation is based on the recognition that lower speed reduces stopping distances and generally reduces severity of accidents. Traffic moving at fairly uniform speed also flows more smoothly, with resultant improvements in both capacity and safety.

Speed regulations and speed limits are intended to supplement motorists' judgment in determining speeds that are reasonable and proper for particular traffic, weather, and roadway conditions. Speed limits are imposed in order to promote lower relative speed conditions and better traffic flow and to reduce accidents. However, if drivers do not consider speed regulations to be reasonable, the limits will be disobeyed and lose much of their value.

High speeds and/or large variations in speeds may be caused by improper speed regulations or the lack of effective enforcement of speed laws, especially if known to motorists. Analyses of speed zoning proposals should always consider whether such problems as have been identified can be solved by better enforcement rather than by engineering speed-zoning methods.

Factors affecting speed regulations

Public attitude. Transportation officials receive many requests for establishing new speed regulations or for altering existing limits. Such requests often reflect the opinion that a particular section of a street or highway is improperly zoned or that the operation of traffic thereon is unsafe. A request for a revised speed limit, usually lower than the limit posted, is sometimes the only immediate solution that many people can envision. Such requests often are based on the public misconception that reducing the speed limit will in fact reduce vehicle speeds and accidents. Citizens, acting as individuals or in groups, will frequently request lower speed limits for their own neighborhood streets than they, as drivers, would consider reasonable in similar neighborhoods elsewhere.

Public reaction to the imposition of speed limits varies. In 1971, West Germany proposed the imposition of a 100-km/hour (62-mph) speed limit on two-lane rural roads where previously no speed limit had been posted. The purpose was to reduce West Germany's high accident rate. The general public reaction was one of anger.[29] In other instances (e.g., residential neighborhoods), speed limits have been welcomed.

---

[28] *Manual for Maintenance and Inspection of Bridges,* American Association of State Highway and Transportation Officials, Washington, DC, 1974.

[29] A. Siegert, "Speed Limit Irks Germans," *Chicago Tribune,* October 11, 1971, Sec. 1-A, p. 3.

In 1974, the United States adopted a national law establishing a maximum speed limit of 55 mph. This law, originally passed to conserve energy, has proved controversial, and a high level of enforcement action has been initiated in many states to attempt to obtain obedience to the limit. Violation of the limit has been (and continues to be) widespread, with more than two-thirds of all vehicles exceeding 55 mph on rural interstate highways during most of the years since the limit went into effect. In 1986, this figure rose to 76%, according to the Federal Highway Administration Summary of Highway Statistics.[30] Some states have imposed nominal fines, and enforcement has appeared to some to be highly selective in practice. Opinion is divided among traffic engineers as to the effectiveness of the 55-mph speed limit.

Although the law did result in energy conservation, it also had a significant secondary benefit in accident reduction. However, in the 1980s, the public no longer perceived an energy crisis and demonstrated increased desires to travel faster than 55 mph, particularly on the interstate highway network.

In 1987, federal legislation allowed for increases of the maximum speed limit to 65 mph on rural segments of the interstate system and other facilities with full access control. Although sufficient data do not exist to judge fully the advisability of these revisions, there are indications that both driver compliance and number of accidents have increased.

Accident frequency and severity as related to speed. Excessive speed is a frequent contributor to the incidence and severity of accidents. For example, the primary cause of 22% of all fatal accidents in California in 1985 was judged to be violation of the basic speed law (that no person shall drive a vehicle upon a highway at a speed greater than is reasonable or prudent having due regard for the weather, visibility, the traffic on, and the surface and width of, the highway, and in no event at a speed which endangers the safety of persons or property).[31] or exceeding the posted speed limit. This was second only to the percentage of accidents directly attributable to alcohol. One-fifth of all injury accidents were listed as involving excessive speed.[32]

Statistics have generally shown that the imposition of appropriate speed limits will lead to a reduction in the serious injury rate in urban areas and in the overall accident rate on specific highway sections.

Figure 11–5, taken from a study made by the Federal Highway Administration, reveals some interesting findings regarding accident involvement and speed on main rural highways, not including freeways. Accident-involvement rates were found to be highest at very low speeds, are lowest at about the average speeds, and high again at very high speeds. A principal conclusion is that the more a driver deviates from the average speed of traffic, the greater is his or her chance of being involved in an accident.

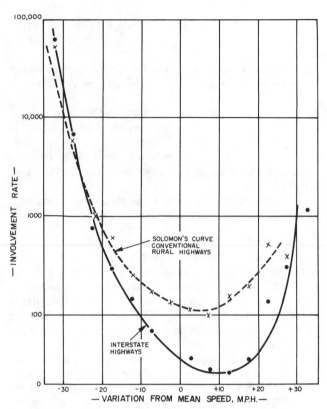

Figure 11–5. Accident involvement rate by variation from mean speed on study units.

SOURCE: "Ramifications of the 55 mph Speed Limit," Committee 4M-2, Institute of Transportation Engineers, Arlington, VA, March 1977.

Effect of environment on speeds. Although much information directed at the driver involves use of the term "safe speed," the term is relative and depends on many conditions and the situation involved. A safe speed in one location may not be safe in another, and a safe speed at a specific time at one location may not be safe under other conditions at the same location.

Roadway type and condition. Higher speeds are relatively safer on roadways with high design standards—wide lanes, absence of sharp curves, adequate sight distance, and clear roadsides—conditions such as exist on freeways. Average speeds for all vehicles by highway type in the United States are shown in Table 11–5. Roadway surface conditions are also a significant factor affecting safe speed, especially surface characteristics that make one surface more slippery than another when wet.

Adjacent land use and access. Safe driving speeds are also affected significantly by intersecting streets and driveways. Speeds on urban streets tend to be much lower than on rural highways because of houses, businesses, and other kinds of development and increased traffic friction. In rural and suburban areas without signals, speed limits may be needed to provide for safe stopping within the distance

[30] U.S. Department of Transportation, Federal Highway Administration, Washington, DC, *Highway Statistics Summary to 1975*, pp. 85–86, *Highway Statistics Summary to 1985*, pp. 257–260, and *Highway Statistics Summary to 1986*, pp. 176–178.

[31] *Uniform Vehicle Code and Model Traffic Ordinance*, Sec. 11-801.

[32] Homburger and Kell, *Fundamentals of Traffic Engineering*, p. 23-1.

TABLE 11-5
Average Speeds of Free-Moving Vehicles, and Percent Exceeding Various Speeds by Type of Highway

Highway System	Average Speed, All Vehicles (mph)				Percent of Vehicles Exceeding							
					55 mph				65 mph			
	1973	1976	1981	1986	1973	1976	1981	1986	1973	1976	1981	1986
Interstate												
Rural	65.0	58.2	57.9	59.7	89	69	68	76	50	10	9	18
Urban	57.0	56.1	55.5	57.4	58	57	51	65	16	5	4	12
Other principal and minor arterials[a]												
Rural	57.1	54.5	54.1	55.3	58	46	43	52	19	6	4	7
Urban	41.8	NA	51.8	53.5	13	NA	30	41	2	NA	2	4

[a]Figures shown for 1973 are for "Rural Primary" and "Urban Primary." Figures are shown for "2-Lane Rural" in 1976. (No figures were available for urban arterials in 1976.)

SOURCE: Highway Statistics Summary for 1985, 1986, U.S. Department of Transportation, Federal Highway Administration.

available after detecting pedestrians crossing the road at crosswalks.[33]

Weather conditions. Weather is also an important factor affecting safe speed. The most significant condition is the presence of snow and ice on the pavement. Rain and fog appear to have less influence. Data obtained at selected sites on California freeways and expressways in both day and night conditions indicate that the effects of fog on traffic flow are small, with mean speeds being reduced by only 5 to 8 mph.[34] In extremely dense fog, of course, traffic may be slowed to crawl speeds. Even heavy rain does not appear to have as much influence because sight distance is usually not as greatly reduced.

Establishment of speed limits

The Uniform Vehicle Code[35] contains the following provision:

> Whenever the (State Highway Commission) shall determine upon the basis of an engineering and traffic investigation that any maximum speed herein before set forth is greater or less than is reasonable or safe under the conditions found to exist at any intersection or other place or upon any part of the (State) highway system, said (Commission) may determine and declare a reasonable and safe maximum limit thereat, which shall be effective when appropriate signs giving notice thereof are erected.

Most states permit local officials to establish speed regulations. This is usually done on the basis of a traffic engineering investigation, and the revision usually takes the form of modifying the basic speed limits set by law or ordinance. (See ITE report on speed zoning.)

There are two basic types of speed controls: (1) regulatory limits that have the effect of law and are enforceable, and (2) advisory maximum speed indications that are not enforceable but which advise or warn motorists of suggested safe speeds for specific conditions at a specific location.

Regulatory controls. Speed regulations may be classified as (1) regulations established by legislative authority and generally applicable throughout a political jurisdiction; and (2) zoned speed regulations established by administrative action on the basis of engineering studies.

There are two basically different types of numerical maximum speed limits: (1) an absolute limit, and (2) a prima facie limit. An *absolute* speed limit is a limit above which it is unlawful to drive regardless of roadway conditions, the amount of traffic, or other influencing factors. A *prima facie* speed is a limit above which drivers are presumed to be driving unlawfully but where, if charged with a violation, they may contend that their speed was safe for conditions existing on the roadway at that time and, therefore, that they are not guilty of a speed limit violation. Enforcement officials prefer the absolute limit because it is much easier to prove guilt in a court of law.

Approximately two-thirds of U.S. states have absolute-type speed limits and one-third have prima facie limits or some of each. The vehicle codes of most states include different types of speed limits for different situations and driving conditions. For example, in the California *Vehicle Code,* absolute limits include the statewide general limit and limits applying to trucks, buses transporting school children, and farm labor vehicles transporting passengers; prima facie limits include speeds at uncontrolled blind intersections, in business and residential districts, posted speed zones, and speed in construction zones.[36]

Advisory controls. Advisory speed signs warn motorists of safe speeds for specific conditions on a highway. They are posted in the form of advisory speed plates (generally supplementary panels) with warning signs or ramp exit speed signs. In some court jurisdictions, driving above the posted advisory speeds may be admitted as evidence that the driver was operating at an unreasonable and unsafe speed.

[33] Ibid., p. 23-4.

[34] *Highway Fog,* National Cooperative Highway Research Project Report 95, Highway Research Board, Washington, DC, 1970, p. 3.

[35] *Uniform Vehicle Code and Model Traffic Ordinance,* Sec. 11-803.

[36] California Department of Motor Vehicles, *Vehicle Code.*

Speed limit studies. Establishment of speed limits should be based on proper engineering and traffic data. Traffic officials are often called upon to testify in court cases regarding speed limits, and they must support their testimonies with data accumulated prior to the establishment of safe speed limits. This information should be of sufficient quantity and of proper quality to justify the value of the speed limit.

At least one state requires that a speed study be conducted at least every 5 years if radar enforcement is to be used, with the 85th percentile used as a basis for the speed zone, unless engineering reasons indicate otherwise.

The following factors should be considered, and appropriate data gathered, in establishing speed limitations.

I. Prevailing vehicle speeds
   A. 85th percentile speed (the speed below which 85% of motorists travel)
   B. Average test-run speeds
   C. Speed distribution data
II. Physical features
   A. Design speed
   B. Measurable physical features
      (1) Maximum comfortable speed on curves
      (2) Spacing of intersections
      (3) Number of roadside businesses per mile
      (4) Restricted sight distances or view obstructions
      (5) Long, steep downgrades or hills
   C. Roadway surface characteristics and conditions
      (1) Slipperiness or roughness of pavement
      (2) Presence of transverse dips and bumps
      (3) Presence and condition of shoulders
      (4) Presence and width of median
III. Accident experience
IV. Traffic characteristics and control
   A. Traffic volumes
   B. Parking and loading vehicles
   C. Commercial vehicles
   D. Turn movements and control
   E. Traffic signals and other traffic control devices that affect or are affected by vehicle speeds
   F. Vehicle-pedestrian conflicts

A Speed Zone Survey Sheet should be prepared to document the above data. An example, published by the Automobile Club of Southern California, is shown in Figure 11–6.[37] Table 11–6 provides a guideline for evaluation of minimum length of the proposed speed zone, distance between intersections, number of roadside businesses, and distribution of 85th percentile speed, pace speed (10-mph range within which most vehicle speeds are observed), and average test-run speeds. The table indicates that locations with higher speed limits should have longer zones, greater distance between intersections, and smaller numbers of roadside businesses. Table 11–6 also suggests that observed speed characteristics should match closely with the proposed speed limit.

In the study of prevailing speeds, observations should be restricted to free-flowing vehicles having generous

headways from those ahead and making no apparent effort to overtake and pass them. The 85th percentile speed as determined by speed studies is a principal factor to be used in the determination of proper speed limits. It is generally assumed that 85% of drivers operate at speeds that are reasonable and prudent for the conditions present in each situation. Hence, the 85th percentile speed of a spot-speed distribution is a *first approximation* of the speed zone that might be imposed, subject to consideration of other factors, as listed above.[38] Additional factors relating to speed studies, sample size, and presentation of data are covered in Chapter 3.

A graphical representation of speed data will usually show that the 85th percentile speed value is the point at which speed values become dispersed. Although collecting speed data is highly satisfactory on streets and highways with moderate to heavy volumes of traffic, it is difficult to do on low-volume roads because of the time consumed in gathering the necessary number of observations. In such cases, trial runs may serve as a satisfactory substitute.

Signing for speed limits. Signing for speed limits should be consistent with the appropriate sections of the latest edition of the MUTCD (used in the United States) or its equivalent in other countries.

Determination of advisory speed indications

Two basically different methods are available for determining advisory speeds on horizontal curves: (1) trial speed runs with an instrument-equipped vehicle, or (2) office calculation. Either method is satisfactory, but field runs to check the office calculations are desirable in any event.

The trial-speed-runs method involves a vehicle equipped with a ball-bank indicator to show the combined effect of the body roll angle, the centrifugal force angle, and the superelevation angle. This method is described in detail in Chapter 3.

In using the office method for the determination of advisory speeds, the appropriate speed to be indicated may be calculated from the simplified formula[39] for relating speed and horizontal curvature:

$$e + f = \frac{V^2}{15R} \qquad (11.1)$$

where: e = rate of roadway superelevation, ft per ft
f = side friction factor
V = vehicle speed, mph
R = radius of curve, ft

Transposing to solve for speed, the formula becomes

$$V = \sqrt{15R[e+f]} \qquad (11.2)$$

---

[37] *Realistic Speed Zoning, Why and How,* Automobile Club of Southern California, Los Angeles, 1976, p. 8.

[38] Homburger and Kell, *Fundamentals of Traffic Engineering,* p. 23-3.

[39] *A Policy on Geometric Design of Highways and Streets,* American Association of State Highway and Transportation Officials, Washington, DC, 1990, p. 141.

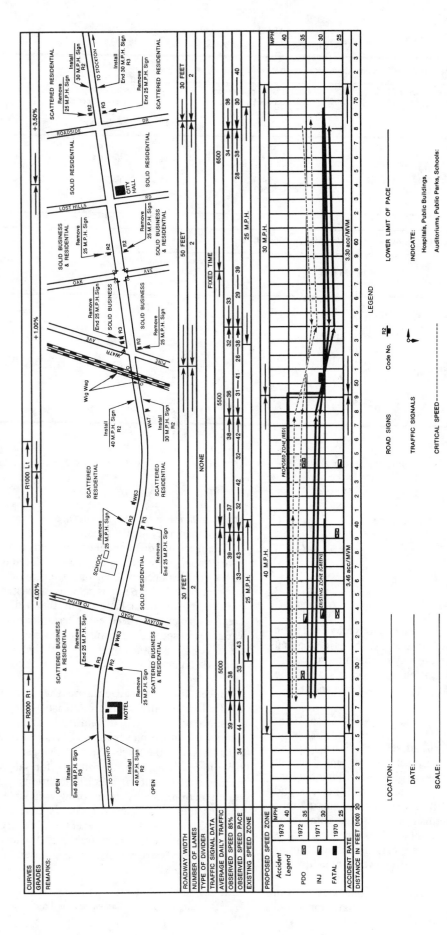

Figure 11–6. Speed zone survey sheet.
SOURCE: *Realistic Speed Zoning, Why and How,* Automobile Club of Southern California, Los
Angeles, 1976, p. 8.

TABLE 11–6

Highway Conditions (Three or More Must Be Satisfied)				Preliminary Estimate of Maximum Speed (mph)
Design Speed (mph)	Minimum Length of Zone Equals or Exceeds (mi)	Average Distance between Intersections Equals or Exceeds (ft)	Number of Roadside Businesses Does Not Exceed (per mi)	
20	0.2	No min.	No max.	20
30	0.2	No min.	No max.	30
40	0.3	125	8	40
50	0.5	250	6	50
60	0.5	500	4	60
70	—	1000	1	70

Speed Characteristics (Two or More Must Be Satisfied)			
85th Percentile Speed (mph)	Limits of 10-mph Pace (mph)	Average Test Run Speed Equals or Exceeds (mph)	Maximum Proposed Speed Limit (mph)
Under 22.5	Under 25	17.5	20
22.5–27.5	11–29	22.5	25
27.5–32.5	16–34	27.5	30
32.5–37.5	21–39	32.5	35
37.5–42.5	26–44	37.5	40
42.5–47.5	31–49	42.5	45
47.5–52.5	36–54	47.5	50
52.5–57.5	41–59	52.5	55
57.5–62.5	46–64	57.5	60
62.5–67.5	51–66	62.5	65
67.5 or over	over 55	67.5	70

*1 mi, 1.61 km; 1 ft, 0.305 mi.

SOURCE: INSTITUTE OF TRANSPORTATION ENGINEERS, *Transportation and Traffic Engineering Handbook,* 2nd ed., © 1982, p. 817. Reprinted by permission of Prentice Hall, Inc., Englewood Cliffs, New Jersey.

For metric units, the formula becomes

$$V = \sqrt{127R[e+f]} \qquad (11.3)$$

where: e is expressed in m/m
V is expressed in km/h
R is expressed in meters

Safe speeds determined by these methods may need to be modified by other factors. For example, the safe stopping sight distance around the curve may require a more restrictive speed than the curvature itself. In this case, it would be advisable to post the advisory speed at the lesser speed.

Special problems

Differential limits by type of vehicle. Some jurisdictions have laws or follow the practice of posting different speed limits for different types of vehicles. Differential limits are most common for (1) passenger cars, (2) trucks, and (3) buses. Special speed limits at levels below general limits are used for trucks in more than half the states. Some jurisdictions also post a reduced limit for towed vehicles, such as trailers, wrecked vehicles, or race cars. Differential limits are likely to occur on at-grade rural highways, rather than on freeways and urban streets.

Special speed limits (lower than for automobiles) exist in about five states for buses and in about 30 states for school

buses. In some cases these limits are the same as for trucks, and in others they are different.[40]

The merits of differential speed limits are subject to debate. Proponents contend that reduced speed is desirable for larger vehicles because their operating characteristics (e.g., stopping distance) are not as good as for passenger cars. Opponents, on the other hand, argue that a differential limit creates variances in speed and a hazardous condition. Such variances in speed are apparently undesirable, as evidenced by the results of a study by the Federal Highway Administration. (see Figure 11–5).

Speed limits for adverse weather conditions. Basic traffic laws usually require motorists to adjust their vehicle speed to existing road conditions. The primary responsibility for adjusting speed to adverse weather conditions thus rests with the driver. Nevertheless, some jurisdictions have found it desirable, primarily for safety reasons, to reduce speed limits at specific locations during adverse weather conditions by means of signs capable of displaying various messages. Such practice is generally limited to freeways or expressways, where speeds are higher and headways between vehicles are typically shorter. Warnings to lower speed have specifically been used in areas of high crosswinds.

---

[40] Homburger and Kell, *Fundamentals of Traffic Engineering,* p. 23-2.

Variable speed limits by lanes on freeways. In order to improve the quality and safety of traffic flow, the use of different speed limits for various lanes of a highway has been tried, principally on freeways and expressways. Where used, the practice is to post the higher limits on lanes closer to the median during peak traffic periods. One study reports that using changeable speed limit signs during the off-peak period produced negligible benefits and that the use of these devices during the peak period was judged to have produced essentially no effect.[41] Separate speed limits are sometimes used for designated truck lanes or grades.

School speed limits. Many jurisdictions establish special speed limits on streets adjacent to or in the vicinity of schools during certain hours of the day. A West Virginia study indicated that the most significant factors influencing speeds in school zones were the approach speed limit, the distance of school buildings from the roadway edge, traffic volumes, and the length of the school zone.[42] To be effective, school speed limits should be based on a documented need.

Minimum speed limits. On limited-access highways designed for high-speed driving, minimum speed regulations are often used to eliminate the differential in speeds, which is often a contributing factor in accidents. Such regulations also prevent motorized farm and construction equipment and other slow-moving vehicles from using high-speed facilities. In Canada, where the maximum legal speed on rural freeways is usually 100 km/h (62 mph), the minimum legal speed is 70 km/h (43 mph). In the U.S., about 16 states have such limits for most or all freeways or interstate highways. Minimum limits are generally 40 mph, but one state (Arkansas) has a minimum limit of 50 mph.[43] Many states have laws that prohibit driving at such slow speed as to impede or block the normal and reasonable movement of traffic.

Speed limits as related to time of day. The fact that fatal accident rates at night are usually greater than during daylight hours has led some jurisdictions to reduce maximum legal speeds at night. Such regulations are now being used with less frequency because they are often considered to be arbitrary and difficult to enforce and there is more and better highway lighting.

Passing and no-passing regulations

The overriding distinction between traffic operations on urban streets and on two-lane, two-way roadways in rural areas lies in the passing maneuver, which must utilize a lane assigned to traffic traveling in the opposite direction.

The need to pass in order to sustain a desired speed is a function of the amount of traffic and the distribution of speeds. The duration of the gap in opposing traffic required

for the passing maneuver is a function of the speed and the relative speed of passed and passing vehicles. The ability to pass is controlled further by the existence or nonexistence of a sight distance sufficient to detect and utilize an available gap with safety. When passing sight distance is not available, all vehicles as they overtake will form into a queue traveling at the speed of the slowest vehicle in the stream.

Passing sight distance

Passing sight distance is the length of highway ahead necessary for one vehicle to pass another before meeting an opposing vehicle that might appear after the pass begins. Passing sight distances used for design, given in Table 11–7, are based on various traffic behavior assumptions.

Passing sight distances for purposes of pavement marking are also given in Table 11–7. No-passing zone markings, given in the *Manual on Uniform Traffic Control Devices* (MUTCD),[44] are based on different assumptions, which result in lower values. No-passing zones are based on the 85th percentile speed during low-volume conditions, which is slightly less than the design speed.

Sight distance adequate for passing should be provided frequently on two-lane highways. The percentage of the highway where passing can take place affects not only capacity but also the safety, comfort, and convenience of all highway users. For purposes of design, passing sight distance for both horizontal and vertical restrictions is measured from a "seeing" height of 3.5 ft to an object height of 4.25 ft. For purposes of marking pavement, it is measured from a "seeing" height of 3.75 ft to an object height of 3.75 ft.

Passing sight distance may be measured graphically from plans or by field methods described in Chapter 3. Sight distance design records are useful for determining the percentage of highway length on which sight distance is restricted to less than the passing minimum—an important criterion in evaluating the overall design and the capacity.

TABLE 11–7
Minimum Passing Sight Distances

Used for Design				Used for Pavement Marking			
Design Speed		Minimum Passing Sight Distance		85th Percentile Speed		Minimum Passing Sight Distance	
mph	km/h	ft	m	mph	km/h	ft	m
20	30	800	245	—	—	—	—
30	50	1100	335	30	48	500	152
40	64	1500	457	40	64	600	183
50	80	1800	549	50	80	800	244
60	97	2100	640	60	97	1000	305
65	105	2300	701	—	—	—	—
70	113	2500	762	70	113	1200	366
75	121	2600	793	—	—	—	—
80	129	2700	823	—	—	—	—

SOURCE: Institute of Transportation Engineers, *Transportation and Traffic Engineering Handbook.* 2nd ed., © 1982, p. 591. Reprinted by permission of Prentice Hall, Inc., Englewood Cliffs, New Jersey.

[41] G.C. Hoff, *A Comparison between Selected Traffic Information Devices,* Chicago Area Expressway Surveillance Project Report 22, Illinois Division of Highways, Chicago, 1969, p. 7.

[42] *Establishing Criteria for Speed Limits in School Zones,* Engineering Experiment Station, West Virginia University, 1967, p. 61.

[43] Homburger and Kell, *Fundamentals of Traffic Engineering,* p. 23-3.

[44] Federal Highway Administration, U.S. Department of Transportation, *Manual on Uniform Traffic Control Devices for Streets and Highways,* Washington, DC, Government Printing Office, 1978, p. 3B-8.

Where a computer is used in the design, it can be programmed to determine the proportion of passing sight distance.

No-passing zones

Both pavement markings and signs are used to mark no-passing zones, as described in the MUTCD. These are most often locations where there is not adequate passing sight distance, although other passing-prohibited zones may be established approaching intersections, bridges, or tunnels, and in residential, business, or school zones.

Stop and yield controls

All right-of-way regulation is based on provisions of vehicle codes that prescribe the procedures to be followed by drivers on approaching and entering intersections. At uncontrolled intersections, the *Uniform Vehicle Code* sets forth the standard right-of-way rule: "The driver of a vehicle approaching an intersection shall yield the right-of-way to a vehicle which has entered the intersection . . . the driver of the vehicle on the left shall yield the right-of-way to the vehicle on the right."[45]

Drivers of vehicles entering an intersection through a stop or yield sign are to yield the right-of-way to other vehicles that have entered the intersection from an intersecting street, or that are approaching so closely as to constitute an immediate hazard, and continue to yield the right of way until such time as they can enter with reasonable safety.

In general, regulatory signs should be used only where needed and warranted, but this is especially true of stop and yield signs since they establish right-of-way and have a pronounced effect on traffic flow and capacity.

Right-of-way assignment

The most generally used methods of assigning the right-of-way at an intersection are the following controls:

Yield sign. The yield sign is used to protect traffic on one of two intersecting streets without requiring traffic on the other street to come to a complete stop. Instead, cross traffic is merely required to yield, as described above. It is thus less restrictive for traffic on a minor street than a stop sign, but does not provide the same degree of protection to traffic on the uncontrolled street. It is not an adequate protection for urban arterials or state highways.

Two-way stop. Where one of the streets at the intersection is more heavily traveled, a designated through street, or a state highway, the stop sign is placed against the traffic on the other street or streets. The sign should be placed against the minor flows of traffic in order to delay as few vehicles as possible.

Four-way (multiway) stop. Four-way stop control has the great disadvantage of delaying all vehicles entering an intersection, whereas it should never be necessary to delay more than half of these vehicles. However, it does provide for safe and orderly movement at intersections where higher volumes of traffic must be accommodated without benefit of traffic signals.

Traffic signals are appropriate under certain circumstances, with specific warrants for installation, as discussed in detail in the MUTCD.

Warrants have also been developed for the use of stop and yield signs. Specific warrants for the use of stop and yield signs are provided in the MUTCD[46] as general policy statements rather than as absolute warrants. Warrants provide a guide to sound sign application and serve as an aid in preventing the overuse of the signs.

Stop sign warrants

A stop sign may be warranted at an intersection where one or more of the following conditions exist:

1. Intersection of a less important road with a main road, where application of the normal right-of-way rule is unduly hazardous.
2. Intersection of a county road, city street, or township road with a state highway.
3. Street entering a through highway or street.
4. Unsignalized intersection in a signalized area.
5. Unsignalized intersection where a combination of high speed, restricted view, and serious accident records indicate a need for control by the stop sign.

Stop signs cannot be erected at intersections where traffic control signals are present because the signals should be operated continuously, either in normal or flashing operation. Moreover, stop signs should not be installed for the sole purpose of controlling the speeds of motorists.

"Multiway" (four-way or all-way) stop installations can be used as a safety measure, or in lieu of traffic signals, at some locations where the volumes on the intersecting roads is approximately equal and the following conditions have been established:

1. Where traffic signals are warranted and urgently needed, the multiway stop is a measure that can be installed quickly and provide for traffic control until signals or other improvements can be provided.
2. An accident problem is indicated by five or more reported accidents in a 12-month period of a type susceptible to correction by a multiway stop installation.
3. Minimum traffic volume:
   a. The total vehicular volume entering the intersection from all approaches must average at least 500 vehicles per hour for any 8 hours of an average day, and
   b. The combined vehicular and pedestrian volume from the minor street or highway must average at least 200 units per hour for the same 8 hours, with an average delay to minor street vehicular traffic of at least 30 sec per vehicle during the maximum hour, but

---

[45] *Uniform Vehicle Code and Model Traffic Ordinance,* Sec. 11-401.

[46] *Manual on Uniform Traffic Control Devices,* 1989, Secs. 2B-5, 2B-6.

c. When the 85th percentile approach speed of the major street traffic exceeds 40 mph, the minimum vehicular volume warrants shall be 70% of the foregoing requirements.

Yield sign warrants

Yield sign warrants are established as follows:

1. On a minor road at the entrance to an intersection when it is necessary to assign right-of-way to the major road, but where a stop is not necessary at all times, and where the safe approach speed on the minor road exceeds 10 mph.
2. On the entrance ramp to an expressway where an adequate acceleration lane is not provided.
3. Within an intersection with a divided highway, where a stop sign is present at the entrance to the first roadway and further control is necessary at the entrance to the second roadway, and where the median width between the two roadways exceeds 30 ft.
4. Where there is a separate or channelized right-turn lane, without an adequate acceleration lane.
5. At any intersection where a special problem exists and where an engineering study indicates the problem to be susceptible to correction by the use of the yield sign.

Yield signs should not be used to control the major flow of traffic, approaches of more than one of the intersecting streets, or at intersections where there are stop signs on one or more approaches. The state of the art in the use of yield signs has been published in the ITE informational report, "Yield Sign Usage and Applications."[47]

Sight obstruction regulations

Obstructions to sight distance must be controlled in order to minimize operational impediments and hazards at intersections and driveways. It is also important to keep intersections clear of view obstructions from parked vehicles on streets and signs located on public or private property. A reduction in accidents at intersections can be achieved by providing a clear view at all intersections.

Clear sight distance areas should be established, where possible, to ensure that obstructions do not infringe upon the lines of sight needed among motorists, pedestrians, bicyclists, and others when approaching potential conflict points. These areas usually take the form of triangles, with the sides along the intersecting pathways being approximately equal to the stopping sight distance for vehicles traveling at design speeds.

For urban conditions, which usually include many uncontrolled intersections, it is typical to assume a design speed of 0 to 5 mph over the posted speed limit, and a stopping distance composed of the vehicle braking distance and the distance traveled during a minimum perception-reaction time of about 1 sec. This assumes reasonable alertness of drivers to

potential conflicts in such an environment. The resultant stopping distance for 25-mph design speeds is on the order of 100 to 120 ft.

In analyzing sight obstruction problems at intersections, a detailed calculation of *safe approach speed* is often advisable.

Safe approach speed

The safe approach speed is the maximum speed at which traffic on one intersection approach can avoid colliding with cross traffic. It is a function of the speed of traffic on the intersecting street, the intersection geometry (angle of intersection, approach alignment, street widths), and the location of view obstructions.

A method of calculation is described in *Fundamentals of Traffic Engineering* by Homburger and Kell.[48]

In calculating safe approach speed, vehicles on the "major" street are assumed to maintain a constant speed, "side" street vehicles are assumed to be able to decelerate at a rate of 10 ft/sec/sec (about the maximum rate of comfortable deceleration), and driver reaction time is assumed to be 1 sec. The analysis is first performed for one side street approach, then repeated for the other. If the two intersecting streets are of about equal importance, one is first designated "major" and the other "side" street; the analysis is then repeated with these labels reversed.

Removal of sight obstructions

Many jurisdictions have enacted ordinances that permit them to clear all obstructions on private property at intersections for a specified distance down the right-of-way line, as shown in Figure 11–7, which is from a typical county ordinance.

Such ordinances prohibit the installation or maintenance of any sign, hedge, shrubbery, natural growth, fence, or other obstruction to view that is more than a specified height (e.g., 2 ft 6 in in the example shown) above the nearest pavement surface or nearest traveled way within the indicated triangular areas. Definitions and exceptions (such as permanent buildings, utility poles, official signs and signals) are usually included, and trees trimmed to the trunk for a specified distance above the ground (typically 6 to 8 ft) are usually exempted. Some jurisdictions also exclude saplings and signs that are mounted to provide a clear, open space above the ground and which has supports that do not constitute an obstruction.

Enforcement is usually the responsibility of the public works department, building inspector, or other official authorized to initiate abatement proceedings. Following investigation of violations, notices are given as may be necessary to carry out the provisions of the ordinance. Any obstruction maintained in violation of the ordinance is deemed a public nuisance. A notice to remove the obstruction may be posted (some jurisdictions send notices by registered mail), and if compliance is not forthcoming, officials may enter the premises and remove or eliminate the obstruction.

---

[47] ITE Technical Council Committee 4A-A, "Yield Sign Usage and Applications," *Transportation Engineering,* 48 (October 1978), 37–43.

[48] Homburger and Kell, *Fundamentals of Traffic Engineering,* pp. 21-6–21-9.

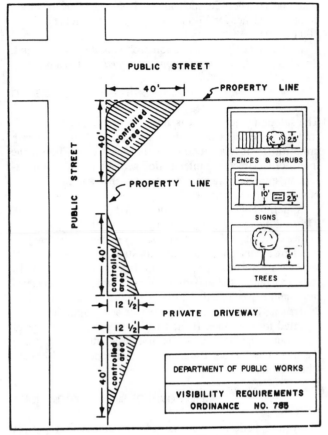

PUBLIC STREET

40'

PROPERTY LINE

controlled area

40'

PROPERTY LINE

controlled area

40'

PUBLIC STREET

12 1/2'

PRIVATE DRIVEWAY

12 1/2'

controlled area

40'

25'

FENCES & SHRUBS

10' 25'

SIGNS

6'

TREES

DEPARTMENT OF PUBLIC WORKS

VISIBILITY REQUIREMENTS
ORDINANCE NO. 785

Figure 11–7.
SOURCE: Perry R. Lowden, Jr., and James C. Ray, *Highway Safety,* Institute of Transportation and Traffic Engineering, University of California, Berkeley, September 1968, p. 91.

Cost of such abatement proceedings can usually be assessed to the owner of the property, who may be guilty of a misdemeanor punishable by fine or imprisonment or both. In practice, the effectiveness and workability of such an ordinance lies in the ability of the governmental agency to use it to obtain voluntary compliance.

Emergency condition regulations and other restrictions

Unexpected events

Events frequently occur on a particular highway (an accident) or in an entire city or area (snow, hurricane, etc.), which create emergency traffic situations that must be handled immediately. If there is snow or a similar weather emergency, advance preparations can often be made based on weather forecasts. However, accidents happen intermittently and advance knowledge is impossible. To ensure that traffic is handled properly during these periods, considerable preplanning is desirable and should include:

1. Arranging with police for traffic control and ambulance or wrecker services.

2. Determining alternative streets to be used in case a street is closed, making police assignments for directing traffic, and having available temporary signs, channelizing devices, flashers, and/or arrow panels.
3. Designing signals and signal systems to let emergency vehicles proceed with minimum delay.
4. Installing variable message highway signs and roadside radio broadcasts to warn motorists of roadway conditions ahead, of alternative routes to use, and similar information. Use of such signs and radios is particularly important on freeways and other high-volume, high-speed highways.
5. Enacting appropriate legislation or directives and procedures to cover action to be taken during the emergency.

An example of legislation relating to emergency or special conditions is Kansas City, Missouri's "Emergency Snow Ordinance," which reads as follows:

Sec. 34.150 Snow Emergency Regulations
1. *Driving emergency.* When snow, sleet or freezing rain is causing slippery or hazardous conditions which might lead to serious traffic congestion, the City Manager may declare a traffic emergency, and until such emergency is terminated, no person shall operate a motor vehicle on any street in such manner or in such condition as to allow or permit such vehicle to become stalled by reason of the fact that the driving wheels of such vehicle are not equipped with effective tire chains or snow tires.
2. *Parking emergency.* Whenever snow has accumulated or there is a possibility that snow will accumulate to such a depth that snow removal operations will be required, the City Manager may declare a parking emergency, and until such an emergency is terminated, no vehicles shall be parked on any streets designated as snow routes by appropriate signs. All vehicles parked on such streets must be removed within two (2) hours after declaration of an emergency or be considered in violation of this section.

Such legislation must be administered properly in order to obtain compliance from motorists. Street and traffic conditions must be closely monitored and the emergency regulation terminated as soon as possible. It is essential that accurate and timely information be provided.

Planned events

Other events, such as parades, motorcades, sports events, concerts, large conventions, and funerals, have the potential for being very disruptive to normal traffic flow unless proper planning and administration are provided so that steps can be taken to minimize and control the event's impact. There should be clear delineation of responsibility for issuing permits, coordinating events, and supplying police and other emergency services, as may be needed.

In some jurisdictions, this responsibility is placed with the police department; in others, it rests with transportation, public works, public services, or a combination of departments. Since traffic flow is only one of a number of concerns (e.g., crowd control, security, and provision of emergency services) related to such events, overall supervision and administration is usually provided by the agency's chief executive officer or perhaps by the chief of police.

Residential street controls

Public uses of residential streets

Residential streets and sidewalks are public spaces for many activities and functions. They offer opportunities for providing landscaped vistas, trees and shrubs, paths for walking, places for talking, rights-of-way for utilities, and—among these and other activities—facility for the movement, stopping, and storage of motor vehicles.

Conflict between traffic and residential uses. The first function of residential streets[49] is to serve the land that abuts them. They provide for access to homes by all who enter and leave, make deliveries, and provide services. A secondary function is to provide routes for those who wish only to pass through the area. It is here that conflict arises, due to the basic discrepancy between the impact of vehicular traffic and the tranquility of a residential street.

Three specific forms of "unwanted traffic" are recognized on residential streets: (1) traffic using the streets as shortcuts, detours (such as going around a residential block because of a left-turn prohibition on an abutting major street), or overflow from a congested arterial; (2) vehicles traveling at excessive speeds; and (3) vehicles parked at curb spaces (with related movements in search for and leaving such spaces) by drivers whose destinations are outside the neighborhood.

Often such intrusion is possible because the geometry of street networks was fixed long before such conflicts had been visualized, although even "modern" residential streets may experience neighborhood traffic problems. The pervasive conflict between traffic and residential uses has been tolerated to a greater or lesser degree throughout the world. However, demands of residents for amelioration of their environment have been growing; since 1960, considerable literature in the planning and transportation fields has been devoted to analyzing the problem and outlining possible solutions.

The traffic engineer will recognize the multiple role of the street to include local access use by vehicles, bicycles, and pedestrians. The resolution of residential street problems is relevant to the basic goals of traffic control, which include safety, the efficient movement of people and goods, and environmental goals. The latter include accessibility for local traffic, minimization of unnecessary traffic, and encouragement of changes in modal choice where this promotes efficiency, quality of life, or other specific objectives of a neighborhood.

Participation by residents in management of street use. The residents must be part of the management process; their contribution is the articulation of values and priorities, their response to proposed plans and designs—perhaps offering

ingenious alternatives of their own—and their ultimate approval and willingness to assist in eventual implementation.

Neighborhood traffic management often represents a proprietary initiative by residents and requires ways of thinking that may be new to both planners and engineers. Traditional emphases on maximizing capacity and offering fast, direct routes for the movement of vehicles need to give way to consideration of the effects of transportation operations on affected land uses.

Neighborhood values concerning streets. A study in San Francisco by Appleyard and Lintell[50] showed a strong relationship between traffic volumes and such values as safety, security, identity, comfort, neighborliness, privacy, and sense of "home." Three streets selected for this analysis were identical in appearance but quite different in the traffic volumes they supported. One street had "light" traffic, some 2,000 vehicles per day; another "medium" traffic, 8,200 vehicles per day; and the third "heavy" traffic volumes, 15,750 vehicles per day. The results of the study, based on extensive interviews and field observations, revealed that residents on all streets were primarily concerned with the dangers of traffic. Excessive speed on the "heavy" street was frequently cited as dangerous. Other factors taken into account in the study included noise, vibration, fumes, soot, and trash. In each case an inverse correlation existed between the adverse environmental impact measured and the degree of livability experienced.

Many of the findings of Appleyard and Lintell's study seem obvious, especially the inverse correlation between traffic flows and levels of livability. However, the implications of these findings were far-reaching with proposed "protected residential areas" throughout the city. These areas were not implemented because of public opposition outside of the neighborhoods, but the study documented the perceptions that residents have of traffic hazards and exposure to possible accidents.

Traffic service and other needs. Provision for the safe and efficient movement of traffic is a guiding factor in street design. The problem of this view, however, is rooted in the conflicts that exist among users of street space. The traffic function—moving traffic streams efficiently—competes with land service functions, such as providing access to properties and parking in the street. Residents may also perceive the street space as a part of an overall system of neighborhood amenities—a view that precludes all but a minor amount of traffic, for either access or movement purposes. In order to identify goals for residential street design, it is important to understand the principal conflicts over the use of street space:

1. Conflict between travelers and neighbors.
2. Conflict among travelers—drivers of cars, trucks and buses, pedestrians, cyclists, etc.
3. Conflict between neighbors, especially between residents and neighboring merchants, institutions, or industries.

---

[49] Based on *Residential Street Design and Traffic Control,* W.S. Homburger, Elizabeth A. Deakin, Peter C. Bosselmann, Daniel T. Smith, Jr., and Bert Beukers, eds. (Englewood Cliffs, NJ: Institute of Transportation Engineers, Prentice-Hall, 1989).

[50] Donald Appleyard and Mark Lintell, "The Environmental Quality of City Streets: The Residents' Viewpoint," *Journal of the American Institute of Planners,* 38, (March 1972), pp. 84–101.

4. Conflict between the public agencies that manage and maintain streets and protect neighborhoods (such as public works, police, and fire services) and the neighbors.

5. Conflicts among the professionals who plan, design, and manage streets, chiefly between engineers and designers.

This list of conflicts suggests the importance of developing a sound public policy for designing and managing streets with all users in mind. It is essential to be willing to compromise, consider trade-offs, and plan the use of each street to its "environmental capacity" appropriate to its particular purpose and function.

Transportation functions of streets

Streets perform two transportation functions: provision of *access* to individual parcels of land and provision of an infrastructure for *movement* between various origins and destinations.

"Access" can be interpreted to include the existence of driveways connecting the street with private property and the availability of parts of the street for parking and loading.

"Movement" comprises both the *capacity* to move quantities of vehicles or people and the ability to do so at a reasonably high *speed*.

Although residents frequently perceive access as a function that primarily serves those within the neighborhood and movement as one that primarily serves those outside the neighborhood, in actuality both functions are necessary to both classes of users, since travel (movement) invariably involves departure and arrival (access) from an origin and to a destination.

Design standards. Agencies in many nations and states have developed design standards for streets. Such standards generally emphasize safe and efficient vehicle operation but may be silent on the relationship of street use to abutting land developments and its users; it generally is assumed that other regulations (such as zoning and building codes) will address these concerns. Thus, although they may make provision for minimizing vehicle conflicts and vehicle-pedestrian conflicts (e.g., offset intersections, continuous sidewalks, etc.), the standards in current use in many jurisdictions do not effectively address the potential for conflicts between needs for residential access and amenities and the needs for traffic movement into and beyond a given neighborhood.

Many neighborhoods also predate current standards, and their streets may have other design problems, such as inadequate driveway spacing, limited setback and space for off-street parking, a "grid" pattern of streets that facilitates the incursion of through traffic and that may result in vehicular conflicts and speeds that are unsuitable for the neighborhood.

Another concern is that design standards usually are applied at the level of a subdivision or local jurisdiction, at least in the United States. There are many cases in which adjacent jurisdictions fail to coordinate their networks, and streets change designation and character as they cross jurisdictional boundaries.

Residential traffic controls. Local residential streets should be protected from through traffic. Residential streets should be linked to traffic-carrying streets in a way that simultaneously provides good access to other parts of the community and minimizes the chance of the residential streets' use by through traffic. These goals should be a part of the planning for new residential areas. In some communities, these objectives have been achieved in older neighborhoods (in Montgomery County, Maryland; Berkeley and Richmond in California; Seattle, Washington; and other locations) through the installation of traffic diverters and barriers.

Residential streets should also be protected from vehicular traffic moving at excessive speed (greater than 25 to 30 mph) and from parking unrelated to residential activities. Figure 11–8 illustrates several types of treatments that are designed to reduce speed and discourage through traffic at minor intersections in residential areas.

A variety of treatments has been devised to accomplish the above objectives—ranging from speed "humps" and "chokers" for speed control to the Dutch "Woonerf" concept, which

Figure 11–8. Treatment of minor intersections in residential areas to reduce speed and discourage through traffic.
SOURCE: W.S. Homburger and J.H. Kell, *Fundamentals of Traffic Engineering*, 12th edition, University of California, Institute of Transportation Studies, Berkeley, 1988, p. 20–7.

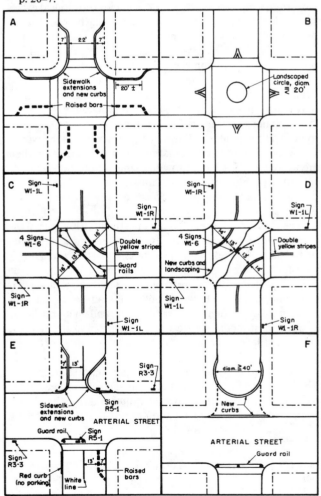

essentially convert neighborhood blocks into pedestrian-oriented precincts.

Speed "humps" are carefully designed undulations in the pavement surface that have been shown to control speed effectively without the risk of interfering with the driver's control of the vehicle. These humps are typically installed across the width of the roadway, with a longitudinal circular section 3 to 4 in in height and approximately 12 ft in length. These should be distinguished from speed bumps, which are much shorter (6 to 12 in long) and which have been associated with maintenance, safety, and liability concerns.

A study of the use of speed humps[51] recommended that the pavement undulations should be used only under the following conditions:

1. The street serves a purely local access function.
2. There is no more than one lane per direction.
3. The street is not a transit or truck route.
4. The street is not zoned above 25 mph.
5. The 85th percentile speed exceeds 30 mph.
6. There is evidence of a speed-related accident problem.

Undulations should be located no less than 200 ft from an intersection or sharp horizontal curve, and each undulation should be visible for at least 200 ft. Standard warning signs (e.g., Sign W8 in the MUTCD) should be used. The use of pavement stripes has also been suggested,[52] but these might give the appearance of crosswalks, which could mislead motorists.

"Chokers" may consist of landscaped bulbs between the sidewalk and the street, widened sidewalk areas, or points where street entrances are necked down. Raised or brick crosswalks may also be used in combination with pavement undulations. In addition to numerous applications of these devices in European and Australian cities, speed humps have been used extensively in Pasadena, California.

The Woonerf requires motorized traffic and bicycle traffic to adapt to pedestrian behavior, and it has become very popular in European countries. Extensive installations and utility relocation are required, and initial costs and maintenance costs may be high. Modified forms of this type of control have been implemented in Boulder, Colorado; San Francisco, and other United States cities.

Neighborhood parking permit programs, to limit long-term parking to those living in the area, have also been implemented in a number of North-American cities and upheld by court decisions. These usually involve standard time-limit parking for the general public (1, 2, or 4 hours) with exemption for vehicles displaying a permit available only to residents. The parking restrictions may also exclude vehicles with no permit at certain hours of the day.

Other traffic controls frequently requested by residents include stop signs, speed limits, turn prohibitions, and one-way street designations. In general, the application of these devices may be expected to have the same effects in residential neighborhoods as they do elsewhere, and their indiscriminate use should be avoided. Stop signs are persistently requested by citizens with the expectation that they will control speeds or reduce traffic volumes and accidents in residential neighborhoods. Although there may be some effect on volume and accidents in certain instances, there is little evidence of effect on traffic speeds attributable to stop sign placement except within about 200 ft of the intersection controlled. Some cities use special warrants with reduced minimum volume requirements for residential neighborhood locations.

Most residential streets are covered by "blanket" speed limits of 25 or 30 mph, though it is not uncommon for 85th percentile speed of traffic to be considerably higher, particularly on collector streets. In such cases, signs merely remind drivers of the general limits applicable to the residential area, unless an overriding problem indicates the need to establish a new speed zone on the street. Studies evaluating the effect of speed limit signs on speed have been largely confined to major streets and have generally shown that signs have very little impact on driver speed on major streets. Drivers consistently drive at speeds which they perceive as reasonable, comfortable, convenient, and safe under existing conditions, regardless of posted speed limits. Consistent enforcement is essential to obtain any measurable effect from posting of speed limits. Alternatives to speed zoning include pavement undulations (described above), traffic circles (discussed below), and podium intersections, where the entire intersection is raised a few inches above the normal grade level with ramps to conform to the grades of the adjacent streets.

Turn prohibitions and one-way streets can have a very significant effect on traffic volumes, if their use is accepted by the affected drivers. Enforcement is essential, particularly when the regulations are first enacted. These measures are low-cost alternatives, and they provide minimum impedance to emergency vehicles, which can travel the "wrong way" when necessary. Speeds tend to be higher on one-way streets. In residential neighborhoods, this can be counteracted by limiting the number of blocks with one-way continuity. One-way streets tend to be inherently safer than two-way streets, but in residential neighborhoods, where irregular patterns of one-way streets are used, careful treatment is essential at intersections. Traffic circles tend to have higher violation rates and may represent a risk for increased accidents because of the tendency of some motorists to violate the one-way pattern, especially in making left turns.

Other regulatory devices, such as traffic signals, yield signs, truck restriction signs, and access regulation signs ("Do Not Enter," "Not a Thru Street," "Dead End," "Local Access Only," and "Thru Vehicles Prohibited"), have also been used in residential settings. The latter signs are used primarily in conjunction with one-way streets (i.e., "Do Not Enter") or as informational signs, although the use of "Local Access Only" signs in the regulatory black-on-white format could conceivably be effective in reducing traffic volume on residential streets if accompanied by enforcement.

Warning signs in residential neighborhoods have limited uses, and drivers and pedestrians usually need to be warned only of special hazards. The attention of the driver is drawn to the location of schools (especially elementary schools) and playgrounds, to pavement undulations, to the fact that traffic

---

[51] Wayne Tanda and others, *Subcommittee Report on Pavement Undulations (Road Bumps)*, Report to the California Traffic Control Devices Committee, November 1983.

[52] Burton W. Stephens, "Road Humps for the Control of Vehicular Speeds and Traffic Flow," *Public Roads,* 50, (December 1986), 82–90.

barriers or diverters are located ahead, and to stop and yield signs ahead if they may not be readily visible because of curves or shrubbery. "Slow" signs, "Children at Play" signs, and novelty signs are vague and unenforceable. The result may, therefore, have little more than a placebo effect on residents. The novelty of a new sign wears off quickly and then no longer attracts the attention of regular passers-by. Nonstandard signs usually have no legal meaning or established precedent; their use is discouraged because of both the lack of proven effectiveness and undesirable liability exposure. Further, driver respect for signs and other traffic controls may be eroded through the use of nonstandard and unneeded devices.

Geometric design features. Geometric design features may be used to restrict access and/or reduce speeds in residential settings. These include median barriers and cul-de-sacs at intersections with major streets, and semi-diverters, diagonal diverters, and midblock cul-de-sacs on local residential streets. Some of these design features are also illustrated in Figure 11–8. These are features that physically restrict and prevent vehicle movement as well as reduce speed. Their common characteristic is that by their physical form they force or prohibit a specific action. Geometric features have the advantage of being largely self-enforcing and of creating a visual impression that a street is not intended for through traffic. The disadvantages relative to other devices are their cost, the potentially negative impact on emergency and service vehicles, and the imposition of inconvenient access on some parts of a neighborhood. They are also static and must be appropriate at all hours of the day and night.

Rumble strips, formed with patterned sections of rough pavement or raised pavement markers, have no effect on traffic volumes and little on speed though they do appear to cause an increase in driver attention. Studies conducted on major streets show that the strips have had a noticeable effect in reducing accidents when placed in advance of a stop sign. Effects in lower-speed residential areas have not been determined.

Introducing curvatures on a previously straight alignment has been discussed as a physical speed control device, but this has produced considerable public controversy and warnings of possible associated safety problems. Use of various designs in Australia is reported to have a very subtle effect on driver behavior.

Valley gutters and rough pavements are two existing devices that tend to control traffic as an unintended by-product of their presence. In neither case can it be suggested that streets should be designed to include valley gutters and rough pavement in order to reduce speed; however, the effect may be an argument for delaying repaving of purely residential streets—an argument that should be carefully weighed against any noted indications of a hazardous or deteriorating structural condition.

Play streets and private streets

Both play streets and private streets are common in some areas of the United States. Traffic flow may be restricted on such streets through the use of temporary barricades, signs, or gates. Such streets can be temporarily closed during certain hours, and permanent closure may also be considered

in areas where vehicular access to homes and garages is provided for by alternate means such as alleys. In Vancouver, British Columbia, a number of blocks have been closed to vehicular traffic and converted to exclusive use by pedestrians, cyclists, strollers, and—in one case—an outdoor cafe. Residents must park their vehicles on a cross street or alley or—where available—in off-street spaces. Provision for emergency vehicle entry into these blocks is provided by use of traversable barriers in some, but not all, cases.

Implementing neighborhood traffic controls

The way in which neighborhood traffic controls are implemented can be as important to their eventual success or failure as the substance of the strategies themselves. Implementation should be considered not as a step but as a *process* requiring careful planning and documentation, public notice, evaluation, and possibly refinement of the strategies. Such a process calls for the same attention to detail and for the same thorough consideration as the initial planning effort.

The implementation of neighborhood traffic control schemes may raise issues about the responsible jurisdiction's legal authority to take such actions. For example, the measures or devices used to effectuate traffic control may be subject to state requirements as to design and/or application. Legal questions may also be raised about restrictions of access caused by the plan, its environmental impact, or concerns for tort liability. It is beyond the scope of this chapter to report or advise on the legal requirements of various United States states or foreign jurisdictions. However, legal counsel is usually advisable in developing and implementing neighborhood traffic control strategies.

Even when there is no question of authority, compliance with standards, or other legal requirements, neighborhood traffic control actions are sometimes challenged by opponents on grounds of denial of access or discrimination against nonresidents. In general, challenges to otherwise authorized traffic control schemes on the grounds that they cause incidental inconvenience to some parties are likely to fail; a community may divert traffic and partially restrict access, but still successfully withstand a legal challenge. Tests of sufficient police power and reasonable exercise of such power must still be met, of course.

Evaluating the impact of neighborhood traffic controls

Evaluation of technical performance and community perceptions is needed to provide a reasonable basis for decisions to keep or abandon a plan. A formal evaluation can clarify issues, bring the more stabilized long-term performance characteristics into focus, and spotlight hidden gains and losses that may be significant. Evaluation can point to opportunities for modifying a traffic control plan to make it perform its intended function better or to lessen adverse impacts. It can also be used to determine whether the plan should be expanded both in terms of devices and geographical area. Finally, evaluation can advance the state of knowledge about neighborhood traffic control and identify problems that might be avoided in future applications.

Effectiveness of controls. An evaluation should start with this question: Do the controls fulfill their intended purposes? Some effects are easily evaluated through a "before-after" traffic study. Other intended purposes involving public perceptions and reactions are best evaluated with public input. Evaluation should go beyond the question of effectiveness in fulfilling the plan's primary intentions, however. In particular, any negative impacts of the plan as implemented should be identified. Technical staff can then follow up on such matters (e.g., increased emergency response time) so as to develop modifications to offset negative effects. Involvement of emergency and service personnel can also help to minimize such adverse effects.

Minor adjustments to a neighborhood plan are a common occurrence. Observation during the period immediately following implementation is critical in order to identify problems that could easily be eliminated by minor adjustments. Additional police surveillance also helps discourage erratic or illegal driving behavior and vandalism.

Maintenance and enforcement issues. Maintenance and enforcement are important to the continued effectiveness of the traffic control scheme and to continued public acceptance. While the physical maintenance of the plan probably will require the greater amount of attention, in a broader sense, maintenance also requires attention to the need for continued driver respect and public support. It may be necessary to remind police officials of the need for enforcement, or to ask them to do "focused enforcement" in areas that appear to have a high violation rate.

REFERENCES FOR FURTHER READING

American Association of State Highway and Transportation Officials, *A Policy on Geometric Design of Highways and Streets,* Washington DC, 1990, pp. 163, 212.

American Association of State Highway and Transportation Officials, *Manual for Maintenance and Inspection of Bridges,* Washington, DC, 1974.

APPLEYARD, DONALD AND MARK LINTELL, "The Environmental Quality of City Streets: The Residents' Viewpoint," *Journal of the American Institute of Planners,* 38 (March 1972), pp. 84–101.

Automobile Club of Southern California, *Realistic Speed Zoning, Why and How,* Los Angeles, 1976, p. 8.

BRUCE, J. A., "One-Way Major Arterial Streets," *Improved Street Utilization through Traffic Engineering,* Highway Research Board Special Report 93, Washington, DC, May 1967.

California Department of Motor Vehicles, *Vehicle Code,* Sacramento, 1986.

California Department of Transportation, *California Ridesharing Facilities,* Sacramento, January 1984.

Establishing Criteria for Speed Limits in School Zones, Engineering Experiment Station, West Virginia University, 1967.

GILLARD, Q., "Los Angeles Diamond Lanes Freeway Experiment," *Traffic Quarterly,* ENO Foundation for Transportation, Westport, CT, April 1978.

HABIB, P. A., AND OTHERS, *Analysis of Pedestrian Crosswalk Safety on One-Way Street Networks,* Report DOT-OS-70057, U.S. Department of Transportation, Washington, DC, September 1978.

Highway Research Board, *Highway Fog,* National Cooperative Highway Research Project Report 95, Washington, DC, 1970.

HOFF, G. C., *A Comparison between Selected Traffic Information Devices.* Chicago Area Expressway Surveillance Project Report 22, Illinois Division of Highways, Chicago, 1969.

HOMBURGER, W. S., ELIZABETH A. DEAKIN, PETER C. BOSSELMANN, DANIEL T. SMITH, JR., AND BERT BEUKERS, *Residential Street Design and Traffic Control,* Institute of Transportation Engineers, Englewood Cliffs, NJ, Prentice Hall, 1989.

HOMBURGER, W.S., AND J.H. KELL, *Fundamentals of Traffic Engineering,* 12th Edition, University of California, Institute of Transportation Studies, Berkeley, 1988.

ITE Technical Council Committee 4A-A, "Yield Sign Usage and Applications," *Transportation Engineering,* 48(10), (October 1978), 37–43.

JOHNSON, R., AND B. McKINLEY, unpublished paper submitted to Technical Award Committee, Missouri Valley Section, Institute of Transportation Engineers, 1971.

MAYER, P. A., "One-Way Streets," *Traffic Control and Roadway Elements—Their Relationship to Highway Safety,* Highway Users Federation for Safety and Mobility, Washington, DC, 1971, Chap. 10.

Michigan Department of State Highways, *The Economic and Environmental Effects of One-Way Streets in Residential Areas,* Lansing, 1969.

National Committee on Uniform Traffic Laws and Ordinances, *Uniform Vehicle Code and Model Traffic Ordinance,* Evanston, IL, 1987.

NIELSEN, E., "Experience from 10 Years Fight against Traffic Congestion," 36th International Congress, International Union of Public Transport, Brussels, Belgium, 1965, p. 15.

Roads and Transportation Association of Canada, *Manual of Geometric Design Standards for Canadian Roads,* Ottawa, Ontario, 1986.

SHOAF, R. T., "Traffic Signs," *Improved Street Utilization through Traffic Engineering,* Highway Research Board Special Report 93, Washington, DC, May 1967.

SIEGERT, A., "Speed Limit Irks Germans," *Chicago Tribune,* October 11, 1971, Sec. 1-A, p. 3.

SINEMUS, HERMAN, *Neighborhood Traffic Controls,* paper presented at the 25th Annual Meeting, Western Section, Institute of Traffic Engineers, Berkeley, CA, July 1972.

STEPHENS, BURTON W., "Road Humps for the Control of Vehicular Speeds and Traffic Flow," *Public Roads,* 50, (December 1986), 82–90.

TANDA, WAYNE, AND OTHERS, *Subcommittee Report on Pavement Undulations (Road Bumps),* Report to the California Traffic Control Devices Committee, November 1983.

U.S. Department of Transportation, *Manual on Uniform Traffic Control Devices,* Washington, DC, 1989.

U.S. Department of Transportation, Federal Highway Administration, *A Bikeway Criteria Digest,* Washington, DC, 1977.

U.S. Department of Transportation, Federal Highway Administration, *Design and Construction Criteria for Bikeway Construction Projects,* Notice of Proposed Rulemaking, *Federal Register,* 45, (151), Washington, DC, August 4, 1980.

U.S. Department of Transportation, Federal Highway Administration, *Highway Statistics Summary to 1975,* pp. 85–86, *Highway Statistics Summary to 1985,* pp. 257–260, and *Highway Statistics Summary to 1986,* pp. 176–178, Washington, DC.

U.S. Department of Transportation, Federal Highway Administration, *Planning and Design Criteria for Bikeways in California,* Washington, DC, 1979.

U.S. Department of Transportation, Transportation Systems Center, *Priority Techniques for High Occupancy Vehicles—State of the Art Overview,* Washington, DC, November 1975.

12

TRAFFIC MANAGEMENT

Herman E. Haenel, P.E., *Consultant**

Introduction

Traffic management may be defined as the utilization of personnel (traffic operations and enforcement), materials, and equipment along freeways, city streets, and rural highways to achieve safe and efficient movement of people, services, and goods. Traffic management draws together the various design elements set forth in other chapters for the purpose of achieving orderly and safe flow and combining the cooperation of the numerous agencies involved in traffic operations and law enforcement. One agency cannot carry out all the operational components of traffic management. For example, freeway incident management cannot be carried out only by the state highway department. Many agencies are involved in this area of traffic management (e.g., city traffic engineering, police, and fire departments). Also, planning is needed in advance to implement most of the traffic management activities so that personnel from all of the agencies involved know what needs to be done when the event occurs.

Traffic management addresses the following areas:

o peak period and other recurring congestion
o accidents and other incidents
o special events
o construction and maintenance work zones
o inclement weather
o catastrophic events (earthquakes, tornados, explosions)

As examples, traffic management is needed in moving traffic around an accident on a freeway, through a construction work zone, and to and from a sports event. Cooperation of many agencies is needed.

An important aspect of traffic management is that it does not end when the peak period hour ends. Because vehicle delay is found to occur both day and night for various reasons, traffic management is vital during both off-peak and peak periods. This is especially true when it is considered that a vast amount of goods movement, service activities, and business meetings occur during off-peak hours. Delay to vehicles involved in these activities directly increases the cost of services and goods to the public.

Traffic management applies to the urban freeways, city streets, and suburban streets adjacent to cities. An incident or work zone activity along a freeway could make it necessary to route traffic along the parallel city and/or suburban street(s). By the same token, an incident or work zone along a city street could necessitate routing of traffic along the freeway.

Traffic management applies to rural areas and to cities of all sizes and to metropolitan areas. A special event in a small city necessitates managing traffic within the city street network, and a major accident on a rural highway may necessitate rerouting traffic along a county road.

The installation of traffic control, surveillance, and motorist information, as well as freeway and street improvements (e.g., pavement widening, geometric changes), access control, and improved utilization of signs, signals, and markings go hand in hand with traffic management. The design of geometric and traffic engineering improvements must keep the future management of traffic in mind as they are carried out. For example, special traffic patterns can be developed

---

* Retired from Texas State Department of Highways and Public Transportation, Field Operations Engineer in Traffic Engineering.

and installed within a signal system computer during the design and implementation of a new traffic signal system. These special patterns will be used for traffic diversion during an incident, special event, or maintenance activity along a freeway or city street once the traffic signal system is installed.

To be effective, traffic management must involve the communication, cooperation, and coordination of applicable city, county, and state traffic operations and law enforcement personnel together with representatives of public transportation and trucking organizations. These personnel must work as a team. The makeup of the team will vary depending on the city or rural location, but the three elements of communication, cooperation, and coordination are very important in achieving effective traffic management.

Need for traffic management

Although congestion has existed in some larger cities for many years, it has not been of significant importance in many of these cities until recent years. The doubling of the number of vehicles between 1960 and 1980, coupled with a decline in new highway construction, has stretched many urban freeways beyond capacity.[1] A study by the Federal Highway Administration has shown that recurring congestion was responsible for over 700 million vehicle hours of congestion along urban freeways during 1987 and that nonrecurring congestion caused over 1.2 million vehicle-hours of delay. Based on this study, as much of 61% of the urban freeway delay could be expected to be caused by non-recurring events.[2] The interstate highway system, which was designed for the rapid movement of goods and military equipment in case of a national emergency and to link cities, has become a major arterial within cities. Although urban freeways make up less than 3% of the total arterial mileage, they carry approximately 30% of the traffic.[3]

Congestion occurs both along freeways and on city streets. An average of 34% of the Daily Vehicle Miles Traveled (DVMT) along freeways and 43% of the DVMT along principal nonfreeway and nonexpressway arterial streets in 29 major urban cities are congested during the peak period.[4] Further, the total vehicle hours of delay (peak period plus incident delay) along nonfreeway and nonexpressway principal arterials within these cities are approximately the same as exists along freeways.[5] Traffic management is needed along arterials as well as along freeways.

Delay on urban freeways is expected to increase 360% from 1985 to 2005 and 433% in outlying areas during the next 20 years unless substantial improvements are made. This includes an increase of more than 300% in urban areas of over 1 million population and 1,000% in urban areas with populations of less than 1 million.[6] The resulting cost of delay along urban freeways can be expected to increase 450% between 1984 and 2005 unless substantial improvements are made.[7] Although significant improvements can be obtained through freeway and arterial street construction and reconstruction, funds are not expected to be sufficient to overcome the anticipated increase in traffic demand and resulting increased delay.

Significant reduction in vehicle delay can be obtained through traffic management applications. For instance, a study of a 42-mile freeway control and surveillance project in Los Angeles showed that traffic management reduced motorist delay by 65%. Other cities have reported reductions in delay by 50% particularly when service patrol use is included. The utilization of incident management teams in Los Angeles reduced delay due to lane blockages from 42 minutes to 21 minutes.[8]

Basic traffic management activities

In order to reduce and eliminate congestion there is a need to:

o Detect and respond promptly to incidents and accidents
o Keep traffic moving as smoothly as possible by improving existing conditions, rerouting traffic, and/or using real-time control strategies
o Provide traffic control plans and monitor special events and construction/maintenance zones
o Deal with snow, flooding, and disasters such as tornados, earthquakes, and explosions.

These basic traffic management activities require varying amounts of study, planning, and action. They also require physical improvements in some instances. Above all, each one of these activities requires teamwork of many agencies involved in improving the operation and safety of freeways and streets. Trying to improve traffic movement seems an impossible task at first glance. By breaking the problems down into basic activities and then merging them into one overall plan through the organized effort of many agencies working together, it is possible to overcome or at least greatly improve the various traffic problems as they occur during the year.

Prompt incident detection and response

A major element of traffic management is to detect incidents and promptly respond to them. This applies to incidents along city streets and on freeways. Prompt detection and response to an incident reduces vehicle delay and can also save lives and

[1] Institute of Transportation Engineers, *A Tool Box for Alleviating Traffic Congestion,* Washington, DC, 1989, p. 5.

[2] J.A. Lindley, *Quantification of Urban Freeway Congestion and Analysis of Remedial Measures,* Oct. 1986 (Federal Highway Administration, Washington, DC), p. 19.

[3] Institute of Transportation Engineers, *Tool Box for Alleviating Congestion,* p. 33.

[4] Texas Transportation Institute, *The Impact of Declining Mobility in Major Texas and Other US Cities,* Report No. 431-1F (College Station: Texas A&M University, Texas Transportation Institute, April 1989), Table B-2, p. B-5.

[5] Ibid., Table 13–3, p. 13-7.

[6] Institute of Transportation Engineers, *Tool Box For Alleviating Congestion,* p. 3.

[7] J.A. Lindley, *Technical Summary, Quantification of Urban Freeway Congestion and Analysis of Remedial Measures* (Washington, DC: U.S. Department of Transportation, Federal Highway Administration, Released February 1988).

[8] Institute of Transportation Engineers, *Tool Box for Alleviating Congestion,* p. 34.

reduce the consequences of an injury. Incident detection, and response along freeways is discussed in Chapter 13.

Delay due to incidents constitutes an average of 55% of the total delay along freeways and an average of 52% of the total delay along major streets in 29 cities in the United States. The results of another nationwide study, which are given in Table 12–1, show that incidents cause 61% of the delay on a freeway. Incidents reduce freeway capacity and in turn cause delays. A vehicle stopped on the shoulder of a six-lane freeway in Houston resulted in a 33% reduction in capacity while the blockage of one lane reduced capacity by 50%.[9] A similar study in Detroit found a 42.9% reduction in volume when one of three lanes was blocked.[10]

The rapid determination of where the incidents occur and the rapid response will significantly reduce delay for all motorists. A study in Los Angeles showed that 4 to 5 minutes of delay are saved by motorists for each minute sooner that vehicles are moved from the freeway.[11] Although the amount of reduced delay varies with the traffic volumes on the freeway, the California study shows that significant savings can be achieved with rapid detection and removal. It is also important to move accident vehicles away from intersections and major city streets as rapidly as possible to eliminate congestion and return the street to free-flow movement.

Another important aspect is to encourage motorists involved in property damage accidents along freeways and major city streets to move their vehicles off of, and away from, the freeway main lanes and off of major streets as quickly as is practical. Doing so reduces congestion that might occur and also makes movement less hazardous for all motorists. However, motorists are reluctant to move their vehicles until a police officer arrives. Thus, information should be provided through broadcasts and publications to encourage voluntary removal of vehicles to a safer location after an accident when this is physically possible.

TABLE 12–1
Urban Freeway Congestions Statistics, 1984

Freeway miles	15,335
Vehicle-miles of travel (millions of miles)	276,645
Recurring congested vehicle-miles of travel (millions of miles)	31,486
Recurring delay (million vehicle-hours)	485.0
Excess fuel consumption due to recurring delay (millions of gallons)	531.6
Delay due to incidents (million vehicle-hours)	766.8
Excess fuel consumption due to incidents (millions of gallons)	845.9
Total delay (million vehicle-hours)	1,251.8
Total excess fuel consumption (millions of gallons)	1,377.5

SOURCE: J.A. LINDLEY, *Quantification of Urban Freeway Congestion and Analysis of Remedial Measures,* Report No. FHWA/RD-87/052 (Washington, DC: Department of Transportation, Federal Highway Administration, October 1986).

---

[9] M.E. Goolsby, "Influence of Incidents on Freeway Quality of Service," *Highway Research Record 349* (Washington, DC: Highway Research Board, 1971).

[10] Herbert L. Crane, "Formation of Detroit Freeway Operations Unit," Report TSD-TR-119-69 (Department of State Highways, Traffic and Safety Division, State of Michigan, June 1969).

[11] David H. Roper, "Manage Traffic—Get Congestion Relief," unpublished paper (Los Angeles: California Department of Transportation, June 1986), p. 3.

Smooth movement of traffic

It is important to keep traffic moving as smoothly as possible during daily recurring congestion and during incident conditions. Studies can be made of existing peak period traffic along freeways and streets to determine where present bottlenecks are and the extent of the problem at these locations. Addition of estimated growth factors permits the engineer to determine to what extent the problem will increase and where additional problem locations will occur. Similar studies can be made using existing traffic volume data to determine what problems will develop when an incident occurs during peak and off-peak periods. Decisions need to be made on the various modifications and approaches required to overcome present and anticipated congestion.

It has been found that motorists make a trip earlier and/or take a different route in going to and from work when freeway entrance ramp control is implemented. The same could occur if motorists know of a major incident before starting a trip or far enough in advance of the incident location to alter their route.

The free flow of traffic also includes the movement of trucks. Since so much commerce today includes freight movement by truck, it may be best to provide special truck routing around a congested area.

Special events and construction/maintenance work zones

Special events and construction/maintenance work zones have certain similarities that are known ahead of time (where and when the event will occur) and the type and extent of the problems that can be expected. This provides time to develop ahead of time a traffic control plan for the upcoming event or work zone. It is also possible to monitor the conditions during the special event or along the work zone and make needed changes to improve operations.

Special events. Special events occur in just about all cities (large or small). These include festivals and parades in smaller towns as well as in larger municipalities. Each of these events require planning so that each agency can carry out its responsibilities when the event occurs. This will generally include advanced information published in the local paper for a festival and the installation of signs directing motorists along selected routes to the event and to parking areas.

Plans for annual events should be reviewed yearly. This review will assure that each agency and person involved is familiar with agency and individual duties and that conditions have not changed since the previous event. Also, where special routing is required to the event and to parking areas, it is important that the route(s) be checked ahead of time to assure that this routing can be used when the event occurs. It is also desirable to have a designated command post during the event.

There should be a specified location to store signs for use during annual festivals and parades and a file to provide instructions to all involved for the next yearly event. The file should include traffic signal timing changes and instructions on where to install signs.

Construction and maintenance work zones. Traffic management in construction and maintenance work zones is important to the safety of both workers and motorists. Traffic control is basically the same along freeway, highway, and city street work sites. Traffic management at work sites can be divided into that needed for long-term construction work zones and short-term maintenance work zones. The issuance of utility permits and the review and monitoring of traffic control plans for utility work should be included as part of work zone management.

It is important to provide good traffic control during construction and maintenance work. Time is required to develop and implement the traffic control properly. No one sequence of traffic control devices can be designed for all situations. An example of the development of a traffic control plan to route traffic during the off-peak period closure during maintenance along U.S. 101 in Los Angeles is given in Figures 12–1—12–4. Note that diversion routes were developed and signed with changeable message signs. A flow chart, which could be used in developing traffic control plans, is shown in Figure 12–5. Where alternate routes are not available or advisable, the different methods of handling traffic around construction and maintenance activities are given in Appendix A.

Static signs are usually adequate for directing traffic in advance of and within the work zone. However, changeable message signs and arrow boards provide additional target value and suitable messages that attract motorists' attention as they approach a work zone. Highway Advisory Radio (HAR) can also be used to provide more information, including route diversion. Newspaper articles and traffic broadcasts can also be employed to alert motorists to construction and maintenance work zone delays.

A discussion on traffic control for construction maintenance and utility work (with illustrations) is presented in the *Manual on Uniform Traffic Control Devices for Streets and Highways.*[12]

Construction work zones. Construction work is generally of a long-term nature. Highway department personnel develop Traffic Control Plans (TCP), which provide plan layout sheets and other information on the management of traffic during different stages of construction. Planning and close cooperation with the contractor is needed to assure the management plans are properly implemented in the field. Close cooperation is also required when changes are made during stage construction since these are the times of possible confusion and hazard to the motorist. It is often desirable on high-volume highways to advise motorists ahead of time of the planned changes in the TCP. If concrete traffic barriers (CTB) are installed (to separate traffic from

---

[12] Federal Highway Administration, *Manual on Uniform Traffic Control Devices for Streets and Highways* (Washington, DC: U.S. Department of Transportation, Federal Highway Administration, 1988), pp. 6.13-4–6.13-12.

Figure 12–1. Development of a traffic control plan for freeway construction with lane closure during off-peak period.
SOURCE: Federal Highway Administration, "Proceedings of Symposium on Work Zone Traffic Control, FHWA Report DOT-1-86-05 (Washington, DC: U.S. Department of Transportation, Federal Highway Administration), 1985.

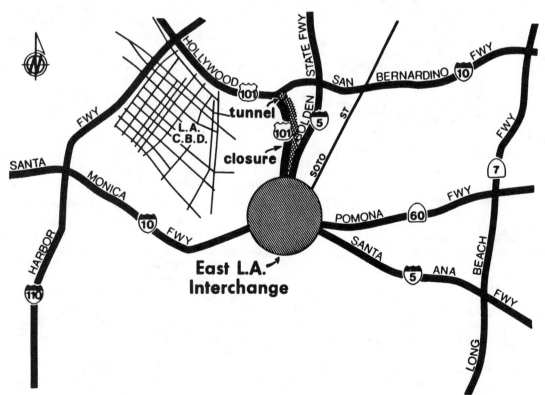

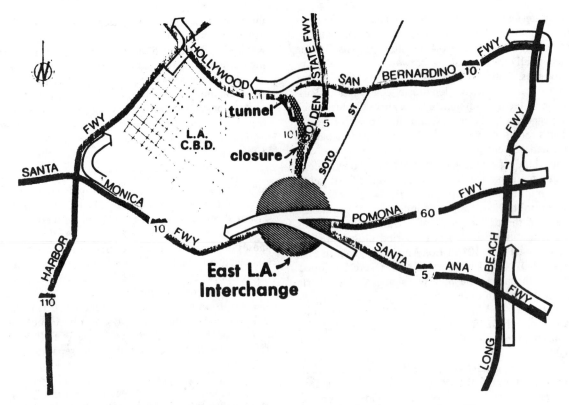

Figure 12–2. Development of a traffic control plan for freeway construction with lane closure during off-peak period.

SOURCE: FEDERAL HIGHWAY ADMINISTRATION, "Proceedings of Symposium on Work Zone Traffic Control, FHWA Report DOT-1-86-05 (Washington, DC: U.S. Department of Transportation, Federal Highway Administration), 1985.

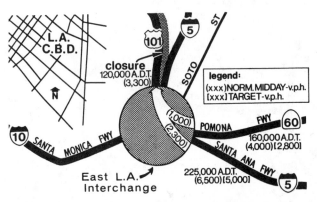

Figure 12–3. Development of a traffic control plan for freeway construction with lane closure during off-peak period.

SOURCE: FEDERAL HIGHWAY ADMINISTRATION, "Proceedings of Symposium on Work Zone Traffic Control, FHWA Report DOT-1-86-05 (Washington, DC: U.S. Department of Transportation, Federal Highway Administration), 1985.

the work area and to protect the construction workers), special signing and temporary delineation and/or route diversion may be needed when these CTB are moved and traffic is shifted. Changes in the TCP should be made during low-volume conditions.

The TCP should be designed to provide the same number of lanes of traffic during construction as before and designed for the same free-flow traffic speeds that existed before freeway construction began. In some cases it may be desirable to close an entrance ramp to achieve safe flow of traffic during construction.

Street reconstruction is complicated because of driveways for homes and businesses. Access must be maintained and at times special signs may be needed for the motorist to find a business driveway. Speeds are slower on streets than along freeways, but signing and delineation are equally as important as along freeways.

A street or highway may need to be closed to through traffic temporarily while changes are made to the TCP. This should be accomplished during low-volume conditions with business inconvenienced as little as possible.

Maintenance work zones. Maintenance work zones are generally of short-term duration—less than a week in length. The same principles apply for maintenance work zones as for construction projects. As with construction projects, care and thought must be given to ensuring worker protection while still providing suitable traffic movement.

Maintenance work should be carried out during off-peak hours. This is generally between 9:00 A.M. and 4:00 P.M. in urban areas. Some freeways and city streets have such heavy volumes, however, that maintenance work must be done on the weekend and/or at night. As many maintenance activities as possible should be conducted at one time so as to reduce the need for frequent interruption of traffic along the

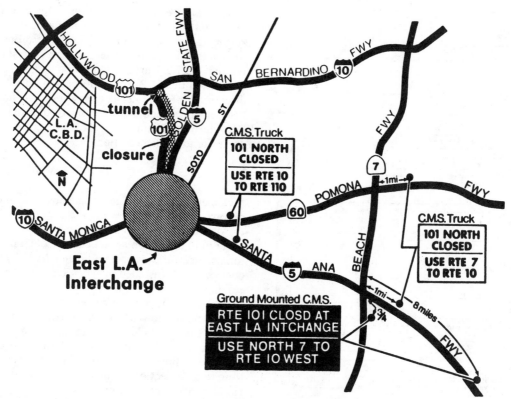

Figure 12–4. Development of a traffic control plan for freeway construction with lane closure during off-peak period.

SOURCE: FEDERAL HIGHWAY ADMINISTRATION, "Proceedings of Symposium on Work Zone Traffic Control, FHWA Report DOT-1-86-05 (Washington, DC: U.S. Department of Transportation, Federal Highway Administration), 1985.

facility due to maintenance. This requires close coordination with utility companies and the management of utility permits to perform work in the roadway.

Close cooperation is also needed between the maintenance and traffic engineering personnel to assure that the TCP is safe and practical to implement. A specially trained traffic handling team is beneficial for managing traffic on high-volume freeways.[13] The team has properly trained flagpersons to direct and control traffic. The team can also temporarily close and open entrance ramps (by using barricades) as a means of both controlling the flow and speed of traffic through the work zone and preventing unnecessary delays.

When ramps are closed, the team members assure that motorists are provided with designated alternate routes (utilizing signs) and that traffic signals are properly timed along the alternate route. Team members can also direct traffic to utilize the shoulder as an added lane. When this is practical, the motorist should be directed to use the shoulder through means of special signing and flagging. The team can also be of special assistance in traffic management as the maintenance work moves along the freeway and signs and channelizing devices are moved.

<hr />

[13] Steven Z. Levine, "Real Time Traffic Control of Urban Freeway Work Zone Operations, *Transportation Research Board Circular 344* (Washington, DC: Transportation Research Board, January 1989).

Traffic volume data and capacity information are needed to determine times during which maintenance can be carried out. A map should be prepared of the city freeways and streets to show when maintenance is permitted along each facility.

Utility maintenance work is similar to highway and street maintenance work. In order to assure prior knowledge of planned utility work, a permit needs to be obtained from a designated agency by the utility company before beginning work. This will give the traffic engineer time to coordinate with the utility company on time and location for the work and review the traffic control plan to use. A permit should be required both on and off the right-of-way anytime traffic is to be interrupted or a potential hazard is expected to exist (e.g., placing an elevated power line across the roadway).

For major maintenance and utility operations, changeable message sign(s) should be placed at the planned work site ahead of time to advise motorists of the dates and time for maintenance and whether ramps are subject to closure. This will enable motorists to make plans to alter their routes while the maintenance or utility work is underway. It is also desirable to publish information on maintenance and utility work zones in the newspapers and broadcast information on radio and television stations. A daily report on maintenance activities is of value to motorists. If it is published on the same page (or same section) of the

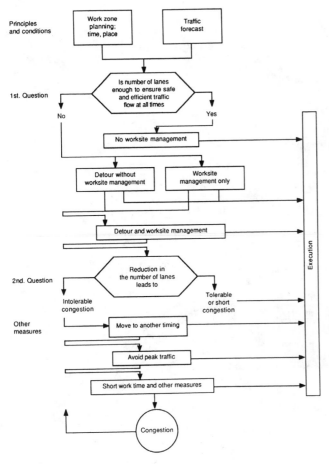

Principles and conditions

Work zone planning; time, place

Traffic forecast

1st. Question

Is number of lanes enough to ensure safe and efficient traffic flow at all times

No — Yes

No worksite management

Detour without worksite management

Worksite management only

Detour and worksite management

2nd. Question

Reduction in the number of lanes leads to

Intolerable congestion — Tolerable or short congestion

Other measures

Move to another timing

Avoid peak traffic

Short work time and other measures

Congestion

Execution

Figure 12–5. Outline of possible planning approach for development of work zone traffic control.

SOURCE: ROAD TRANSPORTATION RESEARCH, *Traffic Management and Safety at Highway Work Zones* (Paris: Organization for Economic Co-operation and Development), 1989.

newspaper each day or scheduled at the same time on the radio or television, motorists will be more apt to use the information.

Night work. There are times when construction work and maintenance work must be carried out at night. Night work presents unique conditions both to the workers and the motorists. Motorists do not expect to encounter night work. Special advanced signing may be needed to advise motorists of the work zone. Because visibility is not as good at night, special attention should be given to delineate a safe path through the work zone. Special care is also needed in placing illumination for the work zone so that it does not interfere with the motorist's vision. There is a higher percentage of impaired drivers at night requiring increased visual cues to offset slower reaction times.

Special attention must also be paid to protecting the worker as well as the motorist at night. It might be desirable to close the freeway or city street when maintenance work is carried out at night and divert the traffic along another route. Further, vehicles should be placed across the lanes in advance of the work area to assure that a motorist breaking through the barricades and/or signs at the beginning of the work zone will not encroach on the work zone.

Law enforcement participation. Police officers should be an integral part of traffic management within work zones. Their presence in itself is a controlling factor since their authority assures better motorist compliance when directing traffic during the initial implementation and changes to the TCP. Police officers are also invaluable in directing traffic at intersections along diversion routes. Police personnel should also be involved in the review of the TCP during preparation of the construction project plans and prior to implementation.

Inclement weather conditions

Inclement weather includes fog, snow, ice, and flood conditions.

Although fog is not common in most areas, it is of major concern in certain areas and can present a problem to motorists. Electric signs operated manually or by fog-detection devices can be installed in specific areas where fog is a recurring event.

Although there are unusual conditions such as major rainstorms or snowfalls that are difficult to plan for, preparations can be made to manage traffic during inclement weather conditions. When heavy rainfall is expected, for example, maintenance personnel can be prepared ahead of time to close sections of highway or city streets subject to flooding at the proper time and provide for predetermined alternate routes and appropriate changes in the timing of traffic signals. There is also a need to gain prior agreement from agencies responsible for work activities. Questions such as who is responsible for deciding when to close a section of freeway or city street, furnishing the barricades, signing alternate routes, and providing the signs and making traffic signal timing changes should be addressed beforehand.

Specific routes for snow clearance are usually predetermined and given priority based on a citywide plan. Parking restrictions should apply during historical periods of snow probabilities (i.e., December 1 to March 1). "No Parking" signs should be posted along the streets where snow is to be removed. Provisions also need to be made to remove snow for both pedestrians and buses (at bus stops).[14]

Shelters may need to be opened for motorists whose cars cannot make it to their destination during a snow or ice storm. Information must be provided via car radios as to the weather conditions and the locations of shelters. A traffic management center should be established and activated during unusual inclement weather conditions to assure that all activities are coordinated. The center can also provide for emergency services.

Making plans for anticipated inclement weather events will go a long way in assuring proper response when unusual situations occur. As with special events, plans need to be reviewed immediately after the plan is used and discussed at a meeting each year by the agencies to assure that everyone involved knows each agency's task and is prepared to carry it out.

---

[14] Institute of Transportation Engineers, "Recommended Practice for Traffic and Parking Control for Snow Emergencies," Pub. NORp-OI2A (Washington, DC: Institute of Transportation Engineers, 1983).

Disasters

Both natural disasters and those caused by humans do occur, and there is a need to be prepared for these unusual events. These calamities include earthquakes, tornados, hurricanes, and manufacturing plant explosions. Some types of transportation accidents (rail, truck) such as oil and chemical spills and gas leaks also have to be considered. Except for hurricanes, such events occur unexpectedly. Agencies within a hurricane area must have time to prepare for these storms and to evacuate residents from low-lying areas and to establish shelters.

Overall planning is needed by cities and communities on the responsibilities of each of the various agencies involved in the disaster. This includes decisions on highway and street routes to be designated for emergency vehicle use. Another determination is where bulldozers and other equipment will be available if needed and the location of operating personnel. Also, a list of agencies and firms specializing in hazardous waste control and cleanup must be readily available including information on the effects of hazardous material emissions and the need to evacuate people in the surrounding area.

The traffic management plan, including the location of the traffic management center, is an important part of the disaster plan. The same traffic management center utilized for daily traffic management and/or inclement weather conditions could be used for handling traffic during disasters. Because a plan may never be used, there is a need to keep agencies abreast of their responsibilities and provide training sessions periodically through mock events.

Traffic management strategies

Improving traffic operations will reduce delay, accidents, vehicle emissions, and fuel consumption. It is important to operate both streets and highways at acceptable levels of service. The following is a discussion of traffic management strategies and Transportation System Management (TSM) improvements available for improving operations along freeways and streets. Transportation System Management projects are those that make the best use of the existing facilities.

Freeways

Operation of a freeway under increasing traffic volumes is shown in Figure 12–6. As shown, an acceptable operation exists until the traffic volumes reach approximately 1,800 vehicles per hour per lane, and freeway occupancy reaches approximately 25% to 30% (capacity varies with the freeway). Beyond these values, congestion can occur. Congestion can also occur, however, at a lower freeway occupancy should an unforeseen event occur that causes the freeway occupancy to increase rapidly to 30% and beyond.

Congestion actually robs the capacity of the freeway when the capacity is needed the most. An example of this is shown by a traffic count (converted to an hourly rate) made along a three-lane section of freeway in Southern California.[15]

---

[15] David H. Roper, "Manage Traffic—Get Congestion Relief," unpublished paper (Los Angeles: California Department of Transportation, June 1986), p. 1.

Figure 12–6. Generalized traffic flow relationships.
SOURCE: Joseph M. McDermott and Others, "Chicago Area Expressway Surveillance and control: Final Report." Report No. FHWA-1L-ES-27 (Washington, DC: U.S. Department of Transportation, Federal Highway Administration, March 1979).

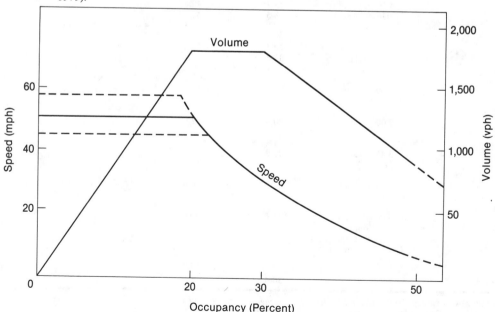

7:15 a.m. 1,767 vehicles per lane per hour Free-flow
7:30 a.m. 1,600 vehicles per lane per hour Slowing
7:45 a.m. 1,467 vehicles per lane per hour Slow-and-go
8:00 a.m. 1,233 vehicles per lane per hour Stop-and-go

The capability of the freeway to carry vehicles decreased from 1,767 vehicles per hour per lane to 1,233 vehicles per hour per lane—a loss of approximately 30% within 45 min.

Since congestion reduces capacity, providing means to maintain traffic volumes at a free-flow level would "increase" capacity. Traffic management strategies permit this to occur. Several traffic management strategies for freeway traffic surveillance and incident management along urban freeways are discussed in Chapter 13. These include:

1. Freeway surveillance to locate incidents, determine steps needed to be taken, and monitor the results of the steps taken. Freeway surveillance includes use of vehicle detectors, closed-circuit television, cellular phones, and CB monitoring.

2. Traffic control to assure a monitored input of traffic onto the freeway. Ramp meters and ramp gates serve to provide this control. As is shown in Table 12–2, ramp meter control has been found to increase main-lane speeds by an average of 9 miles per hour. Based on the information given in Table 12–3, the amount of congested freeway travel measured by minute-miles of travel (minutes of travel multiplied by miles of freeway) has been reduced an average of 68% through the installation of freeway ramp-control systems.[16] As shown in Table 12–4, meter control has also been found to reduce accidents by an average of 31%. Ramp meter control can either be of isolated pretimed or traffic-responsive control or interconnected as part of a freeway control and/or surveillance system.

3. Freeway-to-freeway control has been applied with success. The application of ramp control on freeway-to-freeway connections permits additional control where ramp meters are located along the freeway. Freeway-to-freeway control is best utilized as part of an interconnected traffic control system.

4. Motorist information systems include the utilization of changeable message signs, lane control signals, and highway advisory radio. Commercial radio stations provide traffic information during the peak periods. Although the benefits of motorist information are not easily measured, this information is an essential part of the overall freeway traffic management system. The information system advises motorists of congestion, incidents, and work zones and roadway conditions along with information on alternate routes. Lane-control signals advise the motorist in advance which lane(s) to use during maintenance operations and incidents.

---

[16] Federal Highway Administration, *Traffic Control Systems Handbook*, FHWA Report-1P-85-11 (Washington; DC: U.S. Department of Transportation, Federal Highway Administration, April 1985).

TABLE 12–2

Impacts of Freeway Ramp-Control Systems on Average Speed

Location (References)	Length (Miles)	Time of Day	Average Speed, mph				
			Before Ramp Control	After Ramp Control	Percent Improvement	After Including Ramp Delays	Percent Improvement
Minneapolis, I-35 W (44)							
Inbound	16.6	7:15–8:15 am	33.8	45.5	34	43.0	27
		6:30–9:00 am	43.9	50.1	14	48.5	10
Outbound	12.7	4:30–5:30 pm	33.7	40.1	19	38.6	15
		3:30–6:30 pm	38.5	45.7	19	44.4	15
Chicago, Eisenhower Expressway (45)							
Inbound	9.4	2 hr. am Peak	30.3	33.0	9		
		4 hr. am Peak	37.7	39.7	5		
Los Angeles, Santa Monica Freeway (46)							
Inbound	13.5	6:30–9:30 am	36.2	50.6	40	41.4	14
Houston, Gulf Freeway (47)							
Inbound	6	7:00–8:00 am	20.4	32.6	60		
Los Angeles, Harbor Freeway (48)							
Inbound	4	3:45–6:15 pm	25.9	40.3	55	37.4	44
Detroit, Lodge Freeway (49)							
Inbound	6	2:30–6:30 pm	27.3	36.4	33	32.6	19
Toronto, Queen Elizabeth Way (50)							
Inbound	3.9	7:00–9:00 am (Good Conditions)	21.1	30.9	45	26	21
	13.4	7:00–9:00 am (Poor Conditions)	13.4	21.4	59	16.7	24
Average, All Data			30.2	38.9	29		
Averge, Including Ramp Delay			30.4			36.5	20

SOURCE: Frederick A. Wagner, *Traffic Control System Improvements: Impacts and Costs,* Report No. FHWA-PL-80-005 (Washington, DC: U.S. Department of Transportation, Federal Highway Administration, March 1980).

TABLE 12–3

Summary of Reported Benefits of Entrance Ramp Metering Systems

Location	Length (Mi)	Time Period (Hr)	Travel Time Improvement (%)	Congested Freeway Travel, Minute-Mile, Reduction (%)
Minneapolis (Minnesota) I-35W				
Inbound	16.6	7:15–8:15 a.m.	34	95
		6:30–9:30 a.m.	14	100
Outbound	12.7	4:30–5:30 p.m.	19	59
		3:30–6:30 p.m.	19	92
Chicago (Illinois) Eisenhower Expressway				
Inbound	9.4	2-hr a.m. peak	9	22
		4-hr a.m. peak	5	23
Los Angeles (California) Santa Monica Freeway				
Inbound	13.5	6:30–9:30 a.m.	40	100
Houston (Texas) Gulf Freeway				
Inbound	6	7:00–8:00 a.m.	60	66
Los Angeles (California) Harbor Freeway				
Outbound	4	3:45–6:15 p.m.	55	80
Inbound				
Detroit (Michigan) Lodge Freeway				
Inbound	6	2:30–6:30 p.m.	33	89
Toronto (Canada) Queen Elizabeth Way (QEW)				
Inbound	3.9	7:00–9:00 a.m. (good conditions)	45	45
		7:00–9:00 a.m. (poor conditions)	59	40
Average				65

SOURCE: FEDERAL HIGHWAY ADMINISTRATION, *Traffic Control Systems Handbook,* FHWA Report-1P-85-11 (Washington, DC: U.S. Department of Transportation, Federal Highway Administration, April 1985).

TABLE 12–4

Safety Impact of Freeway Traffic Management Projects

	Reduction in Peak Period Freeway Traffic Accidents (%)
Dallas, North Central Expressway	18
Houston, Gulf Freeway	27
Chicago expressways	35
Minneapolis, I-35W	
a.m. peak	58
p.m. peak	18
Los Angeles, 7 freeway projects	30
San Jose, 29 metered ramps	30
San Diego, I-8 Freeway	30
Average	31

SOURCE: FREDERICK A. WAGNER, *Traffic Control System Improvements: Impacts and Costs,* Report No. FHWA-PL-80-005 (Washington, DC: U.S. Department of Transportation, Federal Highway Administration, March 1980).

5. An analysis of the existing freeway guide signs can result in the installation of improved signing. New concepts such as guide sign spreading, and use of sequence signs, which are shown in Figure 12–7, provide for improved motorist information.[17] Sequence signs name the next two or three exits along the freeway. This allows the motorist to begin changing lanes well in advance of the desired exit ramp. Guide-sign spreading reduces the amount of information at freeway exits

[17] Federal Highway Administration, *Manual on Uniform Traffic Control Devices.*

by showing only the exit information at the exit gore and providing an additional sign for the following exit immediately downstream of the exit. Improved signing reduces motorist confusion, permits smoother traffic flow, and reduces the chances of accidents. In addition, it may increase the capacity of a section of freeway.

6. It is essential to integrate the surveillance, traffic control, and motorist information into one traffic management system so as to provide for optimum operation along the freeway. This can be accomplished through the development of a traffic management center. The center provides one location where traffic management can be carried out for all freeways having surveillance, traffic control, and motorist information within a metropolitan area.

The traffic management center has one or more operators who monitor the operation along the freeway(s) and take steps to dispatch services and implement messages on the changeable message signs when an incident is located. Closed-circuit television permits the operator to verify that an incident exists and the type of assistance that is needed. The traffic management center also provides a location to analyze the freeway operations and to make changes if necessary. The data base within the computer at the center provides a vast amount of information that can be analyzed. This data base can be used to do an analysis of peak-period traffic control patterns, and develop and test new traffic control patterns for implementation. The traffic management center also provides a location for all agencies to meet to coordinate their traffic management activities. It is desirable

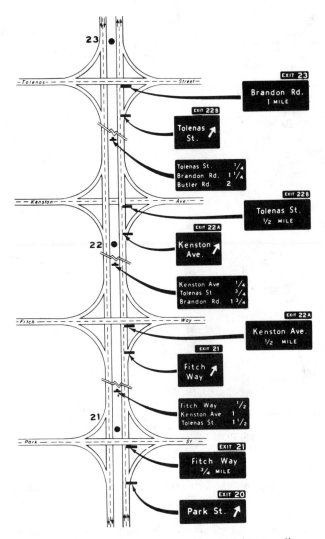

Figure 12–7. Interchange sequence signs and sign spreading.
SOURCE: FEDERAL HIGHWAY ADMINISTRATION, *Manual on Uniform Traffic Control Devices for Streets and Highways* (Washington, DC: U.S. Department of Transportation, Federal Highway Administration, 1988 edition).

to have one joint traffic management center for city, state, and public transportation (where applicable) agencies within a metropolitan area. An alternate would be to have separate central locations with sufficient communications to permit all agencies to work together in providing effective traffic management during periods of congestion. In addition, communications must be provided to the personnel and agencies (tow-truck operators, fire department, law enforcement) that will carry out traffic management in the field. Also, a knowledgeable police officer should be stationed at the traffic management center to be in contact with the police officers in the field involved in carrying out management of the incident.

7. Temporary utilization of a shoulder as an added lane is especially beneficial in providing added capacity past a freeway bottleneck location. For instance—a freeway with three 12-ft lanes could be marked to provide four 10.5 ft lanes by utilizing 6 ft of the left shoulder area. Also, the weaving capacity between an entrance ramp and an exit ramp could

be increased by marking the right shoulder as an auxiliary lane. These modifications should be considered temporary and new construction scheduled to provide the additional lane on a permanent basis and restore the shoulder as soon as possible.

8. Planning is needed to properly provide for traffic management during incidents. Alternate routes should be developed and improvements made to these routes as needed. The California Department of Transportation (Caltrans) prepares incident-management instructions and maps with designated routes from freeway interchanges for use when an incident occurs that creates congestion for at least 2 hours. Vehicles equipped with adequate signs and channelizing devices are designated for traffic-management use. Once such an incident occurs, a command center is established, the alternate route is implemented, and warning signs and channelizing devices are installed. The detour route is checked out by a team member prior to beginning the rerouting to assure that it is still usable before signs are placed and traffic routed along the detour.

9. Studies have shown that approximately 95% of freeway incidents are due to such items as vehicle breakdowns, flat tires, and vehicles running out of gas. The remaining 5% are caused by accidents. Since incidents cause approximately 60% of freeway congestion, it is important to provide motorist aid (service) patrols to get stranded motorists underway as soon as possible.

A study by the Harris County Sheriff's Department[18] (Houston, Texas) showed that a 17:1 benefit to cost ratio has been obtained from its service patrol during peak hours.

10. The Accident Investigation Site (AIS) is an effective low-cost traffic management improvement for reducing congestion and improving safety along the freeway main lanes. The AIS is a designated location out of sight of the freeway where vehicles involved in accidents can be driven or moved by a tow truck. The AIS can be located beneath an overpass structure where the freeway passes over the cross street, along a cross street, or along the frontage road (service road). The AIS can either be similar in design to a bus bay or designated with signs and pavement markings. Blue-and-white-colored information signs are located along the freeway lanes directing motorists to the AIS. Telephones can be provided at the AIS to summon a police officer. The AIS provides a safe location for the motorists to exchange information and for the police officer to conduct the investigation. Moving vehicles to the AIS either voluntarily or by tow truck before the police officer's investigation is made reduces the time that congestion exists on the freeway. A study in Texas showed that AIS provided a 28:1 cost-benefit ratio with a potential ratio of 35:1.[19]

11. High-occupancy vehicle (HOV) provisions should be implemented. These provisions, which include designated freeway lanes and ramp bypass lanes at metered ramps, give priority treatment to motorists using public transportation,

[18] Johnny Klevenhagen, "Motorist Assistance Program: Program Performance Report—Appendix B: Cost-Effectiveness of Harris County Sheriff's Department Motorist Assistance Program During Peak Hours" (Houston: Harris County Sheriff's Department, January 1987).

[19] Conrad Dudek and others, "Promotional Issues Related to Accident Investigation Sites," Transportation Research Board Record 1173, *Urban Freeway Operations* (Washington, DC: Highway Research Board, 1988).

van pools, and car pools. Designated HOV lanes include concurrent flow lanes along the left shoulder or left lane in the same direction of travel as the main lanes, contraflow lanes along the left shoulder or lane in the opposite direction of the main lane of travel, and dedicated lanes within the median area. The HOV lanes can either be the full length of the freeway or only around major bottleneck areas along the freeway. The concurrent and contraflow lanes are generally used only during the peak periods and are separated from the non-HOV vehicle lanes by pylons.

City streets

A city street has unique features and capacity characteristics. As is shown in Table 12–5 and Figure 12–8, little is gained in the capacity of a signalized intersection at cycle lengths above 90 sec. Also, delay generally increases significantly when the critical intersection lane volumes per hour of green exceed volumes between 1,200 and 1,500 vehicles per hour of green.

TABLE 12–5

Capacity Values

Cycle Length (Sec)	Two-Phase Operation		Four-Phase Operation	
	N_c	Percent Increase	N_c	Percent Increase
40	1318.0	–	936.0	–
50	1454.4	10.3	1108.8	18.5
60	1512.0	4.0	1224.0	10.4
70	1553.1	2.7	1306.3	6.7
80	1584.0	1.9	1368.0	4.7
90	1608.0	1.5	1416.0	3.5
100	1627.2	1.1	1454.4	2.7
110	1642.9	0.9	1485.8	2.1
120	1656.0	0.8	1512.0	1.7
130	1667.1	0.7	1534.2	1.5
140	1676.6	0.6	1553.1	1.2
150	1684.8	0.5	1569.6	1.1

SOURCE: P.H. TARNOFF and O. PARSONS, "Selecting Traffic Signal Control at Individual Intersections," *National Cooperative Highway Research Program Report 233* (Washington D.C.: Transportation Research Board, 1981).

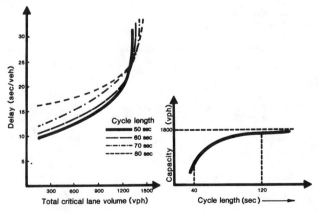

Figure 12–8. Cycle length relationship to capacity and delay.

SOURCE: FEDERAL HIGHWAY ADMINISTRATION, *Traffic Control Systems Handbook*, FHWA Report 1p-85-11 (Washington DC: U.S. Department of Transportation, Federal Highway Administration, April 1985).

As with the freeway, there is a need to improve the capacity and Level of Service of the street system when congestion exists within a corridor. These include:

o Traffic signal retiming.
o Interconnection and/or coordination of traffic signals.
o Application of protected/permissive left-turn operation at signalized intersections with separate left-turn movements.
o Utilization of leading/lagging left turns as well as dual left turns.
o Elimination of protected left-turn phases where they are not needed.
o Relocation or removal of traffic signals to improve progression.
o Installation of a traffic-responsive traffic signal system.
o Installation of closed-circuit television at selected intersections where major streets cross.
o Construction of a traffic management center for operating the traffic responsive traffic signal system(s). The traffic management center also provides a location for dispatching maintenance vehicles to repair traffic signal malfunctions located by the operator at the center and to dispatch emergency vehicles to an accident or other incident reported along a city street. The traffic management center could beneficially be located at the same place as the freeway traffic management center.
o Provisions for channelized left-turn lanes at major intersections.
o Changing four lanes plus parking on each side to five lanes (center lane for left turns) with no parking.
o Use of cellular phones and/or CB monitoring to assist in locating accidents.
o Parking restrictions during peak hours and restriping to provide four lanes in place of two lanes with parking.
o Utilization of a two-way left-turn lane as a reversible lane during peak hours and as a left-turn lane during off-peak hours.
o Construction of an added curbside lane in advance of and beyond a signalized intersection. Such a lane, which could be approximately 300 ft each side of the intersection, would provide for right turn and straight through traffic. The straight through traffic in the right lane would merge beyond the intersection with the adjacent lane. The added curb lane could be installed if there is sufficient right-of-way for the lane and for a sidewalk.
o Provisions for HOV lanes along city streets, installation of traffic signal preemption for buses, and/or an advance green interval for buses at signalized intersection.
o Construction of bus bays. A far-side or midblock bus bay is preferable to a near-side bay.
o Development and implementation of access control along major streets and highways.
o Designation and construction of loading zone bays.
o Implementation of a one-way network of streets. Construction of a mini-freeway utilizing a major street or expressway. This involves the installation of a median or concrete traffic barrier between opposing flows of traffic and grade separations at major street locations. Signalized cross streets located between grade separated

major streets need to provide only two phase operation with no left turns permitted from the major street. This design is discussed more fully in the section in this chapter involving access control.

The improvements gained by some of these applications are shown in Table 12–6. A study of a 180-intersection installation of traffic management improvements including a traffic-responsive computer-controlled traffic signal system in Los Angeles provided the following results:[20]

---

[20] Mobility 2000, *Proceedings of a Workshop on Intelligent Vehicle/Highway Systems,* San Antonio, TX. (Hosted by Texas State Department of Highways and Public Transportation and Texas Transportation Institute, February 1989), pp. 23–24.

TABLE 12–6

Evaluation of City Street Improvements

| TEM Measures | Application | Quantitative Impacts | | Energy Consumption | Emissions |
		Vehicle Delay	Average Speed		
Traffic operations and geometrics					
Widening	Intersection Arterial	40% reduction on arterials	30% increase on arterials	0–1% reduction areawide	Decrease
One-way network	Network	30% reduction	18–25% increase	1–4% reduction areawide	Decrease
Turning lanes	Intersection Arterial	—	16 km/h increase	—	—
Turning and lane-use restrictions	Intersection Arterial	16% reduction	10% increase	—	Decrease
Reversible lanes	Arterial	6% increase (high flow in off-peak direction)	Approximately 4–7% increase	3% reduction	7% decrease
Bus bays	Network	7% reduction	4–5% increase	—	—
Traffic control					
Replace signs and signals	Intersection	—	—	0–2% reduction areawide	Decrease
Remove or replace crosswalks	Arterial	—	—	—	—
Improve signal timing	Intersection Arterial Network	—	6–12% increase	1–4% reduction areawide	Decrease
Coordination of signals	Arterial Network	—	1.5–25% increase	Up to 6% reduction	Decrease
Traffic responsive operations	Intersection	30–40% reduction	22–25% increase	10–30% reduction	Decrease
Flashing of signals	Intersection	—	—	Reduction	Decrease
Parking management					
Curb restrictions	Arterial Network	1% reduction (a.m. peak)	25% increase (no change in a.m. peak)	—	—
Truck loading	Arterial Network	—	1–2% change	0–1% reduction	—
Other tactics					
Railway crossings	—	—	—	—	—
Ramp metering	Freeway	—	2–100% increase on freeway, 0.5% increase areawide	0–1% reduction areawide	HC, CO decrease NO_x increase
Freeway control	Freeway	—	Increase	Reduction	—
Idling control	Arterial Network	—	—	—	—
Improve street signing	Arterial Network	—	—	—	—
Improve route identification	Arterial Network	—	—	—	—
Lighting intensity of signals	Intersection	—	—	—	—

SOURCE: Transportation Energy Management Program, *Traffic Management Measures to Reduce Energy Consumption* (Toronto, Ontario: Ministry of Transportation and Communications, November 1981).

- 20% reduction in delays.
- 35% reduction in stops.
- 10% reduction in emissions.
- 12% reduction in fuel consumption.
- 13% reduction in travel time.
- Estimated 16:1 cost-benefit ratio.

As can be seen, significant improvements can be obtained by implementing traffic management strategies along city streets.

Freeway corridor

Each of the strategies listed and discussed for freeways and city streets will reduce congestion. Where a freeway corridor is involved, a combination of the strategies along the freeway main lanes and city streets will improve the throughput of traffic within the transportation corridor. The city street improvements together with freeway entrance ramp control will encourage motorists making shorter trips to use the street systems rather than the freeway. Improvements along the city streets will also ensure that an alternate route is available during ramp metering and detouring of traffic during incidents, construction/maintenance activities, and special events. Similarly, improvements along the freeway will provide additional capacity for diverting traffic when an incident occurs along a city street. Additional information on freeway corridor control and operation is given in Chapter 13.

The city and the state should work closely to coordinate improvements within freeway corridors. If city and state traffic management centers are separated, close communications is needed between the two centers to assure that the freeway and city streets within the corridor operate as one unit during peak periods, incidents, special events, and construction/maintenance activities.

Coordination of traffic management strategies

Traffic management needs to be carried out through a coordinated effort of individuals involved in traffic operations and law enforcement. Traffic management in urban areas should be carried out both on a corridor basis and on a metropolitan or areawide basis. This requires coordination among many city, county, and state agencies. Traffic management in smaller cities is usually carried out on a network basis and generally involves the city engineer, police, emergency medical service, and fire department personnel, county engineer, sheriff, and state maintenance personnel. In rural areas, traffic management may involve only one or more highways. This generally involves coordination between the highway personnel (traffic, maintenance, construction) and law enforcement people (highway patrol, sheriff).

Traffic delay not only affects motorists going to and from work but also those individuals involved in business, service, and goods movement in larger cities. These businesses use the freeway and city street system during both the peak and the off-peak periods of the day. Delays to commercial vehicles and those involved in business activities increase the costs of the products being moved and services being provided. The faster an incident is taken care of, the less the cost to both the motorist and citizens of the community. Because of business delay costs, it is also desirable in larger cities to coordinate with the Chamber of Commerce and an organization representing trucking firms within the city.

Traffic management also needs to include communication and coordination among city, county, state, and transportation agencies involved in traffic operations and law enforcement in order to assure that optimum traffic operation is provided within the metropolitan area. A major accident along a freeway involves the law enforcement, highway department, and tow truck service personnel. At times it can also include personnel from the fire department and/or emergency medical service. If it is necessary to reroute traffic to adjacent streets, communication, cooperation, and coordination are needed among city and state traffic engineering and police personnel to reroute the traffic effectively. Development of bus routes, park-and-ride lots, and bus schedules generally involves the city, state, and public transportation personnel. Moving goods to and from a harbor or distribution point by truck and railroad can involve a number of organizations involved in shipment of goods as well as the city and state traffic engineering personnel. In order to be effective, these and other types of transportation considerations require teamwork as an organized and ongoing activity.

Traffic management triad

The Traffic Management Triad (Triad) provides a means for organizing a stable and ongoing approach to traffic management. The Triad consists of three components:

- Traffic Management Team (TMT)
- Active Traffic Management
- Design Improvements

The Triad is shown in Figure 12–9. Each one of these components is bound together through communication, cooperation, and coordination, and will be discussed in later sections.

The keystone of the Triad is the Traffic Management Team. This is an organization of personnel from agencies and transportation organizations involved in traffic operations and law enforcement. These personnel meet on a regular basis to study and discuss problems and formulate plans for improving traffic operations within the urban areas. Active traffic management involves city, county, and state engineering and law enforcement personnel working together in managing and directing traffic on a real-time basis. These personnel are from member agencies of the Traffic Management Team. Design improvements include surveillance, control, and motorist information systems and the street and highway improvements needed to eliminate bottlenecks and accident conditions. The interconnection or bonding of these three components through communications, cooperation, and coordination can provide a successful approach to managing

TRAFFIC MANAGEMENT TRIAD

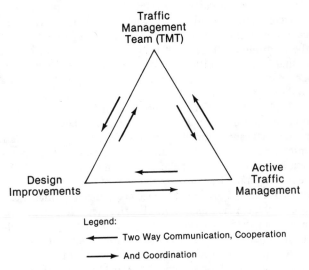

Legend:
◄──── Two Way Communication, Cooperation
────► And Coordination

Figure 12–9. Traffic management triad.
SOURCE: HERMAN E. HAENEL, "The Traffic Management Triad," unpublished paper (Austin: Texas State Department of Highways and Public Transportation, November 1989).

geometrics, traffic signal retiming, traffic signal interconnection, utility work and construction work. Larger cities have the same conditions but to a larger scale and include problems along freeways as well as city streets.

What is unique about the Triad is its simplicity, interlocking activities, and Transportation System Management applications. The personnel involved have common interests that will benefit both their parent organization and the other agencies.

The following sections discuss the activities and benefits of the three components of the Triad.

Traffic management team. The primary purpose of the Traffic Management Team is to permit engineering and law enforcement personnel involved in traffic operations and enforcement from city, county, and state agencies to meet and work together with personnel from public transportation and other transportation organizations in improving the safety and operation of the freeways, city streets, and/or rural roads.

The Traffic Management Team (TMT) can take many forms depending upon the size of the city or metropolitan area, number of governmental agencies (city and/or county) within the area, organization of the governmental agencies, and transportation system. The team can be utilized in a small city of 20,000 population, a large city (100,000 and above), or two or more adjacent cities. TMT's have been established in Texas and Florida. Examples of TMT organization in Texas are shown in Table 12–7.

The TMT provides many opportunities for achieving improved traffic operations and safety. These include:

traffic. With increasing traffic volumes and congestion mentioned earlier, there is a need to develop and implement the Triad in all cities and metropolitan areas. The development of automated freeways and highways and motorist information and guidance systems within vehicles will require even closer city/county/state cooperation in the future if these automated systems are to be successful.

The Triad applies to both small cities and large ones. Smaller cities have festivals and other special events, work zones, and suffer from inclement weather, accident locations, and bottlenecks. These situations require organization for handling traffic and design improvements such as

1. Coordination with planning agencies (where applicable) in developing a program for near-term and long-range improvements.
2. Coordination among city, county, and state traffic engineering agencies to assure that improvements affecting

TABLE 12–7

Disciplines Represented in Selected TMT's in Texas.
Sample Based on Population Size: 1986 Federal Census

Agency	Kerrville (19,890)	Tyler (75,440)	Laredo (117,060)	Midland and Odessa (199,270)	Corpus Christi (263,900)	Fort Worth and Arlington (679,320)	San Antonio (914,350)	Houston (1,698,200)
City								
Traffic	X	X	X	X X	X	X X	X	X
Police	X	X	X	X X	X	X X	X	X
Public works	X		X		X		X	
Fire			X					X
Transit			X			X	X	X
State								
Traffic	X	X	X	X	X	X	X	X
Design	X		X			X		X
Maintenance	X		X		X	X		
Highway patrol	X	X	X	X	X	X		X
County								
Engineer	X	X		X X	X	X	X	X
Sheriff	X	X	X				X	X
Other								
Military base					X			
Airport								X
Railroad assoc.								X

SOURCE: HERMAN E. HAENEL, "The Traffic Management Triad," unpublished paper (Austin: Texas State Department of Highways and Public Transportation, November 1989).

two or more agencies are developed and installed in an organized manner. As an example, improvements along the freeway should be coordinated with improvements along the adjacent street system.

3. Consideration for the needs of public transportation and the effects public transportation will have on traffic.
4. Study of high-accident locations and agreement on improvements and determination on where funds can be obtained to make these improvements.
5. Review of traffic control plans through work areas along freeways and city streets.
6. Review of proposed improvements from different points of view (for example, police officers, traffic engineers, transit operators).
7. Cooperation in planning for traffic management during accidents, special events, and unusual weather conditions (ice storms, snow, hurricane, etc.) and determination how traffic directed from one facility to another during these events will affect traffic within the corridor or network.
8. Cooperation in carrying out the daily operations of the traffic management center(s) for city, county, state, and public transportation organizations.

Cities with populations below 100,000 population are generally not involved with all of the team functions listed, but they have enough activities to make the TMT worthwhile with periodic meetings. Similarly, personnel from several adjacent or closely spaced smaller cities can work with the county and state personnel through a TMT in planning for special events, accidents, and inclement weather conditions. The same applies to a smaller town working with the state and county in providing motorist assistance within the city and rural area adjacent to the city.

The TMT should be kept as small as possible. It should include representation from the principal agencies and transportation organizations. This could include the local railroad and/or trucking association(s), port authority, military base, and public transportation. Smaller cities adjacent to a large city (over 100,000 population) and transportation planning agencies are generally not included as permanent members on the team. There is usually not enough interest on a meeting-by-meeting basis to have the smaller city and planning agency serve as a permanent member. However, the smaller city and planning organization representatives are invited to those meetings where topics that involve them are discussed.

The TMT members are not the top echelon (e.g., mayor, city manager, district engineer) but one or two ranks below (traffic engineer, city engineer, assistant city manager, police captain in charge of traffic enforcement). The team members should be able to speak for their organizations and commit personnel and/or other resources.

A number of factors contribute to making the team a success.

o Team members always attend the meetings and do not send substitutes or alternates.
o Meetings are generally held every month or every other month. However, some of the teams in smaller cities meet only every 3 months unless there is a special need to meet in-between these scheduled meetings. The

meetings are held during the same day of the month and hour of the day so that each member can schedule the meetings for the entire year.
o The chair assures that decisions are carried out. Each team member must also follow through with his or her commitment.
o The chair must also make the team meetings varied and interesting. This can be achieved by contacting team members ahead of the meeting and assuring that various types of problems are included in the agenda.

It has been found that making decisions on a consensus basis rather than on a formal vote basis is preferable.

Teams have enabled the city, county, and state personnel to work closely together in daily activities as well as in active traffic management and improvement programs.

Active traffic management. The second component of the Traffic Management Triad consists of personnel who carry out the traffic management activities on a daily basis and are the heart of the Triad. Included are personnel from the police, fire, and emergency medical service departments, motorist aid patrols, traffic handling organizations, and the traffic management center. Decisions made by the Traffic Management Team must be communicated to these active (real time) traffic management personnel. In this way, cooperation and coordination can be maintained. By the same token, problems in traffic handling and recommendations for improving traffic operations in the field need to be reported back to the Traffic Management Team. The Traffic Management Team members can then obtain and provide the support needed to assist the field personnel and improve the management of traffic on a real-time basis.

Design improvements. The results of traffic studies and recommend improvements by one agency that could affect another agency should be reported at the TMT meeting. The second agency may be planning its own improvements. These improvements can be coordinated and properly scheduled. If the results of the study show that it would be desirable for two or more agencies to make improvements, the study conclusions need to be discussed and acted upon by the TMT members.

The personnel involved in active traffic management will need to have suitable operating systems and geometric design features to assure proper operation. These include:

o Proper traffic control devices (freeway control, traffic signals, stop signs). Traffic signals need to be properly timed and freeway entrance ramp control equipment operated properly.
o Equipment for incident detection and management (loop detectors, television, cellular phone and CB monitoring, call boxes, etc.).
o Motorist information equipment (changeable message signs, highway advisory radio, lane control signals).
o Proper channelization, access control, and parking controls along streets and improved geometrics along freeways.
o Traffic management center located in larger cities for real-time analysis of congestion problems and

approaches for improved operation. The personnel at the center serve as part of the team providing active traffic management. If a city traffic signal control center exists or is planned, this facility could serve as the basis for a traffic management center.

Traffic management can be carried out to a certain degree without the appropriate design improvements, but it is not as easy to do or as effective. Needed improvements in geometrics and traffic control equipment for improved traffic management should be discussed at the Traffic Management Team meeting and steps taken by the appropriate agency to improve traffic operations. Communications, cooperation, and coordination are needed among the various agencies to provide for and/or properly utilize the physical design improvements on a real-time basis.

The Triad is a good approach to apply in analyzing problems, providing proper geometrics and traffic controls, organizing for emergencies, and carrying out the work in the field.

Transportation planning and construction coordination

It is essential to include transportation planning and arterial reconstruction in development of traffic management strategies and plans. Anticipated increase in delay over the next 20 years requires the development of a planning program that determines: (1) where and when growth will take place within urban areas and (2) what construction and improved traffic control systems are required during this period. Commercial and residential development occurs to a great extent as a result of arterial improvements. Zoning, land development restrictions, and environmental strictures also have a significant effect on traffic growth within an urban area. By applying these planning restrictions, it is possible to help develop the total urban roadway network. The combined utilization of available arterial capacity and management of traffic demand (amount of traffic desiring to utilize a freeway or principal arterial) within an area provides feedback for achieving a balance between traffic demand and improvements.

Planning techniques should include a specific network of traffic corridors. Each corridor includes parallel streets and/or freeway(s). Very often, one facility cannot handle all of the demand during peak periods, but a corridor of streets or a combination of freeway(s) and streets can handle the traffic. Since several alternate routes are available within a corridor, it is possible to better manage traffic during recurring congestion, incidents, roadway construction and maintenance work, and special events.

The study of what can be carried out to achieve the optimum operation of a corridor over its design life at the lowest cost includes an analysis of the reconstruction of freeways and streets, development of mini-freeways, and application of Transportation Systems Management (TSM) projects.

In addition, vehicle guidance and other advanced technology need to be incorporated into planning of a system for a 10- to 20-year period. Traffic and planning engineers need to work together in planning future facilities. Traffic engineers provide much of the knowledge about benefits obtained from electronic traffic control and motorist information systems needed to provide for the development of freeway corridors. Traffic engineers can also explain their traffic management needs to the planning engineers to assure that planned improvements will fit traffic management needs. As with traffic management, planning should involve joint city, county, and state cooperation. Certain improvements can be made by one agency (i.e., city) while the remainder will need to be implemented as part of an agreed plan by another agency (i.e., state).

The installation of all or a portion of a traffic management system as part of construction will reduce costs. For example, the installation of vehicle detector loops should be planned for and incorporated as part of a roadway construction project for a freeway or city street. This will reduce cost of installation and future vehicle delay since traffic will not need to be rerouted if the loops are installed during initial roadway reconstruction. Installing the loops is cost-effective since their installation will permit traffic counts to be made using portable traffic counters until a full data collection and surveillance system is installed. The traffic counts can be used for determining the time of day to carry out maintenance operations, when additional traffic control (e.g., ramp meter control) needs to be installed, and when urban traffic patterns change within an area.

Another example is the installation of conduit. This again is cheaper if planned early and installed as part of a construction project. Where the freeway is to be widened and there is limited right-of-way and/or competing utilities, it may be essential to install the conduit to assure that it can be installed or that utility lines will not be damaged at a later date.

When the conduit and vehicle loops are put into place, it will be relatively easy and inexpensive to install detector lead-in wire to the nearest freeway interchange. It will then be possible to make a ground line connection from the detectors to a computer at a central location or to utilize telephone lease lines until a permanent ground line connection can be installed to a traffic management center.

Freeway operations can be improved in the future if the shoulders have the same base and pavement surface as do the adjacent main lanes. This will permit the temporary use of one shoulder for the application of narrow lanes past bottleneck locations, the installation of HOV lanes along the left shoulders during peak periods, and the utilization of the right shoulder as an auxiliary lane between entrance and exit ramps. It will also be worthwhile to construct one lane freeway-to-freeway connections wide enough to permit two narrow lanes at a later date because of changes in traffic patterns. The construction of a one-way reversible or a two-way HOV median lane as part of a freeway reconstruction project along a high-volume freeway can significantly increase the person capacity of the freeway.

The inclusion of corridors in urban network planning for capacity and improved traffic operations and the use of TSM projects are vital to continued mobility. Likewise, incorporating traffic management improvements in freeway and street construction projects—coupled with the capability of future expansion—is needed for the application of traffic management as traffic volumes increase and traffic patterns change.

Access management

Access management provides for the safe movement of traffic along arterials, collectors, and local streets as part of the arterial street and highway plan for the metropolitan area. Access management should include the development and implementation of state and/or city codes. These codes protect the rights of property owners while assuring the best optimum operation of the roadway. The State of Colorado, for example, has developed *The State Highway Access Code,* which provides for "uniform procedures and standards to guide the public, local governments and Department of Highways in the administration of permitting access approaches."[21] Other states and cities also have access codes.

There are at least three reasons why access management is needed. First, there is a need to provide an organized movement of traffic within an urban area. The requirement for safe and efficient movement of traffic at reasonable speeds can be in conflict with access needs. An urban arterial plan, such as that shown in Figure 12–10,

- permits the orderly location of traffic controls;
- provides for the movement of through traffic along designated streets;
- provides residential property owners protection from their local street becoming a thoroughfare;
- assures that all property owners (and potential property owners) know what their access rights are; and
- creates a defined approach to development and improvement of the total street system.

The functional street system concept shown in Figure 12–11 provides different combinations of movement and access for each classification of street shown in Figure 12–10. As depicted, primary and secondary arterials are dedicated to through-traffic movement whereas local streets provide access to abutting property. Collector streets are designed to serve local street traffic. Local and collector streets are not designed or intended to move through traffic. The development and enforcement of the street system shown in Figure 12–10 necessitates the adoption of a zoning plan, subdivision requirements, and geometric and pavement design standards, as well as the access codes.

The second reason why access management is needed is that research has shown that access control provides acceptable capacity and safety. It has been found that the street capacity in one direction along a two-way, four-lane arterial with a 45-mph speed limit is reduced 1% for every 2% per mile of traffic that turns into and out of a driveway.[22] For example, if the arterial carries 1,500 vehicles per hour in one direction, with 150 vehicles turning into driveways along a 1-mile section and 150 turning out of driveways (20% turns), the one-way capacity will be reduced by 10%. It has also been reported that each commercial driveway adds between 0.1 and 0.5 accidents per year along the street. The effects of access points and Average Daily Traffic (ADT) volumes on accident rate are shown in Figure 12–12. Figure 12–13 illustrates the effects of access control on accidents and fatalities. Both show an increased accident rate as access points increase.

Finally, access management reduces the chance that a highway or arterial will need to be relocated or reconstructed. When an arterial street is first constructed, there is little or no commercial development. As traffic volumes increase, however, commercial development occurs. Without adequate access management and zoning the arterial will become degraded in operational capacity and safety. As a result, it may be necessary either to widen the facility or provide a one-way pair. In the case of a highway route, it may be necessary to relocate the route.

Development of an access management program in conjunction with a citywide street plan and zoning preserves the integrity, safety, and capacity of a street and highway.

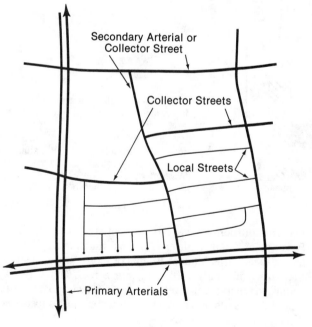

Figure 12–10. Schematic street configuration based on functional classification.

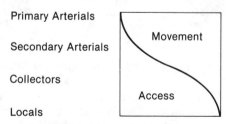

Figure 12–11. Movement and access for street classification.

SOURCE: V.G. Stover, *Texas Engineering Experiment Station Technical Bulletin 81-1,* January 1981.

---

[21] *The State Highway Access Code Amended by the Colorado Highway Commission, August 15, 1985* (Denver: Colorado State Department of Highways).

[22] Brian S. Bochner, "Regulation of Driveway Access to Arterial Streets," *Public Works Magazine* (October 1979), p. 115.

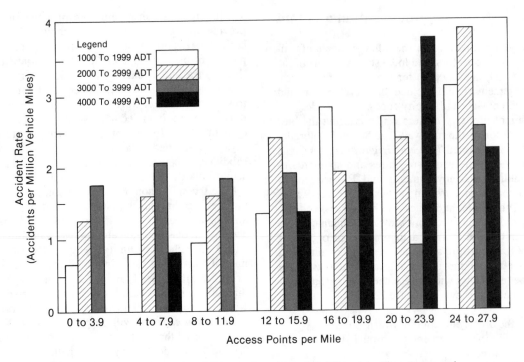

Figure 12–12. Accident rates for road sections with different traffic volumes and access point frequencies.
SOURCE: P.R. STAFFIELD, "Accidents Related to Access Points and Advertising Signs in Study," *Traffic Quarterly* (January 1953).

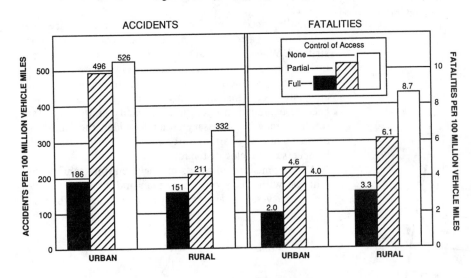

Figure 12–13. Effect of control of access on accidents and fatalities in urban and rural areas.
SOURCE: P.C. BOX, "Traffic Control and Roadway Elements—Their Relationship to Safety," Chapter 5, "Driveways" (Washinton, DC: Highway Users Federation for Safety and Mobility, 1970).

Access management considerations

Several factors should be considered in implementing access management strategies. These include: (1) speed differential between turning vehicles and through vehicles, (2) adequate protection for turning traffic, and (3) location of traffic signals.

Vehicle speed differential. Speed differential between the turning vehicle and through traffic is a significant factor in accidents related to access spacing along arterials. As shown in Figure 12–14, 60% of vehicles involved in two-car

rear-end collisions had a speed differential of 10 mph or more. Fewer than 40% of vehicles in normal traffic were found to have a speed differential of 10 mph or more. Research shows that with commonly used curb return radii and driveway widths, the right-turning vehicle enters a driveway at about 10 mph. This results in high-speed differentials at a substantial distance from the driveway.

Consideration must also be given to the acceleration of a vehicle turning into traffic from a driveway. The minimum spacings of driveways based on the acceleration and deceleration considerations are given in Tables 12–8 and 12–9. Table 12–10 provides similar values to those shown in Table 12–8

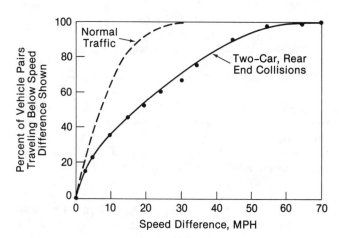

Figure 12–14. Speed difference between passenger cars involved in two-car, rear end collisions compared with normal traffic, day and night combined.

SOURCE: DAVID SOLOMAN, "Accidents on Main Rural Highways Related to Speed, Driver and Vehicle," *Bureau of Public Roads*, (Washington, DC: U.S. Department of Transportation, July, 1964).

TABLE 12–8

Minimum Spacing of Driveways and Unsignalized Access Points Necessary to Reduce Collision Potential Due to Right-Turn Conflicts[a]

Speed (mph)	Minimum Spacing (ft)		
	Author		
	Preferable[b]	Limiting[c]	Glennon et al.[d]
30	185	100	125
35	245	160	150
40	300	210	185
45	350	300	230

[a]Condition: Spacing is minimum so as to allow through vehicle to decelerate to avoid collision with a vehicle entering the through traffic lane from an unsignalized access point.

[b]Measured center-to-center of driveways based on the vehicle in the right-hand through lane cannot change lanes and decelerates at 6.0 fps[2] after a 2.0 sec perception-reaction time; the vehicle entering the traffic stream from a driveway completes the 90° right-turn as it accelerates, from 0 mph to speed equal to that of the decelerated vehicle, at an average acceleration of 2.0 fps[2]; no additional clearance is provided between the through vehicle and the driveway vehicle. The implied speed differentials which result between the driveway vehicle and through vehicle(s) are:

| arterial speed (mph) | 30 | 35 | 40 | 45 |
| maximum speed differential (mph) | 20 | 24 | 28 | 32 |

[c]Measured center-to-center of driveways based on the vehicle in the right-hand through traffic lane cannot change lanes and decelerates at an average of 6.0 fps[2] after a 2.0 sec perception-reaction time; this driveway vehicle completes the 90° right-turn while it accelerates from 0 mph to a speed equal to that of the decelerated through vehicle, at an average of 3.1 fps[2]; no additional clearance is provided between the driveway vehicle and the through vehicle. The implied speed differentials which result between the driveway vehicle and the through vehicle(s) are:

| arterial speed (mph) | 30 | 35 | 40 | 45 |
| maximum speed differential (mph) | 14 | 19 | 24 | 29 |

[d]Measured near curb to near curb based on: 8.5 fps[2] deceleration for vehicle in right-hand through traffic lane, 2.1 fps[2] acceleration from 0 mph start for 30 mph arterial speed and 1.7 fps[2] acceleration of all higher arterial speeds.

TABLE 12–9

Minimum Spacing of Driveways and Unsignalized Access Points Necessary to Allow Vehicle Exiting Access Point to Accelerate to Through Traffic Speed Without Creating Speed Differential in Excess of That Specified[a]

Through Traffic Speed (mph)	Speed Differential[b]			
	−10 mph		−15 mph	
	Spacing (ft)	In Traffic Stream (sec)	Spacing (ft)	In Traffic Stream (sec)
30	210	9.5	160	7.1
35	300	11.9	240	9.4
40	420	14.2	350	11.9
45	550	16.7	470	14.2

[a]Condition: Speed differential is that caused in the through traffic stream; speed change of through vehicle is approximately linear; average acceleration of 3.1 fps[2].

[b]The difference between normal traffic speed and the reduced speed caused by a vehicle entering traffic from a driveway. For example: with a 10 mph differential, a vehicle entering traffic from a driveway would cause an up-stream vehicle(s) in the through traffic lane to decrease speed from a normal speed of 40 mph to a speed of 30 mph.

SOURCE: VIRGIL G. STOVER, "Guidelines for Spacing of Unsignalized Access to Urban Arterial Streets," *Texas Engineering Experiment Station Technical Bulletin 81-1* (College Station: Texas A&M University System, January 1981).

TABLE 12–10

Minimum Driveway Spacing[1,2]

Arterial Speed (mph)	Minimum Separation (ft)
20	85
25	105
30	125
35	150
40	185
45	230

[1]Between two-way driveways. Distances between adjacent one-way driveways with the inbound drive upstream from the downstream drive can be one-half the distances shown above.
[2]Near edge to near edge of adjacent driveways.

SOURCE: GLENNON, J.C., ET AL., "Technical Guidelines for the Control of Direct Access to Arterial Highways," Volume II, *Detailed Description of Access Control Techniques*, Report No. FHWA-RD-76-87, Federal Highway Administration, August 1975.

for speeds of 30 mph and above, and also provides minimum spacings for 20 and 25 mph through-traffic speeds.

Protection for turning traffic. The same speed differentials that apply to right-turn traffic also apply to left-turn traffic. Consideration must be given to providing adequate protection to all turning traffic where necessary. Where a combination of right- and/or left-turning volumes and through traffic are high, added protection can be provided through the construction of right- and/or left-turn lanes.

The results of one study have shown that vehicles turning left into and out of driveways account for 70% of all driveway accidents. Median areas must be considered in developing designs for arterial streets. Medians remove the left-turn vehicle from the through traffic, result in a reduction of delay to through traffic, and have been found to provide a 35% reduction in accidents.[23] The results of a study on the effects of conflict control for commercial driveways with regard to a reduction in left-turn accidents are provided in Tables 12–11, 12–12, and 12–13. The left-turn protection strategies include:

TABLE 12–11

Estimated Annual Accident Reduction (per mile) by Installing Raised Median Divider with Left-Turn Deceleration Lanes

| Level of Development | Driveways per Mile | Highway ADT (vehicles per day) | | |
		Low <5,000	Medium 5–15,000	High >15,000
Low	<30	2.2	4.1	6.3
Medium	30–60	5.8	11.2	17.2
High	>60	10.7	20.7	31.2

SOURCE: FEDERAL HIGHWAY ADMINISTRATION, "Evaluation of Techniques for Control of Direct Access to Arterial Highways," Report FHWA-RD-76-86; Volume II, "Detailed Description of Access Control Techniques," Report No. FHWA-RD-76-87 (Washington, DC: U.S. Department of Transportation, Federal Highway Administration, August 1975).

TABLE 12–12

**Estimated Annual Accident Reduction (per mile) by Installing:
Two-Way Left-Turn Lane or
Continuous Left-Turn Lanes (for each direction of traffic)**

| Level of Development | Driveways per Mile | Highway ADT (vehicles per day) | | |
		Low <5,000	Medium 5–15,000	High >15,000
Low	<30	4.4	8.8	13.3
Medium	30–60	7.1	13.9	20.9
High	>60	9.7	19.0	28.6

SOURCE: FEDERAL HIGHWAY ADMINISTRATION, "Evaluation of Techniques for Control of Direct Access to Arterial Highways," Report FHWA-RD-76-86; Volume II, "Detailed Description of Access Control Techniques," Report No. FHWA-RD-76-87 (Washington, DC: U.S. Department of Transportation, Federal Highway Administration, August 1975).

TABLE 12–13

Estimated Annual Accident Reduction (per driveway) by Offsetting Opposing Driveways

| Driveway ADT | Vehicles per Day | Highway ADT (vehicles per day) | | |
		Low <5,000	Medium 5–15,000	High >15,000
Low	<500	0.4	0.7	1.0
Medium	500–1500	0.9	1.7	2.3
High	>1500	1.6	2.6	3.6

SOURCE: FEDERAL HIGHWAY ADMINISTRATION, "Evaluation of Techniques for Control of Direct Access to Arterial Highways," Report FHWA-RD-76-86; Volume II, "Detailed Description of Access Control Techniques," Report No. FHWA-RD-76-87 (Washington, DC: U.S. Department of Transportation, Federal Highway Administration, August 1975).

[23] Federal Highway Administration, *Access Management for Streets and Highways,* Implementation Package FHWA-1P-82-3 (Washington, DC: U.S. Department of Transportation, Federal Highway Administration, June 1982).

o Table 12–11—Installing a raised median divider with left-turn deceleration lanes
o Table 12–12—Installing two-way continuous left-turn lanes
o Table 12–13—Offsetting opposing driveways

A comparison of Tables 12–11 and 12–12 shows that a raised median can be expected to result in fewer accidents than a two-way left-turn median where there is a combination of a large number of driveways and high-traffic volumes; however, the following benefits may be provided by a two-way left-turn lane:

o Protection for motorists turning left from a driveway on a busy street (shadowed by the left-turn lane while waiting to complete the left turn)
o Utilization where opposing driveways are adequately offset
o Increased maneuverability in case of an emergency condition (permitting vehicle in left lane to swerve to avoid an accident or go around an existing accident)
o Increased storage during peak left-turn conditions
o Ability to utilize the center lane as a reversible lane for car pools and/or buses during peak periods.
o Ability to utilize the center lane as reversible lane for all traffic during peak periods

The last two items listed above reduce the efficiency of through movement in the lane(s) where left turns are not protected when the center lane is utilized as a reversible lane. This needs to be considered where left-turn traffic is heavy. It may be desirable or necessary to prohibit left turns when the center lane is used as a reversible lane.

Another possible left-turn design involves the use of a raised median divider with no openings between intersections and with left-turn lanes at the intersections. This approach can work well along streets having six lanes for through traffic. The design allows traffic desiring to make left turns between intersections an opportunity to make U-turns at the intersections. This approach can be successfully implemented where the U-turns and left turns are not high at the intersection. An alternative approach is to prohibit left turns at major signalized intersections with channelized left turn provided at midblock locations.

Since both the raised median divider and continuous left-turn lane designs have both advantages and disadvantages, consideration must be given to trade-offs in determining whether to provide a two-way left-turn lane or a raised median divider.

Another way of providing protection to motorists as part of access management is the provision for circulation between adjacent commercial sites. Parallel streets that provide adequate access at each commercial site can also provide circulation away from the major street. This will keep circulating local vehicle trips out of the major traffic stream on the arterial.

Location of traffic signals. The number and location of traffic signals is an important part of access management. Traffic signals along arterials should be limited to half-mile intervals for providing progression in both directions at 35 to

45 mph. For collector streets, traffic signals may be installed in multiples of 1,500 to 2,000 ft for two-way progression at 25 to 35 mph. These signal spacings reduce the number of stops and accidents, reduce fuel consumption and vehicle pollution, and increase capacity. They will also limit the installation of traffic signals for commercial driveways. Zoning is needed to assure that major traffic generators are located at designated traffic signal locations. The System Warrant for traffic signals found in the *Manual on Uniform Traffic Control Devices* was developed to provide for the installation of traffic signals at the proper spacing along a developing artery or collector street. Motorists leaving shopping centers (malls) and other high-volume traffic generators should be directed to exits that place them on the signalized cross street. The location of the driveway along the cross street should be

dictated by the probability that the driveway will not be blocked by traffic waiting at a traffic signal. This probability is shown in Table 12–14. The same applies to driveways upstream of traffic signals on the major street (arterial).

Because of the limited funds and ability to provide sufficient right-of-way for construction and widening of freeways, there has been a need to develop "mini-freeways" on narrower right-of-way widths of 110 ft to 200 ft. These major thoroughfares can have grade separations at approximately one-mile intervals with special design for two-phase traffic signal control such as those shown in Figures 12–15—12–20 at suitable locations for optimum movement of thoroughfare traffic at 40 to 45 mph. Access management is a necessity in assuring that these facilities will operate as designed with practical control of access.

TABLE 12–14

Percent of Signal Cycles that Queues Do Not Block Driveway

Arterial Traffic Vehicle/Hour	Red Phase (sec)	Percent of Cycles Driveway Not Blocked, by Distance from Cross Street (in ft)									
		25′	50′	75′	100′	125′	150′	175′	200′	225′	250′
200	15	80%	95%	99%	100%	100%	100%	100%	100%	100%	100%
	25	60	84	95	99	100	100	100	100	100	100
	35	42	69	87	95	98	100	100	100	100	100
	45	29	54	76	89	96	99	100	100	100	100
400	15	50	77	91	97	99	100	100	100	100	100
	25	23	47	70	85	94	98	99	100	100	100
	35	10	25	45	65	80	90	95	98	99	100
	45	4	12	26	44	62	76	87	93	97	98
600	15	29	54	76	89	96	99	100	100	100	100
	25	8	21	40	60	76	87	94	97	99	100
	35	2	7	17	31	47	63	77	86	93	96
	45	0	2	6	13	24	38	52	66	78	86
800	15	15	35	57	76	88	95	98	99	100	100
	25	2	8	19	35	51	68	80	89	94	97
	35	0	2	5	11	21	34	48	62	74	86
	45	0	0	1	3	7	13	22	33	46	58

SOURCE: TRANSPORTATION RESEARCH BOARD, "Guidelines for Medial and Marginal Access Control on Major Roadways," NCHRP Report 23 (Washington, DC: Transportation Research Board, 1970).

Figure 12–15. Geometric design for two-phase traffic signal operation.
SOURCE: RICHARD P. KRAMER, "Access Management via Geometrics & Signalization to Maximize At-Grade Arterial Capacity and Safety," paper presented to the Florida Section, Institute of Transportation Engineers (April 1989).

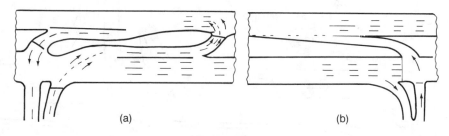

(a) (b)

Figure 12–16. Indirect left turn through a crossover.
SOURCE: FEDERAL HIGHWAY ADMINISTRATION, "Access Management for Streets and Highways," Implementation Package (Washington, DC: U.S. Department of Transportation, Federal Highway Administration, June 1982).

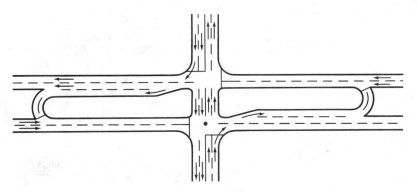

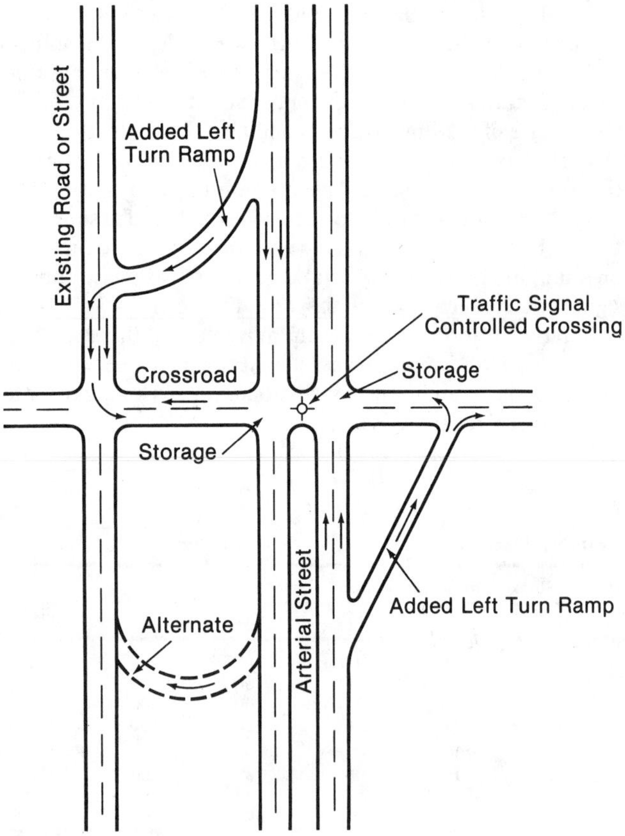

The information given in Tables 12–8 through 12–10 can be utilized in determining the minimum distances between driveways. There are a number of sets of guidelines available for use, and these are given in the various footnotes and source notes throughout the Access Management Section in this chapter. These guidelines should be reviewed by the engineer in developing an access code.

Corner clearance distances to the first driveway vary. One study found that 37 states have a minimum standard for corner clearances, which are typically about 20 to 25 ft for urban areas and about 40 to 50 ft for rural areas.[24] It has also been stated that the minimum tangent curb length between an intersection and a driveway (for a 35 mph arterial) should be 50 ft.[25] A third source recommends a 105- to 150-ft distance on a collector street and a 550- to 600-ft distance (with a 400- to 450-ft limiting distance) along a freeway arterial (with an apparent street speed of 45 mph).[26] Table 12–14 shows the percentage of traffic signal cycles during which vehicles do not block the driveway where the driveway is located at varying distances from the stop line to the driveway of 25 to 250 ft.

Design information for median and right-turn deceleration lanes and openings is given in Figures 12–21—12–25. These are provided for guidance in developing designs for various classifications of streets and vehicle speeds. The longitudinal dimensions shown in Table 12–14 and Figures 12–21

Figure 12–17. Special left-turn designs for traffic leaving highway with narrow median.
SOURCE: FEDERAL HIGHWAY ADMINISTRATION, "Access Management for Streets and Highways," Implementation Package (Washington, DC: U.S. Department of Transportation, Federal Highway Administration, June 1982).

[24] Ibid.

[25] Bochner, "Regulation of Driveway Access to Arterial Streets."

[26] Virgil G. Stover, "Guidelines for Spacing of Unsignalized Access to Urban Arterial Streets," *Texas Engineering Experiment Station Technical Bulletin 81-1* (College Station: Texas A&M University System, January 1981).

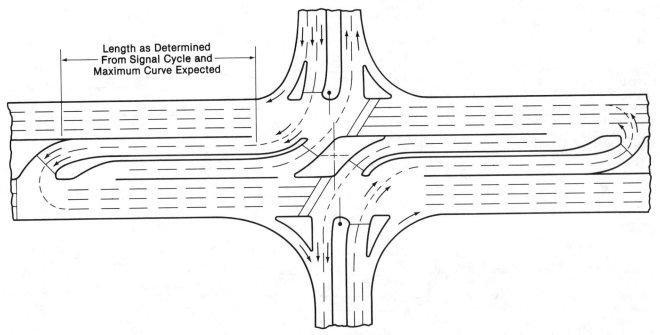

Figure 12–18. Two opposite shopping centers of up to 800,000 sq ft G.L.A. each. Major intersection treatment.

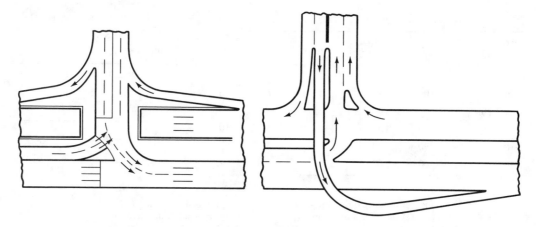

Figure 12–19. Partial grade-separation designs.
SOURCE: Richard P. Kramer, "Access Management via Geometrics & Signalization," paper presented to the Florida Section, Institute of Transportation Engineers, April 1989.

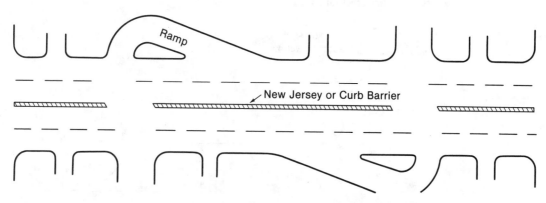

Figure 12–20. Median barrier with indirect left-turn ramps (jug-handle).
SOURCE: Federal Highway Administration, "Access Management for Streets and Highways," Implementation Package (Washington, DC: U.S. Department of Transportation, Federal Highway Administration, June 1982).

through 12–25 are often minimum distances. Where possible, these distances should be increased for improved operations.

Design speed for through traffic must be considered when developing access and turning movement design. Motorists should not have to reduce their speed unduly as they approach the access point and complete their movement. They should be able to enter the street from an access location without interrupting traffic on the street more than is necessary.

As with geometric design along a street or freeway, the design of access points and medians is very important in the overall operation of a facility.

Conclusions

The results of studies have demonstrated the need for a combination of arterial planning, zoning, and access management to assure that all urban streets carry out their designated role. Since suburban facilities eventually become urban arteries, county, city, and state personnel must work together to assure that proper access control is carried out on a metropolitan area basis. Proper access management

planning, implementation, and enforcement will provide great dividends in the safe and efficient movement of traffic.

Evaluation of traffic management systems

A number of measures can be used to assess the effectiveness of traffic management improvements. These include:

o Reduced accidents
o Improved speed and/or reduced travel time
o Reduced delay
o Reduced number of stops
o Increased capacity
o Reduced fuel consumption
o Reduced vehicle emissions
o Improved management of bus route schedules

Studies to quantify these measures can be carried out with minimal training of personnel. The analysis typically

Arterial speed (mph)	Approach geometry (feet)					Departure geometry (feet) all speeds			
	L_1	$L_2{}^a$	$W_1{}^b$	$W_2{}^c$	R_1	$W_3{}^c$	R_2	L_3	0
45 desirable^d	120	250				14	20	10	5
limiting^e	100	130	12	16	20	16	15	15	5
40 desirable	120	190	12	18	15	16	20	0	0
limiting	90	100	12	20	10				
35 desirable	90	120							
limiting	60	60							

^{a} Maneuver distance only; does not include storage for more than one vehicle.

^{b} Excluding curb and gutter.

^{c} Stripe to back of curb.

^{d} 10 mph speed differential, 7 fps^2 average maximum deceleration.

^{e} 15 mph speed differential, 9 fps^2 average maximum deceleration.

^{f} Any set of $L_1 L_2$ values may be used with any set of $W_1 W_2 R_1$ values.

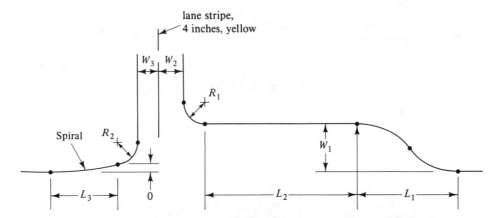

Figure 12–21. Suggested design for undivided driveway access to arterial streets.
SOURCE: V.G. STOVER, *Texas Engineering Experiment Station Bulletin 81-1.*

necessitates a "before" study of the conditions prior to making the improvement and an "after" study after the improvement has been made and traffic conditions have returned to normal. Where accident studies are involved, it is desirable to utilize "control" sections or locations with similar design features for comparison with the improved segment.

One important indicator of the effectiveness of a project is the cost-benefit ratio. As the name implies, calculation of the ratio requires estimating the benefits and the costs of the project. Determining the cost of the project is done with conventional estimating techniques. Estimating the benefits is slightly more difficult.

The first step in estimating the benefits of a project is to estimate the benefits for each year of the project life. This will usually require estimating future traffic volumes. These volumes can then be used to calculate some measure of effectiveness (MOE), which can readily be converted into dollar savings. Delay is the most common MOE used in benefit analysis, with a value of $8 to $10 per vehicle hour, but many others can be used, including gasoline usage, pollution, and maintenance costs.

After the benefits for each year have been estimated, they must be converted to base year values. This requires setting a discount rate, i. The discount rate reflects the value of money and is commonly about 8%. The basic present value equation is then used.

$$PV = FV (1 + i) - n \qquad (12.1)$$

where:

PV is the present value of the future year benefits
FV is the benefit value in the future year
i is the discount rate
n is the number of years from the base year to the future year.

After converting all of the benefits to base-year dollars, they can be added together to obtain the total benefits over the life of the project. The cost-benefit ratio is then found by dividing the benefits by the costs.[27]

---

[27] B. Ray Derr, unpublished memorandum (Austin: Texas State Department of Highways and Public Transportation, January 1990).

Arterial Speed (mph)	Approach Geometry (feet)						Departure Geometry (feet)			
	L_1	L_2		$W_1{}^a$	$W_2{}^b$	R_1	$W_3{}^d$	R_2	L_3	0
45	120	250		12	30	20	28	20	10	5
40	120	200	c	12	32	15	30	15	15	5
35	90	120		12	34	10	30	25	0	0

[a] Excluding gutter.

[b] Back-to-back of curb's total width, entrance width, W_2, divided by solid white line, lane line to back of curb, as follows:

W_2	Right lane	Left lane
30	16.5	13.5
32	18.5	13.5
34	20.5	13.5

[c] Any set of $L_1 L_2$ values may be used with any set of $W_1 W_2 R_1$ values.

[d] Total width back to back of curb; divided solid line as follows:

W_3	Right lane	Left lane
28	14.5	13.5
30	16.5	13.5

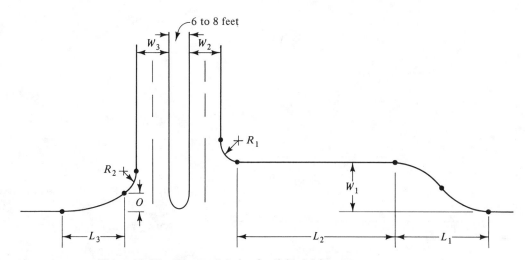

Figure 12–22. Suggested design for divided driveway access.
SOURCE: V.G. Stover, *Texas Engineering Experiment Station Bulletin 81-1.*

Dimension Feet

R_1	R_2	W
15	15	35
20	15	30

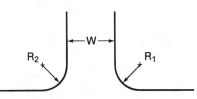

Figure 12–23. Suggested designs for access to collector streets without parking.

The above measures are typically easy for administrators to understand. Cost-to-benefit ratio analysis should also be performed based on the estimated life for the improvement (i.e., 10 years for traffic signal and freeway control improvements; 20 years for access management and geometric improvements; and 1 year for active traffic management techniques). For instance, if it costs $350,000 annually to operate a service patrol unit and the estimated benefits in services to motorists and savings in delay come to $6 million during the year, the cost-benefit ratio is 1:17.

Where a freeway and/or traffic signal system is installed and operating, it should be possible to carry out data collection and analysis at the traffic management center. Before-and-after data for many of the traffic management improvements (i.e., traffic signal retiming) can be collected at the center without having to make field studies.

Improvements should also be studied and analyzed to determine whether further improvements are needed and to determine the costs for these improvements. Studies should also be made periodically with reports provided to the administration on the results of improvements being made.

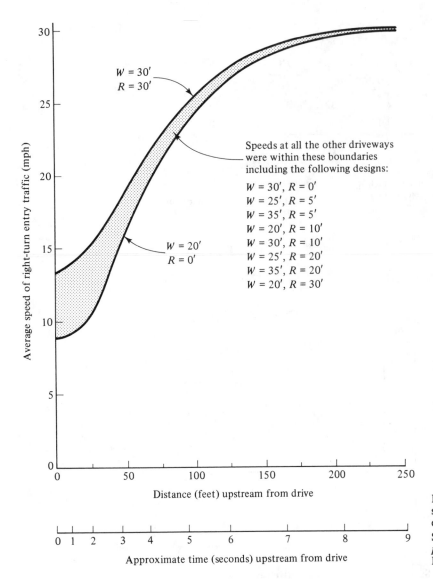

Figure 12–24. Turning vehicle speed at one second intervals upstream from the driveway entry.
SOURCE: V.G. Stover & F. Koepke, *Transportation and Land Development*, I.T.E., Prentice-Hall, 1988, p. 150.

(Figure labels)

W = 30'
R = 30'

Speeds at all the other driveways were within these boundaries including the following designs:

W = 30', R = 0'
W = 25', R = 5'
W = 35', R = 5'
W = 20', R = 10'
W = 30', R = 10'
W = 25', R = 20'
W = 35', R = 20'
W = 20', R = 30'

W = 20'
R = 0'

Summary

Traffic management is not a new concept. Traffic engineers have performed these functions for many years. However, there is a need to reinforce and coordinate all aspects of traffic management on a metropolitan area basis. Additionally, there is an increasing traffic congestion problem with diminishing revenues to construct new facilities and improve existing systems. This will require greater communication, cooperation, and coordination among public transportation officials and city, state, and county traffic engineering personnel and law enforcement agencies. Together they must provide the optimal use of the existing arterial system, and assure that improvements will have the best payoff possible in terms of safety and operations.

Appendix A
Traffic management techniques

Technique*

A. Up to 10 m from carriageway there is no need for warning signs, speed reduction, or overtaking prohibition. The obstacle (such as bridge falsework) need only to be perceived through a proper delineation.**

B. If encroachment of activity area is slight, narrow lanes are the most usual strategy. Only delineation and warning signs are necessary for these lanes. Speed control depends on

---

* Technique designation refers to Table A–1.

** Figure numbers are referenced in the applicable technique.

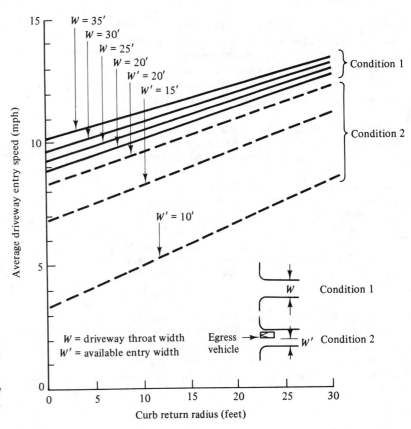

Figure 12–25. Speed of auto entering driveway as a function of curb return radius and driveway width.

SOURCE: V.G. STOVER, *Guidelines for Spacing of Unsignalized Access to Urban Arterial Streets,* Bulletin 81-1, Texas Engineering Experiment Station, Jan. 1981.

TABLE A–1

Traffic Movement Techniques Within Work Zones
Classification of Possible Recommended Techniques
According to the Type of Facility and the Work Zone Location

Work Zone Location (encroachment)		Type of Highway				
		Undivided			Divided	
		2-Lane	2 + 1-Lane	4-Lane	2 × 2-Lane	2 × 3-Lane
External				A		
Outer shoulder				B		
Median shoulder			–		C	
In median			–		D	
In carriageway:	0			B		
number of	1	E	F	G		H
necessary lane	>1	I	E	J	K	L
closures	all			M		

Note: Work zone techniques shown above are described below.

SOURCE: ROAD TRANSPORTATION RESEARCH, *Traffic Management and Safety at Highway Work Zones* (Paris: Organization for Economic Co-operation and Development, 1989).

available width, and overtaking prohibition is not normally needed.

Where encroachment increases, maintaining total number of lanes needs use of available width, through number of reduced width lanes along the main lanes and hard shoulders. Along divided roads, a contra-flow lane, and even temporarily paved medians or parallel strips may be provided. Speed control depends on available width and altered lane geometry. Overtaking prohibition may be necessary for traffic entering a single lane.

On divided highways, use of hard shoulder opposite to work zone lane is preferable to near shoulder, to avoid leaving work zone as an island in the middle of traffic, and related access problems. For this reason, this strategy is not recommended. Thus, this strategy is not recommended for 2 + 1-lane, 4-lane undivided, or 2 × 3-lane divided highways with work zone in the middle lane, F, G, or H below being preferable. Contra-flow lanes need narrowing and/or use of hard shoulder in the carriageway opposed to work zone to maintain the number of its lanes.

Temporarily paved medians or parallel strips can be used economically only for short work zones of long duration.

Different situations are summarized in Figure A–1.

C. This case is similar to B, except that it can be necessary to close the fast lane temporarily to allow access of equipment and agents to the work zone: then it will be similar to G or H.

D. This case is similar to A or C, according to extent of encroachment. Temporary closure of fast lane may be required for access purposes, and then it will be similar to G or H.

E. In this case it will normally be necessary to use shuttle operation. Warning signs, speed limitation up to complete stop, and overtaking prohibition are needed (Figure A–2).

F. One lane per traffic direction can be maintained, closing one lane and altering the others, with proper delineation and signing. Overtaking prohibition is necessary, and speed limitation cannot usually be avoided. Three situations are possible (Figure A–2):

1. Work zone occupies outer (climbing) lane: preferably fast lane should be first closed then provisionally diverted back to its original position. Opposite lane is not modified;
2. Work zone occupies (descending) lane. Fast climbing lane should be closed in order to allow opposite lane to be temporarily diverted on it;

Figure A–1. Technique B.

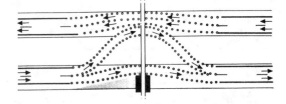

Major encroachment in divided highways: altered and/or narrow lanes plus contra-flow lane (example)

Temporary paving

Median Paving

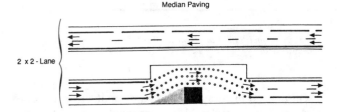

2 x 2 - Lane

Parallel Strip

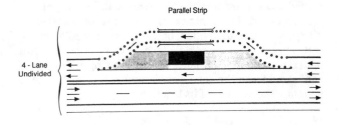

4 - Lane Undivided

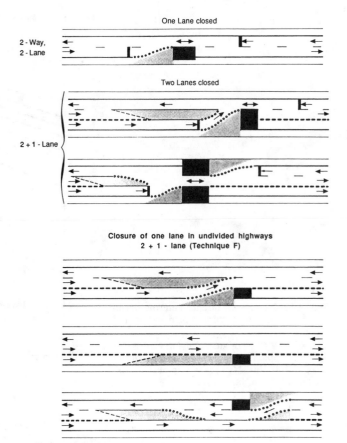

Shuttle operation in undivided highways (Technique E)

One Lane closed

2 - Way, 2 - Lane

Two Lanes closed

2 + 1 - Lane

Closure of one lane in undivided highways
2 + 1 - lane (Technique F)

Figure A–2. Techniques E and F.

3. Closure of middle lane only is not advisable because of access problems; in this case it is better to close two lanes, as in E above.

G. Where one outer lane is occupied by work zone, allocation of remaining 3 lanes should be made as follows (Figure A–3):

1. If traffic volume is low or balanced, 1 provisional lane to traffic affected by work zone—closing first the inside lane—and 2 to opposing traffic;
2. If traffic volume is high and unbalanced, 2 lanes (1 may be contra-flow) should be allotted to direction of higher volume. This may change with time of day or day of week), thus needing reversal of strategy. Problems may arise at divergence of contra-flow lane.

Where one inside lane is occupied by the work zone (Figure A–3) usually only that lane is closed; when traffic is low, both center lanes of an undivided highway may be closed, facilitating access to work zone and additional lateral buffer zone.

In both cases, overtaking prohibition and warning signs are mandatory for traffic direction for which only one lane is left. Speed limitation cannot usually be avoided.

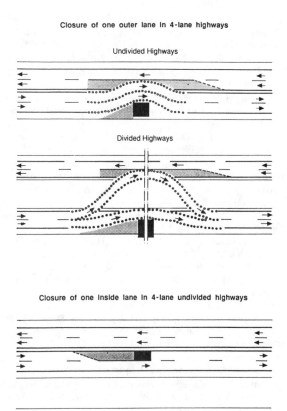

Figure A–3. Technique G.

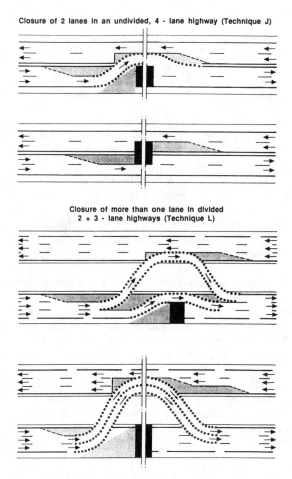

Figure A–4. Techniques J and L.

H. This case is similar to G. Where one lane is occupied by work zone, allocation of remaining 2 lanes for that direction should be made as follows:

1. Normally, fast lane shall be closed, either because affected by work zone or because all traffic is concentrated on the outer lanes before being diverted according to work zone location;

2. If traffic volume is high and unbalanced, and a reduction of number of lanes for traffic direction affected by work zone is not contemplated, even when that means a reduction in the number of those available for the opposite direction, 3 lanes should be allotted to the former; fast lane shall be diverted contra-flow to opposite carriageway (in which fast lane shall have been previously closed), and remaining 2 lanes diverted to inner lanes. This may change with time of day or week, thus needing reversal of strategy.

Overtaking prohibition and advance warning signs are mandatory for concerned traffic. Speed limitation cannot usually be avoided.

I. This case is extrapolation of E, in which the opposite shoulder is used. It can be used only in low-volume roads and for short-duration work.

J. Two situations can be contemplated (Figure A–4):

1. Activity area occupies both lanes for one direction: traffic for that direction shall be diverted to a contra-flow single lane in the rest of the carriageway, which will operate as a 2-way, 2-lane facility;

2. Activity area occupies the 2 fast lanes: both fast lanes shall be closed.

Overtaking prohibition and advance warning signs are mandatory for both directions. Speed limitation cannot usually be avoided.

If more than 2 lanes are to be closed, a special study is necessary since it is a bad solution for the high traffic volumes normally associated with 4-lane roads; adaptation to F or M is possible.

K. Where work zone affects either both outer or both inner lanes, situation can be similar to G or J.

Where closure of affected carriageway is contemplated, the other carriageway shall be operated as a 2-way, 2-lane facility, with one contra-flow lane. Advance warning, overtaking prohibition, and speed limitation are necessary.

1. In the affected carriageway only one lane (either outer or inner, but not the middle one) remains open to traffic, and the opposing carriageway is not impaired. This strategy is only possible

a. For low-traffic volumes, which is seldom the case, thus restraining its use to off-peak hours;

b. For very unbalanced traffic volumes (more in the opposing carriageway);

c. In short-duration work for which the traffic management techniques used for mobile operation are deemed inadequate;

2. On the affected carriageway only one lane (either outer or inner, but not the middle one) remains open to traffic, and another lane is carried contra-flow to the opposite carriageway; a strategy similar to H.

Where full closure of the affected carriageway is necessary, even if this entails a reduction in the number of those available for the opposite direction, two lanes of the former are diverted contra-flow to the inner lanes of the latter (previously closed to opposite traffic); a strategy also similar to H.

In all cases, overtaking prohibition and warning signs are mandatory for concerned traffic. Speed limitation cannot usually be avoided. Traffic-demand balance may change with time of day or day of week, thus needing reversal of strategy.

M.   Where no detour is possible, this case is also an extrapolation of E in which the undivided carriageway, or one of the divided ones, is blocked and therefore traffic has to come to a complete stop. In divided highways, closure of both carriageway at the same time is hardly possible because of capacity problems when completely shutting such facilities even for short-duration work; keeping some lanes open to traffic (through K or L strategy) is always necessary. Multilane carriageways should be reduced to single lane via lane closure to have an orderly stop.

Resumption of traffic can be made through an E situation, if resumption in both directions at the same time is difficult or impossible.

L.   Normally, fast lane shall be closed and all traffic concentrated on the outer lanes before being diverted according to work zone location. Three situations can be contemplated (Figure A–4).

REFERENCES FOR FURTHER READING

A Tool Box for Alleviating Traffic Congestion, Institute of Transportation Engineers, 1989.

"Proceedings of Symposium on Work Zone Traffic Control," FHWA Report DOT-1-86-05, U.S. Department of Transportation, Federal Highway Administration, 1985.

Traffic Management and Safety at Highway Work Zones, Organization for Economic Cooperation and Development, 1989.

Manual on Uniform Traffic Control Devices for Streets and Highways, U.S. Department of Transportation, Federal Highway Administration, 1988 edition.

Traffic Control Systems Handbook, FHWA Report 1p—85-11, U.S. Department of Transportation, Federal Highway Administration, April 1985.

Reconstruction, Conference Proceedings, Transportation Research Board, 1987.

ROTHENBERG, M.J., AND D.R. SAMDAHL, *High Occupancy Vehicle Facility Development—Operation and Enforcement,* Report No. FHWA-IP-82-1, U.S. Department of Transportation, Federal Highway Administration, April 1982.

LANCASTER, A., AND T. LOMAX (Eds.), *Proceedings of the Second National Conference on High Occupancy Vehicle Lanes and Transitways,* Texas Transportation Institute, Texas A&M University, October 1987.

TURNBULL, K.F. (Ed.), *Proceedings of the Third National High Occupancy Vehicle Facilities Conference,* Minneapolis–St. Paul Regional Transit Board, October 1988.

BATZ, T.M., *High Occupancy Vehicle Treatments, Impacts and Parameters—A Synthesis,* U.S. Department of Transportation, Federal Highway Administration, August 1986.

The Effectiveness of High-Occupancy Vehicle Facilities, Institute of Transportation Engineers, Technical Council Committee 6A-37, 1987.

SUMNER, R., AND OTHERS, *Freeway Management Handbook,* U.S. Department of Transportation, Federal Highway Administration, May 1983.

PARVIAINEN, J.A., AND W.M. DUNN, JR., *Freeway Management Systems for Transportation Efficiency and Energy Conversation—Practical Planning Guide for Traffic Engineers,* Report No. TP 622OE, Transportation Development Center, Transport Canada, September 1985.

DUDEK, CONRAD, AND OTHERS, "Promotional Issues Related to Accident Investigation Sites," Transportation Research Board Record 1173, *Urban Freeway Operations,* 1988.

TARNOFF, P.H., AND OTHERS, "Selecting Traffic Signal Control at Individual Intersections," National Cooperative Highway Research Program Report 233, Transportation Research Board, 1981.

FLORA, J.W., AND K.M. KEITT, *Access Management for Streets and Highways,* U.S. Department of Transportation, Federal Highway Administration, Report No. FHWA-IP-82-3, June 1982.

The State Access Code Amended by the Colorado Highway Commission August 15, 1985, Colorado State Department of Highways.

STOVER, VIRGIL T., "Guidelines for Spacing Unsignalized Access to Urban Arterial Streets," *Texas Engineering Experiment Station Technical Bulletin 81-1,* Texas A&M University System, January 1981.

BOCHNER, BRIAN S., "Regulation of Driveway Access to Arterial Streets," *Public Works Magazine,* October 1979.

Access Management for Streets and Highways, Implementation Package, Flt WA 1P-82-3, U.S. Department of Transportation, Federal Highway Administration, June 1982.

"Evaluation of Techniques for Control of Direct Access to Arterial Highways" Report No. FHWA-RD-76-85, "Technical Guidelines for the Control of Direct Access to Arterial Highways," Volume I: "General Framework for Implementing Access Control Techniques," Report No. FHWA-RD-76-86; Volume II; "Detailed Description of Access Control Techniques" Report No. FHWA-RD-76-87 (Washington, DC: Department of Transportation, Federal Highway Administration, August 1975).

Freeway Modifications to Increase Traffic Flow, U.S. Department of Transportation, Federal Highway Administration, January 1980.

WAGNER, FREDERICK A., *Traffic Control System Improvements: Impacts and Costs,* Report No. FHWA-PL-80-005, U.S. Department of Transportation, Federal Highway Administration, March 1980.

Traffic Management Measures to Reduce Energy Consumption, Transportation Energy Management Program, Canadian Ministry of Transportation and Communications, November 1981.

Traffic Management for Freeway Emergencies and Special Events, Transportation Research Circular 344, Transportation Research Board, January 1989.

JUDYCKI, D.C., AND J.R. ROBINSON, "Freeway Incident Management," *Compendium of Technical Papers,* 58th Annual Meeting of the Institute of Transportation Engineers, September 1988.

Urban and Suburban Highway Congestion, Working Paper No. 10 of the Future National Highway Program 1991 and Beyond Task Force, U.S. Department of Transportation, Federal Highway Administration, December 1987.

Recommend Practice for Traffic and Parking Control for Snow Emergencies, Publication NORP-OIZA, Institute of Transportation Engineers, 1983.

"Mobility 2000, Proceedings of a Workshop on Intelligent Vehicle/Highway Systems (Hosted by Texas State Department of Highways and Public Transportation and Texas Transportation Institute), February 1989.

13

FREEWAY SURVEILLANCE
AND CONTROL

JEFFREY A. LINDLEY, P.E., *Urban Transportation Specialist*

Federal Highway Administration

AND

DONALD G. CAPELLE, P.E., *Vice President*

Parsons Brinckerhoff

Urban freeways form the backbone of the transportation system in most urban areas. An effective network of freeways is essential to provide desired levels of mobility among housing, employment, and recreation centers, and to allow efficient movement of commercial goods. In the past several years, traffic demand on urban freeways has risen rapidly, and prevailing traffic patterns have significantly changed as more and more employment centers have moved to suburban locations. This has led to critical freeway traffic congestion problems in many urban areas.

The urban freeway system is essentially completed in most metropolitan areas. New freeways or major expansions in capacity for existing freeways are typically not feasible because of right-of-way, environmental, political, or cost constraints. More and more attention is being focused on maximizing the efficiency of existing freeways to serve increasing traffic demands and reduce traffic congestion levels. This chapter considers the application of surveillance systems and traffic control on urban freeways to achieve this goal.

Urban freeway congestion: Causes and impact

The major operational problem on urban freeways is congestion, which is characterized by slower-than-desired travel speeds, increased and unpredictable travel times, increased accident frequencies, erratic stop-and-go driving, increased vehicle operation costs, and other undesirable conditions resulting in user dissatisfaction. Clearly, a congested freeway is an inefficient one. Figure 13–1 shows a typical relation between speed and volume on an urban freeway. At point A on the curve, flow is uncongested and the mean travel speed is high. As traffic volume increases toward point B, flow becomes more congested and the speed falls to a lower value. Passing from point B to point C, however, results in a lower travel speed and also a reduced volume of traffic on the freeway. Point C is representative of stop-and-go driving conditions during which the volume being accommodated by the freeway is less than capacity, and average speeds are low.

Congestion on urban freeways is of two types: recurring and nonrecurring. Congestion that occurs regularly at particular locations during certain time periods is said to be recurring in nature. Conversely, congestion caused by such random irregular events as accidents, disabled vehicles, and other special situations is referred to as nonrecurring congestion. Both recurring and nonrecurring congestion lead to user dissatisfaction as previously described. There is, however, a difference. With recurring congestion, users can plan their trips according to the expected occurrence and severity of the congestion. On the other hand, nonrecurring congestion can have detrimental effects on a trip that is normally satisfactory. It is bad enough to know that a certain trip takes 10 min longer during the peak period than during off-peak hours, but it is worse if it takes 10 min longer on one day and 20 min longer on another. Predictability is very important to most users.

In nationwide studies of urban freeway congestion, it has been reported that delay from nonrecurring congestion caused by incidents accounts for more than half of the urban

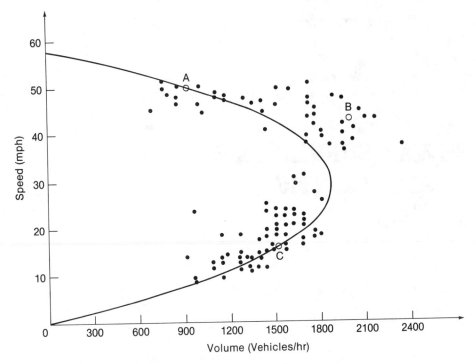

Figure 13–1. Typical speed-volume relationship on an urban freeway.
SOURCE: P.F. EVERALL, *Urban Freeway Surveillance and Control-The State of the Art,* rev. (Washington, DC: U.S. Department of Transportation, Federal Highway Administration, June 1973).

freeway congestion problem.[1, 2] Nonrecurring congestion should thus be considered a serious problem for the public agencies responsible for managing freeway operations.

The occurrence of freeway congestion at any point along its length is essentially the result of traffic demand exceeding the capacity of the freeway at that point, which can be due to excessive demand or some reduction in normal freeway capacity. This is illustrated by the graph in Figure 13–2, which represents the number of vehicles passing a point on the road as a function of time.

The slope of the straight line in Figure 13–2 represents the capacity of a section of freeway at a particular time (i.e., the number of vehicles getting past the point under the prevailing roadway conditions). As long as the traffic demand, or the number of vehicles arriving at that point (shown by the cumulative demand line), is less than or equal to the capacity of that section of the freeway, there is little congestion. However, once the arrival rate begins to exceed the capacity at time T_a, a "bottleneck" is formed and vehicles begin to accumulate upstream of the bottleneck until time T_b when the demand once again falls below the capacity. Congested conditions continue until time T_c when the accumulated traffic at the bottleneck dissipates. The area between the capacity and demand curves during congested conditions is the delay resulting from the congestion.

[1] J.A. Lindley, *Quantification of Urban Freeway Congestion and Analysis of Remedial Measures,* Report No. FHWA/RD-87/052 (Washington, DC: U.S. Department of Transportation, Federal Highway Administration, October 1986).

[2] *Urban and Suburban Highway Congestion,* Working Paper No. 10, The Future National Highway Program 1991 and Beyond Task Force (Washington, DC: U.S. Department of Transportation, Federal Highway Administration, December 1987).

Factors contributing to excessive demand

Excessive demand contributes to the overloading of a facility, which creates turbulence in the traffic stream and leads eventually to a system breakdown if the overloading continues. The capacity of a multilane freeway is discussed in Chapter 5 and is normally considered to be 2,000 to 2,200 passenger vehicles per lane per hour under ideal conditions (straight, level alignment free of lateral obstructions). When the travel demand approaches this maximum value, congestion will result. This situation usually occurs regularly during the peak periods when there are high commuter trip demands. It may also occur as a result of special occasions such as sporting events and holiday travel.

Excessive demand is frequently due to unrestrained access. If the combined volume of a freeway on-ramp and the main freeway lanes exceeds the freeway capacity downstream of the ramp entrance, congestion will develop on the freeway mainline and on the ramp, with queuing upstream of the bottleneck. This type of congestion can be predicted with a fair degree of accuracy in both time and space.

Freeway congestion is sometimes caused by traffic signals and other bottlenecks on city streets that prevent traffic from leaving the freeway. Eventually, the exit ramp queues back onto the freeway where there is inadequate ramp storage. Congestion due to exit ramp queuing can also develop because of heavy exit ramp demands caused by special events.

Factors causing reductions in freeway capacity

Capacity is not uniform along the length of a freeway, but is dependent on many factors, among which the geometric and

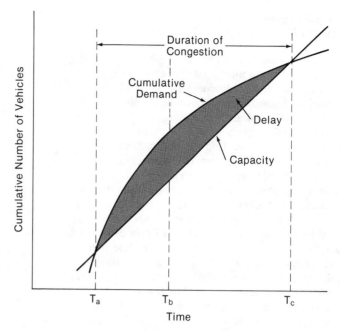

Figure 13–2. Relationships among demand, capacity, and congestion.

SOURCE: Adapted from: D.G. Capelle, *Freeway Traffic Management,* NCHRP Project 20-3D (Washington, DC: Transportation Research Board, September 1979).

When traffic demand exceeds the capacity of a section of freeway a bottleneck is formed and vehicles will accumulate upstream of the bottleneck. The amount of delay is represented by the shaded area.

physical features of the freeway are most important. Certain physical features result in capacity restrictions where they occur. These isolated sections, whose capacity is lower than those of adjacent sections, are referred to as "geometric bottlenecks." When the demand upstream of the bottleneck exceeds the capacity of the bottleneck, congestion develops with queuing in the upstream freeway lanes. Congestion resulting from geometric bottlenecks is of the recurring type, being predictable with reasonable accuracy. Some of the most common causes of geometric bottlenecks and thus of recurring congestion are included in the following discussion.

Lane drops. Congestion may occur at locations where the number of lanes is reduced. Even if the lane drop occurs at an exit ramp, congestion sometimes results if the through-traffic demand exceeds the downstream capacity. In addition, weaving maneuvers out of the dropped lane sometimes create turbulence and result in a drop in operating speed.

Horizontal curvature. A moderately sharp horizontal curve can reduce capacity. During conditions of heavy flow, vehicles weaving into an adjacent lane can cause hesitation and speed reduction, which often creates turbulence.

Ramp design. The design of ramps can have a direct effect on freeway capacity. A large merge angle, short or nonexistent acceleration and deceleration lanes, sharp ramp curvature, and poor sight distance on the ramp all tend to reduce freeway capacity.

Weaving sections. When a freeway on-ramp is closely followed by an off-ramp, the capacity of the intervening section may be significantly below the ideal because of the impact of vehicles exiting and entering the freeway mainline.

These effects can occur for substantial distances upstream and downstream of the weaving area.

Vertical alignment. The grades found on urban freeways generally have negligible effect on passenger-car capacity.[3] However, the effects are considerably more pronounced on the capacity of a freeway with a traffic mix of trucks, buses, and passenger cars. Severe vertical alignment causes reductions in capacity, depending on the percentage of trucks, the steepness of the grade, and the length of grade.[4]

Several other physical features tend to reduce freeway capacity. These include unconventional interchanges, inadequate shoulders, narrow medians, poor surface quality, and poor signing. In addition to physical features, freeway capacity may also be reduced if the traffic stream contains a significant proportion of unfamiliar drivers, such as might be found near recreational areas (see Chapter 5).

Factors contributing to nonrecurring congestion

Freeway congestion is frequently the result of an incident on the freeway that reduces normal capacity by blocking a lane or lanes. Almost any unusual occurrence will have an effect on freeway traffic, but the most common ones encountered on urban freeways include:

○ Accidents
○ Disabled vehicles
○ Spilled vehicle loads

---

[3] "Highway Capacity Manual," *Special Report 209* (Washington, DC: Transportation Research Board, 1985).

[4] Ibid.

o Presence of emergency vehicles
o Vehicles or people on the freeway shoulder

The occurrence of incidents during peak periods can cause congestion on normally uncongested sections and further increase delay on the already congested sections. Even in off-peak periods, congestion may develop, depending on the number of lanes affected. Even incidents involving vehicles completely on the shoulder tend to reduce capacity, and the capacity in the opposite direction also diminishes because of drivers slowing to view the incident (rubbernecking).[5] These random events are not predictable in either time or space, and thus they contribute to nonrecurring congestion.

Maintenance and construction work on freeways also results in a reduction in freeway capacity and can trigger congestion. Major work can result in significant delays and, consequently, efforts should be made to schedule maintenance and construction work during periods when traffic demand will not exceed the reduced capacity provided in the maintenance or construction zone. For this reason, major freeway work is generally conducted only during midday periods or at night in most urban areas. Design of work zone traffic control for maintenance and construction projects is also important, for poorly designed work zone traffic controls can cause traffic congestion problems.

Adverse weather is another unpredictable event that can effectively reduce freeway capacity and cause congestion. Rain, snow, fog, and ice typically reduce freeway capacity by 10% to 20%, with greater reductions possible for unusually severe conditions.

Impact of freeway congestion

Different freeway congestion problems have different degrees of impact. However, the severity of a particular problem generally depends on time of day, location, and duration of occurrence.[6] The severity is also dependent on the interaction between different problems that occur simultaneously. For example, a temporary hazard such as a disabled vehicle on the shoulder of a tangent section of a freeway may have only a negligible impact during off-peak conditions. However, the same incident on a horizontal curve during a peak period with traffic demand approaching capacity may compound the basic overloading of the facility and cause severe congestion.

Several studies have been conducted to measure the effects of freeway problems related to capacity. A study assessing the influence of incidents on freeway quality of service[7] showed that blockages of freeway lanes have a more severe effect than just the obvious reduction in number of lanes. The results, summarized in Table 13–1, show that a one-lane blockage on a three-lane roadway caused by a minor accident or a stalled vehicle reduced capacity by 50%, even though the physical reduction was only 33%. Presence of vehicles on the shoulder

TABLE 13–1

Capacity Reduction Resulting from Lane Blockages on the Gulf Freeway (Houston) (3 lanes in each direction)

Number of Lanes Blocked	Capacity Reduction (%)
Shoulder only	33
One lane	50
Two lanes	79

SOURCE: M.E. Goolsby, "Influence of Incidents of Freeway Quality of Service," *Highway Research Record 349* (Washington, DC: Highway Research Board, 1971).

that had been involved in an accident caused up to a 33% reduction in capacity, although there was no physical reduction of available roadway.

A recent Federal Highway Administration study estimated that recurring urban freeway congestion caused over 700 million vehicle-hours of delay, and nonrecurring congestion was responsible for some 1.2 billion vehicle-hours of delay in the United States in 1987 alone. This equates to approximately $16 billion in excess user costs due to lost time and wasted fuel.[8] It is clear that urban freeway congestion is a serious problem, but some of this problem can be alleviated through the implementation of surveillance and control systems as described in this chapter.

Methods of freeway control

Several methods of freeway traffic control have been developed to improve the operating efficiency of urban freeways and to minimize the occurrence and impact of congestion. These methods can be divided into five main categories:

o Entrance ramp control
o Exit ramp control
o Mainline control
o Priority control
o Corridor control

This section deals with a description of the basic concept of each type of control, and then considers, in some detail, each of the available techniques for implementing the control. Specific details related to the hardware needs are discussed in a later section, which is a combined description of the hardware requirements for surveillance as well as for control.

Entrance ramp control

The most common technique for freeway traffic control is *entrance ramp control*. Entrance ramp control systems are now in place in more than 20 urban areas in North America.[9] The primary objective of any freeway control technique is to improve the safety and efficiency of freeway operations

---

[5] P.F. Everall, *Urban Freeway Surveillance and Control—The State of the Art,* rev. (Washington, DC: U.S. Department of Transportation, Federal Highway Administration, June 1973).

[6] D.G. Capelle, *Freeway Traffic Management,* NCHRP Project 20-3D (Washington, DC: Transportation Research Board, September, 1979).

[7] M.E. Goolsby, "Influence of Incidents on Freeway Quality of Service," *Highway Research Record 349* (Washington, DC: Highway Research Board, 1971).

[8] J.A. Lindley, "Urban Freeway Congestion: An Update," *ITE Journal,* 59, (December 1989).

[9] J. Robinson and M. Doctor, *Ramp Metering Status in North America—Final Report* (Washington, DC: U.S. Department of Transportation, Federal Highway Administration, September 1989).

through the elimination, or at least the reduction, of the factors contributing to congestion. The underlying principle of entrance ramp control is to limit the number of vehicles entering the freeway so that the demand on the freeway will not exceed its capacity. Maximum flow rates will thus be achieved by ensuring that the freeway traffic moves at or near optimum speeds. Entrance ramp control forces motorists desiring to use the freeway to wait on the entrance ramp before being allowed to enter. The resultant queuing on the entrance ramp presents the motorist with four choices:

1. Wait in the queue in hope that the improved freeway speed will more than compensate for ramp delay.
2. Choose another point of entry or another time when the demand is lower.
3. Choose an alternative route to the freeway.
4. Choose another mode of transportation such as transit or some form of ride sharing such as carpools or vanpools.

Some of the benefits of entrance ramp control are clearly apparent. It provides a higher and more predictable level of service on the freeway, while concurrently improving the overall safety of operation, both on the freeway itself and on the entrance ramp. On the other hand, the diverted traffic may create operational problems on the alternative facilities in the corridor if care is not exercised in the implementation of entrance ramp control. Entrance ramp control can achieve improved efficiency of operations and safety on the freeway as well as a more effective utilization of all travel facilities in the corridor if certain conditions are satisfied. These conditions are described in the guidelines for freeway entrance ramp control signals, which can be found in Section 4E of the *Manual on Uniform Traffic Control Devices* (MUTCD).[10] They are as follows:

> There are too many variables that influence freeway capacity (number of lanes, trucks, gradients, merging, weather, etc.) to permit developing numerical volume warrants that are applicable to the wide variety of conditions found in practice. However, general guidelines have been identified for successful application of ramp control.
>
> The installation of ramp control signals should be preceded by an engineering analysis of the physical and traffic conditions on the highway facilities likely to be affected. The study should include the ramps and ramp connections and the surface streets which would be affected by the ramp control, as well as the freeway section concerned. Types of traffic data which should be obtained include, but are not limited to traffic volumes, traffic accidents, freeway operating speeds, travel time and delay on the freeway and on alternative surface routes.
>
> Capacities and demand/capacity relationships should be determined for each freeway section. The location and causes of capacity restrictions and those sections where demand exceeds capacity should be identified. From these and other data, estimates can be made of desirable metering rates, probable reductions in delay of freeway traffic, likely increases in delay to traffic on ramps, and the potential impact on surface streets. The analysis should include an evaluation of storage capacities

on the ramp for vehicles delayed at the signal, the impact of queued traffic on the local street intersection, and the availability of suitable alternative surface routes having adequate capacity to accommodate any additional traffic volume.

> Before installing ramp control signals, consideration should be given to public acceptance potential and enforcement requirements of ramp control, as well as alternate means of increasing the capacity, reducing the demand, or improving characteristics of the freeway.
>
> Installation of freeway entrance ramp control signals may be justified when the total expected delay to traffic in the freeway corridor, including freeway ramps and local streets, is expected to be reduced with ramp control signals and when one of the following instances occurs:
>
> 1. There is recurring congestion on the freeway due to traffic demand in excess of the capacity; or there is recurring congestion or a severe accident hazard at the freeway entrance because of inadequate ramp merging area. A good measure of recurring freeway congestion is freeway operating speed. An early indication of a developing congestion pattern would be freeway operating speeds less than 50 mph, occurring regularly for a period of half an hour. Freeway operating speeds less than 30 mph for a half-hour period would be an indication of severe congestion.
> 2. The signals are needed to accomplish transportation system management objectives identified locally for freeway traffic flow, such as:
>   (a) maintenance of a specific freeway level of service, or
>   (b) priority treatments with higher levels of service, for mass transit and carpools.
> 3. The signals are needed to reduce (predictable) sporadic congestion on isolated sections of freeway caused by short-period peak traffic loads from special events or from severe peak loads of recreational traffic. (pp. 4E-11–4E-12)

Besides the general information provided by these guidelines and a statement of recommended practice on freeway entrance ramp displays prepared by the Institute of Transportation Engineers,[11] the National Cooperative Highway Research Program has produced guidelines for the design, implementation, and operation of entrance ramp control systems.[12] In addition, the Transportation Research Board's Committee on Freeway Operations provides a continuing forum for the review of these guidelines.

Entrance ramp control is implemented through one of the following means:

- ○ Closure
- ○ Pretimed metering
- ○ Traffic-responsive metering
- ○ Integrated system control
- ○ Freeway connector control

Closure. As the name implies, this control technique consists of physically closing the ramp to traffic on either a permanent or a short-term basis. It is the simplest and also the most restrictive form of entrance ramp control and

[10] *Manual on Uniform Traffic Control Devices for Streets and Highways* (Washington, DC: U.S. Department of Transportation, Federal Highway Administration, 1988).

[11] ITE Technical Committee 4M-11, "Recommended Practice: Displays for Metered Freeway Entrance Ramps," *ITE Journal,* 54, (April 1984).

[12] Stanford Research Institute, *Guidelines for the Design and Operation of Ramp Control Systems,* NCHRP Project 3-22 (Washington, DC: National Cooperative Highway Research Program, 1975).

hence is applied sparingly. Permanent closure is generally implemented only as a last resort in situations where there are severe weaving problems on the freeway. Temporary closure of ramps during peak periods is generally considered as a solution only in the following situations:

o The storage area on the ramp for vehicles waiting to enter the freeway is not adequate. These vehicles thus cause a disruption in the traffic on the surface streets.
o The freeway traffic on the section before the entrance ramp is already running at or critically close to capacity, and there is a good alternative route with adequate capacity available for potential freeway users.
o The ramp design does not allow traffic to merge into the freeway traffic without considerable hazard.

Ramp closure is effected through the use of signs, manually placed barriers, and automated barriers. Signs usually tend to be ineffective, with a large number of violations. Manually placed barriers are effective, but they are labor-intensive and are only suited for short-term or trial periods, or where toll booths are located on the entrance ramps, as in Japan.[13] Automated barriers offer the greatest flexibility since entrances can be closed and opened automatically, enabling greater responsiveness to changes in traffic conditions. By and large, however, ramp closure is usually avoided because of strong adverse public reaction that is typically encountered.

Pretimed metering. Ramp metering is a widely used method for limiting vehicular access to the freeway, either to improve the operating conditions on the freeway or to improve the safety of the merging operation. Normally, a single lane ramp may have a capacity of 800 to 1,200 vehicles per hour, provided that the geometric design is adequate.[14] Ramp metering rates for a single lane ramp will generally be

less than this, but greater than a practical minimum of 180 to 240 vehicles per hour.

The simplest form of ramp metering is *pretimed metering,* whereby the metering rate is fixed according to clock time (the time of day and/or the day of week). The basic requirements for pretimed metering are minimal: at least two standard three-section (red-yellow-green) or two-section (red-green) traffic signals located on the ramp, a controller actuated by a time clock, and a warning sign with flashing beacon to indicate to approaching traffic that the ramp is being metered. A typical layout for a pretimed entrance ramp metering system is shown in Figure 13–3.

The metering rates to control the ramp traffic are calculated using historical data. The method of calculation depends on the purpose for which metering is being used—to restrict the volume of traffic entering the freeway or to improve safety of the merging operation by breaking up platoons of merging vehicles. If intended to restrict entering traffic, the metering rate calculation is straightforward, as shown in Figure 13–4. The metering rate is simply set equal to the difference between upstream freeway demand and the downstream freeway capacity—in this case, 300 vehicles per hour (vph). Some additional practical considerations, however, need to be taken into account. These are the conditions addressed by the MUTCD guidelines for entrance ramp control (e.g., is there adequate storage space for the stopping vehicles on the ramp? Are alternative routes and sufficient capacity available in the corridor?). Also, all gaps (up to capacity) may not be candidates for filling since downstream conditions may require some gaps be saved beyond the immediate merge section. Additionally, if the calculated metering rate of 300 vph is deemed too restrictive, consideration should be given to the metering of other ramps upstream of the ramp in question.

If pretimed metering is used only as a means of improving the safety of the merging area, the metering rate used is to enforce single-vehicle entry and must be set to ensure that each vehicle has time to merge before the following vehicle approaches the merging area. The goal is to break up platoons of vehicles, which otherwise lead to rear-end and lane-change

[13] *Traffic Control System of Metropolitan Expressway,* (Tokyo: Metropolitan Expressway Public Corporation, 1988).

[14] Everall, *Urban Freeway Surveillance.*

Figure 13–3. Layout of pretimed, entrance-ramp metering system.
SOURCE: Federal Highway Administration, U.S. Department of Transportation, *Traffic Control Systems Handbook,* Report No. FHWA-IP-85-11, Washington, DC, 1985.

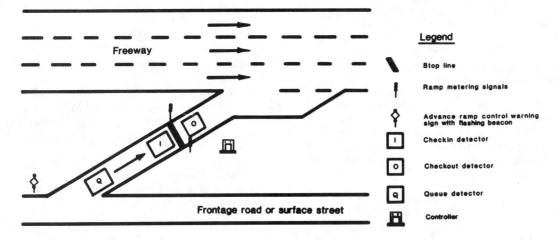

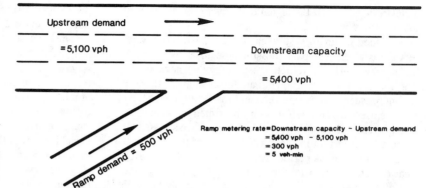

Figure 13–4. Example of pretimed, entrance-ramp metering calculation.
SOURCE: Federal Highway Administration, U.S. Department of Transportation, *Traffic Control Systems Handbook,* Report No. FHWA-IP-85-11, Washington, DC, 1985.

collisions while competing for gaps in the freeway traffic stream. The time to merge for a stopped vehicle depends on the distance from the freeway, ramp geometry, type of vehicle, and availability of acceptable gaps in the freeway traffic. For example, if the average time to merge, taking all these factors into account, is 8 sec, the metering rate would be 450 vph.

The traffic signal in a pretimed metering system operates with a constant cycle according to the metering rate as calculated above. The selected metering rate typically stays in effect for periods of 30 min to an entire peak period. Depending on the magnitude of the metering rate, the metering is either single-entry (for rates up to 900 vph) or platoon-entry (for rates above 900 vph). As the name implies, single-entry metering is accomplished by timing the signal so as to permit only one vehicle to enter the freeway for each green interval. The green-plus-yellow (or just green if yellow is not used) interval is just long enough (usually 3 sec) to allow one vehicle to proceed past the signal. A two-detector layout, as shown in Figure 13–3, terminates the green when the vehicle reaches the checkout detector—this is generally preferable to a constant green time. For the metering rate of 300 vph, or 5 vehicles/min, a green-plus-yellow interval of 3 sec and a red interval of 9 sec is typically used.

Platoon metering allows the release of two or more vehicles per cycle and is used to accommodate high metering rates (greater than 900 vph). Platooning can be achieved by releasing vehicles one behind the other (tandem) or two-abreast in two parallel lanes.[15] For tandem platooning, the green time has to be sufficiently long to allow all vehicles in the platoon to accelerate and pass the signal. A yellow interval is required to prevent sudden braking problems and attendant rear-end collisions. As an example, consider tandem metering of two vehicles per cycle to achieve a metering rate of 1,080 vph or 18 vehicles/min[16] Nine cycles are required per minute, or a cycle of 6.7 sec, which could be split into a 4.7-sec green-plus-yellow interval and a 2-sec red interval. In the United States, experience with tandem metering indicates that two-vehicle platoons can be handled satisfactorily and three-vehicle platoons are a practical maximum, with 1,100 vph the maximum metering rate that can be expected in either case.[17]

In the United Kingdom, however, platoons of up to nine vehicles have been successfully metered onto the freeway.[18]

Similar flow rates can also be achieved by releasing two vehicles in parallel for each cycle. This, of course, requires that two parallel lanes be available for a sufficient distance beyond the ramp-metering signal for the two vehicles to achieve a single-file orientation before merging with the freeway traffic. In one location in Minneapolis, an entrance ramp has been widened to four lanes upstream of the meters, and vehicles are released in pairs on either side of the ramp.[19] The cycle timing for parallel platoon metering is similar to signal-entry metering; that is, each green-plus-yellow interval is just long enough (about 3 sec) to permit one vehicle in each lane to proceed past the ramp-metering signal. For the example of 1,080 vph, requiring nine cycles/lane/min for two parallel lanes, the cycle time of 6.7 sec could be split into a 3-sec green-plus-yellow interval and a 3.7-sec red interval.

Both forms of platoon metering exhibit several disadvantages compared to single-entry metering: greater driver confusion, greater probability of rear-end collisions (tandem metering), and greater disruption of freeway traffic. Of the two platoon-metering schemes, two-abreast is generally preferable, as it tends to create less driver confusion and provides for a safer operation. Platoon metering should only be considered for achieving higher metering rates or when the ramp becomes an added lane on the freeway.[20]

While a pretimed ramp-metering system requires no detectors, some refinements in operation are possible using the detectors indicated in Figure 13–3 as optional equipment. A check-in (demand) detector is highly recommended so that the signal turns green only when a vehicle is waiting. Thus, the signal stays red until the detector is actuated, at which time it turns green provided that the minimum red time has elapsed. During peak periods there will often be a queue at the signal, which will then be continuously metered at the set rate. Painting a stop line at the signal decreases the chance of vehicles not moving forward far enough to actuate the detector with a consequent all-red condition. As a provision for such an occurrence, it is desirable for pretimed metering systems with check-in detectors to have a minimum metering rate (say, 3

[15] Stanford Research Institute, *Guidelines.*

[16] *Traffic Control Systems Handbook,* Report No. FHWA-IP-85-11 (Washington, DC: U.S. Department of Transportation, Federal Highway Administration, April 1985).

[17] Stanford Research Institute, *Guidelines.*

[18] D. Owens and M.J. Schofield, "Access Control on the M6 Motorway: Evaluation of Britain's First Ramp-Metering Scheme," *Traffic Engineering and Control,* 29 (12) (1988).

[19] Robinson and Doctor, *Ramp Metering Status.*

[20] Stanford Research Institute, *Guidelines.*

vehicles/min) at which the signal operates in the absence of detector actuation.

A pretimed metering system can use a checkout (passage) detector installed just past the stop line to ensure single-vehicle entry.[21] When a vehicle is sensed by the checkout detector, the green interval is terminated immediately, thereby ensuring passage of only one vehicle. Finally, a queue detector can be used in conjunction with a pretimed metering system to prevent ramp traffic from blocking surface streets.[22] It is especially important in the initial turn-on of the system at the beginning of a peak period. The queue detector is situated so that its rate of actuation or occupancy indicates that the queue of vehicles waiting on the ramp has possibly become long enough to interfere with the surface street traffic. A higher metering rate can then be used to shorten the queue length.[23] This can be accomplished by having different metering rates programmed in the controller with queue detection transferring signal control to the higher rate. A merge area override detector can also be used with pretimed metering where the merge geometrics are substandard. Finally, when the metering system is not operating (e.g., during off-peak periods), the signals should display a continuous green indication or be turned off.

A discussion of cost-effectiveness of freeway control systems appears in a later section. It suffices here to say that pretimed metering entrance ramp control can be extremely cost-effective.[24] Its major operational advantage is that it provides motorists with a regular metering rate to which they can readily adjust. The principal drawback of pretimed metering is that such a system cannot respond to significant changes in demand or take into account unusual traffic conditions resulting from incidents on the freeway. Because of this insensitivity to changes in traffic conditions, pretimed metering rates are usually set such that the resultant downstream mainline freeway operation is below capacity.

Traffic-responsive metering. Traffic-responsive metering is based on the same principle as is pretimed metering—the demand-capacity constraint. The difference lies in the method of selecting metering rates. Pretimed metering rates are computed using historical measures of traffic conditions, whereas traffic-responsive metering rates are selected using real-time measurements of traffic conditions. Traffic-responsive metering thus overcomes the inherent inability of pretimed metering to respond to short-term changes in traffic conditions.

The basic requirements for implementing traffic-responsive metering are considerably more complex than those for pretimed systems. Such systems require at least two traffic signals; an advance ramp control sign and flashing beacon; a controller with the capability of monitoring traffic variable measurements and using these to select or calculate a metering rate; and various traffic detectors to measure—in real time—the variables used in selecting

the metering rate. A typical layout for a traffic-responsive ramp-metering system is shown in Figure 13–5.

The basic strategy of traffic-responsive metering is to obtain real-time measurements of traffic conditions on the freeway, use these to examine how the freeway is operating with respect to its capacity, and then determine the maximum number of ramp vehicles that can be permitted to enter the freeway without causing congestion. Volume and occupancy (the proportion of time a detector is actuated) are the traffic variables generally used in determining traffic-responsive metering rates as they can be measured by the direct processing of freeway detector data. The traffic-responsive metering selection is based principally on one of two strategies: demand-capacity and occupancy control.

Demand-capacity control. For this type of control strategy, the metering rate is determined using real-time comparison of upstream volume and downstream capacity. The upstream volume is measured in real time and compared either to a preset historical value of downstream capacity or a current value calculated from downstream volume measurements. The effectiveness of demand-capacity control is enhanced if the downstream capacity accounts for the effect on capacity of weather, traffic composition, and incidents. Additionally, since volume alone is insufficient in determining whether the freeway is congested or free-flowing, an occupancy measurement is usually made from at least one of the upstream detectors. The allowable ramp entrance metering rate for the next control period (usually 1 min) is the difference between the upstream volume and downstream capacity. A minimum metering rate of 3 to 4 vehicles/min is used if upstream volume is equal to or greater than downstream capacity, or if the occupancy measurement is above a preset value that—determined from past experience—can be taken as an indicator of congestion.

Occupancy control. For this control strategy, one of several predetermined metering rates is selected for the next control period (usually 1 min) on the basis of occupancy measurements made during the current period. The real-time occupancy measurements are made upstream, downstream, or both upstream and downstream of the ramp in question, and the metering rate is typically based on a plot of historical volume-occupancy data collected at the same locations.[25] Such a plot establishes an approximate volume-occupancy relationship and also determines a value of occupancy at capacity. An example of such a plot is shown in Figure 13–6. Using the historical plot and the measured current value of occupancy, the metering rate is determined as the difference between the historical estimate of capacity and the real-time estimate of volume. If measured occupancy is greater than the (preset) occupancy corresponding to capacity, a minimum metering rate (3 to 4 vehicles/min) is used. Occupancy control offers the advantage of using fewer detectors than demand-capacity control. However, demand-capacity control uses more precise estimates of upstream volume and consequently may bring about more accurate control.[26]

[21] *Traffic Control Systems Handbook.*

[22] Ibid.

[23] Some jurisdictions, such as San Diego, allow metered vehicles to queue onto surface streets. See Robinson and Doctor, *Ramp Metering Status.*

[24] Everall, *Urban Freeway Surveillance.*

[25] *Traffic Control Systems Handbook.*

[26] Stanford Research Institute, *Guidelines.*

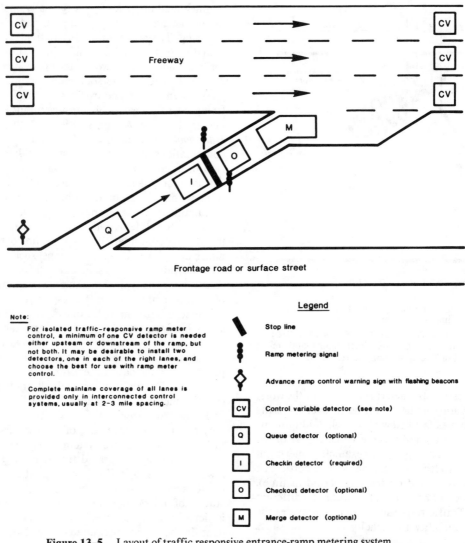

Figure 13–5. Layout of traffic responsive entrance-ramp metering system.

SOURCE: Federal Highway Administration, U.S. Department of Transportation, *Traffic Control Systems Handbook,* Report No. FHWA-IP-85-11, Washington, DC, 1985.

Operation of a traffic-responsive metering system is essentially similar to that of a pretimed metering system, except for the differences in the methods used for selecting the metering rate. Single-entry metering is most often used for implementing traffic-responsive systems, although high metering rates (greater than 800 to 900 vph) may necessitate the use of platoon entry. Both methods of metering (single and platoon entry) should not be used at the same ramp (to avoid driver confusion) as a result of changing from one method to the other between control periods.

Traffic-responsive metering systems generally include override features that tend to enhance system operation.[27] These features are implemented through the use of the optional detectors shown in Figure 13–5. A continued actuation of the queue detector indicates that the queued vehicles on the ramp are likely to interfere with the traffic on the frontage road or surface street, a situation that can then be corrected by selecting a higher metering rate. A merge detector can be used to determine whether a vehicle is still in the merging area. For single-entry metering, such a detection can then be used to preempt subsequent green intervals until the vehicle merges. A checkout (passage) detector can be used to determine whether a vehicle missed the green signal, since in that case there would be no actuation of the checkout detector. The override logic in this case can be set to return the signal to or leave it in green, thus allowing the vehicle to proceed. For systems that rest in red until a check-in detector is actuated, a queue detector can be employed to determine whether a vehicle has stopped short of the check-in detector. This would be indicated by a continued actuation of the queue detector concurrent with no actuation of the check-in detector. The override logic can be set to turn the signal to green and begin the metering cycle.

The principal operational advantage of traffic-responsive metering over pretimed metering is that is tends to utilize current available capacity to the fullest, whereas pretimed

---

[27] *Traffic Control Systems Handbook.*

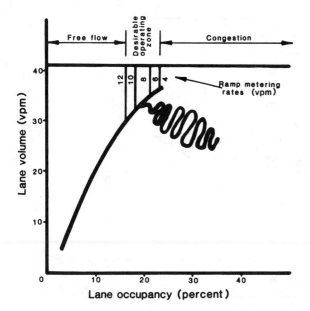

Figure 13–6. Volume-occupancy plot for calculation of entrance-ramp metering rates in Chicago.
SOURCE: J.M. McDermott and others, *Chicago Area Expressway Surveillance and Control,* Final Report, State of Illinois, Department of Transportation, March 1979.

control must, of necessity, use metering rates set for the worst case of peak upstream volume, resulting in over-metering when upstream volumes fall below this level. Additionally, traffic-responsive metering reduces the vulnerability of the control system to certain changes.[28] For example, metering at an entrance ramp with a large volume may solve a bottleneck problem at one location (and during one period of time), but it may trigger another bottleneck to appear at a different location and time. Traffic-responsive metering provides the same degree of adaptability to capacity changes or shifts in bottlenecks. Capacity reductions attributable to weather can also be somewhat accommodated by traffic-responsive metering.[29] The single most important factor, from an operational point of view, in choosing between pretimed and traffic-responsive metering is the variation in demand and capacity likely to be experienced for the freeway segment in question. If demand and capacity variations are essentially insignificant, pretimed metering will meet most needs. However, traffic-responsive metering should be considered when there are frequent occurrences of significant demand and capacity changes.

Integrated system control. All the methods of ramp control discussed so far deal with the control of vehicular flow at an individual ramp. However, it is often necessary to apply ramp control to several entrance ramps in an integrated fashion so that better utilization of the entire freeway system can be achieved. The control strategy to be employed is then based on the demand-capacity considerations for the entire system rather than on the demand-capacity constraints at each individual ramp.

Integrated ramp control is typically applied to a series of ramps that are subject to either pretimed or traffic-responsive metering. The operation of any individual ramp in an integrated system is similar in all respects to the operations described under these two metering schemes *except* for the calculation and selection of metering rates.

A systematic procedure for calculating the pretimed metering rates for a series of entrance ramps is described in the Federal Highway Administration's *Traffic Control Systems Handbook,* cited earlier in this chapter. The metering-rate computations are based on historical data for the time period of interest. Freeway mainline demands and capacity, entrance ramp demands, and descriptions of the traffic patterns within the segment to be controlled are utilized for calculating individual ramp-metering rates.

Integrated systems control using traffic-responsive metering involves coordinating the application of traffic-responsive metering to several entrance ramps, such that the control at each individual ramp is subject to both systemwide and local constraints. The typical method for integrating the operation of several ramps with traffic-responsive metering is to first calculate, usually through a linear programming formulation, sets of integrated metering rates for each ramp for the range of demand-capacity conditions expected. During operation, the rate at each individual ramp is selected from these precomputed sets on the basis of real-time measurements of freeway conditions. Details of other methods of integrated traffic responsive metering are contained in the literature.[30]

Integrated system operation with a series of ramps employing pretimed metering offers the advantage of better utilization of the entire system than does independent pretimed metering. Integrated traffic-responsive metering allows variations in traffic demands to be optimized over the entire freeway system. These advantages are borne out by studies of independent and integrated ramp control, which indicate lower travel time, higher total travel, and fewer accidents in the case of integrated operations.[31]

Freeway connector control. Metering of freeway-to-freeway connectors is a special case of ramp metering that has been applied successfully in several areas. Uncontrolled flow on freeway-to-freeway connectors can lead to merging and congestion problems on the freeway similar to those caused by unrestricted access from surface street ramps. In an area where ramp metering has been implemented for surface street ramps, allowing a freeway connector to remain unmetered can offset much of the benefits of ramp metering. In general, freeway-to-freeway connector metering includes the following benefits:[32]

o Relocation of queuing from the freeway mainline to the connector.

[28] Stanford Research Institute, *Guidelines.*

[29] *Traffic Control Systems Handbook.*

[30] H.J. Payne and others, *Demand Responsive Strategies for Interconnected Freeway Ramp Control Systems* (3 vols.), Report Nos. FHWA/RD-85/109 through FHWA/RD-85/111 (Washington, DC: U.S. Department of Transportation, Federal Highway Administration, March 1985).

[31] *Traffic Control Systems Handbook.*

[32] H.J. Payne and A.D. May, *Control of Mainline Freeway and Freeway-Freeway Connectors for Bottleneck Alleviation,* Report No. FHWA/RD-82/030 (Washington, DC: U.S. Department of Transportation, Federal Highway Administration, June 1982).

o Improved flow through the bottleneck area on the freeway mainline, resulting in increased throughput, reduced delay, and reduced accidents.
o Elimination of the need for less efficient metering on ramps upstream of the connector.
o Diversion of connector traffic to alternate routes, if available.
o Improved overall equity for motorists in the corridor.

Metering techniques used for connector control are basically the same as those used for metering from surface streets, but owing to typically high volumes and potentially high speeds on the connector, careful attention must be given to providing adequate ramp storage space (so that metered traffic does not queue onto the freeway) and advance warning of the need to stop on the connecting ramp. Cities where connector metering has been successfully applied include Minneapolis, San Diego, San Antonio, Seattle, and San Jose.[33]

For further information on all of the above entrance ramp control strategies, a recent Federal Highway Administration report contains details on configuration and benefits of strategies that have been implemented in 20 U.S. and Canadian cities.[34]

Exit ramp control

Control of traffic on exit ramps can be achieved either through closure or some form of metering. However, neither is a universal strategy for freeway control. Metering of exit ramps may offer some relief of congestion on the adjacent surface streets, but it also increases the likelihood of queuing of vehicles onto the freeway, which could cause rear-end collisions on the ramp and on the freeway mainline. Closure of exit ramps is a viable strategy for exit ramp control in the sense that it is applicable as a safety measure in situations where closure would drastically reduce weaving, which often occurs between closely spaced ramps. Also, exit ramp closure can be used at a lane drop location by closing downstream exit ramps to encourage more traffic to leave the freeway prior to the lane drop and thereby reduce freeway demand beyond the lane drop.

The principal disadvantage of exit ramp closure is the adverse public reaction stemming from increased travel time and restriction of access, as well as an increase in the potential for rear-end collisions at such locations. Additionally, if the exit ramp closure is to be instituted only at specific times (e.g., peak periods, special events), considerable operational costs may be incurred. For these reasons, exit ramp closure is generally not considered to be a workable control strategy.

Mainline control

Another approach in improving the safety and efficiency of freeway operations is the control of traffic on the freeway mainline itself. This involves the regulation, warning, and guidance of traffic on the freeway mainline so that some or all of the following objectives are addressed:

o To achieve a more uniform and more stable traffic flow as the freeway demand approaches capacity, leading to improved utilization of the facility and prevention of the onset of congestion.
o To reduce the probability of rear-end collisions caused when motorists unexpectedly encounter congested conditions.
o To distribute total delay in a more equitable manner, by saving some freeway capacity for downstream segments.
o To increase the efficiency of operation under restricted-capacity conditions caused by incidents or maintenance operations.
o To divert some freeway traffic to alternate routes in order to make better use of corridor capacity.

The general indications from the extensive application of various mainline control strategies in West Germany, Japan, and the United Kingdom are that these techniques are workable methods for improving the safety and operating efficiency of freeways. Although there has been relatively little usage of mainline control in the United States, there is interest in the concept, and increased use of these techniques as essential components of freeway traffic control systems is being actively pursued in some areas.

Mainline control is typically implemented through one of the following means:

o Driver information systems
o Variable-speed control
o Mainline metering
o Lane control
o Reversible lane control

Driver information systems. Driver information systems are employed to advise motorists of freeway conditions so that appropriate actions can be taken to enhance the efficiency and safety of freeway operations. The design philosophy is to provide drivers with useful, real-time information about freeway conditions so that they can choose to divert from the freeway onto an alternative route or continue on the freeway under some form of control. Effective design of driver information systems requires adherence to certain principles:[35]

o Observance of primacy (i.e., providing the more important information with more impact);
o Preventing overloading of information;
o Providing information well in advance of decision points;
o Spreading out of information to keep the driver's attention;
o Avoiding unexpected information.

---

[33] Robinson and Doctor, *Ramp Metering Status.*
[34] Ibid.

[35] T.M. Allen, H. Lunenfield, and G.J. Alexander, *Driver Information Needs,* NCHRP Report No. 123 (Washington, DC: National Cooperative Highway Research Program, 1971); and *Proceedings of the International Symposium on Traffic Control Systems,* Vol. 1 (Berkeley: Institute of Transportation Studies, University of California, and U.S. Department of Transportation, Federal Highway Administration, August 1979).

Evaluation studies have determined that drivers prefer information sign displays that are simple and have a unique design. This allows them to be clearly distinguished from other types of freeway signs and also provides a marked difference between displays for normal and abnormal freeway conditions.[36] The driver information systems most deployed in mainline control are single and changeable message signing, and roadside and commercial radio broadcasts.

Signing consists of real-time visual information displays in the form of single or changeable messages. Single-message signs are used to warn drivers of a particular hazardous situation ahead, such as weather, environmental conditions, or congestion; these signs are only turned on when the particular hazard occurs. Clearly, these single-message signs are limited in their application, but they can be very useful at those locations where a hazard is well defined and occurs periodically. Because the signs must be responsive to the occurrence of the particular hazard, some means of the hazard detection is required. This is a primary function of freeway surveillance.

Changeable message signs have the same intended purpose as the single-message signs; but because of their ability to convey a variety of information, they are more effective in presenting motorists with current information on changing freeway conditions. The types of information that are typically presented by changeable message signs include traffic conditions (e.g., location and duration of congestion), weather and environmental conditions (e.g., ice, fog), and routing (e.g., alternative routes available). As with single-message sign, the operation of changeable message signs requires inputs from the surveillance system to select the appropriate sign message. If conditions reported by the surveillance system warrant the selection of two or more messages, a priority system determines the message to be displayed. Although changeable message sign control is done manually in some instances (based on qualitative evaluations of visual observations), the trend is toward automatic control on the basis of quantitative analysis of electronic surveillance data.[37] Even though there is limited empirical data on the effectiveness of changeable message sign systems, it has been determined that motorists are less frustrated and aggravated if they are provided with information on the location and length of the congested area and the expected length of the delay to be encountered.[38] This in itself tends to increase both the safety and comfort of their trips and improve the overall efficiency and safety of the system. However, with either method of signing, it is important to ensure that the driver is presented with accurate, timely information.

If the information is not current and motorists are given reasons to believe it to be erroneous, credibility will suffer and the system will not serve its intended purpose.

Commercial radio is a common means of providing freeway traffic information to drivers; local broadcast stations in many major urban areas provide this type of information on a routine basis during the peak periods of traffic flow. The effectiveness of these broadcasts depends on their timeliness, accuracy, and reliability. Cooperative arrangements between radio stations and freeway operating agencies have now been established in several urban areas whereby detailed traffic information is furnished to the stations at regular intervals, and the stations are notified immediately of traffic problems resulting from accidents or other major incidents. In Chicago, for example, freeway traffic information reports are provided to over 50 media outlets, primarily commercial radio stations.[39] Similar programs are underway in other major cities.

In Minneapolis, for instance, a unique arrangement exists between the Minnesota Department of Transportation (MnDOT) and a local FM public radio station. This agreement allows MnDOT to specify the information to be broadcast and the timing and number of broadcasts. During a major event, such as a heavy snowfall or major accident, MnDOT may specify that the broadcasting of traffic advisory information be continuous.[40] These cooperative arrangements have resulted in a more reliable system for providing real-time driver information. The degree of success of commercial radio as a reliable driver information system is thus clearly dependent on the cooperation of commercial radio stations and the ready access by these stations to accurate and timely traffic information. The potential for communicating motorist information through such links is enormous because more than 80% of the commuters listen to commercial radio during peak travel periods.

Roadside radio, commonly known as Highway Advisory Radio (HAR), is a communication system in which messages can be transmitted directly to the driver on stations at the extreme ends of the AM band from low-power local roadside transmitters. Its major advantage over commercial radio is that the messages broadcast to drivers can be more specific to the conditions at a particular location. Experimental studies have revealed that HAR can be used effectively, and that the system is acceptable to motorists. An HAR system operating in the I-35 freeway corridor in Minneapolis yielded the following findings:[41]

o System operator time was minimal, requiring only 5% of the operators' overall time.
o Over 90% of the commuters were aware of the system. Of the commuters who had tried the system, 80% used it occasionally, frequently, or always.

[36] C.L. Dudek and H.B. Jones, "Evaluation of Real-Time Visual Information Displays for Urban Freeways," *Highway Research Record 366* (Washington, DC: Highway Research Board, 1971); and C.L. Dudek, "Real-Time Displays," *Proceedings of the International Symposium on Traffic Control Systems*, Vol. 1 (Berkeley: University of California, Institute of Transportation Studies, and U.S. Department of Transportation, Federal Highway Administration, August 1979).

[37] D.H. Baxter, "Changeable Message Signs for Freeway Operations—Preliminary Results of Sign Usage in the Long Island Freeway Corridor," *Compendium of Technical Papers* (Washington, DC: Institute of Transportation Engineers, 1989).

[38] Dudek and Jones, "Evaluation of Real-Time Visual Information Displays."

[39] J.M. McDermott, *Advanced Technology Applications in the Chicago Area Freeway Traffic Management Program*, Illinois Department of Transportation, 1989.

[40] *Freeway Operations Program, Status Report*, Minnesota Department of Transportation, June 1990.

[41] G.C. Carlson and others, *Evaluation of Highway Advisory Radio in the I-35 W Traffic Management Network* (Washington, DC: Federal Highway Administration, March 1979).

○ System limitations were that 20% of the commuters never tuned to advisory radio, another 17% had serious tuning difficulties, 3% were not aware of the system, and 3% had no AM radio in their car.

○ Public acceptance of the system was very high, with more drivers liking it than disliking it by a margin of 6:1, and more people favoring expansion to other freeways than not favoring it by a margin of 7:3.

○ The suggestions for system improvement made most frequently were to upgrade the broadcast quality and to provide more detailed traffic information.

○ Traffic diversion studies indicated that commuters divert from the freeway in response to a broadcast when (1) congestion is heavy, (2) they are given sufficient detailed information, and (3) they have an adequate alternative route.

○ System maintenance experience was very favorable.

○ System design, installation, operation, administration, and maintenance costs were low.

Major problems associated with these systems concern the dependency on drivers to have radio receivers in their vehicles, and that they must tune such a radio to the required frequency each time they require information. In the United States, there is the additional problem of Federal Communication Commission (FCC) restrictions on low-range amplitude-modulated (AM) transmission. Finally, such radio systems are opposed by some commercial radio stations who feel that the service provided is competitive with their own ongoing programs.

An emerging technology with potential useful application in motorist information systems is the in-vehicle route guidance system. A number of these systems—which are designed to provide motorists with an in-vehicle, areawide electronic map and information on their current location and routes to their destination—have been developed both in the United States and Europe. An enhancement to these systems that would greatly increase their usefulness is the capability to receive and display information on real-time traffic conditions from existing traffic control systems. This would allow motorists to make better-informed decisions on which route to take to reach their destination. A cooperative experiment, known as the Pathfinder Project, was recently conducted in a 12-mile freeway corridor in Los Angeles to evaluate the effectiveness of providing traffic congestion information to motorists in vehicles equipped with modified in-vehicle route guidance devices.[42]

Variable-speed control. Variable-speed control of the freeway mainline is implemented through changeable message signs that display variable speed limits. This is intended to control the speed of traffic on a freeway at a level that achieves the maximum safe volume of traffic flow. The underlying principle is that, as the demand on the facility increases to optimum density, speed control can help improve the stability of flow and help achieve maximum volumes.

This, of course, is true only if the demand is equal to or less than the capacity of the facility; otherwise, variable-speed control can, at best, only delay the onset of congestion but not prevent it. However, improving the uniformity of speeds does reduce the probability of rear-end collisions as congestion builds up.

In using variable-speed control, traffic flow conditions are defined as a function of one or more traffic flow variables (e.g., volume, occupancy, density, speed) and the speeds to be posted are typically selected on the basis of real-time measurements and historical values of these variables. Variable-speed control can also be used during off-peak periods as an advance warning to alert motorists to such traffic hazards as accidents and traffic obstructions. The very limited experience of the United States in the use of variable-speed control has not been very successful. In some earlier systems,[43] it was found that motorists did not consider the variable speed limits credible and they did not decrease their speeds to coincide with the posted limits unless there was an apparent reason to do so. The control was also not successful in increasing the flow through critical bottlenecks. In Albuquerque, New Mexico, a prototype variable speed limit system in which posted speeds are directly related to real-time traffic stream speed measurements has been operational since 1989 but has had limited effectiveness because, by law, speed limits above 55 mph cannot be posted.[44] In Europe, variable-speed control signs on the freeway have been found to be very effective and have yielded favorable results in terms of improving the speed distribution and reducing the frequency and severity of accidents.[45, 46]

Mainline metering. Mainline metering is a form of mainline control whereby the traffic entering a freeway control section via mainline lanes is controlled according to the input demand and downstream capacity. Although such a tactic can create congestion on the mainline upstream of the control section, it can help maintain uncongested flow on the mainline downstream through the control section. The major objective of such a strategy is to control traffic demand at a mainline control point in a manner that would maintain a desired level of service on the freeway downstream of this control location. The desired level of service may be set to meet various objectives: to maximize the flow through a downstream bottleneck; to provide a high level of service to all vehicles downstream of the control point, especially for buses and carpools; to distribute total delay on the freeway system more equitably; or to encourage diversion of traffic from the freeway because of normal traffic congestion or because of incidents that reduce the effective capacity.

---

[42] F. Mammano and R. Sumner, "Pathfinder System Design," *Conference Record of Papers Presented at the First Vehicle Navigation and Information Systems Conference* (Washington, DC: Institute of Electrical and Electronics Engineers, 1989).

[43] F. DeRose, "Lodge Freeway Traffic Surveillance and Control Project," *Highway Research Record 21* (Washington, DC: Highway Research Board, 1963).

[44] R.L. Sumner and C.M. Andrews, *Variable Speed Limit System,* Publication No. FHWA-RD-90-001 (Washington, DC: U.S. Department of Transportation, Federal Highway Administration, March 1990).

[45] J.T. Duff, "Accomplishments in Freeway Operations Outside the United States," *Highway Research Record 368* (Washington, DC: Highway Research Board, 1971).

[46] K. Russam, "Motorway Signals and the Detection of Incidents," *Transportation Planning and Technology,* 9 (October 1984).

An application of mainline metering has been documented in Japan, where it has been made effective by regulating the number of toll booths open on the mainlines of expressways, thereby controlling the traffic entering the freeway via the mainline according to the available freeway demand and downstream capacity.[47] Another application is the San Francisco–Oakland Bay Bridge mainline metering in Oakland, where controls are used after vehicles pass through the toll plaza at a point prior to entering the bridge.[48]

Other applications of mainline metering on such controlled-access facilities as terminals, bridges, and toll roads have been found to be effective, but the concept is yet to be applied to a typical urban freeway system. Another potentially effective use of mainline metering that has never been applied is in a construction zone where traffic demand greatly exceeds the capacity available and some reserve capacity for vehicles entering at downstream ramps needs to be provided.

Lane control. This method of mainline control operates through the closure of one or more main lanes on the freeway. It is usually implemented during periods of reduced capacity to improve the operating safety and efficiency of the freeway system. Lane control is typically used in mainline control for the following purposes: advance warning of lane blockages, traffic diversion during peak flow periods, tunnel control, and for improvement of merging operations.

Lane control can be implemented through the use of lane-control signs, which indicate green arrows for open lanes and red *X*'s for closed lanes. Downstream lane blockages due to maintenance operations or incidents can be indicated by displaying the red *X* over the affected lane. In the United Kingdom, lane-control signals are used in conjunction with variable-speed control to warn drivers of lane blockages ahead, and they can be used to stop all traffic on the mainline if an incident warrants such action. The system has been found to be effective in reducing the occurrence and severity of accidents on the freeways.[49]

Lane control can be used to divert traffic from the freeway mainline onto alternative routes during peak periods to alleviate congestion. It can also be employed to reduce congestion and hazardous conditions caused by merging operations during heavy traffic flow on the freeway. This can be done by closing—at a point upstream of the merging section—the freeway lane into which the vehicles merge. While such a strategy is likely to reduce the difficulties of the merging operation, some increase in delay to traffic on the mainline can be expected.

Finally, lane control has found application in the control of tunnels on the freeway mainline. An example is a system installed in Italy, where entrance to a tunnel on the freeway mainline is denied by conventional red-yellow-green traffic signals if an incident or very slow traffic is detected in the tunnel through checks of speed and occupancy.[50] A similar technique is being used by the Port Authority of New York and New Jersey in the Lincoln Tunnel.

Reversible lane control. If peak-hour traffic volumes on a section of the freeway exhibit a significant directional imbalance (e.g., 70%–30%), which is predicted to continue for several years, the attendant problems of congestion in the heavier-flow direction may be alleviated through the use of reversible freeway lanes. The directions of flow on one or more lanes in the low-volume direction is reversed, using signs to inform the motorist of the direction of flow, and gates and barriers to prevent wrong-way movement.

If designed for new freeways, it is best to construct a three-roadway facility and separate such lanes from the conventional lanes for safety reasons. It is also advisable to schedule the lane reversal for the same time each day to allow motorists to be accustomed to the system. An application of this technique is Interstate 5 on the northern approach to Seattle, with a 7.5-mile reversible-lane section in the median.[51] Safe reversible lane control can also be effected by using recently developed concrete barrier systems that can be quickly relocated.

The principal advantage of reversible lane control is that it is a more economical and efficient use of road space and right-of-way if traffic flow imbalance is present and is predicted to continue.

Priority control

Priority control of freeways provides preferential treatment for buses, vanpools, and carpools using the freeway, allowing them to bypass congested areas. The essence of this concept is to encourage greater use of high-occupancy vehicles by providing shorter and more reliable travel times, thereby reducing the total vehicle demand on the freeway while serving higher person demand. The different methods of priority control used on freeways are physically separated facilities, contra-flow facilities, concurrent-flow facilities, and priority access control.[52]

Physically separated facilities. Physically separated facilities are limited-access roadways specifically constructed for the exclusive use of high-occupancy vehicles (HOV). They are often used to provide express service between activity centers, or for short bypasses at major freeway bottlenecks. In most cases, the separated roadways are constructed in the median of an existing freeway with some form of barrier providing the physical separation from the general-purpose freeway lanes. The principal advantages of this method are that the priority vehicles can operate safely at high speeds, the efficiency of the existing freeway is not reduced, the separate

[47] *Traffic Control Systems Handbook.*

[48] M.S. MacCalden, "A Traffic Management System for the San Francisco–Oakland Bay Bridge," *ITE Journal,* 54, (May 1984).

[49] Duff, "Accomplishments in Freeway Operations."

[50] R. LePera and R. Nenzi, "TANA–An Operating Surveillance System for Highway Traffic Control," *Proceedings of the Institute of Electrical and Electronics Engineers,* 61 (May 1973). The effectiveness of this system is discussed in P. Ferrari and F. Giannini, "An Aspect of the Determination of the Most Useful Real-Time Traffic Condition Information for Urban Motorway Users," *Proceedings of the International Symposium on Traffic Control Systems* (Berkeley: University of California, Institute of Transportation Studies, and U.S. Department of Transportation, Federal Highway Administration, December 1979).

[51] Everall, *Urban Freeway Surveillance.*

[52] M.J. Rothenberg and D.R. Samdahl, *High Occupancy Vehicle Facility Development: Operation and Enforcement,* Report No. FHWA-IP-82-1 (Washington, DC: U.S. Department of Transportation, Federal Highway Administration, April 1982).

roadways have long service lives and low operating costs, and the enforcement of priority control is simplified. There are, however, some disadvantages, such as high capital costs, long construction time, and potentially low usage during off-peak periods.

Contra-flow facilities. The use of contra-flow facilities essentially consists of using a freeway lane in the off-peak direction for peak direction traffic. Clearly, this requires the existence of highly directional peak-period flow with sufficient excess capacity in the off-peak direction. Movable barriers are normally used to delineate the lane, while access is limited to specific slip ramps or crossover points at each end of the system. A buffer lane may also be used to separate the traffic on the contra-flow facility from the freeway traffic moving in the opposite direction if continuous barriers are not used. Such facilities are typically reserved exclusively for buses and vanpools; carpools are usually not allowed to use non-barrier-separated contra-flow lanes because of potential hazard and liability problems. The principal advantages of contra-flow lanes are the increase in capacity in the peak direction and the fact that the method can be implemented quite rapidly at relatively low capital costs. In addition, violations tend to be minimal because enforcement is effective at the access points or the terminal areas. Disadvantages of contra-flow lanes are the high operating costs, the potential for very serious head-on collisions if continuous barriers are not used, the need for median modification to provide crossover points, and the need for special procedures for handling incidents in the priority lane.

Concurrent-flow facilities. The concept of concurrent-flow reserved lanes for priority vehicles consists of creating an additional lane for use exclusively by priority vehicles, either full-time or during peak periods. The advantages of such a concept are that the high-occupancy vehicles are given a distinct travel-time advantage over other vehicles, capital costs are lower than those for separated facilities, operating costs are low, and the scheme can typically be implemented in a relatively short period of time. The principal disadvantages are the potential hazards of having a high speed differential between priority vehicles and traffic in the general-purpose lanes, and the increased weaving required at ramps for vehicles getting to and from the reserved lane. In addition, provisions for handling breakdowns and enforcement become much more critical and special enforcement programs are normally required to prevent violating vehicles from interfering with the operations of the reserved lane. The general experience with concurrent-flow facilities indicates that the number of ride-sharing vehicles can be expected to rise significantly, although the operation may be less safe than physically separated facilities.

Priority access control. Another technique used to improve the level of service for high-occupancy vehicles is priority control at entrance ramps. This method of control is used in conjunction with ramp metering control by providing the high-occupancy vehicles a ramp bypass lane or an exclusive ramp for their own use. The underlying principle is that the high-occupancy vehicles can avoid the delay in queue caused by ramp metering, and thus can enter an uncongested freeway with minimum disruption. There are several advantages of priority access control: Both capital and operating costs are relatively moderate, underutilization of freeway capacity is avoided, the absence of a speed differential between priority vehicles and other vehicles on the freeway itself enhances the safety of operations, and enforcement is relatively easy.

Corridor control

The purpose of corridor control is to provide an optimum utilization of all available facilities in the freeway corridor. A freeway corridor comprises, in addition to the freeway and its ramps, freeway frontage roads (where they exist), parallel arterial streets that can be used as alternative routes, and cross streets that are links between the freeway ramps and the parallel arterials. Corridor control is thus an integration of urban street control with freeway control toward achieving the optimum utilization of the corridor capacity.

Implicit in the concept of corridor control is that it must be traffic-responsive to be meaningful. The underlying philosophy of corridor control is twofold: restriction and diversion. Restriction consists of limiting demand on corridor links to less than their individual capacities through the coordination of such regulatory controls as ramp, mainline, and arterial-street controls. Diversion of traffic from corridor links with excess demand to links with excess capacity is basically achieved through driver information systems. The essential components of driver information systems for corridor control are:

- A detection system to provide real-time information on how the routes in the corridor are operating.
- A control center that transforms this information into a control strategy.
- A communication system that advises drivers of the decisions of the control center.

The purpose of corridor control is to integrate the operation of the various control and driver information systems in the corridor so that optimum utilization of corridor capacity is achieved. This integration entails application of the following techniques:[53]

- Coordination of traffic signals on frontage roads and on parallel arterials.
- Coordination of ramp control and frontage road operations so as to provide alternative routes on frontage roads during peak-period congestion on the freeway or during unexpected freeway incidents.
- Coordination of the ramp control queue override feature with the frontage-road/cross-street intersection control to prevent queuing across these intersections.
- Provision of turning phases at parallel alternative route intersections with cross streets leading to freeway ramps, possibly in coordination with driver information displays.

---

[53] Everall, *Urban Freeway Surveillance.*

o Coordination of traffic signals at freeway interchanges with arterial cross streets (e.g., at diamond interchanges).[54]

Several efforts are under way that will result in the implementation of corridor control systems. The *Information for Motorists* (INFORM) system on Long Island is one such effort. This system performs surveillance, control, and motorist information activities in a 5-mile-by-35-mile corridor consisting of two parallel freeways, a parallel major arterial, and several crossing roadways.[55] In Los Angeles, the California Department of Transportation (Caltrans) and the City of Los Angeles are jointly developing the Smart Corridor, in which traffic surveillance, control, and driver information system activities will be centrally managed for a 12-mile corridor. Similar projects are planned throughout the states of California and Texas, and in the cities of Seattle and Phoenix.[56]

Traffic surveillance and incident management

Traffic surveillance is an integral and essential part of freeway traffic management systems. Surveillance entails the status monitoring of traffic conditions and of control system operation as well as the collecting of information for implementing controls and for incident detection. The surveillance system provides data on the system operating conditions, upon which appropriate decisions and control actions are taken, and whose effects on the system operations are then monitored by the surveillance system. There is thus a closed loop of information—decision, control, and impact. The concept of surveillance is not new; the effectiveness of implemented controls, as reflected by prevailing traffic conditions and the status of control system operation, has always been of interest to traffic engineering agencies.

These aspects of surveillance are common to both urban street and freeway traffic control systems, whose effectiveness is clearly dependent on the reliability and accuracy of the surveillance system, especially in the case of traffic-responsive control. For freeways, however, perhaps the most important aspect of surveillance is the detection of incidents. Conversely, problems caused by incidents on urban streets are generally less severe than those on freeways, since emergency and repair services, along with alternative routes, are usually more readily available. Thus, the provision of surveillance for incident detection and servicing on urban streets is less common than on freeways. The various methods of surveillance presented in this section are most typically applied to freeways for incident detection. Each method is thus discussed principally in terms of such a

perspective. However, some of these techniques also serve the purpose of measuring traffic performance and control effectiveness, applicable to urban street systems. Such applications are indicated where appropriate.

Incident detection

The earliest traffic surveillance techniques used for incident detection were field observations, police surveillance, and citizen calls. Today, surveillance for incident detection is carried out through the deployment of a wide variety of methods:

o Electronic surveillance
o Closed-circuit television
o Aerial surveillance
o Emergency motorist call systems (including cellular and 911 systems)
o Citizen-band radio
o Police and service patrols

Electronic surveillance. Electronic surveillance for incident detection is accomplished by the real-time computer monitoring of traffic data collected by detectors installed at critical locations. Measurement of traffic performance and control effectiveness, for both urban streets and freeways, can also be carried out by such an installed system. The use of detectors in urban street traffic control systems is described in Chapter 9 and is not repeated here. Also, the discussions of different freeway control strategies in earlier sections of this chapter have indicated the essential manner in which electronic detection is used to implement these strategies. It remains, therefore, to describe how incident detection is accomplished through electronic surveillance.

When a delay causing incident occurs on the freeway, the capacity of the freeway is reduced at the point of occurrence and, if it is reduced to a value less than the prevailing demand, the traffic flow upstream of the incident is also affected. Most freeway incident detection algorithms involve the determination of changes in certain traffic-flow variables (e.g., volume, occupancy, speed), which are believed to be caused by, or correlated with, the occurrence of incidents. If detected changes in the traffic flow variables are greater than some predetermined values, the possible occurrence of an incident is indicated. Thus, incidents are detected by logically evaluating the variations in traffic flow characteristics.[57]

[54] C. Pinnell and D.G. Capelle, "Operational Study of Signalized Diamond Interchanges," *Highway Research Board Bulletin No. 324* (Washington, DC: Highway Research Board, 1962).

[55] P. Zove and others, *Integrated Motorist Information System (IMIS) Feasibility and Design Study, Phase I: Feasibility Study,* Report No. FHWA-RD-77-47 (Washington, DC: U.S. Department of Transportation, Federal Highway Administration, April 1977).

[56] G.W. Euler and J.A. Lindley, "Integrated Traffic Management Concepts and Considerations," *Compendium of Technical Papers* (Washington, DC: Institute of Transportation Engineers, 1989).

[57] For further information on specific methods of incident detection, see *Highway Traffic Detectors and Detection—January 1970–September 1988* (Bibliography) (Springfield, VA: National Technical Information Service, undated); R. Sumner and others, *Freeway Management Handbook* (4 vols.) (Washington, DC: U.S. Department of Transportation, Federal Highway Administration, May 1983); B.N. Persaud and F.L. Hall, "Catastrophe Theory and Patterns in 30-Second Freeway Traffic Data—Implications for Incident Detection," *Transportation Research, Part A: General,* 23A (March 1989); S.A. Ahmed and A.R. Cook, "Application of Time Series Analysis Techniques to Freeway Incident Detection," *Transportation Research Record 841* (Washington, DC: Transportation Research Board, 1982); J. Tsai and E.R. Case, "Development of Freeway Incident Detection Algorithms by Using Pattern-Recognition Techniques," *Transportation Research Record 722* (Washington, DC: Transportation Research Board, 1979); M. Levin and G.M. Krause, "A Probabilistic Approach to Incident Detection on Urban Freeways," *Traffic Engineering and Control,* 20, (March 1979).

Typical of how incidents are detected is the INFORM system operating on Long Island.[58] In this system, changes in lane occupancy between adjacent detectors are used to indicate the occurrence of an incident. At the end of each sampling period, a computer calculates the percent difference in occupancy between adjacent detector stations spaced at approximately half-mile intervals. When the difference between the occupancy of upstream and downstream locations exceeds a predetermined value, and a persistence check is satisfied, the computer automatically signals an alert. Additional information on traffic conditions in the vicinity of the incident can then be obtained, and judgment decisions are made with regard to what other response is needed.

The principal advantages of electronic surveillance for incident detection are: It is the only system that provides a continuous networkwide monitoring capability at relatively low costs; the installed system can be used for many other tasks, such as the establishment of metering rates for traffic-responsive, ramp-metering systems. The main disadvantage is that the nature of the incident cannot be determined by the system, so that some means of verification is required to determine the response needed. Also, electronic surveillance incident detection strategies have not been perfected and can be subject to high false-alarm rates.

Closed-circuit television (CCTV). Closed-circuit television (CCTV) allows operators in a central control room to monitor traffic conditions visually at the locations where CCTV cameras have been placed. When an incident occurs, the operator can readily determine the nature of the incident and the type of response that is required. CCTV generally is applicable to those locations where delay-causing incidents are a chronic occurrence and fast response is essential. In such use, CCTV normally serves as a follow-up to electronic surveillance where incidents are detected automatically and the operator is alerted by an alarm. CCTV can also be applied to urban street systems for surveillance of critical intersections.

The main advantage of CCTV is that it is the only system in which a qualified operator can visually detect incidents over a large section of a road and hence determine the remedial action to be taken. CCTV has also been found to be very useful in the initial evaluation and adjustment phases of freeway management projects or in monitoring critical links on a system where a fast response is often required to maintain an acceptable level of service.

There are, however, several significant disadvantages of CCTV surveillance. These are:

○ It is expensive to install and maintain; recent technological advancements are, however, reducing these costs.
○ Adequate coverage for surveillance may require a large number of cameras.
○ It can be difficult and expensive to obtain monitor pictures of acceptable quality under conditions of adverse weather, darkness, and bright sunlight.
○ Monitoring the TV screens is a tedious task, and operators may not notice incidents immediately for lack of interest, unless there is an automatic alarm.

○ Continuous monitoring of TV screens by qualified operators is quite labor-intensive.

A recent technological advance in the use of CCTV for surveillance and incident detection involves using the video images from CCTV cameras to collect typical traffic data (e.g., volume, occupancy, and speed). The major advantage of this method over conventional electronic surveillance is that several lanes and both directions of the freeway may be covered with a single camera. Development of incident detection algorithms using this technology are also planned.[59]

Aerial surveillance. Aerial surveillance is undertaken by some private organizations (e.g., radio stations) and a few public agencies to get a general overview of traffic in a particular area or corridor. The use of helicopters and light aircraft enables observations of congestion locations and the determination of whether incidents are contributing to the problem. This information can then be broadcast to motorists and used to dispatch aid to the scene of the incident. However, because the pilot is usually covering a wide area, typically an entire city, there is often a considerable delay in relaying the incident information and the subsequent delay in the remedial response.

In general, aerial surveillance for incident detection has yet to be shown to be cost-effective compared to other techniques. The equipment and labor-intensive aspects of the system are expensive, and its effectiveness is sometimes limited by bad-weather conditions, when incident occurrences tend to be highest. Aerial surveillance is, however, gaining widespread support for use by public agencies to monitor response activities and traffic conditions when incidents occur.

Emergency motorist call systems. Emergency call boxes and emergency telephones installed along freeways were among the earliest incident-detection systems used. With either of these systems, a motorist experiencing an incident, or witnessing one, proceeds to the nearest call box or telephone and informs the operating agency of the nature of the incident. With a telephone, the motorist uses voice communication directly with the agency. With a call box, the motorist pushes one of several buttons, each of which sends out a pulse-coded message to the responsible agency requesting a particular service. Telephones are generally preferred because the motorist can explain the nature of the incident and be certain of the appropriate service. However, the cost of installing telephones is considerably more than the cost of installing call boxes. Recent technological advances have reduced this price difference by permitting the development of solar-powered, cellular radio-based telephones.

The main advantages of a motorist call system are that it is an efficient means for signaling a motorist's need for service and provides reassurance that aid will be provided. If emergency telephones are used, other communication requirements can be met (e.g., maintenance control and directions for motorists). The major disadvantage of such systems

---

[58] P. Zove and others, *Integrated Motorist Information System.*

[59] P. Michalopoulos and others, *Wide Area Detection System (WADS) Image Recognition Algorithms,* Publication No. FHWA-RD-89-171 (Washington, DC: U.S. Department of Transportation, Federal Highway Administration, September 1989).

is the time delay inherent in the operation; the motorist determines the occurrence of an incident, decides that the proper action involves an emergency call system, locates the nearest call box or telephone, and then phones the responsible agency. Thus, the detection delay in such a system can be significant. Additionally, a significant number of "gone-on-arrival" calls are generated; the motorist remedies the problem through some other means and when the service agency arrives, the disabled vehicle or accident is no longer there. Finally, motorist call systems present a safety hazard if located on the edge of freeway shoulders; on the other hand, if located off the freeway, their usefulness is considerably diminished. Typically, motorist call systems are installed within 2 ft of the edge of the freeway shoulder, and with safe breakaway terminal posts.[60]

The increased popularity and use of the cellular telephone has solved many of the problems related to conventional emergency motorist call systems noted above. Motorists with cellular telephones can call to report incidents directly from their vehicle. In many urban areas, motorists reporting a freeway incident simply dial 911 as for any other emergency. In Chicago, a recently implemented system allows motorists on the expressway system to reach the Illinois Department of Transportation (IDOT) by dialing *999. An IDOT dispatcher then contacts the appropriate service agency. There is no charge to the motorist for the cellular call. This service handles several thousand calls per month, nearly all of which request service for another motorist.[61]

Citizens-band radio. Citizens-band (CB) radio is another means by which stranded motorists can report their problems directly without leaving their vehicles, or if they do not possess the necessary equipment, fellow motorists can call for them. A CB-radio traffic surveillance system is equally applicable to urban streets and freeways. In a typical system, drivers of CB-radio-equipped vehicles report incidents to a monitoring center, which in turn transmits the information to the appropriate agency that ultimately dispatches the required assistance.

In most systems, requests for assistance are broadcast over Channel 9, which was designated as an official emergency channel by the Federal Communication Commission in 1970. Requests for aid are received directly on this channel and, depending on the geographic characteristics of an area, calls may be effective up to 20 miles.

Some systems make use of CB base station transceivers at unattended remote station locations. This allows the control monitor to receive Channel 9 alarms throughout an entire urban area. The operator can then dial up the CB transceiver near the alarm site and talk with whoever initiated the call. Such a system requires the use of a dial-up phone network and CB units placed at strategic locations to provide major roadway coverage. This type of system can also be used as a follow-up surveillance to verify the nature of a suspected incident following incident detection by electronic surveillance. The operator dials up the nearest roadside CB station once there is an indication of an incident and can listen to a special channel to verify the nature of the suspected incident.

The key elements of this system are properly equipped motorists who are knowledgeable about the system and who are willing to report the incidents they observe. Clearly, the detection capability of the system is directly dependent on the number of such motorists on the road at the time. The major advantage of CB-radio surveillance is that the determination of the nature of the incident and its required response are readily accomplished. On the other hand, because of the voluntary nature of the system, rapid detection is not always assured.

Police and service patrols. The oldest and most common form of incident detection on urban streets and freeways has been the use of patrol systems, notably police patrols that circulate in the traffic stream for the purpose of identifying incidents and determining the response required. The main advantage of patrols is that the response to an incident is immediate once the incident has been detected. The principal disadvantage is the large number of patrols required to cover a freeway or urban street system effectively. The costs are high and, in addition, many incidents do not require police response.

Another commonly used system for freeways is the service patrol. This entails the use of light-duty service vehicles to detect incidents as well as to provide such services as fuel, oil, water, and minor mechanical repairs. As in the case of police patrols, service patrols provide rapid response but are also expensive because of the highly labor-intensive nature of the system.[61A]

Incident servicing

Once an incident is detected, the key to minimizing congestion is the speed with which an appropriate response is made and the incident removed. The longer the duration of the incident, the more severe the resulting congestion and delay for a given level of demand. Consequently, an effective incident-management program must include resources and strategies that provide for the rapid removal of the incident completely off the freeway and out of sight of traffic in both directions.

One important consideration in an incident-management system is the cooperation of the agencies responsible for providing the needed response. Normally, more than one department of an agency or more than one agency are involved. As the priorities within each agency are often different, it is sometimes difficult to achieve the full cooperation of all parties. Matters involving multiple jurisdictions can also complicate the management process. To overcome these differences, it is desirable to create incident-management teams composed of representatives of the major agencies and governmental entities. Responsibility for initiating and coordinating all incident-management activities is given to this team; working together, it can effectively formulate and implement strategies for response to freeway incidents.

[60] Everall, *Urban Freeway Surveillance.*

[61] "Star 999, A Friend on the Expressway," Illinois Department of Transportation Press Release, August 2, 1989.

[61A] For further information on service patrols, see J.D. McDade, "Freeway Service Patrols: A Versatile Incident Management Tool," *Compendium of Technical Papers* (Washington, DC: Institute of Transportation Engineers, 1990).

Some specific strategies to facilitate the quick removal of freeway incidents and minimize disruption of traffic flow are described below.

Tow truck contracts. Agreements with a single operator or with several tow truck operators to respond to incidents in a freeway corridor can significantly reduce response and clearance times for incidents requiring use of a tow truck. Several agencies have developed procedures for qualifying tow truck operators and rotating calls to these operators. A few agencies have an exclusive contract with one tow truck operator to provide towing services in a corridor. Tow truck contracts generally specify a target response time that is not to be exceeded, which minimizes the time spent at an incident scene waiting for a tow truck to arrive. Some agencies also contract with tow truck operators to stand-by at high incident sites (e.g., bridges and tunnels) to minimize response time.

Freeway incident-management manual. This is a consolidated manual available to all response personnel outlining the responsibilities and procedures to be followed in the event of an incident. This manual typically includes a key personnel resource list, copies of interagency and other contractual agreements, locations of key personnel, equipment and facilities, and step-by-step procedures to be followed. All response personnel should have ready access to a copy of the manual and should be familiar with its contents.

Preplanned diversion routes. During an incident, particularly a major incident, much time can be wasted while response personnel decide how to detour or reroute traffic. This time could be saved if establishing alternate routes was a preplanned activity. In Los Angeles, for instance, preplanned alternate routes exist for every mile of freeway. These alternate routes are displayed on color-coded maps easily accessible to response personnel.[62] To plan alternate routes, a lead agency, usually the highway department, selects initial routes considering roadway and intersection capacity, pedestrian volumes, signal timing, and neighborhood impacts. Approval of the alternate routes from agencies responsible for the streets along the alternate route is then obtained.

Flashing lights policy. It has been well established that the presence of an emergency vehicle with activated flashing lights has the effect of significantly slowing traffic, even if it is only parked on the shoulder. Adoption of a policy of using flashing lights only in emergency situations can eliminate some of the congestion resulting from minor incidents, such as disabled vehicles, which make up a majority of all incidents.

Equipping response vehicles with push bumpers. During many minor incidents, lanes are blocked simply because a disabled vehicle cannot be moved out of a traffic lane until a tow truck arrives. Equipping police and other public agency vehicles with push bumpers is a good way to expedite this process. A key to the effective use of push bumpers is a clear procedure for when they should be used, which is usually only

for disablements and minor accidents not involving injuries or complicated investigations. One concern over the use of push bumpers has been liability problems for damage to the vehicles being pushed. Agencies using push bumpers, however, have experienced little or no problems in this area.

Accident investigation sites. A substantial portion of the delay associated with a freeway accident occurs after the involved vehicles are removed from the freeway lanes to the shoulder. Motorists passing the accident scene tend to slow when passing the accident scene, particularly if the vehicles are heavily damaged. One way to overcome this "rubber-necking" problem is to remove the involved vehicles quickly to an off-freeway location to conduct the accident investigation. These accident investigation sites can be specially constructed areas, designated areas of existing streets or parking lots, or a nondesignated off-freeway location selected by the police officer on the scene. Accident investigation sites have been used in Texas, Florida, and Minnesota, and they are being designed into freeway construction and reconstruction projects in several other states. The major obstacle to more widespread implementation of these sites has been perceived legal and insurance problems, but recent research indicates that these concerns can be easily overcome.[63]

Hardware requirements

Previous sections of this chapter described the application of various concepts of freeway control and traffic surveillance. This section discusses the hardware components needed for the implementation of these concepts. The hardware components consist essentially of detectors located on the freeway and on surface streets, communication links to and from a control center, information and control displays in the field, and control and display equipment in the control center. (The basic hardware requirements of traffic control systems for urban streets are described in Chapter 9 and will not be repeated here. Detailed discussions of signs and markings are found in Chapter 8.) This discussion is not intended to be comprehensive; rather, it is aimed at providing some essential hardware considerations required in the implementation of surveillance and control concepts. Clearly, the type of hardware and installation configuration is dependent on the functional purpose that needs to be served in the surveillance and control system.

Detectors

The various types of detectors used for traffic surveillance and control for urban streets, discussed in Chapter 9, also apply to freeways. Three types of detectors have been most widely used in freeway applications—ultrasonic, magnetometer, and inductive loop—of these, the inductive loop detector is the most widely used. The general location of detectors for various freeway ramp control strategies is

---

[62] D.J. Roper, "Alternate Route Planning: Successful Traffic Management," *Special Report 153* (Washington, DC: Transportation Research Board, 1975).

[63] C.L. Dudek and others, *Promotional Issues Related to Off-Site Accident Investigation,* Report No. FHWA/RD-87/036 (Washington, DC: U.S. Department of Transportation, Federal Highway Administration, January 1987).

described in an earlier section, the exact location being dependent on the particular strategy employed.[64]

Closed circuit television equipment

Closed circuit television cameras are available from several manufacturers. Both high-resolution black-and-white and color cameras have been used in existing systems. Cameras typically include remote control pan, tilt, focus, and zoom features and can be rotated at least 180° so that the freeway both upstream and downstream of the camera location can be viewed. Cameras are typically pole mounted so that the view of the entire segment of the freeway is not obstructed. Further details on CCTV equipment can be found in the *Freeway Management Handbook* cited earlier in this chapter.[65]

Ramp metering installations

The equipment for a specific ramp control system depends on the control strategy to be employed, but in general such systems have the following common requirements: warning and regulatory signs, pavement markings, traffic signals, and local controllers.

Signs. Ramp metering signs are used to warn the driver that the ramp is under control and to provide instructions as to what action to take. Advance warning signs, with such a legend as "Ramp Signal Ahead," alert approaching traffic to the presence of a ramp signal and should be located at the head of the ramp at a sufficient distance (around 200 ft) in advance of the ramp signal to allow for adequate evaluation and reaction time.[66] The effectiveness of warning signs is enhanced when used in conjunction with flashing yellow beacons. A suitable sign legend in such cases is "Ramp Metered When Flashing."

Signs are also normally used at the signal to inform drivers of what actions are expected of them. It is a fairly standard practice to mount a sign on the same pedestal as the signal head, bearing such legends as "Stop Here on Red" or "Wait Here for Green," with an arrow pointing to the stop line. Supplementary signs may be used at the signal to indicate number of vehicle departures during the green. Unfortunately, such signs can create confusion if a steady green display is used during unmetered periods.

Pavement markings. Pavement markings for ramp control systems are used to augment signs in positioning the vehicle over the check-in detector and facilitate single-vehicle entry. They normally consist of a stop line and striping to guide the vehicle as desired. The stop line should extend across the whole ramp, be 12 in wide, and be reflectorized. Also, the majority of ramps require some pavement striping to ensure single-lane entry. A typical example of pavement markings on an entrance ramp with a ramp signal and check-in detector is shown in Figure 13–7.

---

[64] Specific guidelines to freeway ramp detector location and configuration are given in the *Traffic Control Systems Handbook*.

[65] R. Sumner and others, *Freeway Management Handbook*.

[66] Everall, *Urban Freeway Surveillance*.

Traffic signals. Various configurations of ramp signals have been used in ramp control applications. As a result of the experience to date, certain standards have now been recommended and are outlined in the most recent edition of the *Manual on Uniform Traffic Control Devices*. They are:

o The standard display for freeway entrance ramp control signals shall be either a two-lens signal face containing red and green lenses, or a standard three-indication signal face containing red, yellow, and green lenses.
o There shall be a minimum of two signal faces per ramp, facing entering traffic.
o On entrance ramps having more than one lane there shall be a signal face mounted on the left side and on the right side.
o The required signal faces should be mounted such that the height to the bottom of the housing of the lowest signal face is between 4 1/2 and 6 ft. The height of any supplemental signal faces should be consistent with sound design principles and engineering judgment.
o All ramp control signals shall utilize vertically aligned lenses with minimum nominal diameter of 8 in.
o Ramp control signals need not be illuminated when not in use.

There has been some controversy with regard to the choice of two or three lenses for ramp control signals. The relative advantages and disadvantages are summarized in Table 13–2. The use of two-lens signals with 8-in lenses is generally recommended, wherever they are legal. For platoon metering applications, the standard three-lens head should be used.

Local signal controllers. The various types of local signal controllers for urban street control are described in Chapter 9. The same controllers are suitable for use in freeway control systems. Three principal modes of operation of local controllers are usable in freeway control: pretimed control, central control, and actuated control.

Pretimed control, the simplest of the three, is typically implemented using a standard solid-state or microprocessor-based controller units. Control is typically activated, deactivated, or changed based on time-of-day.

In the central control operation, the controller functions are incorporated within a central computer and the need for controller hardware at the ramp site is virtually eliminated. Thus, timing and control operations are carried out by power relays with instructions to the computer performed by software or by manual inputs. Such a system offers greatly increased flexibility by virtue of the computer's ability to store a large number of timing plans in its memory. The disadvantages of central control are the need for a fairly elaborate communication system between field equipment and the central computer, and the possible loss of system operation due to computer or communications malfunction. The latter problem can be avoided by using a field controller unit to provide backup service during an equipment failure.

Although a central computer can be utilized for both systemwide and individual traffic-responsive ramp control, often it may only be necessary or desirable to provide individual traffic-responsive control at selected ramps. Locally

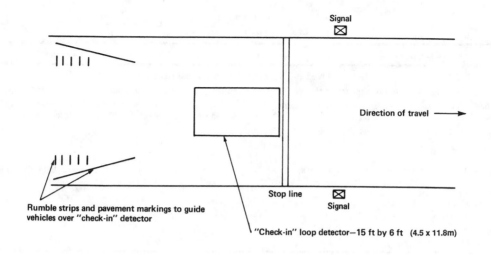

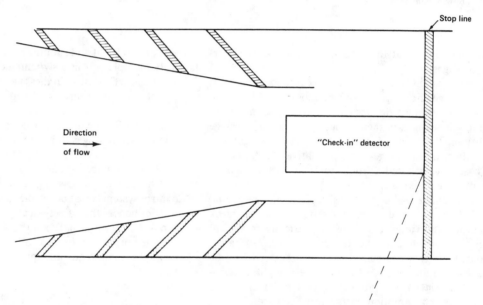

Figure 13–7. Location of "check-in" loop and pavement markings for freeway ramp metering.
SOURCE: P.F. EVERALL, *Urban Freeway Surveillance and Control—The State of the Art,* Revised (Washington, DC: United States Department of Transportation, Federal Highway Administration, June 1973.

actuated controllers can be used to accomplish such objectives. These controllers operate in stand-alone fashion by selecting metering rates based on real-time traffic information gathered by ramp and mainline detectors. Microprocessor-based controller units are most suitable for such operations as they can be programmed not only to perform all local control functions, based on detector data, but they can serve as the building blocks for a comprehensive centrally controlled operation.

Changeable message signs

Several different types of changeable message signs are available for applications in freeway surveillance and control

systems. The simplest type is the *blank-out sign.* This type of sign is either internally or externally illuminated and contains one message that can be turned off or blanked out when the message is not applicable. The major disadvantage of this type of sign is obviously its limited message flexibility. However, such signs are a relatively inexpensive way to advise motorists of regularly recurring conditions.

Rotating drum signs contain one or several multisided drums. Each drum consists of a polygon shape, so that, depending on the shape, up to six predetermined messages (or blanks) can be stored on each drum. The messages are typically externally illuminated, though internal illumination can be used if the characters are formed with translucent material. Rotating drum signs offer greater flexibility than

TABLE 13–2

Three-Lens or Two-Lens Signal Systems

Advantages	Disadvantages
Three Lens, Red/Yellow/Green Signal Head	
Familiar to motorist; therefore he or she knows what action to take	Minimum cycle length is increased, thereby reducing rate possible
Amber may be flashed to convey special messages, such as instigation of control each peak period	Is not suitable for single metering since motorists may try to "beat the amber," with possible poor safety consequences
Amber is necessary if bulk or platoon metering is to be employed on a single-lane ramp	
Two-Lens, Red/Green Signal Head	
Ramp control is essentially different from intersection control, and therefore it is advantageous not to use the same signal for both purposes; the ramp-metering signal is strictly for stop-and-go past a point, whereas an intersection traffic control signal assigns right-of-way	More inflexible than three-lens signal in that bulk or platoon metering cannot be employed on a single-lane ramp
Gives clearer indication that only one vehicle at a time is allowed	
With no amber phase, a faster maximum metering rate can be employed in uncongested conditions	

SOURCE: P.F. EVERALL, *Urban Freeway Surveillance and Control—The State of the Art,* rev. (Washington, DC: U.S. Department of Transportation, Federal Highway Administration, June 1973).

do blank-out signs, but they are still limited to predetermined messages.

Bulb matrix signs consist of an array of incandescent bulbs for each message line. Various combinations of these bulbs are illuminated to produce the desired message. The major advantage of these signs is their virtually unlimited message flexibility and their good visibility for both day and night conditions. Major disadvantages include relatively high maintenance costs (necessitated by bulb replacement) and power consumption requirements.

Electromagnetic disk signs consist of several individual matrices of disks that are reflectorized on one side and black on the other. Individual disks are magnetically rotated to produce the characters for the desired message. A separate light source, either contained in the sign or external to the sign, is required for messages to be visible at night. Electromagnetic disk signs offer unlimited message flexibility and reduced power consumption costs when compared to bulb matrix signs. However, they do not typically provide comparable message visibility, particularly at night or under poor weather conditions.

Light emitting diode (LED) signs use clusters of LEDs to form the characters desired for the sign message. Message flexibility for this type of sign is unlimited, power requirements are very low, and reliability of the LEDs is very high. Reported message visibility problems with early versions of this relatively new technology have been solved in recent versions. To date, this technology has primarily been used overseas.[67]

Fiber optic signs use bundles of glass fibers in a matrix arrangement similar to those used in the other sign technologies. Message flexibility for this type of sign is unlimited and message visibility is very good. And since an individual bulb in the sign can feed several hundred fiber bundles, power consumption and maintenance requirements are lower than for bulb matrix signs.

For blank-out and rotating drum signs, the local sign controller simply acts as a switching device. For sign types that allow more flexibility in message content, the local sign controller normally consists of a microprocessor unit that can store preset messages and receive instructions from the control center for special messages. A flashing capability, which allows two messages to be alternately displayed, is also usually included, as well as verification of the message displayed back to the control center.

Changeable message signs for freeway surveillance and control systems must be visible to several lanes of high-speed traffic; hence, they are typically large and are mounted on support structures that span the width of the freeway. A catwalk is normally provided so that maintenance can be performed without requiring equipment to block a portion of the freeway.[68]

Highway Advisory Radio (HAR) equipment

Highway advisory radio (HAR) systems may operate on one of two frequencies, 530 and 1610 kHz on the AM broadcast band. Implementation of a HAR system first requires a license from the Federal Communications Commission (FCC). Procedures for applying for this license are contained in the *HAR Operational Site Survey and Broadcast Equipment Guide.*[69] In some areas, the FCC prohibits the licensing of the 530 or 1610 kHz frequency for HAR applications because of potential interference with the broadcast signal of other nearby AM stations. In a few limited areas of the United States, *no* HAR system can be licensed because of these potential conflicts.

---

[68] For more information on changeable message sign technology, see R. Sumner and others, *Freeway Management Handbook* and *Technology Evaluation for Changeable Message Signs.*

[69] W.F. Dorsey, *Highway Advisory Radio Operational Site Survey and Broadcast Equipment Guide,* Report No. FHWA-RD-79-87 (Washington, DC: U.S. Department of Transportation, Federal Highway Administration, April 1979).

---

[67] *Technology Evaluation for Changeable Message Signs* (Downsview, Ontario: Ontario Ministry of Transportation, May 1989).

Figure 13–8 shows the two basic types of highway advisory radio systems. One uses a monopole antenna to radiate a signal for a diameter of approximately 1 mile from the antenna. The other type uses a cable antenna installed along the roadway, to radiate a signal for approximately 200 ft from the cable. For either system, a modulating source, a communications link, and a transmission system are required. These may be located in a single cabinet, with messages updated by visiting the cabinet, or from a remote location via telephone or cellular radio.

Data transmission systems

All surveillance and control systems require some form of data transmission. The transmission system is defined as the equipment needed to transmit detector data to the control

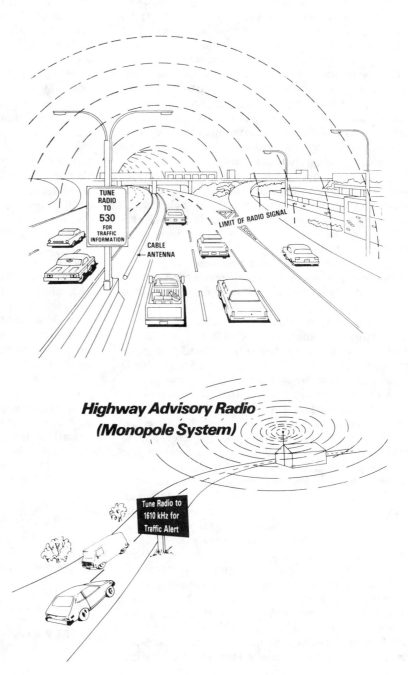

Figure 13–8. Types of Highway Advisory Radio (HAR) antennas.
SOURCE: W.F. DORSEY, *Highway Advisory Radio Operational Site Survey and Broadcast Equipment Guide,* Report No. FHWA-RD-79-87, Washington, DC: U.S. Department of Transportation, Federal Highway Administration, April 1979.

equipment being used, to transmit CCTV images to the central control center if necessary, and to send control commands to the ramp control equipment, cameras, and changeable message signs. The type of system used will depend on both the type of control being employed (i.e., local or central control) and the type of information to be transmitted. For local control alone, direct wire connections are most feasible because of the short distances between detectors, controllers, and displays. On the other hand, central control typically necessitates data transmission over long distances, and different means of data transmission must be considered.

Essentially, two types of carriers can be used in long-distance data transmission for urban and freeway traffic control systems: cable media and air-path media such as radio, microwave, satellite, and laser technologies. Air-path media have been tried as the primary communications carriers for some surveillance and control systems, but they are not used extensively because of several disadvantages: high cost and complexity of terminal equipment, problems of maintenance and reliability, and need for high transmitter power for remote locations. Some form of physical interconnection using cable has been generally found to be the most feasible for the primary means of data transmission in freeway surveillance and control systems. Radio, microwave, satellite, and laser systems are, however, used for specialized applications such as crossing obstructions (e.g., a river) and linking remote locations to the central control system.

The four basic types of cable commonly used for data transmission in freeway surveillance and control systems are conventional interconnect cable, twisted pair cable, coaxial cable, and fiber optics cable. Each type has different characteristics of cost and performance. The conventional interconnect cable, the least expensive, is basically an open-wire conductor used to complete a circuit between two points or a series of points. However, it suffers from a limitation on the amount of information that can be reliably transmitted using such a system.

Significant improvements in performance, in terms of transmitting more information reliably, can be achieved through the use of systems employing twisted-pair cable, coaxial cable, or fiber optics cable. Twisted-pair cable is made up of sets of two wires that are wrapped around each other, which reduces interference from external sources, since the pair of conductors carrying the data are always immediately adjacent to each other in the cable. However, if not properly designed and located, interference from electrical sources can still be a problem with twisted-pair cable systems. Twisted-pair cable is used primarily for data and voice applications (e.g., highway advisory radio).

Coaxial cable consists of a central conductor within an outer cylindrical conductor separated by an insulator. This makes the cable virtually immune to external noise except at connection points. It also has a large bandwidth that can accommodate CCTV image transmission as well conventional data and voice applications. However, coaxial cable is generally more expensive and more complex to install properly than twisted-pair cable.

Fiber optics cable consists of one or more strands of glass fiber constructed to reflect light in a highly efficient and predictable fashion. This design completely eliminates potential interference from electrical sources. Fiber optics cable has a very large bandwidth and can accommodate data, voice, and video transmissions. Like coaxial cable, fiber optics cable is relatively expensive and complex to install properly.

The cable facilities for freeway surveillance and control systems, whether comprised of twisted-pair, coaxial, or fiber optics cable, can either be owned by the operating agency or leased from local telephone or cable television companies. A user-owned cable system typically offers higher reliability than a leased system, but it also requires a significant initial investment and considerable continuing maintenance expenditures. Leased-line systems are generally less expensive than agency-owned systems and accommodate incremental expansions of the system more easily. However, leasing rates are subject to substantial change and satisfactory priority in maintaining the cable system may not be provided by the leasing agency. The magnitude and importance of these advantages and disadvantages will, of course, vary depending on the particular leasing agency and public agency involved.

Since the communications subsystem can easily be the most costly component of a freeway surveillance and control system, performance of a detailed trade-off analysis is essential to identify the most appropriate system configuration. This trade-off analysis typically involves an examination of the information required to be transmitted, an inventory of existing communications resources and constraints, and an evaluation of the available communications alternatives, in terms of effectiveness, reliability, and cost. More details on the communication trade-off analysis process as well as communications technology can be found in *Communications in Traffic Control Systems.*[70]

Central control and display equipment

For large-scale freeway traffic management systems, the control center is the focal point for all communications and control. It receives information on traffic conditions, analyzes and evaluates it, and makes decisions as to what controls should be used. The equipment normally found in control centers consists of the computer and its related peripheral accessories, communication consoles, CCTV monitors and other display components, and equipment for dispatching emergency and maintenance vehicles to the problem locations.

A digital computer is the nucleus of most urban and freeway traffic management systems. Some of its features particularly suited to surveillance and control are its ability to do the following:[71]

○ Perform various functions according to a preestablished priority order.
○ Gather information, calculate parameters, make decisions, and output control commands in the same system at high speeds.
○ Operate in real time so that events, as they occur, initiate preestablished priority-level programs.

[70] L.G. Neudorff and D.C. Terry, *Communications in Traffic Control Systems,* Publication No. FHWA-RD-88-012 (Washington, DC: U.S. Department of Transportation, Federal Highway Administration, August 1988).

[71] P.J. Athol, "Traffic Surveillance and Control," *Transportation and Traffic Engineering Handbook* (Englewood Cliffs, NJ: Prentice-Hall, 1976), 19.

o Operate automatically with minimal human intervention on a 7-day, 24-hour/day basis.
o Report system performance in printed form for permanent records as well as in display form to allow operator interaction with the system.
o Control various kinds of systems, including signing and television subsystems, by turning on, turning off, or selecting equipment, based on control and decision logic programmed into the computer.
o Perform technical analysis in the free time of the control system.

Rapid advances in digital computer technology in the last decade have produced significant reductions in the cost and size of computers, while their computing power has increased noticeably. The majority of computers installed in existing control systems are either minicomputers or medium-scale computers. Minicomputers are attractive for smaller control and surveillance applications in that they are high-speed devices with reasonably powerful input/output capability and low capital cost. Medium-scale computers are required for applications requiring complex, high-speed processing tasks or transferring large amounts of data.[72]

The peripheral equipment normally used in traffic control systems includes interactive peripherals and storage devices. The interactive peripherals are those devices that enable the operator to interact either with the computer or the system. They provide a means of obtaining information as well as to input commands or alter performance. Microcomputers, workstations, and printers are typically used for these functions.

Magnetic tapes or hard disks are typically used for storing large quantities of data at low cost. These devices are extremely useful for logging system operations, for they allow subsequent playback for additional processing. The hard disk has the advantage of higher-speed random access to data than with tape. Thus, the disk is typically used for storing programs that are not resident in the computer but must be called upon for special situations, and for storing data relating to operating strategies that must be recalled according to some predetermined sequence.[73]

For systems using CCTV, monitors that display the CCTV transmissions and are easy to view by the system operators are a necessary feature of the control room. For small systems, a separate monitor and controls for remote operation can be provided for each camera. For larger systems, where it is impractical to provide a separate monitor for each camera, a bank of several monitors with switching capability between cameras is usually provided.

Many existing systems use a display map to provide a visual indication of how the system is working. The display map is usually a scaled replica of the traffic control system with colored lights to depict traffic conditions on the freeway and urban streets. In recent years, static display maps have gradually been replaced or supplemented by graphics systems that provide greater flexibility in the information that can be displayed. These graphics systems can typically provide display information for a number of measures of effectiveness (e.g., speed, occupancy, volume) and can "window-in" on selected portions of the system.

Operations and maintenance requirements for freeway surveillance and control systems

Unlike many types of freeway improvements, such as widening, the implementation of a freeway surveillance and control system requires a significant continuing expenditure for operations and maintenance. Many agencies, particularly those implementing systems for the first time, have underestimated system operations and maintenance requirements, leading to resource shortages or cost overruns, and poor system performance. To avoid these problems, operations and maintenance considerations must be addressed in the earliest stage of the system planning process.

During peak traffic periods, freeway surveillance and control systems typically require continuous monitoring and intervention by system operators. Their duties typically include:[74]

o Monitoring system peripherals (CCTV monitors, status displays, CB radio monitors) and analyzing traffic flow status.
o Operating changeable message signs and updating HAR messages.
o Initiating and monitoring ramp controls.
o Reporting incidents to police and other emergency services.
o Reporting malfunctions to maintenance personnel.
o Providing traffic information to the media.
o Keeping logs of system operation and incidents.

Several existing systems have found it useful to share the operations center for the freeway surveillance and control system with police dispatch personnel. This reduces the time necessary for communication between operations and dispatch personnel and reinforces the notion that system operations and police personnel share a common goal.

During off-peak periods, system operators can devote time to such activities as updating hard-copy records, performing preventative maintenance for computer hardware, and evaluating the effectiveness of control strategies.

An effective maintenance-management program is also critical to the successful operation of a freeway surveillance and control system. Lack of proper maintenance can degrade system performance and drastically reduce equipment service life. Required system maintenance extends to both system hardware and software. Hardware maintenance involves preventative procedures and responding to known malfunctions. Software maintenance typically involves fixing flaws in system programs and enhancing software performance by adding system features and reports.

---

[72] *Traffic Control Systems Handbook.*

[73] Everall, *Urban Freeway Surveillance.*

[74] L.G. Neudorff, *Guidelines for Successful Traffic Control Systems,* Publication No. FHWA-RD-88-014 (Washington, DC: U.S. Department of Transportation, Federal Highway Administration, August 1988).

Many agencies that operate freeway surveillance and control systems use maintenance contracts for all or some of the critical components of the systems. Use of maintenance contracts frees the agency from hiring additional staff and stocking specialized spare parts. Contract maintenance is used most commonly for high-technology hardware, such as computers, peripherals, and communications units, but has also been used for routine hardware and software maintenance in several systems. A recent Federal Highway Administration report contains a list of sample provisions typically contained in contract maintenance agreements.[75] In addition to contracting for system maintenance, a few systems, most notably the INFORM system on Long Island, also use service contracts with private engineering firms to staff and operate the system.[76]

Evaluation of freeway surveillance and control systems

Traffic control system investments are usually justified to decision-makers on the basis of some forecast of the improvements or benefits that may accrue. The real need for evaluation, then, is either to confirm or to reject a hypothesis that expected results will occur. Such evaluations also become useful tools in the identification of ways to increase benefits further and/or justify system expansion. Further, the dissemination of the evaluation results provides user experience that is often beneficial to other agencies considering similar improvements.

Measures of effectiveness

The evaluation procedure must be able to measure the effects of surveillance and control on traffic flow and safety. Although there are many possible measures of performance, several basic parameters are typically used in measuring the performance of most surveillance and control systems:[77]

Total travel: the total amount of traffic expressed in vehicle-miles per unit of time. Its primary advantage is that it takes into account the length of trip as well as the number of vehicles served.

Total travel time: the sum of individual vehicle travel times along a specified section of roadway, expressed in vehicle-hours per unit of time. It is representative of the amount of time spent by all system users. Both "total travel" and "total travel time" can be expressed in terms of "persons" rather than "vehicles" if vehicle classification and vehicle occupancies are known. The variance of travel times for individual trips can also be use as a measure of performance for trip predictability.

Safety: the number of accidents on a freeway or portion thereof, categorized by severity and by type. These accident numbers are usually related to the amount of travel and the length of roadway under consideration in determining rates useful for evaluation and comparison.

Speed: often cited as the most descriptive variable of traffic flow. Point samples of the average stream speeds or the speed traces of individual vehicles can be used to locate problem areas and also to provide useful data for developing other performance measures such as fuel consumption and air pollution. The inverse equivalent of speed—travel time—is useful to quantify the amount of delay experienced when compared with normal operation.

Diversion: some diversion of trips from the freeway, particularly shorter trips, is an expected and desirable result of implementing a freeway surveillance and control system. However, diversion of too many trips to alternate routes can be undesirable.

Public acceptance: freeway surveillance and control systems, particularly those that include ramp metering, can be controversial, and it is important to communicate to the public the benefits of the system. Public acceptance is typically measured through opinion surveys. Total public acceptance of the system is not necessary, but acceptance by a majority of the system users is usually required to retain the critical support of political figures and decision-makers.

Specific procedures and techniques for performing studies that develop these measures of performance are given in Chapter 3.[78]

Cost-effectiveness analysis

With most freeway transportation projects, administrators are constrained principally by budgetary limitations. They thus strive for a strategy allowing them to provide the greatest benefits with available funds. The measures of performance described in the previous section are useful in this decision process by:

- o Determining which of the proposed alternative solutions are most effective from an economic perspective;
- o Allowing a "before" and "after" comparison of an implemented solution and thus giving an overall indication of the effectiveness of the project

The logical framework for both these analyses is the same (the difference being that the first one uses predicted "after" values, whereas the second uses measured "after" values) and therefore the steps for both are similar. The success of both forms of analyses is also clearly dependent on collection of good "before" data for the various measures of effectiveness, a step that has been overlooked in some system evaluations. Collection of a sound set of "before" data requires a well-developed data collection plan to ensure that all information necessary for the subsequent analyses are collected prior to system implementation.[79]

---

[75] Ibid.

[76] D.H. Baxter and R.N. Russo, "Contracting Maintenance for Traffic Signal Systems," *Transportation Research Record 1021* (Washington, DC: Transportation Research Board, 1985).

[77] Stanford Research Institute, *Guidelines.*

[78] See also Everall, *Urban Freeway Surveillance; Traffic Control Systems Handbook;* and J.A. Wattleworth and W.R. McCasland, "Study Techniques for Planning Freeway Surveillance and Control," *Highway Research Record 99* (Washington, DC: Highway Research Board, 1965).

[79] See Everall, *Urban Freeway Surveillance;* and R. Winfrey, *Economic Analysis for Highways* (Scranton, PA: International Textbook Co., 1969).

The basic technique employed in the cost-effectiveness analysis is one in which the benefits, or favorable changes in a number of measures of performance, are evaluated and, in an economic assessment, compared with the estimated costs of implementation.

The total traffic and safety benefits are evaluated by adding the dollar value of the traffic benefits to the value of the changes in congestion or accidents. In an evaluation of an implemented solution, the total monetary benefits of the implementation are assessed by comparing the traffic and safety effects before with those afterward. However, in an economic assessment of proposed alternative solutions, there is the problem of comparing the costs and benefits of these to determine which is the most effective. There are standard methods for doing this, some of which are very detailed.[80] A somewhat simpler method has the following steps:

1. Decide over how many years the capital and installation costs are to be amortized.
2. Decide at what interest rate the costs should be amortized
3. Estimate the total capital and installation cost.
4. Calculate the equivalent annual cost of this total capital and installation cost, and to this add the annual maintenance and operating costs to give the total equivalent annual costs.
5. Estimate the traffic and safety benefits over the first year.
6. Calculate the ratio of the benefits to the costs.

Those projects with a ratio greater than unity are worth undertaking, and in general those with the biggest ratio are worth doing first. However, some overriding practical factors are always present in the evaluation and selection of a major improvement project. Obviously, there are investment constraints that often rule out some projects, regardless of the extent of the benefits. Also, the ability of the operating agency to staff and maintain the project will be a significant factor.

Expected future developments in freeway surveillance and control technology

The most significant future development in freeway surveillance and control system technology is likely to be the expanded integration of these systems with other computerized traffic control systems (e.g., traffic signal systems). This will permit an entire freeway corridor or even an urban area to be controlled by a single system, which will result in more effective allocation of available roadway capacity. Another resulting benefit of this level of integration will be improved equipment and procedures for surveillance, control strategy evaluation and selection, and provision of information to motorists. Improvements in the technology for the provision of in-vehicle congestion and route guidance information will also tend to increase the effectiveness of freeway surveillance and control systems.

Finally, another area with great potential for growth is the use of expert systems to assist system operators and maintenance personnel in performing their tasks. The development of prototype systems for areas such as incident management is currently underway. As these systems are further developed and perfected, they should prove to be a cost-effective supplement to system operations and maintenance staff.

REFERENCES FOR FURTHER READING

ANDERSON, R., AND OTHERS, *Alternate Designs for CCTV Traffic Surveillance Systems,* Texas Transportation Institute, August 1976.

BANKS, J.H., *Performance Measurement for Centrally Controlled, Traffic Responsive Ramp Metering Systems,* Report No. FHWA/CA/SDSU/CE87141, San Diego State University, August 1987.

CHU, J., "I-66/I-395 Traffic Management System," *ITE Journal,* 54, (September 1984).

CIMA, B., AND OTHERS, *Evaluation of the Dallas Freeway Corridor System—Final Report,* Report No. FHWA/RD-81/058, Federal Highway Administration, May 1981.

DELSEY, M.J., AND S.E. STEWART, "Queen Elizabeth Way—Burlington Skyway Freeway Traffic Management System," *Transportation Forum,* 2, (September 1985).

DUDEK, C.L., AND OTHERS, *San Antonio Motorist Information and Diversion System,* Report No. FHWA/RD-81/018, Federal Highway Administration, September 1981.

FAMBRO, D.B., AND OTHERS, "Cost-Effectiveness of Freeway Courtesy Patrols in Houston," *Transportation Research Record 601,* (1976).

GOLDBLATT, R.B., "Investigation of the Effect of Location of Freeway Traffic Sensors on Incident Detection," *Transportation Research Record,* 773, Transportation Research Board, 1980.

HALL, F.L., AND D. BARROW, "The Effect of Weather on the Relationship Between Flow and Occupancy on Freeways," *Transportation Research Record 1194,* Transportation Research Board, 1988.

IWASAKI, M., AND OTHERS, "Improvement of Congestion Detection on Expressways," *ASCE Journal of Transportation Engineering,* III, (July 1985).

JHK & ASSOCIATES, *San Francisco Bay Area Traffic Operations Center Study—Final Report,* March 1987.

JU, R.S., AND OTHERS, "Techniques for Managing Freeway Traffic Congestion," *Transportation Quarterly,* 41, (October 1987).

KENTON, E., *Highway Ramp Control, 1964–June 1980 (Bibliography),* National Technical Information Service, July 1980.

KOBLE, H.M., AND OTHERS, *Control Strategies in Response to Freeway Incidents* (4 vols.), Report Nos. FHWA-RD-80-004 through 007, Federal Highway Administration, November 1980.

LARI, A.D., *I-35W Incident Management and Impact of Incidents on Freeway Operations,* Report No. FHWA/MN/TE-82/04, Federal Highway Administration, January 1982.

DE LASKI, A.B., AND P.S. PARSONSON, *Traffic Detector Handbook,* Report No. FHWA-IP-85-1, Federal Highway Administration, April 1985.

MCCASLAND, W.R., AND R.G. BIGGS, *A Study of Microwave Television for Traffic Surveillance in Texas,* Texas Transportation Institute, Research Report 165-9, August 1975.

MORALES, J.M., "Analytical Procedures for Estimating Freeway Traffic Congestion," *ITE Journal,* 57, (January 1987).

NIHAN, N.L., *Impact of Freeway Surveillance and Control on Eastbound SR 520,* Report No. WA-RD-99.1, Washington State Department of Transportation, February 1987.

NIHAN, N.L., AND G.A. DAVIS, "Estimating the Impacts of Ramp Control Programs," *Transportation Research Record 957,* Transportation Research Board, 1984.

---

[80] See Winfrey, *Economic Analysis for Highways,* for one such procedure.

PARVIAINEN, J.A., AND W.M. DUNN, JR., *Freeway Management Systems for Transportation Efficiency and Energy Conservation—Practical Planning Guide for Traffic Engineers,* Report No. TP 6220E, Canadian Government Publishing Centre, September 1985.

PRETTY, R.L., "Detection of Congestion on an Urban Freeway for the Purpose of Access Control," *ITE Journal,* 54, (September 1984).

RAJAN, S.D., AND OTHERS, *On-Ramp Traffic Control on the Black Canyon Freeway,* Report No. FHWA/AZ-86/211, Federal Highway Administration, April 1986.

RITCH, G.P., *An Application of RF Data Transmission in Freeway Ramp Metering,* Report No. FHWA/TX-82/6+210-9, Federal Highway Administration, September 1981.

RITCH, G.P., *State-of-the-Art of Motorists Aid Systems,* Texas Transportation Institute, Research Report 165-17, June 1975.

ROPER, D.H., "The Commuter Lane: A New Way to Make the Freeway Operate Better," *Transportation Research Record 1081,* Transportation Research Board, 1986.

TEAL, R.F., *Estimating the Full Economic Costs of Truck Incidents on Urban Freeways,* Report No. UCI-ITS-RR-88-3, University of California-Irvine, November 1988.

TRANSPORTATION RESEARCH BOARD, *Traffic Management for Freeway Emergencies and Special Events,* Transportation Research Circular 344, January 1989.

TRANSPORTATION RESEARCH BOARD, *Research Problem Statements: Freeway Operations,* Transportation Research Circular 354, 1990.

TSUI, D., *Highway 400/417 Cellular Emergency Callbox Pilot Project,* Ontario Ministry of Transportation, August 1988.

URBANEK, G.L., AND OTHERS, *Alternative Surveillance Concepts and Methods for Freeway Incident Management,* (6 vols.), Report Nos. FHWA-RD-77-58 through -63, Federal Highway Administration, March 1978.

YAGAR, S., Metering Freeway Access, *Transportation Quarterly,* 43, (April 1989).

14

TRAFFIC ADMINISTRATION

WILLIAM G. VAN GELDER, P.E., *Traffic Engineer*

City of Pleasanton, California

AND

WOODROW W. RANKIN, P.E., *Engineering Consultant*

AND

DR. JOHN E. BAERWALD, P.E., *President*

John E. Baerwald, P.C.

Highway and street systems that operate efficiently and safely are essential for continued national and local economic growth and improved quality of life. Effective administration and management of the agencies responsible for traffic engineering at all levels of government is a key element in achieving and maintaining safe and efficient traffic operations.

The basic principles of administration and management that apply to commercial and industrial operations also apply to the administration of traffic engineering agencies. This chapter does not cover all those principles. Rather it focuses on the application of ones that are needed for the effective operation of a traffic engineering agency. Part A looks at the public relations aspect of traffic engineering agency operations. Part B examines the legislative authority, functions, organization, personnel requirements, and funding requirements of traffic engineering agencies, and how management controls are used in the operations of agencies at various levels of government. Part C discusses the scope of the legal liability problem for traffic engineering agencies and actions an agency can take to reduce its liability risks.

Part A:
Public Relations And Program Implementation Methods

W.G. VAN GELDER

Public relations and program implementation

A successful traffic engineering program is based on the ability to communicate to the motorist, to the general public, to elected officials, and to the various employees working for governmental agencies.

The essential element of effective communication is to discern the various needs of all concerned parties prior to communication. Communications must fill a need, and must be timely and accurate. The traffic engineer must develop a sense of timing to know when and how much to communicate. This chapter focuses on the following subject areas of concern to traffic engineers:

1. Developing a positive public image and public confidence;
2. Adopting reasonable traffic policies;
3. Securing staff assistance and cooperation;
4. Obtaining an adequate planning design, operational and maintenance budget, and staffing; and
5. Funding capital improvement projects related to traffic facilitation or control.

All of the program implementation items listed above relate to portions of a greater governmental process and the competing demands on people's time or dollars. They also relate to convincing people that the needs of the traveling public have a high enough priority to deserve current attention. This requires salesmanship, a good knowledge of the various clients, and the clients' perspective regarding transportation. In all cases, it also requires a sound technical knowledge of the transportation needs and reasonable solutions.

Developing a positive public image

Public perception and image of the agency and the agency's professionals can be improved with the right approach. First it is necessary that individuals within the agency maintain the appropriate attitude, be visible, and be responsive to the public.

Attitude

Chuck Haley, while he was traffic engineer of Phoenix, Arizona, kept the following sign over the door of the traffic engineering office where employees could see it as they left the office: "What have you done to help move traffic today?" The traffic engineer and engineering staff must first believe that theirs is a service function that can help the public. If the staff attitude is "Look out, here comes another *complaint!*" then this attitude will be detected by the citizen and a lack of public confidence will result. Staff must keep clearly in mind that there is a difference between a "complaint" and a *request for service*. The job of the traffic engineer is to help the requester define the problem and then explain and implement possible solutions.

Visibility

Are you accessible? People must know you exist before you can gain their confidence. Be sure you are listed in the phone book where people can call "Traffic Engineering Department; Traffic Signs and Signals."

First impressions are important. If the telephone is answered "Transportation Services, may I help you?," then the caller who believes that "*They* don't care about the roads!" may begin to have a change of attitude. Keep in mind that we owe the public a better conception of who we are and what we do.

Be responsive

The success of transportation services, like any other business, depends upon how the service is perceived by the public. As with any other business, word of mouth is the most effective advertisement. "How was the service?" "I don't know. I never got a call back," is bad advertising. Preferably the response will be, "They sounded really interested in the problem. They're going to look into the problem and see what can be done." Or better yet: "They mentioned several things that might work. They're going to start with the simplest and see if that works."

Not all public contact requires a follow-up, but when one is promised be sure to make the call, even if it is not the answer the citizen wants.

Principles of public contact

There are a number of principles that should be followed when dealing with the public.

Understand the person's point of view

Try to put yourself in the other person's position. To do this (so that you can appreciate the "problem") takes emotional maturity and communication skill. Find out what the person's concern is. Talk in terms of his or her experiences, what he or she expects, and what brings that individual to you. Do not react personally, even if the complaint takes the form of a personal attack.

Let the person tell his or her story

The best medicine for angry citizens is to let them get things off their chests, *without getting you upset!* Give them a chance to express their feelings. Draw them out with questions or noncommittal remarks like "I know how you feel." Let people know that you understand what they are saying without interrupting their flow of thoughts. This will help them calm down. It may also reveal some points of agreement or settlement that are important in leading to a solution. You can count on some people not listening to you until they tell their story anyway. So plan for it. Even if the matter seems to be trivial or ridiculous, remember—it is important to that person.

Learn to listen

It's not enough to sit passively by while the other person talks. One has to listen with an active mind, looking for the paths that lead to understanding and problem solving. This involves leading the speaker with apt and timely questions. It requires the ability to turn the speaker's question back to him or her so that the individual will tell the story fully and not be given answers before he or she is ready. Don't give a solution until a clear-cut, mutually understood problem definition is agreed to. Listening is a skill!

Speak their language

It won't help to use the terms and nomenclature common to your profession or your specialty when dealing with the public. You have to translate. You should never embarrass your listeners or put them on the defensive by making them ask "What do you mean?" Find the words they will understand when talking about your product or service. Aim at communicating, not showing off your knowledge.

Say it with respect

Courtesy, respect, and consideration are all shown in little ways: a friendly tone of voice (but not a honeyed one); a manner that shows the person that he or she is considered worthy of respect and courtesy; a controlled volume to your voice—don't speak too loud, or too soft. Choose words that will be meaningful to your visitor.

Make people feel important

You may see 50 or 100 people every day. But they see you only once. When you have other work, besides meeting people, set it aside when a visitor comes to see you. Try not to make a person wait until you have finished your work. If you are pressed, say so; let people know that you would rather attend to their problem but must finish this one task and will return to them as soon as possible. Do so quickly and return immediately. Learn the person's name right away and use it. A person's name is important to that person; thus, people should feel that their names are important to you.

Be prepared

When you know a specific individual is coming to see you, review his or her file in advance. Should you not know of his or her arrival, you can still prepare your interview by drawing out the person in a warm, interested, and friendly way. When you are dealing with facts or data that you give to your clients, you must keep information current. If it is necessary, plan to arrive at work 15 min early and review the information you must have during the day. Read Dale Carnegie's *How to Win Friends and Influence People.* [1]

Be honest with yourself

Bluffing may be all right in poker, but it won't do in public contacts. When you don't have the information, don't fake it. If you don't know the answers, refer the client elsewhere or admit that you don't know the answer. Don't make excuses and don't argue.

Be presentable

There are limits to presentability, but they should be set by necessity, not sloth. Many times a public contact can be rendered ineffective because the person at the desk or counter is carelessly groomed, or his or her desk is untidy. Often people are offended by such an image. Sloppiness suggests a lack of interest in oneself and therefore a lack of interest in others.

Know when to terminate an interview

Don't lose the effectiveness of your discussion by letting it drag on. When you feel that the problem has been solved, you can courteously end the contact. This takes tact, but it should be done firmly and pleasantly. Often simply rising from your seat, extending your hand, and saying, "Thank you for coming in; we hope to see you again soon," is all that is needed. If more time is required to resolve the problem, explain that it is too important to be handled by a snap judgment and that some additional information and data are required. Let the person know about how long this will take. This will help ease the citizen's concern and extend his or her patience. A good sense of closure—however you apply it—is essential.

Developing positive relations with various client groups

In addition to individual citizens, the traffic engineer has a number of support groups that help determine the community character and how funds are to be spent. The basic principle of our democratic form of government is that "the will of the people will be done." Community organizations carry a lot of weight in determining what programs should be given emphasis. School and neighborhood groups are always interested in child-safety programs. Obtaining representation from these groups on a "Traffic Safety Committee" may improve coordination of education, engineering,

and enforcement needs and also establish the needed priority for staffing and materials necessary to do the job.

Business groups and chambers of commerce are always interested in adequate parking and traffic circulation and access. The traffic engineer can help provide answers on what improvements will work and what they will cost. Meeting these needs builds public confidence. You will be asked many questions about what is going on in the town or city. Be sure you have a good general knowledge of the status of current projects or programs and a good speaking knowledge of the funding picture in your locale. Keep in mind that while funding decisions rest with elected officials, the citizenry also look to you for answers. Your job is to make it clear who decides and at the same time not make people look bad.

The citizen group

It has often been heard at transportation seminars that the three old *E*'s of Engineering, Enforcement, and Education have been replaced by Energy, Environment, and Environmental Impact Statement (EIS). The simple fact is that democracy demands communication with the public, and communication requires a sender and a receiver. EIS laws came about partly because engineers and builders are notorious noncommunicators. By law, we are mandated to do written communication.

Today there are quite adequate formats prescribed for how to write an EIS. But how do we communicate with other groups? We generally don't get to choose the format or forum.

Meet anywhere, anytime

Be willing to meet with citizen groups at anytime and anyplace with or without the media present. If you are backed into a corner and have a little information, do not hesitate about meeting; but make it clearly known that you are there to gather information in order that you can more thoroughly analyze the problem. In other words, defuse the initial contact, support the more rational leaders of citizen interest groups, but don't ignore the radicals completely.

Citizen interest groups may contain a number of people who have developed strongly charged feelings in opposition to the subject matter being presented. These feelings are frequently brought on either by lack of information or the wrong information. Thus, it is critical to separate the emotion from the concern, and to identify what the concerned people would like to see happen and why they feel that way. Don't assume anything with a citizen group. Make certain that you are communicating at basic levels. Try to avail yourself of interpersonal-relations skills, and attempt to gain skills in the identifying styles of, for example, "driver," "expressive," "amiable," and "analytical." Being able to identify these styles will afford you an opportunity for more efficient communication, in that you must communicate with each faction differently so as to be effective.

Bring out the problem statement

Get training in leading group discussions. Without that ability, most group discussions will not result in positive progress.

---

[1] Dale Carnegie, *How to Win Friends and Influence People,* revised edition (New York: Simon & Schuster, 1981).

If you are not comfortable in this role, there may be times when a "nonbiased facilitator" can be employed. The key is to "be sure all viewpoints are allowed to surface." When you see people are not participating in the group discussion, draw them out. Make liberal use of the *why, how,* and *what* questions. Learn the proper easel-and-blackboard techniques.

It is very effective in a community meeting to have staff members write down citizen comments on a flip chart. This graphic portrayal of "See, we are listening" will promote a lot of confidence. Learn to identify opinion leaders. Get names, phone numbers, and addresses of participants. Prior to subsequent meetings, double-check with these people—"Here's what we heard, right?" "This sounds like what is wanted, and what we should do is. . . ." This process should identify those whom you can trust and those who have hidden agendas.

Never talk down to individuals or to group members, or undersell the person high on an ego trip. The individual who must lead the attack on you may just be filling a need for self-worth among peers. Such a person, if coached as an expert or soothsayer, can become your best ally. Don't worry about the obnoxious person. Frequently, a statement like "That's an interesting viewpoint; there may be others here who agree with you" will suffice. If the person keeps up with the "dominant obnoxious" role, others in the audience almost always will tell such a person to shut up and sit down. This is one of the joys of the democratic process.

Never try to accomplish too much at one meeting

Just remember if everything were simple, there probably would be no reason for a meeting. With most community groups, consider it a huge success if at the first meeting a general agreement on "the problem" can be reached together with a date for the next meeting. Perhaps a week or two is needed for trauma to subside; that is, "The world did not end; perhaps the problem wasn't as bad as we first thought." The second meeting may then proceed with a more reasonable look at alternatives.

Forming a "Citizens Advisory Group" may be advisable to clarify concerns and desirable actions. A sign-up sheet should be made available at public meetings to allow citizens to write their names, addresses, and phone numbers and if they would be willing to participate in a Citizen Transportation Committee. Try to accomplish as much as possible with the citizen committee before arranging another open meeting with the entire neighborhood. Much time can be saved in this manner.

Also, at a public meeting, have forms available for the public to list comments and suggestions. Many citizens are uncomfortable speaking in public, but their input could be valuable. Moreover, have them indicate their names, addresses, and phone numbers. It is helpful to know from which section of the neighborhood the particular comments originate.

Most transportation engineers work under a great deal of pressure with limited staff. Thus, there is a desire to resolve each problem with a minimum of work. But many neighborhood studies require extensive consultation and data collection. Hence, the engineer should recruit as much free labor as possible from within the community. Frequently, this is an excellent buffer—that is, if the magnitude of the problem is not sufficient to warrant citizen use of their time. Then perhaps a government of the people and by the people should also go easy. It is a fascinating dynamic to watch the fires of indignation subside when the possibility of self-help surfaces. Citizens may also find when they collect data that the problem is not as major as they first thought.

The Federal Highway Administration's "RD-80/092, State of the Art Report: Residential Traffic Management 1980" is an excellent resource document for group relations. Two very strong statements are made in the Introduction of the report.

> The planning process is more important than selection of the right device; more important than design, or more important than implementation technique.

Sound like a planner? You bet. The point, however, is: Surveys of many cities showed

> The failure of a program can be traced directly to either a breakdown in the planning process or the failure to have a structure at all.

Some examples of failure were total lack of action, leaping to "obvious" solutions, too limited a focus, and lack of community involvement. All of the above are obviously communication problems.

Simple success formula

- ○ Assess the problem or needs.
- ○ Develop alternative plans.
- ○ Evaluate alternative plans and plan selection.
- ○ Implement selected plan.
- ○ Modify the plan if necessary.

As stated earlier, "defining the problem" is a key element. The system is frequently circumvented by citizens demanding a solution and the desire to do something even though it may not be the right solution.

The problem with applying solutions to nonproblems is that one never knows when to quit; that is, how do you know when you have solved the "nonproblem"?

Dealing with the media

It is not possible to control the media.

It is possible to control media relations.

It is, as a matter of fact, imperative.

Who is this "fourth estate" that has constitutional rights and freedoms? When things are running well, the press is never around, but once something goes wrong—watch out! The media have the freedom to show us at our worst.

Ongoing friendly relations with the media (newspapers, TV, radio, etc.) can make a big difference. Keep in mind that most "make or break" decisions are subject to public opinion, and local press coverage makes the difference. There is obviously no such thing as "the media." There are individuals who work in TV, radio, and for newspapers. Each medium has unique needs, and each reporter and writer has an individual style, assignment, and temperament. However, there are some things common to all media and some things unique to each.

Charles M. Rossie, Jr., in *Media Resource Guide*[2], points out that "thoughtful editors are genuinely interested in their communities and the voluntary and nonprofit groups that serve them. This fact alone should encourage representatives of community groups to take the steps to identify themselves to the news people in their area."[3]

The transportation professional should "develop a list of all news vehicles that reach the people you are trying to reach. Most good libraries have directories of publications and broadcasters. Many communities have lists that are kept reasonably up to date by Chamber of Commerce or United Way agencies,"[4] Mr. Rossie states.

"Checking through the list of media outlets, you should determine the ones most likely to be interested in you or your activities. Not all news media will be interested. Nor will you consider all media outlets your best targets,"[5] Mr. Rossie contends.

Develop empathy

The current trend in the electronic medium is for more news coverage, longer news periods, and more competitive news gathering. Reporters are always looking for a story. Of concern to traffic engineers is turning a possible negative situation into a positive asset. The best way is empathy—empathy with the public and empathy with the reporter. The public has a right to know. What is it that the reporter needs to know in order to tell the story? "No comment" is usually taken as an "admission of guilt" or "Keep out, we have things to hide."

Be outraged

The public has a right to be outraged if there is an occurrence involving senseless loss. If the timing calls for outrage, be outraged also, and then return to the cool logic of the professional.

For example:

"How many do we have to kill here, Mr. Transportation Engineer?"

"The accident was appalling, with senseless injury to the child. I can understand how the parents and friends must feel! If it was my son, I would be beside myself. What we have to do is reconstruct what caused the accident, and then see what we can do to prevent similar ones. We will have our investigators begin immediately."

The three guidelines for surviving bad news are:

1. Don't make it worse;
2. Get it over with; and
3. Remember, you may be dealing with the same news agencies long after the present item is forgotten.[6]

Priority to the press

Hank Barnes, one of our profession's most successful and renowned transportation engineers, had a simple rule:

Always give first priority to the press. Drop everything, listen, and respond quickly.

The press always has a deadline to meet. The press can present the situation as it sees it, and unlike other forums, you do not get equal time. Keep in mind the following:

1. You should know the general subject matter much better than the press does;
2. You should be interested; that is, if it is worth the press's time, it is certainly worth yours;
3. You have to be honest.

Also bear in mind that there is no such thing as "off the record." If something can't be repeated in public, don't say it in private. The reporter's job is to spread information. Your off-the-record remark may end up in print.

Every journalist also knows that if you (the traffic engineer) have it written down, it is public record, unless you can prove otherwise. The law is clear that if you are the custodian of the public record, you must make it available to others and provide it in a timely fashion.

Develop empathy with the reporter

Be honest, and never mislead

The reporter relies on the information you give to him or her. If the reporter accepts this information as fact and is later made to look silly, he or she will never forget it. As one reporter said, "I don't keep score, but I always get even." Not knowing is forgivable, as is refusing to answer if a credible reason is offered.

Always call back as soon as possible

Timing is what makes a reporter competitive. Empathize with this need. If a reporter "enterprises" a story, that is, begins research and writes articles on something he finds interesting that you are doing, don't call the reporter's competitor. Old news is not news; it's history.

You should have a central contact for press information in your organization. This is especially true for major events

---

[2] Foundation for American Communications, *Media Resource Guide: How to Tell Your Story,* 4th ed. (Washington, DC: Author). This publication can be ordered through the foundation at 1627 K Street N.W., Suite 403, Washington, DC 20006.

[3] Ibid., p. 2.

[4] Ibid., p. 5.

[5] Ibid., p. 7.

[6] Ibid., p. 45.

or disasters. It is always better to say, "Call Ruth on this one; she has all the latest information." If the press talks to three people in your organization and gets three separate and different answers, you know what you look like. A good rule is always to report all press contacts to your boss. Keep surprises to the boss to a minimum.

Get the facts

Do your homework, especially if you are going "live." If you are not accustomed to give-and-take public debate or free-style interviews, then practice talking to yourself. As with cooking, you need to have a pretty good idea of the end product before you start tossing ingredients together. Ask yourself: Why is the public interested in this subject? What will the reporter be likely to ask me? Plan your answers, then practice saying them aloud. Remember, written language is often different from spoken language. When you have what you believe is a concise answer, your confidence will rise; you will see yourself as knowledgeable and give a smooth answer. When you are "on," your answers will flow smoothly and concisely.

Misquotes and hostile reporters

Reporters may have an assignment or an angle. If so, they are out to collect the facts that fill in the assignment or angle. The chance of misquotes is reduced if you know the angle. A reporter asking about auto accident figures may have little or no interest in transportation matters but just be looking for background to fill in a hypothesis about the behavior of today's youth.

Pick your fights sparingly and check your won-loss record. You cannot fight the press. You can, however, obtain accuracy. If you believe you are receiving shabby reporting and inaccurate press coverage, you can appeal to the editorial board or the editor. You may not get a retraction; but if it can be shown that there has been a flagrant violation of facts, you will usually get fairer treatment in the future.

Keep in mind that reporters make mistakes too. Many are young and inexperienced, but few are stupid. Most are trying to do an honest job. Find out where the reporter wants to go on a particular story. The trouble with most people is that they talk when they should be listening. The trouble with the rest is that they don't talk at all. Just as the reporter is trying to draw you out, you must do the same. What does the reporter think? For a moment, set aside your ego and desire to be the expert and learn what the newsperson has to say. For example, find out who will be on the proposed panel. This may make a statement of the weighted answer— that is, four environmentalists vs. one engineer.

Television

The emphasis is on the *vision*. Fame used to be through written media or being known as a "person of letters." Now it is how often you are invited to TV talk shows. In TV, the emphasis is on the "visual" as much as the "tell." If it's action news, you can get it on the tube; down with the old, up with the new:

crews installing a new sign system, new bus lanes, new school signs, new paint striper, etc.

> "Look, citizens, you are getting your money's worth for once."
> "See, someone at city hall actually works."
> "See the action."

Dan Jones, public information officer for the City of Houston Public Works, offers training in dealing with the media. He makes some excellent points:

1. Interviews are three-sided; the third being the public. You are making a statement while you are listening to a question;
2. You will be watched for a reaction while the question is still unfolding.
3. Don't be a cold fish; demonstrate interest, lean forward, nod understanding, express shock, and be animated in response.
4. Look and sound like you care about what you are saying.

Dan also makes the excellent point that you should "Bridge to the positive." Yes, it's terrible that the accident happened, but we have taken the worst of these locations and proposed a new bond issue to rehabilitate the signals. Not only will they be safer, but they will reduce pollution and save everyone a lot of valuable time by being more responsive . . . etc. Yes, it's terrible, but here's what we have been doing with what we have. We are proud of these accomplishments. We plan to do the following . . . in the next several months. . . .

The public service announcement (PSA)

The public service announcement (PSA) is a good way to get a message across. These PSAs are free of advertising cost. Professionally done, PSAs on seat belts, school crossings, ridesharing, and bicycle safety will get air time. Most homemade PSAs get the late, late, late night shot or none at all. If you want a 30- to 60-sec message produced professionally, you can expect to spend from $5,000 to $10,000 minimum, providing you get some donated services. Professional actors want royalties if subsequent showings are run.

Written media

The press presentation can be topical (newsworthy) or of general interest (Sunday supplement). Christopher Little, publisher of the *Everett-Herald* and former attorney for the *Washington Post,* while talking to a Washington–Oregon and British Columbia ITE meeting made this statement:

> A column and one-half of newsprint usually has more material than a one-hour TV newscast. You can cover a lot of ground in the written media.

Items relative to new traffic control installations, modernizations, or removals are newsworthy and should be released as expeditiously as possible. Remember to credit the local politician, if practical. This type of information is generally best released to newspapers because they usually have reporters on the government beat who, if properly treated, will find it to their advantage to drop by the transportation engineer's office at least daily.

If the jurisdiction in question has more than one daily newspaper, it is best to share "topical" news on an equitable basis. This sharing may be on a rotational basis if the item is of human interest vs. press releases, that is, immediate need to know. Other media can be handled in a like manner. It is best for the professional to prepare written releases covering the event fully.

> Keep in mind the newsrooms of America are in constant danger of being buried in a blizzard of paper news releases. Anyone reading your news release will decide on the basis of what they can see in the first 60 seconds whether to read on let alone print it. Any ideas you present must jump off the page if you are to survive the first screening.[7]

These written releases, after being read by the reporter, will stimulate questions. The written release, plus answers to these questions, will permit the reporter accurately to paraphrase the information in his or her own news story. If accuracy is critical, it is proper to request that the news story be reviewed for accuracy before publication.

Items of general interest, such as the traffic control plans for a proposed large shopping center, a proposed coordination system for the central business district (CBD) signal system, or a new one-way street system for the CBD are best presented in the Sunday paper or in a local TV special.

If the professional believes that it is important that the information be disseminated, it is his or her obligation to prepare the necessary diagrams, charts, and maps needed by the media. Many agencies have weekly news bulletins that go to all media outlets.

As with topical releases, if there is more than one channel for the release of the information, they should be rotated with meticulous equality. If such a course is not followed, the governmental agency affected may end up with one or two friendly media contacts and a number of unfriendly contacts. Nobody needs enemies, particularly when it can be easily avoided with a little care. And as with topical press releases, wherever possible an elected official should be given appropriate credit.

Give the press a source person

Press releases for major highway or transportation facilities, or detours, must be timely, or there will be no coverage. Too soon can be as bad as too late. Contractors frequently miss deadlines. If the facility is to open in 48 hours, be certain this is accurate. If the detour is to begin tomorrow, be sure it begins tomorrow. Keep your credibility intact. Three axioms to remember: (1) Include the name and phone number of a person a reporter can call for more information; (2) accept as a

given that what you release will be rewritten; and (3) newspapers are always looking for good pictures, but they usually want to take their own. As for the last point, send the photo, but don't feel insulted if it's not used.

Finally, consider this word of caution given by a transportation engineer who proceeded through the school of hard knocks:

> Press releases will not take the place of personal contact with the reporters of the TV stations and with the major newspapers. As an example, I once decided to make our press releases more efficient by sending them directly to the AP and UP wire services instead of the daily newspaper. I was absolutely flabbergasted when the City Manager called and said the City Editor was foaming at the mouth and was out for my hide. I hightailed over to the City Editor's office and he read me up one side and down the other because he was damned if he was going to read news in his town off a wire service. The lesson learned was that reporters take a tremendous pride in discovering and writing various news stories. They are exceedingly offended if treated like parts of a vast machine to be used only when it is to the advantage of the public official. Use press releases for the routine matters but don't expect press releases to take the place of personal contact.[8]

Developing reasonable traffic policies

The traffic engineer has a major role in creating certain public policy. History repeats itself. A number of situations will confront the community and the traffic engineer and require a decision. Installing stop signs or not installing them, painting crosswalks or not painting them, installation of traffic signals, speed bumps, speed limits, and intersection turn restrictions are a few of these examples. To the extent possible, the traffic engineer must develop a framework of policy and/or ordinance support for making rational decisions on such repetitive requests. A considerable amount of staff work is needed on each policy development. You must make sure you have done your homework and talked to all concerned. You need to determine what information is needed to make the decision: Who will make the decision, who cares about the decision, and how will they react? The timing of when to take the policy for adoption or review is critical. The engineer must recognize that there are emotional times when sound engineering judgment and logic are irrelevant.

On the other hand, in a less heated environment, if we have a good solution and methodology to treat common situations in a common logical manner then reasonable elected officials should be able to reach a logical conclusion and adopt a logical policy. Pick your time to adopt policy carefully and only if you have counted the votes ahead of time. If you have been consistently providing good answers to service requests referred to you by the elected officials, then credibility is established and you have earned the right to be heard.

---

[7] Ibid, p. 8.

[8] Chuck Haley, *Effective Communications for the Transportation Professional* (Washington, DC: ITE Journal, 1984) 25.

Securing staff assistance and cooperation

Interjurisdiction coordination

In most urban areas no single jurisdiction—whether city, county, or state—is solely responsible for the total traffic operations on all the major streets and highways in the region. Thus, interjurisdiction coordination of the traffic regulations and controls on major streets and highways is essential for efficient and effective areawide traffic operations. Traffic signal system operations, freeway HOV lanes and ramp metering, major highway incident management, and major highway construction traffic management plans are all activities that can require interjurisdiction coordination. Developing and participating in programs to achieve that coordination are important urban traffic engineering activities.

Interjurisdiction coordination can be arrived at in a number of ways. A 1988 Transportation Research Board study found that in most urban areas, staff members from jurisdictions concerned with specific coordination problems work out solutions together. In some areas, coordination is achieved through regularly scheduled meetings of senior traffic engineering staff members from all the jurisdictions in the region. At these meetings, current and possible future problems are addressed. The meetings are sponsored by either one jurisdiction or jointly rotated among several jurisdictions.

Elected officials also have an important need for interjurisdictional coordination and cooperation. They are involved in many interagency committees that plan and fund needed transportation activities and projects. The basis for the informed public official decision is timely information. It is up to the staff not just to respond to requests from elected officials but to anticipate the needs and provide information that will afford elected officials the best opportunity to influence decisions in favor of their jurisdiction. This means not just accurate information, but, more important, timely information.

There is also a need to examine the impact of projects or plans from other jurisdictions on the needs or quality of life of your own jurisdiction. The enlightened staff person who can make an elected official look good by providing accurate and timely help in getting a better project or a better share of the pie will gain the respect of that official and others. Project needs and project construction cut across political boundaries. City needs are different from those of the county, as are the county needs different from those of the State. A strong tendency exists both at the staff and elected official level to have tunnel vision with regard to the needs of other jurisdictions. We tend to see our own needs as paramount and the needs of others as immaterial. Local jurisdictions resent having to go to higher levels of government for funding requests and design approval. Yet all parties must guard against developing an adversarial relationship, or mind-set. The attitude that "the state and feds are inflexible and overly bureaucratic" will not advance good working relations.

Similarly, a "holier than thou" attitude of "the state highway and interstate system are the most important roads, and our standards must be followed no matter what" will not sit too well with the local officials who also have highway needs. We must continue to pursue the answer to our various needs and be discussing joint needs, something along the lines of: "If the current system does not provide the solution, then what can we jointly do to change things so that we both benefit?"

Intrajurisdiction coordination

Traffic engineering expertise is needed at all levels of government and across various departmental lines. Providing service across these lines and among the various governmental levels is what builds credibility and recognition that there is a useful activity to be performed in traffic and transportation engineering.

Once this credibility has been developed, established procedures can then be set in place for review and comment on the potential traffic impact and design needs of new urban development in the private sector or the need for new public transportation facilities.

Procedures are also required to assure that the agency reviews and comments on the scheduling of construction projects and their needed traffic control plans, major maintenance work, and utility work on major arterials. The development of sound traffic control plans and clearly written traffic control special provisions regarding hours of work, lane closures, and detour routes are mandatory for a local area to maintain a minimum quality of life, business, and safety. Notifying the private sector of our own plans is essential.

In government there is always more work to be done than can ever be completed. Traffic engineering is also an activity that can be data-intensive. This means that a person can't do all the bean counting and still devote large amounts of time to policy matters. The work needs to be shared with those needing the answers. The spirit of cooperation may have you doing the complete job on the first request, and then having the requester providing the needed data on subsequent occasions once the requester knows what is required. This is the typical situation on environmental assessments. Much of the required data collection can be required of the developer. The traffic engineer can then spend his or her limited time in scoping the requirements and reviewing alternatives.

Training

The best public relations tool that the traffic engineer can have is a well-trained staff that is competent to do the required job. This is not done without proper selection and training. A number of self-help opportunities are available to local agencies. A call to ITE headquarters staff will put you in touch with the opportunities in your area. Moreover, local universities and colleges normally have professional development courses that are beneficial to practicing professionals.

Police

Cooperation with the local police department is also key to a successful traffic and transportation program. Most transportation management systems require some enforcement. Traffic signing for parking control or turn restrictions and speed limits are ineffective without reasonable enforcement. In a similar vein, the police are the first to pick up on situations that seem to create a high violation problem. These situations are frequently due to some facility inadequacy or possibly a traffic engineering need. Keeping clear lines of communication open will not only help you stay on top of both problems and ideas but also provide you with some insights into solutions that would not be considered without this small quality-circle approach.

Inviting police officials to attend meetings of community groups who are working on a traffic problem helps clarify both benefits and limitations of signing and enforcement. It also builds teamwork, which keeps each party from passing the buck: "The traffic engineers should put up some signs." "All we need is some enforcement," comes the reply.

A well-coordinated traffic control management program will keep the police up to date on all new and proposed traffic control devices requiring enforcement. Similarly, a sound program will have the police feeding back information on maintenance needs, such as down or missing signs, burned out street lights, etc. We should keep in mind that transportation cuts across all walks of life and governmental agencies. We need constantly to ask ourselves, "Who would also like to know about this project, activity, or concern?" Coordination with our counterparts not only improves working relations but also improves the end product.

Planning and funding standard traffic engineering programs

The difference between the professional traffic engineer and the politician is that the professional is supposed to decide based upon engineering facts and information. The politician has license to use "gut" reactions and, in fact, this is what got him or her elected (i.e., the politico's ability, as a generalist, is to make enough of the right choices that people agree with).

Traffic needs will be brought to the elected official's attention by citizens and groups as they see the need for service arise. As a traffic engineer, you owe elected officials the evaluation of how much safety or efficiency they can buy for the desired service proposed versus using the same amount of money on some other location. Always keep in mind that there may be other factors that go into any project or service that are beyond traffic safety or mobility concerns.

Standard needs lists

The traffic engineer needs to keep a relative ranking of problem areas by solution categories. The most common elements on anyone's needs list are: "traffic signal"; "school beacons"; widening "bottleneck intersections"; "channelization needs"; and "pedestrian overpasses."

There are a number of ways of arraying each of these needs, and the choice is up to each local jurisdiction. As a professional, it is important that you set forth all of the pertinent facts in any ranking. For example, you may wish to show for each location on the needs list the "Past 3-year accident history," "Traffic volumes being served," etc. You should check to see that there is a clear statement of how bad the

traffic problem is and how well it will be relieved by the solutions being proposed. Bear in mind the basic concept of safety per dollar spent. The elected official may be looking at "votes per dollar spent." The final decision regarding which is the best system usually rests with the voter. Remember that if there is no needs list then the citizen need currently being requested of the elected official is his or her highest priority.

Understanding the budget and capital improvement process

The efficiency of any transportation program rests with the ability to define the current needs adequately and to show what can be done to improve the situation.

Few jurisdictions have the luxury of throwing money at a problem in hope that it will go away. Thus, the traffic engineer needs to answer some very basic questions: What are you doing with the money we now give you? What makes you think that what you are asking for will be needed and will solve the problem? Who "sez" we need it?

Depending upon the size of the governmental agency, the budget process can range from "complex" to "monumental." The transportation official must make a concerted effort to understand the process by which decisions get made. This is a high-priority administrative task.

The basics of the budget process are similar regardless of governmental jurisdiction. "Wish lists" are prepared and compared against revenues. Departmental or division managers compare relative needs against their perspective of priorities and forward these upward in the budget decision chain. In larger organizations the Office of Management and Budget may play a key role in cutting back both transportation capital improvement program requests as well as transportation planning, design, and operating budget requests.

As a transportation official, you must get to know who is doing the staff work at each stage of the process and the basics of the decision process. It is critical that your proposals are adequately presented, that they are clear, concise, and that they are not misrepresented or undersold. Rest assured that the primary purpose of budget analysts is to "cut the program down to size."

The budget process should not be approached as a necessary evil where you spend only the minimum amount of effort to complete the forms requested. Wherever possible, "program narrative" should be developed clearly and concisely and supplemented with backup information and pictures where necessary. Historical background might be helpful in showing why the current budget proposal is appropriate. If budgets have been cut in the past, it may be feasible to paint a successful "I already gave" picture. "We are already getting more done with less. If you cut any deeper we won't be able to produce the efficiencies we developed," you might argue.

Conversely, "With the last budget increase we were able to develop the following budget efficiencies. With the requested increase, we can build on these and deliver the following. . . ."

Be sure to review other portions of the total budget and capital improvement program (CIP) submittals. Of special concern is the data-processing bandwagon. Computer processing budgets are increasing, and there is a great need to assure that traffic data gets its fair share as compared to possibly some that do not relate to public safety as strongly. In summary, the secret to good budgeting and capital programs is to know what it costs to do business and what other agencies are paying in staff and budget to do the functions identical to yours. This knowledge coupled with the knowledge of what else is competing for the same dollar that you are will give you a clear advantage over most other managers. Knowledge of the competing needs together with knowledge of the elected officials' feelings toward your needs and others will tell you how hard to push for your current needs.

Organizing the CIP submittal

Most capital improvement programs are a "line item" type of presentation with limited narrative and specific cost information. All too often this limited narrative is all that the policymaker ever sees of your project. To a large degree, the CIP narrative is like an ad in the classified pages. The readers of the ad are the department heads and budget analysts and then the elected officials. Bread-and-butter projects are generally fairly well known to the policymakers because of their cost and impact. Routine programs are another story. Some of the typical annual programs are as follows:

"High Hazard Traffic Signal Replacement" refers to signals from a replacement list that have the highest accident rate of all those on the list.

"Replacement of High Maintenance Cost Locations" is a good seller. What administrator wants to be stuck with "high maintenance costs"? A "Safety Lighting" program improves nighttime pedestrian or vehicular lighting at high-activity locations and locations where the night/day accident rates are higher than normal.

Be sure also to report on successes. The general rule is if the money is well spent then next time you may get what you ask for. ("Did you make us look good, Charlie?")

The benefit of establishing annual programs is that a systematic approach to problems can be taken and more efficiencies of scale can be developed. This is true in design as well as in construction. If the annual work plan can be projected for a number of years then staffing decisions can be made regarding hiring consultants vs. adding city staff.

Annual programs also work well in both the areas of "new" facilities and "major maintenance." The press has been a great asset in the past few years with regards to making a buzz word of "infrastructure." ("We don't know what this is, but it is worn out, needs to be replaced, and we are in bad need.")

The reality of developing a categorical needs list is that by the time you have got the list compiled you probably are doing a good job of analysis of what needs to be done and how it should be done. If the inventory is small, Joe, the maintenance foreman, knows the worst problem area. That is if good old Joe knows what he is doing.

Other typical program areas are: "Bottleneck Improvements," "Safety Lighting" (or Crime Lighting), "Thermoplastic Replacement Program," "Line Marker Replacement," and "Regulatory and Warning Sign Replacement." Signing can be sold as a capital item if the dollar value is high for the total program and if the signs have outlived their

useful life. An assessment needs to be made to determine whether there is more money in the "capital" budget or the general "operating" budget. If competition is keen for "general" fund monies then perhaps "capital" funds should be requested. Other programs to consider for annual funding are "Bicycle Spot Safety," and "Neighborhood Traffic Control."

The traffic engineer should look for ways to incorporate traffic safety items into other projects planned for construction. A good example is the installation of thermoplastic markings as part of any street construction or reconstruction. Similarly, traffic signs, signals, and detector loops should not be overlooked. If these are not included in capital improvements, then operating budgets will have to bear the costs of such work. The capital improvement list helps you provide a choice to elected officials on the transportation priorities.

Transportation officials need to keep alert to the latest grant-funding efforts of both the states and the federal government. There is always a new current pet project and funding scheme in the works. For example, when the oil companies overcharged on petroleum products, these overcharges were rebated to the states. It was stipulated that these funds go toward energy-reduction programs. Programs were set up by the states to spend energy-rebate money on signal timing and fuel-efficiency programs. Windows of opportunity also exist for ride-sharing and other transportation-efficiency programs.

How the budget analyst will get you

Here are some examples of what budget analysts will tell you:

1. We are cutting this year's program because there are funds left in your last year's program that you didn't spend. It must not have been a very high priority need.
2. We cut it because it wasn't clear where the money was to be spent or what locations were to be done.
3. We cut it because your cost estimate was too high.
4. We cut it because the cost estimate was too low. You should just do the work with funds from your operating budget.
5. They didn't think it was a high enough priority. (This goes with "They cut it last year so let's not put it in again this year.")

You should always include your lower-priority needs as well as your higher-priority ones. Remember that the budget analyst's job is to cut things out. If you do not ask for the items that will make for a top-notch transportation system then it will never be achieved. Always be sure that the money will be well spent if you perchance happen to get what you ask for.

Assisting the public to obtain efficient transportation systems is one of the most rewarding jobs for any administrator to undertake. Regardless of the size of the organization, the essentials are the same; providing timely, effective, and affordable solutions requires the support of the public at large as well as that of elected officials and co-workers.

Finally, the selection and training of competent staff must be done in light of the many community needs and various priorities. If both transportation capital improvement and operating budget needs are presented in understandable terms and in a timely manner, then the public will be well served.

Part B:
Traffic Administration

Woodrow W. Rankin

This section covers the administration of city, urban county, and state traffic engineering agencies.

City and urban county agencies

The need for traffic engineering services is essentially the same in cities and urban counties, and the administration of the agencies is similar.

Legislative authority. The *Uniform Vehicle Code* (UVC)[9] is a model set of laws designed to be a comprehensive guide for state motor vehicle regulations. It is based on experience under various state laws in the United States. Its use promotes national uniformity in traffic regulations, which contributes to safe, efficient traffic flow. As a city or county must be authorized by the state to perform traffic control and regulatory functions, the usual powers and duties of local authorities with respect to those functions are provided in Chapters 11, 14, and 15 in the UVC.

Sections in Chapter 15 of the UVC provide the basic authority for local jurisdictions to regulate and control traffic, including the use of traffic control devices, and they establish the state's control over the use of that authority. Other sections, not only in Chapter 15 but also Chapters 11 and 14 of the UVC, authorize local jurisdictions to implement specific traffic controls and regulations including speed limits, truck traffic prohibitions, and parking controls. Most state motor vehicle codes are in substantial compliance with those provisions of the UVC.

It is desirable that local jurisdictions have laws that establish the position of traffic engineer and the authority and responsibilities of the position. Usually this can be done with a local ordinance. The *Model Traffic Ordinance* (MTO),[10] a companion publication to the *Uniform Vehicle Code,* has provisions in Chapter 2 for establishing the traffic engineering position, and designating its authority and its responsibilities. The authority and responsibilities include the installation, operation, and maintenance of all traffic control devices, and the planning and implementing of traffic flow and traffic safety improvement measures. All actions of the traffic engineer in carrying out his (or her) duties are subject to the controls of the state vehicle code.

Functional responsibilities. To provide effective and efficient management of traffic operations in urban areas, a traffic engineering agency must have principal responsibility for four key traffic functions:

- o The development of new and revised traffic laws and ordinances;
- o The design, installation, maintenance, and operation of all traffic control devices;
- o The traffic operations aspects of transportation system planning; and
- o The traffic operations aspects of street improvements.

Each of these functions includes a number of activities, which are given in Table 14–1.

A 1984–1985[11] study by the Institute of Transportation Engineers looked at the functions and administration of traffic engineering agencies in 131 U.S. cities between 50,000 and 1,000,000 population. The study found that 55% of the agencies were responsible for all four key traffic functions. Another 26% were responsible for three of the functions and had substantial responsibility for the fourth. The percentage of the agencies responsible for each function is shown in Table 14–2.

The study also collected information from several major urban counties (population over 250,000) that were responsible for traffic operations on major streets and highways. The study found that their traffic engineering agencies had

TABLE 14–1
Urban Traffic Engineering Agency Functional Activities

1. New and revised traffic laws and ordinances
   Initiate recommendations
   Review and comment on recommendations of others
2. Use of traffic control devices
   Design and application (including design by contract)
   Installation (including by contract)
   Maintenance (including by contract)
   Operation
3. Transportation system planning
   Participation in urban area planning process
   Review and comment on traffic operations aspects of site plans and development proposals
   Provide data for planning process
4. Street improvement
   Traffic operations review of designs for major improvements
   Identification of locations for safety improvements
   Design of safety improvements (including by contract)
   Participation in development of traffic control plans for major street construction projects

SOURCE: "Urban Transportation/Traffic Engineering Agency Functions and Administration Study," *ITE Journal,* 55 (1985), 21–23.

TABLE 14–2
Functional Responsibilities of Urban Traffic Engineering Agencies

Function	Percent of Agencies Responsible
1. New and revised traffic laws and ordinances	85
2. Use of traffic control devices	76
3. Transportation system planning	87
4. Street improvements	89
All four functions	55

SOURCE: "Urban Transportation/Traffic Engineering Agency Functions and Administration Study," *ITE Journal,* 55 (1985), 21–23.

---

[9] *Uniform Vehicle Code,* 1987 ed. (Evanston, IL: National Committee on Uniform Laws and Ordinances, 1987), 58–59, 64, 68, 93–96.

[10] *Model Traffic Ordinance,* 1987 ed. (Evanston, IL: National Committee on Uniform Laws and Ordinances, 1987), 5–6.

---

[11] Woodrow Rankin, "Urban Transportation/Traffic Engineering Agency Functions and Administration Study," *ITE Journal,* 1985, 55 (4), 21–23; 55 (12), 23–26.

substantial responsibility for the four key traffic engineering functions.

Traffic engineering agencies often are responsible for one or more other traffic-related functions. In a 1984–1985 study by the Institute of Transportation Engineers,[12] 70% of the traffic engineering agencies also reported they were responsible for street lighting, and 68% reported they were responsible for off-street parking programs. Other functional responsibilities reported by some cities were transit planning and operations, airport planning and operations, on-street parking enforcement, and high occupancy vehicle programs. In cities using parking meters, the traffic engineering agencies usually were responsible for installation and maintenance of the meters.

Organization. Traffic engineering agencies are located in various positions in the organizational structure of cities and urban counties. Regardless of its organizational arrangement, the traffic engineering agency must have a qualified traffic engineering professional as its chief administrator. It should have authority to perform effectively the essential traffic engineering functions, and have authorized communication channels with the executive officer and the major department heads of the city or county.

In the 1984–1985 study by the Institute of Transportation Engineers, the 131 city traffic engineering agencies furnishing information were located in the organizational structure as follows:

o 61% were a unit of the public works department
o 18% were in a traffic engineering department
o 8% were in the engineering department
o 4% were in a transportation department, and
o 9% were in other departments.

Most urban county traffic engineering agencies were in public works or transportation departments.

---
[12] Ibid.

Figure 14–1 is a typical organization chart of a city or an urban county where the traffic engineering agency is in the public works or some other department. Figure 14–2 is a typical organization chart of a city with a separate traffic engineering department.

The internal organizational structure of traffic engineering agencies primarily varies with the size of the city or county. The larger the jurisdiction the more complex the organization. In smaller cities, the traffic engineer usually directs all activities through traffic technicians or supervisors. Figure 14–3 is a typical organization chart for a small-city traffic engineering agency.

In larger cities and urban counties, usually two or more traffic engineering activities are grouped in divisions or sections of the traffic engineering agency under the direction of a professional traffic engineer. Figure 14–4 is a typical organization chart for a traffic engineering agency in a mid-sized city with two divisions: one for studies, planning, and design, and one for traffic control device installation, maintenance, and operation.

In urban jurisdictions over 400,000 population, usually subunits of divisions or sections also are headed by a traffic engineer, and often there is a deputy traffic engineer to assist in the overall administration of the agency. Figure 14–5 is a typical organization chart for a large city or urban county traffic engineering division or department with a deputy traffic engineer and the principal units and subunits headed by traffic engineers.

Staff levels. To perform the four traffic engineering functions effectively, an urban traffic engineering agency must have an adequate number of qualified professional, technical, and maintenance personnel. In addition to the area's population, a number of factors influence what an urban agency's staff level should be. The principal ones are:

o The area's rate of traffic and population growth;
o The area's traffic density levels;

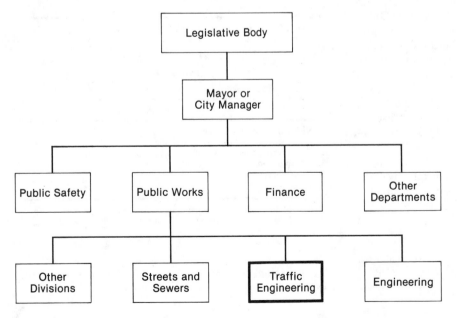

Figure 14–1. Typical organization chart of a city where traffic engineering is a division in the public works department.

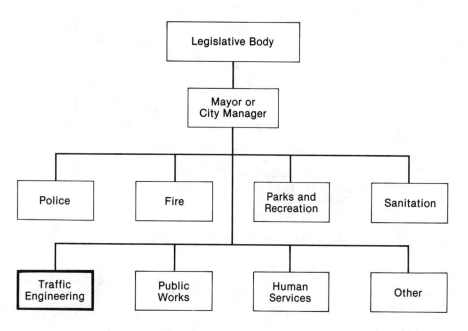

Figure 14–2. Typical organization chart of a city with a separate traffic engineering department.

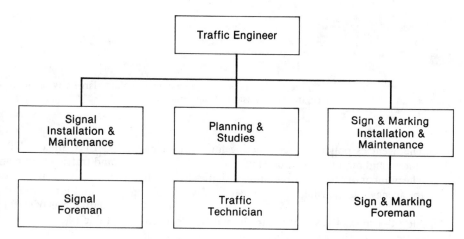

Figure 14–3. Typical organization chart of a small-city traffic engineering division or department.

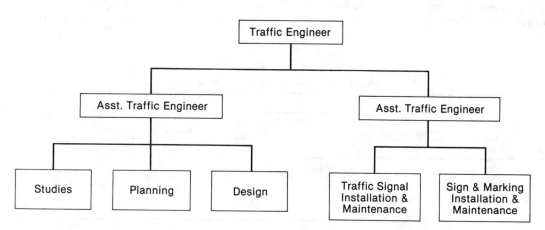

Figure 14–4. Typical organization chart of a traffic engineering division or department in a mid-sized city.

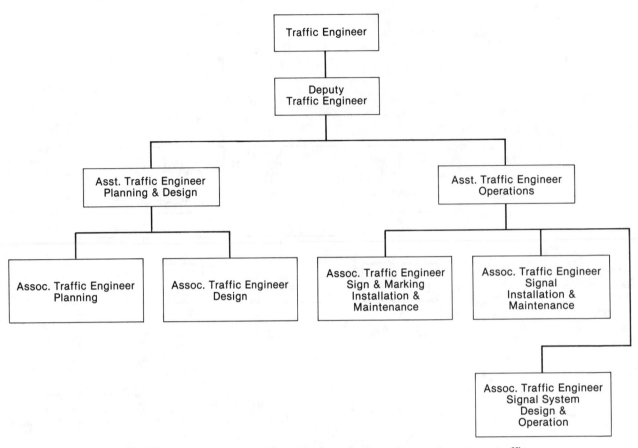

Figure 14–5. Typical organization chart of a large city or urban county traffic engineering division or department.

○ The agency's policies on contracting for engineering studies and maintenance work; and

○ The number of other transportation activities for which the agency is responsible.

Table 14–3 shows the variations in the staff levels of urban traffic engineering agencies in a representative group of cities between 50,000 and 1,000,000 population. Staff levels, in employees per 100,000 populations, are given for professionals, traffic engineering technicians, maintenance workers and supervisors, and all employees.

Ever since 1970, the percapita level of professional and traffic technician staffing in urban traffic engineering agencies has increased, and the level of maintenance workers and total employees has decreased. This change in the relative level of staff in the four employee groups is shown in Table 14–4.

Operating budget levels. To provide all essential traffic engineering services, a city or urban county traffic engineering agency must have adequate funding for its operations, engineering, and maintenance activities. In addition

TABLE 14–3
Urban Traffic Engineering Agency Staffing, 1984

Population Group	No. of Cities	Personnel per 100,000 Population*											
		Professionals			Traffic Engineering Technicians			Maintenance Workers			All Employees		
		Mean	Median	Range	Mean	Median	Range	Mean	Median	Range	Mean	Median	Range
50,000–100,000	33	3.3	3.1	1.1–7.4	2.8	1.6	0–15.7	13.0	12.2	3.4–28.7	21.4	20.3	8.8–38.6
100,000–250,000	31	2.8	2.3	0.6–6.9	2.4	1.9	0–11.8	12.1	10.9	2.5–29.9	19.4	18.2	7.4–47.3
250,000–500,000	12	2.4	2.0	0.8–5.9	2.9	1.9	0.4–11.0	16.3	14.3	7.9–34.9	23.3	22.0	10.4–52.6
500,000–1,000,000	8	2.7	2.8	0.5–5.1	3.2	2.8	0.5–6.5	11.1	10.5	7.2–17.4	18.0	19.0	10.2–26.3
All cities	84	2.9	2.6	0.5–7.4	2.7	1.9	0–15.7	13.0	11.9	2.5–34.9	20.6	19.6	7.4–52.6

*1980 census.

SOURCE: "Urban Transportation/Traffic Engineering Agency Functions and Administration Study," *ITE Journal,* 55 (1985), 23–26.

No. of Cities	Population Range	Year	Personnel per 100,000 Population*			
			Professionals	Traffic Engineering Technicians	Maintenance Workers	All Employees
			Mean	Mean	Mean	Mean
41	100,000 to 1,000,000	1970–1975	2.3	2.1	14.7	20.8
		1984	2.8	2.6	13.7	20.7
Change			+0.5	+0.5	−1.0	−0.1

*1980 census.

SOURCE: "Urban Transportation/Traffic Engineering Agency Functions and Administration Study," *ITE Journal,* 55 (1985), 23–26.

to the area's population, a number of factors affect what an agency's funding level should be. The principal ones are:

o Size of agency staff;
o Miles of street per capita the agency is responsible for;
o Amount of new or replacement traffic control device installations done with operating funds;
o Local wage rates; and
o Local utility costs.

Table 14–5 summarizes the 1984 level of funding per capita for the operating and engineering and the maintenance budgets reported by 54 city traffic engineering agencies in the 1984–1985 study by the Institute of Transportation Engineers. The cities ranged from 50,000 to 1,000,000 population. All the agencies had substantial responsibility for the four key traffic engineering functions.

In Table 14–5, the maintenance budgets fund all the costs for personnel, materials, and operations associated with the routine installation, repair, and replacement of all traffic control devices. The operating and engineering budgets fund the agency's engineering and technical personnel costs and all central office costs. No capital improvement program funding is included in Table 14–5.

State traffic engineering agencies

All state highway agencies are responsible for traffic engineering activities on rural sections of the state highway system, but in urban areas there is a wide range in the level of state responsibility. Most states are responsible for traffic engineering

activities on all urban sections of their highway system. However, a few states have little or no traffic engineering responsibilities inside incorporated cities, and others have limited responsibilities in some cities and full responsibility in others.

Legislative authority. The authority for state highway departments to regulate and control traffic is established by state statute. Chapters 11 and 15 of the Uniform Vehicle Code have model statutes for that purpose. In Chapter 15, Section 15-104 establishes the state highway department's authority over the design and use of traffic control devices on all streets and highways in the state. The section requires the state highway agency to adopt a traffic control device manual and specifications for the use of traffic control devices on all streets and highways in the state.

Section 15-105 of the code assigns to the state highway department the responsibility for use of traffic control devices on its highway system and, consequently, the regulation and control of traffic. The section also prohibits local jurisdictions from placing or maintaining traffic control devices on state-controlled highways without the permission of the state highway agency. Other sections in Chapter 15 as well as in Chapter 11 of the code establish the state highway department's authority to implement specific traffic control measures including speed limits, one-way traffic, parking regulations, and both stop and yield intersections.

Functional responsibilities. There is a wide range in the level of responsibility of state traffic engineering agencies for traffic engineering functions. A 1985 survey of state traffic

Population Group	No. of Cities	Budget Category								
		Maintenance			Operations & Engineering			Both Categories		
		Mean	Median	Range	Mean	Median	Range	Mean	Median	Range
50,000–100,000	17	8.30	5.00	0.50–28.80	4.05	2.05	0.90–18.50	12.35	8.30	2.30–31.40
100,000–250,000	22	5.50	4.70	2.10–17.70	3.15	1.70	0.70–10.40	8.70	6.85	3.10–23.30
250,000–500,000	10	8.70	6.50	2.60–22.10	2.70	2.10	0.60– 7.40	11.40	7.80	3.40–29.40
500,000–1,000,000	5	5.10	4.20	1.30–10.40	3.30	3.50	0.90– 5.00	8.40	8.00	2.20–15.40
All cities	54	6.95	5.10	0.50–28.80	3.40	2.05	0.70–18.50	10.30	7.60	2.20–31.40

*1984 dollars; population as per 1980 census.

SOURCE: "Urban Transportation/Traffic Engineering Functions and Administration Study," *ITE Journal,* 55 (1985), 23–26.

engineering agencies[13] looked at their level of responsibility for 12 traffic engineering functions. The study found that all of the agencies had some level of responsibility for the design of traffic control in construction areas, and 70% had no responsibility for traffic count programs. Table 14–6 lists both the functions looked at in the study and the percent of states with these levels of responsibility for each function: Prime, Review, Assist, and None.

Organization. The status and location of the traffic engineering function in the organization of state highway departments varies from state to state. A 1983 study of organization charts for 39 highway/transportation departments found that an identifiable unit was responsible for traffic engineering in 72% of those states.[14] In the other states, the function was divided between two or more units or there was no identifiable traffic engineering unit. This distribution of the traffic engineering function is given in Table 14–7.

In 59% of the states, the identifiable traffic engineering unit was a section of a major division of the highway department. Figure 14–6 illustrates this type of organization

structure of a department. In 13% of the states, traffic engineering was a major division. Figure 14–7 illustrates this type of organization.

In many states there are traffic engineering units in the district offices of the highway department. A 1985 survey of 39 states[15] found that the authority for major technical decisions given to district traffic engineering units varies between states. These variations are summarized in Table 14–8.

Professional staffing. The number of professional traffic engineers in state highway departments varies substantially among states because of variations in population, urbanization, and other factors. A 1985 survey of 39 states[16] found that the range was from one to 300. On a percapita basis, professionals per million state population, the range was from 0.6 to 15.8. Table 14–9 summarizes this information on the number of professional traffic engineers in state highway departments.

Traffic records

Up-to-date, readily available traffic records are essential for effective traffic engineering. Three types of records are needed:

o Records of traffic operations
o Records of traffic regulations
o Records of traffic control device installation and maintenance.

All the records should have the same location reference system to facilitate looking at all the traffic-related factors at any location. In small-urban jurisdictions and urban counties, manual records—using file cards, copies of written reports, and spot maps—can be used. In larger urban jurisdictions and states, the records should be in compatible computer files. Refer to Chapter 3, "Traffic Studies," and Chapter 4, "Traffic Accidents and Highway Safety," of this text for a more detailed discussion of record systems.

Traffic operations records. Two types of records of traffic operations are needed: traffic accidents and traffic volumes. The accident record system should have information on location, date, time, severity, and the accident report file number for all reportable traffic accidents occurring on streets and highways controlled by the jurisdiction. It is not essential that the traffic engineering agency be responsible for maintaining the system, but the agency should have ready access to it. At a minimum, the traffic volume system should have volume and turning count information at all signalized intersections controlled by the jurisdictions. In urban jurisdictions, the traffic engineering agency should maintain the traffic volume system.

Traffic regulations records. The traffic regulations record system should have the location for the following regulations: one-way streets, through streets, isolated stop signs and yield signs, traffic signals, turn restrictions, speed

TABLE 14–6
State Traffic Engineering Agency Functional Responsibilities

Function	Percent of Agencies with Level of Responsibility			
	Prime	Review	Assist	None
Geometric design	16	62	19	3
Traffic control device design	85	13	0	2
Design site access plans	16	38	28	18
Evaluate requests for traffic signals	95	0	0	5
Maintenance of traffic control devices	26	14	31	29
Design traffic control in construction areas	56	20	24	0
Develop speed zones	82	5	3	10
Maintain accident records system	43	7	10	40
Develop FHWA safety program	54	10	17	19
Manual traffic counts	30	0	0	70
AADT count program	18	0	0	82

SOURCE: "Traffic Engineering Functions in State Government," *ITE Journal,* 58 (1988), 76–78.

TABLE 14–7
Location of Traffic Engineering Function in State Highway Departments

Location of Function	Percent of States
Identifiable unit in:	
Operations	26
Engineering	10
Traffic Engineering	13
Maintenance	5
Maintenance & Operations	8
Other divisions	10
Multiple units	10
No identifiable unit	18

SOURCE: "Traffic Engineering Functions in State Government," *ITE Journal,* 58 (1988), 76–78.

[13] ITE Technical Council Committee 2-30, "Traffic Engineering Functions in State Government," *ITE Journal,* 1988, 58 (6), 76–78.

[14] Ibid.

[15] Ibid.

[16] Ibid.

LOCATION	REQUESTOR NAME	PHONE	DATE RECEIVED	DATE COMPLETE	TYPE	TIME IN SYSTEM	DISPOSITION	ACTION PERSON
OLD SANTA RITA N OF SANTA RITA RD	HAROLD RADTKE	484 3926	07-Jul-88		PARK	66		WVG
HOPYARD RD @ W.LOS POSITAS	HAROLD RADTKE	484 3926	07-Jul-88		MARKING	66		WVG
HOPYARD S OF W.LOS POSITAS	HAROLD RADTKE	484 3926	07-Jul-88	25-Jul-88	MARKING	66		WVG
KOLLN ST	BRIAN SANBORN	846 7898	25-Jul-88		VOLUME	48	REPT CC	WVG
BERNAL AVE @ INDEPENDENCE	PHIL LOWE	846 8384	13-Jul-88		R 1	60		WVG
MAIN STREET @ ABBIE ST.	MRS FOX	462 5903	01-Jun-88		X WALK	102		WVG
ARTHUR @ CHERYL	STACY SWARTY	462 2073	01-Jun-88		R 1	102		WVG
VALLEY @ BUSH	PAUL HELMS	847 8086	19-May-88		TRUCKS	115		WVG
KOLLN @ MOHRE	MS LANEY BROWN		20-May-88	27-May-88	VO	7	NA	WVG
KOLLN N OF VALLEY	SUSAN DAVIS	462 2057	23-May-88	25-May-88	R 1	2	FILE	WVG
W LOS POSITAS @ SANTA RITA	KEVIN BROWN	847 4526	24-May-88	25-May-88	TIMING	1	>TIME	WVG
PASEO SANTA CRUZ @ ALTIMIRA	MRS KRALL	833 0984	27-Apr-88	30-Apr-88	R 1	3	FILE	WVG
VALLEY @ CRESTLINE	DON ADAMS	939 5610	27-Apr-88	17-May-88	X WALK	20	MARK	WVG
SANTA RITA RD @ VALLEY	DAVE MARCHE	847 8195	29-Apr-88		SIGNING	135		WVG
OLD BERNAL @ PETERS	BRIAN SWIFT		27-Apr-88	04-May-88	PARK	7	FILE	WVG
DEL VALLE PKWY @ CHURCH	BILL IRICK	828 1363	10-May-88	11-Oct-88	MEDIAN CU	154	NEW CON	WVG

TRAFFIC SERVICE REQUEST LOG   REPORT DATE 11-Sep-88

Figure 14–9. Sample traffic service request log.
SOURCE: Courtesy of Mr. William Van Gelder, Transportation Engineer, City of Pleasanton, CA.

Legal background

...lawsuit, involves two or more entities or ...st one of which feels that he or she has been ...ed.[17] Usually, none of the participants feel ...lty of any overt act or omission, wrong...nce on their part. Thus, it then becomes ...independent trier of the facts (the court) to ...ll be responsible for the damages incurred ...r form of any remedial actions.

...deral, state, and local government powers

...tes, the authority for the federal govern...t in streets and highways has as its foun...merce Clause" in the U.S. Constitution. ...given the authority to regulate commerce. ...ourts have extended the clause to include ...munications, aviation and airports, and ...ys. It is through this clause that the devel...rstate highway system gains its authority ...nstruction, that the authority to develop ...s and highways occurs, and that the gen...ets, highways, and related matter begins. ...tion reserves certain powers to the states. ...e police powers, which are not limited to ...as the name implies, but include all as...ty, health, and welfare. Thus, the states ...motor vehicle movement, construct and ...and streets, and have a role in pollution ...ronmental improvement. These powers ...two factors: They may not infringe upon ...of interstate commerce granted to it by the Constitution, and state control may not be so overly broad as to approach an abuse of police powers. They must further be applied equally to all classes of citizens so as to meet the constitutional requirement of equal protection under law to all.

It is under these police powers that the states—and the local communities when the authority is delegated to them by the states—have the authority to develop and pass motor vehicle codes and ordinances. In the interest of uniformity of motor vehicle laws, a *Uniform Vehicle Code* and a *Model Traffic Ordinance* have been prepared every 2 to 6 years since 1926. First published as separate booklets, they now appear in a single volume.[18] The publication's foreword indicates that it is not a panacea for all ills and that it does not cover all conceivable legal provisions, but that it does provide the framework for uniformity. And with uniformity come familiarity and safety.

The Highway Safety Act of 1966 (*U.S. Codes,* Chap. 23, Sec. 401–404) directed attention to highway safety. Through the National Highway Traffic Safety Administration, a series of highway safety standards has been promulgated, and the states are required to conform to these standards. One standard deals with vehicle codes and laws.[19] It has the objective that the states

should develop and implement a traffic code and laws program, which includes plans to:

1. Eliminate all major variations in traffic codes, laws, and ordinances among its political subdivisions.
2. Increase the compatibility of these ordinances with a unified, overall State policy on traffic safety codes and laws.
3. Further the adoption of appropriate aspects of the Rules of the Road chapter of the *Uniform Vehicle Code.*

---

[18] *Uniform Vehicle Code,* 1987 ed., and *Model Traffic Ordinance,* 1987 ed. (Evanston, IL: National Committee on Uniform Traffic Laws and Ordinances, 1987).

[19] National Highway Traffic Safety Administration, *Codes and Laws.* Highway Safety Program Manual 6, November 1974.

...ject area (pp. 438–444, up to before section on ...ve Measures") was prepared by Sheldon Pivnik, ...d Signs Division, Metro-Dade County, Public ...i, Florida.

...ng Handbook

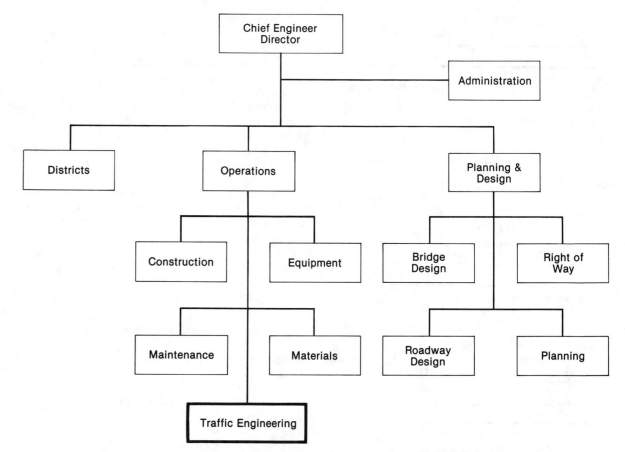

Figure 14–6. Typical organization chart of a state highway department with traffic engineering a section of a major division.

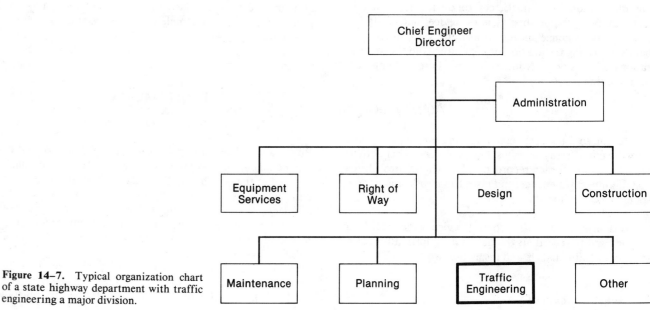

Figure 14–7. Typical organization chart of a state highway department with traffic engineering a major division.

TABLE 14-8

Where Final Major Technical Decisions Are Made

	Percent of States
Percent at central office	
100	20
50–99	62
<50	18
Percent at district office	
100	0
50–99	30
<50	70

SOURCE: "Traffic Engineering Functions in State Government," *ITE Journal,* 58 (1988), 76–78.

TABLE 14-9

Number of Professional Traffic Engineers in State Highway Departments

	Number of States
Number of professionals	
1–10	17
11–20	9
21–50	7
51–100	4
>100	2
Professionals per capita*	
<3.0	11
3.0–6.0	10
6.1–9.0	11
9.1–12.0	3
>12.0	4

*Per million state population, 1980 census.

SOURCE: "Traffic Engineering Functions in State Government," *ITE Journal,* 58 (1988), 76–78.

zones, major parking restrictions, and school area traffic controls.

Traffic control device records. The traffic control device record system, at a minimum, should have the following information for all traffic control devices for which the jurisdiction is responsible: type of device, date installed, dates of maintenance and/or inspections, work order numbers authorizing installation or upgrading, and the identification numbers or other identification of crews doing work or inspecting the device.

Activity records

In addition to usual cost-accounting and personnel records, a traffic engineering agency should keep two other activity records: a work order record of all traffic control device installation, maintenance, and inspection work; and a correspondence/telephone log.

Work orders. A traffic control device work order form can be used to record all the necessary information on installation, maintenance, and inspection work:

- What was done
- Where it was done
- When it was done
- Who authorized the work
- Who supervised the work
- What material, equipment, and labor was used

Information from completed work orders should be used to update the traffic control device records system. The information also can be used to develop budget information on time, equipment, and material costs. To facilitate this in larger cities, the time, equipment, and material use information is put into a computer file.

Correspondence and telephone log. A traffic engineering agency should have a record of all the correspondence and telephone calls it receives relative to its traffic operations and planning activities. These include:

— Request for new or revised traffic regulations and control devices

— Complaints about traffic operations

— Reports of traffic control devices that are missing, damaged, in poor condition, or malfunctioning.

A sample traffic control request form is illustrated in Figure 14-8.

The record can be kept as a written log. At a minimum, the following information should be recorded for each communication:

— Type: letter, phone call, internal memorandum, etc.

— Date and time received

— Name, address, and phone number of individual and/or organization from whom it was sent

— Name of staff person it was assigned to

— Action taken and date.

The agency should have established procedures in place to assure the log is used to monitor the status of the action on all items and to assure the action is completed in an appropriate time period. A sample computer system log is illustrated in Figure 14-9.

Part C: Legal Liability

JOHN E. BAERWALD

Public and private transportation organizations as well as individual employees of these organizations are increasingly exposed to the possibility of litigation. This can range from being a witness whose only function is to testify about certain factual information (signal timing programs, results of traffic data measurements, date and type of equipment and/or site modifications, etc.) to being a defendant in a civil liability lawsuit.

A growing number of transportation and traffic engineers are being called as expert witnesses by both plaintiffs and defendants in liability suits, as well as in hearings before

TRAFFIC CONTROL REQUEST
PLEASANTON DEPARTMENT OF PUBLIC WORKS
TRANSPORTATION DIVISION

Location File ____________

Reported by ____________ Address ____________

LOCATED on E W N S side of ____________ N S E W of ____________

PROBLEM STATEMENT:
(Sketch on back side ☐)

Written by ____________

INVESTIGATOR'S EVALUATION

Investigation by ____________

Information Reviewed:

Collision Records ____ Traffic Volume Counts ____ Field

Location File Check ____ Paint Sketch ____ Phot

Other Traffic Surveys ____ List: ____________

Other Records ____ List: ____________

Recommendation or action ____________

Figure 14–8. Sample traffic control request form.
SOURCE: Form provided courtesy of Mr. William Van G
Engineer, City of Pleasanton, CA.

administrative bodies. This involvement can be time-consuming, expensive, and traumatic, but it can also be professionally beneficial and an important element of the legal process—provided you are qualified and properly prepared.

This part of the chapter is intended to provide basic information for those persons and entities interested in reducing exposure to tort liability, risk-management, and the involvement of professional transportation and traffic engineers in

litigation acti
and only ma
cluded. More
Readings at t
Because n
the widely di
the world, th
in the Unite

A tort action, c
individuals, at le
wronged or inju
that they are gu
doing, or neglig
necessary for an
determine who w
and the amount

Fe

In the United St
ment's involveme
dation the "Com
Here Congress is
Congress and the
radio and TV co
streets and highw
opment of the int
for funding and c
standards for stre
eral funding for st
Yet the Constit
Among these are
police protection,
pects of public sa
are able to regulat
maintain highway
reduction and en
are limited by only
Congress's control

[17]Material in this s
"Organizational Protec
Chief, Traffic Signals
Works Department, Mi

Vehicle codes and ordinances have an important place in liability litigation. By violating a code provision, motorists may find themselves in a position where they become the negligent party or contributed to their own negligence, so that, as explained later in this section, they may be precluded from recovery for injuries sustained on the highway system or may get a reduced award from the courts. Transportation professionals must therefore be familiar with the codes and ordinances of their states and local communities in order to apply their engineering skill and judgment properly for the maximum safety of the users of the road system.

Public liability

The federal government, while a developer and prime mover of the interstate highway system and the *Manual on Uniform Traffic Control Devices* (MUTCD)[20] has been held harmless by the courts for injuries and losses that have as their proximate (legal) cause infractions of standards, changes in design, and so on, occurring with the knowledge of the federal government. The U.S. Supreme Court has held that the role of the federal government is one of financial management and not designer and operator of the street and road system. Under this theory, lawsuits against the federal government for negligence of design and operation of the highway system generally fail.

The courts hold that the state is not an absolute insurer of the safety of the user of the highway system. However, they have indicated that state and local governments are required to maintain streets and roads in a reasonably safe condition for users. Failure to do so and which results in injury or property loss to users will subject the agency having the prime responsibility to liability for that loss.

In the operation of street and road systems, four separate areas are involved: design, implementation, operation, and maintenance. The liability exposure differs in each of these areas.

In the area of design, as long as discretionary authority in the development of the design has not been abused, an agency is held generally immune from tort liability. In the areas of implementation, operations, and maintenance, any breach of the duty required by that activity which results in an injury or loss may expose the agency to liability.

All liability suits are founded in a particular area of law called *tort* law. A tort is a private or civil wrong that results in injury or loss. It is not a criminal act or a contractual breach, both of which have their own remedies. It is an injury that one inflicts upon another by an overt act or omission.

Negligence

The most common of all tort cases is the negligence case. Negligence is the failure to use reasonable care in the dealings that one has with another. In a negligence suit the plaintiff (injured party) must prove the five elements of a negligence suit:

---

[20]Federal Highway Administration, *Manual on Uniform Traffic Control Devices* (Washington, DC: Government Printing Office, 1988).

1. The defendant had a duty to the injured. For example, public entities have a duty to maintain their street and road system in a reasonably safe condition.
2. There was a breach of that duty. For example, there was a failure or an abuse of discretion in the design, or a failure in the operation or maintenance of the system.
3. The breach of duty was the *proximate cause* of the injury. The proximate cause is the legal cause of the injuries or damages that are sustained. Generally, when an accident has occurred it has two causes:
   a. The actual cause: for example, the cause in a right-angle intersectional crash was the fact that the traffic signals were completely out, not operating.
   b. The proximate cause: the proximate (legal) cause in the example was the fact that signals were not maintained properly. The breach of the proximate (legal) cause determines how liability will be assigned to a particular individual or public entity.
4. There was *no* contributory negligence on the plaintiff's part. In some states, if the plaintiffs have contributed to their own injuries, they are precluded from any recovery. A more detailed analysis of contributory negligence and its role in reducing awards to an injured person will follow later in this section.
5. There must have been damages. To recover, the injured party must have suffered some kind of personal injury or property damage. The purpose of the award is to put the injured parties back in the position in which they were prior to the accident, either to rehabilitate them, or to reimburse them for their monetary loss.

Immunity

The best defense to a lawsuit is a preventive defense. Such defenses can preclude the plaintiff from successful pursuit of the lawsuit, and a preventive defense may even result in a summary judgment (i.e., dismissal of the suit prior to hearing any of the substantive testimony).

The major preventive defense is *sovereign immunity.* Sovereign immunity is an example of judge-made law. It evolved from the action of William the Conqueror, King of England during the latter half of the eleventh century (Figure 14–10). William brought to England a new form of government, the feudal system, which included the Common People's Court. In this court, neighbor could bring an action against neighbor for a loss or injury. But the one thing William would not allow was a suit against himself. After all, William ruled by "divine right" and, therefore, could do no wrong.

The first recorded case in English legal history to discuss sovereign immunity was *Russell v. the Men of Devon,* 1788. Russell owned a horse and wagon that traveled the English countryside. The horse and wagon plummeted through a defectively maintained bridge in Devon and the horse was killed. Russell sued the Men of Devon, which today could be equated to a law suit against an entire county.

The court handed down a multipronged ruling. First, it said the Men of Devon governed with the authority of the king and therefore had the same immunity as the sovereign. Thus, they were not required to answer in court for any wrong they did while performing their governmental function.

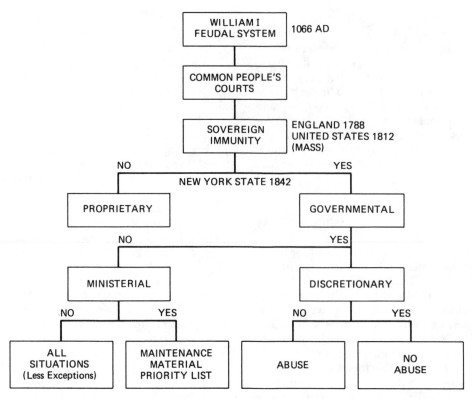

Figure 14–10. History and variations of sovereign immunity.

Second, the court ruled that if Parliament wanted government to be held responsible it would have passed enabling legislation. The court further indicated that it was reluctant to permit a suit because, if governments were required to be responsible for the repair and maintenance of their public works projects, they might not be inclined to provide future projects. Finally, the court said to allow a suit to be successful against the government might open the gates of litigation.

Twenty-four years later, in 1812, in the Massachusetts case of *Mower v. the Inhabitants of Leicester,* a similar event was litigated. Mower owned a stage coach that operated along the Boston Post Road carrying passengers between New York and Boston. One day the coach broke through a hole in a bridge in the town of Leicester. The horse, valued at $120, died from the injuries sustained. In its decision, the court held that Mower could not sue the town because, as in the case of *Russell v. the Men of Devon* (an example of precedent ruling), the town of Leicester could do no wrong because it had the immunity of the ruling sovereign, which in this case was the Commonwealth of Massachusetts.

Subsequently, the doctrine of sovereign immunity became well established in the laws of the United States. The doctrine was two-pronged:

1. No one could sue the government unless the government gave permission to do so.
2. Even if one could sue the government, the government was not responsible for the acts of its employees.

Immunities are the freedom from all tort liability as a favored defendant. Historically, there have been three basic immunities: family, charitable, and governmental immunities. We are concerned with only the last, yet inroads are being made in all three. It is apparent that it is unfair to injured plaintiffs that they should suffer the entire consequences of the loss just because they were injured by the acts of a government or its agents.

Exceptions to immunity

Some courts have done away completely with sovereign immunity, while others have sought to modify it to conform more to the social needs of the day. Where modified, certain breaches of the duty owed to the public by a public entity are permitted to go to trial. A notable area of modification divides the governmental operations into two areas, holding that the government is liable for injuries caused in one area and not the other; this is the governmental-proprietary distinction, developed by the courts of the State of New York in 1842 as a test of municipal tort liability.

The court held that the distinction roughly parallels the dual nature of the great English trade ventures such as the Massachusetts Bay Company. A municipal corporation could not be held liable for injuries resulting from the performance of a governmental function that otherwise would be performed by the state. It could, however, be liable for functions that mainly benefited the proprietors (e.g., the inhabitants of a municipality rather than all the residents of the state), functions frequently performed by private parties, and those for which it charged a fee or through which it made a profit. As rule of thumb, where a city derives revenues from a service, this service can be considered a proprietary function. At the state level it is often held that public authorities are immune for negligence arising in the performance of governmental functions, but the cases do not indicate that the converse is true—that the state authority can be held liable for negligence in the performance of its proprietary functions.

Not being satisfied with the governmental-proprietary dichotomy, the courts developed the discretionary-ministerial doctrine (see Figure 14–10). This is used more uniformly and applies almost equally to both state and local government. In this doctrine, an act that is considered discretionary and that results in an injury or property loss to an individual is generally held immune from tort liability. However, the public entity is held responsible for the injuries that it has caused by an act considered ministerial.

The courts have recognized that all acts performed by an individual or government have discretionary aspects. The courts have, therefore, defined discretionary as having the power and duty to make a decision among valid alternatives, and the exercise of independent judgment in making this selection. It is not expected that government must be an absolute insurer against accidents. Courts have often decided it would be too harsh to impute liability where authorities have carried out a reasonable plan of improvement, especially in the absence of latent as well as patent defects, and that the adoption of a plan for public improvement is a legitimate function involving independent judgment. However, an abuse of discretion does not enjoy the protection of immunity; neither does a complete failure to exercise discretion.

A ministerial act, however, is performed within a narrow set of guidelines, and any injuries that may result from the performance of this act will expose the agency that has the prime responsibility to liability for those injuries. In the case of *Croft v. Gulf and Western* 506 P.2d 541 (Oregon 1972), the Oregon State Highway Commission was adjudged liable for injuries sustained at a defective traffic signal, partly because the signal technician failed to use the test meter given him when performing maintenance checks—clearly a breach of a ministerial duty.

How, then, can individuals maintain a lawsuit against a public entity when some form of sovereign immunity still exists? First, they must seek permission to sue. This requires going before the public entity and claiming that injury has occurred because of negligence and indicating that they wish to file suit for recovery of loss. Many states have given up this permissive requirement by passing blanket laws that authorize suits. Generally, under these laws, government retains a vestige of the permissive doctrine by requiring notification of the government within a specified time period that an injury has been sustained. Second, the insurance waiver doctrine as developed by the courts holds that the public entity has waived any immunity that it may still have if it has purchased liability insurance, at least up to the value of that insurance.

Abrogation by law offers a third way for an individual to bring a case into court. This is done by means of a *tort claim act*. The federal government passed a law in the early 1940s that made itself responsible in a narrowly defined area for the tortious acts of its employees done within the scope of their employment. Many states now have adopted similar acts. Known as *little tort claims acts*, they permit

individuals who fall within the area covered by these acts to bring an action in court for their loss.

Finally, there are the safe street exceptions. Several states, recognizing the importance of commerce and economic growth that derives from the safe and expeditious movement of goods, passed limited statutes known as the "safe street exceptions" to sovereign immunity. They are a form of little tort claims act, except that these laws do not open up the state and its political subdivisions to total tort liability, but create tort liability only in the area of safe operation of the street and road system.

At one time the safe street exceptions applied only to the roadbed itself. For example, if the roadbed was maintained improperly or potholes or breaks in the pavement were allowed to exist, any injuries or damages that resulted therefrom would be covered under the safe street exception. However, the courts, with their ability to expand or narrow laws under the common law doctrine, have increased the exposure that government faces under the safe street exceptions by expanding the coverage to include pavement markings, traffic signs, traffic signals, and roadside hazards placed in the right-of-way.

As noted earlier, a major defense available to a public entity is the doctrine of contributory negligence. In some states, a plaintiff who was injured in an automobile accident and contributed to his own injury, even if the amount was on 1% of the total negligence while the government entity or individual that was primarily responsible was 99% negligent, is precluded from recovery. However, only in extreme circumstances do passengers have the driver's negligence imputed to them. An extreme circumstance may exist if a passenger knowingly gets into a car with a known unsafe driver or with a driver who is totally incapacitated and continues to remain with that driver without attempting to leave the vehicle or stop the driver in one way or another. The driver's own negligence then may be imputed to the passenger. Should injuries result, the passenger as well as the driver may be precluded from recovery.

Several states have recognized the harshness of the contributory negligence doctrine—that persons only minutely responsible for their own injuries cannot recover a single penny. Their legislatures have enacted a new doctrine called *comparative negligence* by which the relative negligence of the parties involved is weighed. For example, if the injured party is found to be 40% negligent and the public entity 60% negligent, any award to the injured party for recovery of loss will be reduced by that percentage. Thus, if a $100,000 judgment is awarded, the injured but partly negligent party will be entitled to only $60,000. Modifications to comparative negligence have resulted in several different forms of this law.

Notice of defect

The courts have uniformly held that a dual duty arises once the public entity that has the responsibility for a highway or street system has notice of a defect; it then has the duty to repair it or, if unable to do so within a reasonable time, to warn the public adequately. The question arises as to what constitutes notice. *Actual notice* is very simple to understand. *Constructive notice,* however, does not require specific notice of the defect. The courts have held that if a defect has existed for an unreasonable period of time, the agency should have discovered the defect, and therefore it has constructive notice of its existence. Also, courts have held that local police departments are considered to be agents of county maintenance organizations, and if the police department received notice of a defect but did not forward this notice to the maintaining agency, the public entity will be considered to have constructive notice that said defect exists.

Notice is not required where the defect has been caused by the public entity's own act. For example, if the public entity has improperly maintained or improperly repaired the highway, it does not have to receive notice that the defect exists. The courts uniformly say that the entity that performed the act is aware of its own act and thus, if it created a defect, it is responsible. Also, once the duty to repair or warn arises, the warning to the driver must be commensurate with the danger and adequate to convey to the driver that the danger exists. Signs must be used that convey to drivers the type of danger they face. They must be adequate, they must be visible, and they must convey to the motorist or the pedestrian the danger to be faced. The signing must be so located as to allow ample decision time. Failure to provide adequate warning is an abuse of discretion and will create liability.

The traffic engineer should be alert to the fact that while the duty to remedy or warn arises when notice is received, a warning sign is not the panacea it appears to be. Courts require the warning when repairs cannot be made within a reasonable period of time. However, the warning sign will be tolerated only for a reasonable period of time, just until repairs can be effected. If the warning sign is allowed to remain for an unreasonable period of time without repairs being made, the courts will consider that the public entity is maintaining a dangerous condition of public property and the public entity will be just as responsible for injury or loss as if it had done nothing once it had notice.

Design immunity

Courts in many of the states that have passed design immunity statutes have held that the development of a design is a discretionary act, particularly if the design was accepted by a lawfully authorized body designated to approve the design. Yet the same courts have held that there is no design immunity if the design itself is arbitrary, unreasonable, or made without adequate care, or if it is dangerous or manifestly unsafe after use and the agency has received notice of that fact. Under those circumstances, to permit continued use of the road system is an abuse of discretion, and any immunity existing could be waived by the court exposing the agency to liability for injuries sustained as a result of that defect.

At one time it was assumed that changed conditions would not create an abuse of design discretion. However, most state courts have revised their view and have said that if changed conditions have created an unsafe or dangerous situation and if the situation is allowed to exist for an unreasonable period of time, liability exposure will exist. An operational review must be made after the installation. It must be determined that changed physical conditions have not produced a dangerous situation. If it has, action must be taken to correct it; the

courts have often said that the public entity must not develop an ostrichlike syndrome, burying its head in the sand and hoping that the defect will go away if ignored. The defect will not go away, and its continued maintenance will result in liability exposure on the part of the public entity.

It is uniformly held that construction and maintenance, as they relate to road and street operations, are considered proprietary or ministerial functions. Hence, there are no immunities to protect the agency that is performing the duty. It means that any time an agency does any kind of construction or maintenance it is as liable to the public as an individual is liable to another individual in the performance of an everyday act. However, the courts have modified this by holding, particularly in the area of maintenance, that there are two subareas where the court will grant discretionary immunity: in materials selection and in the order of making the repairs (the establishment of a priority list). The courts have held that the selection of the material to be used in making the repair is a discretionary one and that the only way that liability can attach because of injuries resulting from the use of a particular material selected is if there was an abuse of discretion. For example, if there was a superior material available and if there was no rational or logical basis for the selection of the inferior material, liability will attach because of abuse of discretion.

"Rational" or "logical" basis varies from court to court. Governmental operations generally require low bid acceptance. The courts recognize this; yet they uniformly hold that dollars must not be the sole reason that a particular material was selected. It can be one of the factors, but in making the decision, one must also look at the quality, availability, and suitability. Under these circumstances the public entity will most likely retain discretionary immunity.

If an established priority list for repairs is not followed, and there is no rational reason for the deviation, there is no immunity, and liability can occur. The courts uniformly hold that as long as a priority list is established on a rational-logical basis and repairs are made in accordance with that priority list, there is no liability because an accident occurred at a location with a low priority where repairs were not yet made.

Personal liability

Various provisions exist for handling the liability of the individual transportation engineer. Many states have indemnification laws under which the state requires the public entity to defend its employees in cases where the employees have been charged with a tortious act—while working within the scope of their employment—that has resulted in an injury. Under those laws the agency is responsible for providing an adequate defense on behalf of its employees. The agency must pay any amounts that the courts may require these individuals to pay to the injured party.

In other states, laws provide that injured individuals who bring an action against the agency itself are precluded from bringing an action against individual employees. Under most circumstances, this provides employees with adequate protection, because under the "deep pocket" practice, attorneys for injured parties usually seek an award against the party who is in the best position to pay. The individual engineer, maintenance crew member, or truck driver is presumed to have limited means, but it is assumed that the public entity has a relatively unlimited source of income and the ability to pay a large award if one is made.

In two cases, *Garcher v. City of Tamarac* (Florida), and *Grubaugh v. City of St. Johns* (Michigan), the respective towns had to raise their property tax rates to pay court-awarded liability judgments. Tamarac, faced with a $2.8-million judgment out of a total award of $4.9 million, raised taxes by 6.7 mills for a year; St. Johns paid off a $565,000 judgment in a 2-year period by raising property taxes by 4.2 mills each year.

In the private sector, individual engineers are most vulnerable. They have none of the government immunities of their colleagues in the public sector. They have no statutes to indemnify them, and employer policy may preclude such a defense. Under these circumstances there is an increasing trend for the private-sector engineer to obtain liability and malpractice insurance. This trend is also evident in the public sector, despite immunities and indemnification statutes.

Acting under orders

There is often concern about the predicament that arises when one is told to do something that can create a dangerous situation with the likelihood that someone will be injured. The courts have held that the defense "acting under orders" is a good defense in this situation. To cite an individual with personal liability when acting under orders would be unfair. As a policy consideration, it would make it harder for public entities to obtain employees if they were exposed to the risk of liability under these circumstances. But courts also say that, in extreme cases, there is a duty for a person acting under orders to attempt to mitigate the act and to bring the danger of that act to the attention of superiors. Professionals cannot escape liability by blindly relying on the fact that they are acting under orders. This amounts to an abuse of discretionary power, and the courts will not tolerate abuse of any of the protective devices an individual has.

Respondeat superior

Another doctrine that can create individual liability is *respondeat superior*. This doctrine is concerned with both the personal liability of an individual supervisor and of the employer in the public or private sector for the tortious acts of subordinates or employees, respectively. Generally, supervisors have no individual liability if their subordinates commit a tortious act. Again, the courts have carved out exceptions to that rule. For example, allowing subordinates to perform their duties with knowledge that they are of such a temperament or such a violent nature that they are prone to injure persons they deal with possibly exposes the superior to joint liability together with the party causing the injury. If a supervisor has knowledge that a subordinate is intoxicated and yet allows that individual to operate a city vehicle, the supervisor may face personal liability for injuries or property loss caused by that operator.

Courts generally hold that the employer (public or private) is responsible only for tortious acts committed by employees within the scope of their employment. A classic example of being outside one's scope of employment is the employee who uses a company vehicle during lunch hour to visit a friend. The courts have held that if the employee has an accident while enroute to the friend's house it is outside the scope of employment and the employer is not responsible. Courts have refined this approach to the extent that once the employee has left the friend's house and has commenced the return trip to the place of employment, the employee has returned to the scope of employment. Any accidents then may be the employer's responsibility.

Another example is the maintenance person who gets into an argument with a motorist at a traffic signal and punches the motorist. Courts will generally hold that the maintenance person acted outside the scope of employment and the employer is not responsible. On the other hand, if a bartender has an altercation with a patron and injures the patron, the bar owner may be held responsible because the courts may consider the nature of the bartender's duty to require contact with patrons. In summary, if the act is done within the scope of employment, the employer *is* responsible; if the act is done outside the scope of employment, the employer *is not* responsible.

Recovery from negligent employees

One of the major concerns to a transportation professional is "recovery from negligent employees." Can the employer recover from the employee any award that must be paid to an injured party as a result of an employee's negligent act? Under the legal theory of "agency," employers are permitted to recover from employees any awards that they must pay out because of the employees' negligent acts. In the United States today this seldom occurs. Union agreements, civil service rules, social and political pressures, and many other facets of modern personnel management keep such claims to a minimum. However, the transportation professional should be aware of the fact that this practice is permitted under law. Currently, many employees are indemnified by the employer for losses attributed to them. Yet employees might be in a position where they would be required to seek malpractice or liability insurance to protect themselves from liability exposure caused by changing doctrines of law.

Reducing exposure to liability (risk management)

It is a simple matter to file a lawsuit and, no doubt, it appears to some that litigation has become as popular a national pastime as baseball or football. Succeeding in a lawsuit is another matter, particularly when the government or one of its employees is the defendant. As pointed out earlier, lawsuits of this kind are a two-tier project. To get the case before a court, any immunity that may exist must first be overcome. This can happen by a judicial or legislative abrogation of the Doctrine of Sovereign Immunity, the passage of a Tort Liability Act, the Safe Street Doctrine, and even the defendant's purchase of liability insurance. If the plaintiff can surmount the defenses of immunity, it still must be proven that the defendant was the cause of the injuries or property loss.

Lawsuits arising out of accidents, or resulting from an alleged defect in a highway or road system, generally involve three principal issues as well as the previously discussed five elements of the negligence suit:

1. Did a potentially dangerous defect exist?
2. Was the defect the proximate cause of the accident?
3. Did the defendant have any actual or constructive knowledge of the hazardous condition?

In seeking answers to these questions the courts have established some guidelines that are helpful in determining responsibility.

1. The state is not an insurer of the roads or guarantor of absolute safety.
2. Motorists have the right to presume, and act upon the presumption, that the highway is safe for usual and ordinary traffic in the daytime and at night. They are not required to anticipate extraordinary dangers, impediments, or obstruction to which their attention has not been directed or of which they have not been warned.
3. The public highway must be maintained in a way that is reasonably safe for travel, within accepted and understood criteria, under generally promulgated engineering standards or practices.
4. In maintaining the highway in a manner that is reasonably safe for travel, there is wide latitude in the exercise of administrative discretion, but continual supervision and inspection are of the utmost importance.
5. The courts do recognize factors in establishing what is reasonably safe and include among them the terrain that is encountered, weather conditions, and the materials used in construction.
6. Recovery is based upon more than the mere presence of a hazardous condition; such presence must be due to negligence.
7. Negligence stems from knowledge or notice of the existence of a dangerous or defective condition and a subsequent failure to safeguard against such a condition.

General duties are the most important guidelines in protecting against liability suits. Basically, there is a duty to maintain the roadway in a reasonably safe condition. This duty involves the inspection, anticipation of defect, and conformity with generally accepted standards and practices. There is no requirement for perfect condition or repair, or for actions beyond the limits of human ingenuity. The key term is *reasonability.* There are many factors upon which determination of what is reasonable may be based, among them the character of the roadway in question, the width and construction of the road, characteristics of the traffic, the turning requirements, the side friction, and the development of traffic generators alongside the roadway. These all must be taken into account when trying to determine the guidelines for protection against the liability suit.

Organizational protective measures

Following currently accepted technical procedures and practices, keeping proper records and decision documentation, having professional decisions made by competent professionals, and implementing an effective risk-management program are the primary means of reducing organizational liability exposure.

There are many ways in which organizations can reduce their liability risks, including the following:

1. There should be a clear definition and understanding of the duties, responsibilities, and authority of the institution, its subunits, and each individual in the organization. Administrators and professional employees should know the laws and ordinances relating to their jurisdiction and the application of traffic control devices.
2. Officials and employees should clearly understand and subsequently perform their general duties in a satisfactory manner.
3. Decisions concerning technical plans and/or programs, such as the physical and geometric design of traffic facilities and the application of traffic control devices and regulations, should be either made by competent professionals or be based on the advice of such persons. Possession of a valid driver's license and years of accident and arrest-free driving are not the basic qualifications for making traffic engineering decisions.
4. Public highway units, within the limits of their staff and budgetary constraints, should establish and maintain adequate record systems to provide current *facts* about existing conditions. These systems should include:
   a. Traffic accident records and procedures for identifying high-accident locations.
   b. Inventory procedures that will provide reasonably current information about the physical features and conditions of existing transportation facilities (i.e., photo logging and condition ratings) and traffic control devices (such as the location, model and/or type and size, date installed and/or repaired, condition, function, reliability, and operational criteria like traffic signal timing and sequence).
   c. Complete and current maintenance records can provide information about type and character of repair and/or replacement activity including what trouble was found, what repairs were made, and what materials were used.
5. A system of regular inspection should be established and maintained on a continuing basis within the limits of fiscal and personnel limitations. These inspections should cover the physical conditions of facilities and traffic control devices. If possible, traffic signals should be checked at a maximum of 6-month intervals and traffic signs should be inspected at least twice annually under day and night conditions. They should also be checked during adverse weather conditions. Traffic markings should be checked as needed, but special attention should be directed in the late summer and early fall while there are favorable weather conditions for remedial efforts. Inspections in late winter and early spring will identify locations that should be upgraded as soon as the weather conditions become favorable.

Temporary traffic control devices (such as those placed in construction or maintenance areas) should be checked on a daily basis (day and night), including workdays, weekends, and holidays. More frequent inspections should be made in major work areas.

A chain of command should be established for the inspection process so that changing conditions can be anticipated, present and potential defects can be reported, and prompt action undertaken on these reports.

An extremely helpful type of inspection is periodic trips to problem areas made by the traffic engineer and his (or her) traffic enforcement counterpart (supervisor of traffic, head of the accident records department, etc.). Not only do these individuals have an opportunity to become better acquainted on a personal basis, but they can also evaluate critical locations on the basis of their individual professional backgrounds.

Another source of inspection capability is to develop an awareness and sense of responsibility on the part of all employees, even nontechnical staff members, so that they will constantly be on the lookout for vandalized or malfunctioning traffic control devices or other hazardous conditions.

6. Traffic control devices should be evaluated for replacement at the end of their warranty life.
7. All traffic control devices should comply with provisions of the MUTCD. Whenever possible, the "should" and "may" alternatives in the manual should be considered with any decision to deviate from these alternative requirements well documented.
8. An established procedure for the handling of complaints and defect reports should be developed and maintained with one person or one office being designated to receive and record all such reports and to take appropriate action. Effective handling of complaints has legal as well as good public relations benefits, as noted earlier in this chapter.
9. All designs of facilities and/or traffic control device installations should be in accordance with currently adopted policies, guidelines, standards, and manual specifications. Geometric designs should be predicated on criteria well above established minimum standards. Field conditions should be correlated with traffic controls (i.e., having a centerline stripe without marking no-passing zones is unsafe and not reasonable).
10. Standards of performance should be adopted in the areas of design, construction, operations, and maintenance. These standards should be realistically attainable and be followed and exceeded whenever possible.
11. Rational procedures for determining improvement priorities and programming should be established and followed. Normally this will include a consideration of the cost-effectiveness of various alternatives.
12. There should be design and operational reviews both before and after any facility and/or traffic control changes are made. Both the basic design and the traffic control elements should be checked in the field. Reviewers should be alert for changing conditions such as increased traffic movements, changes in vehicle types, and so forth. There should be inspections of active and completed projects.
13. All agency employees should be impressed with the importance of reasonable care in the fulfillment of their individual duties as well as the overall group mission.
14. Beware of false economy. The foolish cutting of necessary expenditures in order to appear fiscally responsible inevitably leads to careless and negligent work.

15. Provide liability insurance against claims, if not self-insured or insured and indemnified by others such as contractors and others.
16. Special attention should be directed toward traffic control in work zones because these zones present special and unusual problems. Public agencies cannot delegate "duty"; consequently, they are responsible that contractors and other participants use appropriate measures, comply with provisions of the MUTCD, secure necessary permits, etc.

Individual protective measures

In addition to complying fully with all of the organizational efforts to minimize tort liability as previously described, individual officers and employees can minimize their personal exposure to tort liability claims by:

1. Expending their best effort in every assignment and action. This is the best way to avoid litigation and claims.
2. Exercising their best judgment based upon current practice and knowledge of their profession by:
   a. Reading the latest professional journals and knowing about recent innovations in their field,
   b. Keeping up with revisions and interpretations of professional standards and guidelines, and
   c. Actively participating in professional organizations and attending conferences and workshops.
3. Keeping written records of the basis for decisions. Such records can be invaluable to anyone subsequently reviewing the decision-making process and also tends to ensure a more complete consideration of all relevant factors.
4. Personal liability insurance should be obtained on the part of any individual who is not sure of being fully protected by his or her employer.

Following all of the above recommendations will not guarantee that an organization or an individual will be immune from tort liability suits, but the vulnerability to such suits and possible adverse verdicts will be greatly lessened.

Expenditure of funds should be based on the consideration of safety of existing facilities (including related conditions such as record-keeping, maintenance, and safety improvements) versus other expenditures for new programs and facilities, in order to serve and protect the public and in turn reduce exposure to liability suits.

The engineer as expert witness

Professional traffic engineers may become personally involved in trial activity or in public hearings as fact witnesses or as expert witnesses. In either case, honesty, professional competence, and proper preparation are fundamental considerations.

A *fact witness* is someone called upon to testify concerning information known to the witness. Results of personal observations relative to the problem at issue or the existence and content of certain records (i.e., accident reports and/or summary tabulations or diagrams, traffic control device installation, operational and maintenance records, etc.) are typical examples of traffic and transportation engineers' involvement as fact witnesses.

Fact witnesses are frequently subpoenaed (a court writ that requires an appearance at a specified time and place to give testimony and/or the production of certain records, plans, and/or other documents). Appropriate legal counsel should be consulted upon receipt of the subpoena.

The following discussion concentrates on the engineering expert witness.

Expert witness qualifications

Because an expert witness, when accepted as such by the court or hearing body, is permitted to render opinions on matters beyond the knowledge of most people and thus would be helpful to the jurors or hearing body in forming their conclusions, it is fundamental that the engineering expert witness is qualified to testify about the subject matter at issue.

These qualifications normally include professional education and experience background and demonstrated expertise in the area at interest. Professional publications, lecture presentations, technical committee participation, and unique experience are also considered. Honest representations of this information should be given to the legal counsel who seeks your assistance so that appropriate evaluations can be made.

Pretrial and prehearing involvement

Identification of critical issues, evaluation of strong and weak points of the case or proposition, and research into past and current (time of the subject crash) design and operational criterion (MUTCD warrants and statements, geometric design policies, traffic study procedures, etc.) are possible forms of pretrial and prehearing involvement.

Objective evaluation of all available facts and other relative elements of the case is fundamental to expert witness involvement. Public employee engineers, consulting engineers who are retained by governmental agencies, as well as independent consulting engineers are sometimes pressured to shade their objectivity with threats about future job security and/or employment. These pressures should be resisted with every possible means.

Every effort should be expended to obtain all available factual information, and reasonable assumptions should be made only when necessary. "State of the art" practices and criteria at the time of critical previous decisions should be identified and studied.

Reports to legal personnel are usually given verbally, and written reports are only prepared after being specifically requested.

Replying to interrogatory questions and submitting to one or more discovery depositions are also part of the pretrial process, especially if the engineer is expected to testify during a trial.

Interrogatory questions are written and require written answers. Consequently, there normally is sufficient time to prepare honest and complete answers.

During the discovery deposition, the witness is under oath and is questioned by attorneys representing the opposing side of the case. An expert witness should never participate in a deposition unless associated legal counsel is present.

Interrogatory replies and deposition testimony may be used during trials, especially for possible impeachment of the witness.

Trial and public-hearing involvement

Expert witness testimony in trials and public hearings is intended to provide technical information in a readily understandable manner to a decision-maker (judge, jury, board or committee). In public hearings, the expert's testimony may also be a vehicle for public education.

In all cases, the engineering expert witness must be prepared to discuss honestly, clearly, and thoroughly the expert's opinions and conclusions and the basis thereof. This can only be accomplished after thorough preparation. This preparation can include selection and sequence of testimony subject matter, the production of exhibits and other visual aids, and prior inspection of the court or hearing room.

The expert witness's role in the overall presentation of the case or project must be clearly defined and understood.

In some situations expert witnesses are employed to sit in and evaluate the presentations of other expert witnesses. This evaluation can range from merely listening to critiquing and possibly preparing questions to be directed to the other expert witness.

The expert witness's testimony should be honest, clear, concise, objective, and to the point. One should know the source of all factual information and be absolutely certain as to the proof, explanation, and verification of anything that is presented during testimony. No tables, charts, graphs, diagrams, or any exhibit should be used that cannot be explained or for which the deriving formulas are unknown.

Question response guidelines

The following guidelines should be followed when answering all questions relative to court or public-hearing testimony:

1. Tell the truth. Nobody can cross up an expert witness if the truth is told and told accurately.
2. Do not guess. If the answer is not known, admit it with no apologies.
3. Be sure you understand the question. If there is any doubt, ask to have it repeated and/or clarified or rephrased.
4. Before answering oral questions, first pause to allow time for possible objections and also to give the necessary thought to your answer. Formulate the answer in your mind, give the answer, and then stop. Do not volunteer additional information.
5. Answer questions with a "yes" or "no" whenever possible.
6. Do not equivocate an answer. The answer should be definite and positive.
7. All oral testimony should be presented in a clear and audible manner that is loud enough for all to hear.
8. Be sure that answers to technical questions are readily understandable to the court and the jury or other decision-makers. Avoid technical jargon.
9. The expert witness should present a professional and neutral attitude.

10. Always be courteous to everyone, including opposing attorneys; look at the jury or hearing body when answering questions; do not become intimidated or lose your temper.

The engineering expert witness fills a very important role in the legal and public-administration process. The testimony of a properly qualified and prepared expert witness can be a major help to the decision-making body.

Therefore, serving as an engineering expert witness requires the highest degree of professional competence and commitment.

REFERENCES FOR FURTHER READING

BAKER, J. STANNARD AND LYNN B. FRICKE, *The Traffic Accident Investigation Manual,* 9th edition, Northwestern University, Evanston, IL, 1986.

Forensic Procedures in Transportation Engineering, an ITE Informational Report, Institute of Transportation Engineers, Washington, DC, 1987.

FRICKE, LYNN B., *The Traffic Accident Reconstruction Manual,* Volume 2, Northwestern University, Evanston, IL, 1990.

JUDGE, JAY S. "A Positive Approach to Defending Municipalities," Judge & Knight, Brochure No. 119. Park Ridge, IL, 1984.

KUHLMAN, RICHARD S., "Killer Roads: From Crash to Verdict," The Michie Company, Charlottesville, VA, 1986.

LEWIS, RUSSELL M., *Practical Guidelines for Minimizing Tort Liability,* N.C.H.R.P. Synthesis of Highway Practice No. 106, Transportation Research Board, Washington, DC, December 1983.

LUCK, REBECCA JENNATHAN AND JACK B. HUMPHREYS, *The Transportation Engineer as a Court Witness,* Institute of Transportation Engineers, Washington, DC, 1984.

PAGAN, ALFRED R., "Ten Commandments (More or Less) for the Expert Witness," Reprinted from and published by *Better Roads,* Rosemont, IL, 2nd Edition, 1987.

Tort Liability and Risk Management, Transportation Research Circular, No. 361, Transportation Research Board, Washington, DC, July 1990.

15

INTELLIGENT VEHICLE–HIGHWAY SYSTEMS

GARY W. EULER, *Chief*

Program Management & Systems Engineering Branch
Federal Highway Administration

Introduction

This chapter of the text, in contrast to the preceding ones, which dealt primarily with traffic engineering techniques and practices that are widely applied, discusses systems that are only beginning to be developed. It does not present a picture of state-of-the-art traffic engineering practice so much as a vision of what that state-of-the-art could be in the year 2000 and beyond.

The term *Intelligent Vehicle–Highway Systems* (IVHS) applies to transportation systems that involve integrated applications of advanced surveillance, communications, computer, display and control process technologies, both in the vehicle and on the highway. The essence of IVHS is to make significant improvements in mobility, highway safety, and productivity by building transportation systems that draw upon advanced electronic technologies and control software. IVHS will also cause reassessment of how transportation services are provided. The communication of reliable, accurate information in real-time among users, vehicles, and transportation management centers will require partnership arrangements between the private and public sectors and among local and state government agencies to develop the applications, and to plan, design, build, operate, and maintain the types of systems needed.

While the technologies and software used will be the most advanced available, the ideas behind the applications will not necessarily be new ones. The next section provides a brief background on the history of IVHS and the reasons for the current strong interest in the subject.

Background

In the late 1960s and early 1970s, the U.S. Federal Highway Administration (FHWA) and a number of other agencies and governments around the world sponsored research in a number of areas considered part of the umbrella of IVHS today. These included early research and development of computerized traffic signal control systems,[1] freeway ramp-metering systems,[2] and integrated, corridor traffic control systems, notably the Dallas Corridor Project in Texas[3] and the Integrated Motorist Information System (IMIS) in Long Island, New York.[4]

A system called the Electronic Route Guidance System (ERGS) was also developed and tested.[5] The ERGS system used loop detectors to locate properly equipped vehicles in a network. Information on the vehicle's destination was transmitted through the loops to computers located at the roadside. Instructions on which way to go at the next major

---

[1] J. MacGowan, and I. Fullerton, "Development and Testing of Advanced Control Strategies in the Urban Traffic Control System," *Public Roads,* 43 (September 1979); 43 (December 1979); and 43 (March 1980).

[2] J.A. Wattleworth, and C.E. Wallace, "Evaluation of the Operational Effects of an On-Freeway Control System," *Highway Research Record 368,* Washington, DC, 1971.

[3] J.D. Carvell, Jr., "Dallas Corridor Study," Texas Transportation Institute, Texas A&M University, College Station, 1976.

[4] P. Zove, and others, "Integrated Motorist Information System (IMIS) Feasibility and Design Study," Federal Highway Administration, FHWA-RD-78-23, -24, May 1978.

[5] D.A. Rosen, F.J. Mammano, and R. Favout, "An Electronic Route Guidance System for Highway Vehicles," *IEEE Transactions on Vehicular Technology,* VT-19 (February 1970).

intersection was transmitted back to the vehicle, and a display device in the vehicle showed an arrow that directed the driver. These ideas formed the basis for many of the electronic route guidance systems being tested in the early 1990s, but the technology was 20 years more advanced. The FHWA and others also sponsored research and testing of automated control techniques such as lateral control, merging strategies, alternative guidance systems, and hazard warning systems.[6]

The strong interest in IVHS has been sparked by the serious congestion problem that many metropolitan areas around the world are facing and the dramatic increases predicted for the future as discussed in earlier chapters of this text. The possibility of alleviating the congestion problem through technological means while preserving individual mobility is very appealing. The issue of national economic competitiveness has also been instrumental in igniting interest in IVHS. For example, the Conference Report on the Fiscal Year 1989 United States Department of Transportation Appropriations Act directed the U.S. Secretary of Transportation to report to the Congress on ongoing European, Japanese, and U.S. IVHS research initiatives, the potential impacts of foreign programs on the introduction of advanced technology for the benefit of U.S. highway users and on U.S. vehicle manufacturers and related industries, and to make appropriate legislative and/or programmatic recommendations. That report to the Congress recommended that the federal government take a more active role in IVHS research and operational demonstrations in the form of public/private cooperative partnerships.[7]

Development of IVHS in the late 1980s was proceeding in Europe and Japan at perhaps an even faster pace than in the United States. The next section summarizes the international scene as of 1990.

IVHS programs

Europe

A number of programs are underway in Europe to research, develop, and demonstrate IVHS technology. Included in EUREKA, a $5 billion, 19-country program designed to stimulate cooperative research and development among industries and governments in Europe, are the following IVHS-related programs:

- EUROPOLIS—a $150 million, 7-year research project to design automated road systems and to develop technologies to automate driver functions
- CARMINAT—a 4-year research project to develop in-vehicle electronic navigation and communications systems
- ATIS—an $8.5 million, 5-year project to provide pre-trip information on traffic conditions

- ERTIS—a $2.7 million, 3-year project to develop a common road information and communications system for motor carriers across Europe.[8]

The two most important European IVHS programs are PROMETHEUS, which stands for *PRO*gra*M*me for *Eu*ropean *T*raffic with *H*ighest *E*fficiency and *U*nprecedented *S*afety,[9, 10] and DRIVE (*D*edicated *R*oad *I*nfrastructure for *V*ehicle Safety in *E*urope).[11] PROMETHEUS, which is also a EUREKA program, began in 1986 as an 8-year, $800 million program. It involves a consortium of 14 European automobile manufacturers, electronic supply companies, and approximately 40 research institutes and state authorities, led by the automobile manufacturers. Its general objectives are to improve traffic safety, enhance vehicle operating efficiency, and reduce the adverse environmental effects of automobile travel by using the latest advances in electronics and information technology in order to shape computer-aided driving. Safety is a major aspect of the program. A target of reducing European traffic fatalities by 50% by the year 2000 has been established.

DRIVE is a European Community program of collaborative research and development to find ways to alleviate road transportation problems through the application of advanced information and telecommunications technology. The stated goal of DRIVE is to improve road safety, promote transport efficiency, and reduce environmental pollution. The program was initiated in 1988 and had a total committed funding level of almost $150 million over a 3-year period, half from the public and half from the private sector. A second multiyear DRIVE program was being planned. The DRIVE program brings together road users, research institutions, providers of broadcasting and telecommunications services, industry, and road transportation authorities. A major objective of the DRIVE program is to develop standardized technology and common functional specifications so that any products or services developed can be used throughout the European Community.

These two programs, PROMETHEUS and DRIVE, are different in the sense that PROMETHEUS is vehicle oriented and is largely a private-sector effort, whereas DRIVE is traffic-management oriented and has been organized by the European Community.

Two major European IVHS projects are ALI-SCOUT and Autoguide. ALI-SCOUT, developed in the Federal Republic of Germany, is a route-guidance system that uses infrared transmitters and receivers to transfer routing information between roadside beacons and in-vehicle displays.[12] The routing information is computed at a systems center

[6] J.G. Bender, and others, "System Studies of Automated Highway Systems," General Motors Transportation Systems Center Report EP-810141A, August 1981; FHWA Report No. FHWA/RD-82/003, 1982.

[7] U.S. Department of Transportation, Office of the Secretary of Transportation, *Report to the Congress on Intelligent Vehicle–Highway Systems*, DOT-P-37-90-1, March 1990.

[8] Ibid.

[9] Tage Karlsson, "PROMETHEUS, The European Research Program," paper presented at the Transportation Research Board Annual Meeting, January 1988.

[10] Statement of Karl-Heinz Faber before the Motor Vehicle Safety Research Advisory Committee, February 1988.

[11] Dr. ir. A. Vits, "DRIVE Activities and Areas for Future Work," paper prepared for the Vehicle Navigation and Information Systems Conference, Toronto, September 1989.

[12] Dr. J.M. Sparmann, "LISB Route Guidance and Information System: First Results of the Field Trial," paper prepared for the Vehicle Navigation and Information Systems Conference, Toronto, September 1989.

and is based on current travel times as provided by the equipped vehicles. A pilot test of ALI-SCOUT began in Berlin in June 1989 involving 700 vehicles and 260 beacons. Autoguide is a similar British system that is being tested in a corridor between London and Heathrow Airport.[13] A pilot Autoguide system is being planned for the London area, involving 1,000 vehicles and up to 300 roadside beacons. This system is being designed to be capable of being upgraded to become the first commercial system in Britain. (See Figure 15–1.)

Japan

Japan also has two major IVHS programs: the Advanced Mobile Traffic Information and Communication System (AMTICS) and Road-Automotive Communication System (RACS). Both of these programs emphasize communications and traffic control and have been given a high priority by the Japanese government because of the severe traffic

[13] "Autoguide, Pilot Stage Proposals, A Consultation Document," issued by the Department of Transport, London, April 1988.

Figure 15–1. The Autoguide System Concept.

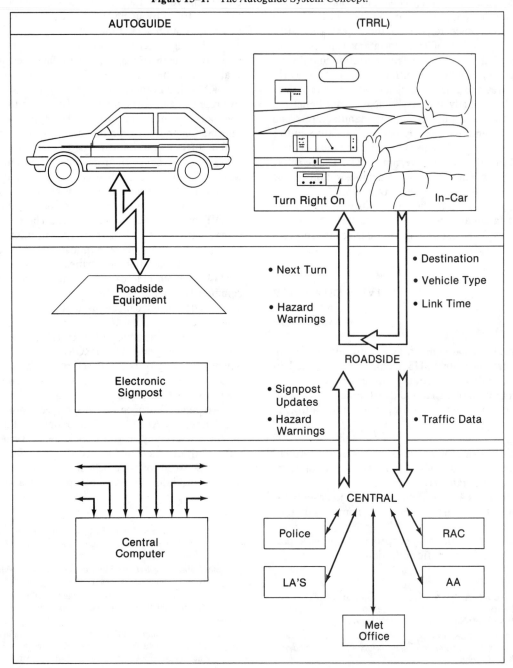

congestion problem in Japan, and because the Japanese government wants to encourage technological innovation in the Japanese automotive and electronics industries.[14]

AMTICS transmits traffic-congestion information from a traffic control center to an in-vehicle display through a two-way digital cellular communications system.[15] It provides static information using in-vehicle compact discs (CD-ROM) and display terminals to show road maps, local traffic regulations, and the location of parking lots, hospitals, gas stations, and other useful information. A dynamic component provides real-time information on traffic conditions, weather and accident warnings, and parking space availability. AMTICS can also be modified to allow communication among vehicles, making full use of its communication system. The AMTICS project was launched in 1987, and pilot experiments were started in 1988. Wider implementations in Tokyo and Osaka are planned.

RACS is a parallel project that uses a different communications technology.[16] RACS consists of roadside communication beacons, in-vehicle units, and a system center. System functions are classified into navigation, traffic information, and message systems. Real-time traffic information is collected at the system center, sent to the roadside beacons, and disseminated between the roadside beacons and vehicles using microwave technology. This includes information on congestion, construction work, accidents, traffic control, parking availability and estimated travel time. The roadside beacons also transmit digital map information, which corrects accumulated errors in the in-vehicle navigation units. The in-vehicle map shows the vehicle's current location and recommended routes and provides information on traffic patterns. Because the microwave communications system is two-way, it is also possible for the system to monitor and locate specific vehicles and collect real-time traffic data. Messages, facsimile, and other types of information may also be distributed. The RACS project was started in 1984. Two field tests of individual RACS functions were conducted in 1987 and 1988. An integrated road test for RACS began in 1989.

United States

Before 1990, the United States did not have a formal IVHS program similar to the ones in Europe. IVHS activities in the United States were being coordinated through a group called Mobility 2000, which was an ad hoc coalition of industry, university, and federal, state, and local government participants convinced that IVHS technology would make highways more productive and vehicles safer and more efficient. This group was involved in the initial formulation of a national program of IVHS research, development, and operational testing of activities.

In May 1990, the Highway Users Federation for Safety and Mobility (HUFSAM) sponsored a National Leadership Conference on Implementing Intelligent Vehicle-Highway Systems. This conference recommended establishment of a formal national organization to coordinate and set priorities for U.S. IVHS research, development and testing activities, and asked HUFSAM to take the necessary steps to form such an organization. That organization, called IVHS America, was incorporated later in 1990 as a non-profit organization, funded by the federal government and through membership fees, that would serve the functions recommended by the National Leadership Conference. IVHS America would serve both the public and private sectors, and would also be used in an official advisory capacity to the federal government.

A number of IVHS activities in the United States have already begun under the sponsorship of the federal and state governments in partnerships with the private sector. For example, a 10-mile-long, 2-mile-wide section of the Santa Monica Freeway corridor in Los Angeles is called the Smart Corridor.[17] This project involves a partnership among the federal government and state and local government traffic operations and police agencies in the Los Angeles area. It is scheduled to become operational in late 1992. Sensors on the Santa Monica Freeway and five parallel arterials will feed traffic flow data to the CALTRANS Semi-Automated traffic Management System (SATMS) and the city of Los Angeles' Automated Traffic Surveillance and Control (ATSAC) computers. The computers will make use of "expert system" algorithms to evaluate traffic conditions and make response decisions. A central Smart Corridor computer will oversee the incident detection function of the above systems, evaluate traffic conditions and operator inputs, and recommend coordinated responses. Current traffic information will be made available to motorists for pre-trip planning through various media such as telephone dial-up, personal computers, radio, and television. Motorists will also be provided with traffic advisories through changeable message signs, highway advisory radios, and commercial radio. (See Figure 15–2.)

Another feature of the Smart Corridor is the Pathfinder project.[18] Pathfinder is a cooperative venture among state and federal government agencies and the private sector. It began as a field evaluation of an in-vehicle urban freeway navigation and information system. The experiment initially involved 25 vehicles equipped with electronic navigation systems that were supplemented with information on current traffic conditions. Pathfinder provided the first assessment of in-vehicle information technology in the United States.

Other industry and government agencies are developing additional cooperative operational tests. The second such test planned is the TravTek project in Orlando, Florida.[19] This cooperative project initially involves 100 vehicles and will

[14] U.S. Department of Transportation, Office of the Secretary of Transportation, *Report to the Congress on Intelligent Vehicle–Highway Systems.*

[15] H. Kawashima, "Integrated System of Navigation and Communication in Japan," paper presented at the midyear meeting of the Transportation Research Board's Communications Committee, July 1990.

[16] H. Kawashima, "Future Research Plans in RACS Project," paper prepared for the panel discussion "Future Directions," IEEE International Conference on Vehicle Navigation and Information Systems, Toronto, September 1989.

[17] JHK & Associates, "Smart Corridor Demonstration Project Conceptual Design Study," final report prepared for the city of Los Angeles, October 1989.

[18] F. Mammano, and R. Sumner, "PATHFINDER System Design," Conference Record of Papers, Vehicle Navigation and Information Systems Conference, Toronto, September 1989.

[19] "TravTek—Travel Technology—An Advanced Motorist Information Demonstration," brochure prepared by the Federal Highway Administration, May 1990.

Figure 15–2. The Smart Corridor Concept.

provide real-time route guidance information as well as information on motorist services. (See Figure 15–3.)

A project different in orientation is the Heavy Vehicle Electronic License Plate Program's (HELP) Crescent Project.[20] This project plans to test whether heavy trucks operating along a major truck corridor in six states will be able to use transponders to communicate weigh-in-motion and automatic vehicle classification data collected on the roadway to state regulatory and law enforcement officials. This information will go into a central data base, the entire content of which is illustrated in Figure 15–4. The goal is to reduce costs associated with motor carrier regulatory requirements. The HELP program involves highway agencies, associated motor carrier groups, and the federal government.

A number of states are also sponsoring IVHS research, development, and testing programs. Perhaps prime among these is California's Program on Advanced Technology for the Highway (PATH).[21] This program is aimed at evaluating whether advanced highway technology can increase the capacity of California's highway system. Primary focuses of this program are automatic vehicle control systems and clean propulsion systems.

A number of research and development activities are also underway. For example, the use of a collision avoidance radar system as a headway control system on the San Diego Freeway is being evaluated.[22] Human factors guidelines and evaluation methodologies are also being developed for in-vehicle information systems.[23]

---

[20] "Heavy Vehicle Electronic License Plate (HELP) Program," Executive Summary, Castle Rock Consultants, April 1989.

[21] Robert E. Parsons, "Program on Advanced Technology for the Highway, Phase I Final Report," Institute of Transportation Studies, University of California, Berkeley, December 1987.

[22] Ibid.

[23] "Assessment of the Effects of In-Vehicle Display Systems on Driver Performance," Federal Highway Administration Contract No. DTFH61-89-C-00044, Transportation Research Institute, University of Michigan.

Figure 15–3. The TravTek System.

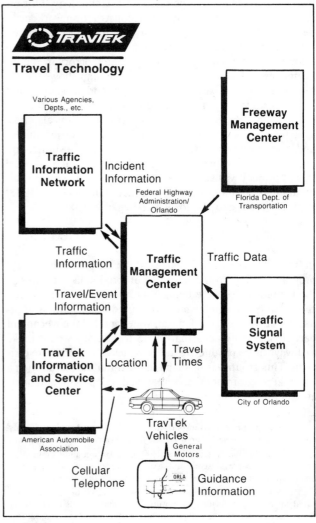

Crescent Data Base Inputs

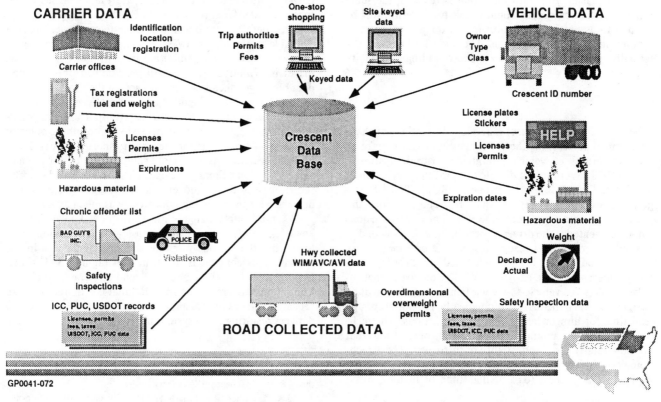

CARRIER DATA

Carrier offices — Identification location registration

Tax registrations fuel and weight

Licenses Permits
Expirations

Hazardous material

Chronic offender list

Safety Inspections

ICC, PUC, USDOT records
Licenses, permits fees, taxes USDOT, ICC, PUC data

One-stop shopping
Trip authorities Permits Fees

Site keyed data

Keyed data

Crescent Data Base

Violations

Hwy collected WIM/AVC/AVI data

ROAD COLLECTED DATA

Overdimensional overweight permits

VEHICLE DATA

Owner Type Class

Crescent ID number

License plates Stickers

Licenses Permits

HELP

Expiration dates

Hazardous material

Weight

Declared Actual

Safety Inspection data
Licenses, permits fees, taxes USDOT, ICC, PUC data

GP0041-072

Figure 15–4. Improving commercial vehicle productivity; the Crescent Project.

Role of IVHS in transportation systems of tomorrow

Implementation of IVHS technology will change the way transportation engineers manage transportation services. The transportation engineer of tomorrow will have to be knowledgeable in areas such as information management, communications technology, control software algorithms, and systems engineering. The effects that IVHS will have in the areas of transportation operations, safety, and productivity are briefly described in the following sections.

Operations

The essence of IVHS as it relates to transportation operations in the future will be the improved ability to manage transportation services because of availability of accurate, real-time information. Decisions that individuals make as to time, mode, and route choices will be influenced by information that is either not available today, is not available when it is needed, or is incomplete, inconvenient, or inaccurate. For example, advanced IVHS technology will enable operators to detect incidents more quickly; to provide information immediately to the public on where the incident is located, its severity, effect on traffic flow and expected duration; to change traffic controls to accommodate changes in flow brought

about by the incident; and to provide suggestions on better routes to take and information on alternative means of transportation. This could include information on transit alternatives, a carpool-matching data base, and information on congestion pricing that would be implemented through automated toll-collection systems. The information would be provided to people through computer or cable television networks at home or places of employment, at transit stops and other transportation terminals, and in the vehicle itself through both visual displays and audio means.

The availability of this information will also enable the development of new transportation control strategies. For example, to obtain recommended routing information, drivers will have to specify their origins and destinations. Knowledge of origin and destination information in real-time will enable the development of traffic assignment models that will be able to anticipate when and where congestion will be occurring. Control strategies that integrate the operation of freeway ramp-metering systems, driver information systems and arterial traffic signal control systems, and meter flow into bottleneck areas can be developed to control traffic better.

Eventually, perhaps, toward the third decade of the twenty-first century, fully automated facilities may be built on which vehicles would be totally controlled by electronics in the highway. This would enable vehicles to move along a crowded urban facility at very high speeds and at very close headways, leading to large-capacity increases. Long-distance travel along interstate highways could also be made faster, safer, and easier. But until that point is reached, the same types of technologies that would enable total control will be applied to assume partial control; for example, to assist in maintaining the appropriate speed behind a lead vehicle, potentially smoothing traffic flow and also leading to capacity increases.

Safety

Whereas many safety measures developed over the years have been aimed at lessening the consequences of accidents (e.g., crashworthiness of vehicles; forgiving roadside features), many IVHS functions are directed toward the *prevention* of accidents. A premise of the European PROMETHEUS program, for example, is that 50% of all rear-end collisions and accidents at cross roads and some 30% of head-on collisions can be prevented if the driver is given another half-second of advanced warning and reacts correctly. Over 90% of these accidents could have been avoided had the drivers taken countermeasures 1 sec earlier.[24] IVHS technologies that involve sensing and vehicle-to-vehicle communications will be designed to warn the driver automatically, providing enough lead time for him or her to take evasive actions. The technologies may also assume some of the control functions that are now totally the responsibility of the driver. These technologies—which include hazard warning systems; vision enhancement systems (e.g., at nighttime or in fog); speed

control, braking, and steering assistance systems—will be designed to compensate for some of the limitations of human drivers, and enable them to operate their vehicles closer together and safer at the same time.

Even before these crash-avoidance technologies become available to the public, IVHS holds promise for improving safety by providing for smoother traffic flow. For example, driver information systems will provide warnings on incident blockages ahead, which may soften the shock wave that propagates because of sudden and abrupt decelerations triggered by unanticipated slowdowns. Transportation information systems may also cause diversion onto other routes or modes away from incidents, leading to shorter queues, fewer abrupt decelerations, and more rapid return to normal and safer conditions. Finally, navigation systems should serve to reduce excess travel and erratic maneuvers made by lost drivers, perhaps reducing fatalities by as much as 7%.[25]

Potential safety dangers must also be acknowledged. A key issue involves driver distraction and information overload from the various warning and display devices in the vehicle. This concept is illustrated in Figure 15–5, which hypothesizes that the amount of road and traffic information retained decreases as an individual is required to make repeated glances at a display device.[26] Other issues include dangers resulting from system unreliability (e.g., a warning

---

[24] Statement of Karl-Heinz Faber before the Motor Vehicle Safety Research Advisory Committee.

[25] G.F. King, and T. Mast, "Excess Travel: Causes, Extent and Consequences," paper presented at 66th Annual Meeting of the Transportation Research Board, Washington, DC, January 1987.

[26] J.A. Parviainen, R.L. French, and H.T. Zwahlen, "Mobile Information Systems Impact Study," report prepared for the Ontario Ministry of Transportation and Transport, August 1988.

Figure 15–5. Effect of repeated glances on information retention.
SOURCE: *Mobile Information Systems Impact Study,* Ontario Ministry of Transportation.

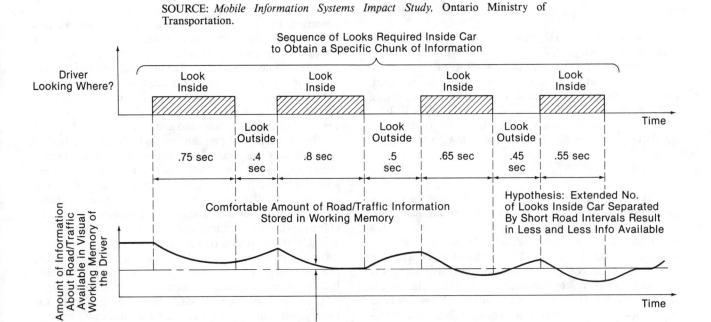

[1]When obtaining a specific chunk of information in four intermittent looks from a sophisticated in-vehicle display *or* When operating a sophisticated in-vehicle control requiring four intermittent looks.

or driver-aid system that fails to operate), and the incentive for risky driving that IVHS technologies may provide.[27] These are important research issues that must be addressed before these types of systems are widely implemented.

Productivity

The availability of accurate, real-time information will especially be useful to operators of vehicle fleets, including emergency vehicles, fire, police, transit vehicles, and truck fleets. Here, quick response is essential. Time lost is lost productivity and income. Operators will know where their vehicles are, how long a trip can be expected to take, will be able to advise on best routes to take, and, thus, will be better able to manage their fleets.

Great potential exists for productivity improvements in the area of regulation of commercial vehicles. Automating and coordinating regulatory requirements through application of IVHS technologies can, for example, reduce delays currently incurred at truck weigh stations, reduce labor costs to the regulators, and minimize the frustration and costs of red tape to long-distance commercial vehicle operators. There is also potential to improve coordination among freight transportation modes (e.g., use of the same electronic container identifiers by the maritime and trucking industries would greatly improve freight-handling efficiencies.)

IVHS categories

As part of the process of planning a national IVHS program, the following categories have been defined by Mobility 2000 and popularly accepted in the United States: Advanced Traffic Management Systems (ATMS), Advanced Traveler Information Systems (ATIS), Commercial Vehicle Operations (CVO), and Advanced Vehicle Control Systems (AVCS). These are defined below, and a vision of what these systems will be and how they will evolve is presented.

Advanced Traffic Management Systems

Advanced traffic management systems have six primary characteristics that differentiate them from the typical traffic management systems of the early 1990s.[28]

1. An ATMS works in real time.
2. An ATMS responds to changes in traffic flow. In fact, an ATMS may be one step ahead, predicting where congestion will occur based on collected origin-destination information.
3. An ATMS includes areawide surveillance and detection systems.
4. An ATMS integrates management of various functions, including transportation information, demand management, freeway ramp metering, and arterial signal control.

5. An ATMS implies collaborative action on the part of the transportation management agencies and jurisdictions involved.
6. An ATMS includes rapid response incident-management strategies.

The ultimate aim of the IVHS program in the ATMS area is to develop what is called an *interactive traffic control system*. In these systems, access to transportation information systems will be available in homes, at places of employment, at transportation terminals, and in vehicles. Accurate information on current and predicted traffic conditions and the best modes and routes to take will be provided to the public, commercial vehicle operators, and transportation managers. Accurate real-time origin-destination information will be available. Advanced, integrated transportation management strategies will be implemented in transportation operations centers in urban areas across the country. Major delays caused by incidents will be avoided, goods will be delivered on time, and people will be able to make informed choices about whether and when to travel, and the best modes and routes to take. (See Figure 15–6.)

To implement these types of systems, fundamental research will have to be conducted into driver behavior and travel information needs. One key issue will be how people respond to suggestions to take alternate routes, and whether this can be predicted in advance. Real-time traffic monitoring and data management capabilities will have to be developed, including advanced detection technology such as image-processing systems, automatic vehicle location and identification techniques, and the use of vehicles themselves as probes. Software and expert systems techniques to manage the plethora of real-time information that will be transmitted to an operations center from multiple sources must also be developed. New traffic models will have to be created, including real-time dynamic traffic assignment models, real-time traffic simulation models, and corridor optimization techniques. Applicability of artificial intelligence and expert systems techniques will have to be assessed and applications such as rapid incident detection, congestion anticipation, and control strategy selection developed and tested. Responsive demand

Figure 15–6. The INFORM Control Center on Long Island, New York, illustrates the concept of integrated traffic control.

[27] D.K. Willis, "Future Technologies: What Impact on Highway Safety?" ITE Conference on Meeting the Transportation Challenges of the 1990s: Land Development, Traffic Congestion, and Traffic Safety, Garden Grove, CA, March 1990.

[28] Mobility 2000, "Final Report of the Working Group on Advanced Traffic Management Systems," February 1990.

management strategies will need to be evaluated, including such things as HOV or transit incentives, parking restrictions, and congestion pricing during periods when heavy congestion is predicted.

Advanced Traveler Information Systems

Advanced Traveler Information Systems (ATIS) will provide drivers with information on congestion and alternate routes, navigation and location, and roadway conditions through audio and visual displays in the vehicle.[29] This information might include incident locations, location of fog or ice on the roadway, alternate routes, recommended speeds, and lane restrictions. ATIS will provide information that would assist in trip planning at home, at work, and by operators of vehicle fleets. ATIS will provide information on motorist services (e.g., restaurants, tourist attractions, and the nearest service stations and truck and rest stops). This has been called the "Yellow Pages" function. ATIS could also include on-board displays that replicate warning or navigational roadside signs that may be obscured during inclement weather or when the message should be changed—for example, to lower speed limits approaching congested freeway segments or fog areas. In concert with advanced vehicle control systems, ATIS

___
[29] Mobility 2000, "Final Report of the Working Group on Advanced Driver Information Systems," February 1990.

could also provide information that would warn of potentially dangerous situations. An automatic MAYDAY feature might also be incorporated that would provide the capability to summon emergency assistance and provide vehicle location automatically. (See Figure 15–7.)

The vision of the ultimate ATIS is similar to that described above for ATMS. The role of ATIS in an interactive traffic management system would be to enable vehicles acting as traffic "probes" to transmit information on origins and destinations and real-time travel times to a transportation operations center where it would be combined with all other sources of information (e.g., from freeway surveillance and control and computerized traffic signal systems, police and commercial radio surveillance, etc.) using data-fusion models. This combined information would be analyzed and the location of congestion anticipated. Information would then be transmitted on the best routes and modes to take, and traffic management strategies developed that optimize flow on an areawide basis. This advisory function would be integrated with those described above, in essence creating a comprehensive, real-time transportation information system.

To get to this system, research and development (R&D) is required to define the communications technology, architecture and interface standards that will enable two-way communication of information between vehicles and a management center in real time. Possibilities include radio data communications, cellular telephone, roadside beacons

Figure 15–7. General concept of Advance Traveler Information Systems (ATIS).

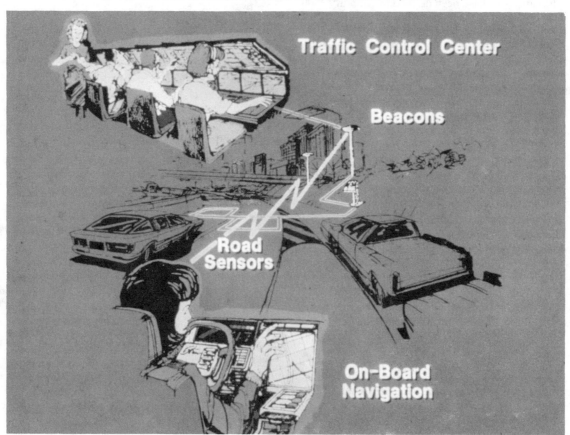

used in conjunction with infrared or microwave transmissions or low-powered radio signals, and satellite communications. Software methods to fuse the information collected at the management center and format it for effective use by various parties must also be developed. These parties include commuters, other trip-makers and commercial vehicle operators, both before they make a trip and en route; operators of transportation management systems, and police, fire, and emergency response services. A number of very critical human factors issues must also be investigated. These include looking at how individuals make travel, mode, and routing decisions, accounting for the requirements and characteristics of special groups such as commercial vehicle operators and the elderly, and identifying the critical pieces of information and best way of conveying these. Only by understanding the behavior and needs of all users will systems be implemented that meet those needs. Also required is a critical examination of in-vehicle display methods. Among the issues here are: What are the roles of audio messages and in-vehicle displays? Where should visual displays be located? (Included would be an assessment of heads-up displays that project an image onto the windshield, theoretically reducing the time a driver has to glance away from the road scene.) What should be shown on visual displays? Possibilities include directional arrows to guide vehicles through intersections and interchanges, text messages, electronic maps which display information on traffic conditions, perhaps using color graphics, and critical roadway signs. (See Figures 15–8 and 15–9.)

Commercial vehicle operations

Application of IVHS technologies holds great promise for improving the productivity, safety, and regulation of all commercial vehicle operations, including large trucks, local delivery vans, buses, taxis, and emergency vehicles.[30] These will make possible faster dispatching, efficient routing, and more timely pickups and deliveries, which will have a direct impact on the quality and competitiveness of business and industries at both the national and international levels. IVHS technologies such as weigh-in-motion scales, automated vehicle identification transponders and classification devices, automatic vehicle location and tracking, routing algorithms, in-vehicle text and map displays, and two-way communications will reduce the time spent at weigh stations, improve hazardous material tracking, reduce labor costs required to administer government truck regulations, and minimize costs to commercial vehicle operators. Advanced vehicle control technologies, such as anti-lock brakes, blind-side and near-obstacle detection, and collision-avoidance control systems, will serve

---

[30] Mobility 2000, "Final Report of the Working Group on Commercial Vehicle Operations," February 1990.

Figure 15–8. An in-vehicle display unit.

Figure 15–9. A heads-up display.

to reduce public concern about the mix of smaller cars and larger trucks on highways.

IVHS technology will manifest itself in numerous ways in commercial vehicle operations. For example, for long-distance freight operations, on-board computers will not only monitor the other systems of the vehicle, but may also function to analyze driver fatigue and provide communications between the vehicle and external sources and recipients of information. Applications will include automatic processing of truck regulations (e.g., commercial driver license information, safety inspection data, and fuel tax and registration data), avoiding the need to prepare redundant paperwork and leading to "transparent borders"; provision of real-time traffic information through advanced driver information systems; proof of satisfaction of truck weight laws; and communication with fleet dispatchers. Regulatory agencies will be able to take advantage of computerized record systems and will be able to target their weighing operations and safety inspections at those trucks that are most likely to be in violation. Local delivery trucks will benefit from receiving real-time route guidance information. Automatic toll debiting will also be possible through automatic vehicle identification technology. Trucks and buses will be able to signal when there is an emergency or mechanical problem, and operators will be able to identify where the vehicle is located and send help promptly.

Applications of IVHS technologies could also lead to significant improvements in bus and paratransit operations in urban and rural areas. Dynamic routing and scheduling could be accomplished through on-board devices, communications with a fleet management center, and public access to a transportation information system containing information on routes, schedules, and fares. Automated fare collection systems could also be developed, which would enable extremely flexible and dynamic fare structures and relieve drivers of fare collection duties.

Research and development and initial testing of many of these applications have already been completed. Truck drivers are communicating in real-time with dispatchers and receiving route guidance information. The missing link is real-time traffic information to supplement the route guidance system, and ties with state truck regulatory requirements. There is also need for research into the

human factors aspects of information processing by commercial vehicle drivers, whose driving task is already complicated. Displays must be simple, but clear, and driver overload must be avoided. Another promising area of research is that of vehicle dynamics. Systems that warn of impending problems related to vehicle roll, and collision-avoidance systems in both commercial vehicles and private vehicle fleets, could prevent many accidents from happening.

Advanced Vehicle Control Systems

Whereas the other categories of IVHS primarily serve to make traveling more efficient by providing more timely and accurate information about transportation, AVCS serves to improve safety greatly and potentially make dramatic improvements in highway capacity by providing information about changing conditions in the vehicle's immediate environment and sounding warnings and assuming partial or total control of the vehicle.[31] The technologies involved in AVCS are very extensive, including driver warning, vision enhancement and assistance systems, automatic headway control (platooning), obstacle avoidance and automatic braking, automatic trip routing and scheduling, control merging of streams of traffic, and transitioning to and from automatic control.

Early implementation of AVCS technologies might include a number of systems to aid with the driving task. These include hazard warning systems that sound an alarm or actuate a light when a vehicle moves dangerously close to an object—for example, when backing up or moving into the path of another vehicle in its blind spot when changing lanes. Infrared imaging systems might also be implemented that enhance driver visibility at night. (See Figure 15–10.) AVC technologies may also include adaptive cruise control and lane-keeping systems that automatically adjust vehicle speed and position within a lane—for example, through radar systems that detect the position and speed of a lead vehicle, or possibly through electronic transmitters in the pavement that detect the position of vehicles within the lane and send messages to a

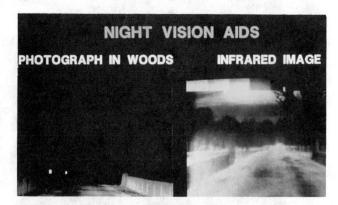

Figure 15–10. Infrared imaging system.

[31] Mobility 2000, "Final Report of the Working Group on Advanced Vehicle Control Systems," February 1990.

computer in the vehicle that has responsibility for partial control functions. As technology advances, lanes of traffic may be set aside exclusively for automated operation. These are called *platooning highway systems*. Small groups of vehicles, perhaps up to 12 per platoon, will travel together at high speeds, maybe 65 mph with very short headways, controlled through obstacle-detection and automatic speed control and braking systems. These automated facilities have the potential to greatly increase highway capacity, while at the same time providing for safer operation. Eventually, AVCS technologies will provide for complete control of the driving function for vehicles operating on specially equipped freeway facilities. Roadway electrification might also assist in the recharging of electric vehicle batteries, enabling, for example, small delivery, transit, and other electric vehicles to be easily integrated into an electric roadway network. Totally automated facilities could also be implemented in high-speed intercity corridors to make business travel more productive, and to ease the burden on tourists.

There is, of course, much R&D work and testing to be done before these types of systems can be built and implemented. Applications involving alternative technologies will have to be developed and evaluated. Complicated software will have to be written and tested. Perhaps the most important issues, though, relate to the role of humans in the system—for example, public acceptability and how it is likely to affect system effectiveness. Other human factors issues include driver reaction to partial or full control; for instance, will it cause them to lose alertness; will it cause them to drive more erratically? Another important area of research is the assessment of AVCS technology performance, reliability, and cost-effectiveness. A final important issue is the effect that the threat of liability may have on the willingness of potentially creative and innovative private developers of AVCS applications to get involved.

A scenario for the year 2020

Even though IVHS is a collection of different technologies and systems, the vision is that these would all eventually work together in a single, integrated advanced transportation system. To illustrate how this system might operate, the Advanced Vehicle Control Systems Working Group of Mobility 2000 described the following scenario for an automated vehicle-highway system of tomorrow, perhaps by the year 2020. (See Figure 15–11.)

> For commuters living in the suburbs, the morning drive is now a realization of the many benefits of a state-of-the-art transportation system, not a mind numbing lesson in confusion and frustration. From the moment Norm turns the ignition key, he is aware of the advances in the transportation system. He pushes the "office" button on his destination selection console to indicate that he wants to head toward his office and the computer starts working out the fastest route to get there in this morning's traffic. The car completes its systems maintenance check and the journey begins with Norm driving toward the nearest freeway on-ramp. Norm controls the steering, but receives assistance from the "situation awareness" electronics equipment in the car which warns him of hazards ahead and to the side. Having avoided the neighborhood

Figure 15–11. Automated control systems of tomorrow.

fender bender, he leaves his subdivision and pulls onto a major street.

The dramatic improvements in freeway capacity and efficiency have eased the bottleneck that used to occur just a few blocks away from his home in the suburbs. Cars from the residential subdivision can move along the surrounding streets and toward the freeway on-ramps with greater ease now that freeway congestion has been drastically reduced. The driver information and warning systems do not indicate any problems this morning, so Norm knows he can proceed directly to the nearest freeway ramp.

At the freeway entrance ramp, Norm responds to the electronic signal and pushes the button on his dashboard that initiates automatic control of his vehicle and directs his car to the off-ramp closest to his office. He can now take his hands off the wheel and his feet away from the pedals, since these no longer affect the movement of his car. He notices a brief slowdown of the car as the roadside diagnostic unit interrogates his car to ensure that the necessary on-board equipment is operational and that the vehicle condition is satisfactory to complete the trip. It was a good thing he had remembered to refill his tank last night so that he didn't have to repeat his experience of last week, when his car was directed into the reject lane and back to the local street because it didn't have enough fuel to complete the trip.

Once past the brief diagnosis period, the car accelerates smoothly and merges into a platoon of fast-moving cars. It's such a relief, not having to worry through the tricky merge maneuver the way he used to in the old days, when he never knew where he would be able to find a space in the heavy traffic stream and was always worried about getting rear-ended or sideswiped.

Norm pulls out his newspaper and catches up with the latest developments in the world, until he remembers the memo he forgot to write last night in preparation for his morning meeting. He calmly retrieves his laptop computer from the back seat and types out the memo. When he finishes, he notices that he is not on Highway 580, where he normally travels, but is on 880 instead. His curiosity aroused, he turns on his map display unit and discovers that there is a resurfacing project underway on 580, which has taken a lane out of service, reducing its capacity. The central computer enabled him to avoid this by rerouting his trip onto the alternate route. Even though this would be a couple of miles longer, it would save some time this morning.

With the integration of electronic signposts, highway signs, and road markers, the cars in the platoon stay together for much of the journey to the city, with the faster-moving platoons in the left lane and the slower platoons in the right lane. As Norm's car approaches its exit, the trip computer beeps to warn him that it is preparing to exit the freeway. Although Norm's car will remain in the fully automated mode until he has safely reached the exit ramp, he is alerted prior to the exit so that he is prepared to resume control of the car. The detection of the road markers and the electronic signs prevent the car from running the signal at the bottom of the off-ramp. Norm punches in the right numerical code to assure the computer that he is awake and alert and ready to resume control of the car (just like one of those old sobriety check devices his boss had once told him about).

The traffic signals at the intersections near the freeway exits are integrated into the new transportation system to manage the flow of cars that have exited the freeways. The traffic signal integration system allows Norm and his fellow commuters to move away from the freeway and into the commercial and business district of the city with much greater ease and much less of the stop-and-go traffic patterns that were present before the new traffic signal system.

Although more commuters like Norm have chosen to drive to work, the traffic management capabilities of the advanced traffic network integrated with AVCS have kept the morning and afternoon commuting times to a minimum. Rush-hour commuters are not the only ones that have benefitted from the changes that have taken place. Sports and concert fans, along with theater patrons, and shopping center customers cherish the simplicity with which they drive to the stadium, theater, and mall. The horrendous traffic jams that used to occur with regularity have given way to a much more orderly exit from these places.

As cars leave the parking areas in the semiautomatic mode, they enter automatic merging lanes leading to the adjacent freeways. In the merging lanes, small groups of cars form platoons and accelerate and merge into the traffic flow as space permits. The newly formed platoons travel down major inner-city expressways, allowing the traffic to exit the parking areas with greater efficiency and speed than many had imagined. The parking lot that used to take a half hour to empty after the basketball game is now deserted within 10 minutes.

Elsewhere on the West side of town, the roadway electrification program has proven its usefulness with the postal service and the public transit authorities. For vehicles such as buses and trucks running along fairly consistent routes, the electricity available through the supplemental power supply in the streets has allowed the use of electric vehicles to be integrated into the city's transportation network. Electric vehicles, operating primarily on batteries, extend their range using these recharging facilities in the downtown area. The electric roadway program has helped the city keep its commercial vehicle pollution under control while providing a major reduction in the noise level in many downtown areas.

Commuters to and from nearby cities continue to enjoy the benefits brought about by the modernization of existing highways. With the complete integration of the automated vehicles and the advanced highways, businessmen and workers travel between cities in much less time than required just 20 years ago. Instead of traveling at average speeds of 30–40 mph, cars operating in the fully automated mode now travel at the 75 mph speed limit for the entire trip. Tourists visiting

Washington, D.C., can now routinely visit nearby cities like Baltimore and Annapolis as the difficulty of traveling on unfamiliar roads has been eliminated and the speed of automobile travel between these urban areas has increased."[32]

Benefits and costs of IVHS

Benefits

Attempting to quantify the benefits of widely deployed IVHS technologies at this stage must be similar to what planners of the U.S. interstate highway system tried to do in the 1950s. There is no way to anticipate all of the ways that applications of IVHS technology may affect society, just as planners of the interstate system could not have anticipated all of its effects on American society. Recognizing the importance of the issue, however, one of the working papers prepared by the Mobility 2000 Group was on the potential benefits of applying IVHS technology in the United States.[33] One of the interesting findings was that IVHS was not just for urban areas. Numerous benefits were also found for rural areas, and for targeted groups such as the elderly and disadvantaged travelers, and fleet operators. Positive benefits were also found in regard to safety and the environment. A complete list of potential IVHS benefit measures and groups is shown in Tables 15–1 and 15–2. Some of the specific findings that were reported include:

TABLE 15–1
Categories of IVHS Benefits

A. Travel	B. Economic
Travel Time • Average speed • Reliability • Predictability	• Product innovation • Sales (revenue) • Productivity • Improved skills (human capital) • Competitiveness
Travel Safety • Fatalities • Injuries • Property damage	• Supplier industries • International • Standard setting
Travel Availability/Accessibility • Routes • Modes (e.g., ride-sharing) • Services	Economic Development • Economic growth • Improved standard of living • Enhanced land use
Travel Comfort • Stress • Stops	Environmental • Air pollution • Noise • Amenities
Travel Security • Emergency services	• Guideway • Flow
Travel Cost • Operating cost • Fuel/Energy • Parking • Insurance	• Meeting environmental standards Information • Efficiency • Traffic enforcement
Distance Traveled • Vehicle miles of travel	• Improved transportation investment and operating decisions

SOURCE: Mobility 2000, "Final Report of the Working Group on Operational Benefits," February 1990.

---

[32] Ibid.

[33] Mobility 2000, "Final Report of the Working Group on Quantification of Benefits," February 1990.

TABLE 15–2
Groups Participating in IVHS Benefits

Users (Groups)	Organizations
People • Urban • Suburban • Rural • Car-less • Commuter • Elderly • Handicapped • Tourists/visitors • Military • Regions **Freight (Consumers)** • Manufacturers • Retailers and other private firms • Defense **GENERAL POPULATION** **Other Transportation System Users** • All categories **Nonusers** • System abutters • Enjoying living standard changes • Changing settlement patterns, etc.	**Public-Sector Operators** • State DOTs • Traffic departments • Transit agencies • Department of Defense • Authorities (airports/toll roads/ports) • Police/Emergency • Environmental • Planning • School systems • Human service agencies **Private-Sector Operators** • Trucking companies • Bus companies • Taxis • Limousines • Small package delivery • Terminals • Freight • Air • Water • Emergency services • Railroads • Airlines **Industry** • Auto manufacturers • Electronics/communications • Auto suppliers • Construction • Academic • Research • Energy • Financial • Other Industry

SOURCE: MOBILITY 2000, "Final Report of the Working Group on Operational Benefits," February 1990.

○ Fully deployed combinations of advanced traffic-management systems and advanced traveler information systems can produce congestion-cost decreases in urban areas ranging from 25% to 40%.

○ It was estimated that the cost of delay in the United States in 1990 was approximately $100 billion annually. The value of time saved alone, therefore, would be at least $25 billion in 1990, and would grow substantially since total travel was expected to increase by about 50% by the year 2005.

○ Unchecked traffic congestion is the single largest contributor to poor air quality and wasted fuel consumption. Reductions in traffic congestion will lead to improvements in these areas.

○ By 2010, annual savings of approximately 11,500 lives and $22 billion in accident costs could be realized.

○ By 2020, annual savings of 33,500 lives and $65 billion in accident costs could be realized, as advanced vehicle control strategies achieve a large market penetration.

○ Rural areas have the most to gain in relation to safety improvements, since 57% of fatal accidents occur in rural areas where collision speeds are likely to be higher.

○ Older and disadvantaged drivers can benefit by having specific devices available to offset the slowing down of their capabilities—for example, infrared imaging, obstacle-detection and warning systems, radar braking

and steering override, and on-board replication of maps and signs.

○ Motor carrier productivity can be significantly increased and fuel costs decreased through automated collection of tolls, through the provision of real-time routing information and "Yellow Pages" services, through automated processing of permits and licenses, and through on-board computers that provide information on vehicle performance.

○ Vehicle congestion and crashes could virtually be eliminated on automated facilities.

Finally, assuming that the benefits and cost-effectiveness are proven and the public exhibits a willingness to pay for these systems, it is estimated that the U.S. market alone for automotive electronics will amount to $28 billion annually by the year 2000, and that the U.S. highway infrastructure costs for these systems would be around $30 billion by the year 2010.[34, 35] There will be a very substantial international market for IVHS products and services that can be supplied by the private sector, contributing to economic growth. There is further potential growth from spinoff products developed as a result of IVHS research and development.

Costs

The costs of developing, testing, and deploying IVHS technologies may more appropriately be viewed as investments that will produce better transportation for the public and create markets for the private sector. The Mobility 2000 Group has estimated the costs of researching, developing, testing, and deploying IVHS technologies in the United States, but the estimating process is a very difficult one. It is impossible to predict how technology will advance over, say, a 10-year period, and how quickly transportation applications of that technology will be developed. It is also hard to estimate accurately how much private industry will spend on researching and developing IVHS technology, and even what should be included within the scope of the definition of IVHS research. Nevertheless, at a March 1990 workshop in Dallas, the Mobility 2000 Group produced the following rough estimates of the costs of a national U.S. IVHS program, including both private and public funds:[36]

1. The total cost of research and development over a 20-year period would be almost $1.4 billion. Over half of this would be in the area of automated vehicle control strategies. Three quarters of the total would be invested in the first 10 years of the program.

2. The total cost of operational testing over the same 20-year period would be just over $3 billion. Again, roughly half

---

[34] U.S. Department of Transportation, Office of the Secretary of Transportation, *Report to the Congress on Intelligent Vehicle–Highway Systems.*

[35] Mobility 2000, "Draft Report of the Breakout Group on Program Funding Requirements," prepared for the proceedings of a workshop held in Dallas, March 1990.

[36] Mobility 2000, "Draft Report of the Breakout Group on Program Funding Requirements," prepared for the proceedings of a workshop held in Dallas, March 1990.

of this would be in the area of automated vehicle control strategies, where the private sector would likely contribute substantially. Over 80% of the operational testing costs would be in the last 15 years of the program.

3. The total cost to deploy IVHS highway technology (i.e., not including the cost of producing vehicle electronics) was estimated to be around $30 billion. Approximately 60% of this cost would be invested to pay for the basic surveillance infrastructure that would also allow early implementation of advanced traffic management and traveler information systems. Over 80% of the cost of deployment would occur in the last 15 years of the program.

4. The cost to the individual to purchase the vehicle technology was estimated to be around $800 to $1,000 per vehicle.

What would this investment buy? To illustrate, the Mobility 2000 Group described the following scenario:

○ Instrumentation of 18,000 miles of freeway and integration of the operation of these freeway surveillance and control systems with the operation of approximately 200,000 signals in 250 of the largest metropolitan areas in the United States.
○ Establishment of advanced traveler information systems integrated with the advanced traffic management systems in these 250 metropolitan areas, and establishment of an advanced driver information systems center for rural areas in every state.
○ Instrumentation of the 42,500-mile interstate highway system and the remainder of the National Truck Network for commercial vehicle operations applications.
○ Sixteen platooning highway systems for headway, speed, and merge control, and 44 electric propulsion systems (in 25-mile increments) in areas over 1 million population.

To provide some scale to these estimates, the total investment (for research, development, testing, and deployment of IVHS highway technology) of just over $34 billion *for the 20-year period* is roughly half of what was being spent *annually* on highways in the United States in 1990 considering all sources. If the total annual highway investment remained constant, the *annual* investment in IVHS would translate to an equivalent of 2.5% per year of all highway expenditures.

Institutional Issues

Administrative

Deployment of Intelligent Vehicle–Highway Systems will cause changes in the way public agencies provide transportation services to the public. Accurate and timely information on traffic conditions on all facilities will be needed, necessitating a cooperative approach to collecting the information and sharing it among responsible operating agencies (as discussed in Chapter 12). Control decisions will have to be made jointly among agencies; for example, recommendations to divert traffic off a freeway onto an arterial should be accompanied by changes in signal timing along the arterial.

Providing information to the public through computer networks, cable television, or in-vehicle devices will require cooperation among the agencies providing the information and the operators of information networks, including vehicle fleets (e.g., transit systems, emergency vehicle fleets, or trucking companies). Organizational structures and attitudes will have to adapt to the change from being operators of single systems (e.g., traffic signal systems, freeway surveillance and control systems, transit systems) to being players in the operation of a signal areawide transportation management system in which information and management decisions are shared and made jointly. Transportation management centers, staffed by multiple agencies, will be developed to serve as the focal point of transportation operations in urban areas. Eventually, some urban areas may choose to combine operating agencies, perhaps even to create a single transportation operating agency or authority.

Government agencies involved in the regulation of commercial vehicles, especially long-distance carriers, will also have to act cooperatively, sharing information data bases so that carriers can truly benefit from "one-stop shopping" and "transparent borders." Commercial vehicle operators will also have to be convinced that the benefits of using IVHS technology exceed their perceived risks of closer surveillance that IVHS technologies will provide. Safeguards will have to be established to avoid abuses of truck and driver location and performance data that violate the privacy rights of individuals.

The private sector will also have an increasingly larger role to play. Certainly much of the research and development work that has to be done, especially in the commercial vehicle and advanced control areas, will be done in the private sector. Private firms will also provide much of the talent that will plan, design, and build IVHS systems. It is also probable that some urban areas will hire private companies to operate and maintain their IVHS systems, perhaps in a franchise arrangement. Private capital may also be used to fund the operation of information networks. For example, owners of restaurants or other services might pay to advertise through these networks. Finally, the establishment of communications networks will require that producers of IVHS products develop national standards for communications interfaces and protocols so that vehicles will be able to interact with transportation information networks or automated facilities in any locale across the country. Indeed, these standards may well be international standards.

Given the expected changes in how public transportation agencies operate and the increasing role of the private sector, it is very likely that public/private partnerships will flourish in an IVHS environment. This will cause changes in the traditional "hands-off" relationship between private companies and public agencies. Private firms will have to become more sensitive to the demands of the public, and public agencies will have to become more entrepreneurial in their approach to providing services. Cooperation will be of the essence, and agencies will have to be willing to take calculated risks to install new IVHS technologies.

Support

Successful deployment of IVHS systems is not only dependent on the successful development of transportation applications of advanced electronics technology and control strategies, but also on the successful recruitment of talented people into the transportation engineering profession, and the political success the profession has in convincing policymakers and the public of the need for and cost-effectiveness of deploying IVHS systems. In order to be a planner, designer, or operator of an IVHS system, people with new skills and new disciplines will be needed in addition to transportation engineering. These will obviously include electrical engineering, computer programmers, systems analysts, and data management experts, but will also include experts on human factors, liability, and public relations. Transportation engineering programs will have to teach about these issues, and efforts to recruit bright young people into our profession will have to be continued and perhaps heightened. Increased funding for IVHS research programs should also serve to attract students to graduate work in this area, providing a training ground for future professionals.

Decision-makers and the public will have to be convinced of the social and individual benefits of IVHS systems. The systems and benefits will have to be marketed because the cost will be high. Innovative ways of financing deployment of IVHS systems will have to be sought, involving mixes of federal, state, or province, and local funds, private investment, dedicated taxes and bond issues, and developer fees and other tax mechanisms. Experience has already shown that public support will evaporate quickly if early trials fail; thus, care must be taken to ensure that implemented systems are reliable, well-maintained, and effectively operated.

The budget problem will not only be an initial capital budget problem; the cost of operating and maintaining IVHS systems will be substantial, estimates ranging on the order of 10% to 15% of the initial capital costs annually.[37] Programs will need to be committed to in the project planning stages to provide for adequate operations and maintenance funding and staff resources. The practice of waiting to address operation and maintenance considerations until after the project is implemented will not suffice for IVHS.

---

[37] Mobility 2000, "Final Report of the Working Group on Advanced Traffic Management Systems."

Key considerations

Intelligent Vehicle–Highway Systems hold great potential for revolutionizing surface transportation around the world. Commuting, business, freight, and social travel could be made substantially easier, faster, and safer. Economic growth could be enhanced not only because of decreases in the cost of transportation, but also because of the creation of a substantial international IVHS market. In summary, three key ingredients are needed to realize this ambition:

1. *People.* This human resource must be enhanced by bringing more people with new and different skills into the profession. These individuals must not only be technically excellent but must also be bold and innovative thinkers.

2. *Entrepreneurship.* Public agencies and private firms, in partnership with one another, must be willing to take risks and must provide the leadership and action necessary to develop and sell IVHS systems to the public.

3. *Money.* Capital must be made available to research, develop, and test promising IVHS technology and strategies, to plan, design, and build IVHS systems, and to operate and maintain them once they are built. Innovative financing mechanisms, involving both public and private funds, must be established.

The opportunity to make significant improvements in surface transportation systems is on the horizon. It is up to the profession to take advantage of that opportunity.

REFERENCES FOR FURTHER READING

"Assessment of Advanced Technologies for Relieving Urban Traffic Congestion," National Cooperative Highway Research Program Report No. 3-38(1), Castle Rock Consultants.

Conference Record of Papers, Vehicle Navigation & Information Systems Conference, IEEE, Toronto, Ontario, September, 1989.

Executive Summary, Mobility 2000 Workshop on Intelligent Vehicles and Highway Systems, March 19–21, 1990 Dallas, Texas Transportation Institute. (Full proceedings were not available at the time this was written, but were also expected to be available from the Texas Transportation Institute.)

"Proceedings of the National Leadership Conference on Intelligent Vehicle–Highway Systems," Highway Users Federation for Safety and Mobility, Orlando, May 1990.

INDEX

operating costs, 42–45
   rolling resistances of, 34
   travel speeds of, 54
   weight/power ratios of, 36
Passenger zones parking, 233
PASSER-II Model (Progression
      Analysis and Signal System
      Evaluation Routine), 298
Passing and no-passing regulations,
      351–52
Passing sight distance, 162, 163,
      351–52
Passing zones on two-lane highways, 39
Passive restraints, 105
Passive signs, 252
Pathfinder Project, 403, 451
Paths, bicycle, 342
Patrols, police freeway, 408
Pavement markings. *See* Marking(s)
Pavement reflectance lighting,
      312–13
Peak hour factor (PHF), 77, 120
Peak hour volume, 49–50, 156
Pedestals, movable, 336–37
Pedestrian(s), 19–28
   accidents, 19–20
   children, 24–27
   elderly, 27
   fatalities, 22, 23
   handicapped, 24
   nighttime conditions and, 23
   safety countermeasures for, 27–28
   signals, 20–22
   social factors and, 23–24
   studies of, 78–79
   walking speeds, 19, 20
   walkway capacity, 148
Pedestrian facilities, design of,
      198–201
Pedestrian-only streets, regulations
      for, 343–44
Pedestrian signals, 289
Pedestrian walkway:
   defined, 314
   recommended average maintained
      illuminance levels, 321
People mover, 46
Percent time delay, 56
Perception:
   of drivers, 4–6
   of own driving ability, 13
Perception model of driver behavior,
      2, 3
Perception-reaction time (PRT), 6–7
Performance, drive:
   positional, situational, and
      navigational, 313
Periodic-check parking studies, 84
Permanent International Association of
      Road Congresses (PIARC), 115
Personality of drivers, 13–14

Personal liability, 442–43
Personal liability insurance, 445
Phase sequence, 291, 297
Phasing, 291–93, 295–96
Photography method survey, 69
Physical condition diagrams, 279
Physically separated HOV roadways,
      186–88
Planning:
   of bikeways, 342–43
   for one-way streets, 334
   traffic engineering, 427–29
Planning analysis capacity, 123
*Planning and Design Criteria for
      Bikeways in California,* 342
Planning application worksheet, 139
Planning procedure, defined, 138
Plastic signs, 271
Platooning, vehicle, 56, 296
Platooning highway systems, 459
Platoon ramp metering, 397
Play streets, 358
Plywood signs, 271
Point of capacity, 120
Poisson distribution, 57
Pole and mast arm inspection, 305
Police, 426–27
Police patrols, 408
Policy development, 425
*Policy on Geometric Design of
      Highways and Streets, A*
      (AASHTO, 1990), 32–33, 37,
      77, 154, 155, 156, 165, 166
Polyester lane-striping materials, 273
Positional performance, 313
Positive Guidance Process, 4,
      243–44
Postcard survey, 85–86
Power vehicle, 35
Preemptor signal, 284
Preliminary engineering studies, 176
Preplanned diversion routes, 409
Press releases, 425
Pretimed controller operation, 280,
      281
Pretimed ramp metering, 396–98,
      410–11
Prévention Routière Internationale
      (PRI), 115
Preventive maintenance signals, 305
Prima facie speed limit, 347
Prime parking area, 224
Priority access control, 405
Priority control freeway, 404–5
Prismatic (cube-corner)
      retroreflection, 249
Private streets, 358
Productivity, IVHS's role in, 455
Program on Advanced Technology for
      the Highway (PATH), 452
Progression factor, 77

PROMETHEUS (PROgraMme for
      European Traffic with Highest
      Efficiency and Unprecedented
      Safety), 449, 454
Protected-only left-turn phasing, 295
Protected/permissive left-turn
      phasing, 295–96
Proximate cause, 439
Public acceptance of freeway
      surveillance and control
      systems, 416
Public agency parking, 233
Public attitude, speed regulations and,
      345–46
Public education, accidents and,
      102–3
Public health measures, 95
Public hearings, 177
Public liability, 439–42
   immunity, 439–41, 442
   negligence, 439, 441
   notice of defect, 441–42
Public relations, 419–25
   client relations, developing positive,
      421–22
   media, dealing with, 422–25
   positive public image, developing,
      419–21
Public service announcement
      (PSA), 424
Public transit service inventories, 79
Public transit studies, 79
Pulsed flow, 296

Q

Queue studies, 76–77
   length, 73
Queuing (waiting) pedestrian areas,
      148, 149

R

Radar speed studies, 64
Radio, 285
   citizens-band (CB), 408
   commercial, 402
   highway advisory (HAR), 412–13
   signal interconnect, 285
Railroad car, commuter, 46
Railroad crossing:
   markings, 261, 262, 270
   signs and signals, 339
Railroad-highway grade-crossing
      signals, 306
Rail station change-of-mode facilities,
      parking facilities for, 226–28
Rail transit car, 45, 46, 47
Ramp, garage, 216, 217, 218

Traffic management (*cont.*)
  special events and
    construction/maintenance
    work zones, 362–66, 387–90
  defined, 360
  evaluation of, 383–86
  law enforcement participation in,
    366
  need for, 361
  strategies, 367–76
    city streets, 371–73
    coordination of, 373–76
    freeway corridor, 373
    freeways, 367–71
  techniques, 386–90
  transportation planning and
    construction coordination, 376
Traffic management center, 369–70
Traffic Management Team, 373–75
Traffic Management Triad (Triad),
  373
Traffic records, 434–36
Traffic regulations, 329–59
  for bicycle lanes and bikeways,
    342–43
  for emergencies, 354–55
  observance of, 87
  one-way streets, 330–35
    benefits of, 334–35
    roadway requirements, 335
    termini design, 335
    usage criteria, 333–34
  for passing and no passing, 351–52
  for "pedestrian-only" streets,
    343–45
  purpose and scope of, 329–30
  records, 434–36
  for residential streets, 355–59
  reversible lanes and roadways,
    335–37
  sight obstruction, 353–54
  speed, 345–51
    determination of advisory speed
      indications, 347, 348–50
    establishment of speed limits,
      347–48
    special problems of, 350–51
    stop and yield controls, 352–53
  for transit and carpool lanes,
    340–42
  for turns, 337–40
Traffic Research Corporation, 300
Traffic-responsive metering, 398–400
Traffic safety, 95
Traffic service request log, 436–37
Traffic Signal Installation and
  Maintenance Manual, 305
Traffic Signal Optimization Model,
  Version III, 304
Traffic studies, 59–93
  of conflicts, 74

engineering study techniques,
  91–93
gap, 74, 280
of highway operations, 73–74
of intersections, 74–77
inventories, 59–60
observance of control devices,
  86–87
for one-way streets, 333–34
parking, 79–86
  duration and turnover studies,
    83–84
  occupancy studies, 80–83
  parking demand and generation
    studies, 84–86
  requirements of, 80
  special parking studies, 84
  truck loading studies, 84
pedestrian, 78–79
preliminary engineering, 176
public transit, 79
for reversible lanes and roadways,
  336
of speed, 64–67
  on curves, 66–67
  data analysis and presentation,
    65–66
  speed limit, 348
  spot-speed studies, 64
traffic impact studies, 87–91
  framework for, 88–89
  impact analysis, 89
  mitigation measures, 89–90
  need for, 87–88
  reports, 90–91
of travel time and delay, 67–73,
  280
  intersection delay studies, 70–73
  presenting results, 69–70
  route studies, 68–69
of volume, 60–64, 279
Traffic volume, 46–51
  accident rates and, 159
  annual average daily traffic
    (AADT), 46, 48, 49–51
  daily variations, 48, 49, 63
  design controls and criteria, 156
  design-hour, 156
  design hourly volume (DHV),
    50–51, 156
  growth of, 51, 52
  hourly variations, 48–51, 63
  neighborhood values concerning
    streets and, 355
  peak-hour, 49–50, 156, 335
  seasonal and monthly variations,
    47–48, 63
  studies of, 60–64, 279
Trail, bicycle, 314
Trailers, 30–31, 32
Training, 426

Tranquilizers, 10
Transformer base, frangible, 324
Transitions, unspiraled, 166
Transit lanes, regulations for, 340–42
Transit load checks, 79
Transit vehicles, 45–46, 47, 48
Transportation Demand Management
  (TDM), 90
Transportation Planning Handbook,
  64, 79
Transportation Research Board, 117
  Committee on Freeway Operations,
    395
Transportation Research Circular, No.
  212, 152
Transportation Systems Management
  (TSM), 90, 340, 367, 376
Transverse pavement markers,
  260–61, 270
TRANSYT-7F Model, 303–4
Traveled way, 169–70
Travel speed (space-mean speed), 53,
  66
  average, 119
Travel time and delay studies, 67–73,
  280
  intersection delay studies, 70–73
  presenting results, 69–70
  route studies, 68–69
Travel-time study, 67–68
TravTek project, 451–52
Tri-Level Study of the Causes of
  Traffic Accidents, 94
Trip assignment, 88–89
Trip distribution, 88–89
Trip generation, 88
Trip Generation Handbook (ITE), 51
Trolley bus, 45
Truck(s). *See also* Light trucks
  aerodynamic drag coefficient of, 35
  combination, 30, 31, 32, 33
  double-trailer, 31
  driveways for, 235
  IVHS technology for, 457–58
  offtracking, 33–34
  parking facilities for, 235–36
  regulations, 345
  rolling resistances of, 34
  single-trailer, 31, 32
  single-unit, 30, 32
  travel speeds of, 54
  triple-trailer, 31
  weight/power ratios of, 35–36
Truck loading studies, 84
Truck operators, 15–17
  traffic control devices relevant to,
    16–17
Trumpet interchange types, 182
Tubes, movable, 336–37
Tunnels, lighting in, 325
Turning movement counter, 61